Frontiers in High Pressure Biochemistry and Biophysics

Frontiers in High Pressure Biochemistry and Biophysics

Editors:

Claude Balny
CNRS
Montpellier, France

Patrick Masson
Department of Toxicology
Enzymology Unit
La Tronche, France

Karel Heremans
Department of Chemistry
University of Leuven,
Leuven, Belgium

Reprinted from *Biochimica et Biophysica Acta (BBA)*, volume 1595 1/2

ELSEVIER

2002
Amsterdam – Boston – London – New York – Oxford – Paris – San Diego – San Francisco – Singapore – Sydney – Tokyo

ELSEVIER SCIENCE B.V.
Sara Burgerhartstraat 25
P.O. Box 211, 1000 AE Amsterdam, The Netherlands

First edition 2002

Library of Congress Cataloging in Publication Data
A catalog record from the Library of Congress has been applied for.

British Library Cataloguing in Publication Data
Frontiers in high pressure biochemistry and biophysics
1. High pressure biochemistry 2. Biophysics 3. High pressure (Science)
I. Balny, Claude II. Masson, Patrick III. Heremans, Karel
IV. Biochimica et biophysica
660.6

ISBN 0444510990

Reprinted from *Biochimica et Biophysica Acta (BBA)*, volume 1595 1/2
ISBN: 0-444-51099-0

⊗ The paper used in this publication meets the requirements of ANSI/NISO Z39.48-1992 (Permanence of Paper).
Printed in The Netherlands

CONTENTS

ELSEVIER

Biochimica et Biophysica Acta 1595 (2002) 1–2

BIOCHIMICA ET BIOPHYSICA ACTA
BBA
www.bba-direct.com

Preface

Richard R. Ernst [1],*

ETH Zürich, Laboratorium für Physikalische Chemie, ETH Hönggerberg HCI, 8083 Zurich, Switzerland

The BBA Special Issue 'Frontiers in High Pressure Biochemistry and Biophysics' is a veritable treasure chest! Perhaps, it will remove the very last doubts that pressure is indeed an invaluable tool for exploring and comprehending biological function. Action and reaction or input–output relations are the approaches which science is using to elicit the secrets from nature.

While for small molecules and simple systems the pressure effects are often disappointingly small, they can become quite significant for larger systems, especially for biosystems. This is hardly astonishing as intermolecular non-covalent interactions dominate and determine structure and function. In extended molecular complexes, a subtle equilibrium between the various interacting components is established which is very susceptible to volume and pressure effects. Often, cavities are important for permitting substrate binding and biological functionality, and pressure can significantly alter cavity properties.

True, pressure is apparently a non-biological parameter in the frame of our daily experiences, but this view disregards the existing internal pressures in liquids, solids, and molecules which are often much larger than external perturbations by ambient pressure. In many situations, pressure is an ideal parameter for reversible changes of biomolecular assemblies. Sometimes, it has to be combined with thermal and chemical conditioning for selecting a range of optimum pressure sensitivity. In some situations, also irreversible pressure effects are encountered and provide further insight into the inner workings of biomolecules.

When studying pressure effects, it is most important to select the proper analytical tools for monitoring the induced changes. Naturally, spectroscopy is usually the most convenient approach. The articles in this issue present a full arsenal of tools and spectroscopic approaches, ranging from fluorescence to infrared spectroscopy, and NMR. X-Ray diffraction and neutron small angle scattering can also be revealing. The technical aspects of using advanced tools can become quite demanding when combined with high pressure equipment, but the present issue demonstrates that these problems are solvable also for routine-type applications.

The present issue contains a wealth of applications of high pressure studies to biological systems. Naturally, the effects on individual molecules and molecular complexes are in the foreground. They can be rationalized by physico-chemical arguments and lead to further understanding. But also fascinating and important pressure effects on bioorganisms are discussed. Obviously their explanation is slightly more demanding and sometimes perhaps preliminary.

Certainly one of the most intriguing fields of high pressure studies are folding and unfolding processes of proteins. Here science is still at the beginnings and much progress is expected in the next few years. Surely, pressure investigations will prove to be particularly revealing, but also protein–nucleic acid in-

* Fax: +41-1-632-1257.
E-mail address: ernst@nmr.phys.chem.ethz.ch (R.R. Ernst).

[1] Nobel Prize in Chemistry.

0167-4838/02/$ – see front matter
PII: S0167-4838(01)00330-2

teractions, lipid structures, and the amyloid formation can fruitfully be explored by high pressure investigations as is demonstrated in this issue.

The editors, Claude Balny, Patrick Masson, and Karel Heremans are to be congratulated for initiating a fascinating collection of papers. It will open new avenues for further revealing high pressure studies.

ELSEVIER

Biochimica et Biophysica Acta 1595 (2002) 3–10

BIOCHIMICA ET BIOPHYSICA ACTA
BBA
www.bba-direct.com

Introduction

High pressure effects on biological macromolecules: from structural changes to alteration of cellular processes

Claude Balny [a,*], Patrick Masson [b], Karel Heremans [c]

[a] *INSERM Unité 128, IFR 24, CNRS, 1919, route de Mende, 34293 Montpellier Cedex 5, France*
[b] *CRSSA, Department of Toxicology, Enzymology Unit, BP 87, 38702 La Tronche Cedex, France*
[c] *Department of Chemistry, Katholieke Universiteit Leuven, B-3001 Leuven, Belgium*

Received 12 December 2001; accepted 12 December 2001

1. Historical background

The development of high pressure bioscience is not documented in a single volume. Several books contain chapters that discuss some topics related to bioscience. Bridgman himself always kept a lively interest in the field as may be judged from the last chapter in his book 'The physics of high pressure' [1] and in the many reviews that are collected in his 'Collected experimental papers' [2]. The most extensive source on the older literature is probably to be found in 'The kinetic basis of molecular biology' by Johnson et al. [3] or the newer edition 'The theory of rate processes in biology and medicine' [4]. Books that concentrate only on pressure effects are: 'High pressure effects on cellular processes' by Zimmerman [5] and 'The effects of pressure on organisms', a symposium volume edited by Sleigh and Macdonald [6].

Over a period of about a century the emphasis in our understanding of the effect of pressure on organisms shifted from the question of how organisms can survive under those conditions to the recent discovery that some organisms actually thrive under these extreme conditions [7].

The present volume is strongly biased to the behavior of biomolecules, a story that started with the observations of Bridgman (1914). He reported 'a fact of possible biological interest': 'If the white of an egg is subjected to hydrostatic pressure at room temperature, it becomes coagulated, presenting an appearance much like that of a hard boiled egg'. Being interested in phase transitions of compounds as a function of pressure and temperature, he notes: 'The effect of temperature, which is not large, seems to be such that the ease of coagulation increases at low temperatures, contrary to what one might expect' [8].

In France, Basset and Macheboeuf reported their results on the effect of high pressure on the inactivation of enzymes, viruses, antigens, antibodies, some of them being quite pressure stable, microorganisms, tissues and cells [9]. In the USA, Johnson and coworkers at Princeton studied various biosystems at low pressures in order to get insight in the physicochemical mechanisms [3,4]. In Japan, the research concentrated on the denaturation of proteins. Here we see the first report on the pressure and temperature effects on ovalbumin and hemoglobin with the data reported as a phase diagram [10]. At low temperatures and high pressures, negative activation energies are reported and interpreted as the pressure-induced penetration of water into the protein as the first step in the denaturation process according to the theory of protein denaturation of Wu [11].

$$\text{Protein} + H_2O \rightarrow \text{protein}(H_2O) \rightarrow \text{unfolding}$$

* Corresponding author. Fax: +33-4-67-52-36-81.
E-mail address: balny@u128.crbm.cnrs-mop.fr (C. Balny).

PII: S0167-4838(01)00331-4

Recent high pressure computer simulations support this mechanism [12]. It is of interest to note the similarities between pressure-induced unfolding and unfolding by chemical denaturants:

$$\text{Protein} + \text{urea} \rightarrow \text{protein(urea)} \rightarrow \text{unfolded protein}$$

Thus, it is now possible to make a unified phenomenological description of the cold, heat, pressure and chemical denaturation of proteins [13–16].

About a decade ago, the use of high pressure found new applications and industrial realizations, following the suggestions made by Hayashi and coworkers in Japan. Although Hite in the USA [17] and Basset and coworkers in France [18] had studied the possible applications of high pressure for the treatment of milk and production of vaccines respectively, it was in Japan that the first products were put on the market. This was followed by an explosion of new papers exploring various possibilities mostly in food science [19] but also in medical and pharmaceutical applications [20,21].

The developments made in Japan are recorded in the volumes (in Japanese with one page abstracts in English) edited by R. Hayashi under the title: High Pressure Bioscience Conferences in Japan [22]. A total of eight volumes have been published up to now.

The first international conference on High Pressure Bioscience and Biotechnology was organized in 1992 in La Grande Motte–Montpellier (France). The publication of the proceedings proved to be a commercial success [23]. Other conferences were held, both in Japan and Europe, the proceedings of which are available [24–26]. There is also the long tradition of NATO Advanced Study Institutes; the proceedings of the most recent one contain a number of chapters on bioscience [27].

Except for one paper, pressure effects on cells are not discussed explicitly in this volume but numerous studies indicate that the effect of pressure on proteins is quite relevant for our understanding of the effect of pressure on organisms.

2. Pressure effects compared with temperature effects

The quantitative description of the effects of temperature and pressure on chemical processes originated at the time of the development of chemical thermodynamics especially by Gibbs. But since his papers were rather mathematical, he was understood by only a few. The contributions of van't Hoff, Helmholtz and Planck were more accessible for chemists. In 1884, van't Hoff derived an equation for the effect of temperature on a chemical equilibrium,

$$\partial \ln K/\partial T = \Delta H/RT^2$$

based on the Clapeyron–Clausius equation (1834). A few months later, Le Châtelier stated the *Le Châtelier principle* based on the equation proposed by van't Hoff. Van't Hoff also derived an equation for the effect of temperature on a reaction rate, but the equation became known as the Arrhenius (1889) equation:

$$\partial \ln k/\partial T = E_a/RT^2$$

The derivation of the equations describing the effect of pressure on the equilibrium was then a trivial matter except for the fact that it is much more difficult to verify experimentally. In 1887 Planck proposed the equilibrium equation

$$\partial \ln K/\partial p = -\Delta V/RT$$

followed by the kinetic equation

$$\partial \ln k/\partial p = -V_a/RT$$

proposed by van't Hoff (1901) and Evans and Polanyi (1935). The last equation is often found in high pressure articles or books as it gives for the first time a molecular interpretation of the activation volume in terms of intrinsic volume effects and solvent effects.

Bridgman first discussed the combined effects of pressure and temperature on physical equilibrium in 1915. Brandts and Hawley used the application of these equations in the early seventies of last century [28]. The treatment makes use of the change of the Gibbs free energy change as a function of pressure and temperature:

$$d(\Delta G) = (\Delta V)dp - (\Delta S)dT$$

With the relation between ΔG and the equilibrium constant K, an expression is obtained, assuming that ΔV and ΔS are pressure and temperature dependent, that is used to analyze thermodynamic as well as kinetic data for the denaturation of proteins, the inactivation of enzymes and microorganisms, and many other biochemical and microbiological phenomena. The most salient feature of the equation is that it describes the cold, the heat as well as the pressure denaturation of proteins [13].

A molecular interpretation of kinetic and thermodynamic data is usually based on data obtained from low molecular weight model systems [29,30]. This is done in terms of intrinsic volume and solvent effects as first proposed by Evans and Polanyi, but the separation of these effects can only be based on molecular models. These models, again, start from some basic assumptions about the pressure and temperature behavior of matter.

For proteins and other biopolymers there is the additional factor of uncertainty: the packing of the atoms which is not always perfect. In order to take this effect into account, Kauzmann [31] proposed to consider the partial molar volume (V_{i}) of a protein in solution as being composed of three contributions: the atoms, the cavities and the hydration.

$$V_{\mathrm{i}} = V_{\mathrm{atoms}} + V_{\mathrm{cavities}} + \Delta V_{\mathrm{hydration}}$$

It is clear that pressure as well as temperature effects give rise to changes in the cavities as well as the hydration. Independent data on the cavities and the hydration are hard to get and most investigators are forced to make assumptions on the contributions of these factors. One possible technique that could be used to probe the role of cavities in proteins is positron annihilation lifetime spectroscopy. It is a powerful technique for determining the size of cavities in synthetic polymers.

In order to make the role of the solvent water (W) more explicit in biochemical processes, it is useful to consider the following reaction of a macromolecule (P):

$$\mathrm{P} + \mathrm{W} \rightarrow \mathrm{PW}$$

The replacement of water by a ligand or another cosolvent (X) can be explicitly included:

$$\mathrm{P} + \mathrm{W} + \mathrm{X} \rightarrow \mathrm{PWX}$$

The effect of the binding of the ligand on the equilibrium can be expressed with linked functions as developed by Wyman:

$$\partial \ln K / \partial \ln x = \Delta X$$

Here x refers to the activity of the ligand X and ΔX to the change in moles of X. A similar equation can be written for the effect of the change in the activity of water (w).

$$\partial \ln K / \partial \ln w = \Delta W$$

At low concentrations of the macromolecule, the activities of the water and that of the ligand are related via the Gibbs–Duhem equation. Putting all the effects into one equation yields:

$$\mathrm{d}(\ln K) = (\partial \ln K / \partial p)_{T,w}\, \mathrm{d}p + (\partial \ln K / \partial T)_{p,w}\, \mathrm{d}T + (\partial \ln K / \partial \ln w)_{p,T}\, \mathrm{d}w$$

3. Molecular interpretation of thermodynamic and kinetic data

For a molecular interpretation of thermodynamic and kinetic data, it is important to keep in mind that the above-presented equations are expressed in terms of partial molar quantities.

This may be illustrated with the partial molar volume. It is not equal to the volume of the molecule or the ions since it includes also the interaction with the solvent. This may be seen from the fact that the partial molar volume of the uncharged glycolamide is larger (56.2 ml/mol) than that of the zwitterionic amino acid glycine (43.5 ml/mol). The difference is interpreted as being due to the electrostriction of the dipolar water molecules by the charges present in glycine. These considerations are also applicable to enthalpy, entropy and free energy changes.

Studies on various small molecule model systems have given information on the effect of pressure on noncovalent intermolecular interactions [29,30]. One of the important results is that hydrogen bonds are stabilized by high pressures. Hydrophobic interactions that play a substantial role in the stabilization of the tertiary structure and in protein–protein interactions are destabilized under high pressure. On the other hand, stacking interactions between aromatic rings show negative volume changes and are stabilized by pressure.

3.1. Thermodynamic modeling

Data for the unfolding of proteins under the influence of temperature, at constant pressure, are fitted with the Gibbs–Helmholtz equation, an equation well known in the biochemical literature:

$$\Delta G = \Delta G^0 - \Delta S^0 (T - T_0) + \Delta C_p \left[(T - T_0) - T \ln \frac{T}{T_0} \right]$$

The following equation describes, at constant temperature, the pressure-induced protein unfolding:

$$\Delta G = \Delta G^0 + \Delta V^0 (P - P_0) + \frac{\Delta \beta}{2} (P - P_0)^2$$

One can see that the role of the change in heat capacity (ΔC_p) that is fundamental for the heat-induced unfolding is now replaced by the change in the compressibility factor ($\Delta\beta = \partial \Delta V / \partial p$). But as the changes in volume are not only pressure but also temperature dependent, a complete thermodynamic description also contains the change in the thermal expansivity factor ($\Delta\alpha = \partial \Delta V / \partial T$):

$$\Delta V = \Delta V^0 + \Delta\alpha (T - T_0) - \Delta\beta (P - P_0)$$

These equations give a consistent thermodynamic description of the heat, pressure and cold denaturation [13]. There are two limitations for the application of these equations. First, they apply at constant composition of the solvent as discussed by Kornblatt and Kornblatt [32]. Second, they provide only the differences in the thermal expansivity factor, compressibility factor, and heat capacity between the denatured and the native state. Absolute quantities must be obtained from density, ultrasound velocity and calorimetric measurements.

The relevance of the compressibility to the biological function can be understood through the statistical mechanical relation between the isothermal compressibility (β_T) and the volume fluctuations of biomolecules as discussed by Taulier and Chalikian [33]:

$$\langle \partial V^2 \rangle = kTV\beta_T$$

This equation relates the volume fluctuations of the system to the isothermal compressibility. Because of the small size of most biomolecules, the volume fluctuations are relatively large. In a thermodynamic system with constant T and P, the heat capacity can be regarded as a measure of the entropy fluctuations of the system which can be visualized from changes in conformation between ordered and less ordered structures. The thermal expansivity is related to the cross-correlation of the volume and entropy fluctuations. This agrees with the intuition that the thermal expansivity is a coupling between the thermal and the mechanical parameters.

3.2. Computer modeling

Molecular simulations can also assist molecular interpretation of experimental data. Computer modeling starts from a physical mechanical model of molecules and considers the possible changes in conformation and interaction with the solvent as a result of changes in energy and volume. The role of the size and the shape of molecules are especially clear in these model calculations. Molecular dynamic (MD) simulations are still restricted to small proteins, but MD simulation attempts on large proteins with their hydration shell have made progress in the past years [34]. In the future one may expect that an increased interaction between modeling and experiment will stimulate new ideas and progress [35].

4. Temperature and pressure effects on biopolymers

The pressure behavior of lipids, nucleic acids and polysaccharides (except starch) may be summarized as a pressure-induced ordering of the structure, i.e. pressure has an effect that can be compared with lowering the temperature. This is clear from the diagram of lipids [36]. Proteins (and starch), on the other hand, show a pressure-induced disordering in their structure, i.e. the effect of pressure induces effects that are similar, but not identical to those induced by temperature [13]. A phase diagram has now been proposed for nucleic acid stability on the basis of new experimental findings [37].

In proteins the various processes that play a role in the pressure- and temperature-induced effects are unfolding, phase separation and coagulation. Although papers in this volume concentrate on the unfolding aspect, processes like phase separation, coagulation, and crystallogenesis are not only of tremendous practical relevance but also at the same time a challenge from a physico-chemical point of view.

Of particular importance are the intermolecular interactions in proteins that follow after a partial unfolding and that give rise to the formation of fibrils [38]. On the other hand it is of interest to note that some proteins do not form fibrils unless under well defined conditions [39]. Whether this is related to the shape of the stability diagram remains to be investigated.

Many of these observations can be described with thermodynamic based equations. The study of model systems, e.g. synthetic polymers [40], may assist in the interpretation of thermodynamic parameters in terms of molecular mechanisms. Molecular engineering may then be used to affect the unfolding, the phase separation and the interaction (coagulation) steps of the process. Protein crystallization, a process of considerable theoretical and practical interest, is particularly sensitive to pressure [41].

5. Temperature and pressure effects on cellular processes

Cellular processes are characterized by a wide variety of reactions, including ligand–protein interactions, redox reactions, protein–protein interactions, interactions between proteins and nucleic acids, etc. The difference in complexity between interactions involving macromolecules and those between molecules of small molar mass is nicely illustrated by the papers of Macyk and van Eldik [42] and those from Hui Bon Hoa et al. [43] and Jung [44].

The sensitivity for pressure and temperature of supramolecular assemblies such as microtubuli and flagellin suggests that we may expect that cells are quite sensitive to pressure. This is supported by the observations on viruses. Very small pressures induce changes that make viruses biologically inactive although the structure may be little affected [45]. The molecular adaptation mechanisms that play a role in the composition of bacterial membranes may be interpreted along the same lines [46], more in particular on the ion channels [47].

The observed role of water in the stability of macromolecules is also found in cells. It is well known that bacterial spores are very difficult to inactivate by high pressures. On the other hand the germination of spores takes place at pressures of the order of 100 MPa [48]. An extreme example may be seen with tardigrades [49]. Tardigrades are composed of about 40 000 cells, become immobile and shrink into a special state when the humidity of the surroundings decreases. In such a state they can survive temperature from −253°C up to 151°C and pressures up to 600 MPa. In the normal state they are killed at 200 MPa. This behavior is quite similar to that of bacterial spores and dry proteins where pressures of more than 1 GPa are not able to provoke any changes. The fact that tardigrades can undergo a transformation to an extreme dry state may therefore be much more exceptional than the fact that they are resistant to extremes of temperature and pressure in the dry state.

Pressure may have delicate effects on the metabolism and the gene expression of cells [46]. However, the description of these effects assumes a complete knowledge of the metabolic pathways and the genetic information of the cell.

A number of observations point to the possibility of an understanding of pressure effects on the biology of the cell from an understanding of the effects of pressure on the interactions between macromolecules and their reactions. New data on the interactions between proteins and nucleic acids reveal quite unexpected results as shown by the contributions from Silva et al. [45] and Lynch and Sligar [50], but it is also obvious that cell biology may need more theory to make sense of the complex networks of metabolism. Rate and equilibrium data are needed on reactions and interactions between molecules. Computational modeling of cell biology may stimulate new experiments and help in the understanding of the present results.

The final step in our understanding should be the explanation of the sensitivity of different microorganisms submitted to treatment by hydrostatic pressure. This may be a problem of understanding not only the metabolism, the gene expression but also more complex phenomena such as unfolding of critical enzymes followed by phase separation that the cell cannot cope with anymore by the available repair mechanisms. It would be of interest to look into the differences between eukaryotes and prokaryotes. It is of interest to note that many extremophilic bacteria are of the latter type. The biotechnological relevance of extremophilic enzymes has attracted considerable attention [51].

6. Major developments in experimental methodology

Progresses in biophysical techniques adapted to high pressure measurements and in molecular biology methods have been permitting refined analysis of the effects of pressure on biomacromolecules at the submolecular level. For instance, recent developments in derivative spectroscopy allow detecting discrete pressure-induced conformational changes in the vicinity of aromatic amino acid residues [52]. The developments in fluorescence [53,54] and phosphorescence [55] spectroscopy that provide information on local changes together with Fourier transform infrared spectroscopy [56] are promising approaches. Whereas high resolution NMR spectroscopy [57] gives an overwhelming amount of details, techniques such as synchrotron X-ray and neutron small angle scattering give details on the overall shape of a protein as a function of pressure and temperature [58]. While the classical approach still leads to interesting new insights into enzymatic activity [59], it seems that studies on the effect of pressure on structure and activity of mutant proteins generated by site-directed mutagenesis provide unique information on compressibility of enzymes and thus on the role of protein flexibility on the activity [60]. More recently, the double-mutant thermodynamic cycle analysis at varying pressure has been successfully applied to investigate interactions between two amino acids along the enzyme reaction pathway [61].

New methodological developments have made possible the use of high pressure techniques in a number of biotechnological applications [62,63].

7. Conclusions: specificity in pressure effects?

There are several reasons to make it clear that pressure effects on biological macromolecules reveal a specificity that is different from those of temperature and from chemical effects.

First there is the thermodynamic or energetic point of view. In contrast to increasing temperature, which focuses on energy and volume effects due to thermal expansivity, pressure effects are mainly on the volumetric aspects via the compressibility of the system. Apart from this equilibrium aspect, information is also obtained on the dynamics of the system in view of the relation between the compressibility and the volume fluctuations. Secondly, from a mechanistic point of view these volume fluctuations, which are of the order of 1% or less of the total volume of a protein, arise from contributions of the internal cavities and the interactions with water. Recent computer simulations suggest that the contribution of the cavities to the compressibility may be much larger than assumed up to now. Pressure-induced effects in protein may then be largely attributed to the penetration of water into the protein structure and this results in unfolding. In the case of heat-induced denaturation, unfolding may take place first followed by water penetration. The fact that unfolding is the first step in the heat unfolding may explain the observation that aggregation results more frequently from heat than from pressure treatment.

The reduction of living systems to molecules has proved to be a fruitful approach provided one keeps in mind the simplifications that are introduced. Do we then need new strategies for the study of the complexity of life? As expected, opinions are divided, but the observation that stability diagrams for bacteria, viruses and other complex systems are similar to those obtained for proteins suggests that proteins are the primary targets in the pressure and temperature behavior of organisms.

A better understanding of the behavior of proteins and other biomolecules under extreme conditions may be expected from a concerted application of a number of existing and new experimental approaches. New developments in science usually imply fresh views at existing experimental data. These new views stimulate new experiments. And the best experiments in science are those that lead to the least expected results!

Acknowledgements

This collective work has partially been done in the framework of the European Cooperation in the field of Scientific Technical Research, COST D 10. The authors thank the contributors and referees for making this volume a source of stimulating ideas and reliable information for those who want to enter the field of high pressure bioscience. They are also grateful to Prof. R. Ernst for his support to this project.

References

[1] W. Bridgman, The Physics of High Pressure, Dover Publications, New York, 1970.
[2] P.W. Bridgman, Collected Experimental Papers, Harvard University Press, Cambridge, MA, 1964.
[3] F.H. Johnson, H. Eyring, M.J. Polissar, The Kinetic Basis of Molecular Biology, Wiley, New York, 1954.
[4] F.H. Johnson, H. Eyring, B.J. Stover, The Theory of Rate Processes in Biology and Medicine, Wiley, New York, 1974.
[5] A.M. Zimmerman, High Pressure Effects on Cellular Processes, Academic Press, New York, 1970.
[6] M.A. Sleigh, A.G. Macdonald, The Effects of Pressure on Organisms, Cambridge University Press, Cambridge, 1972.
[7] K. Horikoshi, W.D. Grant, Extremophiles. Microbial Life in Extreme Environments, Wiley-Liss, New York, 1998.
[8] P.W. Bridgman, J. Biol. Chem. 19 (1914) 511–512.
[9] J. Basset, M. Macheboeuf, E. Wollman, Ann. Inst. Pasteur 58 (1937) 58–77.
[10] K. Suzuki, Rev. Phys. Chem. Jpn. 29 (1960) 91–97.
[11] H. Wu, Chin. J. Physiol. 5 (1931) 321–344 (reprint in: Adv. Prot. Chem. 49 (1995) 6–26.
[12] B. Wroblowski, J.F. Diaz, K. Heremans, Y. Engelborghs, Proteins Struct. Funct. Genet. 25 (1996) 446–455.
[13] L. Smeller, Biochim. Biophys. Acta 1595 (2002) 329–344.
[14] S. Kunugi, N. Tanaka, Biochim. Biophys. Acta 1595 (2002) 11–21.
[15] S. Perrett, J.M. Zhou, Biochim. Biophys. Acta 1595 (2002) 210–223.
[16] G. Pappenberger, C. Saudan, M. Becker, A.E. Merbach, T. Kiefhaber, Proc. Natl. Acad. Sci. USA 97 (2000) 17–22.
[17] B.H. Hite, W. Virginia, Agric. Exp. Station Bull. 58 (1899) 15–35.
[18] J. Basset, P. Lépine, L. Chaumont, Ann. Inst. Pasteur 90 (1956) 575–596.
[19] D. Knorr, Curr. Opin. Biotechnol. 10 (1999) 485–491.
[20] P. Masson, C. Tonello, C. Balny, J. Biomed. Biotechnol. 1 (2001) 85–88.
[21] L.P. Gaspar, A.C.B. Silva, A.M.O. Gomes, M.S. Freitas, A.P.D. Ano Bom, W.D. Schwarcz, J. Mestecky, M.J. Novak, D. Foguel, J.L. Silva, J. Biol. Chem. (2002) in press.
[22] R. Hayashi, Use of Pressure in Food, San-Ei Shuppan Co., Kyoto, 1989, and subsequent volumes.
[23] C. Balny, R. Hayashi, K. Heremans, P. Masson (Eds.), High Pressure and Biotechnology, INSERM/Libbey, Paris, 1992.
[24] R. Hayashi, C. Balny (Eds.), High Pressure Bioscience and Biotechnology, Elsevier, Amsterdam, 1996.
[25] K. Heremans (Ed.), High Pressure Research in the Biosciences and Biotechnology, Leuven University Press, Louvain, 1997.
[26] H. Ludwig (Ed.), Advances in High Pressure Bioscience and Biotechnology, Springer, Heidelberg, 1999.
[27] R. Winter, J. Jonas (Eds.), High Pressure Molecular Science, Kluwer Academic Publishers, Dordrecht, 1999.
[28] S.A. Hawley, Biochemistry 10 (1971) 2436–2442.
[29] B.B. Boonyaratanakornkit, C.B. Park, D.S. Clark, this volume.
[30] C.A. Royer, Biochim. Biophys. Acta 1595 (2002) 201–209.
[31] W. Kauzmann, Adv. Protein Chem. 14 (1959) 1–63.
[32] J.A. Kornblatt, M.J. Kornblatt, Biochim. Biophys. Acta 1595 (2002) 30–47.
[33] N. Taulier, T.V. Chalikian, Biochim. Biophys. Acta 1595 (2002) 48–70.
[34] V. Lounnas, S.K. Lüdemann, R.C. Wade, Biophys. Chem. 78 (1999) 157–182.
[35] E. Paci, Biochim. Biophys. Acta 1595 (2002) 185–200.
[36] R. Winter, Biochim. Biophys. Acta 1595 (2002) 160–184.
[37] R.B. Macgregor, Biochim. Biophys. Acta 1595 (2002) 266–276.
[38] T.W. Randolph, M. Seefeldt, J.F. Carpenter, Biochim. Biophys. Acta 1595 (2002) 224–234.
[39] M. Fandrich, M.A. Fletcher, C.M. Dobson, Nature 410 (2001) 165–166.
[40] S. Kunugi, N. Tanaka, Biochim. Biophys. Acta 1595 (2002) 11–21.
[41] Y. Suzukietal, Biochim. Biophys. Acta 1595 (2002) 345–356.
[42] J. Macyk, R. vanEldik, Biochim. Biophys. Acta 1595 (2002) 283–296.
[43] G. HuiBonHoa, M.A. McLean, S.G. Sligar, Biochim. Biophys. Acta 1595 (2002) 297–308.
[44] C. Jung, Biochim. Biophys. Acta 1595 (2002) 309–328.
[45] J.L. Silvaetal, Biochim. Biophys. Acta 1595 (2002) 250–265.
[46] D.H. Bartlett, Biochim. Biophys. Acta 1595 (2002) 367–381.
[47] A.G. Macdonald, Biochim. Biophys. Acta 1595 (2002) 387–389.
[48] H. Ludwig, Biochim. Biophys. Acta 1595 (2002) 390–391.
[49] K. Seki, M. Toyoshima, Nature 395 (1998) 853–854.
[50] T.W. Lynch, S.G. Sligar, Biochim. Biophys. Acta 1595 (2002) 277–282.
[51] E. Mombelli, E. Shehi, P. Fusi, P. Tortora, Biochim. Biophys. Acta 1595 (2002) 392–396.
[52] R. Lange, C. Balny, Biochim. Biophys. Acta 1595 (2002) 80–93.
[53] K. Ruan, C. Balny, Biochim. Biophys. Acta 1595 (2002) 94–102.

[54] P. Tauc, C. ReyesMateo, J.C. Brochon, Biochim. Biophys. Acta 1595 (2002) 103–115.
[55] P. Cioni, G.B. Strambini, Biochim. Biophys. Acta 1595 (2002) 116–130.
[56] W. Dzwolak, M. Kato, Y. Taniguchi, Biochim. Biophys. Acta 1595 (2002) 131–144.
[57] J. Jonas, Biochim. Biophys. Acta 1595 (2002) 145–159.
[58] R. Winter, Biochim. Biophys. Acta 1595 (2002) 160–184.
[59] D.B. Northrop, Biochim. Biophys. Acta 1595 (2002) 71–79.
[60] K. Gekko, Biochim. Biophys. Acta 1595 (2002) 382–386.
[61] P. Masson, M.-T. Froment, S. Fort, F. Ribes, N. Bec, C. Balny, L. Schopfer, Biochim. Biophys. Acata, in press.
[62] P. Lemay, Biochim. Biophys. Acta 1595 (2002) 357–366.
[63] R. Hayashi, Biochim. Biophys. Acta 1595 (2002) 397–399.

ELSEVIER

Biochimica et Biophysica Acta 1595 (2002) 11–29

BIOCHIMICA ET BIOPHYSICA ACTA
BBA

www.bba-direct.com

Review

Pressure–temperature phase diagrams of biomolecules

László Smeller *

Department of Biophysics and Radiation Biology, Semmelweis University, Puskin u. 9, PF. 263, H-1444 Budapest, Hungary

Received 24 September 2001; received in revised form 16 November 2001; accepted 19 November 2001

Abstract

The pressure–temperature phase diagram of various biomolecules is reviewed. Special attention is focused on the elliptic phase diagram of proteins. The phenomenological thermodynamic theory describing this diagram explains the heat, cold and pressure denaturations in a unified picture. The limitations and possible developments of this theory are discussed as well. It is pointed out that a more complex diagram can be obtained when the intermolecular interactions are also taken into account. In this case metastable states appear on the pressure–temperature (p–T) diagram due to intermolecular interactions. Pressure–temperature phase diagrams of other biopolymers are also discussed. While the p–T diagrams of helix–coil transition of nucleic acids and of gel–liquid crystal transition of lipid bilayers are non-elliptical, those of gelatinization of starch and of phase separation of some synthetic polymers show an elliptic profile, similar to that of proteins. Finally, the p–T diagram of bacterial inactivation is shown to be elliptic. From the point of view of basic science, this fact shows that the key factor of inactivation should be the protein type, and from the viewpoint of practical applications, it serves as the theoretical basis of pressure treatment of biosystems.

Keywords: Protein; Phase diagram; Pressure; Unfolding; Biopolymer

1. Introduction

Although in biochemical/biophysical experiments temperature is a much more widely varied parameter than pressure, there is no doubt about their theoretically equivalently important role as thermodynamic variables [1]. Only the experimental difficulties can explain the relative rareness of pressure studies. In the last decades the number of high pressure experiments performed on biosystems has considerably increased [2–8]. There are several reasons to measure the effect of pressure on a wide variety of thermodynamic systems. Perhaps the most important argument is that one can separate the effects of volume and thermal energy changes, which appear simultaneously in temperature experiments [9]. Molecular processes, which alter the volume of the system, are especially sensitive to pressure. Volume changes associated to (bio)chemical processes can therefore be revealed using pressure as an experimental variable.

The above arguments make clear that several phenomena are worthwhile to be studied not only in temperature but also in pressure experiments, or both. Proper presentation of these data needs a phase space spanned by the pressure and temperature axes. In this paper we discuss the different thermodynamic processes (denaturation, phase transition, etc.) on this pressure–temperature (p–T) plane.

This review is dedicated to the p–T diagram of biomolecules. Most part of the paper deals with pro-

* Fax: +36-1-266-6656.
E-mail address: smeller@puskin.sote.hu (L. Smeller).

0167-4838/02/$ – see front matter © 2002 Elsevier Science B.V. All rights reserved.
PII: S0167-4838(01)00332-6

teins, because of their interesting and well characterized elliptic phase diagram. Other biomolecules like DNA and lipids are also shortly discussed, and at the end we turn to simple living systems like viruses and bacteria.

Of course experiments can cover only a small section of the whole *p–T* plane. Even smaller is the region where the behavior of biomolecules is of interest for scientists. Therefore it is useful here to define the region which we deal with. Naturally this region lies around the living conditions of the biosystems. Considering temperature, the interesting section corresponds roughly to the one where the water is liquid, because as a matter of fact biological systems are aqueous solutions. The highest pressure in the biosphere is slightly above 0.1 GPa (1 kbar)[1] at the bottom of the Mariana trench in the Pacific Ocean. Although this pressure is more than 1000 times higher than the atmospheric pressure, it is not able to disturb the conformation or organization of biomolecules significantly. Higher pressures are usually needed for drastic changes in the structure of biomolecules, such as protein denaturation. The range where interesting biophysical transformations occur is typically 0.5–1 GPa and very rarely up to 2 GPa. This limit is in accordance with the pressure induced solidification of water (see Section 2), which occurs in this pressure interval (depending on the temperature). Therefore we restrict our considerations mainly to the pressure–temperature range of liquid water.

2. Water

Most of the systems which are the subjects of this review are aqueous solutions, of one or several kind of biomolecules. Therefore the pressure–temperature phase diagram of water is discussed first. It is well known that water is a very special liquid [10]. Several practical applications utilize the unique features of the phase diagram of water (see e.g. [11,12]).

Most of the studies on the phase diagram of water were performed by Bridgman at the beginning of the last century [13,14]. These data were summarized by Eisenberg and Kauzmann [10]. The whole phase diagram is rather complicated containing several ice phases. Since this issue is dedicated to high pressure behavior of biomolecules, we want to discuss only that part in detail which has importance in experiments on aqueous solutions of biological molecules. Fig. 1 shows the phase diagram of water in the temperature range of −50 to +100°C up to a pressure of 2.2 GPa (22 kbar). The most unusual property of water is that the ice formed at ambient pressure (ice I) has a lower density than water. As a consequence, the pressurizing of water at temperatures slightly below 0°C does stabilize the liquid phase. This phenomenon can be observed below 200 MPa. The lowest temperature at which water can be held liquid is −22°C at 207 MPa [10]. This makes many applications possible, ranging from basic studies on pressure assisted cold denaturation [12,15–18] to the practical applications of storage of food materials under subzero temperature, without freezing, in order to avoid the damaging effect of ice crystal for-

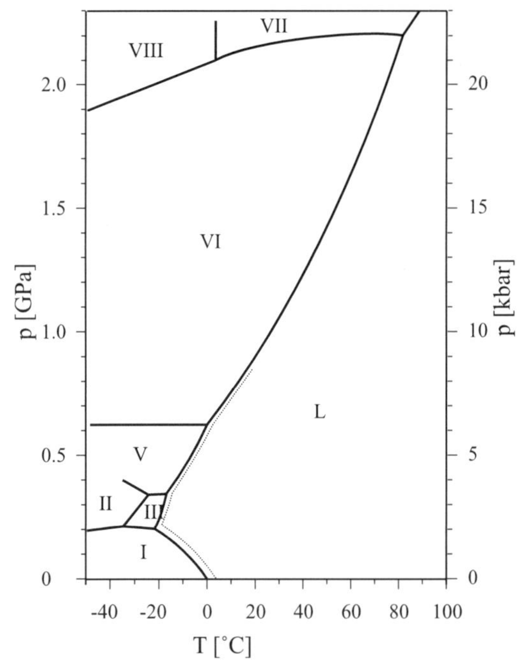

Fig. 1. Phase diagram of water in the temperature range of −50 to +100°C up to a pressure of 2.2 GPa. L refers to the liquid phase; roman numbers (I–VIII) show the different ice phases. The dotted line shows the melting curve of heavy water.

[1] Pressure units: 1 GPa = 10 kbar, 0.1 GPa = 100 MPa = 1 kbar, 0.1 MPa = 1 bar = atmospheric pressure.

mation, or rapid freezing achieved by depressurizing the samples pre-cooled under 0°C [11].

Pressure induced ice formation at room temperature can be found around 1 GPa. In concentrated solutions the freezing temperature decreases, which corresponds to an increase in the freezing pressure at a given temperature. The freezing of water limits the experiments investigating the pressure stability of proteins. However, proteins that are stable up to this pressure in solution are very rare [19,20], therefore this limit is not frustrating from the point of view of protein experiments.

Some experimental techniques, such as Fourier transform infrared (FTIR) and nuclear magnetic resonance (NMR) spectroscopy, require heavy water (^2H_2O) in order to avoid the strong signal of water, which would obscure the spectral features of solute molecules such as proteins. Therefore it is worthwhile to observe the phase diagram of ^2H_2O as well. In its main features this is similar to that of water, with small temperature and pressure shifts. The dotted line in Fig. 1 shows the solidification curve of heavy water. As can be seen, this broken line is practically parallel to that of normal water, but shifts slightly to higher temperatures.

3. Proteins

3.1. Protein unfolding

It has been known for a long time that proteins unfold at high temperature, as well as at high pressure (see e.g. [1] and references therein). Later, cold denaturation of proteins was also predicted [21], but it was only measured for a relatively small number of proteins [15,16,18,22]. The picture which emerged from all these experimental facts assumed a stable native state of the protein in a closed range of the p–T plane. Crossing the boundary of that region, the native conformation looses its stability, and the protein unfolds. The consistent thermodynamic description of the phase boundary of protein unfolding was developed by Hawley [23]. The theory explained his measurements on the unfolding of chymotrypsinogen and the data from experiments by Brandts on ribonuclease A [24]. Hawley calculated the Gibbs free energy difference (ΔG) of the denatured and the native state of the protein and obtained a second order curve for the denaturation conditions ($\Delta G = 0$). This curve turned to be elliptical in case of proteins. The experimental data on chymotrypsinogen and ribonuclease A [23,24] also confirmed that the shape of the phase diagram of protein unfolding is elliptical. In the frame of this theory, heat, pressure, and cold denaturations are only three very special ways of unfolding (Fig. 2).

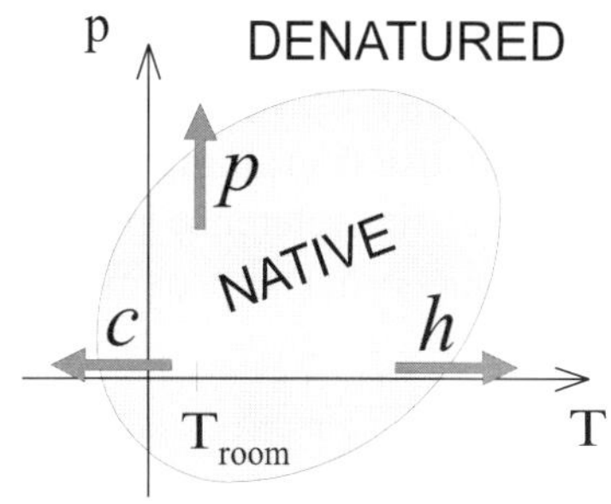

Fig. 2. Schematic representation of the elliptic phase diagram of proteins. The arrows marked by the letters p, h, c show the specific denaturation ways known as pressure, heat and cold denaturation.

3.1.1. Hawley theory for the elliptical phase diagram

Hawley [23] started from the assumption that there are only two distinct states of the protein (native and denatured) and the transition between them is a two-state process. The Gibbs free energy difference between these states is defined as:

$$\Delta G = G_{\text{denatured}} - G_{\text{native}} \tag{1}$$

$$\mathrm{d}\Delta G = -\Delta S \mathrm{d}T + \Delta V \mathrm{d}p \tag{2}$$

Upon integration of this equation from an arbitrarily chosen reference point T_0,p_0 to T,p one obtains:

$$\Delta G = \frac{\Delta\beta}{2}(p-p_0)^2 + \Delta\alpha(p-p_0)(T-T_0) - \Delta C_p\left[T\left(\ln\frac{T}{T_0}-1\right)+T_0\right] + \Delta V_0(p-p_0) - \Delta S_0(T-T_0) + \Delta G_0 \tag{3}$$

where Δ always means the change of the corresponding parameter during denaturation (i.e. the value in the denatured state minus that in the native state). β is the compressibility factor: $\beta = (\partial V/\partial p)_T$; α is the thermal expansivity factor $\alpha = (\partial V/\partial T)_p = -(\partial S/\partial p)_T$; C_p is the heat capacity $C_p = T(\partial S/\partial T)_p$; all the other symbols have their usual meaning.

In the vicinity of the reference point one can use the following second order approximation:

$$T\left(\ln\frac{T}{T_0}-1\right)+T_0=\frac{(T-T_0)^2}{2T_0} \tag{4}$$

Substituting it into Eq. 3 one gets:

$$\Delta G=\frac{\Delta\beta}{2}(p-p_0)^2+\Delta\alpha(p-p_0)(T-T_0)-\frac{\Delta C_p}{2T_0}(T-T_0)^2+\Delta V_0(p-p_0)-\Delta S_0(T-T_0)+\Delta G_0 \tag{5}$$

The transition line, where the protein denatures (or refolds depending on the direction of the crossing), is defined by $\Delta G=0$. Substituting this into Eq. 5 one gets the equation of a second order curve in the T–p plane. Mathematically this is a general equation for a conic section, which can be either a hyperbola or an ellipse, but actually it was found to be always elliptic for proteins. The mathematical constraint ensuring the elliptic shape is:

$$\Delta\alpha^2>\Delta C_p\Delta\beta/T_0 \tag{6}$$

which was satisfied by the different signs of ΔC_p and $\Delta\beta$ in case of chymotrypsinogen and ribonuclease studied by Hawley [23].

The function on the right hand side of Eq. 5 gives an elliptic paraboloid for ΔG as shown in Fig. 3a. Fig. 3b is the projection of this function to the p–T plane, showing the origin of the elliptic phase diagram. The region where the protein is more stable in the native phase ($\Delta G>0$) is shown in gray. The gray ellipse shows the conditions for the native state. Curves with similar elliptic shape, but different actual values of thermodynamic parameters have been determined for several proteins [39,68,111,112].

It should be mentioned that Eq. 5 is formally the second order Taylor series of ΔG expanded with respect to p and T around T_0p_0. This approach was used by Clark [25], who developed a similar theory for the explanation of the so called re-entrant phase diagram of certain liquid crystals [26]. This re-entrant p–T phase diagram shows elliptic features like the one of proteins, and it can be explained by the same phenomenological thermodynamic theory.

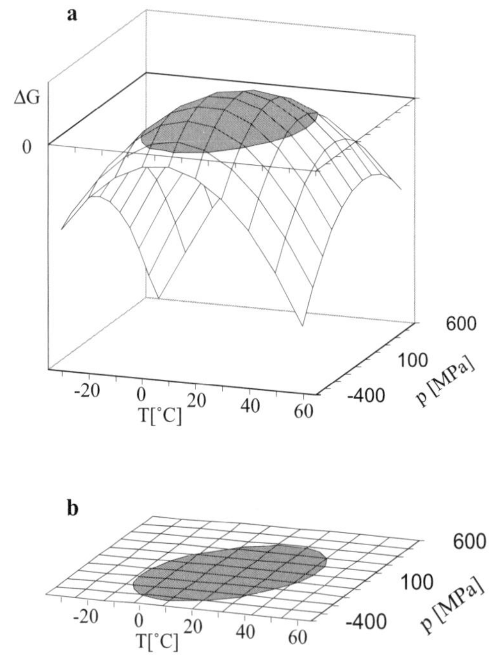

Fig. 3. (a) ΔG as a function of temperature and pressure for chymotrypsinogen, using the parameters determined by Hawley. (b) The elliptical phase diagram shown as the horizontal projection of the above function. The gray area shows the region where the native state is more stable.

From a pure mathematical point of view the reason why one gets an elliptic shape for the phase boundary curve is the fact that one truncates the series after the second order term. Taking only the first order terms into account would give a linear phase boundary. The question arises: what happens if one goes further with the series expansion, taking the higher order terms into account [27]?

Physically, truncating the Taylor series at the second order terms means that the second derivatives of the Gibbs free energy difference (ΔG) do not change with temperature and pressure:

$$\left(\frac{\partial\Delta C_p}{\partial T}\right)_p=0,\left(\frac{\partial\Delta\beta}{\partial T}\right)_p=\left(\frac{\partial\Delta\alpha}{\partial p}\right)_T=0$$
$$\left(\frac{\partial\Delta\beta}{\partial p}\right)_T=0,\left(\frac{\partial\Delta\alpha}{\partial T}\right)_p=\left(\frac{\partial\Delta C_p}{\partial p}\right)_T=0 \tag{7}$$

Although the conditions of Eq. 7 are usually assumed to be valid, there are some indications that this is not always the case. Yamaguchi et al. [28] reported pressure dependence of ΔC_p in the case of ribonuclease A. An extended analysis of the $\Delta G(T,p)$

function is thus necessary, where the third order terms are also involved [27]. Fig. 4 shows the distortion of the phase diagram when one introduces terms proportional to T^3, T^2p, Tp^2, and p^3. The physical meaning of these terms is the temperature and pressure dependence of ΔC_p, $\Delta\beta$ and $\Delta\alpha$. The main form of the ellipse is clearly not abolished, but the actual shape of the curve is distorted. As one would expect, the most distorted part of the ellipse is at high pressure and/or temperature, where these terms become dominant. Such distorted phase boundaries were already observed, and associated to the limitation of the theory behind the elliptic phase diagram [29].

There are several specific points on the elliptic diagram. The heat denaturation temperature T_h (at atmospheric pressure), the denaturation pressure p_d (at room temperature) and the cold denaturation temperature T_c could be the three points of highest interest. They can be given by the following equations:

$$T_h = -\frac{\Delta S_0 T_0}{\Delta C_p} + \sqrt{\frac{\Delta S_0^2 T_0^2}{\Delta C_p^2} + 2\frac{\Delta G_0 T_0}{\Delta C_p}} + T_0 \tag{8}$$

$$P_d = -\frac{\Delta V_0}{\Delta\beta} + \sqrt{\frac{\Delta V_0^2}{\Delta\beta^2} - 2\frac{\Delta G_0}{\Delta\beta}} + p_0 \tag{9}$$

$$T_c = -\frac{\Delta S_0 T_0}{\Delta C_p} - \sqrt{\frac{\Delta S_0^2 T_0^2}{\Delta C_p^2} + 2\frac{\Delta G_0 T_0}{\Delta C_p}} + T_0 \tag{10}$$

Since the choice of the reference point T_0,p_0 is arbitrary, we have chosen for the sake of simplicity: p_0 = atmospheric pressure and T_0 = room temperature.

The slope of the phase boundary in an arbitrarily chosen point of the ellipse can be written in the following form:

$$\frac{\partial p}{\partial T} = -\frac{\frac{\partial\Delta G}{\partial T}}{\frac{\partial\Delta G}{\partial p}} = \frac{\Delta S_0 - \Delta\alpha(p-p_0) + \Delta C_p \frac{T-T_0}{T_0}}{\Delta V_0 + \Delta\beta(p-p_0) + \Delta\alpha(T-T_0)} \tag{11}$$

or in the form of the Clausius–Clapeyron equation:

$$\frac{\partial T}{\partial p} = -\frac{\frac{\partial\Delta G}{\partial p}}{\frac{\partial\Delta G}{\partial T}} = \frac{\Delta V_0 + \Delta\beta(p-p_0) + \Delta\alpha(T-T_0)}{\Delta S_0 - \Delta\alpha(p-p_0) + \Delta C_p \frac{T-T_0}{T_0}} \tag{12}$$

At the maximum pressure and maximum temperature, where the native state is stable, the slope of the ellipse is zero and infinite at p_{max} and T_{max} respectively. Using the Clausius–Clapeyron equation, one gets $\Delta V = 0$ for the point of T_{max} and $\Delta S = 0$ for p_{max}. The $\Delta V = 0$ curve is a line, crossing the center of the ellipse. Similarly the $\Delta S = 0$ can also be represented by a line. These lines can be given by:

$$p_{\Delta S=0} = \frac{\Delta C_p}{\Delta\alpha}\frac{T-T_0}{T_0} + \frac{\Delta S_0}{\Delta\alpha} + p_0 \tag{13}$$

$$p_{\Delta V=0} = \frac{\Delta\alpha}{\Delta\beta}(T-T_0) + \frac{\Delta V_0}{\Delta\beta} + p_0 \tag{14}$$

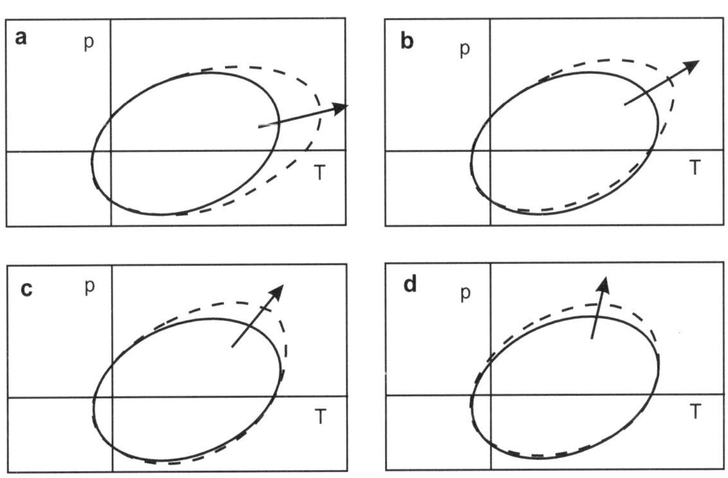

Fig. 4. Effect of higher order terms on the shape of the elliptic diagram. Effect of terms containing (a) T^3; (b) T^2p; (c) Tp^2; (d) p^3. The solid line shows the original phase diagram; the broken line shows the phase diagram if the higher order term is taken into account. The arrows show the direction of the distortion when the higher order term increases.

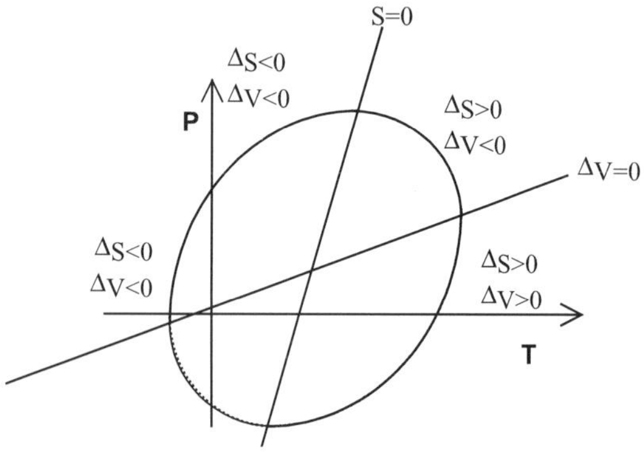

Fig. 5. Relative position of the $\Delta S=0$ and $\Delta V=0$ lines compared to the ellipse.

Fig. 5 shows these lines for a general case of the ellipse. Below these lines the entropy or volume changes associated to the unfolding are positive. Below the $\Delta S=0$ line the order decreases during denaturation. Above the $\Delta V=0$ line, the contribution of decreasing volume to the ΔG promotes unfolding.

The actual shape, size, and orientation of the elliptic boundary are defined by six thermodynamic parameters ($\Delta\beta$, $\Delta\alpha$, ΔC_p, ΔV_0, ΔS_0, ΔG_0). Although the equations for the unfolding points are quite complex, some rules of thumb can be given to predict the effect of the above mentioned parameters. Decrease of ΔC_p broadens the ellipse in the direction of the temperature axis. Similarly does $\Delta\beta$ in the direction of the pressure axis. $\Delta\alpha$ determines predominantly the orientation of the ellipse.

The phase diagrams determined by Hawley for chymotrypsinogen and ribonuclease differ markedly in their shape (Fig. 6). While the phase diagram of chymotrypsinogen shows a stabilizing effect of moderate pressure against heat denaturation (the slope of the curve, $dp/dT>0$ at T_h), pressurizing decreases the unfolding temperature of ribonuclease. The phase diagrams of proteins can be classified using these two forms as a first approximation. We will call the one of chymotrypsinogen 'tongue'-like, and the one of ribonuclease 'hillside'-like.

3.1.2. *Limitations of the Hawley theory*

3.1.2.1. General assumptions. The first remark must be that Hawley's theory is a purely *phenomenological* one, which does not involve any structural knowledge about the actual system (i.e. the structure of the protein). It is also very important to notice that the whole theory is based on the assumption that there are *only two* possible *states* of the protein, and the transition is a two-state one, without any intermediate state. This latter point was questioned recently by several measurements, where an intermediate called 'molten globule' was found during the denaturation of the protein [30–32,35].

3.1.2.2. Reversibility. Hawley used *equilibrium thermodynamics* in the description of denaturation. This implies *reversible unfolding* of the protein, which is not always the case. In several experiments the denaturation was found to be irreversible, especially at high temperatures. Above the heat denaturation temperature gel formation is observed at sufficiently high concentrations [34]. Hawley already noticed that

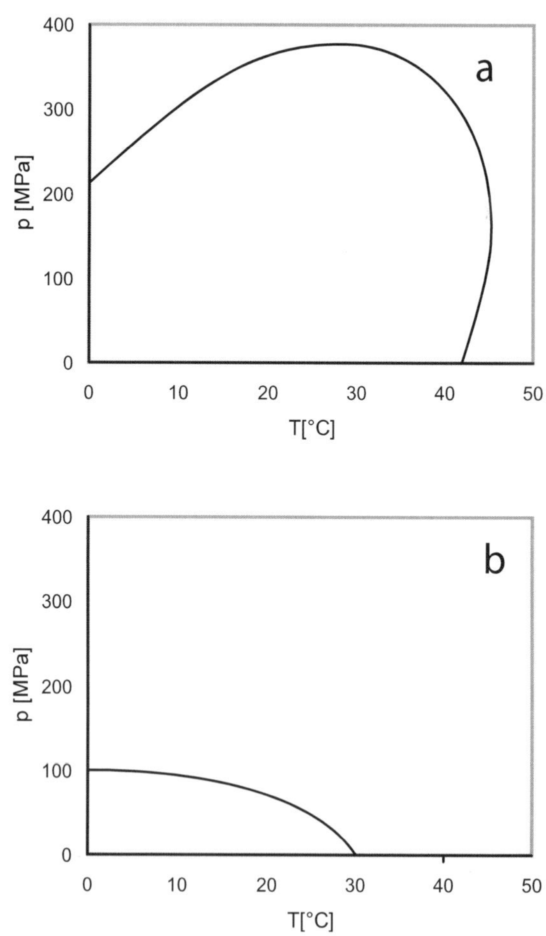

Fig. 6. Types of protein phase diagrams. (a) 'Tongue'-like diagram of chymotrypsinogen (pH 2.07). (b) 'Hillside'-type diagram of ribonuclease (pH 2). Drawn after [23].

above 45°C it was not possible to differentiate between reversible and irreversible denaturation. Zipp and Kauzmann [29] also observed irreversible effects (precipitation) of myoglobin in a certain pH range. Irreversible denaturation can be observed most markedly in the case of infrared experiments [33–35,112], where the temperature unfolded protein usually forms insoluble gel stabilized by intermolecular hydrogen bonds. This hydrogen bonding network remains unchanged after cooling the sample below the denaturation temperature. The appearance of these intermolecular connections is promoted by the high concentration (50–100 mg/ml) used in the infrared spectroscopy. In contrast, the fluorescence and UV absorption experiments require much lower concentration, which reduces the probability of the formation of intermolecular interactions. In this respect the lack of reversibility points to the failure of the basic assumption of two states. Irreversibility is a result of a third phase, an aggregated one, which is not included in the Hawley theory. The intermolecular interactions will be discussed in detail in Section 3.2. There are few proteins which are exceptions to this rule, and do not form intermolecular interactions in the heat denatured state. Ribonuclease A is one of them; therefore it is a favored object of folding–refolding studies [36–38,111,68]. Recently a similar behavior of staphylococcal nuclease (SNase) was reported by Panick et al. [39].

3.1.2.3. Kinetic effects. On the transition line $\Delta G=0$, which means that the folded and the unfolded proteins have the same Gibbs free energy. Assuming equilibrium thermodynamics, both states are equally populated here. Nothing is known, however, about the barrier height between the two states which determines the kinetics of the transition. The early investigators of the elliptic boundary already noticed that in some parts of the p–T plane it took a very long time (hours or days) to reach equilibrium near the unfolding conditions. This indicates a very high barrier between the folded and unfolded states.

3.1.2.4. Role of the experimental methods. Elliptic phase diagrams similar to that of Hawley were determined for several proteins [39,68,111,112]. Different spectroscopic methods were applied to obtain data at high pressure and/or temperature. Care must be taken when one compares results from different methods using different spectroscopic markers. Certain methods provide information about the vicinity of a specific site of the protein, while others reflect the global state of the protein. Fluorescence spectroscopy especially reports the neighborhood of the fluorescent marker (usually tryptophan or a prosthetic group [40,41]). UV absorption spectroscopy (used by Hawley [23]) utilizes the absorption of the aromatic side chains. Fourth derivatives are often used to follow small spectral changes in this region [42,43]. Zipp and Kauzmann [29] followed the denaturation of myoglobin utilizing the visible absorption of the heme group at 536 and 585 nm. Global information is obtained in the infrared (or Raman) spectroscopy, which characterizes the state of the protein using conformation sensitive vibration of the protein backbone [44]. The simplest one-dimensional NMR studies of proteins are based on the signal of histidine residues [45,28], but use of the more sophisticated NOESY spectra allows to follow the hydrogen–deuterium exchange at several sites of a small protein [46,47], revealing detailed information about its flexibility. It is not surprising that the denaturation parameters determined by different methods can deviate. One example is the staphylococcal nuclease, where the elliptic denaturation contour determined by tryptophan fluorescence is slightly smaller compared to the one measured by FTIR and small angle X-ray scattering [39].

3.1.2.5. Role of water in protein unfolding. The role of water in protein unfolding should neither be under- nor overestimated. The importance of the presence of water is clearly proven by the fact that the elliptic diagram can only be observed for protein solutions. If proteins are in the dry state, they are very pressure stable. On the other hand, a model was proposed to explain cold and heat denaturation of proteins on the basis of the dome shaped variation in the density of water with temperature [48]. This curve was correlated with the dome shaped $\Delta G(T)$ function of proteins. As pointed out by Heremans et al. [49], this analogy does not apply to pressure.

3.1.2.6. Effect of pH on the phase diagram. It is well known that the denaturation temperature and

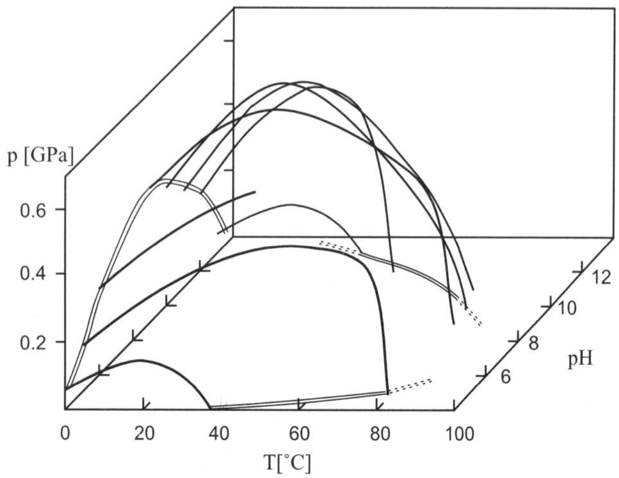

Fig. 7. Three-dimensional representation of the pressure temperature and pH dependence of the unfolding of myoglobin. (Drawn after [29].)

pressure do usually decrease at extreme pH values [50–52]. This results in the reduction of the size of the ellipse. Zipp and Kauzmann [29] measured the phase diagram of myoglobin in a wide pH range. In Fig. 7 the equilibrium transition curves obtained by them are plotted in a three-dimensional plot. In this way the pH dependence of the denaturation pressure and temperature can be visualized (double lines). The three-dimensional phase diagram thus obtained is a more or less ellipsoid shaped surface. The vertical sections containing the p and T axes represent the elliptic phase diagram described mathematically in Section 3.1.1. The reduction of the size of the ellipse at extreme pH will bring the heat and cold denaturation points closer to each other, and lower the unfolding pressure. The lowering of the denaturation pressure at acidic pH is used by several investigators to bring the pressure unfolding into the range of their pressure equipment. A similar effect can be reached by using moderate concentrations of chemical denaturants, such as guanidine hydrochloride and urea. The first studies of the phase diagram were also carried out at pH 2 [23,24]. The same approach can be used to raise the cold unfolding temperature, which for most proteins lies below the freezing point of water at normal conditions. Privalov measured the cold denaturation of metmyoglobin and apomyoglobin around pH 3.8 and 5, respectively [22,53].

3.1.2.7. Effect of protein mutation on the phase diagram. Site directed mutation is nowadays a standard method of biochemistry to produce proteins with altered properties (e.g. increased activity or environmental flexibility). Temperature (and sometimes also pressure) stability of these proteins is an important factor.

In some cases drastic changes in pressure stability can be reached by mutation of only one amino acid. This was observed in the case of ribonuclease P2 from the thermophilic archaebacterium *Sulfolobus solfataricus*. The wild type of this protein showed extreme stability against pressure and temperature treatment. The reason for this is assumed to be the hydrophobic core containing three aromatic acid residues. A drastic destabilization takes place when Phe31 is replaced by Ala31 [54]. At lower pH the cold, heat and pressure denaturations of wild type and two mutants have been measured and compared [55]. The response of the protein was found to be different towards cold, heat and pressure.

Panick et al. compared the phase diagram of wild type and a V66A mutant of staphylococcal nuclease [39]. The elliptic diagrams seemed to have approximately the same center, but the size of the native region was reduced by the mutation.

A systematic study on 10 mutants of ribonuclease was performed by Torrent et al. [37,38]. All the mutations were at the carboxy-terminal site. Although the whole diagram was not determined for all of the mutants, the linear correlation between the heat denaturation temperature and the unfolding pressure is an interesting result of this study. They concluded that pressure and temperature have comparable effects on the hydration of hydrophobic residues. From the viewpoint of thermodynamics, the simplest explanation for this correlation is a simple increase (or decrease) in the stability of the protein (ΔG_0) due to mutations. Lifting the whole elliptic paraboloid (Fig. 3a) increases both the unfolding pressure and temperature, by the increase in ellipse size. However, two points (T_h, $p_{1/2}$) do not allow to draw conclusions about all the thermodynamic parameters which can influence the changes of the p–T diagram during mutations.

Rubens et al. studied three different α-amylases from *Bacillus* species: *Bacillus subtilis* (BS), *Bacillus amyloliquefaciens* (BA), and *Bacillus licheniformis*

(BL) [56]. The first two have quite similar diagrams. The third one has considerably higher temperature stability, which is, however, not accompanied by higher pressure stability. Observing the phase transition diagram, in the case of BL the region of the native phase is either shifted to higher temperatures, or broadened in the direction of the temperature axis. The former can be explained by an increased entropy change, while the latter involves a ΔC_p increase.

A special environment of the proteins can increase their pressure stability. A few examples are the carbohydrate shell [57] and inclusion of the protein into reversed micelles [58]. The presence or absence of a prosthetic group [59] or a stabilizing ion [60] can also change the size and shape of the *p–T* diagram. Glycerol cosolvent is used in several low temperature studies to prevent ice formation [61]. Kaposi et al. measured horseradish peroxidase down to 10 K without observing any cold denaturation [62].

3.1.3. Phase diagram from kinetic measurements

3.1.3.1. Denaturation kinetics. All the above experiments focus on the structural changes in proteins during unfolding. When the kinetic aspects of denaturation are measured as a function of pressure and temperature, isokineticity curves can be drawn, connecting the points on the *p–T* space where the rate constants of denaturation are the same. A pioneering work on the unfolding and coagulation kinetics of ovalbumin and carboxyl hemoglobin was performed by Suzuki [63]. These measurements resulted in isokineticity curves which are similar in shape to the elliptic phase diagram discussed above.

It must be pointed out that the similarity of the elliptic profiles of the isokineticity and equilibrium denaturation curves is far from trivial. While the equilibrium line is determined by the equal G_{native} and $G_{\text{denatured}}$, the kinetics of the unfolding is strongly influenced by the activated state, i.e. the height of the barrier between the native and denatured phases. Mathematical analysis of the isokineticity curves therefore gives the activation parameters of the unfolding [63].

3.1.3.2. Phase diagram obtained from enzymatic activity. Although structure and function are expected to be correlated, functional studies on enzymes are also important to characterize the effect of pressure and temperature on proteins. The enzymatic activity of α-chymotrypsin has been measured by Taniguchi and Suzuki at various pressures and temperatures [64]. In these experiments the activity first increases with pressure because of the negative activation volume [65]. After reaching a maximum, the activity decreases due to the unfolding of the protein. The curve on the *p–T* diagram where 50% of the protein is unfolded corresponding to the $\Delta G = 0$ phase transition line was determined this way [64].

More usually, enzyme kinetics is measured *after* some *p–T* treatment but at ambient conditions [33,66,67]. It must be noted that the important kinetic aspect of these measurements is not the speed of the catalytic process, but the time course of the *p–T* treatment. The rate constant of the enzymatic function after a certain *p–T* treatment gives information about the number of protein molecules which survived the treatment. Surviving means here avoiding an irreversible denaturation, i.e. this fraction was either not denatured, or reversibly refolded. Naturally short treatment can be more drastic (in terms of high pressure and/or temperature), while the phase diagrams obtained from activity measurements after long treatments are closer to the equilibrium curve.

An explanation of these experiments is not possible within the theory of the two-state reversible unfolding/folding. An intermediate state has to be included according to the following scheme:

$$\mathrm{N} \leftrightarrow \mathrm{I} \rightarrow \mathrm{D}$$

where N is the native, I is an intermediate and D is the irreversibly denatured state of the protein. The intermediate state sometimes shows the features of a molten globule. For example, pressure unfolding of carboxypeptidase Y shows more than one transition, involving distinct intermediate states [57]. The intermediate state, however, is not always separated from the native and denatured phases by clear phase transitions.

3.1.4. Comparison of the pressure, temperature, and cold denatured protein structures

This section discusses the question whether the elliptic diagram is a complete description of the pressure–temperature effects on proteins, or whether there are other structural states and corresponding phase transition lines which are necessary for an adequate description.

According to the basic assumption of the Hawley theory, there exist only two states of the protein. This means that all the denatured protein structures are of the same kind. This was, however, questioned by several studies in which the pressure, temperature, and cold denatured phases were compared.

Privalov found the unfolded state as being universal, for both the cold and the heat denatured case [53]. His calorimetric studies, however, could not provide any structural information. More detailed spectroscopic studies, however, did not support his assumption.

Thermodynamic analysis of pressure, heat, and cold denaturation can also predict differences between the denatured structures. Observing the $\Delta V=0$ and $\Delta S=0$ lines in Fig. 5, one can see that the temperature denaturation is entropy driven, while the other two are mainly volume change driven. Therefore both the pressure and cold denatured phases are expected to be more compact than the heat unfolded one.

Jonas and coworkers investigated the heat, cold and pressure unfolding of ribonuclease A by NMR spectroscopy [68]. On the basis of the hydrogen–deuterium exchange rates these authors concluded that the cold and pressure denatured RNase A molecules contain some partial secondary structure, in contrast to the thermally denatured one. A similar conclusion has been obtained by Meersman et al. analyzing the secondary structure sensitive amide I region of the FTIR spectrum of myoglobin [17]. Griko and Kutyshenko found the cold and heat denatured phases of β-lactoglobulin to be different [69]. Their NMR measurements led to the conclusion that the network of residual interactions is more extensive in the cold denatured state than in the heat unfolded one.

On the basis of the above arguments, the temperature unfolded state is different from the others. This, however, does not imply the existence of a phase boundary line between the high temperature and high pressure phases, because the properties of the unfolded state can also change continuously.

Table 1
Pressure, heat and cold denaturation parameters for proteins

Protein	T_h (°C)[a]	p_d (GPa)[b]	T_c (°C)[c]	Ref.
Chymotrypsinogen pH 2.07	42[d]	0.34[d]	−9[d,e]	[23]
Ribonuclease pH 2	30[d]	0.1[d]	−32[d,e]	[23,24]
Ribonuclease p^{2}H 2.4	49[d]	0.38[d]	−25[d]	[68]
Ribonuclease A p^{2}H 5				
FTIR amide I band	61.5	0.74	–	[38][g]
FTIR tyrosine band	63.6	0.78	–	[38]
Ribonuclease p^{2}H 7.0	65[d]	0.75[d]	–	[111]
Lipoxygenase (soybean) p^{2}H 7.8	68[d]	0.6[f]	–	[33]
β-Lactoglobulin A	60[d]	0.15	–	[112]
Metmyoglobin pH 9	78	0.6	–	[29]
Staphylococcal nuclease pH/p^{2}H 5.5	48[d]	0.21[d]	0[d,h]	[39]
Staphylococcal nuclease V66A mutant p^{2}H 5.5	38[d]	0.09[d]	–	[39]

[a]At ambient pressure.
[b]At 20°C if not stated otherwise.
[c]At ambient pressure if not stated otherwise.
[d]Estimated from figures of the given reference.
[e]Determined from the extrapolated region of the fitted ellipse at 0.3 GPa.
[f]Determined from the extrapolated region of the fitted ellipse at room temperature (exact temperature is not given).
[g]These authors also determined T_h and p_d for 10 mutants of RNase; details can be found in the original publication.
[h]At 150 MPa.

3.1.5. *Examples of proteins with known p–T denaturation characteristics*

Although the elliptic diagram for proteins was established more than 30 years ago, the *p–T* behavior of proteins has been measured in detail only for a limited number of proteins. Recently this number has increased considerably [33,39,111,112]. The pressure, heat (and sometimes the cold) denaturation points are determined for many more enzymes. These values are given for a selection of proteins in Table 1.

3.2. *Aggregation: influence of intermolecular interactions on the phase diagram*

As has already been mentioned, the temperature denatured protein has the tendency to aggregate. This can mainly be observed in the infrared spectroscopic experiments [33,34,36,112], for two reasons: (1) the concentration is relatively high with this technique, and (2) a pair of specific absorption bands characteristic for the intermolecular hydrogen bonds can be seen on the two sides of the conformation sensitive amide I band [70,35]. On the other hand, pressure is known to favor dissociation of oligomers [71,72]. These basic facts suggest that description of protein interactions using the pressure–temperature diagram reveal new features of this process. It is, however, clear that the aggregation cannot be described by a simple two-state model. Since the aggregate formed from the temperature denatured protein remains stable (or at least metastable) after cooling down the sample, the phase diagram which takes into account the intermolecular interactions of proteins will contain metastable phases.

The biological/medical impact of aggregation was emphasized recently, when it turned out that several serious diseases are the result of deposition of insoluble protein aggregates in the cell [73,74]. Amyloidosis, Alzheimer disease and transmissible spongiform encephalopathy are such diseases where destabilized proteins form insoluble fibers [75,76]. Kinetic studies on fibril formation suggested that an amorphous aggregate formed immediately after denaturation of the protein can slowly transform into a fibrillar structure [77].

Several pressure works have been performed on

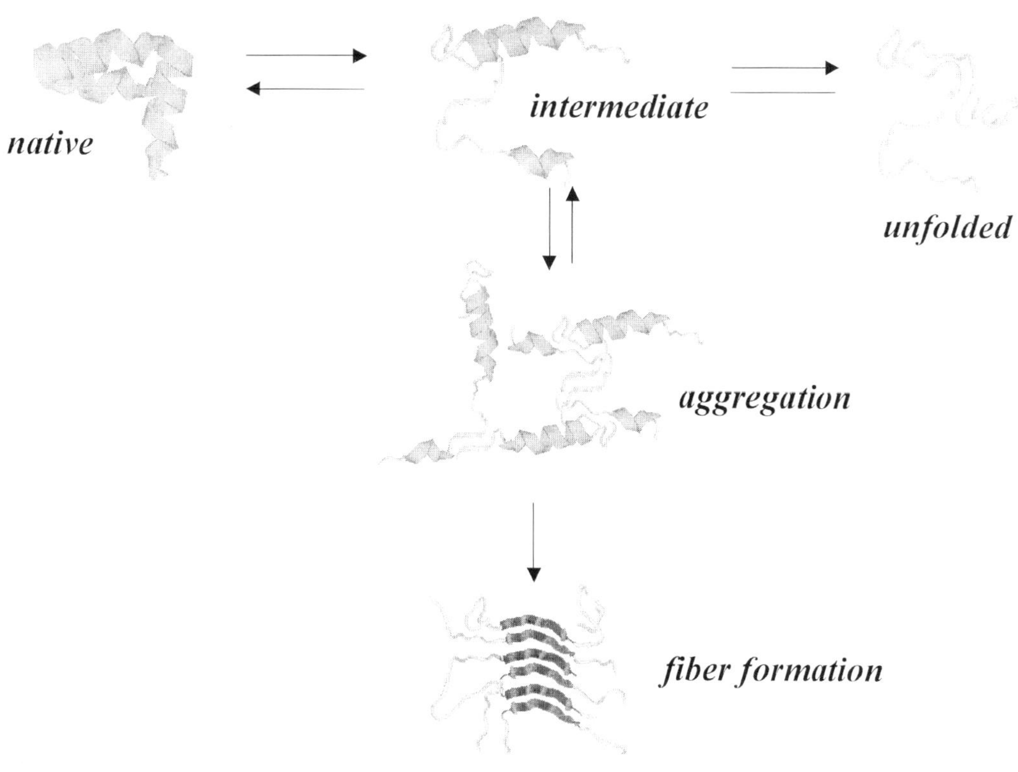

Fig. 8. Scheme of the formation of protein aggregates. (After [73].)

multimeric proteins. One of the common findings was that the dissociation occurs in the range of 100–200 MPa, while the typical unfolding pressure of proteins is around 500 MPa. The relative high pressure sensitivity of the aggregation is the result of the large volume change associated with disaggregation.

Two types of intermolecular organization can be observed among proteins. Some proteins exist in dimer, tetramer [78] or oligomer [79] form in the native state. On the other hand, there are proteins which do aggregate only if their structure is somehow destabilized [76]. Formation of amyloid fibers belongs to the latter type of aggregation. Partially folded intermediates play a key role in this process. It has been shown that the intermediate structure which appears during the refolding of a pressure denatured protein can also be regarded as a destabilized state of the protein which is prone to aggregation [35]. Fig. 8 shows the suggested scheme of the aggregation.

Although there are very few systematic investigations on the pressure–temperature diagram of aggregation, the tendency of the data collected by several pressure studies allows us to draw some conclusions. However, at the present state some details of the following diagram should be treated as hypothetical. From the comparison of the structures of the pressure and heat denatured proteins [17], it is clear that formation of a gel stabilized by the intermolecular hydrogen bonding network is present only after heat denaturation at atmospheric pressure. When the heat denaturation was performed at 3 kbar for myoglobin, no intermolecular interactions were observed (Smeller, Heremans, unpublished). Nor were they observed at pressure denaturation, at high pressure [34]. However, the specific infrared bands characteristic for these intermolecular interactions were seen after pressure denaturation when the pressure was released below 200 MPa, if the pressure denaturation was performed slightly above room temperature [35]. This suggests that above approx. 40°C the aggregated state is more stable than the intermediate, formed during refolding. The intermediate and the unfolded states cannot form an aggregate if the pressure is higher than 200 MPa. Taking into account that the aggregate, if it has already been formed once, e.g. at high temperature, is also stable at room temperature, the phase diagram shown in Fig. 9 can be suggested for myoglobin. Some parts contain metastable states, which are indicated by the letters in parentheses. The most stable state is indicated without parentheses.

It has to be mentioned, however, that not all proteins form an intermolecular hydrogen bonding network even at high temperature and atmospheric pressure. One of the already mentioned examples is RNase [80]. Recent infrared studies enlarged the list of these proteins with SNase [39] and horseradish peroxidase [81]. The phase diagram suggested in Fig. 9 cannot be applied to these proteins. An interesting remark can be made if one compares the shape of the phase diagrams of aggregating and non-aggregating proteins. RNase and SNase have hillside-type phase diagrams, while myoglobin has a rather tongue-like diagram or at least something between these two types. If we suppose that the line separating the denatured and aggregated phases crosses the ellipse at the highest temperature point (maximum temperature stability, $\Delta V=0$), the disappearance of the aggregated phase for the hillside-type phase diagrams is a natural consequence. Whether this can be treated as a general rule is not yet clear; more detailed thermodynamic analysis of unfolding and aggregation would be necessary to confirm this hypothesis.

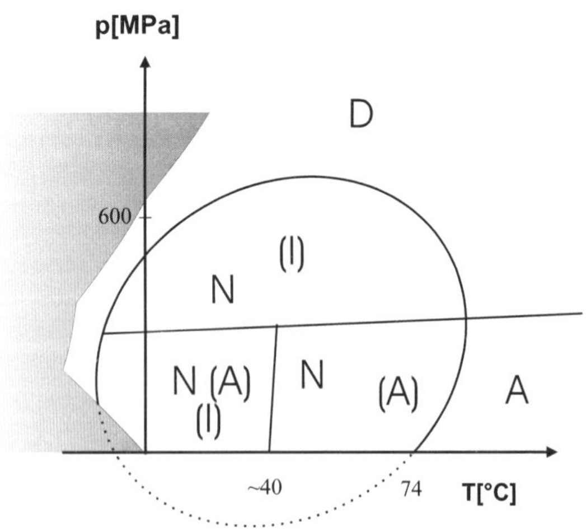

Fig. 9. Extended phase diagram suggested for myoglobin. The diagram includes both the denaturation and aggregation phenomena, and therefore contains several metastable phases. The metastable phases are indicated in parentheses. N, native; D, denatured; A, aggregated; I, intermediate.

Since aggregation is a rather irreversible process in biosystems, protective mechanisms exist in nature. One of these is the synthesis of the so-called heat shock proteins, whose role is to prevent the aggregation of other proteins after unfolding caused by heat or another shock [82]. One of their subfamilies consists of the so-called small heat shock proteins with a molecular mass of ≈20 kDa. Interestingly, they form large oligomers in their 'resting state', and there are several indications that their heat shock functioning is accompanied by changes in their quaternary structure. Recent pressure studies indicate that besides elevated temperature, elevated pressure can also induce an increase in the chaperone activity of these proteins [83].

4. Other biopolymers

Keeping in mind that the elliptic shape of the *p–T* phase diagram of proteins can be explained by a purely phenomenological theory, the question arises whether this shape is unique for proteins. In this section we review shortly the results on biopolymers: nucleic acids, starch, and some non-biological synthetic polymers.

4.1. Nucleic acids

In contrast to proteins, whose secondary, tertiary and quaternary structures are also essential for their function, DNA contains the most important information in its primary structure, i.e. the sequence of the base pairs. It has been known for a long time that the double stranded helical structure of DNA is abolished at high temperature. This 'melting' is an endothermic process. Besides by calorimetry, it can also be detected by the increase in UV absorption above the melting temperature, which is the result of the disappearance of the hypochromic effect due to the breaking of the double helix. Hawley and Macleod [84] studied the effect of pressure on the melting temperature of *Clostridium perfringens* DNA. Unlike the heat unfolding temperature of proteins, the melting temperature of DNA did not show any curvature, but a purely linear pressure dependence in the studied pressure range (Fig. 10). The slope of the curve as well as the melting temperature

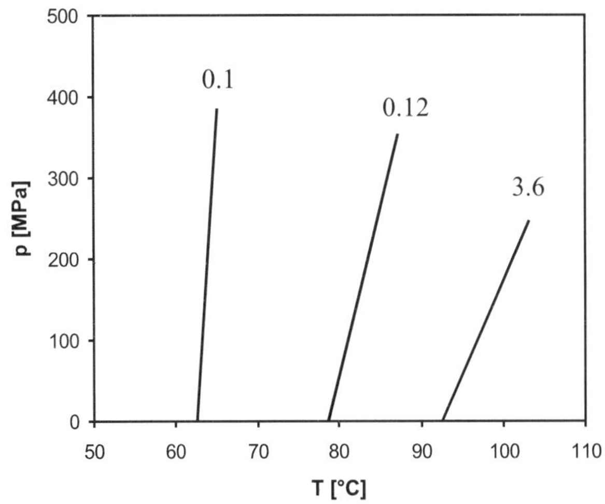

Fig. 10. Pressure–temperature diagram of DNA helix–coil transformation. Unlike the heat unfolding temperature of proteins, the melting temperature of DNA does not show any curvature, but a purely linear pressure dependence. The numbers refer to the molar concentration of neutral salts. Drawn after [84].

at ambient pressure varied after the addition of neutral salts. The volume change determined from the slope of the phase transition lines varied from almost zero to 8.6×10^{-3} ml/g depending on the salt concentration. No additional transitions were detected in spite of an extensive search in the temperature range from −22°C and 200 MPa to 58°C and 0.9 GPa. The conclusion from this study is that melting of DNA does not show an elliptic phase diagram. In the case of proteins the thermodynamic reason of the elliptic shape is that the second order terms in the series expansion of ΔG were important. Because of the absence of the re-entrant behavior in the case of DNA, the contribution of $\Delta\beta$, ΔC_p, and $\Delta\alpha$ should be different from that in the protein case.

The pressure–temperature stability of the double stranded nucleic acids was recently studied by Dubins et al. [85]. The stability diagram calculated by them is consistent with the earlier measurements in the low pressure range [84]. On the calculated diagram one expects a stabilizing effect of pressures below 200 MPa (melting temperature lowered by pressure) if the melting point was above 50°C (at atmospheric pressure), and a destabilizing effect if the melting point was below 50°C. Above 200 MPa this trend reversed, and at high pressure (~1 GPa) the melting temperature was 50°C, independent of the T_{melting} at atmospheric pressure. A low temperature single stranded phase was also obtained from the calculations.

A less drastic transition was reported at high pressures [86]. The right-handed B-form of DNA transforms into the left-handed Z-form at 6 kbar pressure. This conformational change has been interpreted as the effect of different hydration of various DNA forms [87].

4.2. Starch

Starch is an energy reserve of plants. It is organized in granules containing crystalline and amorphous parts. Gelatinization of starch is a process which occurs in daily life during cooking of foodstuff. Increasing temperature leads to swelling of the granules and at the gelatinization temperature the destabilized crystalline regions melt, leading to destruction of the granular structure [88]. Pressure studies on gelatinization temperature showed that the transition line is curved, i.e. after a slight stabilization up to 100–200 MPa, pressure decreases the gelatinization temperature [89,90]. Rubens et al. performed a systematic study on starches from six different sources [91–93]. In all the cases pressure induced gelatinization of starch was observed between 430 and 650 MPa depending on the source of the starch. A pressure–temperature phase diagram was constructed for rice starch. The diagram has a clear re-entrant character; however, the form of the curve is rectangular rather than typical elliptic (Fig. 11). Such a distortion of the elliptic curve can, however, be explained by the contribution of the higher order terms (see Section 3.1.1). The parameters of the thermodynamic analysis were also determined, by fitting the experimental points to Eq. 5.

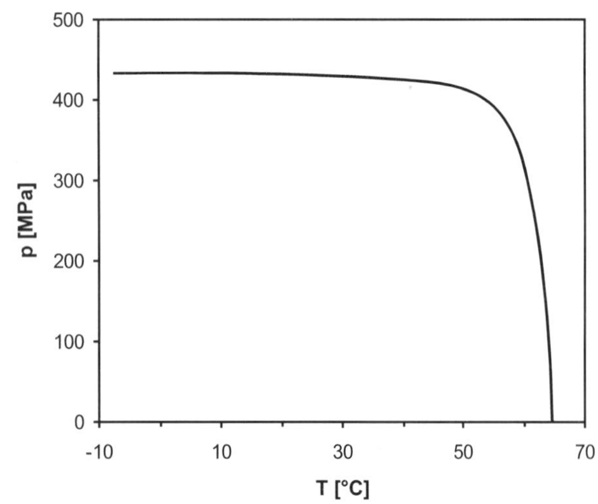

Fig. 11. p–T diagram of rice starch. (Drawn after [93].)

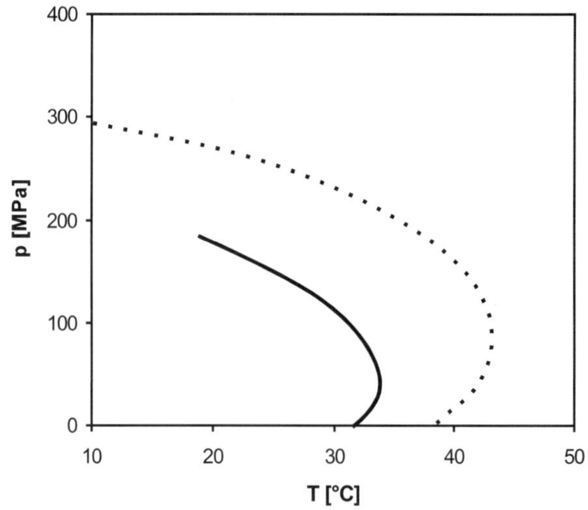

Fig. 12. Pressure–temperature diagram of poly(N-isopropylacrylamide) (solid line) and poly(N-vinylisobutyramide) (dotted line). Drawn after [94] and [96], respectively.

4.3. Synthetic polymers

A number of synthetic polymers have been investigated [94–96] where the phase separation phase diagram follows a similar elliptic shape as described by Hawley for proteins [23]. The native phase of proteins corresponds to the one-phase region of the polymer phase diagram, while the denatured phase corresponds to the phase separated polymer solution. Fig. 12 shows the pressure–temperature diagram of poly(N-isopropylacrylamide) [94] and poly(N-vinylisobutyramide) [96]. The similarity of these polymers to proteins is that they have both hydrogen bonding and hydrophobic interactions [96].

5. Multimolecular systems

5.1. Lipid membranes

Aqueous lipid suspensions are known to form bilayers spontaneously. These structures can be treated as simple models of biological membranes. Such lipid–water systems were intensively studied as a function of temperature and pressure. One-component

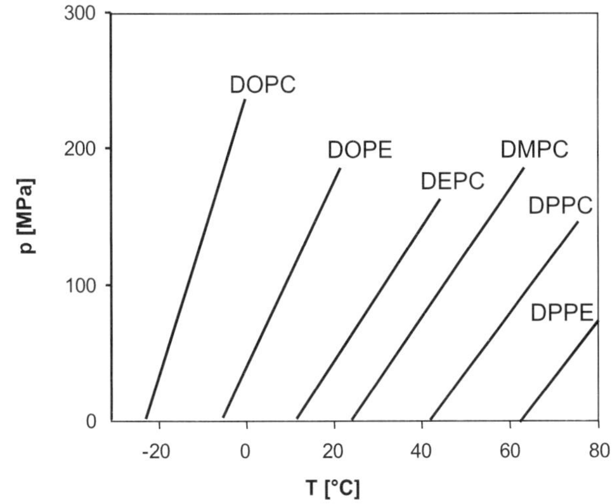

Fig. 13. Phase diagram of the gel–liquid crystal transition for some of the most frequent lipids. After [98].

lipid systems show a so-called main phase transition where the fatty acid chains melt, i.e. the number of gauche conformers increases drastically. It is also referred to as gel to liquid crystal phase transition. Above its temperature the membrane becomes more fluid-like. This transition can be observed in all main lipid types; however, the actual temperatures vary with the head group and the length of the fatty acid. A number of works have been performed on the pressure dependence of the phase transition temperature [97,98]. These can be briefly summarized as follows: the gel–fluid transition temperature increases with a rate of approx. 0.2°C/MPa [99]. Lipids with longer hydrophobic chains are more sensitive to pressure, i.e. the main transition temperature increases slightly more rapidly. Fig. 13 shows the phase transition line for some of the most frequent lipids. It is clear that the phase diagrams have no re-entrant character. This was proven by pressure experiments up to 5 GPa in the case of dimyristoylphosphatidylcholine (DMPC) [100]. Besides the main phase transition, some lipids (such as dipalmitoylphosphatidylcholine (DPPC)) show phase changes with smaller rearrangements. The pressure effect on these transitions was also studied by high pressure calorimetry, and the transition lines were almost linear, similar to the main transition [101].

Biological membranes are mixtures of lipids of several types. The first step from one-component lipid bilayers towards biomembranes is the two-component lipid membrane. These show phase diagram

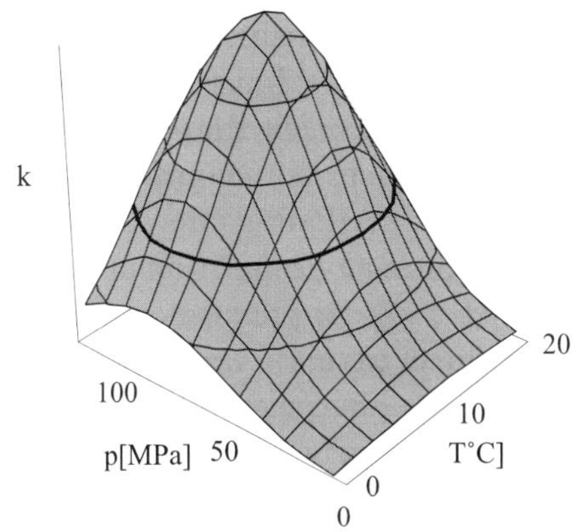

Fig. 14. Schematic *pTk* diagram of bacterial growth. (After [103].) The thick line shows the isokineticity curve, which is similar to the one of protein unfolding.

with various exotic phases, especially if the two lipids have very different character. Detailed analysis of these phases is beyond the scope of this review, and the reader is directed to review papers of this field [102].

5.2. *Microorganisms*

The growth rate constant of a bacterial strain plotted as a function of pressure and temperature gives a mountain-like surface with maximum growth rate [103]. As shown in Fig. 14, the constant growth rate curves are closed loops, similar to the isokineticity curves of proteins. If the pressure of the opti-

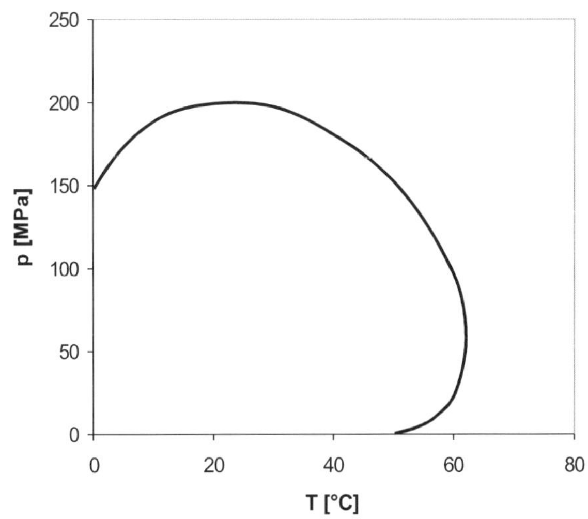

Fig. 15. The stability diagram of *E. coli*, showing a decrease of bacterial count by two orders of magnitude within 5 min, has a clearly elliptic character. (Drawn after [106].)

mal growth conditions is higher than the atmospheric one, the bacteria are called piezophiles. Such bacterial strains were found in the deep sea. On the adaptation mechanism there is a comprehensive review elsewhere in this issue [104].

Another interesting phenomenon is the inactivation of bacteria by high pressure and temperature, because of its direct practical application in sterilization of food and pharmaceutical products [105]. Pressure inactivation of several bacteria was studied by Ludwig et al. [106,107]. The stability diagram of *Escherichia coli*, showing a decrease in bacterial count by two orders of magnitude within 5 min, has a clearly elliptic character (Fig. 15). Since the elliptic diagram is characteristic for proteins, but not for membranes or nucleic acids, these authors conclude that the key role in bacterial inactivation is played by a certain protein (or proteins).

Inactivation of viruses due to pressure treatment was also observed [108]. Virus inactivation was also observed at low temperature, which resembles the cold denaturation of proteins [109,110]. This again underlines the role of the elliptic pressure–temperature phase diagram of proteins in a wide area of complex biological systems.

6. Conclusions

As shown in this paper, the pressure–temperature phase diagrams of different macromolecules are different. Elliptic and linear diagrams are both present among the main components of biosystems. It was also shown that combined pressure–temperature studies on simple systems can reveal the molecular basis of the pressure effect on much more complex biosystems. Practical applications, based on the knowledge of pressure–temperature phase diagrams of molecular processes reviewed here, are underway mainly in the food and pharmaceutical industry, as can be found in other reviews in this issue.

Acknowledgements

This work has been supported by Hungarian Project OTKA 032117.

References

[1] K. Heremans, L. Smeller, Protein structure and dynamics at high pressure, Biochim. Biophys. Acta 1368 (1998) 353–370.

[2] C. Balny, R. Hayashi, K. Heremans, P. Masson (Eds.), High Pressure and Biotechnology, John Libbey Eurotext, Montrouge, 1992.

[3] R. Hayashi, C. Balny (Eds.), High Pressure Bioscience and Biotechnology, Elsevier Science Publishers, Amsterdam, 1996.

[4] J.L. Markley, D.B. Northrop, C.A. Royer (Eds.), High Pressure Effects in Molecular Biophysics, Oxford University Press, New York, 1996.

[5] M. Gross, R. Jaenicke, Proteins under pressure, Eur. J. Biochem. 221 (1994) 617–630.

[6] K. Heremans (Ed.), High Pressure Research in Bioscience and Biotechnology, Leuven University Press, Louvain, 1997.

[7] H. Ludwig (Ed.), Advances in High Pressure Bioscience and Biotechnology, Springer, Heidelberg, 1999.

[8] Y. Taniguchi, H.E. Stanley, H. Ludwig (Eds.), Biological Systems under Extreme Conditions: Structure and Function, Springer, Heidelberg, 2001.

[9] G. Weber, H.G. Drickamer, The effect of high pressure upon proteins and other biomolecules, Q. Rev. Biophys. 16 (1983) 89–112.

[10] D. Eisenberg, W. Kauzmann, The Structure and Properties of Water, Oxford University Press, Oxford, 1969.

[11] J.C. Cheftel, J. Levy, E. Dumay, Pressure-assisted freezing and thawing: principles and potential applications, Food Rev. Int. 16 (2000) 453–483.

[12] S. Kunugi, H. Yamamoto, M. Makino, T. Tada, Y. Uehara-Kunugi, Pressure-assisted cold-denaturation of carboxypeptidase Y, Bull. Chem. Soc. Jpn. 72 (1999) 2803–2806.

[13] P.W. Bridgman, Water in the liquid and five solid forms under pressure, Proc. Am. Acad. Arts Sci. 47 (1911) 441–558.

[14] P.W. Bridgman, The phase diagram of water to 45000 kg/cm^2, J. Chem. Phys. 5 (1937) 964–966.

[15] D. Nash, J. Jonas, Structure of the pressure-assisted cold denatured state of ubiquitin, Biochem. Biophys. Res. Commun. 238 (1997) 289–291.

[16] D. Nash, J. Jonas, Structure of the pressure-assisted cold denatured lysozyme and comparison with lysozyme folding intermediates, Biochemistry 36 (1997) 14375–17383.

[17] F. Meersman, L. Smeller, K. Heremans, Pressure-assisted cold unfolding of proteins and its effects on the conformational stability compared to pressure and heat unfolding, High Press. Res. 19 (2000) 263–268.

[18] S. Kunugi, H. Yamamoto, M. Makino, T. Tada, Y. Uehara-Kunugi, Pressure-assisted cold-denaturation of carboxypeptidase Y, Bull. Chem. Soc. Jpn. 72 (1999) 2803–2806.

[19] K. Goossens, L. Smeller, J. Frank, K. Heremans, Pressure-tuning spectroscopy of bovine pancreatic trypsin inhibitor: a high pressure FT-IR study, Eur. J. Biochem. 236 (1996) 254–262.

[20] B. Wroblowski, J.F. Diaz, K. Heremans, Y. Engelborghs, Molecular mechanics of pressure-induced conformational changes in bovine pancreatic trypsin inhibitor, Proteins Struct. Funct. Genet. 25 (1996) 446–455.
[21] J.F. Brandts, Conformational transitions of proteins in water and in aqueous mixtures, in: S.N. Timasheff, G.D. Fasman (Eds.), Structure and Stability of Biological Macromolecules, Vol. 2, Marcel Dekker, New York, 1969, p. 213.
[22] P.L. Privalov, Y.V. Griko, V.P. Kutysenko, S.Y. Venyaminov, Cold denaturation of myoglobin, J. Mol. Biol. 190 (1986) 487–498.
[23] S.A. Hawley, Reversible pressure-temperature denaturation of chymotrypsinogen, Biochemistry 10 (1971) 2436–2442.
[24] J.F. Brandts, R.J. Oliveira, C. Westort, Thermodynamics of protein denaturation. Effect of pressure on the denaturation of ribonuclease A, Biochemistry 9 (1970) 1038–1047.
[25] N.A. Clark, Thermodynamics of the re-entrant nematic-bilayer smectic a transition, J. Phys. 40 (1979) 345–349.
[26] D.D. Klug, E. Whalley, Elliptic phase boundaries between smectic and nematic phases, J. Chem. Phys. 71 (1979) 1874–1877.
[27] L. Smeller, K. Heremans, Some thermodynamic and kinetic consequences of the phase diagram of protein denaturation, in: K. Heremans (Ed.), High Pressure Research in Bioscience and Biotechnology, Leuven University Press, Louvain, 1997, pp. 55–58.
[28] A. Yamaguchi, T.H. Yamada, K. Akasaka, Thermodynamics of unfolding of ribonuclease A under high pressure. A study by proton NMR, J. Mol. Biol. 250 (1995) 689–694.
[29] A. Zipp, W. Kauzmann, Pressure denaturation of metmyoglobin, Biochemistry 12 (1973) 4217–4228.
[30] O.B. Ptitsyn, V.N. Uversky, The molten globule is a third thermodynamical state of protein molecules, FEBS Lett. 341 (1994) 15–18.
[31] C. Redfield, R.A.G. Smith, C.M. Dobson, Structural characterisation of a highly ordered molten globule at low pH, Nat. Struct. Biol. 1 (1994) 23–39.
[32] C. Clery, F. Renault, P. Masson, Pressure-induced molten globule state of cholinesterase, FEBS Lett. 370 (1995) 212.
[33] O. Heinisch, E. Kowalski, K. Goossens, J. Frank, K. Heremans, H. Ludwig, B. Tauscher, Pressure effects on the stability of lipoxygenase: Fourier transform infrared spectroscopy and enzyme activity studies, Z. Lebensm. Unters. Forsch. 201 (1995) 562–565.
[34] K. Heremans, K. Goossens, L. Smeller, Pressure-tuning spectroscopy of proteins: Fourier transform infrared studies in the diamond anvil cell, in: J.L. Markley, D.B. Northrop, C.A. Royer (Eds.), High Pressure Effects in Molecular Biophysics, Oxford University Press, New York, 1996, pp. 44–61.
[35] L. Smeller, P. Rubens, K. Heremans, Pressure effect on the temperature induced unfolding and tendency to aggregate of myoglobin, Biochemistry 38 (1999) 3816–3820.
[36] G. Panick, R. Winter, Pressure-induced unfolding/refolding of ribonuclease A: static and kinetic Fourier transform infrared spectroscopy study, Biochemistry 39 (2000) 1862–1869.
[37] J. Torrent, J.P. Connelly, M.G. Coll, M. Ribo, R. Lange, M. Vilanova, Pressure versus heat-induced unfolding of ribonuclease A: the case of hydrophobic interactions within a chain-folding initiation site, Biochemistry 38 (1999) 15952–15961.
[38] J. Torrent, P. Rubens, M. Ribo, K. Heremans, M. Vilanova, Pressure versus temperature unfolding of ribonuclease A: an FTIR spectroscopic characterisation of 10 variants at the carboxy-terminal site, Protein Sci. 10 (2001) 725–734.
[39] G. Panick, G.J.A. Vidugiris, R. Malessa, G. Rapp, R. Winter, C. Royer, Exploring the temperature-pressure phase diagram of staphylococcal nuclease, Biochemistry 38 (1999) 4157–4164.
[40] M.C. Marden, G. Hui Bon Hoa, F. Stetzkowski-Marden, Heme proteins fluorescence versus pressure, Biophys. J. 49 (1986) 619–627.
[41] O. Roitel, N. Bec, R. Lange, C. Balny, G. Branlant, Pressure denaturation of phosphorylating glyceraldehyde-3-phosphate dehydrogenase from *Bacillus stearothermophilus*, Biochem. Biophys. Res. Commun. 283 (2001) 347–350.
[42] K.C. Ruan, C.H. Xu, Y. Yu, J. Li, R. Lange, N. Bec, C. Balny, Pressure-exploration of the 33-kDa protein from the spinach photosystem II particle, Eur. J. Biochem. 268 (2001) 2742–2750.
[43] C. Balny, V.V. Mozhaev, R. Lange, Hydrostatic pressure and proteins: basic concepts and new data, Comp. Biochem. Physiol. 116A (1997) 299–304.
[44] W. Dzwolak, M. Kato, Y. Taniguchi, Biochim. Biophys. Acta 1595 (2002) 131–144.
[45] J. Jonas, A. Jonas, High pressure NMR spectroscopy of proteins and membranes, Annu. Rev. Biophys. Biomol. Struct. 23 (1994) 287–318.
[46] D. Nash, J. Jonas, Hydrogen-exchange kinetics in the cold denatured state of ribonuclease A, Biochim. Biophys. Acta 1297 (1996) 40–48.
[47] H. Li, H. Yamada, K. Akasaka, Effect of pressure on individual hydrogen bonds in proteins. Basic pancreatic trypsin inhibitor, Biochemistry 37 (1998) 1167–1173.
[48] I.M. Klotz, Parallel change with temperature of water structure and protein behaviour, J. Phys. Chem. 103 (1999) 5910–5916.
[49] K. Heremans, F. Meersman, H. Pfeiffer, L. Smeller, Water mediated effects in high pressure processing, in: Z. Berk, R.B. Leslie, O.J. Lillford, S. Mizrahi (Eds.), Water Science for Food, Health, Agriculture and Environment, Technomic Publishing Co., Lancaster, 2001.
[50] G. Acampora, J. Hermans Jr., Reversible denaturation of sperm whale myoglobin I. Dependence on temperatures, pH and composition, J. Am. Chem. Soc. 89 (1967) 1543–1552.
[51] D. Puett, The equilibrium unfolding parameters of horse and sperm whale myoglobin, J. Biol. Chem. 245 (1973) 4623–4634.
[52] P. Degraeve, P. Delorme, P. Lemay, Pressure-induced inac-

tivation of *E. coli* β-galactosidase: influence of pH and temperature, Biochim. Biophys. Acta 1292 (1996) 61–68.

[53] P.L. Privalov, Cold denaturation of proteins, Crit. Rev. Biochem. Mol. Biol. 25 (1990) 281–305.

[54] P. Fusi, K. Goossens, R. Consonni, M. Grisa, P. Puricelli, G. Vecchio, M. Vanoni, L. Zetta, K. Heremans, P. Tortora, The extremely heat- and pressure-resistant 7 kDa protein P2 from the archaeon *Sulfolobus solfataricus* is dramatically destabilized by a single aminoacid substitution, Proteins Struct. Funct. Genet. 29 (1997) 381–390.

[55] E. Mombelli, M. Afshar, P. Fusi, M. Mariani, P. Tortora, J.P. Connelly, R. Lange, The role of phenylalanine 31 in maintaining the conformational stability of ribonuclease P2 from *Sulfolobus solfataricus* under extreme conditions of temperature and pressure, Biochemistry 36 (1997) 8733–8742.

[56] P. Rubens, Exploring the Pressure-temperature Phase Diagram of Biopolymers, Ph.D. Thesis, Louvain, 1999.

[57] M. Dumoulin, H. Ueno, R. Hayashi, C. Balny, Contribution of the carbohydrate moiety to conformational stability of the carboxypeptidase Y, Eur. J. Biochem. 262 (1999) 475–483.

[58] R.V. Rariy, N. Bec, J.-L. Saldana, S.N. Nametkin, V.V. Mozhaev, N.L. Klyachko, A.V. Levashol, C. Balny, High-pressure stabilization of α-chymotrypsin entrapped in reversed micelles of Aeroso OT in octane against thermal inactivation, FEBS Lett. 364 (1995) 98–100.

[59] L. Smeller, K. Goossens, K. Heremans, High pressure FTIR studies on hemoproteins, in: W.A. Trzeciakowski (Ed.), High Pressure Science and Technology, World Scientific Co., Singapore, 1996, pp. 863–865.

[60] W. Dzwolak, M. Kato, A. Shimizu, Y. Taniguchi, Fourier transform infrared spectroscopy study of the pressure-induced changes in the structure of the bovine α-lactalbumin: the stabilizing role of the calcium ion, Biochim. Biophys. Acta 1433 (1999) 45–55.

[61] J. Fidy, M. Laberge, A.D. Kaposi, J.M. Vanderkooi, Fluorescence line narrowing applied to the study of proteins, Biochim. Biophys. Acta 1386 (1998) 331–351.

[62] A.D. Kaposi, J. Fidy, E.S. Manas, J.M. Vanderkooi, W.W. Wright, Horseradish peroxidase monitored by infrared spectroscopy: effect of temperature, substrate and calcium, Biochim. Biophys. Acta 1435 (1999) 41–50.

[63] K. Suzuki, Studies on the kinetics of proteins denaturation under high pressure, Rev. Phys. Chem. Jpn. 29 (1960) 91–98.

[64] Y. Taniguchi, K. Suzuki, Pressure inactivation of α-chymotrypsin, J. Phys. Chem. 87 (1983) 5185–5193.

[65] C. Balny, P. Masson, F. Travers, Some recent aspects of the use of high-pressure for protein investigations in solution, High Press. Res. 2 (1989) 1–28.

[66] L. Ludikhuyze, Indrawati, I. Van den Broeck, C. Weemaes, M. Hendrickx, Effect of combined pressure and temperature on soybean lipoxygenase. I. Influence of extrinsic and intrinsic factors on isobaric isothermal inactivation kinetics, J. Agric. Food Chem. 46 (1998) 4074–4080.

[67] L. Ludikhuyze, W. Claeys, M. Hendrickx, Combined pressure temperature inactivation of alkaline phosphatase in bovine milk: a kinetic study, J. Food Sci. 65 (2000) 155–160.

[68] J. Zhang, X. Peng, A. Jonas, J. Jonas, NMR study of the cold, heat, and pressure unfolding of ribonuclease A, Biochemistry 34 (1995) 8361–8641.

[69] Yu.V. Griko, V.P. Kutyshenko, Differences of β-lactoglobulin cold and heat denaturations, Biophys. J. 67 (1994) 356–363.

[70] A.A. Ismail, H.H. Mantsch, P.T.T. Wong, Aggregation of chymotrypsinogen: portrait by infrared spectroscopy, Biochim. Biophys. Acta 1121 (1992) 183–188.

[71] J.L. Silva, G. Weber, Pressure stability of proteins, Annu. Rev. Phys. Chem. 44 (1993) 89–113.

[72] G. Weber, Protein Interactions, Chapman and Hall, New York, 1992, pp. 235–270.

[73] R.W. Carrell, B. Gooptu, Conformational changes and disease – serpins, prions and Alzheimer's, Curr. Opin. Struct. Biol. 8 (1998) 799–809.

[74] T.W. Randolph, M. Seefeldt, J.F. Carpenter, Biochim. Biophys. Acta 1595 (2002) 224–234.

[75] A.L. Fink, Protein aggregation: folding aggregates, inclusion bodies, and amyloid, Fold. Des. 3 (1998) R9–R23.

[76] D.R. Booth, M. Sunde, V. Bellotti, C.V. Robinson, W.L. Hutchinson, P.E. Fraser, P.N. Hawkins, C.M. Dobson, S.E. Radford, C.C.F. Blake, M.B. Pepys, Instability, unfolding and aggregation of human lysozyme variants underlying amyloid fibrillogenesis, Nature 385 (1997) 787–793.

[77] J. Zurdo, J.I. Guijarro, J.L. Jimenez, H.R. Saibil, C.M. Dobson, Dependence on solution conditions of aggregation and amyloid formation by an SH3 domain, J. Mol. Biol. 311 (2001) 325–340.

[78] S. Pin S, C.A. Royer, E. Gratton, B. Alpert, G. Weber, Subunit interactions in hemoglobin probed by fluorescence and high pressure techniques, Biochemistry 29 (1990) 9194–9202.

[79] J. Horwitz, Q.-L. Huang, L. Ding, M.P. Bova, Lens alpha crystallin has chaperone like properties, Methods Enzymol. 290 (1998) 365.

[80] H. Fabian, C. Schultz, D. Naumann, O. Landt, U. Hahn, W. Saenger, Secondary structure and temperature-induced unfolding and refolding of ribonuclease-T(1) in aqueous-solution – a Fourier-transform infrared spectroscopic study, J. Mol. Biol. 232 (1993) 967–981.

[81] I.E. Holzbaur, A.M. English, A.A. Ismail, FTIR study of thermal denaturation of horseradish and cytochrome *c* peroxidases in D_2O, Biochemistry 35 (1996) 5488–5494.

[82] J.P. Hendrick, F.-U. Hartl, Molecular chaperone functions of heat-shock proteins, Annu. Rev. Biochem. 62 (1993) 349–384.

[83] L. Smeller, F. Meersman, F. Tölgyesi, Cs. Böde, J. Fidy, K. Heremans, From aggregation to chaperoning: pressure effect on intermolecular interactions of proteins, High Press. Res. (2002) in press.

[84] S.A. Hawley, R.M. Macleod, Pressure-temperature stability of DNA in neutral salt solutions, Biopolymers 13 (1974) 1417–1426.

[85] D.N. Dubins, A. Lee, R.B. Macgregor Jr., T.V. Chalikian, On the stability of double stranded nucleic acids, J. Am. Chem. Soc. 123 (2001) 9254–9259.
[86] A. Krzyzaniak, P. Salanski, J. Jurzak, J. Barciszewski, B-Z DNA reversible conformation changes effected by high pressure, FEBS Lett. 279 (1991) 1–4.
[87] J. Barciszewski, J. Jurzak, S. Porowski, T. Specht, V.A. Erdmann, The role of water structure in conformational changes of nucleic acids in ambient and high pressure conditions, Eur. J. Biochem. 260 (1999) 293–307.
[88] R.E. Cameron, A.M. Donald, A small-angle X-ray scattering study of starch gelatinisation in excess and limiting water, J. Polym. Sci. 31 (1993) 1197–1203.
[89] J.M. Thevelein, J.A. Van Assche, K. Heremans, S.Y. Gerlsma, Gelatinization temperature of starch as influenced by high pressure, Carbohydr. Res. 93 (1981) 304–307.
[90] A.H. Muhr, R.E. Wetton, J.M.V. Blanshard, Effects of hydrostatic pressure on starch gelatinization, as determined by DTA, Carbohydr. Polym. 2 (1982) 91–102.
[91] P. Rubens, K. Goossens, K. Heremans, Pressure induced gelatinisation of different starch types: a FTIR study, in: K. Heremans (Ed.), High Pressure Research in Bioscience and Biotechnology, Leuven University Press, Louvain, 1997, pp. 55–58.
[92] P. Rubens, J. Snauwaert, K. Heremans, R. Stute, In-situ observations of pressure-induced gelation of starches studied with FTIR in the diamond anvil cell, Carbohydr. Polym. 39 (1999) 231–235.
[93] P. Rubens, J. Snauwaert, K. Heremans, Pressure-temperature gelatinization phase diagram of starch: an in situ Fourier transform infrared study, Biopolymers 54 (2000) 521–530.
[94] K. Otake, R. Karaki, T. Ebina, C. Yokohama, S. Takahashi, Pressure effects on the aggregation of poly(N-isopropylacrylamide) and poly(N-isopropylacrylamide-co-acrylic acid) in aqueous solutions, Macromolecules 26 (1993) 2194–2197.
[95] Y. Yamada, T. Tada, S. Kunugi, Effects of acrylic acid incorporation on the pressure-temperature behavior and the calorimetric properties of poly(N-isopropylacrylamide) in aqueous solutions, Colloid Polym. Sci. 278 (2000) 80–83.
[96] S. Kunugi, K. Takano, N. Tanaka, K. Suwa, M. Akashi, Effects of pressure on the behavior of the thermoresponsive polymer poly(N-vinylisobutyramide) (PNVIBA), Macromolecules 30 (1997) 4499–4501.
[97] P.T.T. Wong, H.H. Mantsch, Pressure effects on the infrared spectrum of 1,2-dipalmitoyl phosphatidylcholine bilayers in water, J. Chem. Phys. 83 (1985) 3268–3274.
[98] R. Winter, A. Landwehr, T. Brauns, J. Erbes, C. Czeslik, O. Reis, High pressure effects on the structure and phase behavior of model membrane systems, in: J.L. Markley, D.B. Northrop, C.A. Royer (Eds.), High Pressure Effects in Molecular Biophysics, Oxford University Press, New York, 1996, pp. 274–297.
[99] K. Heremans, High pressure effects on proteins and other biomolecules, Annu. Rev. Biophys. Bioeng. 11 (1982) 1–21.
[100] P.T.T. Wong, Correlation field splitting of chain vibrations: structure and dynamics in lipid bilayers and biomembranes, in: J.L. Markley, D.B. Northrop, C.A. Royer (Eds.), High Pressure Effects in Molecular Biophysics, Oxford University Press, New York, 1996, pp. 256–273.
[101] S. Utoh, T. Takemura, Phase transition of lipid multilamellar aqueous suspension under high pressure I investigation of phase diagram of dipalmitoyl phosphatidylcholine bimembrane by high pressure DTA and dilatometry, Jpn. J. Appl. Phys. 24 (1985) 356–360.
[102] R. Winter, Effects of hydrostatic pressure on lipid and surfactant phases, Curr. Opin. Colloid Interface Sci. 6 (2001) 303–312.
[103] A.A. Yayanos, The properties of deep-see piezophilic bacteria and their possible uses in biotechnology, in: H. Ludvig (Ed.), Advances in High Pressure Bioscience and Biotechnology, Springer, Heidelberg, 1999.
[104] D.H. Bartlett, Biochim. Biophys. Acta 1595 (2002) 367–381.
[105] N. Merkulow, R. Eicher, H. Ludwig, Pressure inactivation of fungal spores in aqueous model solutions and in real food systems, High Press. Res. 19 (2000) 643–652.
[106] H. Ludwig, W. Scigalla, B. Sojka, Pressure and temperature inactivation of microorganisms, in: J.L. Markley, D.B. Northrop, C.A. Royer (Eds.), High Pressure Effects in Molecular Biophysics, Oxford University Press, New York, 1996.
[107] C. Schreck, G. Layh-Schmitt, H. Ludwig, Inactivation of *Mycoplasma pneumoniae* by high hydrostatic pressure, Pharm. Ind. 61 (1999) 759–762.
[108] J.L. Silva, D. Foguel, A.T. Da Poian, P.E. Prevelige, The use of hydrostatic pressure as a tool to study viruses and other macromolecular assemblages, Curr. Opin. Struct. Biol. 6 (1996) 166–175.
[109] A.T. Da Poina, A.C. Oliveira, J.L. Silva, Cold denaturation of an icosahedral virus. The role of entropy in virus assembly, Biochemistry 34 (1995) 2672–2677.
[110] A.C. Oliveira, D. Ishimaru, R.B. Goncalves, T.J. Smith, P. Mason, D. Sa-Carvalho, J.L. Silva, Low temperature and pressure stability of picornaviruses: implications for virus uncoating, Biophys. J. 76 (1999) 1270–1279.
[111] N. Takeda, M. Kato, Y. Taniguchi, Pressure and thermally-induced reversible changes in the secondary structure of ribonuclease A studied by FT-IR spectroscopy, Biochemistry 34 (1995) 5980–5987.
[112] G. Panick, R. Malessa, R. Winter, Differences between the pressure- and temperature-induced denaturation and aggregation of β-lactoglobulin A, B, and AB monitored by FTIR spectroscopy and small-angle X-ray scattering, Biochemistry 38 (1999) 6512–6519.

ELSEVIER

Biochimica et Biophysica Acta 1595 (2002) 30–47

BIOCHIMICA ET BIOPHYSICA ACTA
BBA

www.bba-direct.com

Review

The effects of osmotic and hydrostatic pressures on macromolecular systems

Jack A. Kornblatt *, M. Judith Kornblatt

Enzyme Research Group, Department of Biology, Concordia University, 1455 de Maisonneuve, Montreal, QC, Canada H3G 1M8

Received 24 July 2001; received in revised form 17 October 2001; accepted 18 October 2001

Abstract

Osmotic pressure and hydrostatic pressure can be used effectively to probe the behavior of biologically important macromolecules and their complexes. Using the two techniques requires a theoretical framework as well as knowledge of the more common pitfalls. Both are discussed in this review in the context of several examples.

Keywords: Protein hydration; Hydrostatic pressure; Osmotic pressure

1. Introduction

It is somewhat ironic that two of the earlier quantitative studies on enzymes and proteins involved manipulation of water, although the importance of that factor was not recognized at the time. Michaelis and Menten [1] worked initially with the enzyme invertase. Bridgman [2] worked with ovalbumin. The former showed that the hydrolysis of sucrose could be accounted for by rate laws that ignored the concentration of water; at 54 'M' in their solutions, it was effectively constant. The latter showed that the application of high hydrostatic pressure (hp) would denature ovalbumin. Many years later, we accept that hp acts at the level of driving a system from a lower density to a higher density; Bridgman's phenomenon results from water at pressure forcing the hydration of a protein's buried surface. This exposure of new surface may lead to aggregation. We have, at the same time, been slow to recognize that if we change the activity of water, increase the osmotic pressure (op), such as Michaelis and Menten did, we can alter the way enzymes behave. Increasing op drives a system from the more hydrated state to the less hydrated according to the number of solute-inaccessible waters that connect the two states. The effects of hp and op on nucleic acids, proteins, enzymes and their assemblies are mainly mediated through the effects of water on those systems. Accordingly, this review deals with water effects on systems and what can be learned about them using hp and op as tools. The two pressures are two of the three parameters that can be varied in a thermody-

Abbreviations: op, osmotic pressure; hp, hydrostatic pressure; T state, the taut conformer of hemoglobin: it has a low affinity for O_2; R state, the relaxed conformer of hemoglobin: it has a high affinity for O_2; $\alpha_2\beta_2$, $\alpha\beta$, α, β, the hemoglobin tetramer, dimer and individual subunits; P450cam^{o,os}, substrate free and substrate bound forms of cytochrome P540; cytox, cytochrome *c* oxidase from ox heart; cyt *c*, cytochrome *c*; cyt, cytochrome; DPG, diphosphoglyceric acid; DOPC, dioleyl phosphatidylcholine

* Corresponding author. Fax: +1-514-848-2881.

E-mail address: krnbltt@vax2.concordia.ca (J.A. Kornblatt).

0167-4838/02/$ – see front matter
PII: S0167-4838(01)00333-8

namic treatment; we have now and then considered the third parameter, T, when it is logical to do so.

Some of the conclusions discussed here will sound quite familiar since they have been expressed in other reviews [3–8].

If water, studied by the application of op and hp, is the real subject of this review, what roles does it play in the lives of enzymes, other proteins and other macromolecular systems? The two most familiar roles are those of solvent and substrate/product. More than 99% of the known biological macromolecules function in an aqueous environment. More than 90% of the known enzymes and ribozymes have water as either a substrate or as a product.

Other roles for water include the following. It will fill voids when it is energetically favorable to do so [9]. It stabilizes conformations and facilitates the interconversions between conformations. It facilitates binding between distant groups thereby functioning as a molecular bridge. Water modulates midpoint potentials, ionization potentials, pKs, and binding coefficients. It acts as a molecular wire to transmit protons from one site to another. Lastly, water can act as a bulk carrier for protons, other ions, uncharged molecules, etc.

These points are important in that the application of our two pressures is going to influence all of them. Even when we are careful, the fact that they are sensitive to changing water activities will confuse the interpretation of results.

For the analysis of how hp and op can give us information about the effects of water on our macromolecular systems we will try to restrict ourselves to 'simple' but real cases; complex cases, enzyme aging or whole cell studies have been recently reviewed [10,11]. The simple cases fall into two categories.

1. Equilibria between two states of a macromolecule in solution in which temperature, hp or op is varied. This constitutes the vast majority of cases. We will also consider equilibria in which hp and temperature, or hp and op, or temperature and op are varied two at a time. This approach has been profitably used by several groups; the original data should be consulted [12–15]. In principle, all three parameters could be varied. There are no data on biological systems here but there are indications as to how they might behave based on non-biological systems [16].
2. Kinetics in which hp or op changes the overall rate by changing the rates of specific steps in the reaction.

We start by considering two relatively simple and well defined cases – the phospholipid bilayer and hemoglobin.

1.1. The influence of temperature, hp and op on the behavior of multilamellar phospholipid bilayers

These are multiple bilayers stacked parallel to each other and separated by layers of water, in much the way that the myelin sheath is composed.

There is a wealth of information on the effects of temperature on the liquid-crystalline, gel transition of phospholipid bilayers [17]. Increasing temperature causes the bilayer lipids to go from the ordered gel state with the hydrocarbon chains crystallized, to the less ordered liquid crystalline state as the hydrocarbon chains disorder and melt. From the structural point of view this means that as temperature increases the fatty acid chains become disordered and are able to sweep out a volume that is larger than the volume swept out in the gel phase. The phosphoderivative head groups have greater mobility in the latter state. With increasing T, the hydrocarbon chains of bilayers melt at T_c, and thin as T increases further.

The inter bilayer or polar group layer between hydrocarbon regions of opposing bilayers can be hydrated to a limited extent if the lipids have no net electric charge. There are about 32 waters per DOPC head group in DOPC membranes at saturation [18–20]. Using the op stress approach, one can reduce the activity of water under tightly controlled conditions. With increasing op, water is removed osmotically, forcing/pushing bilayers together. This results in a strong mutual repulsion. The bilayers thicken, molecular area decreases and the dipoles (of the head groups) become more perpendicular to the plane of the bilayer. T_c increases.

Others have found that the application of hp to phosphatidyl choline multilayer membranes causes (1) a linear increase in T_c along with an interdigitation of acyl chains from apposing leaflets [21], (2) a tilting of the head groups [22] and (3) increases in the bilayer thickness and separation of the liquid crystal to gel transition from the gel to ordered crystalline phase transition [23,24].

In other words, op can cause membrane thickening in some species of phospholipids while hp can cause a *similar* net result in other phospholipid species. If we belabor this example of phospholipids and bilayers, it is because one starts with the preconceived idea that bilayers should be relatively easy to analyze. One also often starts with the preconceived idea that op and hp should have opposite effects on the system. The above points out one significant fact: temperature, op and hp perturbations yield unexpected results because they do not interrogate the same end states. Phospholipid bilayers can exist in more than two states; one is not dealing with a simple two-state equilibrium.

1.2. The influence of temperature, hp and op on the oxygen binding capacity of hemoglobin

The effects of temperature on the equilibrium shown below were studied well before the era of X-ray crystallography and the Monod–Wyman–Changeux model of cooperative interactions [25].

$$\text{(T state) Hb} + 4\text{O}_2 \rightleftarrows [\text{Hb } 4\text{O}_2] \text{ (R state)}$$

Temperature increases influence this reaction by lowering the affinity of Hb for oxygen. The p_{50}, the partial pressure of oxygen required to bring about 50% saturation of Hb, increases as temperature increases. ΔG° changes because reactions are governed either by an enthalpy term, ΔH°, or by an entropy term, $T\Delta S^{\circ}$. Perutz [26] pointed out, based on his crystal structure, that T state hemoglobin contained many salt bridges that were not present in the R state protein. These would have to be broken in order for the T to R transition to take place. Breaking them required heat; the T state of Hb is stabilized by an enthalpy term. hp should also act to break these salt bridges, thereby favoring the T to R transition, whereas op should act to strengthen their interactions. Furthermore, to go from the T to R state requires deprotonating Hb at some of those same residues as well as releasing bound Cl^-. Protons and the organophosphates stabilize the fully deoxygenated T state of Hb. A thorough analysis of the energetics of hemoglobin binding oxygen showed that the above reaction was really a simplification of reality [27]. The binding reaction was better described by the equation:

$$\text{(T state) } [\text{Hb}, \text{Cl}^-, \text{DPG}^{2-}, \text{H}^+] + 4\text{O}_2$$

$$\rightleftarrows [\text{Hb}4\text{O}_2] + \text{H}^+, \text{DPG}^{-2} + \text{Cl}^- \text{ (R state)}$$

in which oxygen bound to hemoglobin in four discrete steps. ΔH and $T\Delta S$ for the individual binding reactions were determined under several conditions [27]. Imai showed several salient features of the reaction. ΔH and $T\Delta S$ *appeared* to vary widely for the individual steps in oxygen binding; once the contributions for ion binding and protonation were subtracted, ΔH and $T\Delta S$ for the actual binding of oxygen were more or less the same for the individual steps. The variation in the two terms resulted and results from the binding of other substances during the binding of oxygen and the T to R transition that accompanies the oxygenation.

With so many different things binding or released, with the known large conformational change that takes place, it should go without saying that hp should have major effects on the oxygen binding curve of hemoglobin and the associated T to R conformational change. The effects of hp on the above equilibrium can be easily summed up. When the various contributions to the equilibrium are examined there is general agreement that hp does not influence the T to R transition [28–30]. hp has either a small or no discernible influence on the equilibrium [31,32]. The volume change for the system must be close to zero. There is a small influence of pressure on the spin state of the hemes but this does not appear until pressures above 400 MPa [28,33]. One place where pressure does have an influence is on the overall equilibrium $\alpha_2\beta_2 \rightleftarrows \alpha\beta \rightleftarrows \alpha+\beta$ [34]. hp drives the equilibrium in the direction of free monomers. Accordingly, if there is a significant concentration of dimer in a sample, hp will displace the overall equilibrium away from the native tetramer.

The hemoglobin story has always been interesting but it became more so when the three dimensional structures of both R and T state hemoglobins were established. The two structures clearly demonstrated the conformational differences between the two states but they also showed that the oxygenated R state contained about 60 waters 'tightly' associated with it. This motivated Colombo et al. [35–37] to ask whether the excess waters were a necessary part of the structure of oxygenated hemoglobin. The waters turned out to be an obligatory component of oxygenated hemoglobin. The approach used relied on lowering the activity of water with cosolvents; the log of the binding constant, p_{50}, was then shown to vary as the op and as $[Cl^{-1}]$. Applying the Wyman linkage relationships [38], the authors showed that concomitant with the binding of four oxygens, approximately 60 waters were also bound as the hemoglobin molecule went from T state to R state. In the absence of anions, only 25 additional waters bound to hemoglobin as it bound oxygen [39]. From hindsight, this means that the equilibrium written above should really be written

$$\text{(T state) } [Hb.Cl^{-1}H^{+}DPG^{2-}] + 4O_2 + 60H_2O \rightleftarrows$$

$$[Hb4O_2 60H_2O] + H^{+} + DPG^{-2} + Cl^{-1} \text{ (R state)}$$

or

$$\text{(T state) } [Hb, H^{+}, DPG^{2-}] + 4O_2 + 25H_2O \rightleftarrows$$

$$[Hb4O_2 25H_2O] + H^{+} + DPG^{-2}] \text{ (R state)}$$

The Colombo experiments were performed at concentrations of Hb between 15 μM (4 heme tetramer) and 45 μM, concentrations where significant fractions of dimer were, in the past, thought to have been present [27]. More recent data [40,41] indicate that the K_d for the transition varies between 0.2 μM and 1 μM depending on [salt], pH, and temperature; the fraction dimer present is small but significant at the lower concentrations. In the unlikely case that the dimer/tetramer equilibrium is contributing to the change in water binding that accompanies oxygen binding and the T to R transition, the analysis would become quite complicated.

The differences between the four above equations point out another unexpected feature. Release of chloride from bound Hb should result in water binding to the chloride. Electrostriction of the chloride does not result in the disappearance of 35 waters. This means that there must be another form of hemoglobin; the above authors refer to this as P state hemoglobin. We have referred to the case of hemoglobin binding oxygen and water as elegant; it also points out the same trap as the phospholipid bilayers pointed out. Increasing hp and op experiments drive systems in the direction of increased density and decreased water associations. If our probes are not specific for our end points, it is possible to interrogate different end points without being aware of doing so. In such cases, the thermodynamic conclusions are not valid. If biochemical conclusions or physiological conclusions are based on invalid thermodynamic conclusions, they too cannot be valid.

What do we vary in hp and op experiments? In an hp experiment we measure changes in equilibria (or rates) as we vary hp. We analyze those changes in terms of changes in volume or changes in density between two states of a system. In an hp experiment we equate those two states of the system with two states of a macromolecule or macromolecular interaction. This is convenient but it is not strictly true. The change in the macromolecule is only a portion of the overall change exhibited by the system. In an op experiment we measure changes in equilibria (or rates) as we vary water activity (op). We analyze those changes in terms of the number of solute-inaccessible waters that must change to go between two states of a system. In this instance, the equations that relate the chemical potential of water to the chemical potential of the macromolecule are exact and require no assumptions other than *everything else is kept constant.* If this were true in hp experiments and op experiments, there would be few problems with the interpretation of results. However, most often, the data from either type of experiment are open to other interpretations. The most important things to keep in mind are the following.

hp acts at the level of density. Atoms are not compressible. Voids can be compressed. Solvent water contains 'void' space compared to the space the actual atoms occupy but protein interiors also contain void space [3]. Solvent can solvate a macromolecular surface where it may occupy less volume (where its density may be increased) but where it may also occupy more volume [42]. Where the void space of a protein interior can be reversibly compressed without disrupting the total energetics of folded state, hp will compress those spaces. op acts at the level of number of waters. Comparing two solute-inaccessible compartments, it is unimportant whether the waters occupy more or less space in the comparison. What is important is that the numbers of waters in the two are different.

This means that predicting whether hp will force a system into one particular direction is not possible without prior knowledge of the densities of the components.

It also means that predicting whether op will force a system into one particular direction is not possible without prior knowledge of the required hydration states of the components.

Crystal structures and NMR structures may be useful in rationalizing the experimental results of the two kinds of pressure but are only useful for prediction 'after the fact'.

What things interfere with our interpretation of pressure data? What kinds of things can be varied in a biochemical experiment and what other things do we vary when we change them? The traditional enzymologist varies [x], pH, T, I and measures a response. In cryoenzymology, for instance, one lowers T while trying to keep other things 'under control'. We include the quotation marks because you cannot change T and not change a large number of other factors at the same time. They are linked. Changing T for instance changes the dielectric coefficient of the solution. If we try to correct for this (using graded concentrations of zwitterions for example) we then have other additions that must be corrected for. Is it important to keep the dielectric coefficient constant? The answer is: 'It all depends' [43]. If one is measuring the equilibrium between two states, denatured and native, the stability of the two states is a function of the sum total of the forces acting on the molecule. If there are non-neutralized charges in the interior of a macromolecule, those charges act to destabilize it. The destabilization energy scales as

$$E_s = q^2/2Dr_s \qquad (1)$$

where q is the charge, D the dielectric coefficient in the vicinity of the charge, and r_s the radius over which the charge is distributed. Increasing T causes D of the solvent to decrease and D of the protein interior to decrease. The net change in stability is, in part, a reflection of the difference in D internal vs. external.

What else changes? The addition of high salt to a solution exerts an op but it also disorders solvent such that the solvent looks as though it is under high hp [44]. From the point of view of a macromolecule in solution, this means that the addition of salt increases the op of the solution – it decreases the chemical potential of water – but it also provides a milieu in which the remaining water looks like water at hp. This means that bifurcated hydrogen bonds are introduced where they did not previously exist.

We have spent a substantial amount of effort to persuade readers that op and hp experiments must be planned and interpreted with caution. Moreover, the cautions that we suggest should be viewed in addition to, rather than instead of, those of other authors [6–8,45]. This does not mean that hp and op experiments should not be done.

What tools does one need to plan and execute an experiment using op as a perturbant? One needs four equations and a practical framework.

The four equations are:

$$d(RT\ln K_M) = (\Delta N V_0)d(op) \qquad (2)$$

where K_M is the equilibrium coefficient for the two states of the macromolecule, ΔN is the number of solute-inaccessible waters associated with one of the states that are absent in the other state, V_0 is the partial molar volume of water and op is determined from

$$op = (RT/V_0)\ln X_{H_2O} \qquad (3)$$

X, the mole fraction, is most easily obtained from either the Handbook of Chemistry and Physics or from one of two WEB sites (http://aqueous.labs.brocku.ca/osfile.html; http://dir.nichd.nih.gov/Lpsb/docs/osmdata/osmdata.html).

The left side of Eq. 3 is the familiar expression for the free energy of a system, in this case, the chemical potential. Inspection of Eq. 3 indicates that the right side of Eq. 3 is also an energy term. We plot $RT\ln K_M$ vs op. The slope of the resulting straight line is equal to the difference in the number of solute inaccessible waters associated with the two forms of the macromolecule.

To carry out an op vs temperature experiment requires knowing that ΔC_p, the heat capacity of the protein, does not vary as a function of the op perturbant. The basic experiment is to measure K_M as a function of op at a fixed temperature and to extrapolate the data to find the op at which $\Delta G = 0$. This experiment is then repeated at a number of different temperatures. One then plots T vs. op at $\Delta G = 0$ to yield a curve in much the same way that one handles hp vs. temperature data [12]. For this curve,

$$\partial T/\partial(\mathrm{op}) = (\Delta N V_0)(\Delta S^{\circ} + \Delta C_p((T - T_o)/T_o))^{-1} \quad (4)$$

To carry out an op vs. hp experiment requires knowing that compressibility does not vary as a function of the osmotic perturbant. The basic design is the same except that we vary op at constant hp. In this case we plot hp and op at $\Delta G = 0$ to yield the curve. The equation is similar to Eq. 4:

$$\partial(\mathrm{hp})/\partial(\mathrm{op}) = (\Delta N \mathrm{V}_0)(\Delta \mathrm{V}^{\circ} + \Delta\beta((\mathrm{hp} - \mathrm{hp}_o)/\mathrm{hp}_o))^{-1} \quad (5)$$

In an op vs. temperature experiment or an op vs. hp experiment, the instantaneous slope (Eqs. 4 and 5) gives information on ΔN, ΔS°, and ΔC_p or on ΔN, ΔV° and $\Delta\beta$. The complete derivations of Eqs. 2–5 are given in the Appendix.

The above analyses are what we, as physical biochemists, should be trying to achieve. As will be evident in the remaining sections of this article, in practice, hp and op have not been used to their full potential.

1.3. Practice

Traps to avoid.

Eqs. 2–5 are totally general and rigorous. Using them to carry out an analysis requires some caution.

All calculations must be based on activities and only activities. The use of molar or molal concentrations leads to false conclusions except in the case of very dilute solutions. Since we work at high op, we are not working in the very dilute regime.

Calculating the chemical potential of the macromolecule usually requires that you know that there are only two end states in question. It requires that the population of intermediates between those states is small.

Finally it requires that the materials used to alter the op are acting only at the level of the activity of water and not by some direct interaction between macromolecule and perturbant. Op effects should be invoked because most other considerations do not explain a set of data.

The reader is directed to [7,8] in order to obtain a full appreciation for the problems that arise during analysis and the level of friction those problems can generate.

1.4. The application of op to perturb equilibria

The past 10 years has seen a large number of papers in which equilibria have been studied as a function of cosolvent activity. We have chosen a very few selected case studies based on our familiarity with those; we apologize to the authors of other studies whose work has not been included here. The omission carries no judgement with it.

2. The case of cytochrome P450cam

Cytochrome P450cam is a heme containing enzyme from *Pseudomonas putida* which catalyzes the hydroxylation of camphor in an ordered reaction as shown in Fig. 1. The enzyme has been studied since the late 1950s [46–49].

The individual steps leading to the conversion of camphor to 5-exo-hydroxycamphor are shown in Fig. 1. There is little formation of the dead-end product cytochrome P420 during steady state turnover. There is little electron transfer from reduced putidaredoxin (Pd^r) in the absence of bound camphor which means that there is little accumulation of activated oxygen products during steady state turnover. P450cam is a well designed enzyme.

From the point of view of op studies, the following points should be kept in mind.

(i) The crystal structures of $P450^o$ and $P450^{os}$ were established many years ago and are shown in Fig. 2 [50,51]. The pocket of $P450^o$ is large, consists of a hydrophobic surface and is filled with about six water molecules; the sixth ligand to the oxidized iron is probably a water molecule. The substrate pocket of $P450^{os}$ is filled with a well ordered camphor molecule; it displaces the waters including that which is ligated to the iron in $P450^o$. The entire cat-

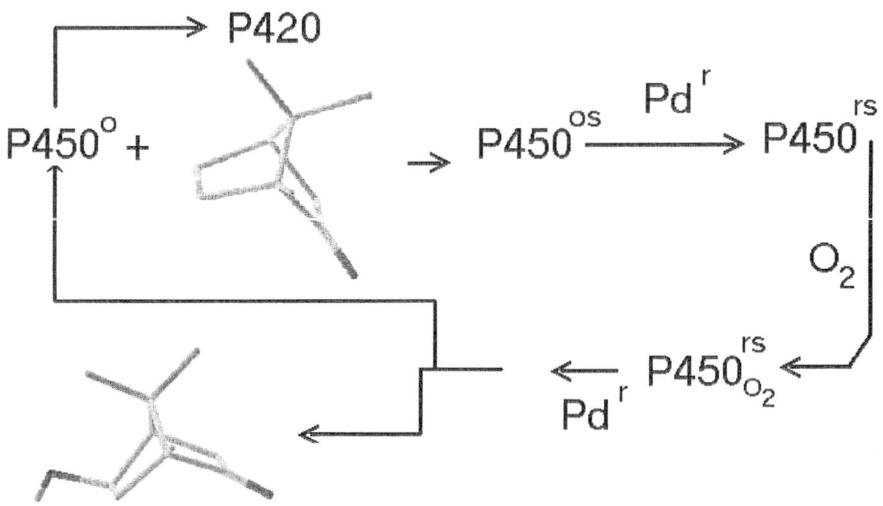

Fig. 1. Cytochrome P450 converts camphor to 5-exo-hydroxycamphor. The enzyme, cytochrome P450, is represented as P450. P420 is a non-native form which is enzymatically inactive. The enzyme form with camphor bound is $P450^{os}$, in which 'o' means oxidized and 's' means that camphor is bound. Putidaredoxin reduces the enzyme to $P450^{rs}$ in which 'r' means reduced. O_2 binds and forms $P450^{os}_{O_2}$. The final electron transfer step comes when the second molecule of putidaredoxin binds.

alytic pathway at atomic resolution has been recently elucidated and summarized [52].

(ii) $P450^{o}$ contains low-spin iron whereas it is high-spin in $P450^{os}$. The iron of P450 undergoes a spin transition [53,54] which is associated with substrate binding [55–57]. The spin transition is susceptible to hp [58,59]. Starting with $P450^{os}$, hp drives the complex low-spin and drives the camphor from and water into the pocket [60]. In addition to hp, KCl, temperature and cosolvents also influence the equilibrium between the two spin states [49].

(iii) Qualitatively, the effects of *T*, hp and op on the spin equilibrium of P450 are consistent. Higher temperatures promote the high spin, camphor containing form. hp promotes expulsion of the camphor from the pocket, op promotes its retention. Quantitatively the situation is somewhat more complex. The volume change for hp expulsion of camphor from the P450 pocket is about 30 ml/mol [61]. This number is (1) consistent with 6–7 water molecules if one assumes that the density of water in the P450 pocket is about 25% greater than that of bulk water [62] and (2) satisfying when one considers that there are about six waters in the pocket of $P450^{o}$. However, this number increases to 17 waters if one assumes only a 10% difference in water density [63]. A second problem arises when one considers the op data. Di Primo et al. showed that the volume change for the spin transition is about 350 ml/mol [61]. In an op experiment volume and number are related by the partial molar volume of water which means that this corresponds to about 20 waters. Clearly, the waters probed by op include too many to be associated with just the camphor pocket. *Is this really surprising*? Of course not! We perturb the equilibrium with cosolvents or other compounds that change the activity of water; in the best of cases, such as the Di Primo studies, it is reasonably clear that the osmotic perturbants are only influencing water activity and are not binding to the protein. Nonetheless, in this system there are waters associated with the P450 equilibrium other than just those in the heme/camphor pocket. Amongst those are the waters that hydrate the released camphor. Until such time as we have a complete picture of the hydration states of all the components of the system, we prefer to take a cautious attitude. All the data point to about 20 osmotically sensitive waters being involved in the spin equilibrium. Thermodynamics cannot tell us where these waters are.

(iv) The spin equilibrium is not the only aspect of the P450 cycle (Fig. 1) that is sensitive to op. The binding of putidaredoxin, the reducing substrate, to the P450 is also perturbed by op [64,65]. The effects of temperature on this system were the first indication that waters might mediate the binding. When two solvated molecules bind with a resulting dehydration of their surfaces, there is a large gain in en-

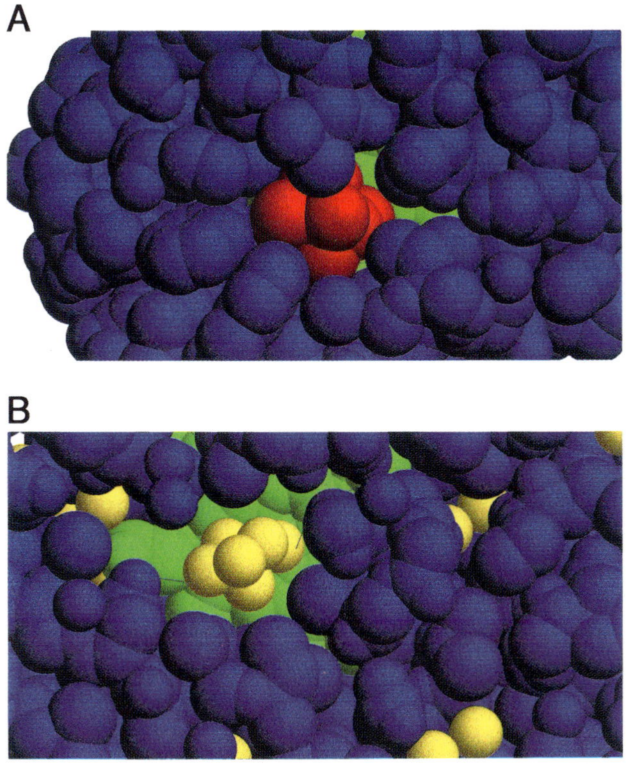

Fig. 2. The crystal structure of cytochrome P450. (A) $P450^{os}$ is shown looking into the camphor pocket at the camphor (PDB2CPP). The amino acids are blue, heme is green and the camphor is shown in red. The camphor does not quite fill the pocket. (B) $P450^{o}$ is shown looking into the same camphor pocket; in this case camphor is not present but has been replaced by about six, well ordered water molecules, shown in yellow ((PDB1PHC). Note that this view is similar to but not exactly the same as that in A.

tropy by the system. Water at the macromolecular surface is more ordered than bulk and its release into the bulk phase must be accompanied by a large gain in entropy. The converse is true. If there is a large decrease in entropy on binding of two macromolecules, there must be net amounts of solvent taken up from the bulk. In the case of some ferridoxin type associations, the reactions are entropy driven [66]. In the case of P450 and putidaredoxin, the binding is enthalpy driven [67]. When the complex was subjected to osmotic stress, it was found that binding was weakened as though there were waters sequestered at the interface which were necessary for binding [64,65]. The crystal structure of a related complex, the P450 BM-3/P450 reductase, shows waters sequestered at the interface [68]. The numbers of waters associated with the P450cam/putidaredoxin case are large: 25 waters in the physiologically relevant case between the oxidized P450 and the reduced putidaredoxin; 13 waters between the non-physiologically relevant partners, oxidized P450 and the oxidized putidaredoxin [65].

(v) To summarize what we know about P450cam: Temperature studies indicated that there was an hydration difference between the camphor containing protein and the ligand free species. This was reinforced by hp studies which showed that expulsion of camphor from the P450 pocket was probably accompanied by water entry. The crystal structures of the ligated and unligated forms substantiated this deduction. The waters associated with the P450 pocket can be perturbed osmotically but additional waters associated with the system are also perturbed. The interfacial area between putidaredoxin and P450 probably contains water. Lastly, the numbers of waters in the physiologically relevant and irrelevant complexes differ. We have mastered the art of applying the three thermodynamic perturbants to the cytochrome P450cam; in retrospect we can rationalize their significance even if we have difficulty with absolute numbers.

3. The case of cytochrome *c* oxidase

Cytox (cytochrome *c* oxidase) is the terminal electron transfer component of the mitochondrial electron transport chain. It is an integral membrane protein that can be solubilized with detergents. It consists of four redox centers: two hemes as cytochromes *a* and a_3, CuA and CuB. It has been crystallized and the crystal structure determined [69,70]. The reactions catalyzed by the protein are shown below. Reduced cyt *c* (cytochrome *c*) must bind transiently to cytox, it must transfer an electron and then dissociate from the complex; this cycle must be repeated four times in order for oxygen to be reduced. The binding and electron transfer reactions occur on the cytosolic side of the mitochondrion; oxygen reduction occurs towards the inner space of the organelle. Four protons are consumed inside the mitochondrion to form water; at the same time, four protons are transferred against the electrochemical gradient from inside to outside the mitochondrion.

$$4\ \text{cyt c(II)} + O_2 + 4H^+ + 4\ H^+ \rightleftarrows 4\ \text{cyt c(III)} + 2H_2O + 4\ H^+$$

The protein has been extensively studied since the 1950s and the studies reviewed on a regular basis. A reasonably current view of the inner workings of electron transfer, proton transfer and oxygen reduction can be found in two recent references [71,72]. There have been innumerable temperature studies on the oxidase. Most have been carried out at low temperature in order to establish the electron transfer path from reduced cytox to oxygen. None to the best of our knowledge allows for the decoupling of individual electron transfer steps, proton transfer steps or cyt *c* binding and release.

Cytox has been the subject of several pressure studies. The fully oxidized protein shows a small spin change when subjected to hp; the volume change is similar to those found in other heme containing proteins [73]. Similarly, the totally reduced protein is impervious to hp [74]. During catalytic turnover, the situation changes and the protein becomes highly sensitive to pressure. A block in electron transfer is introduced between cytochrome *a* and cytochrome a_3 for which the volume change is about 80 ml/mol [74]. This was postulated to be too large a volume change to be associated with anything other than solvent movements into and out of the cytox during turnover. Osmotic perturbation of cytox showed that the two static forms of the protein were relatively impervious to op but that the enzyme during turnover was quite sensitive to inhibition by op. The response to op was consistent with a catalytic cycle in which about 10 waters must enter and leave the molecule with each turnover [75]. Other techniques have indicated that the actual number of waters that must enter and leave during catalysis range from about 4 to 10 [76–78]. At first glance, it is surprising that op and hp both inhibit the activity of the enzyme. However, an enzyme during turnover is quite different from a system in equilibrium. If, during turnover, there is a step which involves an increase in the number of water molecules associated with the protein, then there must be a later step in which those water molecules are lost from the protein. Thus the catalytic cycle would include steps that would be inhibited by op (and activated by hp) and others which exhibit the opposite pattern.

It is not surprising that the degree of solvation should change as macromolecules change conformations and oxidation states. What is surprising is that these changes should be obligatory in a macromolecule whose only roles are electron transfer and proton pumping. What is gratifying is that a water cycle should be predicted on the basis of osmotic perturbation studies and that 6 years later the X-ray crystal structure [69,70] should reveal a cluster of waters in the vestibule of the heme pocket of this protein. This cluster is perfectly positioned in the region where proton transfer is postulated to occur. We envisage a proton pump in which the hydrated proton is transferred across the molecule rather than a proton wire in which there is *transient* dehydration of each proton transferred. What is significant about the work on cytox is that osmotic perturbation predicted water involvement in electron and proton transfer and that this idea has subsequently been supported by several independent lines of evidence.

The catalytic cycle of cytox relies on binding and release of cytochrome *c*. The equilibrium binding has been subjected to the same sort of scrutiny as the catalytic cycle. Early studies from the Margoliash laboratory [79] established quite clearly that the binding of cyt *c* to cytox involved a ring of positive charge on the cyt *c* molecule and that these charges interacted with negative charges on cytox. The ionic strength dependence of interaction indicated that about four charges on each protein should take part in binding. This should have meant that there would be a strong force favoring electrostriction of the individual charges when the complex was subjected to hp. The experimental results were somewhat surprising in that hp had little effect on the interaction between cyt *c* and cytox, nor between cyt *c* and cytochrome *c* peroxidase. There was a substantial influence on the interaction between cyt *c* and cyt b_5 [80–83] which had a volume change of about 50 ml/mol. The binding sites for cyt *c* on all three acceptor proteins were thought to be similar but hp results indicated that the b_5 site differed from the other two. Modeling had shown that the b_5, cyt *c* complex formed a tight fit with little space between the molecules. Osmotic perturbation of the cyt *c*/cyt ox and cyt *c*/cyt b_5 complex showed entirely different results [84,85]. In the cyt *c*/cyt ox couple, op caused the interaction between the two proteins to *weaken*. In the cyt *c*/cyt b_5 couple op caused the interaction between the two proteins to *strengthen*. This indicated that the interface between cyt *c*/cyt ox contained sequestered solvent and that forces that inhibited the sequestration of the solvent would simultaneously inhibit binding. The data indicated that there was no such sequestration of solvent in the cyt *c*/b_5 interface. When this work was published, 1993, the crystal structure of the cytochrome *c* peroxidase/cyt *c* complex had just been published [86] but details of the interaction site were not available. Since then the coordinates have been deposited and the interaction site shows that the contacts between the two partners are made, with few exceptions, through water bridges (Fig. 3). Once again the utility of the *combined hp, op approach* was to reveal details of the interaction site, details that were later confirmed by X-ray crystallography.

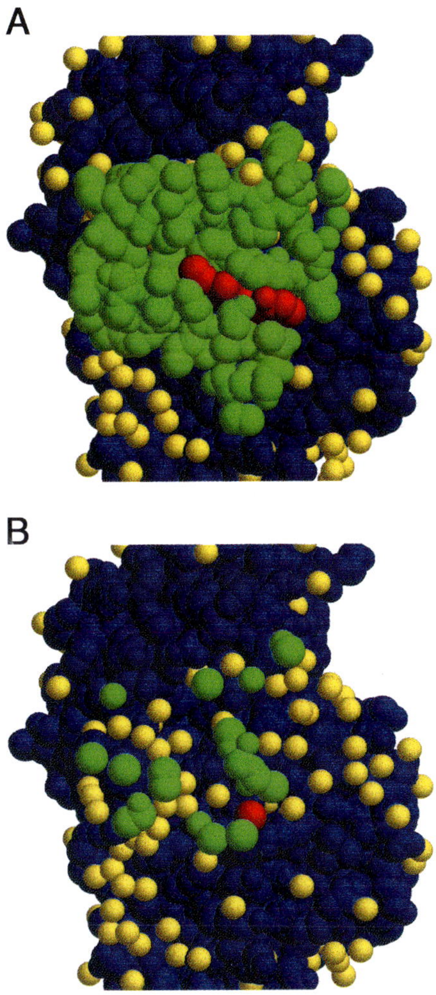

Fig. 3. The crystal structure of the cytochrome *c*, cytochrome *c* peroxidase complex (PDB2PCC). (A) The complex as viewed from cytochrome *c* down onto the peroxidase. On the cytochrome *c*, all atoms further away than 10 Å from the peroxidase have been removed so that one can see the broad outline of the interfacial region. The amino acids of the peroxidase are blue, those of cytochrome *c* are green. Heme is red and water is yellow. (B) The exact same view of the complex removing those atoms of cytochrome *c* that are further away than 4 Å from the peroxidase. The interfacial waters are clearly seen as well as a few atoms of cytochrome *c* that are within the restricted distance. There are more waters close to the peroxidase than there are atoms of cytochrome *c*.

4. The case of enolase

Enolase catalyzes the interconversion of 2-phosphoglycerate and phosphoenolpyruvate. The yeast protein is a homodimer as are known mammalian enzymes with the exception of some brain forms. From the point of view of one surface recognizing another, of one surface making good contacts with another, this poses a problem. There are constraints operating on the structure of homodimeric enzymes that do not operate on heteropolymers.

Under controlled conditions the dissociation of enolase into subunits is a reversible reaction in which intermediates do not accumulate. hp promotes dissociation [87,88] as would be predicted if the subunit surfaces which are buried in the dimer become hydrated upon dissociation. op promotes dimer formation [89]. The volume changes measured by op, hp and dilatometry are consistent (unpublished data). With yeast enolase, ΔV for dissociation, as measured by hp, is -260 ml mol^{-1}. The presence of osmolytes stabilizes the dimeric state and shifts the curve for dissociation to higher hydrostatic pressures. Osmolytes do not, however, change ΔV for dissociation. The op experiments yield a value of about 120 waters that bind to the protein upon dissociation. One hundred and twenty bound waters corresponds to the hydration of about 860 Å^2 of surface. The published X-ray structures report that almost 4000 Å^2 of surface are buried at the subunit interface [90,91]. This means that as much as 3/4 of the dimer interface should contain water molecules.

The X-ray crystallographic structure of yeast eno-

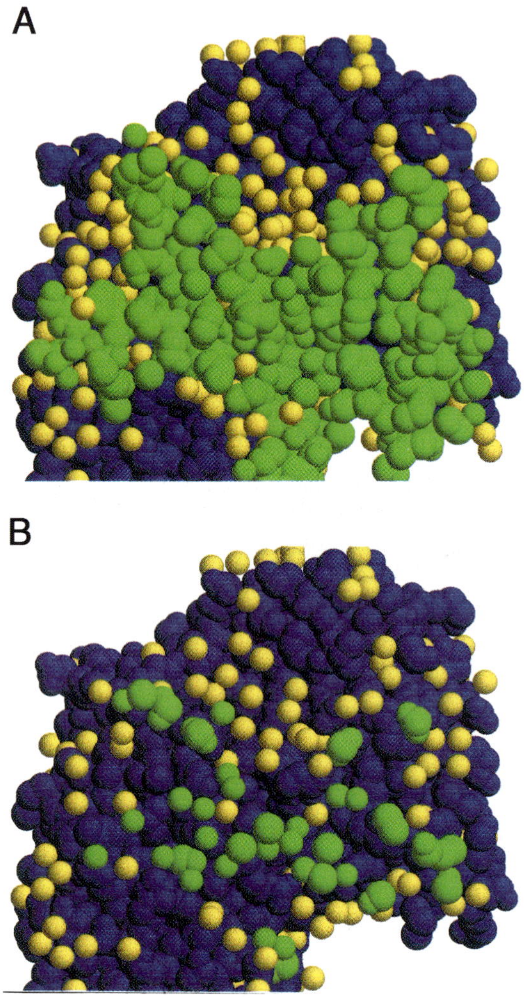

Fig. 4. The crystal structure of yeast enolase dimer (PDB1EBH). (A) The complex as viewed down one monomer onto the other. One subunit is blue, the other green. The waters seen in the complex are yellow. All atoms of the top subunit further than 7.0 Å from the bottom subunit have been removed such that some feel for the interfacial region can be obtained. (B) The exact same view of the complex removing all those atoms of the top subunit that are more than 3.5 Å from the bottom subunit. Note the large number of waters at the interface. Clearly, the water network covers less than 3/4 of the interface but it is still extensive.

lase is shown in Fig. 4 [92]. There is a rugged protein surface that has to make contact with an identical surface on the other monomer. The problem is how to maximize the good contacts and avoid creating large, potentially destabilizing voids. Evolution in the yeast enolase case has solved the problem by having a relatively small surface that makes good contacts on the two monomers and having an auxiliary surface that makes poorer contacts that are linked by water molecules. The interfacial waters are shown in Fig. 4.

What is significant about these waters is that in some instances they make good contacts with monomer A, in other instances with monomer B and in some instances with other water molecules. They are far from the active site. These interfacial waters form a cushion – they take up space, improve van der Waals contacts and hydrogen bonding – that improves otherwise poor contacts; most significantly, the waters are probed by both hp and op. There is also a cleft between the two subunits that is filled with water, including some waters sufficiently immobile to be seen in the X-ray structure.

The measured volume changes and the effects of both hp and op reflect the net balance, upon dissociation, between buried surface that becomes hydrated and interfacial waters and inter-cleft waters that are liberated.

Dissociation of enolase is promoted by removal of the bound Mg^{++}. ΔV for dissociation of apo-enolase is only -110 ml mol^{-1}, much smaller than the 260 ml mol^{-1} for Mg^{2+}-enolase [93]. This difference in ΔV tells us that there must be a significant difference between the two dimeric forms (apo vs. +Mg^{2+}) or between the monomeric forms. In this case, it has been interpreted to mean that pressure dissociation involves loss of the bound Mg^{2+} and that Mg^{2+}-monomers and apo-monomers do not have the same structure.

The dissociation of one of the mammalian enolases (form $\gamma\gamma$ from rabbit) has also been studied. In this case, the rate of dissociation was measured as a function of hp and op [89]. Both pressures act primarily on the rate of dissociation; hp increases the rate and op decreases the rate. $\Delta V^{\ddagger}$, as measured by hp is negative, indicating that the transition state for dissociation is more hydrated than is the ground state.

5. The case of protein–nucleic acid interactions

The specificities of many nucleic acid–protein interactions are mediated by individual water molecules acting as bridges between the two macromolecular components of the system [94]. Does this water do anything?

*Eco*RI is a restriction endonuclease from *Escherichia coli*. It binds to a well defined sequential site (GAATTC) on DNA where it cleaves the DNA. High op causes the enzyme to bind to sites other than those for which it is specific and introduce cleavage sites where they would not otherwise occur [95]; the simultaneous application of hp restored the lost specificity [96] to *Eco*RI. The compounds used as osmolytes in this and subsequent studies covered a wide range of characteristics. The loss of specificity correlated only with op and not with any other property of the solutions. As molecular biologists well know, the presence of osmolytes is not the only factor that can produce this altered specificity; cleavage at the alternate sites is promoted by changes in pH, ionic strength, or divalent cation concentration or identity. Changes in specificity produced this way were not reversed by hp. This suggests that there are multiple mechanisms for altering substrate specificity. The same conclusions applied to the endonucleases *Pvu*II and *Bam*HI but not to *Eco*RV [97].

The binding of this enzyme to non-specific sites, the alternate site (TAATTC) and the cognate (specific) site has been studied [95,98,99]. All three modes of binding involve loss of significant numbers of water molecules and therefore are all promoted by op. What is interesting here is that binding to the three sites involves loss of different amounts of water: 100 for the non-specific site, 150 for the canonical site and 200 for the alternate site [99]. DNA and the protein, free in solution, would both be highly hydrated. The X-ray structure of the complex with the cognate DNA reveals an anhydrous DNA/protein interface [100,101]. Both groups propose that upon forming the non-specific complex, there is dehydration of both partners but retention of a full hydration layer at the interface. The binding at the alternate site is more difficult to explain, since this structure releases about 50 more waters than does the canonical structure, yet both are catalytically competent. If the DNA/protein interface in the specific complex is largely anhydrous, where are these 50 additional waters? Possible explanations include significant changes in the conformation or extent of local folding of the DNA or protein or both.

The story becomes even more complicated when the catalytic activity is studied; op increases k_{cat} at the alternate site but decreases k_{cat} at the specific site. The opposite effects may be the result of a change in the rate-limiting step of the reaction as one changes substrate. The obvious take-home message here is that understanding the effects of hp or op on enzyme activity require being able to look at both binding and catalysis and knowing the identity of the slow steps in the reaction.

op has been applied to other DNA binding proteins, often with similar results. The homeodomains 'ultrabithorax' and 'deformed' bind DNA with the release of water; the equilibrium positions are sensitive to applied op [102]. The gal repressor also brings about the release of water when it binds to its specific sequence [103].

Transcription, the process by which DNA is copied into functional RNA, is a truly 'messy' biological system. It functions during the elongation process with a minimum of eight different components, of which the RNA polymerase and the DNA each contain multiple components. Erijman and Clegg [104] have subjected a transcribing system to high hp. The system stalls. The components do not dissociate; they simply stop doing what they were doing. When the pressure is released they resume doing what they were doing where they were doing it. Erijman and Clegg believe that the effects of pressure here are the result of 'a reduction of the partial specific volume of the RNA polymerase' thereby generating an enzyme which is too rigid to function. hp might also be interfering with hydration cycles as it does in the cytochrome *c* oxidase discussed above. Given that transcription involves bond making, unwinding and rewinding of a small segment of DNA and movement of the polymerase relative to the DNA and RNA, one can imagine that there may be multiple steps that involve water movement. Transcription is a prime candidate for systematic probing with op. It might help shed light on the process. Finally, the Silva group has looked at many DNA or RNA binding systems using hp as a probe. Often hp brings about denaturation or dissociation followed by denaturation of the binding protein. The nucleic acid displaces this protein equilibrium towards stability [105]. This system joins the list of those which might profit from osmotic perturbation.

We, along with numerous other authors, have used hp and op as probes of the thermodynamics of the macromolecular systems we study. The two approaches are powerful in that they can tell us about the involvement of solvent in the catalytic working of proteins (and probably catalytic RNAs) and can tell us about the involvement of solvent at the interfaces between proteins, nucleic acids and phospholipids. The problem, as has been stated hundreds of times, is that thermodynamics does not give information on mechanism, on path, on structure. The effects of hp on protein structures have been determined both by X-ray diffraction [106,107] and by NMR [108]. The technical problems with hp and X-ray diffraction are such that it is highly unlikely that it will ever be used for combined hp/op studies. This type of experiment should, however, be amenable to NMR. With NMR, we can obtain, in principle, both the total involvement of solvent in macromolecular interactions and where exactly the solvent is acting. Currently the laboratories of Akasaka and Markley and a few others are pursuing this approach [109–112,23, 24]. The rest of us would be well advised to join them.

Acknowledgements

We thank the Natural Science and Engineering Research Council of Canada for generous financial assistance. We owe a debt of gratitude to V.A. Parsegian, R.P. Rand, P. Nicholls, M.C. Marden and K. Heremans for *trying* to keep us honest. We extend apologies to all those many workers, not cited in this review, who have contributed to our knowledge of how op and hp might work. We note that the final word on the subject has not been spoken. Lastly, we wish to thank the editors of this volume for the invitation to be amongst the contributors.

Appendix. The theoretical ins and outs of op as a perturbant. The relations between op, temperature and hp

A1. Theory

The relations between free energy, T, op, and hp are easily restated.

$$dG = (\partial G/\partial T)_{P,n}dT + (\partial G/\partial P)_{T,n}dP + \sum(\partial G/\partial n_i)_{T,P}dn \quad \text{(A1)}$$

since

$$dG = dE + VdP + PdV - SdT - TdS \quad \text{(A2)}$$

and

$$dE = TdS - PdV \quad \text{(A3)}$$

substituting Eq. A3 into Eq. A2 yields

$$dG = VdP - SdT \quad \text{(A4)}$$

and

$$dG = -SdT + VdP + \sum \mu_i dn_i \quad \text{(A5)}$$

We can make the following simplifications: at constant T

$$dG = VdP + \sum \mu_i dn_i \quad \text{(A6)}$$

at constant P

$$dG = -SdT + \sum \mu_i dn_i \quad \text{(A7)}$$

at constant n

$$dG = VdP - SdT \quad \text{(A8)}$$

The equations relating changes in hp and changes in temperature have recently been reiterated [12].

At constant T, P

$$dG = \sum \mu_i dn_i \quad \text{(A9)}$$

Eq. A9 can be simplified if we recall that

$$G = \sum n_i \mu_i \quad \text{(A10)}$$

therefore differentiating gives

$$dG = \sum n_i d\mu_i + \sum \mu_i dn_i \quad \text{(A11)}$$

and setting Eq. A9 equal to Eq. A11

$$dG = \sum \mu_i dn_i = \sum n_i d\mu_i + \sum \mu_i dn_i \quad \text{(A12)}$$

ergo

$$0 = \sum n_i d\mu_i \text{ and } n_w d\mu_w + n_s d\mu_s = 0 \quad \text{(A13)}$$

The utility of Eq. A13 is that it relates the chemical potential of water to the chemical potential of the macromolecule. If we change the chemical potential of water, we necessarily change that of the macromolecule. The two are related by the ratio n_s/n_w or ΔN which is the number of waters that are associated with the macromolecule in one state that are not associated with it in the other state. In order to relate Eq. A13 to op, we need to be able to equate the chemical potential of water with the op.

$$d\mu_w = -d(op)/V_0 \quad (A14)$$

where V_0 is the partial molar volume of water. op is obtained from the mole fraction of water, X_{H_2O},

$$op = (RT/V_0)\ln X_{H_2O} \quad (A15)$$

X is most easily obtained from either the Handbook of Chemistry and Physics or from one of two WEB sites (http://aqueous.labs.brocku.ca/osfile.html; http://dir.nichd.nih.gov/Lpsb/docs/osmdata/osmdata.html). Combining Eqs. A13–A15 we obtain

$$d(RT\ln K_M) = \Delta V d(op) = (\Delta N^* V_0)d(op) \quad (A16)$$

where K_M is the equilibrium coefficient for the two states of the macromolecule, ΔV is the volume of water associated with one of the states that are absent in the other state, ΔN is the number of waters associated with the interconversion and op is the osmotic pressure. It is necessary to include the partial molar volume of water in Eq. A16 because of the nature of the derivation of Eq. A15.

Eqs. A15 and A16 are the only two that are needed to carry out an op experiment. We plot $RT\ln K_M$ vs. op and directly obtain the numbers of waters associated with the one equilibrium form that are not associated with the other.

To carry out an op vs. temperature experiment requires knowing that ΔC_p does not vary as a function of the op perturbant. If that is true, one plots T and op at $\Delta G = 0$ to yield a curve in much the same way that one handles hp vs. temperature data [12]. For this curve,

$$\Delta G = \Delta G^o - \Delta S^o(T - T_o) - \Delta C_p(T - T_o)^2/2T_o + (\Delta N V_0)(op - op_o) \quad (A17)$$

and

$$\partial T/\partial(op) = (\Delta N V_0)(\Delta S^o + \Delta C_p((T - T_o)/T_o))^{-1} \quad (A18)$$

To carry out an op vs. hp experiment requires knowing that compressibility does not vary as a function of the osmotic perturbant. In this case

$$\Delta G = \Delta G_o + \Delta V^o(hp - hp_o) + \Delta\beta/2(hp - hp_o)^2 + (\Delta N V_0)(op - op_o) \quad (A19)$$

and

$$\partial(hp)/\partial(op) = (\Delta N V_0)(\Delta V^o + \Delta\beta((hp - hp_o)/hp_o))^{-1} \quad (A20)$$

In ideal cases the macromolecule is not readily compressible and Eq. A20 reduces to

$$\partial(hp)/\partial(op) = \Delta N V_0/\Delta V^o \quad (A21)$$

In an op vs. temperature experiment or an op vs. hp experiment, the instantaneous slope (Eq. A18 and A21) gives information on ΔN, ΔS^o, and ΔC_p or on ΔN, ΔV^o and $\Delta\beta$.

The equation required to evaluate the simultaneous variation of T, op and hp is somewhat complex and because of its complexity we do not present it here. Suffice it to say that it must contain at least 10 terms [113]. No one in biochemistry has yet approached the complete problem of how the free energy of the system varies as T, op and hp are varied.

References

[1] L. Michaelis, M.L. Menten, Biochem. Z. 49 (1913) 333.

[2] P.W. Bridgman, The coagulation of albumen by pressure, J. Biol. Chem. 19 (1914) 511–512.

[3] P.N. Nicholls, Introduction: the biology of the water molecule, Cell Mol. Life Sci. 57 (2000) 987–992.

[4] P.S. Belton, NMR of the hydration of proteins and nucleic acids, Cell Mol. Life Sci. 57 (2000) 993–998.

[5] R.P. Rand, V.A. Parsegian, D.C. Rau, Intracellular osmotic action, Cell Mol. Life Sci. 57 (2000) 1018–1032.

[6] V.A. Parsegian, R.P. Rand, D.C. Rau, Macromolecules and water: probing with osmotic stress, Methods Enzymol. 259 (1995) 43–95.

[7] V. A Parsegian, R.P. Rand, D.C. Rau, Osmotic stress, preferential hydration, and binding; a comparison of perspectives, Proc. Natl. Acad. Sci. USA 97 (2000) 3987–3992.

[8] S.N. Timasheff, In disperse solution, "osmotic stress" is a restricted case of preferential interactions, Proc. Natl. Acad. Sci. USA 95 (1998) 7363–7367.

[9] J.D. Dunitz, The entropic cost of bound water in crystals and biomolecules, Science 264 (1994) 670.

[10] P. Masson, C. Clery, P. Guerra, A. Redslob, C. Albaret, P.-L. Fortier, Hydration changes during the aging of phosphorylated butyrylcholinesterase, Biochem. J. 343 (1999) 361–369.
[11] P. Mentre, G. Hui Bon Hoa, Effects of high hydrostatic pressures on living cells, Int. Rev. Cytol. 201 (2001) 1–84.
[12] K. Heremans, L. Smeller, Protein structure and dynamics at high pressure, Biochim. Biophys. Acta 1386 (1998) 353–370.
[13] C. Balny, F. Travers, Activation thermodynamics of the binding of carbon monoxide to horseradish peroxide. Role of pressure, temperature and solvent, Biophys. Chem. 33 (1989) 237–244.
[14] G. Panick, G.J.A. Vidugiris, R. Malessa, G. Rapp, R. Winter, C.A. Royer, Exploring the temperature-pressure phase diagram of staphylococcal nuclease, Biochemistry 38 (1999) 4157–4164.
[15] A. Zipp, W. Kauzmann, Pressure denaturation of Mb, Biochemistry 12 (1973) 4217–4228.
[16] G.M. Schneider, Phase behavior of aqueous solutions at high pressures, in: F. Franks (Ed.), Water: A Comprehensive Treatise, Plenum Press, New York, 1973, pp. 381–404.
[17] N.A.H.L. Ruthven, I. Winter, M. Kriechbaum, K. Lohner, R.N. McElhaney, Studies of the structure and organization of cationic lipid bilayer membranes: calorimetric,spectroscopic, and X-ray diffraction studies of linear saturated P-O-ethyl phosphatidylcholines, Biophys. J. 80 (2001) 1329–1342.
[18] R.P. Rand, V.A. Parsegian, Hydration forces between phospholipid bilayers, Biochim. Biophys. Acta 988 (1989) 351–376.
[19] Z. Zhou, B.G. Sayer, D.W. Hughes, R.E. Stark, R.M. Epand, Studies of phospholipid hydration by high-resolution magic-angle spinning nuclear magnetic resonance, Biophys. J. 76 (1999) 387–399.
[20] J.F. Nagle, Y. Liu, S. Tristram-Nagle, R.M. Epand, R.E. Stark, Re-analysis of magic angle spinning nuclear magnetic resonance determination of interlamellar waters in lipid bilayer dispersions, Biophys. J. 77 (1999) 2062–2065.
[21] L.F. Braganza, D.L. Worcester, Hydrostatic pressure induces hydrocarbon chain interdigitation in single component phospholipid bilayers, Biochemistry 25 (1986) 2591–2596.
[22] X. Peng, J. Jonas, High pressure ^{31}P NMR study of dipalmitoylphosphatidylcholine bilayers, Biochemistry 31 (1992) 6383–6390.
[23] B. Bonev, M.R. Morrow, Hydrostatic pressure-induced conformational changes in phosphatidylcholine headgroups: a ^{2}H NMR study, Biophys. J. 69 (1995) 518–523.
[24] B. Bonev, M.R. Morrow, Effects of hydrostatic pressure on bilayer phase behavior and dynamics of dilauroylphosphatidylcholine, Biophys. J. 70 (1996) 2727–2735.
[25] J. Monod, J. Wyman, J.P. Changeux, On the nature of allosteric transitions: a plausible model, J. Mol. Biol. 12 (1965) 88–118.
[26] M. Perutz, Stereochemistry of cooperative effects in haemoglobin, Nature 228 (1970) 726–739.
[27] I. Imai, Thermodynamic aspects of the co-operativity in four-step oxygenation equilibria of hemoglobin, J. Mol. Biol. 133 (1979) 233–247.
[28] G.B. Ogunmola, A. Zipp, F. Chen, W. Kauzmann, Effects of pressure on visible spectra of complexes of myoglobin, hemoglobin, cytochrome *c* and horse radish peroxidase, Proc. Natl. Acad. Sci. USA 74 (1977) 1–4.
[29] I. Morishima, M. Hara, High pressure NMR studies on hemoproteins, J. Biol. Chem. 258 (1983) 14428–14432.
[30] I. Morishima, M. Hara, High pressure NMR studies on hemoproteins, Biochem. Biophys. Res. Commun. 121 (1984) 229–236.
[31] F.H. Johnson, F.McK. Schlegel, Hemoglobin oxygenation in relation to hydrostatic pressure, J. Cell. Comp. Physiol. 31 (1948) 421–425.
[32] F.G. Cary, F. Knowles, Q.H. Gibson, Effect of hydrostatic pressure on ligand binding to hemoglobin, J. Biol. Chem. 252 (1977) 4102–4107.
[33] R.G. Alden, J.D. Satterlee, J. Mintorovitch, I. Constantinidis, M.R. Ondrias, B.I. Samson, The effects of high pressure upon ligated and deoxyhemoglobins and myoglobin, J. Biol. Chem. 264 (1989) 1933–1940.
[34] S. Pin, C.A. Royer, E. Gratton, B. Alpert, G. Weber, Subunit interactions in hemoglobin probed by fluorescence and high-pressure techniques, Biochemistry 29 (1990) 9194–9202.
[35] M.F. Colombo, D.C. Rau, V.A. Parsegian, Protein solvation in allosteric regulation: a water effect on hemoglobin, Science 256 (1992) 655–659.
[36] M.F. Columbo, D.C. Rau, V.A. Parsegian, Reevaluation of chlorides's regulation of hemoglobin oxygen uptake: The neglected contribution of protein hydration in allosterism, Proc. Natl. Acad. Sci. USA 91 (1994) 10517–10520.
[37] M.F. Colombo, G.O. Bonilla-Rodrigues, The water effect on allosteric regulation of hemoglobin probed in water/glucose and water/glycine solutions, J. Biol. Chem. 271 (1996) 4895–4899.
[38] J. Wyman, S.J. Gill, Binding and Linkage, Functional Chemistry of Biological Macromolecules, University Science Books, Mill Valley, 1990.
[39] M.F. Colombo, F.A.V. Seixas, Novel allosteric conformation of human Hb revealed by hydration and anion effects on oxygen binding, Biochemistry 38 (1999) 11741–11748.
[40] L. Kiger, A.L. Klinger, L.D. Kwiatkowski, A. De Young, M.L. Doyle, J.M. Holt, R.W. Noble, G.K. Ackers, Thermodynamic studies on the equilibrium properties of a series of recombinant W37 hemoglobin, Biochemistry 37 (1998) 4336–4345.
[41] N. Griffon, V. Baudin, W. Dieryck, A. Dumoulin, J. Pagnier, C. Poyart, M.C. Marden, Tetramer-dimer equilibrium of oxyhemoglobin mutants determined from auto-oxidation rates, Protein Sci. 7 (1998) 673–680.
[42] S.J. Hubbard, P. Argos, Detection of internal cavities in globular proteins, Protein Eng. 8 (1995) 1011–1015.
[43] A.A. Milne, Winnie the Pooh, Dutton, New York, 1926.
[44] R. Leberman, A.K. Soper, Effect of high salt concentrations on water structure, Nature 378 (1995) 364–366.

[45] C.R. Robinson, S.G. Sligar, Hydrostatic and osmotic pressure as tools to study macromolecular recognition, Methods Enzymol. 259 (1995) 395–427.

[46] W.H. Bradshaw, H.E. Conrad, E.J. Corey, I.C. Gunsalus, Microbiological degradation of (+)-camphor, J. Am. Chem. Soc. 81 (1959) 5507–5507.

[47] J. Hedegaard, I.C. Gunsalus, Mixed function oxidation. IV. An induced methylene hydroxylase in camphor oxidation, J. Biol. Chem. 240 (1965) 4038–4043.

[48] J.G. Rheinwald, A.M. Chakrabarty, I.C. Gunsalus, A transmissible plasmid controlling camphor oxidation in *Pseudomonas putida*, Proc. Natl. Acad. Sci. USA 70 (1973) 885.

[49] R. Lange, C. Bonfils, P. Debey, The low-spin/high-spin transition equilibrium of camphor-bound cytochrome P-450. Effects of medium and temperature on equilibrium data, Eur. J. Biochem. 79 (1977) 623–628.

[50] T.L. Poulos, B.C. Finzel, A.J. Howard, High-resolution crystal structure of cytochrome P450cam, J. Mol. Biol. 195 (1987) 687–700.

[51] T.L. Poulos, B.C. Finzel, A.J. Howard, Crystal structure of substrate free *Pseudomonas putida* cytochrome P450, Biochemistry 25 (1986) 5314–5322.

[52] I. Schlichting, J. Berendzen, K. Chu, A.M. Stock, S.A. Maves, D.E. Benson, R.M. Sweet, D. Ringe, G.A. Petsko, S.G. Sligar, The catalytic pathway of cytochrome P450cam at atomic resolution, Science 287 (2000) 1615–1622.

[53] R. Tsai, C.A. Yu, I.C. Gunsalus, J. Peisach, W. Blumberg, W.H. Orme-Johnson, H. Beinert, Spin-state changes in cytochrome P-450cam on binding of specific substrates, Proc. Natl. Acad. Sci. USA 66 (1970) 1157–1163.

[54] M. Sharrock, P.G. Debrunner, C. Schulz, J.D. Lipscomb, V. Marshall, I.C. Gunsalus, Cytochrome P450cam and its complexes. Mossbauer parameters of the heme iron, Biochim. Biophys. Acta 420 (1976) 8–26.

[55] S.G. Sligar, Coupling of spin, substrate, and redox equilibria in cytochrome P450, Biochem. 15 (1976) 5399–5406.

[56] S.G. Sligar, I. Gunsalus, A thermodynamic model of regulation: modulation of redox equilibria in camphor monoxygenase, Proc. Natl. Acad. Sci. USA 73 (1976) 1078–1082.

[57] P.E. Cole, S.G. Sligar, Temperature-jump measurement of the spin state relaxation rate of cytochrome P450cam, FEBS Lett. 133 (1981) 252–254.

[58] G. Hui Bon Hoa, M.C. Marden, The pressure dependence of the spin equilibrium in camphor-bound ferric cytochrome P-450, Eur. J. Biochem. 124 (1982) 311–315.

[59] M.C. Marden, G. Hui Bon Hoa, Dynamics of the spin transition in camphor-bound ferric cytochrome P-450 versus temperature, pressure and viscosity, Eur. J. Biochem. 129 (1982) 111–117.

[60] M.T. Fisher, S.F. Scarlata, S.G. Sligar, High-pressure investigations of cytochrome P-450 spin and substrate binding equilibria, Arch. Biochem. Biophys. 240 (1985) 456–463.

[61] C. Di Primo, E. Deprez, G. Hui Bon Hoa, P. Douzou, Antagonistic effects of hydrostatic pressure and osmotic pressure on cytochrome P-450cam spin transition, Biophys. J. 68 (1995) 2056–2061.

[62] W. Zhang, M.W. Capp, J.P. Bond, C.F. Anderson, M.T. Record Jr., Thermodynamic characterizations of native bovine serum albumine with highly excluded (glycine betaine) and moderately accumulated (urea) solutes by a novel application of vapor pressure osmometry, Biochemistry 35 (1996) 10506–10516.

[63] D.I. Svergun, S. Richard, M.H. Koch, Z. Sayers, S. Kuprin, G. Zaccai, Protein hydration in solution: experimental observation by X-ray and neutron scattering, Proc. Natl. Acad. Sci. USA 95 (1998) 2267–2272.

[64] C. Di Primo, S.G. Sligar, G. Hui Bon Hoa, P. Douzou, A critical role of protein-bound water in the catalytic cycle of cytochrome P-450 camphor, FEBS Lett. 312 (1992) 252–254.

[65] Y. Furukawa, I. Morishima, The role of water molecules in the association of cytochrome P450cam with putidaredoxin, J. Biol. Chem. 276 (2001) 12983–12990.

[66] I. Jelesarov, H.R. Bosshard, Thermodynamics of ferredoxin binding to ferredoxin:NADP+ reductase and the role of water at the complex interface, Biochemistry 33 (1994) 13321–13328.

[67] M. Aoki, K. Ishimori, H. Fukuda, K. Takahashi, I. Morishima, NMR studies of putidaredoxin: associations of putidaredoxin with NADH-putidaredoxin reductase and cytochrome p450cam, Biochim. Biophys. Acta 1384 (1998) 180–188.

[68] I.F. Sevrioukova, H. Li, H. Zhang, J.A. Peterson, T.L. Poulos, Structure of a cytochrome P450-redox partner electron-transfer complex, Proc. Natl. Acad. Sci. USA 96 (1999) 1863–1868.

[69] T. Tsukihara, H. Aoyama, E. Yamashita, T. Tomazaki, H. Yamaguchi, K. Shinzawa-Itoh, R. Nakashima, R. Yano, S. Yoshikawa, The whole structure of the 13-subunit oxidized cytochrome *c* oxidase at 2.8 Å, Science 272 (1996) 1136–1144.

[70] S. Yoshikawa, K. Shinzawa-Itoh, R. Nakashima, R. Yaono, E. Yamashita, N. Inoue, M. Yao, M. Jie Fei, C. Peters Libeu, T. Mizushima, H. Yamaguchi, T. Tomizaki, T. Tsukihara, Redox-coupled crystal structural changes in bovine heart cytochrome *c* oxidase, Science 280 (1998) 1723–1729.

[71] S. Ferguson-Miller, Mammalian cytochrome *c* oxidase, a molecular monster subdued, Science 272 (1996) 1125.

[72] D.A. Mills, L. Florens, C. Hiser, J. Qian, S. Ferguson-Miller, Where is 'outside' in cytochrome *c* oxidase and how and when do protons get there?, Biochim. Biophys. Acta 1458 (2000) 180–187.

[73] J.A. Kornblatt, G. Hui Bon Hoa, The conformations of cytochrome oxidase: thermodynamic evaluation of the 418 nm and 428 nm forms, Biochem. 21 (1982) 5439–5444.

[74] J.A. Kornblatt, G. Hui Bon Hoa, K. Heremans, Pressure induced effects on cytochrome *c* oxidase: the aerobic steady state, Biochemistry 27 (1988) 5122–5128.

[75] J.A. Kornblatt, G. Hui Bon Hoa, A non-traditional role for water in the cytochrome *c* oxidase reaction, Biochemistry 29 (1990) 9370–9376.

[76] J.A. Kornblatt, M.J. Kornblatt, Cytochrome *c* oxidase: the presumptive channel holds at least four water molecules, Biochim. Biophys. Acta 1099 (1992) 182–184.

[77] J. A Kornblatt, M.J. Kornblatt, I. Rajotte, G. Hui Bon Hoa, P.C. Kahn, Thermodynamic volume cycles for electron transfer in the cytochrome *c* oxidase and for binding cytochrome *c*, Biophys. J. 75 (1998) 435–444.

[78] J.A. Kornblatt, The water channel of cytochrome *c* oxidase: inferences from inhibitor studies, Biophys. J. 75 (1998) 3127–3134.

[79] S. Ferguson-Miller, D.L. Brautigan, E. Margoliash, Kinetics of the reaction of carboxydinitrophenyl-cytc with cytochrome *c* oxidase, J. Biol. Chem. 253 (1978) 149–159.

[80] J.A. Kornblatt, G. Hui Bon Hoa, A.M. English, Volume changes associated with the cytochrome *c* oxidase/porphyrin c equilibrium, Biochemistry 23 (1984) 5906–5911.

[81] J.A. Kornblatt, A.M. English, G. Hui Bon Hoa, The effects of pressure on yeast cytochrome *c* peroxidase, Eur. J. Biochem. 159 (1986) 39–43.

[82] J.A. Kornblatt, G. Hui Bon Hoa, L. Eltis, A.G. Mauk, The effects of pressure on porphyrin cytochrome *c*–cytochrome b_5 complex formation, J. Am. Chem. Soc. 110 (1988) 5909–5911.

[83] J.A. Kornblatt, J. Theodorakis, G. Hui Bon Hoa, E. Margoliash, Cytochrome *c*, cytochrome *c* oxidase interactions: the effects of ionic strength and hydrostatic pressure studied with site-specific modifications of cytochrome *c*, Biochem. Cell Biol. 70 (1992) 539–547.

[84] J.A. Kornblatt, M.J. Kornblatt, G. Hui Bon Hoa, A.G. Mauk, Responses of two protein-protein complexes to solvent stress: Does water play a role at the interface?, Biophys. J. 65 (1993) 1059–1065.

[85] P. Nicholls, Y. Sternin, J. Loewen, T. Jennings, B. Tattrie, Hydration as a control mechanism in membrane proteins: the case of cytochrome *c* oxidase, Faraday Discuss. 103 (1996) 313–323.

[86] H. Pelletier, J. Kraut, Crystal structure of a complex between electron transfer partners, cytochrome *c* peroxidase and cytochrome *c*, Science 258 (1992) 1748–1755.

[87] A.A. Paladini, G. Weber, Pressure-induced reversible dissociation of enolase, Biochemistry 20 (1981) 2587–2593.

[88] J. Kornblatt, J. Kornblatt, G. Hui Bon Hoa, The pressure-induced, reversible inactivation of mouse brain enolases, Eur. J. Biochem. 128 (1982) 577–581.

[89] M.J. Kornblatt, J.A. Kornblatt, G. Hui Bon Hoa, The role of water in the dissociation of enolase, a dimeric enzyme, Arch. Biochem. Biophys. 306 (1993) 495–500.

[90] S. Duquerroy, C. Camus, J. Janin, X-ray structure and catalytic mechanism of lobster enolase, Biochemistry 34 (1995) 12513–12523.

[91] B. Stec, L. Lebioda, Refined Structure of Yeast Apo-enolase at 2.25 Å Resolution, J. Mol. Biol. 211 (1990) 235–248.

[92] T.M. Larsen, J.E. Wedekind, I. Rayment, G.H. Reed, A carboxylate oxygen of the substrate bridges the magnesium ions at the active site of enolase: structure of the yeast enzyme complexed with the equilibrium mixture of 2-phosphoglycerate and phosphoenolpyruvate at 1.8 Å resolution, Biochemistry 35 (1996) 4349–4358.

[93] M.J. Kornblatt, R. Lange, C. Balny, Can monomers of yeast enolase have enzymatic activity?, Eur. J. Biochem. 251 (1997) 775–780.

[94] Z. Otwinowski, R.W. Schevitz, R.-G. Zhang, C.L. Lawson, A. Joachimiak, R.Q. Marmorstein, B.F. Luisi, P.B. Sigler, Crystal structure of *trp* repressor/operator complex at atomic resolution, Nature 335 (1998) 321–329.

[95] C.R. Robinson, S.G. Sligar, Molecular recognition mediated by bound water. A mechanism for star activity of the restriction endonuclease *Eco*RI, J. Mol. Biol. 234 (1993) 302–306.

[96] C.R. Robinson, S.G. Sligar, Hydrostatic pressure reverses osmotic pressure effects on the specificity of *Eco*RI–DNA interactions, Biochem. 33 (1994) 3787–3793.

[97] C.R. Robinson, S.G. Sligar, Heterogeneity in molecular recognition by restriction endonucleases: osmotic and hydrostatic pressure effects on *Bam*HI, Pvu II, and *Eco*RV specificity, Proc. Natl. Acad. Sci. USA 92 (1995) 3444–3448.

[98] N.Y. Sidorova, D.C. Rau, Differences in water release for the binding of *Eco*RI to specific and nonspecific DNA sequences, Proc. Natl. Acad. Sci. USA 93 (1996) 12272–12277.

[99] C.R. Robinson, S.G. Sligar, Changes in solvation during DNA binding and cleavage are critical to altered specificity of the *Eco*RI endonuclease, Proc. Natl. Acad. Sci. USA 95 (1998) 2186–2191.

[100] J.A. McClarin, C.A. Frederick, B.-C. Wang, P. Greene, H.W. Boyer, J. Grable, J.M. Rosenberg, Structure pf the DNA–*Eco* RI endonuclease recognition complex at 3 Å resolution, Science 234 (1986) 1526–1541.

[101] J.M. Rosenberg, Structure and function of restriction endonucleases, Curr. Opin. Struct. Biol. 1 (1991) 104–113.

[102] L. Li, K.S. Matthews, Differences in water release with DNA binding by ultrabithorax and deformed homeodomains, Biochemistry 36 (1997) 7003–7011.

[103] M.M. Garner, D.C. Rau, Water release associated with specific binding of gal repressor, EMBO J. 14 (1995) 1257–1263.

[104] L. Erijman, R.M. Clegg, Reversible stalling of transcription elongation complexes by high pressure, Biophys. J. 75 (1998) 453–462.

[105] R. Mohana-Borges, A.B.F. Pacheco, F.J.R. Sousa, D. Foguel, D.F. Almeda, J.L. Silva, LexA repressor forms stable dimers in solution: the role of specific DNA in tightening protein–protein interactions, J. Biol. Chem. 275 (2000) 4708–4712.

[106] C.E. Kundrot, F.M. Richards, Crystal structure of hen egg-white lysozyme at a hydrostatic pressure of 1000 atm, J. Mol. Biol. 193 (1987) 157–170.

[107] C.E. Kundrot, F.M. Richards, Effect of hydrostatic pressure on the solvent in crystals of hen egg-white lysozyme, J. Mol. Biol. 200 (1988) 401–410.

[108] J. Jonas, L. Ballard, D. Nash, High-resolution, high-pres-

sure NMR studies of proteins, Biophys. J. 75 (1998) 445–452.

[109] Y.O. Kamatari, H. Yamada, K. Akasaka, J.A. Jones, C.M. Dobson, L.J. Smith, Response of native and denatured hen lysozyme to high pressure studied by (15)N/(1)H NMR spectroscopy, Eur. J. Biochem. 268 (2001) 1782–1793.

[110] M.W. Lassalle, H. Yamada, K. Akasaka, The pressure-temperature free energy-landscape of staphylococcal nuclease monitored by (1)H NMR, J. Mol. Biol. 298 (2000) 293–302.

[111] H. Li, H. Yamada, K. Akasaka, Effect of pressure on individual hydrogen bonds in proteins. Basic pancreatic trypsin inhibitor, Biochemistry 37 (1998) 1167–1173.

[112] K.E. Prehoda, E.S. Mooberry, J.L. Markley, Pressure denaturation of proteins: evaluation of compressibility effects, Biochemistry 37 (1998) 5785–5790.

[113] G. Illian, H. Kneppe, F. Schneider, Phase diagrams with closed loop smectic-nematic phase boundaries, Ber. Bunsenges. Phys. Chem. 92 (1988) 776–780.

ELSEVIER

Biochimica et Biophysica Acta 1595 (2002) 48–70

BIOCHIMICA ET BIOPHYSICA ACTA
BBA

www.bba-direct.com

Review

Compressibility of protein transitions

Nicolas Taulier, Tigran V. Chalikian *

Department of Pharmaceutical Sciences, Faculty of Pharmacy, University of Toronto, 19 Russell Street, Toronto, ON, Canada M5S 2S2

Received 3 September 2001; received in revised form 5 November 2001; accepted 6 November 2001

Abstract

We review the results of compressibility studies on proteins and low molecular weight compounds that model the hydration properties of these biopolymers. In particular, we present an analysis of compressibility changes accompanying conformational transitions of globular proteins. This analysis, in conjunction with experimental compressibility data on protein transitions, were used to define the changes in the hydration properties and intrinsic packing associated with native-to-molten globule, native-to-partially unfolded, and native-to-fully unfolded transitions of globular proteins. In addition, we discuss the molecular origins of predominantly positive changes in compressibility observed for pressure-induced denaturation transitions of globular proteins. Throughout this review, we emphasize the importance of compressibility data for characterizing protein transitions, while also describing how such data can be interpreted to gain insight into role that hydration and intrinsic packing play in modulating the stability of and recognition between proteins and other biologically important compounds.

Keywords: Globular protein; Conformational transition; Hydration; Intrinsic packing; Compressibility; Ultrasonics

1. Introduction

The problem of protein folding remains a fascinating and much debated challenge of molecular biophysics [1–12]. After Anfinsen demonstrated that proteins fold spontaneously and introduced the hypothesis that the native state of the protein corresponds to the global minimum of free energy [13,14], thermodynamic characterizations of proteins in their native and denatured states have emerged as a major and fruitful approach to tackling the problem of protein folding [15–20]. The native state of the protein is separated from the ensemble of its denatured states by highly cooperative transition of the all-or-none type [21–25]. Elucidation of the thermodynamic nature and relative importance of the forces that govern the cooperativity of the folding/unfolding transitions of proteins is a key to understanding and, ultimately, addressing the protein folding problem [26,27]. In recognition of this fact and to this end, the thermodynamics of protein transitions has been thoroughly studied based on temperature-dependent [15,17,20], pH-dependent [19,24,25], denaturant-dependent [16,28,29], and pressure-dependent [18,30–34] protocols. In these studies, the equilibrium between the native and denatured protein states is perturbed by changing solution conditions (temperature, pressure, pH, salt, co-solvent, and denaturant concentration) and the concomitant change in a thermodynamic state function is monitored.

The effect of pressure on protein stability is determined by changes in volume, expansibility (the first

* Corresponding uthor. Fax: +1-416-978-8511.
E-mail address: chalikan@phm.utoronto.ca (T.V. Chalikian).

0167-4838/02/$ – see front matter
PII: S0167-4838(01)00334-X

temperature derivative of volume), and compressibility (the first pressure derivative of volume) associated with the unfolding transition of the protein at the experimental temperature and pressure [31]. The volumetric properties (volume and its derivatives expansibility and compressibility) of proteins and their low molecular weight analogs have proven uniquely sensitive to solute–solvent interactions (hydration) and intrinsic packing of globular proteins [35–38]. Consequently, any transition-induced change in protein hydration and/or intrinsic packing should be reflected in the accompanying changes in volume, expansibility, and compressibility [38–41]. Therefore, volumetric parameters represent a useful tool for macroscopic characterizations of structural transitions of proteins. It should be noted that thermodynamic information on intra- and intermolecular interactions of proteins provided by pressure-related volumetric observables (such as volume and compressibility) is complementary to that provided by temperature-related calorimetric observables (such as heat capacity and enthalpy). Hence, the combined use of volumetric and calorimetric observables often leads to more complete thermodynamic characterizations of protein systems and their conformational transitions.

The current interest of the biophysical/biochemical community toward compressibility as a tool for protein studies was first triggered several decades ago by progress in development of high precision techniques of sound velocity measurements and high-pressure studies [42–45]. Subsequently, compressibility measurements have begun to find an increasing employment as an independent and effective means for characterizing protein systems in a number of research laboratories across the globe [46–55]. In this paper, we review recent progress in the field of compressibility-based characterizations of structural transitions of water-soluble globular proteins. This review is not meant to be comprehensive with respect to critically evaluating the existing literature data on protein compressibility and microscopic interpretations of these data in terms of protein hydration and intrinsic packing. Instead, we focus on presenting our own results while also describing selected studies from other research groups. We begin by defining the partial compressibility of a solute and describing how it can be interpreted microscopically. Then, we briefly review the data on the partial compressibility of amino acids, peptides, and native globular proteins. Next, we present some theoretical considerations on compressibility changes associated with protein transitions. Finally, we present a survey of published experimental data on changes in compressibility associated with temperature-, pressure-, denaturant-, and pH-induced structural transitions of globular proteins. Throughout the review, we emphasize the significance of compressibility measurements for gaining insight into the forces that govern conformational preferences of polypeptide chains at given solution conditions.

2. Partial compressibility of proteins

2.1. *Definitions*

The most comprehensive and thermodynamically rigorous review of definitions and calculations of solute compressibility and related state functions has been recently presented by Blandamer et al. [56]. The isothermal compressibility, K_T, of a system can be defined as the negative pressure derivative of its volume, V, taken at constant temperature:

$$K_T = \beta_T V = -(\partial V/\partial P)_T \quad (1)$$

where T is the absolute temperature; and $\beta_T = -V^{-1}(\partial V/\partial P)_T$ is the coefficient of isothermal compressibility.

Similarly, the adiabatic compressibility of a system is the negative pressure derivative of its volume, V, taken at constant entropy:

$$K_S = \beta_S V = -(\partial V/\partial P)_S \quad (2)$$

where S is the entropy; and $\beta_S = -V^{-1}(\partial V/\partial P)_S$ is the coefficient of adiabatic compressibility.

Coefficients of isothermal, β_T, and adiabatic, β_S, compressibility are related to each other as follows:

$$\beta_T = \beta_S + T\alpha^2/\rho c_P \quad (3)$$

where ρ is the density; $\alpha = V^{-1}(\partial V/\partial T)_P$ is the coefficient of thermal expansion; and c_P is the specific heat capacity at constant pressure.

The partial molar isothermal/adiabatic compressibility, $K°_{T/S}$, of a solute is defined as the partial derivative of the isothermal/adiabatic compressibility of the solution with respect to the number of moles of the solute, N:

$$K^{\circ}_{T} = (\partial K_{T}/\partial N)_{T,P} \quad (4)$$

$$K^{\circ}_{S} = (\partial K_{S}/\partial N)_{T,P} \quad (5)$$

For proteins, when it is often needed to compare molecules with vastly different molecular weights, M, the values of partial specific isothermal, $k^{\circ}_{T} = K^{\circ}_{T}/M$, and adiabatic, $k^{\circ}_{S} = K^{\circ}_{S}/M$, compressibilities are more convenient for use than the partial molar values. The relationship between K°_{T} and K°_{S} is as follows [56]:

$$K^{\circ}_{T} = K^{\circ}_{S} + (T\alpha_0^2/\rho_0 c_{P0})(2E^{\circ}/\alpha_0 - C^{\circ}_{P}/\rho_0 c_{P0}) \quad (6)$$

where ρ_0 is the density of the solvent; α_0 is the coefficient of thermal expansion of the solvent; c_{P0} is the specific heat capacity at constant pressure of the solvent; E° is the partial molar expansibility of a solute; and C°_{P} is the partial molar heat capacity of a solute.

Sometimes, the so-called adiabatic, $\kappa_S = K^{\circ}_{S}/V^{\circ}$, and isothermal, $\kappa_T = K^{\circ}_{T}/V^{\circ}$, compressibilities of a protein are used in volumetric studies. The use of these characteristics is justified as far as there is a clear understanding of the relationships between κ_S, κ_T, K°_{S}, and K_T. In this respect, it should be noted that the relationship between κ_S and κ_T is given by a modification of Eq. 6 [$\kappa^{\circ}_{T} = \kappa^{\circ}_{S} + (T\alpha_0^2/V^{\circ}\rho_0 c_{P0})\,(2E^{\circ}/\alpha_0 - C^{\circ}_{P}/\rho_0 c_{P0})$] and not by Eq. 3.

2.2. Measurements

Currently, the most accurate method of determining the partial molar adiabatic compressibility, K°_{S}, of a solute is based on the Newton-Laplace equation which relates the coefficient of adiabatic compressibility, β_S, of a medium with its density, ρ, and sound velocity, U [49]:

$$U^2 = (\beta_S \rho)^{-1} \quad (7)$$

Differentiating Eq. 7 with respect to the molar concentration of a solute, C, and performing simple arithmetical rearrangements one obtains the following relationship for infinitely diluted solutions [57,58]:

$$K^{\circ}_{S} = \beta_{S0}(2V^{\circ} - 2[U] - M/\rho_0) \quad (8)$$

where β_{S0} is the coefficient of adiabatic compressibility of the solvent; $V^{\circ} = (\partial V/\partial N)_{T,P}$ is the partial molar volume of a solute; $[U] = (U - U_0)/U_0 C$ is the relative molar sound velocity increment of a solute; and U and U_0 are the sound velocities in the solution and solvent, respectively.

The partial molar volume, V°, of a solute can be calculated based on differential solution-versus-solvent density measurements [59]:

$$V^{\circ} = M/\rho_0 - (\rho - \rho_0)/\rho_0 C \quad (9)$$

where ρ and ρ_0 are the densities of the solution and solvent, respectively.

Inspection of Eqs. 7, 8, 9 reveals that the partial molar adiabatic compressibility of a solute can be calculated from differential solution-versus-solvent measurements of density and sound velocity. Recent developments of highly sensitive densimetric and acoustic apparatuses have enabled differential measurements of solution density and sound velocity in liquid samples with the minimum required volume of less than 1 ml and relative precision on the order of 10^{-4}% [49,60–65]. This precision permits one to carry out reliable compressibility measurements in solutions with solute concentration of ~1 mg/ml or less. The partial molar isothermal compressibility, K°_{T}, of a solute can be recalculated from K°_{S} using Eq. 6 provided that the partial molar expansibility, E°, and heat capacity, C°_{P}, have been determined independently. Significantly, by conducting densimetric and acoustic measurements in a protein solution as a function of temperature, pressure, pH, salt, or denaturant concentration (all of which may perturb the equilibrium between the native and denatured protein species), one can determine the change in compressibility associated with a temperature-, pressure-, pH-, salt-, or denaturant-induced protein transition.

An alternative way of calculating the change in isothermal compressibility, ΔK_T, accompanying the pressure-induced denaturation of a protein is based on a two-state analysis of the transition profile monitored by some physical (often spectroscopic) observable [31,44,45,66]. In this approach, a two-state analysis is used to describe the pressure dependence of the free energy, ΔG, of protein denaturation. The change in compressibility is determined from approximation of the pressure dependence of ΔG by the function [31]:

$$\Delta G = \Delta G_0 + \Delta V(P - P_0) + 0.5\Delta K_T(P - P_0)^2 \quad (10)$$

where ΔG_0 is the free energy of protein denaturation at reference pressure, P_0; and ΔV and ΔK_T are, respectively, the changes in volume and isothermal compressibility accompanying the pressure-induced denaturation of the protein.

Finally, in one recent study, the partial specific isothermal compressibility, $k°_T$, of the globular protein staphylococcal nuclease was determined from direct density measurements as a function of pressure over the range of 1 to 1000 bar [67]. Even though this promising approach represents the most straightforward way of determining $k°_T$, its relative precision is still low. Consequently, high protein concentrations are required for experiments. For example, Seemann et al. have used a protein concentration of ~40 mg/ml for evaluating $k°_T$ of staphylococcal nuclease [67].

2.3. Interpretations

The partial molar adiabatic compressibility, $K°_S$, of a solute can be represented by the sum of intrinsic, K_{SM}, and hydration, ΔK_{Sh}, contributions [37,38,48,49,68]:

$$K°_S = K_{SM} + \Delta K_{Sh} = \beta_{SM} V_M + n_h(K_{Sh} - K_{S0}) \quad (11)$$

where K_{SM} is the intrinsic adiabatic compressibility of a solute; V_M is the intrinsic volume of a solute; β_{SM} is the intrinsic coefficient of adiabatic compressibility of a solute; ΔK_{Sh} is the compressibility effect of hydration; K_{S0} and K_{Sh} are the partial molar adiabatic compressibilities of water in the bulk state and in the hydration shell of a solute, respectively; and n_h is the 'hydration number' which corresponds to the number of water molecules in the hydration shell of a solute.

The partial molar isothermal compressibility, $K°_T$, of a solute can be interpreted in terms of hydration and intrinsic contributions using a similar relationship:

$$K°_T = K_{TM} + \Delta K_{Th} = \beta_{TM} V_M + n_h(K_{Th} - K_{T0}) \quad (12)$$

In Eqs. 11, 12, the intrinsic volume, V_M, represents the geometric volume of the solute that is not penetrable by surrounding water molecules. For a globular protein, V_M is equal to the sum of the van der Waals volumes, V_W, of all constituent atoms plus the volume of intraglobular voids, V_V, resulting from its imperfect internal packing. Consequently, V_M reflects the spatial architecture of protein interior. The intrinsic coefficient of adiabatic, β_{SM}, or isothermal, β_{TM}, compressibility of a globular protein represents a measure of intraglobular interactions and, consequently, reflects the tightness of intrinsic packing. In addition, β_{TM} reflects the dynamic properties of a protein since the mean square volume fluctuation of a solute molecule, $\langle V_M^2 \rangle$, is directly proportional to β_{TM} [69,70]:

$$\langle V_M^2 \rangle = k_B T \beta_{TM} V_M \quad (13)$$

where k_B is Boltzmann constant.

The hydration terms, ΔK_{Sh} and ΔK_{Th}, in Eqs. 11, 12 represent the change in solvent compressibility induced by solute–solvent interactions. Being so, ΔK_{Sh} and ΔK_{Th} are sensitive to the amount (n_h) and strength (K_{Sh} and K_{Th}) of solute–solvent interactions in the vicinity of solute molecule. Consequently, the values of ΔK_{Sh} and ΔK_{Th} depend on the number and chemical nature (charged, polar, and nonpolar) of solvent-exposed protein groups.

Based on the foregoing discussion, it is clear that any transition-induced change in protein hydration, interior packing, or size of the solvent-inaccessible core should be reflected in the values of $K°_S$ and $K°_T$. Hence, compressibility measurements can be used as an effective means for detecting and characterizing protein transitions. Eqs. 11, 12, 13 in conjunction with the accumulated volumetric database on globular proteins and their low molecular weight model compounds form the basis for microscopic interpretations of compressibility data.

3. Partial compressibility of low molecular weight protein analogs and unfolded polypeptide chains

In an attempt to better understand and, ultimately, predict the hydration properties of proteins, their low molecular weight analogs, including amino acids and short oligopeptides, have been widely used as model systems. To this end, the volumetric properties of these relatively simple compounds have been studied as a function of temperature, pressure, and pH [71–91]. One important advantage of small molecules for modeling protein hydration is that their intrinsic compressibility, K_M, which is determined mostly by the compressibility of covalent bonds and external electron shells, is small and can be neglected in Eqs. 11, 12 [37,38,49]. Consequently, as is seen from Eqs. 11, 12, the partial compressibility of a low molecular weight model compound is predominantly determined by its solute–solvent interactions as reflected in the values of ΔK_{Sh} and ΔK_{Th}.

Based on model compound studies, the compressibility contributions of charged, polar, and nonpolar groups have been determined as a function of temperature. Fig. 1A presents the temperature dependence of the adiabatic compressibility contribution of an independently hydrated pair of the charged amino $-NH_3^+$ and carboxyl $-COO^-$ (●) termini in long α,ω-aminocarboxylic acids [79]. Fig. 1B presents our data on the adiabatic compressibility contributions of an independently hydrated methylene $-CH_2-$ group in α,ω-aminocarboxylic acids [79] (○), peptide –CONH- group in oligoglycines [81] (■), hydroxyl –OH group in pentoses and hexoses [92] (□), and benzene $-C_6H_5$ ring in the GlyGlyPhe tripeptide [88] (◇) plotted against temperature. In addition, Fig. 1B presents the literature data on the temperature dependence of the adiabatic compressibility contribution of the hydroxyl –OH group in long α,ω-diols (◆) [77]. The data presented in Fig. 1A,B in conjunction with Eq. 11 enable one to quantitatively characterize the hydration properties of charged, polar, and nonpolar groups and evaluate the differential compressibility of water of hydration and bulk water. Within the entire temperature range studied, waters solvating charged groups are less compressible than bulk water [37,49,77,79]. Water molecules solvating aliphatic groups are less compressible than bulk water below 35°C while exhibiting a greater compressibility at higher temperatures [37,49,77,79]. Hydration of a polar group depends on its position relative to other polar or charged groups in the solute molecule, with the two extreme cases represented by 'closely located' and 'single' polar groups [77]. For closely located polar groups (such as the –OH groups in pentoses and hexoses), the compressibility of water of hydration is reduced compared to that of bulk water and weakly depends on temperature. By contrast, for a single polar group (such as, the –OH groups in alcohols and long α,ω-diols), the compressibility of water of hydration is significantly larger than that of bulk water, but the difference steeply decreases with increasing temperature (see Fig. 1B). Recently, the thermodynamics of water solvating atomic groups of different chemical nature has been

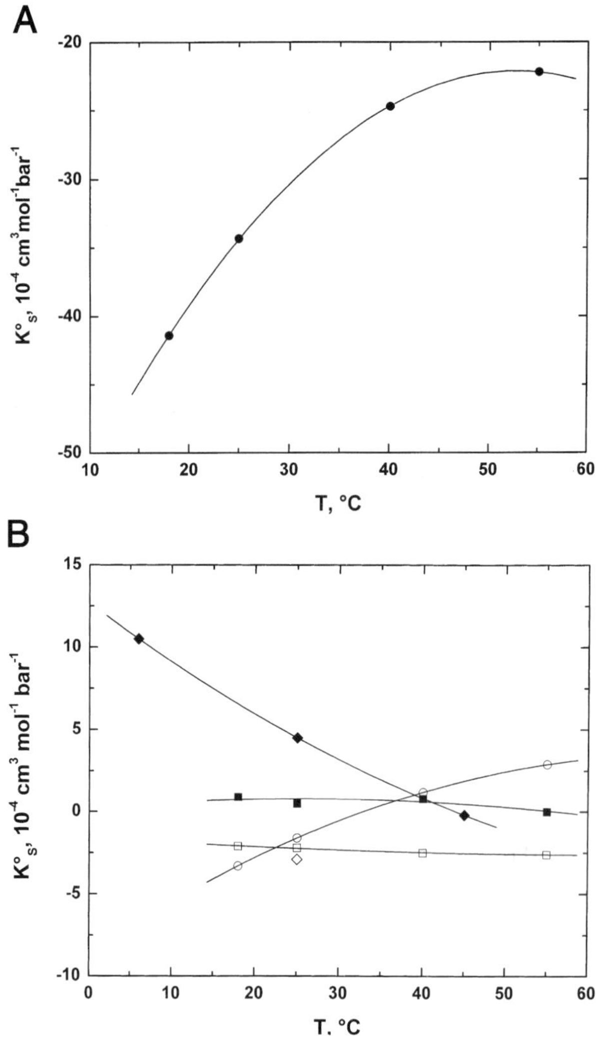

Fig. 1. The temperature dependences of compressibility contributions of different atomic groups: (a) a pair of the charged amino $-NH_3^+$ and carboxyl $-COO^-$ (●) termini in long α,ω-aminocarboxylic acids [79]; (b) an aliphatic $-CH_2-$ group in α,ω-aminocarboxylic acids [79] (○), peptide –CONH- group in oligoglycines [81] (■), hydroxyl –OH group in pentoses and hexoses [92] (□), aromatic benzene $-C_6H_5$ ring in the GlyGlyPhe tripeptide [88] (◇), and the hydroxyl –OH group in long α,ω-diols (◆) [77].

analyzed and rationalized based on the two-state representation of liquid water [93].

Volumetric studies on amino acids and systematically altered oligopeptides have been used to determine the compressibility contributions of glycyl unit (-CH_2CONH-) and different amino acid side chains [71,73,77,79–81,83,84,88,90]. These contributions are required to develope additive algorithms for calculating the partial molar (or specific) adiabatic compressibility of an extended polypeptide chain of a known primary sequence [77,83,94]. However, a survey of the relevant literature reveals that the magnitude and even the sign of the compressibility contribution of a given amino acid side chain strongly depends on the model compound used. For example, the compressibility contribution of the Phe side chain calculated as the difference between the partial molar adiabatic compressibilities of the amino acids Phe and Gly equals -7.54×10^{-4} $cm^3\,mol^{-1}\,bar^{-1}$ [71]. The same contribution determined from di- and tripeptide studies is equal to -1.79×10^{-4} $cm^3\,mol^{-1}\,bar^{-1}$ (GlyPhe/GlyGly) [75], 5.8×10^{-4} $cm^3 mol^{-1}\,bar^{-1}$ (GlyPheGly/GlyGlyGly) [83], 1.2×10^{-4} $cm^3\,mol^{-1}\,bar^{-1}$ (PheGlyGly/GlyGlyGly), and 0.6×10^{-4} $cm^3\,mol^{-1}\,bar^{-1}$ (GlyGlyPhe/GlyGlyGly) [88]. These disparities, which result from unaccounted intramolecular interactions between the side chain and the rest of the solute molecule, make it very difficult to choose a satisfactory set of data for modeling the compressibility contributions of amino acid residues in a polypeptide chain. In this respect, free amino acids are the least reliable models since, in these molecules, the closely located side chain and zwitterionic skeleton may strongly interact with each other via overlap of hydration shells. Hence, the side chain of a free amino acid can hardly be considered independently hydrated. Perhaps, amino acid side chains in Gly-X-Gly tripeptides and uncharged *N*-acetyl amino acid amides represent the best models for mimicking amino acid residues in proteins. Unfortunately, systematic volumetric databases on these promising substances are still lacking.

The compressibility contribution of glycyl unit (-CH_2CONH-) obtained on the basis of similar (but not identical) peptide systems also can be quite different. For example, in two recent studies on the homologous series of oligoglycines (glycine to pentaglycine), similar values for the compressibility contribution of glycyl unit equal to $-(1.1\pm0.5)\times10^{-4}$ $cm^3\,mol^{-1}\,bar^{-1}$ [81] and $-(1.08\pm0.08)\times10^{-4}$ $cm^3\,mol^{-1}\,bar^{-1}$ [90] have been obtained. By contrast, the same contribution obtained from the Ala, AlaGly, $AlaGly_2$, $AlaGly_3$, and $AlaGly_4$ series is significantly more negative and equals $-(2.75\pm0.03)\times10^{-4}$ $cm^3\,mol^{-1}\,bar^{-1}$ [81]. The reason(s) for the observed disparity is (are) yet to be uncovered.

In the aggregate, a wealth of information on the partial molar adiabatic compressibilities of amino acids and oligopeptides has been accumulated to date. Nevertheless, reliable derivation of the compressibilty contributions of amino acid residues required for calculating the partial compressibility of individual polypeptide chains of known primary sequences is not yet possible. It should be noted however that, in a recent publication, an additive method for calculating the partial compressibility of an unfolded polypeptide chain has been put forward [94]. Based on this method, the author calculated the compressibility effects of complete unfolding for a number of globular proteins [94]. In our opinion, the proposed additive method is of limited applicability for experimental analyses and practical predictions. In this method, amino acid residues of polypeptide chains have been somewhat arbitrarily modeled by zwitterionic amino acids. The use of free amino acids may be the source of large, unaccountable calculation errors since, as noted above, in these compounds, the side chain is not independently hydrated. Hence, free amino acids make a poor model for mimicking amino acid residues in polypeptide chains.

Another strategy for estimating the partial compressibility of an unfolded protein is to focus on an 'average' polypeptide chain rather than to attempt discriminating between individual polypeptides differing in primary sequences. The partial compressibility of an average unfolded protein can be obtained based on the following assumption: the average unit hydration (normalized per 1 Å^2) of an unfolded protein, independent of its exact amino acid sequence, is roughly similar to that of short, structureless oligopeptides of mixed composition, such as tripeptides. Based on this assumption, the average unit compressibility contribution (per 1 Å^2) of a polypeptide surface, γ, should be roughly equal to the arithmetic mean of unit contributions of a large number of relatively short oligopeptides [95]. For example, for the X-Gly-Gly and Gly-Gly-X families of tripeptides (where X is an apolar amino acid), the values of γ range from -10×10^{-6} to -14×10^{-6} $cm^3\,mol^{-1}\,bar^{-1}\,Å^{-2}$ as can be calculated by dividing the partial molar adiabatic compressibilities, $K°_S$, of a tripeptide by its solvent accessible surface area, S_A ($\gamma = K°_S/S_A$) [88]. In our analysis below, we will assume that the unit compressibility contribution of an unfolded protein, γ_U, is equal to $(-12\pm2)\times10^{-6}$ $cm^3\,mol^{-1}\,bar^{-1}\,Å^{-2}$. This approach does not permit one to discriminate between specific polypeptides, but rather provides an estimate of the partial specific adiabatic compressibility of an 'average' fully extended polypeptide chain. An implicit assumption of this approach is that the partial *specific* adiabatic compressibility, $k°_S$, of an unfolded polypeptide chain is roughly similar to that of short oligopeptides. Based on our tripeptide data, we estimate the partial specific adiabatic compressibility, $k°_S$, of an 'average' unfolded protein to be $-(18\pm3)\times10^{-6}$ $cm^3\,g^{-1}\,bar^{-1}$ [88].

4. Partial compressibility of native globular proteins

The largest set of partial specific adiabatic compressibilities, $k°_S$, of native globular proteins at 25°C has been reported by Gekko and coworkers [46,47]. Subsequently, Gekko and Hasegawa [96] and Chalikian et al. [97] studied the effect of temperature on the partial compressibility of native globular proteins. Analysis of the available data on protein compressibility reveals that native globular proteins predominantly exhibit positive values of $k°_S$. Microscopic interpretation of data on the partial specific adiabatic compressibility of proteins conventionally relies on an expanded version of Eq. 11 [98]:

$$k°_S = k_{SM} + \Delta k_{Sh} + k_{rel} \quad (14)$$

In this equation, $k_{SM} = K_{SM}/M$ is the specific intrinsic compressibility of a protein; M is the molecular weight of a protein; $\Delta k_{Sh} = M^{-1}\sum_i n_{Shi}(K_{Shi} - K_{S0})$ is the compressibility effect of hydration; K_{Shi} is the partial molar adiabatic compressibility of water solvating the ith solvent-exposed protein group; n_{hi} is the number of water molecules in the hydration shell of the ith solvent-exposed protein group; k_{rel} is the relaxation contribution to protein compressibility, which can be presented as the sum of the pH-dependent, $k_{rel}(pH)$, and pH-independent, k_{rel0}, terms:

$$k_{rel} = k_{rel0} + k_{rel}(pH) \quad (15)$$

The pH-independent term, k_{rel0}, results from fluctuations (subtransitions) within the ensemble of nearly isoenergetic, native-like conformational states of the protein, which are, nevertheless, distinct with respect to their partial volumes. Based on the theory of chemical relaxation [99], it can be shown that the pH-independent relaxation contribution to protein compressibility is given by the following expression:

$$k_{rel0} = M^{-1}\sum_j [\Delta V_j(\Delta V_j - \alpha\Delta H_j/\rho c_P)/RT]$$
$$[K_j/(1+K_j)^2](1+\omega^2\tau_j^2)^{-1} \quad (16)$$

where ΔV_j is the differential volume of the two native-like conformational substates of the jth subtransition; ΔH_j is the enthalpy of the jth subtransition; α, ρ, and c_P are the coefficient of thermal expansion, density, and specific heat capacity of the protein solution, respectively; K_j is the equilibrium constant for the jth subtransition; ω is the angular frequency of ultrasonic waves used for sound velocity measurements; $\tau_j = k_{fj} + k_{rj}$ is the relaxation time of the jth subtransition; and k_{fj} and k_{rj} are the rate constants for the forward and reverse processes of the jth subtransition, respectively. The characteristic range of relaxation times for a native globular protein is on the order of 10^{-9} to 10^{-3} s [98].

The pH-dependent term, $k_{rel}(pH)$, becomes important only at acidic and alkaline pH and is related to proton transfer reactions accompanying protonation/deprotonation of acidic ($-COO^- + H^+ \leftrightharpoons -COOH$) and basic ($-NH_3^+ + OH^- \leftrightharpoons -NH_2 + H_2O$) groups [72,78,98–108]:

$$k_{rel}(pH) = M^{-1}\sum_k [\Delta V_k(\Delta V_k - \alpha\Delta H_k/\rho c_P)/RT]$$
$$\Gamma_k(1+\omega^2\tau_k^2)^{-1} \quad (17)$$

where ΔV_k and ΔH_k are the volume and enthalpy of the kth proton transfer reaction; τ_k is the relaxation time for the kth proton transfer reaction. For acidic pH, the concentration-dependent coefficient Γ_k can be calculated from the relationship:

$$\Gamma_k^{-1} = [(2 + 10^{\mathrm{pH}-\mathrm{p}Kak} + 10^{pKak-\mathrm{pH}})/C_k] + 10^{\mathrm{pH}} \quad (18)$$

where C_k is the molar concentration of the kth atomic group being titrated; and pK_{ak} is the dissociation constant of the kth atomic group being titrated.

For alkaline pH, Γ_k is given by the relationship:

$$\Gamma_k^{-1} = [(2 + 10^{\mathrm{pH}-\mathrm{p}Kak} + 10^{\mathrm{p}Kak-\mathrm{pH}})/C_k] + 10^{14-\mathrm{pH}} \quad (19)$$

For acidic pH, the relaxation time, τ_k, can be calculated from the equation:

$$\tau_k^{-1} = k_{fk}[C_k/(1 + 10^{\mathrm{p}Kak-\mathrm{pH}}) + 10^{-\mathrm{pH}} + 10^{-\mathrm{p}Kak}] \quad (20)$$

where k_{fk} is the rate constant for the forward kth proton-transfer reaction.

For alkaline pH, τ_k is give by the following relationship:

$$\tau_k^{-1} = k_{fk}[C_k/(1 + 10^{\mathrm{pH}-\mathrm{p}Kak}) + 10^{\mathrm{pH}-14} + 10^{\mathrm{p}Kak-14}] \quad (21)$$

Experimentally, the net relaxation contribution to protein compressibility, k_{rel}, can be determined from measurements of the excess ultrasonic absorption per wavelength, $\alpha\lambda$, in the protein solution [72,78,98, 99,108]:

$$k_{\mathrm{rel}} = -\beta_{\mathrm{S0}}/(\pi\omega)\sum_{j,k}[\alpha\lambda]_{j,k}/\tau_{j,k} \quad (22)$$

where $[\alpha\lambda]_{j,k} = [(\alpha\lambda)_{j,k} - (\alpha\lambda)_0]/c$; $(\alpha\lambda)_{j,k}$ is the excess ultrasonic absorption per wavelength in the protein solution caused by the jth (subtransition between the native-like protein states) or kth (proton-transfer reaction) relaxation process; $(\alpha\lambda)_0$ is the ultrasonic absorption per wavelength in the neat solvent; and c is the specific protein concentration. Significantly, $[\alpha\lambda] = \Sigma_{j,k}[\alpha\lambda]_{j,k}$ is the net specific increment of ultrasonic absorption per wavelength which can be measured directly. Based on a careful analysis of ultrasonic data, Sarvazyan and Hemmes [98] have demonstrated that the relaxation contribution, k_{rel}, to the partial specific adiabatic compressibility, $k°_{\mathrm{S}}$, of globular proteins does not exceed ~10%. This contribution is well within experimental error of most $k°_{\mathrm{S}}$ measurements. Therefore, as a first approximation, the relaxation term, k_{rel}, in Eq. 14 can be neglected, and the partial specific adiabatic compressibility, $k°_{\mathrm{S}}$, of a globular protein can be viewed as the sum of the intrinsic, k_{SM}, and hydration, Δk_{Sh}, contributions.

Our ability to interpret protein compressibility data in terms of microscopic events ultimately depends on our ability to resolve $k°_{\mathrm{S}}$ into its k_{SM} and Δk_{Sh} contributions. This separation is model-dependent and can be performed only based on some assumptions. Below, we discuss some current approaches to evaluating the intrinsic and hydration contributions to the partial specific adiabatic compressibility of a globular protein.

4.1. Intrinsic compressibility

The protein material is very tightly packed as can be judged by the value of the packing density of the protein interior [109,110]. From this point of view, the protein interior is solid-like with the mean packing density of ~0.75 which is close to the maximum value for closely packed spheres (0.74) and corresponds to the upper limit of the packing densities of organic crystals [97,109,110]. The amount of the interior empty space is only on the order of 25% of the total protein volume although may vary somewhat from protein to protein [111]. As pointed out by Richards and Lim [110], the packing density is not uniform inside the protein. For example, packing of the backbone of the polypeptide chain is much denser than that of the side chains.

Analogous to packing density, the intrinsic compressibility of a globular protein falls within the range typical for organic crystals and characterizes the protein interior as a solid-like material [38,46, 47,97,112–114]. Compressibility is not uniform inside the protein structure in which rigid (low compressibility) and soft (high compressibility) packing domains can be intermixed [115–122]. In particular, it has been suggested that helices and loops exhibit larger microscopic compressibilities than β-strands [118]. There have been numerous attempts to evaluate the intrinsic compressibility of globular proteins [42,46,47,51,97,112,122–128]. Most of these evaluations yield the values of the coefficient of adiabatic (β_{SM}) or isothermal (β_{TM}) compressibility of the protein interior within the range of 10×10^{-6} to 30×10^{-6} bar^{-1}. In these estimates, the difference between β_{TM} and β_{SM} is thought to be smaller than the calculation uncertainties and, therefore, can be neglected. It is out of the scope of this paper to describe and/or critically analyse all methods and approaches that have been used to estimate the compressibility of the protein material. We believe that the most reliable estimate of the intrinsic compressibility of a globular protein reported to date is 25×10^{-6} bar^{-1}. This value was originally determined by Chalikian et al. [97] based on the first thermodynamic characterization of protein hydration and interior packing that does not depend on model compound data but rather is based exclusively on macroscopic (volumetric) and microscopic (X-ray) measurements on protein molecules themselves. Specifically, acoustic and densimetric techniques were used to measure $k°_S$ for 15 globular proteins over a temperature range of 18–55°C [97]. For the subset of the 12 proteins with known three-dimensional structures, the molecular volumes, V_M, and the solvent-accessible surfaces of the constituent charged, S_c, polar, S_p, and nonpolar, S_n, atomic groups were calculated. Assuming that the partial specific adiabatic compressibility, $k°_S$, of a protein is the sum of its intrinsic compressibility and hydration contributions of charged, polar, and nonpolar groups, one obtains the following linear relationship [97]:

$$k°_S = (\beta_M V_M + B_c S_c + B_p S_p + B_n S_n)/M \quad (23)$$

where B_c, B_p, and B_n are the unit compressibility contributions (per 1 Å^2) of the solvent accessible surfaces of charged, polar, and nonpolar atomic groups, respectively.

This regression analysis allows determination, as a function of temperature, of the intrinsic coefficient of adiabatic compressibility, β_M, and the average hydration contributions to $k°_S$ of 1 Å^2 of the charged, polar, and nonpolar solvent-accessible protein surfaces. The value of β_M was found to be $(25\pm1)\times10^{-6}$ bar^{-1} and practically independent of temperature between 18°C and 55°C [97].

Another way to estimate the intrinsic compressibility of globular proteins was proposed several years ago by Kharakoz and Sarvazyan [112]. It is based on an analysis of the correlation between the partial molar adiabatic compressibility, $K°_S$, and solvent accessible surface area, S_A, of a globular protein. Recall that $K°_S = K_M + \Delta K_{Sh} = \beta_M V_M + \gamma_N S_A$, where γ_N is the average unit contribution (per Å^2) of the surface of a native protein to its partial molar adiabatic compressibility. The partial specific adiabatic compressibility, $k°_S$, of a globular protein is then given by the relationship:

$$k°_S = K°_S/M = k_M + \Delta k_{Sh} = \beta_M v_M + \gamma_N S_A/M \quad (24)$$

where $v_M = V_M/M$ is the specific intrinsic volume of a globular protein (in cm^3 g^{-1}).

Using this method and a spherical approximation of a globular protein (S_A was set to be proportional to $M^{2/3}$), Kharakoz and Sarvazyan [112] calculated the value of β_M equal to 14×10^{-6} bar^{-1} relative to our value of 25×10^{-6} bar^{-1}. However, as emphasized by Miller et al. [129], because of the roughness of the protein surface, the solvent accessible surface area, S_A, of a globular protein cannot be approximated as proportional to $M^{2/3}$. In fact, Miller et al. have shown that S_A is proportional to $M^{0.73}$ [129]. Consistent with this estimate, our own data (not shown) on the solvent accessible surface areas, S_A, for 20 monomeric globular proteins yield a slope of 0.76 ± 0.03 for the dependence of $\log(S_A)$ on $\log(M)$. Specifically, we found that S_A (Å^2) = (4.7 ± 0.2) $M^{0.76\pm0.03}$. Hence, based on Eq. 24, $k°_S$ should be proportional to $M^{-0.24}$ and not to $M^{-1/3}$ as proposed in [112]. Fig. 2 presents the values of $k°_S$ for native globular proteins obtained at 25°C as a function of $M^{-0.24}$. Note that the data presented in Fig. 2 can be approximated reasonably well by a straight line. The intercept of this line on the y-axis (at $M^{-0.24}=0$) yields the value of $\beta_M v_M$ (see Eq. 24) equal to 16×10^{-6} cm^3 g^{-1} bar^{-1}. The intrinsic volume, V_M (Å^3), of a globular protein is related to its molecular weight via V_M (Å^3) = $(1200\pm500)+(1.04\pm0.02)M$ [97]. Consequently, the specific intrinsic volume, v_M (cm^3 g^{-1}) = $N_A V_M/M$ (where N_A is Avogadro's number), of a globular protein is equal to 700/M+0.626. For a very large protein, for which $M^{-0.24}$ is zero, v_M equals 0.626. Using this value, one estimates the intrinsic coefficient of adiabatic compressibility of a globular protein, β_M, of 25×10^{-6} bar^{-1} (16×10^{-6}/0.626). This value is significantly larger than the original value of 14×10^{-6} bar^{-1} calculated by Kharakoz and Sarvazyan [112] and coincides with our previous estimate of 25×10^{-6} bar^{-1} [97]. It should be noted however that, in his subsequent publications, Kharakoz emphasized that 25×10^{-6} bar^{-1} is the most reliable estimate of β_M [113,114].

Importantly, the value of 25×10^{-6} bar^{-1} is in good agreement with the pressure dependent X-ray crystallographic data presented by Katrusiak and Dauter [125], who determined the unit cell dimensions and the volume of hen egg-white lysozyme crystals (orthorhombic form) at pressures from 1 to 1000 bar. A nonlinear exponential approximation of the pressure dependence of the crystal volume yields at 1 bar a coefficient of isothermal compressibility, β_M, of 20×10^{-6} bar^{-1} [38].

In the aggregate, the most reliable estimate of the intrinsic coefficient of protein compressibility is 25×10^{-6} bar^{-1}. This value characterizes the protein interior as a rigid, tightly packed, solid-like substance. For comparison, the coefficients of adiabatic compressibility, β_S, of liquid water, benzene, and hexane are equal to 45×10^{-6}, 96×10^{-6} bar^{-1}, and 165×10^{-6} bar^{-1}, respectively. However, the protein interior is much more flexible (softer) than most of the crystalline solids: for example, β_S of ice is 13×10^{-6} bar^{-1}. In a way, this observation 'makes sense': certain level of dynamic flexibility of the polypeptide chain and amino acid residues is required for the protein molecule to 'breath' and perform its biological function [130].

4.2. Hydration contribution to compressibility

A major question related to protein solvation is whether low molecular weight model compounds can model the hydration properties of native globular proteins. Alternatively, this question can be reformulated as whether the hydration characteristics of atomic groups at the protein surface are identical to those of similar groups in small molecules. One way to answer this question is to compare the average compressibility contribution of 1 Å^2 of protein surface, γ_N, with that of small molecules. As is seen from Eq. 24, the value of γ_N can be obtained if the difference between $K°_S$ and $\beta_M V_M$ is plotted against the protein solvent accessible surface areas, S_A. Fig. 3 shows such plots for 12 globular proteins at 18, 25, 35, 45, and 55°C (the data are taken from our previous work [97]). The data presented in Fig. 3 have been approximated by straight lines with slopes equal to the average values of γ_N at specific temperatures studied.

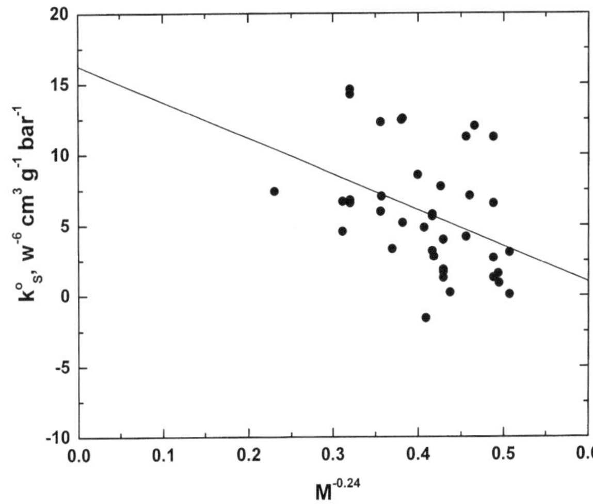

Fig. 2. The dependence of the partial specific adiabatic compressibility, $k°_S$, of globular proteins at 25°C [47,97] on $M^{-0.24}$ (where M is the molecular weight). The straight line represents the best-fit linear approximation of the data.

Fig. 4 shows the temperature dependence of γ_N between 18°C and 55°C. Inspection of Fig. 4 reveals that the average unit compressibility contribution of the protein surface, γ_N, is highly negative and slightly changes with temperature. The temperature dependence of γ_N can be approximated reasonably well by a second order polynomial. At 25°C, γ_N is equal to $(-23\pm3)\times10^{-6}$ cm^3 mol^{-1} bar^{-1} Å^{-2}. Note that this value is almost twice as small (more negative) as the average unit contribution of an unfolded polypeptide chain, γ_U, of $(-12\pm2)\times10^{-6}$ cm^3 mol^{-1} bar^{-1} Å^{-2} (see Section 3). For comparison, this value is even more negative than the unit compressibility contribution of glycine, the hydration shell of which is predominantly dictated by electrostatic solute–solvent interactions! At 25°C, the partial molar adiabatic compressibility, $K°_S$, of glycine is -26.6×10^{-4} cm^3 mol^{-1} bar^{-1} [71,77,79] while its solvent accessible surface area, S_A, is 136 Å^2. Hence, the unit compressibility contribution for glycine is -20×10^{-6} cm^3 mol^{-1} bar^{-1} Å^{-2} (-26.6×10^{-4}/136).

The highly negative value of γ_N suggests that the native state of a globular protein is extensively hydrated with the pattern of hydration being distinct from that of low molecular weight model compounds. We propose that this distinction is related to the differential hydration of polar groups in proteins and small molecules. In fact, a polar group in a protein is characterized by a three times more negative unit compressibility contribution than a similar group in a small molecule [97]. This disparity may reflect the cooperative formation of networks of water molecules adjacent to the rigid matrix of closely located polar protein groups, with these networks involving waters from the second and third coordination spheres. Such cooperative networks have not been observed in small molecules.

Based on the foregoing discussion, the hydration properties of native globular proteins cannot be satisfactorily modeled based on additive calculations using small molecule data. In contrast to this notion, in two recent publications it has been suggested that

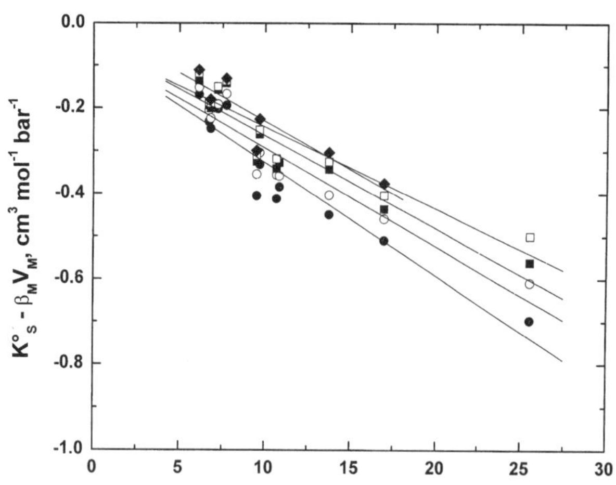

Fig. 3. The dependences of the difference ($K°_S-\beta_M V_M$) on the solvent accessible surface area, S_A, for globular proteins at 18°C (●), 25°C (○), 35°C (■), 45°C (□), and 55°C (♦). The data on the partial molar adiabatic compressibility, $K°_S$, intrinsic volume, V_M, and solvent accessible surface area, S_A, of globular proteins are from [97]. The value of the coefficient of intrinsic compressibility, β_M, of a globular protein is 25×10^{-6} bar^{-1}. The straight lines represent the best-fit linear approximations for each data set.

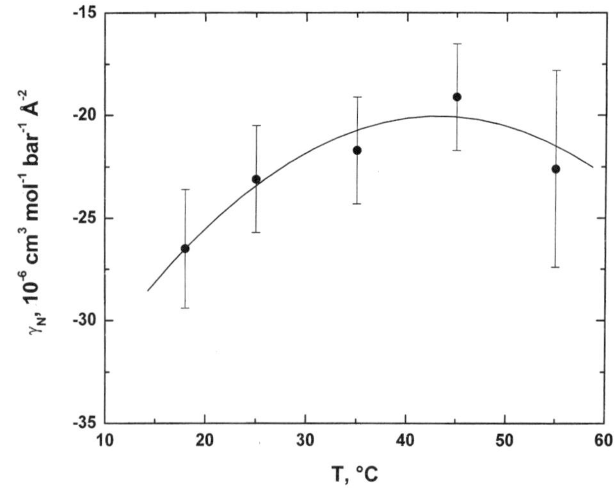

Fig. 4. The temperature dependence of the average unit compressibility contribution, γ_N, of the surface of a native globular protein.

hydration of protein groups is identical to that of similar groups in small molecules [94,112]. Based on this suggestion, additive calculations have been performed to determine the hydration contributions to compressibility, Δk_{Sh}, and intrinsic compressibilities, $k_M=\beta_M v_M$, for some globular proteins [112]. However, these calculations have produced significantly underestimated values of the intrinsic coefficient of protein compressibility, β_M, which undermines the validity of such additive analyses.

In the aggregate, the unit compressibility contribution of the surface of the native globular protein, γ_N, is 50% smaller (more negative) than the average unit contribution of an unfolded polypeptide chain, γ_U. At 25°C, γ_N is equal to $(-23\pm3)\times10^{-6}$ cm^3 mol^{-1} bar^{-1} Å^{-2}. This value is reflective of cooperative water networks adjacent to the polar protein surface and underscores the limitations of additive models of protein hydration based on small molecule data.

5. Compressibility changes associated with protein transitions: Theoretical considerations

5.1. *The magnitude of compressibility changes*

Some time ago, based on analysis of experimental data on compressibility changes associated with protein transitions, Δk_S, we proposed that a general relationship exists between the class of protein conformational transitions and the sign and magnitude of Δk_S [40]. Specifically, we found that native-to-compact intermediate (such as native-to-molten globule) transitions are accompanied by small increases in $k°_S$ of $+(1–4)\times10^{-6}$ cm^3 mol^{-1} bar^{-1}, native-to-partially unfolded transitions are accompanied by small decreases in $k°_S$ of $-(3–7)\times10^{-6}$ cm^3 mol^{-1} bar^{-1}, and native-to-fully unfolded transitions are accompanied by large decreases in $k°_S$ of $-(18–20)\times10^{-6}$ cm^3 mol^{-1} bar^{-1} [40]. In other words, we proposed that for single-domain globular proteins, changes in $k°_S$ correlate with the type of transition being monitored, independent of the specific protein. To rationalize the observed correlation, we offered a qualitative explanation based on the interplay between the changes in intrinsic, k_M, and hydration, Δk_{Sh}, compressibilities associated with each transition type [40]. Below, we present a more quantitative analysis that shows that the observed correlation between Δk_S and the transition type can be obtained theoretically based on the interior packing, size of the solvent-inaccessible protein core, and solvent-accessible surface areas of each of the three protein states (R. Filfil, T.V. Chalikian, unpublished results).

From Eq. 24, the net change in compressibility, Δk_S, associated with a conformational transition of a globular protein can be presented as the sum of the changes in the intrinsic, Δk_M, and hydration, $\Delta\Delta k_{Sh}$, compressibilities:

$$\Delta k_S = \Delta k_M + \Delta\Delta k_h \tag{25}$$

where $\Delta k_M = k_{MD} - k_{MN}$; $k_{MN} = v_{MN}\beta_{MN}$ and $k_{MD} = v_{MD}\beta_{MD}$ are the intrinsic compressibilities of the native and denatured protein states, respectively; v_{MN} and v_{MD} are the intrinsic volumes of the native and denatured protein states, respectively; β_{MN} and β_{MD} are the intrinsic compressibilities of the native and denatured protein states, respectively; $\Delta\Delta k_{Sh} = \Delta k_{hD} - \Delta k_{hN}$; Δk_{hN} and Δk_{hD} are the hydration contributions to compressibility for the native and denatured protein states, respectively.

The denatured protein, while exposing to the solvent a substantial number of previously buried atomic groups, generally retains a water-inaccessible core of loosely packed amino acid residues. The water-inaccessible core may be sizeable, as in molten globules, or insignificant, as in random coil-like, fully unfolded polypeptide chains. Quantitatively, the water-inaccessible core can be described in terms of the degree of unfolding, σ, of a protein. The degree of unfolding, σ, can be defined as the ratio of the number of amino acid residues in the unfolded, random coil-like domains of the polypeptide chain to the total number of amino acids. Clearly, the value of σ equals zero for the native state and unity for the fully unfolded state.

Recall that the intrinsic volume, V_M (Å^3), of a native globular protein of a molecular weight, M, is equal to $1200+1.04M$ [97]. Consequently, the specific intrinsic volume, v_{MN} (cm^3 g^{-1}), is equal to $(N_A/M)(1200+1.04M)$. Since, the molecular weight of the solvent-inaccessible core of the protein in its denatured state is equal to $M(1-\sigma)$, the value of v_{MD} is $(N_A/M)[1200+1.04M(1-\sigma)]$. As discussed above, the intrinsic coefficient of compressibility of the tightly packed, solid-like interior of the native protein state, β_{MN}, is 25×10^{-6} bar^{-1}. By contrast, the loosely packed solvent inaccessible core of a denatured protein is liquid-like. Its intrinsic compressibility should be close to that of organic liquids ($\sim100\times10^{-6}$ bar^{-1}). With this notion in mind, we assume that the intrinsic compressibility of the denatured state, β_{MD}, linearly changes with increasing σ from 25×10^{-6} bar^{-1} at $\sigma=0$ to 100×10^{-6} bar^{-1} at $\sigma=1$: $\beta_{MD}=25\times10^{-6}+75\times10^{-6}\sigma$. Consequently, one obtains the following relationship for the denaturation-induced change in the intrinsic compressibility of a protein, Δk_M:

$$\Delta k_M = k_{MD} - k_{MN} = (N_A/M)\{[1200 + 1.04M(1-\sigma)](25\times10^{-6} + 75\times10^{-6}\sigma) - (1200 + 1.04M)25\times10^{-6}\} \tag{26}$$

Further, we assume that the hydration contributions to the partial compressibility of both the native, Δk_{hN}, and denatured, Δk_{hD}, protein states correlate to the solvent-exposed surface area, S_A. Recall that, for the native state, S_{AN} (Å^2) is equal to $4.7\ M^{0.76}$. Consequently, the hydration contribution to compressibility for the native state is given by the expression:

$$\Delta k_{hN} = M^{-1} S_{AN}\gamma_N = 4.7M^{-0.24}\gamma_N \tag{27}$$

where γ_N is equal to -23×10^{-6} cm^3 mol^{-1} bar^{-1} Å^{-2}.

For the denatured state, the total solvent accessible surface area, S_{AD}, can be broken down into two contributions. The first contribution comes from the unfolded part of the chain. The solvent accessible surface area (in Å^2) of an extended polypeptide chain is equal to 1.45 M [109]. Hence, the solvent-accessible area of the unfolded domain of the polypeptide chain can be set equal to 1.45 σM. The hydration contribution to compressibility of the unfolded domain is equal to 1.45 $\sigma M \gamma_U$ (where γ_U is equal to -12×10^{-6} cm^3 mol^{-1} bar^{-1} Å^{-2}). The second contribution to S_{AD} comes from the solvent accessible surface area of the compact domain of the polypeptide chain which can be set to be equal to 4.7 $[M(1-\sigma)]^{0.76}$. In this representation, we assume that the unit compressibility contribution of the compact domain linearly changes with increasing σ from γ_N at $\sigma=0$ to γ_U at $\sigma=1$. Consequently, for the denatured state, Δk_{hD} is given by the relationship:

$$\Delta k_{hD} = M^{-1}\{4.7[M(1-\sigma)]^{0.76}[\gamma_N + (\gamma_U-\gamma_N)\sigma]+ 1.45M\sigma\gamma_U\} = 4.7M^{-0.24}(1-\sigma)^{0.76} [\gamma_N + (\gamma_U-\gamma_N)\sigma] + 1.45\sigma\gamma_U \tag{28}$$

Combining Eqs. 27, 28, one derives the following relationship for the change in the hydration contribution to compressibility, $\Delta\Delta k_h$, associated with protein denaturation:

$$\Delta\Delta k_h = \Delta k_{hD}-\Delta k_{hN} = 4.7M^{-0.24}(1-\sigma)^{0.76} [\gamma_N + (\gamma_U-\gamma_N)\sigma] + 1.45\sigma\gamma_U-4.7M^{-0.24}\gamma_N \tag{29}$$

Finally, substituting Eqs. 26 and 29 into Eq. 25, one obtains the expression for calculating the change in adiabatic compressibility, Δk_S, associated with the protein transition. We used Eqs. 25, 26, 29 to calculate the change in compressibility, Δk_S, associated with protein denaturation as a function of the degree of unfolding, σ. Fig. 5 shows results of these calculations for proteins with molecular weights, M, of 10, 15, 20, 30, 50, and 80 kDa. Inspection of Fig. 5 reveals a number of significant observations.

Firstly, for any molecular weight, M, the σ-dependence of Δk_S is parabolic-like. At small values of σ (when σ is smaller than $0.5\div0.6$), Δk_S is positive while becoming negative at larger values of σ.

Secondly, recall that, in molten globules, 40–60% of previously solvent inaccessible protein groups be-

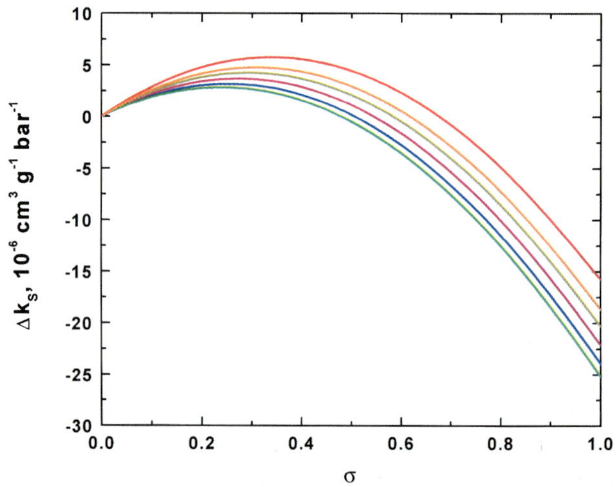

Fig. 5. The calculated dependences of the change in adiabatic compressibility, Δk_S, accompanying protein denaturation on the degree of unfolding, σ, for proteins with molecular weights, M, of 10 (red), 15 (orange), 20 (dark yellow), 30 (magenta), 50 (blue), and 80 kDa (green). Calculations have been performed using Eqs. 25, 26, 29.

come exposed to the solvent. This estimate was made by comparing the changes in hydrodynamic and partial specific volumes accompanying native-to-molten globule transitions as well as from differential scanning calorimetric measurements [131–133]. By extension, the value of σ for the molten globule state can be set to be $0.4\div0.6$. Hence, inspection of Fig. 5 reveals that the range of Δk_S values for native-to-molten globule transitions is -3×10^{-6} to 6×10^{-6} cm^3 g^{-1} bar^{-1}. These values are in excellent agreement with the experimentally observed range of Δk_S values for native-to-molten globule transitions of 1×10^{-6} to 4×10^{-6} cm^3 g^{-1} bar^{-1} [40]. Note that in this treatment, there is no need to assume that water penetrates the intrinsic core of the molten globule as proposed in ref. [134].

Thirdly, the solvent accessible surface area of a protein in its partially unfolded state is about 70–80% of the value expected for the fully extended conformation [38,95,108,135–138]. Thus, the degree of unfolding, σ, for the partially unfolded state can be set to be $0.7\div0.8$. With this estimate, inspection of Fig. 5 reveals that, for native-to-partially unfolded transitions, the values of Δk_S are between -2×10^{-6} and -12×10^{-6} cm^3 g^{-1} bar^{-1}, in agreement with the experimentally observed range of -3×10^{-6} to -7×10^{-6} cm^3 g^{-1} bar^{-1} [40].

Fourthly, for native-to-fully unfolded transitions ($\sigma = 1$), the values of Δk_S are between -16×10^{-6} and -25×10^{-6} cm^3 g^{-1} bar^{-1}, in excellent agreement with experimental values of Δk_S on the order of -20×10^{-6} cm^3 g^{-1} bar^{-1} [40].

In the aggregate, the observed correlation between Δk_S and the transition type can be accounted for by simple theoretical calculations based on Eqs. 25, 26, 27, 28, 29. The excellent numerical agreement between our calculated values and the observed range of experimental values of Δk_S for each transition type lends credence to the assumptions and approximations employed in our analysis (including the specific choice of the values of the γ_N and γ_U).

5.2. Transition profiles

Recently, Rösgen and Hinz [139,140] presented a comprehensive theoretical analysis of the thermodynamics of conformational transitions of proteins. For the simplest case of a monomeric protein that may exist only in two conformational states (native, N, and denatured, D), the partial specific volume, $v°$, at any given solution condition can be presented as follows:

$$v^\circ = v_N + f_D \Delta v \quad (30)$$

where $\Delta v = v_D - v_N$; v_N and v_D are the partial specific volumes of the native and denatured states, respectively; $f_D = K/(1+K)$ is the fraction denatured; and $K = [D]/[N]$ is the equilibrium constant.

The partial specific isothermal compressibility, $k°_T$, of the protein can be found by differentiating Eq. 30 with respect to pressure:

$$k^\circ_T = -(\partial v^\circ/\partial P)_T = k_{TN} + f_D \Delta k_T - (\partial f_D/\partial P)_T \Delta v \quad (31)$$

where $k_{TN} = -(\partial v_N/\partial P)_T$ is the partial specific isothermal compressibility of the native state; $\Delta k_T = k_{TD} - k_{TN}$; $k_{TD} = -(\partial v_D/\partial P)_T$ is the partial specific isothermal compressibility of the denatured state.

It can be shown that the pressure slope, $(\partial f_D/\partial P)_T$, is given by the relationship:

$$(\partial f_D/\partial P)_T = -(\Delta v/RT) f_D (1 - f_D) \quad (32)$$

Substituting Eq. 32 into Eq. 31 and taking into account that $f_D = K/(1+K)$, one obtains the following relationship for $k°_T$:

$$k^\circ_T = -(\partial v^\circ/\partial P)_T = k_{TN} + [K/(1+K)]\Delta k_T + (\Delta v^2/RT)[K/(1+K)^2] \quad (33)$$

Using a similar line of reasoning, it can be shown that the partial specific adiabatic compressibility, $k°_S$, of a protein determined from ultrasonic velocity measurements is given by the equation:

$$k^\circ_S = k_{SN} + [K/(1+K)]\Delta k_S + [\Delta v(\Delta v - \alpha \Delta h/\rho c_P)/RT] [K/(1+K)^2](1+\omega^2\tau^2)^{-1} \quad (34)$$

where $\Delta k_S = k_{SD} - k_{SN}$; k_{SN} and k_{SD} are the partial specific adiabatic compressibilities of the native and denatured states, respectively; Δh is the differential specific enthalpy (per gram of protein) of the denatured and native states; ω is the angular frequency of ultrasonic waves; and τ is the relaxation time of the native-to-denatured protein transition.

Since the coefficient of thermal expansibility, α, of an aqueous solution is small while its specific heat capacity, c_P, is large, Eq. 34 can be simplified to the form:

$$k^\circ_S = k_{SN} + [K/(1+K)]\Delta k_S + (\Delta v^2/RT)[K/(1+K)^2](1+\omega^2\tau^2)^{-1} \quad (35)$$

Inspection of Eq. 35 reveals that the transition-induced change in protein compressibility, $k°_S - k_{SN}$, consists of the two terms. The structural term, $[K/(1+K)]\Delta k_S$, is frequency-independent and reflects the transition-induced change in the equilibrium distribution between the native and denatured protein species. The relaxation term, $(\Delta v^2/RT)[K/(1+K)^2](1+\omega^2\tau^2)^{-1}$, is frequency-dependent and reflects the pressure-induced change in the distribution between the two species. It should be noted that, the structural term in Eq. 35 can be either positive or negative (depending on the sign of Δk_S), while the relaxation term is always positive. If the ultrasonic velocity measurements were performed at very low frequencies ($\omega \approx 0$), the relaxation term in Eq. 35 would exhibit its maximum value. An increase in frequency, however, will bring about a decrease in the relaxation term, which, at very high frequencies ($\omega\tau \gg 1$), subsides to zero.

Relaxation times, τ, for protein folding/unfolding transitions are, typically, between milliseconds and days. On the other hand, due to technical difficulties, frequencies of ultrasonic velocimetric measurements are usually limited to the range of 1 to 10 MHz (i.e., $\omega\tau \gg 1$). Consequently, as emphasized by Rösgen and Hinz [140], in protein compressibility studies performed by means of ultrasonic measurements, the relaxation term in Eq. 35 is negligible and cannot be used for gaining practical information about the kinetics of protein folding/unfolding transitions. However, should a high precision method for direct measurements of isothermal compressibility in protein solutions be developed, the relaxation term can be determined based on Eq. 33.

Fig. 6 presents our calculated high-frequency and low-frequency profiles for the heat-induced denaturation of a hypothetical protein which exhibits either a positive (Fig. 6A) or a negative (Fig. 6B) value of Δk_S. Inspection of Fig. 6 reveals that the high-frequency transition profiles have a classical sigmoidal shape expected for a cooperative, two-state transition. In contrast, the low-frequency profiles deviate from the sigmoidal shape with the deviation being proportional to Δv^2. Thus, in principle, the difference between the low-frequency and high-frequency profiles can be used to evaluate the transition volume. In addition, inspection of Eq. 35 reveals that frequency-dependent studies may yield kinetic information about the relaxation time of the transition, τ. However, such low-frequency ultrasonic studies are not yet possible due to unresolved technical difficulties. Therefore, all reported ultrasonically detected transition profiles for temperature-, pH-, denaturant-, salt-, and cosolvent-induced conformational changes of proteins are of the high frequency, sigmoidal type.

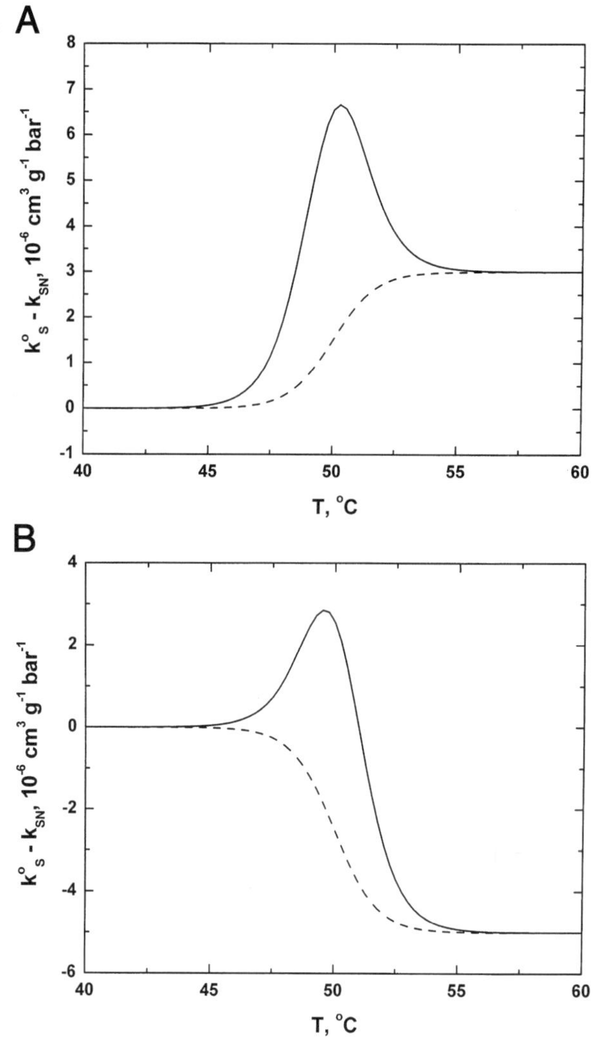

Fig. 6. The temperature dependence of the partial specific adiabatic compressibility, $k°_S$, of a hypothetical globular protein as calculated from Eq. 35: (a) $\Delta k_S = 3\times10^{-6}$ cm^3 g^{-1} bar^{-1}; (b) $\Delta k_S = -5\times10^{-6}$ cm^3 g^{-1} bar^{-1}. The transition volume, Δv, is 0.005 cm^3 g^{-1}. The equilibrium constant, K, is given by $(T/T_M)^{\Delta C_P/R}\exp[(T_M\Delta C_P-\Delta H°)(T^{-1}-T_M^{-1})/R]$, where the denaturation temperature, T_M, is 50°C; the heat capacity change, ΔC_P, associated with protein denaturation is 1 kcal mol^{-1} K^{-1}; and the transition enthalpy, ΔH, is 100 kcal mol^{-1}. The solid and dashed lines represent the calculations performed for low ($\omega\tau \approx 0$) and high ($\omega\tau \approx 1$) frequencies of ultrasonic waves, respectively.

6. Compressibility changes associated with protein transitions: Experimental observations

Proteins may exist in their native, molten globule (compact intermediate), partially unfolded, or fully unfolded states. Transitions between these states can be induced by altering temperature, pressure, pH, salt, denaturant, and cosolvent concentration. Compressibility measurements have been applied to characterizing transition profiles, although such measurements are still relatively scarce. For the simplest case of two-state, sigmoidal transitions (with the exception of pH-induced transitions), the change in compressibility can be determined by extrapolating the 'native' and 'denatured' baselines to the transition midpoint and calculating the difference.

For a pH-induced transition, the situation is more complex since an alteration in the solution pH results in a change in the state of ionization of protein titrable groups with a concomitant change in hydration. By itself, a change in the state of ionization of titrable groups may or may not result in a global change in protein conformation. However, independent of the conformational state of the protein, the change in protein hydration due to protonation/deprotonation of ionizable groups will bring about a change in compressibility. The overall pH-induced change in protein compressibility can be presented as the sum of the three terms:

$$\Delta k_S(\mathrm{pH}) = \Delta k_{\mathrm{ion}} + \Delta k_{\mathrm{conf}} + \Delta k_{\mathrm{rel}} \quad (36)$$

where the relaxation term, Δk_{rel}, is mostly related to proton-transfer reactions as discussed above in Section 4.

The ionization term, Δk_{ion}, in Eq. 36 is due to the change in hydration of ionizable groups upon their protonation/deprotonation. For protonation reactions, which predominantly occur at acidic pH, Δk_{ion} is given by the expression:

$$\Delta k_{\mathrm{ion}} = M^{-1}\sum_i \Delta K_{si}(1 + 10^{\mathrm{pH}-\mathrm{p}K a i})^{-1} \quad (37a)$$

For deprotonation reactions, which predominantly occur at alkaline pH, Δk_{ion} is given by the expression:

$$\Delta k_{\mathrm{ion}} = M^{-1}\sum_i \Delta K_{Si}(1 + 10^{\mathrm{p}K a i-\mathrm{pH}})^{-1} \quad (37b)$$

where $\mathrm{p}K_{ai}$ is the dissociation constant of the ith titrable group; ΔK_{Si} is the molar change in adiabatic compressibility upon complete protonation/deprotonation of the ith group; and M is the molecular weight of the protein.

Assuming a two-state transition for protein unfolding, the pH-dependence of the structural term, Δk_{conf}, in Eq. 36 can be expressed by a sigmoidal function:

$$\Delta k_{\mathrm{conf}} = \Delta k_S K/(1 + K) \quad (38)$$

where K is the equilibrium constant that can be calculated if the $\mathrm{p}K_a$ values of all ionizable groups are known for the native and denatured protein states [24,25]; Δk_S is the pH-independent 'conformational' change in adiabatic compressibility equal to the difference in the partial specific adiabatic compressibility, k°_S, between the denatured and native protein states determined at the same pH.

6.1. Atmospheric pressure experiments

6.1.1. Native-to-molten globule transitions

Table 1 presents compilation of literature data on changes in adiabatic compressibility, Δk_S, accompanying native-to-molten globule (compact intermediate) transitions of globular proteins. There are only five proteins, for which the changes in compressibility upon the native-to-molten globule (N-to-MG) transition have been determined. Specifically, the following transitions have been studied: the acid-induced N-to-MG transitions of cytochrome c [54,95], human [134] and bovine [141] α-lactalbumin; the sorbitol-induced N-to-MG transition of cytochrome c [142]; the heat-induced N-to-MG transitions of ribonuclease A [143] and α-chymotrypsinogen A [136]; and the base-induced N-to-MG of bovine α-lactalbumin [141].

Table 1
Changes in adiabatic compressibility associated with native-to-molten globule transitions of globular proteins

Protein/transition	Δk_S, 10^{-6} cm^3 g^{-1} bar^{-1}	Ref.
Cytochrome c (acid-induced in 200 mM NaCl at 25°C)	0.3[a]	[54]
Cytochrome c (acid-induced in 200 mM CsCl at 25°C)	1.7[a]	[95]
Human α-lactalbumin (acid-induced in 50 mM KCl at 25°C)	1.8[a]	[134]
Bovine α-lactalbumin (acid-induced in 2 mM $CaCl_2$ at 25°C)	−0.6[a]	[141]
Bovine α-lactalbumin (base-induced in 2 mM $CaCl_2$ at 25°C)	1.6[a]	[141]
Cytochrome c (sorbitol-induced in water at 20°C)	11.2	[142]
Ribonuclease A (heat-induced in ~10 mM HCl at pH 1.9; T_M = 26°C)	4.0	[143]
α-Chymotrypsinogen A (heat-induced in 10 mM glycine buffer at pH 2.0; T_M = 41°C)	3.6	[136]

[a] Ionization effects have been taken into account according to Eqs. 36, 37a, 37b, 38.

Table 2
Changes in adiabatic compressibility associated with native-to-partially unfolded transitions of globular proteins

Protein/transition	Δk_S, 10^{-6} cm^3 g^{-1} bar^{-1}	Ref.
Cytochrome *c* (acid-induced in water at 25°C)	−3.8[a]	[95]
Cytochrome *c* (base-induced in water at 25°C)	−3.9[a]	[108]
Staphylococcal nuclease (acid-induced in 20 mM NaCl at 25°C)	−6.3[a]	[137]
β-Lactoglobulin (base-induced in 10 mM NaCl at 25°C)	−7.0[a]	[144]
Myoglobin (acid-induced in 100 mM acetic buffer at 20°C)	−4.8	[145]
Equinatoxin II (acid-induced in water at 25°C)	−9.5[a]	[138]

[a]Ionization effects have been taken into account according to Eqs. 36, 37a, 37b, 38.

Inspection of the data presented in Table 1 reveals that compressibility changes associated with N-to-MG transitions of globular proteins are generally positive (or slightly negative) ranging from -0.6×10^{-6} to 11.2×10^{-6} cm^3 g^{-1} bar^{-1}. It should be noted that that the outstanding value of 11.2×10^{-6} cm^3 g^{-1} bar^{-1} obtained for the sorbitol-induced N-to-MG transition of cytochrome *c* [142] may be an overestimate, since this value was determined by a linear extrapolation to zero of very 'noisy' compressibility data obtained at high concentrations of sorbitol. As discussed above (see Section 5.1), experimental values of Δk_S accompanying N-to-MG transitions of globular proteins can be accounted for by a relatively simple analysis which takes into account transition-induced changes in the intrinsic and hydration contributions to protein compressibility. The mostly positive values of Δk_S for N-to-MG transitions reflect the increase in the intrinsic compressibility, k_M, which prevails over the decrease in the hydration compressibility, Δk_h.

6.1.2. Native-to-partially unfolded transitions

Table 2 presents compilation of literature data on changes in adiabatic compressibility, Δk_S, accompanying native-to-partially unfolded (N-to-PU) transitions of globular proteins. Specifically, Table 2 presents the data for the base-induced N-to-PU transition of cytochrome *c* [108] and β-lactoglobulin [144] and the acid-induced N-to-PU transitions of cytochrome *c* [95], staphylococcal nuclease [137], equinatoxin II [138], and myoglobin [145].

Inspection of the data presented in Table 2 reveals that the compressibility changes associated with N-to-PU transitions of globular proteins are all negative ranging from -3.8×10^{-6} to -9.5×10^{-6} cm^3 g^{-1} bar^{-1}. As discussed above (see Section 5.1), the observed values of Δk_S for N-to-PU transitions of globular proteins can be accounted for by decreases in the intrinsic and hydration contributions to protein compressibility.

6.1.2. Native-to-fully unfolded transitions

Table 3 presents the data on changes in adiabatic compressibility, Δk_S, accompanying the GuHCl-induced native-to-fully unfolded (N-to-FU) transitions of ribonuclease A [143] and lysozyme [146]. These changes are equal to -11×10^{-6} to -18×10^{-6} cm^3 g^{-1} bar^{-1}, respectively. As discussed above (see Section 5.1) such large decreases in compressibility are anticipated for N-to-FU transitions of globular proteins when the intrinsic compressibility diminishes practically to zero and the majority of protein groups becomes exposed to solvent.

6.2. Pressure-induced transitions

Table 4 presents compilation of literature data on changes in isothermal compressibility, Δk_T, accompanying pressure-induced denaturation of chymotrypsi-

Table 3
Changes in adiabatic compressibility associated with native-to-fully unfolded transitions of globular proteins

Protein/transition	Δk_S, 10^{-6} cm^3 g^{-1} bar^{-1}	Ref.
Ribonuclease A (GuHCl-induced at pH 2.0 and 15°C)	−18	[143]
Lysozyme (GuHCl-induced at pH 4.0 and 25°C)	−11	[146]

Table 4
Changes in isothermal compressibility associated with pressure-induced denaturation of globular proteins

Protein/transition	Δk_T, 10^{-6} cm^3 g^{-1} bar^{-1}	Ref.
Ribonuclease A (pressure-induced at pH 2.0–4.0 and temperatures of 20–50°C)	0.7–1.5	[45]
Chymotrypsinogen (pressure-induced at pH 2.07 and 0°C)	1.2	[44]
Metmyoglobin (pressure-induced at pH 4.0–6.0 and 20°C)	−0.2	[147]
Ribonuclease A (pressure-induced at pH 2.0 and 22°C)	1.1	[66]
Staphylococcal nuclease (pressure-induced at pH 7.0 and 45°C)	2.4	[67]
Staphylococcal nuclease (pressure-induced at pH 5.3 and 24°C)	0.9	[148]

nogen [44], ribonuclease A [45,66], metmyoglobin [147], and staphylococcal nuclease [67,148]. Inspection of the data presented in Table 4 reveals that pressure-induced conformational transitions of globular proteins are generally characterized by positive values of Δk_S. In other words, the partial compressibility of the pressure-induced denatured state of a protein is generally larger than that of the native state. It is difficult to unequivocally rationalize this observation. Four explanations can be hypothesized. Firstly, in all of these studies (except [67]), the values of Δk_T have been calculated based on a two-state analysis of the denaturation profiles monitored by spectroscopic observables. Hence, should the transitions deviate from the two-state behavior (in other words, if intermediates are accumulated at some stage of the transition), calculations based on a two-state analysis may yield erroneous results.

Secondly, the pressure-induced denatured state of a globular protein may retain a sizeable core of loosely packed, water-inaccessible amino acid residues. In this respect, the pressure-induced denatured state of a protein may resemble a molten globule thereby exhibiting an enhanced compressibility.

Thirdly, as proposed by Kharakoz [94], the partial compressibility of the unfolded state may increase with increasing pressure more steeply than that of the native state. Consequently, at some elevated pressure, the compressibility difference between the denatured and native states will become positive even if the initial difference at atmospheric pressure was negative. In this scenario, the pressure-induced denatured state of a protein may be either unfolded or molten globule-like, in both cases exhibiting a partial compressibility greater than that of the native state.

Finally, it would not be unexpected if, in future studies (especially when Δk_T is measured directly), some proteins will be found to exhibit negative values of Δk_T for their pressure-induced conformational transitions. Clearly, further pressure-related studies on a larger set of globular proteins are required to determine the validity of each of the above outlined possibilities/explanations.

7. Concluding remarks

Recent advances in the development of highly sensitive, small volume densimetric, acoustic, and high-pressure spectroscopic instrumentation have facilitated compressibility studies on protein systems. This experimental capability is significant since compressibility data are uniquely sensitive to the hydration properties and intrinsic packing of proteins, while providing complementary information to other techniques for defining the nature of the forces that control stability and conformational transitions of proteins. In this article, we have reviewed recent progress in volumetric investigations of conformational transitions of proteins, with particular attention given to the partial compressibility of the native and denatured protein states.

Despite the limited number of compressibility measurements that have been conducted on proteins and their low molecular weight analogs, it already is clear that this characteristics is exquisitely sensitive to intrinsic packing and hydration properties. This sensitivity is particularly important considering the role of hydration and intraglobular interactions in stabilizing protein structure and in modulating protein recognition events. We believe that compressibility measurements represent a relatively untapped yet powerful means of probing and characterizing all interactions stabilizing/destabilizing the native and denatured protein states, particularly, as they relate to the problem of protein folding.

Acknowledgements

We would like to thank Dr. Armen P. Sarvazyan for his critical comments and many stimulating discussions and Mr. Arno G. Siraki for his careful reading of the manuscript.

References

[1] M. Levitt, M. Gerstein, E. Huang, S. Subbiah, J. Tsai, Protein folding: the endgame, Annu. Rev. Biochem. 66 (1997) 549–579.

[2] J.N. Onuchic, Z. Luthey-Schulten, P.G. Wolynes, Theory of protein folding: the energy landscape perspective, Annu. Rev. Phys. Chem. 48 (1997) 545–600.

[3] C.M. Dobson, A. Šali, M. Karplus, Protein folding: a perspective from theory and experiment, Angew. Chem. Int. Ed. 37 (1998) 868–893.

[4] D.V. Laurents, R.L. Baldwin, Protein folding: matching theory and experiment, Biophys. J. 75 (1998) 428–434.

[5] R.L. Baldwin, G.D. Rose, Is protein folding hierarchic? I. Local structure and peptide folding, Trends Biochem. Sci. 24 (1999) 26–33.

[6] R.L. Baldwin, G.D. Rose, Is protein folding hierarchic? II. Folding intermediates and transition states, Trends Biochem. Sci. 24 (1999) 77–83.

[7] B. Honig, Protein folding: from the Levinthal paradox to structure prediction, J. Mol. Biol. 293 (1999) 283–293.

[8] R. Jaenicke, Stability and folding of domain proteins, Prog. Biophys. Mol. Biol. 71 (1999) 155–241.

[9] K.A. Dill, Polymer principles and protein folding, Protein Sci. 8 (1999) 1166–1180.

[10] R. Jaenicke, H. Lilie, Folding and association of oligomeric and multimeric proteins, Adv. Protein Chem. 53 (2000) 329–401.

[11] O.V. Galzitskaya, D.N. Ivankov, A.V. Finkelstein, Folding nuclei in proteins, FEBS Lett. 489 (2001) 113–118.

[12] J.-E. Shea, C.L. Brooks III, From folding theories to folding proteins: a review and assessment of simulation studies of protein folding and unfolding, Annu. Rev. Phys. Chem. 52 (2001) 499–535.

[13] C.B. Anfinsen, Principles that govern the folding of protein chains, Science 181 (1973) 223–230.

[14] C.B. Anfinsen, H.A. Scheraga, Experimental and theoretical aspects of protein folding, Adv. Protein Chem. 29 (1975) 205–300.

[15] G.I. Makhatadze, P.L. Privalov, Energetics of protein structure, Adv. Protein Chem. 47 (1995) 308–425.

[16] C.N. Pace, K.L. Shaw, Linear extrapolation method of analyzing solvent denaturation curves, Proteins 4 (Suppl) (2000) 1–7.

[17] E. Freire, Thermodynamics of partly folded intermediates in proteins, Annu. Rev. Biophys. Biomol. Struct. 24 (1995) 141–165.

[18] M. Gross, R. Jaenicke, Proteins under pressure. The influence of high hydrostatic pressure on structure, function, and assembly of proteins and protein complexes, Eur. J. Biochem. 221 (1994) 617–630.

[19] M.R. Eftink, R. Ionescu, Thermodynamics of protein folding: questions pertinent to testing the validity of the two-state model, Biophys. Chem. 64 (1997) 175–197.

[20] A.D. Robertson, K.P. Murphy, Protein structure and the energetics of protein stability, Chem. Rev. 97 (1997) 1251–1267.

[21] R. Lumry, R. Biltonen, J.F. Brandts, Validity of the 'two-state' hypothesis for conformational transitions of proteins, Biopolymers 4 (1966) 917–944.

[22] P.L. Privalov, Stability of proteins. Small globular proteins, Adv. Protein Chem. 33 (1979) 1–71.

[23] W. Kauzmann, Some factors in the interpretation of protein denaturation, Adv. Protein Chem. 14 (1959) 1–63.

[24] C. Tanford, Protein denaturation. Parts A and B, Adv. Protein Chem. 23 (1968) 121–282.

[25] C. Tanford, Protein denaturation. Part C, Adv. Protein Chem. 24 (1970) 1–95.

[26] K.A. Dill, Dominant forces in protein folding, Biochemistry 29 (1990) 7133–7155.

[27] W. Kauzmann, Thermodynamics of unfolding, Nature 325 (1987) 763–764.

[28] C.N. Pace, Determination and analysis of urea and guanidine hydrochloride denaturation curves, Methods Enzymol. 131 (1986) 266–280.

[29] G.I. Makhatadze, Thermodynamics of protein interactions with urea and guanidinium hydrochloride, J. Phys. Chem. B 103 (1999) 4781–4785.

[30] K. Heremans, High pressure effects on proteins and other biomolecules, Annu. Rev. Biophys. Bioeng. 11 (1982) 1–21.

[31] K. Heremans, L. Smeller, Protein structure and dynamics at high pressure, Biochim. Biophys. Acta 1386 (1998) 353–370.

[32] C.A. Royer, Application of pressure to biochemical equilibria: the other thermodynamic variable, Methods Enzymol. 259 (1995) 357–377.

[33] C.R. Robinson, S.G. Sligar, Hydrostatic and osmotic pressure as tools to study macromolecular recognition, Methods Enzymol. 259 (1995) 395–427.

[34] J.S. Silva, G. Weber, Pressure stability of proteins, Annu. Rev. Phys. Chem. 44 (1993) 89–113.

[35] H. Durchschlag, Specific volumes of biological macromolecules and some other molecules of biological interest, in: H.-J. Hinz (Ed.), Thermodynamic Data for Biochemistry and Biotechnology, Springer, Tokyo, 1986, pp. 45–128.

[36] A.A. Zamyatnin, Amino acid, peptide, and protein volume in solution, Annu. Rev. Biophys. Bioeng. 13 (1984) 145–165.

[37] T.V. Chalikian, A.P. Sarvazyan, K.J. Breslauer, Hydration and partial compressibility of biological compounds, Biophys. Chem. 51 (1994) 89–109.

[38] T.V. Chalikian, K.J. Breslauer, Thermodynamic analysis of

biomolecules: a volumetric approach, Curr. Opin. Struct. Biol. 8 (1998) 657–664.
[39] T.V. Chalikian, K.J. Breslauer, On volume changes accompanying conformational transitions of biopolymers, Biopolymers 39 (1996) 619–626.
[40] T.V. Chalikian, K.J. Breslauer, Compressibility as a means to detect and characterize globular protein states, Proc. Natl. Acad. Sci. USA 93 (1996) 1012–1014.
[41] H.-J. Hinz, T. Vogl, R. Meyer, An alternative interpretation of the heat capacity changes associated with protein unfolding, Biophys. Chem. 52 (1994) 275–285.
[42] B. Jacobson, On the adiabatic compressibility of aqueous solutions, Ark. Kemi 2 (1950) 177–210.
[43] G.R. Anderson, A study of the pressure dependence of the partial specific volume of macromolecules in solution by compression measurements in the range 1–8000 arm, Ark. Kemi 20 (1963) 513–571.
[44] S.A. Hawley, Reversible pressure-temperature denaturation of chymotrypsinogen, Biochemistry 10 (1971) 2436–2442.
[45] J.F. Brandts, R.J. Oliveira, C. Westort, Thermodynamics of protein denaturation. Effect of pressure on the denaturation of ribonuclease A, Biochemistry 9 (1970) 1038–1047.
[46] K. Gekko, H. Noguchi, Compressibility of globular proteins in water at 25°C, J. Phys. Chem. 83 (1979) 2706–2714.
[47] K. Gekko, Y. Hasegawa, Compressibility-structure relationship of globular proteins, Biochemistry 25 (1986) 6563–6571.
[48] A.P. Sarvazyan, Ultrasonic velocimetry of biological compounds, Mol. Biol. (USSR) 17 (1983) 739–749.
[49] A.P. Sarvazyan, Ultrasonic velocimetry of biological compounds, Annu. Rev. Biophys. Biophys. Chem. 20 (1991) 321–342.
[50] D. Eden, J.B. Matthew, J.J. Rosa, F.M. Richards, Increase in apparent compressibility of cytochrome *c* upon oxidation, Proc. Natl. Acad. Sci. USA 79 (1982) 815–819.
[51] B. Gavish, E. Gratton, C.J. Hardy, Adiabatic compressibility of globular proteins, Proc. Natl. Acad. Sci. USA 80 (1983) 750–754.
[52] F.J. Millero, G.K. Ward, P. Chetirkin, Partial specific volume, expansibility, compressibility, and heat capacity of aqueous lysozyme solutions, J. Biol. Chem. 251 (1976) 4001–4004.
[53] D. Valdez, J.-Y. Le Huérou, M. Gindre, W. Urbach, M. Waks, Hydration and protein folding in water and in reverse micelles: compressibility and volume changes, Biophys. J. 80 (2001) 2751–2760.
[54] B. Nölting, S.G. Sligar, Adiabatic compressibility of molten globules, Biochemistry 32 (1993) 12319–12323.
[55] M. Iqbal, R.E. Verrall, Implications of protein folding. Additivity schemes for volumes and compressibilities, J. Biol. Chem. 263 (1988) 4159–4165.
[56] M.J. Blandamer, M.I. Davis, G. Douhéret, J.C.R. Reis, Apparent molar isentropic compressions and expansions of solutions, Chem. Soc. Rev. 30 (2001) 8–15.
[57] S. Barnatt, The velocity of sound in electrolytic solutions, J. Chem. Phys. 20 (1952) 278–279.
[58] B.B. Owen, H.L. Simons, Standard partial molal compressibilities by ultrasonics. 1. Sodium chloride and potassium chloride at 25°C, J. Phys. Chem. 61 (1957) 479–482.
[59] D.W. Kupke, Density and volume change measurements, in: S.J. Loch (Ed.), Physical Principles and Techniques of Protein Chemistry, Part C, Academic Press, New York, 1973, pp. 1–75.
[60] O. Kratky, H. Leopold, H. Stabinger, The determination of the partial specific volume of proteins by the mechanical oscillator technique, Methods Enzymol. 27 (1973) 98–110.
[61] F. Eggers, Th. Funck, Ultrasonic measurements with millilitre liquid samples in the 0.5-100 MHz range, Rev. Sci. Instrum. 44 (1973) 969–978.
[62] A.P. Sarvazyan, Development of methods of precise ultrasonic measurements in small volumes of liquids, Ultrasonics 20 (1982) 151–154.
[63] F. Eggers, Ultrasonic velocity and attenuation measurements in liquids with resonator, extending the MHz frequency range, Acustica 76 (1992) 231–240.
[64] F. Eggers, U. Kaatze, Broad-band ultrasonic measurement techniques for liquids, Meas. Sci. Technol. 7 (1996) 1–19.
[65] A.P. Sarvazyan, T.V. Chalikian, Theoretical analysis of an ultrasonic interferometer for precise measurements at high pressures, Ultrasonics 29 (1991) 119–124.
[66] K.E. Prehoda, E.S. Mooberry, J.L. Markley, Pressure denaturation of proteins: evaluation of compressibility effects, Biochemistry 37 (1998) 5785–5790.
[67] H. Seemann, R. Winter, C.A. Royer, Volume, expansivity, and isothermal compressibility changes associated with temperature and pressure unfolding of staphylococcal nuclease, J. Mol. Biol. 307 (2001) 1091–1102.
[68] T.V. Chalikian, K.J. Breslauer, Volumetric properties of nucleic acids, Biopolymers 48 (1998) 264–280.
[69] A. Cooper, Thermodynamic fluctuations in protein molecules, Proc. Natl. Acad. Sci. USA 73 (1976) 2740–2741.
[70] A. Cooper, Protein fluctuations and the thermodynamic uncertainty principle, Prog. Biophys. Mol. Biol. 44 (1984) 181–214.
[71] F.J. Millero, A. Lo Surdo, C. Shin, The apparent molal volumes and adiabatic compressibilities of aqueous amino acids at 25°C, J. Phys. Chem. 82 (1978) 784–792.
[72] A.P. Sarvazyan, D.P. Kharakoz, P. Hemmes, Ultrasonic investigation of the pH-dependent solute–solvent interactions in aqueous solutions of amino acids and proteins, J. Phys. Chem. 83 (1979) 1796–1799.
[73] M. Iqbal, R.E. Verrall, Partial molar volumes and adiabatic compressibilities of glycyl peptides at 25°C, J. Phys. Chem. 91 (1987) 967–971.
[74] J.V. Leyendekkers, Solutions of organic solutes. 1. Volume and compressibility of amino acids, J. Phys. Chem. 90 (1986) 5446–5455.
[75] G.R. Hedwig, H. Høiland, Thermodynamic properties of peptide solutions: 7. Partial molar isentropic pressure coefficients of some dipeptides in aqueous solutions, J. Solution Chem. 20 (1991) 1113–1127.
[76] G.R. Hedwig, H. Høiland, Thermodynamic properties of peptide solutions. 8. Isentropic pressure coefficients

$(\partial V_{2,\Phi}/\partial p)_S$ of the apparent molar volume $V_{2,\Phi}$ for each of the aqueous solutes: diglycine, triglycine, and tetraglycine, J. Chem. Thermodynamics 23 (1991) 1029–1035.

[77] D.P. Kharakoz, Volumetric properties of proteins and their analogues in diluted water solutions. 2. Partial adiabatic compressibilities of amino acids at 15–70°C, J. Phys. Chem. 95 (1991) 5634–5642.

[78] T.V. Chalikian, D.P. Kharakoz, A.P. Sarvazyan, C.A. Cain, R.J. McGough, I.V. Pogosova, T.N. Gareginian, Ultrasonic study of proton-transfer reactions in aqueous solutions of amino acids, J. Phys. Chem. 96 (1992) 876–883.

[79] T.V. Chalikian, A.P. Sarvazyan, K.J. Breslauer, Partial molar volumes, expansibilities, and compressibilities of α,ω-aminocarboxylic acids in aqueous solutions between 18 and 55°C, J. Phys. Chem. 97 (1993) 13017–13026.

[80] G.R. Hedwig, H. Høiland, Thermodynamic properties of peptide solutions. 9. Partial molar isentropic pressure coefficients in aqueous solutions of sequence isomeric tripeptides with a single -CH_3 side-chain, J. Chem. Thermodynamics 25 (1993) 349–354.

[81] T.V. Chalikian, A.P. Sarvazyan, Th. Funck, K.J. Breslauer, Partial molar volumes, expansibilities, and compressibilities of oligoglycines in aqueous solutions at 18–55°C, Biopolymers 34 (1994) 541–553.

[82] T.V. Chalikian, A.P. Sarvazyan, Th. Funck, C.A. Cain, K.J. Breslauer, The partial molar characteristics of glycine and alanine in aqueous solutions at high pressures calculated from ultrasonic velocity data, J. Phys. Chem. 98 (1994) 321–328.

[83] G.R. Hedwig, H. Høiland, Thermodynamic properties of peptide solutions. Part 11. Partial molar isentropic pressure coefficients in aqueous solutions of some tripeptides that model protein side-chains, Biophys. Chem. 49 (1994) 175–181.

[84] G.R. Hedwig, H. Høiland, Partial molar isentropic pressure coefficients of some N-acetyl amino acid and peptide amides at infinite dilution in aqueous solutions at the temperature 298.15 K, J. Chem. Thermodynamics 27 (1995) 745–750.

[85] G.R. Hedwig, H. Høiland, Partial molar isothermal and isentropic compressibilities of glycine, alanine, and glycylglycine in aqueous solutions at 25°C, J. Phys. Chem. 99 (1995) 12063–12064.

[86] G.R. Hedwig, J.D. Hastie, H. Høiland, Thermodynamic properties of peptide solutions: 14. Partial molal expansibilities and isothermal compressibilities of some glycyl dipeptides in aqueous solution, J. Solution Chem. 25 (1996) 615–633.

[87] G.R. Hedwig, H. Høiland, E. Høgseth, Thermodynamic properties of peptide solutions. Part 15. Partial molar isentropic compressibilities of some glycyl dipeptides in aqueous solution at 15 and 35°C, J. Solution Chem. 25 (1996) 1041–1053.

[88] T.V. Chalikian, V.S. Gindikin, K.J. Breslauer, Hydration of diglycyl tripeptides with non-polar side chains: a volumetric study, Biophys. Chem. 75 (1998) 57–71.

[89] O. Likhodi, T.V. Chalikian, Partial molar volumes and adiabatic compressibilities of a series of aliphatic amino acids and oligoglycines in D_2O, J. Am. Chem. Soc. 121 (1999) 1156–1163.

[90] A.W. Hakin, H. Høiland, G.R. Hedwig, Volumetric properties of some oligopeptides in aqueous solution: partial molar expansibilities and isothermal compressibilities at 298.15 K for the peptides of sequence Ala(gly)$_n$, $n=1$–4, Phys. Chem. Chem. Phys. 2 (2000) 4850–4857.

[91] O. Likhodi, T.V. Chalikian, Differential hydration of α,ω-aminocarboxylic acids in D_2O and H_2O, J. Am. Chem. Soc. 122 (2000) 7860–7868.

[92] T.V. Chalikian, Ultrasonic and densimetric characterizations of hydration properties of polar groups in monosaccharides, J. Phys. Chem. B 102 (1998) 6921–6926.

[93] T.V. Chalikian, Structural thermodynamics of solute hydration, J. Phys. Chem. 105 (2001) 12566–12578.

[94] D.P. Kharakoz, Partial volumes and compressibilities of extended polypeptide chains in aqueous solution: additivity scheme and implication of protein unfolding at normal and high pressure, Biochemistry 36 (1997) 10276–10285.

[95] T.V. Chalikian, V.S. Gindikin, K.J. Breslauer, Volumetric characterizations of the native, molten globule, and unfolded states of cytochrome *c* at acidic pH, J. Mol. Biol. 250 (1995) 291–306.

[96] K. Gekko, Y. Hasegawa, Effect of temperature on the compressibility of native globular proteins, J. Phys. Chem. 93 (1989) 426–429.

[97] T.V. Chalikian, M. Totrov, R. Abagyan, K.J. Breslauer, Hydration of globular proteins as derived from the volume and compressibility measurements: Cross correlating thermodynamic and structural data, J. Mol. Biol. 260 (1996) 588–603.

[98] A.P. Sarvazyan, P. Hemmes, Relaxational contributions to protein compressibility from ultrasonic data, Biopolymers 18 (1979) 3015–3024.

[99] J. Stuehr, E. Yeager, The propagation of ultrasonic waves in electrolytic solutions, in: W.P. Mason (Ed.), Physical Acoustics. Vol. II, Part A, Academic Press, New York, 1965, pp. 351–462.

[100] M. Hussey, P.D. Edmonds, Proton transfer reactions. A mechanism for the absorption of ultrasound in aqueous solutions of proteins, J. Phys. Chem. 75 (1971) 4012–4019.

[101] R. Zana, C. Tondre, Ultrasonic studies of proton transfers in solutions of poly(lysine) and poly(ornithine). Implications for the kinetics of the helix-coil transition of polypeptides and for the ultrasonic absorption of proteins, J. Phys. Chem. 76 (1972) 1737–1743.

[102] H. Kanda, N. Ookubo, H. Nakajima, Y. Suzuki, M. Minato, T. Ihara, Y. Wada, Ultrasonic absorption in aqueous solution of lysozyme, Biopolymers 15 (1976) 785–795.

[103] K.C. Cho, W.P. Leung, H.Y. Mok, C.L. Choy, Ultrasonic absorption in myoglobin and other globular proteins, Biochim. Biophys. Acta 830 (1985) 36–44.

[104] P.K. Choi, J.-R. Bae, K. Takagi, Ultrasonic spectroscopy in bovine serum albumin solutions, J. Acoust. Soc. Am. 87 (1990) 874–881.

[105] W.D. O'Brien Jr., F. Dunn, Ultrasonic absorption mechanisms in aqueous solutions of bovine hemoglobin, J. Phys. Chem. 76 (1972) 528–533.
[106] L.J. Slutsky, L. Madsen, R.D. White, Acoustic absorption and proton-exchange kinetics in aqueous bovine pancreatic ribonuclease A, J. Phys. Chem. 88 (1984) 5679–5683.
[107] T. Yamashita, K. Tanaka, H. Yano, S. Harada, Micellar effect on the kinetics of the base equilibrium of amino acids studied by the ultrasonic absorption method, J. Chem. Soc. Faraday Trans. 87 (1991) 1857–1861.
[108] T.V. Chalikian, V.S. Gindikin, K.J. Breslauer, Spectroscopic and volumetric investigation of cytochrome *c* unfolding at alkaline pH: characterization of the base induced unfolded state at 25°C, FASEB J. 10 (1996) 164–170.
[109] F.M. Richards, Areas, volumes, packing, and protein structure, Annu. Rev. Biophys. Bioeng. 6 (1977) 151–176.
[110] F.M. Richards, W.A. Lim, An analysis of packing in the protein folding problem, Q. Rev. Biophys. 26 (1994) 423–498.
[111] A.A. Rashin, M. Iofin, B. Honig, Internal cavities and buried waters in globular proteins, Biochemistry 25 (1986) 3619–3625.
[112] D.P. Kharakoz, A.P. Sarvazyan, Hydrational and intrinsic compressibilities of globular proteins, Biopolymers 33 (1993) 11–26.
[113] D.P. Kharakoz, Nonlinear elasticity and dynamics of globular proteins, Biophysics 45 (2000) 603–614.
[114] D.P. Kharakoz, Protein compressibility, dynamics, and pressure, Biophys. J. 79 (2000) 511–525.
[115] C.E. Kundrot, F.M. Richards, Crystal structure of hen eggwhite lysozyme at a hydrostatic pressure of 1000 atmospheres, J. Mol. Biol. 193 (1987) 157–170.
[116] K. Akasaka, T. Tezuka, H. Yamada, Pressure-induced changes in the folded structure of lysozyme, J. Mol. Biol. 271 (1997) 671–678.
[117] H. Li, H. Yamada, K. Akasaka, Effect of pressure on individual hydrogen bonds in proteins. Basic pancreatic trypsin inhibitor, Biochemistry 27 (1998) 1167–1173.
[118] K. Akasaka, H. Li, H. Yamada, R. Li, T. Thoresen, C.K. Woodward, Pressure response of protein ackbone structure. Pressure-induced amide ^{15}N chemical shifts in BPTI, Protein Sci. 8 (1999) 1946–1953.
[119] H. Li, H. Yamada, K. Akasaka, Effect of pressure on the tertiary structure and dynamics of folded basic pancreatic trypsin inhibitor, Biophys. J. 77 (1999) 2801–2812.
[120] H.R. Kalbitzer, A. Görler, H. Li, P.V. Dubovskii, W. Hengstenberg, C. Kowolik, H. Yamada, K. Akasaka, ^{15}N and ^{1}H NMR study of histidine containing protein (HPr) from *Staphylococcus carnosus* at high pressure, Protein Sci. 9 (2000) 693–703.
[121] Y.O. Kamatari, H. Yamada, K. Akasaka, J.A. Jones, C.M. Dobson, L.J. Smith, Response of native and denatured hen lysozyme to high pressure studied by $^{15}N/^{1}H$ NMR spectroscopy, Eur. J. Biochem. 268 (2001) 1782–1793.
[122] M. Iwadate, T. Asakura, P.V. Dubovskii, H. Yamada, K. Akasaka, M.P. Williamson, Pressure-dependent changes in the structure of the melittin α-helix determined by NMR, J. Biomol. NMR 19 (2001) 115–124.
[123] J. Zollfrank, J. Friedrich, J. Fidy, J.M. Vanderkooi, Photochemical holes under high pressure: compressibility and volume fluctuations of a protein, J. Chem. Phys. 94 (1991) 8600–8603.
[124] E. Paci, M. Marchi, Intrinsic compressibility and volume compression in solvated proteins by molecular dynamics simulation at high pressure, Proc. Natl. Acad. Sci. USA 93 (1996) 11609–11614.
[125] A. Katrusiak, Z. Dauter, Compressibility of lysozyme protein crystals by X-ray diffraction, Acta Cryst. D 52 (1996) 607–608.
[126] V.M. Dadarlat, C.B. Post, Insights into protein compressibility from molecular dynamics simulations, J. Phys. Chem. B 105 (2001) 715–724.
[127] E. Paci, M. Marchi, On the volume of macromolecules, Biopolymers 41 (1997) 785–797.
[128] K. Gekko, Flexibility of globular proteins in water as revealed by compressibility, in: H. Levine, L. Slade (Eds.), Water Relationships in Food, Plenum, New York, 1991, pp. 753–771.
[129] S. Miller, J. Janin, A.M. Lesk, C. Chothia, Interior and surface monomeric proteins, J. Mol. Biol. 196 (1987) 641–656.
[130] J.M. Vanderkooi, The protein state of matter, Biochim. Biophys. Acta 1386 (1998) 241–253.
[131] O.B. Ptitsyn, The molten globule state, in: T.E. Creighton (Ed.), Protein Folding, Freeman, New York, 1992, pp. 243–300.
[132] O.B. Ptitsyn, Molten globule and protein folding, Adv. Protein Chem. 47 (1995) 83–229.
[133] M. Arai, K. Kuwajima, Role of the molten globule state in protein folding, Adv. Protein Chem. 53 (2000) 209–282.
[134] D.P. Kharakoz, V.E. Bychkova, Molten globule of human α-lactalbumin: hydration, density, and compressibility of the interior, Biochemistry 36 (1997) 1882–1890.
[135] B. Lee, Isoenthalpic and isoentropic temperatures and the thermodynamics of protein denaturation, Proc. Natl. Acad. Sci. USA 88 (1991) 5154–5158.
[136] T.V. Chalikian, J. Völker, D. Anafi, K.J. Breslauer, The native and the heat-induced denatured states of α-chymotrypsinogen A: thermodynamic and spectroscopic studies, J. Mol. Biol. 274 (1997) 237–252.
[137] R. Filfil, T.V. Chalikian, Volumetric and spectroscopic characterizations of the native and acid-induced denatured states of staphylococcal nuclease, J. Mol. Biol. 299 (2000) 829–844.
[138] N. Poklar, J. Völker, G. Anderluh, P. Macek, T.V. Chalikian, Acid and base induced conformational transitions of equinatoxin II, Biophys. Chem. 90 (2001) 103–121.
[139] J. Rösgen, H.-J. Hinz, Statistical thermodynamic treatment of conformational transitions of monomeric and oligomeric proteins, Phys. Chem. Chem. Phys. 1 (1999) 2327–2333.
[140] J. Rösgen, H.-J. Hinz, Response functions of proteins, Biophys. Chem. 83 (2000) 61–71.

[141] V. Gindikin, Charge Effects in Proteins: Volumetric Study, Ph.D. Thesis, Rutgers, The State University of New Jersey, USA, 2000.

[142] T. Kamiyama, Y. Sadahide, Y. Nogusa, K. Gekko, Polyol-induced molten globule of cytochrome *c*: an evidence for stabilization by hydrophobic interaction, Biochim. Biophys. Acta 1434 (1999) 44–57.

[143] Y. Tamura, K. Gekko, Compactness of thermally and chemically denatured ribonuclease A as revealed by volume and compressibility, Biochemistry 34 (1995) 1878–1884.

[144] N. Taulier, T.V. Chalikian, Characterization of pH-induced transitions of β-lactoglobulin: ultrasonic, densimetric, and spectroscopic studies, J. Mol. Biol. 314 (2001) 873–889.

[145] W.P. Leung, K.C. Cho, Y.M. Lo, C.L. Choy, Adiabatic compressibility of myoglobin. Effect of axial ligand and denaturation, Biochim. Biophys. Acta 870 (1986) 148–153.

[146] T. Kamiyama, K. Gekko, Compressibilty and volume change of lysozyme due to guanidine hydrochloride denaturation, Chem. Lett. (1997) 1063-1064.

[147] A. Zipp, W. Kauzmann, Pressure denaturation of metmyoglobin, Biochemistry 12 (1973) 4217–4228.

[148] M.W. Lassale, H. Yamada, K. Akasaka, The pressure-temperature free energy-landscape of staphylococcal nuclease monitored by ^{1}H NMR, J. Mol. Biol. 298 (2000) 293–302.

ELSEVIER

Biochimica et Biophysica Acta 1595 (2002) 71–79

BIOCHIMICA ET BIOPHYSICA ACTA

BBA

www.bba-direct.com

Review

Effects of high pressure on enzymatic activity

Dexter B. Northrop *

Division of Pharmaceutical Sciences, School of Pharmacy, 777 Highland Avenue, University of Wisconsin-Madison, Madison, WI 53705, USA

Received 28 September 2001; received in revised form 2 November 2001; accepted 6 November 2001

Abstract

Effects of high pressure on enzymatic reactions are poised to revolutionize enzyme kinetics. The reason for this is that experimental designs are at hand to separate effects on equilibria between reactant states from effects on catalytic transition states and both yield new information. The first of the former runs contrary to Pauling's hypothesis that substrates are bound more tightly in the transition state, while the latter penetrates the 'black box' of catalysis, the stabilized transition state itself, and returns a precise measure of a physical parameter, $\Delta V^{\ddagger}$. This in turn opens the door to new forms of structure–activity relationships. The first of these has been described, the effect of pressure on isotope effects, with the surprising finding that the entire isotope effect comes from a transition state phenomenon such as quantum mechanical hydrogen tunneling.

Keywords: Steady-state kinetics; Pauling's hypothesis; Transition state; Structure–activity relationship; Isotope effect; Quantum mechanical hydrogen tunneling; Substrate capture; Commitments to catalysis; Yeast alcohol dehydrogenase; Yeast formate dehydrogenase; Coupled motion; Substrate oscillation; Vibrationally enhanced ground state; Torsional motion

1. Introduction

When confronted with theoretical considerations of enzyme kinetics at high pressure, most people ask: 'what happens to the enzyme protein?' The answer is, 'it doesn't matter'. The physical origins of changes in reaction rates is completely irrelevant to their kinetic analysis. All pressure does is change the distribution between equilibria that are already present at atmospheric pressure, according to Le Chatelier's principle. It is in this sense that pressure is non-invasive; it adds nothing new to the system. These changes in distribution obey the function[1]:

$$K_p = K_0 e^{-\Delta V_K p/RT} \tag{1}$$

where K_0 is an equilibrium constant under zero or atmospheric conditions (the difference is insignificant), ΔV_K is the volume difference between the two states at equilibrium, p is the pressure, R is the gas constant, and T is the temperature. When multiple equilibria are present, data as a function of pressure are fitted to the appropriate algorithm consisting of multiple representations of Eq. 1 combined in different ways. The final fit is, in effect, an extrapolation back to atmospheric pressure, and the result-

* Fax: +1-608-262-3397.
E-mail address: dbnorthrop@pharmacy.wisc.edu
(D.B. Northrop).

[1] A correct fit at the level of Eq. 1 returns the equivalent of a pair of asymptotes in the log plot, not tangents.

0167-4838/02/$ – see front matter
PII: S0167-4838(01)00335-1

ing parameters describe the system under normal conditions. What happens and why at extremes of pressure are of no consequence. The only matters that are of consequence are: that some changes (plural) do occur as a function of pressure; that the results of these changes be measured precisely; and that the changes fit one algorithm representing a unique chemical system and do not fit other algorithms depicting other systems.

Experimental effects of pressure on enzymatic activity have a long history, beginning with Eyring's study of invertase in 1946 [1]. In 1981, Morild tabulated positive and negative effects on apparent activity for 135 enzymes, with only eight showing a zero volume change [2]. Recent interest has focused on hyperthermophilic enzymes from extremophiles [3], but most kinetic studies yield ambiguous results because of uncertainties with regard to possible protein denaturation, what step or steps of a catalytic turnover are being affected, and complexities of data reduction when multiple effects are expressed. All that has changed. The breakthrough in overcoming these uncertainties required at least six of eight separate developments.

2. Recent developments in theory and experimental design

2.1. Analyzing non-linear functions

According to the absolute rate theory of Glastone et al. [4] a rate constant, k, can be considered a quasi-equilibrium constant between the reactant state and the transition state, $K^{\ddagger}$, multiplied by some physical constants. Hence, Eq. 1 may be recast for kinetics:

$$k_p = k_0 e^{-\Delta V^{\ddagger} p/RT} \tag{2}$$

where k_0 is the rate constant for the chemical transformation under atmospheric conditions, $\Delta V^{\ddagger}$ is the activation volume or volume difference between the reactant state and transition state. Physical chemists have a long tradition of taking the logarithm of Eq. 2 in order to reformulate it into a *monophasic* or linear form:

$$\ln(k_p) = \ln(k_0) - \Delta V^{\ddagger} p/RT \tag{3}$$

This latter construct made graphical analysis possible; plots of $\ln(k_p)$ versus p generate a straight line with an intercept of $\ln(k_0)$ and slope of $-\Delta V^{\ddagger}/RT$ in simple systems. Whenever a non-linear or *biphasic* plot resulted within such plots (called 'strange phenomena' by Morild [5]), physical chemists were stumped as to how one should proceed with any meaningful interpretation. If a log plot of activity is non-linear, which is the rule rather than an exception in enzyme kinetics, it is mathematically impossible to move in the opposite direction, from a curved logarithmic form of Eq. 3 back to something analogous to Eq. 2. Instead, the tradition from chemistry has been to fit such curves to an arbitrary quadratic equation:

$$\ln(k_p) = A + Bp + Cp^2 \tag{4}$$

Term A is obviously $\ln(k_0)$ and B is $-\Delta V^{\ddagger}/RT$ but the third term, C, dubbed 'compressibility' in thermodynamics, has no physical meaning in kinetics, either chemical or enzymatic. Moreover, it was recently shown that fittings of kinetic data to Eq. 4 are clearly incorrect as they lead to nonsensical isotope effects [6]. In contrast, by *not* taking logarithms and instead evaluating primary data by non-linear regression using modern computers rather than graphical analysis using pen and paper, it is now possible to escape the tyranny of the straight line and the shackles of archaic analytical geometry.

2.2. Changing paradigms

Throughout the history of the kinetics of enzyme-catalyzed reactions, the dominant paradigm was a simplistic formulation in which one transition state presented the highest activation energy barrier, and the chemical transformation of substrate to product was considered to be the rate-limiting step. A smaller barrier for substrate binding was consistent with the *rapid equilibrium binding hypothesis* employed by Michaelis and Menten [7]. Despite the demonstration by Briggs and Haldane [8] that this restriction was unnecessary and that their alternative *steady-state hypothesis* could account equally well for the kinetic behavior of enzymes, the simpler paradigm often persisted, quite satisfactorily in many applications. During the past two or more decades, however, a renaissance occurred in the use of isotope effects to study enzyme mechanisms. This was triggered by simply replacing the rapid equilibrium hypothesis with the steady-state hypotheses along with multiple reactive steps [9–12]. A similar paradigm shift has now been introduced to pressure effects, with dramatic results [13].

2.3. *Placing pressure effects fully in the context of modern steady-state kinetics*

With a few exceptions, most physical chemists who attempted to study enzymes at high pressures had only a textbook understanding of enzyme kinetics, within the old paradigm. They tried to interpret pressure effects on the maximal velocity, V_{max}, or k_{cat}, believed to be a single chemical transformation, and on the Michaelis–Menten constant, K_m, believed to be a binding phenomenon. Both are actually complex mixtures of reactant states and transition states that lay beyond the reach of pressure functions [14]. When multiple effects were addressed, a 'weighted mean' of the two activation volumes was sometimes computed to give a 'composite volume' which could then be addressed in a traditional logarithmic equation of pressure dependence [15]. The composite volume turned out to be the zero tangent to the resulting biphasic logarithmic curve [13] and there was no way back from the slope of this tangent to the operative rate equation nor even to values for either of the two volumes responsible for the curve. However, if reaction velocities are plotted as a function of substrate concentrations in the traditional double-reciprocal form (non-logarithmic) and pressure is treated as an inhibitor or activator, then useful kinetic *patterns of straight lines* can be obtained and interpreted [5,13].

2.4. *Addressing the kinetics of iso-mechanisms*

Everyone who considered the effects of pressure on enzymatic reactions has worried about the possibility of denaturing the enzyme protein, either as unfolding of the protein or by dissociation of subunits. Indeed, Balny and co-workers have specifically documented both phenomena with several enzymes, for example carboxypeptidase [16] and enolase [17], respectively, and attempted to isolate changes in catalytic rate constants by performing presteady-state kinetics at high pressure [14]. Others have devised indirect means to reduce this uncertainty in kinetic studies, such as conducting HPLC analyses immediately following high pressure treatment [18]. It turns out that gross phenomena which change chromatographic behavior, plus much more subtle conformational changes that do not, subsume nicely to the kinetics of iso-mechanisms [19]. The experimental design is based upon that of Parmentier et al. [20] who used isotope effects to distinguish between alternative models of allosteric regulation: the two-state concerted model of Monod et al. [21] and the multi-state sequential model of Koshland et al. [22]. Because an isotope effect reports back only from active enzyme, it can distinguish between a loss of activity due to *removal* of enzyme from participation in catalytic turnovers (the Monod model, which will not change the isotope effect on the remaining activity), and a loss due to *modulation* of rate constants contributing to catalytic turnovers (the Koshland model, which may change the expression of intrinsic isotope effects). This experimental design is elegant in its simplicity and power, and not only distinguishes protein thermodynamics from enzyme kinetics but also returns precise quantitative information on both.

To illustrate, the following Scheme 1 shows a Cleland line diagram of an iso-mechanism with two steps in the isomerization segment following the formation of a product form of enzyme, F: the conversion to an intermediate form, G (which could be the release of a second product or a second half-reaction), and isomerization of the intermediate back to the substrate form, E; The conversion of G to E is shown twice: first, as a reversible process in the beginning of the diagram which is how it appears at low [S], and second as an irreversible process at the end which is how it appears at high or infinite [S]. At

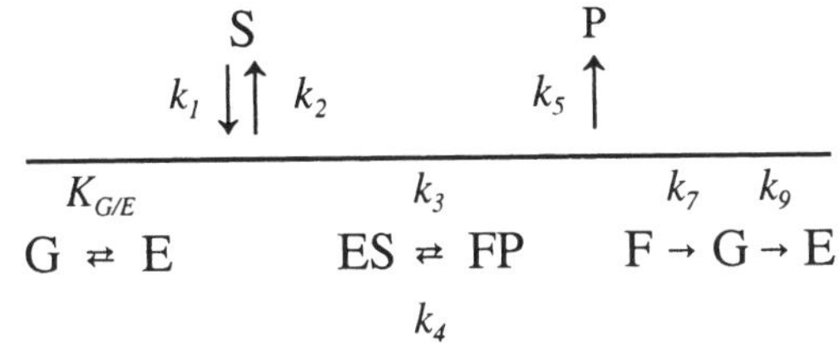

Scheme 1.

high pressure, G may stand for an obligate intermediate, as in Scheme 1, or any other form or forms that may co-exist with E, including partially unfolded or dissociated enzyme. Changes in the equilibrium between G and E encompass the Monod phenomena and exclude Koshland phenomena. Conversely, changes in the rate of conversion of ES to G follow the Koshland model and exclude the Monod model.

2.5. Understanding the kinetics of capture

In contrast to the various complexities of V_{max} and K_m, V_{max}-over-K_m is relatively simple algebraically. Contrary to popular belief, V/K is not an alternate measure of enzymatic catalysis but only a measure of substrate *capture* [23]. Particularly significant is the silence of all intermediate reactant states contributing to substrate capture, which makes possible the interpretation of complex pressure effects in fewer terms. These are: (a) an iso-mechanism equilibrium, (b) a single transition state, and (c) particular relationships between multiple transition states called *commitments to catalysis* [19]:

$$|V/K|_p = \left(\frac{k_1}{1 + K_{G/E}e^{-\Delta V_{G/E}p/RT}}\right) \cdot \left(\frac{R_0 e^{-\Delta V^{\ddagger}p/RT}}{1 + C_f e^{-\Delta V^{\ddagger}p/RT} + C_r e^{-(\Delta V^{\ddagger} - \Delta V_{Keq})p/RT}}\right) \quad (5)$$

The first expression in parentheses contains the diffusion-controlled rate constant for collisions between substrate and enzyme, k_1, which is least affected by changes in viscosity due to pressure at temperatures near 30°C [24]. The expression also contains the iso-mechanism equilibrium constant, $K_{G/E}$, between the form of enzyme, G, that is unable to participate in substrate capture and that which can, form E, governed by the volume change, $\Delta V_{G/E}$. In iso-mechanism kinetics, the G form is generated within a catalytic turnover, but as used here it includes small conformational changes and ionizations that are not necessarily part of a turnover, and may extend to include unfolding of the enzyme protein and even subunit dissociations; the algebra for the two scenarios is the same. The numerator of the second term contains a collection of rate constants, R_0, equal to k_3k_5/k_2k_4 in Scheme 1, governed by a *single* activation volume, $\Delta V^{\ddagger}$. The nature of this volume change is illustrated by the displacement of the vertical arrow in the activation energy diagram in Fig. 1, governed by k_1R_0. It is independent of intermittent changes that occur, in ES for example, because it is a state function, like entropy or equilibria, i.e. twists and turns in a garden hose do not affect the water levels at its two ends. This is the volume difference between free enzyme (or an enzyme–coenzyme complex, see below) and substrate, E+S, and the transition state of an isotopically sensitive step, here $EX^{\ddagger}$. The denominator of the second term contains the forward and reverse *commitments to catalysis*, C_f and C_r, equal to k_3/k_2 and k_4/k_5, respectively, in Scheme 1. Commitments signify only differences between various transition states contributing to substrate capture, meaning that concentrations of intermediates such as ES or EP cancel out, which is not true for V_{max}. Commitments may change with pressure, most likely as a simple function of $\Delta V^{\ddagger}$ as shown in Eq. 5 (with a correction for the volume difference between the last substrate and first product, ΔV_{Keq}), which keeps the number of variables to a minimum despite kinetic complexity.

2.6. Combining substrate isotope effects and pressure effects

Because every rate constant, enzyme complex and transition state could potentially be affected by pressure, interpreting pressure effects on complex enzymatic reactions seemed doomed to ambiguity [14]. However, because substrate isotope effects usually arise from a single transition state and are usually pressure-independent [25], they can provide an anchor or reference point and thereby facilitate in the sorting out of multiple reaction components that change with pressure. This is exemplified in the equation [26]:

$$|V/K|_p = \left(\frac{k_1}{1 + K_{G/E}e^{-\Delta V_{G/E}p/RT}}\right) \cdot \left(\frac{R_0 e^{-\Delta V^{\ddagger}p/RT}}{1 + C_i[^{D}k-1] + C_f e^{-\Delta V^{\ddagger}p/RT} + C_r e^{-(\Delta V^{\ddagger} - \Delta V_{eq})p/RT}}\right) \quad (6)$$

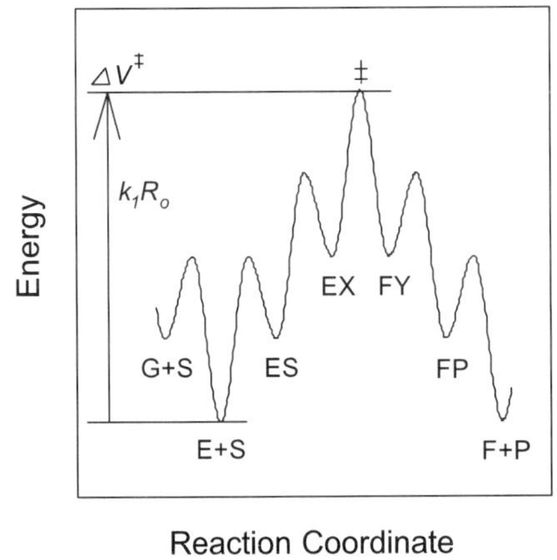

Fig. 1. Activation energy diagram for substrate capture within an iso-mechanism with small commitments to catalysis. The volume change represented by $\Delta V^{\ddagger}$ encompasses the portion of the mechanism represented by the vertical arrow. As such it includes any and all volume changes that occur between the limits of E+S and $EX^{\ddagger}$ and, being a system function, any volume changes in the solvent associated with the binding of substrate.

where C_i is the decimal fraction of isotopic labeling and ${}^{D}k = k_H/k_D$ is the intrinsic isotope effect. When commitments are very small and the intrinsic isotope effect is fully expressed on substrate capture (which sometimes obtains with poor substrates) Eq. 6 reduces to:

$$|V/K|_p = \left(\frac{k_1}{1 + K_{G/E}e^{-\Delta V_{G/E}p/RT}}\right)\left(\frac{R_0 e^{-\Delta V^{\ddagger}p/RT}}{1 + C_i[{}^{D}k-1]}\right) \quad (7)$$

This is as simple as the general equation can get; on the left is the Monod component, on the right the Koshland component. This bifurcation of pressure effects on protein folding and enzyme kinetics, respectively, is cleanly separated and further simplification may occur with enzymes that express only one and not the other.

2.7. Combining solvent isotope effects and pressure effects

Similar separations are even more important with respect to solvent isotope effects because these are complicated by the fact that multiple primary effects, secondary effects, and medium effects are multiplied together, as in the following equation [27]:

$$\left(\frac{k_{H_2O}}{k_{D_2O}}\right)_{\text{total}} = \left(\frac{k_H}{k_D}\right)_{\text{pri}} \times \left(\frac{k_H}{k_D}\right)_{\text{sec}} \times \left(\frac{k_{H_2O}}{k_{D_2O}}\right)_{\text{medium}} \quad (8)$$

In addition, particularly in enzymatic reactions, there may be multiple primary and secondary effects. Attempts to factor experimental data into multiple components center around the *proton inventory experiment* in which solvent isotope effects are plotted against the atom fraction of deuterium in the solvent. A linear plot represents a single effect whereas a nonlinear plot reveals multiple effects. To evaluate proton inventories arising from multiple states, first-order rate constants are examined by fittings to the equation of Kresge [28]:

$$k_n = k_H \frac{(1-n+n\phi^T)^p}{(1-n+n\phi^R)^p} \quad (9)$$

where k_n represents apparent rate constants in mixtures of D_2O and H_2O, k_H is an apparent rate constant in H_2O, n is the fraction of deuterium in mixed isotopic waters, ϕ^T is the fractionation factor for a transition state, p is the number of protons in flight and ϕ^R is the fractionation factor for a reactant state. Hence, interpretations are difficult not only because of the products of multiple effects of Eq. 8, but also because of the quotients of multiple reactant and transition state fractionation factors of Eq. 9. Further difficulties arise in solvent isotope effects on enzyme-catalyzed reactions because of their inherent kinetic complexity, which can distort proton inventories [29]. What is needed to resolve these difficulties is some means to perturb the diverse components differently as a continuous function of something else. High pressure has been shown to be such a perturbant [30]. Eq. 6 becomes:

$$|V/K|_p = \left(\frac{\dfrac{k_1}{1+0.24C_i}}{1 + \dfrac{K_{G/E}e^{-\Delta V_{G/E}p/RT}}{1 + C_i[{}^{D2O}K-1]}}\right) \cdot \left(\frac{R_0 e^{-\Delta V^{\ddagger}p/RT}}{1 + C_i[{}^{D2O}k-1] + C_f e^{-\Delta V^{\ddagger}p/RT} + C_r e^{-(\Delta V^{\ddagger}-\Delta K_{eq})p/RT}}\right) \quad (10)$$

where the 0.24 term corrects for medium viscosity

effects of D_2O [27], ${}^{D2O}K$ is the solvent isotope effect on an iso-mechanism equilibrium, and ${}^{D2O}k$ is an intrinsic solvent isotope effect. When commitments are very small and the intrinsic solvent isotope effect is fully expressed on substrate capture, Eq. 10 reduces to:

$$|V/K|_p = \left(\frac{\frac{k_1}{1+0.24C_i}}{1+\frac{K_{G/E}e^{-\Delta V_{G/E}p/RT}}{1+C_i[{}^{D2O}K-1]}}\right)\left(\frac{R_0 e^{-\Delta V^{\ddagger}p/RT}}{1+C_i[{}^{D2O}k-1]}\right) \quad (11)$$

For the first time, solvent isotope effects arising from binding and catalysis can be separated, assigned and quantified. But more is still possible; fitting pressure data to Eqs. 10 or 11 combined with Eqs. 8 or 9 may make it possible to tease the latter two apart as well.

2.8. Incorporating the effects of pressure on hydrogen tunneling

Stumbling serendipitously upon an isotope effect that was extremely sensitive to pressure [26] led to an examination of the only eight chemical reactions reported by Bell [31] that show a similar phenomenon [25]. This, in turn, led to deriving an equation to address and quantify the effect of pressure on hydrogen tunneling in chemical reactions [32]:

$$|{}^{D}k|_p = {}^{D}k({}^{D}Q-1)\cdot e^{-\Delta V_Q p/RT} + {}^{D}k \quad (12)$$

where ${}^{D}k$ is the semi-classical isotope effect, ${}^{D}Q = Q_H/Q_D$ is the Bell tunneling correction and ΔV_Q is the apparent volume difference between a hydride and deuteride ion. It is important to note that while formulated in the language of hydrogen tunneling, this equation is actually a generic one (a hyperbolic function) that can address any pressure-sensitive transition state phenomena and distinguish them from reactant state phenomena so long their volumes are different. Incorporating Eqs. 12 into 7 yields the triphasic pressure function for hydrogen tunneling in an enzyme reaction [26]:

$$|V/K|_p = \left(\frac{k_1}{1+K_{G/E}e^{-\Delta V_{G/E}p/RT}}\right)\cdot \left(\frac{R_0 e^{-\Delta V^{\ddagger}p/RT}}{1+C_i[{}^{D}k({}^{D}Q-1)e^{-\Delta V_Q p/RT}+{}^{D}k-1]}\right) \quad (13)$$

This new and unique key is destined to unlock secrets of transition state chemistry in both enzyme-catalyzed and uncatalyzed reactions. Extensive review of the literature reveals that there are in fact two types of so-called 'hydrogen tunneling' – but never formally identified as such. There is *Tunneling* (capital T) which originates in the duality of matter and is dependent upon DeBroglie wavelengths of small particles and barrier widths as described by Bell [31]. Then there is *tunneling* (small t) which is a catch-all for any large deuterium isotope effect of unknown origin or a proton transfer reaction with anomalous Arrhenius plots. The latter type will not be dependent necessarily upon DeBroglie wavelengths, etc., and therefore responses to pressure may vary, and that variance may be used to distinguish between the two. The former has been modeled theoretically leading to the common conclusions that Tunneling has a dynamic component, termed variously as 'coupled motion', 'substrate oscillation', and a 'vibrationally enhanced ground state' [33–36]. It is this dynamic component that must be responsible for the effects of pressure on isotope effects, and on the sign and magnitude of ΔV_Q.

3. Applications

The oxidation of benzyl alcohol by yeast alcohol dehydrogenase was chosen first as a candidate for high pressure kinetics because it fully expressed a substrate deuterium isotope effect and was suspected of hydrogen tunneling. Fig. 2A shows the biphasic effect of pressure expressed on *V/K*, for both protium and deuterium transfers [26]. Moderate pressure increases the capture of benzyl alcohol by activating the hydride transfer step [19]. This means that the transition state for hydride transfer has a smaller volume than the free alcohol plus the capturing form of enzyme, E*-NAD$^+$, with a $\Delta V^{\ddagger}_H = -38 \pm 1$

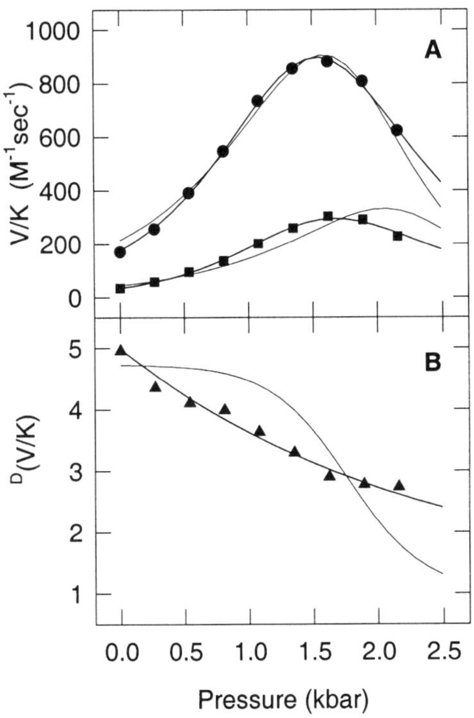

Fig. 2. Effect of pressure on the capture of benzyl alcohol (●), dideutero benzyl alcohol (■) by yeast alcohol dehydrogenase and on deuterium isotope effects (▲). The benzyl alcohol and NAD^+ were varied from 0.83 mM to 12.4 mM (K_m = 3.1 mM) and from 0.75 mM to 15.1 mM (K_m = 1 mM), respectively. The concentration of alcohol dehydrogenase was 3.19 μM. Tris–HCl (pH 8.5) buffer was used to prevent a pressure-dependent change in pH at high pressure due to electrostriction.

ml/mol. Pressures above 1.5 kbar decrease capture of benzyl alcohol by favoring a conformation of enzyme which binds NAD^+ less tightly [19]. This means that the ground state for tighter binding, E*-NAD^+, has a larger volume than the collision complex, E-NAD^+, with a $\Delta V^* = +73 \pm 2$ ml/mol. The equilibrium constant of the conformational change is $K_{eq}^* = 75 \pm 13$ at one atmosphere of pressure. Fig. 2B shows the ratio of pressure effects on protium and deuterium capture, expressed as monophasic decrease in the deuterium isotope effect [26]. It follows, therefore, that the volume of activation for the transition state of deuteride transfer must be even more negative, at $\Delta V^{\ddagger}_D = -49 \pm 1$ ml/mol.

What was most surprising of the fit of the data in Fig. 2 to Eq. 13 was the return of a value for the intrinsic isotope effect not significantly different from one, meaning that there is no semi-classical isotope effect arising from vibrational differences in a reactant state being expressed. Instead, all of the effect arises from a transition state phenomenon, expressed as $^DQ = 4.99 \pm 0.37$ with an apparent volume change of $\Delta V_Q = 10.4 \pm 1.5$ ml/mol. These numbers are consistent with the vibrationally enhanced ground state as described by Bruno and Bailek [34] and match a $Q_H = 11.7 \pm 1.1$ with a reaction frequency of $\nu^{\ddagger}_H = 1220 \pm 11$. This volume change is a very strange number. According to the Born–Openheimer approximation, which describes the energy surface of a reaction diagram as being isotopically independent and only the occupation of certain vibrational states differ with mass differences, this cannot be a real volume difference. The latter two numbers are curious, but cannot have any physical meaning either because the data are essentially telling us that Bell's conception of Tunneling does not apply. Instead, these numbers say something about the dynamic aspects of enzymatic catalysis, but just what is not clear.

Similar experiments were performed with yeast formate dehydrogenase and produced similar biphasic curves, characterized by $\Delta V^{\ddagger}_H = -9.3 \pm 1.1$ ml/mol, $\Delta V^* = -84 \pm 6$ ml/mol, $K_{eq}^* = 1000 \pm 500$ ml/mol and $^Dk = 1.02 \pm 0.03$ [37]. However, the isotope effect on capture *increased* with pressure, dependent upon an apparent volume change of $\Delta V_Q = -2.5 \pm 2.2$ ml/mol. The lack of a decrease is not consistent with the Bruno and Bailek model, and underscores the conclusion that this form of tunneling is something new that we do not understand.

The effect of pressure on the capture of benzyl alcohol by yeast alcohol dehydrogenase was measured in D_2O and compared to data in Fig. 2 using Eq. 11 [30]. The result is a smooth sigmoidal transition from the Koshland portion dominant at low pressure to the Monod portion dominant at high pressure. The absence of the latter in Fig. 2B but present here generates confidence that these assignments are indeed correct. The solvent isotope effect on proton transfer is small but significantly inverse, $^{D2O}k = 0.901 \pm 0.075$ while the effect on the conformational change of E-NAD^+ is large and also inverse, $^{D2O}K^* = 0.38 \pm 0.04$. The former provides evidence in support of a proposed mechanism in which the breaking of a low barrier hydrogen bond is concerted with hydride transfer [38], while the latter provides evidence in support of the proposal that torsional motions of exchangeable protons are restricted when enzymes 'crunch' down on substrates [39].

4. Significance

Escaping from the tyranny of the log plot allows for new applications and interpretations of high pressure as a system variable in thermodynamic as well as kinetic studies, because tangents to extremes of non-linear log plots have been misinterpreted in the past. Because pressure is non-invasive, can be varied as a continuous function and can generate positive, null or negative effects, a systematic study of its effects on enzymatic activity holds promise of becoming a much more powerful tool than any other system variable. It has more versatility and applicability than can be found within the disciplines of temperature effects, isotope effects or pH effects, for example. All rate constants are sensitive to temperature but in the same way, while most rate constants are insensitive to isotopes or changes in pH. In contrast, nearly all enzyme-catalyzed reactions respond to pressure, but in different ways. More specifically, separating and quantifying effects on binding from those on catalysis opens the doors to new questions and already challenges fundamental principles of modern enzymology. The first of the former runs contrary to Pauling's hypothesis that substrates are bound more tightly in the transition state, while the first of the latter runs contrary to classical, semi-classical and even quantum mechanical rate theory. Retrieving quantitative physical data from a transition state opens the door to new forms of structure–activity relationships. The first of these has been described, the effect of pressure on isotope effects, but many others are possible. An enticing example would be to determine the volume changes associated with a series of transition state analog inhibitors and compare them to $\Delta V^{\ddagger}$; the closest numerical match should identify the most analogous inhibitor which could then be used as a guide in the design of new inhibitors.

References

[1] E. Eyring, F.H. Johnson, R.L. Gensler, J. Phys. Chem. 50 (1946) 453.
[2] E. Morild, Adv. Protein Chem. 34 (1981) 93–166.
[3] M.M.C. Sun, D.S. Clark, Methods Enzymol. 334 (2001) 316–327.
[4] S. Glastone, K.J. Laidler, H. Eyring, The Theory of Rate Processes, McGraw-Hill, New York, 1941.
[5] E. Morild, Biophys. Chem. 6 (1977) 351–362.
[6] D.B. Northrop, J. Am. Chem. Soc. 121 (1999) 3521–3524.
[7] L. Michaelis, M.L. Menten, Biochem. Z. 49 (1913) 333–369.
[8] G.E. Briggs, J.B.S. Haldane, Biochem. J. 19 (1925) 338–339.
[9] D.B. Northrop, Biochemistry 14 (1975) 2644–2651.
[10] D.B. Northrop, in: W.W. Cleland, M.H. O'Leary, D.B. Northrop (Eds.), Isotope Effects on Enzyme-Catalyzed Reactions, University Park Press, Baltimore, MD, 1977, pp. 122–152.
[11] D.B. Northrop, Biochemistry 20 (1981) 4056–4061.
[12] D.B. Northrop, Methods Enzymol. 87C (1982) 607–625.
[13] D.B. Northrop, in: J.L. Markely, C.A. Royer, D.B. Northrop (Eds.), Effects of High Pressure on Molecular Biology and Enzymology, Oxford University Press, New York, 1996, pp. 122–241.
[14] C. Balny, in: J.M. Markely, C.A. Royer, D.B. Northrop (Eds.), Effects of High Pressure on Molecular Biology and Enzymology, Oxford University Press, New York, 1996, pp. 196–210.
[15] K.L. Laidler, P.S. Bunting, The Chemical Kinetics of Enzyme Action, 2nd edn., Clarendon, Oxford, 1973, pp. 220–232.
[16] M. Dumoulin, H. Ueno, R. Hayashi, C. Balny, Eur. J. Biochem. 262 (1999) 475–483.
[17] M.J. Kornblatt, R. Lange, C. Balny, Eur. J. Biochem. 251 (1998) 775–780.
[18] S. Dallet, M.-D. Legoy, Biochim. Biophys. Acta 1294 (1996) 15–24.
[19] Y.-K. Cho, D.B. Northrop, Biochemistry 38 (1999) 7470–7475.
[20] L.E. Parmentier, M.H. O'Leary, H.K. Schachman, W.W. Cleland, Biochemistry 31 (1992) 6598–6602.
[21] J. Monod, J. Wyman, J.-P. Changeux, J. Mol. Biol. 12 (1965) 88–118.
[22] D.L. Koshland Jr., G. Nemethy, D. Filmer, Biochemistry 5 (1966) 365–385.
[23] D.B. Northrop, J. Chem. Ed. 75 (1998) 1153–1157.
[24] K.E. Bett, J.B. Cappi, Nature 207 (1965) 620–621.
[25] N.S. Isaacs, in: E. Buncel, C.C. Lee (Eds.), Isotope Effects in Organic Chemistry, Vol. 6, Elsevier, Amsterdam, 1984, pp. 67–105.
[26] D.B. Northrop, Y.-K. Cho, Biochemistry 39 (2000) 2406–2412.
[27] K.B.J. Schowen, in: R.D. Gandour, R.L. Schowen (Eds.), Transition-states of Biological Processes, Plenum Press, New York, 1978, pp. 225–283.
[28] A.J. Kresge, Pure Appl. Chem. 8 (1964) 243–258.
[29] D.M. Kiick, J. Am. Chem. Soc. 113 (1991) 8499–8504.
[30] D.B. Northrop, Y.-K. Cho, Biophys. J. 79 (2000) 1621–1628.
[31] R.P. Bell, The Tunnel Effect in Chemistry, Chapman and Hall, New York, 1980.

[32] D.B. Northrop, J. Am. Chem. Soc. 121 (1999) 3521–3524.
[33] W.P. Huskey, R.L. Schowen, J. Am. Chem. Soc. 105 (1983) 5704–5706.
[34] W.J. Bruno, W. Bialek, Biophys. J. 63 (1992) 689–699.
[35] D. Antoniou, S.D. Schwartz, Proc. Natl. Acad. Sci. USA 94 (1997) 12360–12365.
[36] J. Rucker, J.P. Klinman, J. Am. Chem. Soc. 121 (1999) 1997–2006.
[37] D.J. Quirk, D.B. Northrop, Biochemistry 40 (2001) 847–851.
[38] S. Ramaswamy, D.H. Park, B.V. Plapp, Biochemistry 38 (1999) 13951–13959.
[39] W.W. Cleland, Bioorg. Chem. 15 (1987) 283–302.

ELSEVIER

Biochimica et Biophysica Acta 1595 (2002) 80–93

BIOCHIMICA ET BIOPHYSICA ACTA
BBA
www.bba-direct.com

Review

UV-visible derivative spectroscopy under high pressure

Reinhard Lange [1], Claude Balny *

INSERM, Unité 128, IFR 24, CNRS, 1919, route de Mende, 34293 Montpellier Cedex 5, France

Received 21 September 2001; received in revised form 12 November 2001; accepted 13 November 2001

Abstract

High hydrostatic pressure affects proteins, changing their intra- or intermolecular interactions, conformation and solvation. How to detect these changes? In this paper, via some selected examples, we show the potentiality (but also the limits) of the ultraviolet derivative spectroscopy specially adapted to high pressure experiments.

Keywords: High pressure; Absorbance spectroscopy; Derivative; Protein inactivation; Folding; Unfolding

1. Introduction

High hydrostatic pressure induces changes in protein conformation, solvation and enzyme activities via reversible and non-reversible effects on intra- and intermolecular interactions (non-covalent bonds) [1]. To have access to these pressure-induced structural modifications, spectroscopic investigations are useful methods when biological compounds have chromophores. It is a rather simple and easy method adaptable to high pressure which is complementary to NMR or FTIR spectroscopy, for example. However, to work under high pressure requires special technical adaptations. One recent improvement concerns the use of the derivative modes to treat the optical signal.

Whatever the pressure value, it is relatively easy to characterize conformational equilibria and elementary steps of proteins when chromophores absorbing in the visible region are present. However, it is more difficult to obtain protein structural information from bands absorbing in the ultraviolet region, since here many overlapping individual electronic transitions broadened by their vibronic structure are found. Not only the three aromatic amino acids (phenylalanine, tyrosine and tryptophan) but also cysteine, histidine and UV components of visible chromophores such as the heme group contribute to the ultraviolet absorbance of proteins resulting in rather structureless spectra. Yet, a better resolution can be achieved, by calculating the second or the fourth derivative of the spectra [2–4]. Second derivative spectroscopy is the more widely used method [5–10], in particular for the detection of structural changes in the environment of tyrosine residues [11]. Fourth derivative spectroscopy has the advantage that a maximum in the original spectrum corresponds to a maximum in the derivative spectrum. Furthermore, the fourth derivatives are more selective for narrow bands than the second derivatives,

Abbreviations: CD, circular dichroism; FTIR, Fourier transform infrared spectroscopy

* Corresponding author. Fax: +33-4-67-52-36-81.

E-mail address: balny@u128.crbm.cnrs-mop.fr (C. Balny).

[1] Also corresponding author.

0167-4838/02/$ – see front matter
PII: S0167-4838(01)00336-3

and they enable a more detailed study of the environment of all three aromatic amino acids [12].

To work under high pressure, some problems arise: for a protein chemist or enzymologist who wishes to study the effects of pressure on a protein, he/she must: (a) observe and quantify the effects and (b) interpret them. As mentioned above, the majority of proteins do not contain easily studied chromophores but they do contain substantial amounts of the aromatic amino acids phenylalanine, tryptophan and tyrosine which are characterized by strong but overlapping absorption bands in the ultraviolet region. A protein that contains five to ten of each of the aromatics presents a spectrum that is a composite of the contributions of each. A change in a protein, such as a conformational change, hydration variation, dissociation, or denaturation, may result in a change in the environment of one or more of these residues. However, the UV spectrum of a protein is broad and featureless: the changes of interest are masked by the contributions from all the other aromatics whose environment has not changed.

2. Methods

Two spectroscopic methods have been recently developed and adapted to high pressure experiments: the second and the fourth ultraviolet derivative spectroscopies.

2.1. Second derivative spectroscopy

Second derivative spectroscopy is a means to extract information from the spectrum. In contrast to the relatively featureless UV spectra, the second derivatives of UV spectra are characterized by two sharp peaks and troughs [9]. It is an approach which has been developed for high pressure experiments by the Kornblatt group and Fig. 1 shows the ultraviolet spectra of yeast enolase at 0.28 and 200 MPa (zero order). Fig. 2 shows the second derivative of those spectra [13].

Ragone [9] has determined the effects of solvent polarity on the second derivative spectra of model compounds. For both tyrosine and tryptophan, as the solvent polarity goes from non-polar (as when the residue is buried inside a protein) to polar (exposed to solvent), the peak and trough positions shift to the blue (shorter wavelengths). There are also changes in amplitude, which are best described by calculating the ratio ($r = a/b$) of the two peak-to-trough values marked in Fig. 2 [9]. For tyrosine, the value of r decreases as solvent polarity decreases, while the value of r for tryptophan is almost independent of solvent polarity. Thus, in the second derivative spectrum of a protein, the ratio of the peak to trough values and the positions of the peaks and troughs are a function of the relative amounts of the two amino acids and of the average polarity of the

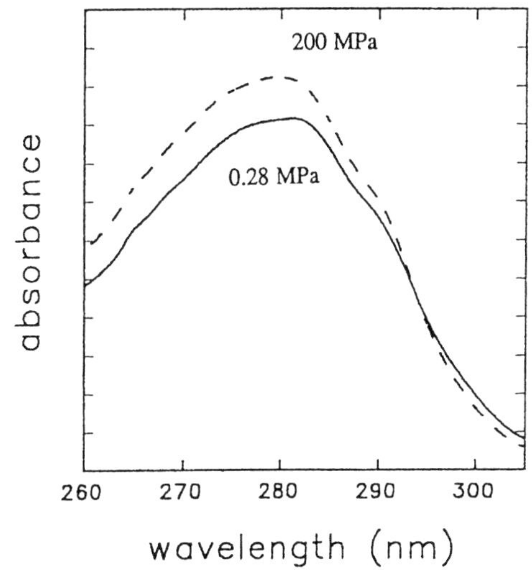

Fig. 1. UV spectra of yeast enolase at 0.28 and 200 MPa. (From [14], with permission.)

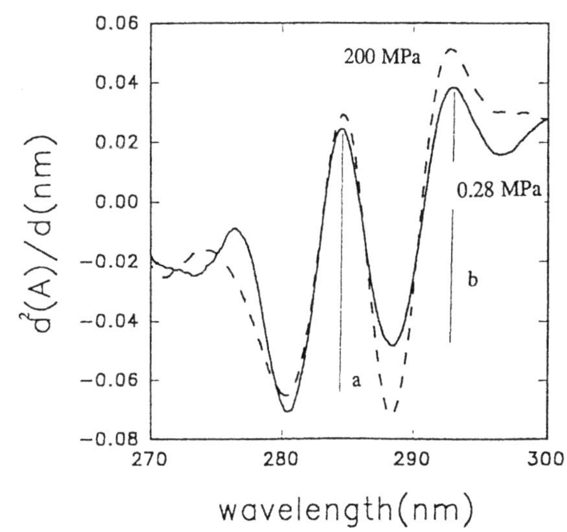

Fig. 2. Second derivative of the UV spectra of the yeast enolase shown in Fig. 1. (From [14], with permission.)

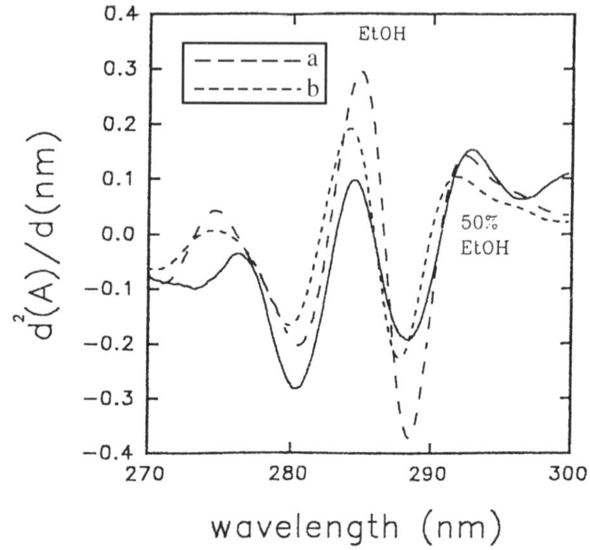

Fig. 3. Simulated spectra of enolase. ——, native enolase; (a) tyr in water and trp in 100% ethanol; (b) tyr in water and trp in 50% ethanol. (From [14], with permission.)

environments of the tyrosines as well as that of the tryptophans. For the protein shown in Fig. 2, increasing pressure from 0.28 to 200 MPa has decreased the average polarity around the tyrosine residues, while not changing the environment around the tryptophan residues.

Extracting the physical meaning of the changes in second derivative spectra relies on knowing the amino acid composition of the protein and using the constituent spectra to simulate the observed spectra. Simulations were performed by combining the second derivative spectra of tyrosine- and tryptophan-ethyl ester in solvents of various dielectric constants, in a molar ratio of 9 tyr to 5 trp (that of enolase). Fig. 3 shows the observed spectra of native (dimeric) yeast enolase and that of tyr in water and trp in 100 and 50% ethanol. The simulations produce average values for changes in polarity. When an X-ray structure is also available, it may be possible to assign the changes in polarity to changes in the environment of specific tyrosines or tryptophans [14].

2.2. *Fourth derivative spectroscopy*

The selective resolution enhancement in derivative spectroscopy is pushed even further in the fourth derivative mode. From a methodological point of view, the derivatives are usually computed by shifting a spectrum for a given wavelength difference (called the derivation window) and subsequent subtraction from the original spectrum. The derivation window is a very important feature of this technique, since it determines the spectral amplitude and resolution of the derivative [19]. Its judicious choice therefore allows one to study selectively certain electronic transitions, according to their spectral band widths. As in the case of second derivative spectroscopy, the amplitude and the position of the derivative spectral bands of the aromatic amino acids are related to the polarity of the medium. A detailed description of the method adapted to high pressure has been published elsewhere [2,3,15,16]. In this laboratory, we have undertaken a systematic investigation of these spectral features of the *N*-acetyl *O*-ethyl esters of tyrosine and tryptophan in various solvents of different polarity (from cyclohexane to water). Astonishingly, a simple relationship between the spectral parameters of the fourth derivatives and the dielectric constant was found [2]. As shown in Fig. 4, for tyrosine it is the position of λ_{max}, and for tryptophan it is the derivative amplitude which depends linearly on the dielectric constant ε_r. Since the fourth derivative spectra of these model compounds do not depend significantly on pressure (at least up to 500 MPa), these spectral features may be used as an intrinsic probe to sense the dielectric constant in the vicinity of tyrosine and tryptophan.

3. Technological requirements

3.1. *The high pressure cell adapted for UV/vis spectroscopy*

The design developed in our laboratory is an im-

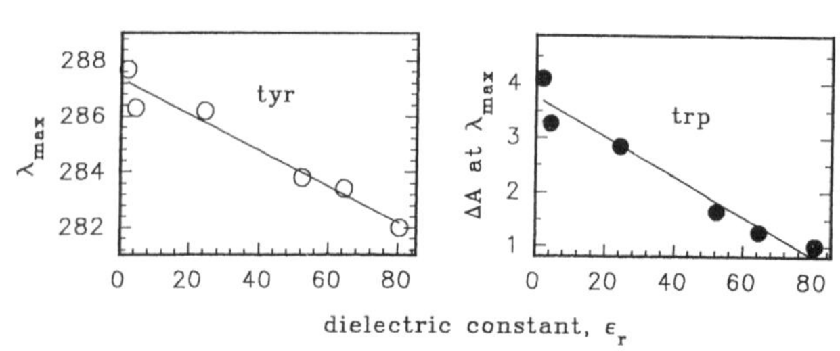

Fig. 4. Amplitude and λ_{max} of the fourth derivative spectra of tyrosine (left) and tryptophan (right) as a function of the dielectric constant ε_r. (From [14], with permission.)

provement of devices already described for room temperature experiments at high pressure. In addition to the classical characteristics of a device working at room temperature, a high pressure optical bomb for biochemical sub-zero temperature measurements should permit accurate measurements in absorbency (and in fluorescence) mode, with, in addition, the possibility of carrying out photochemistry, in a temperature range from −50 to 90°C. The design should include a device that will avoid any ice formation at the windows or at the sample cell. This high pressure optical bomb should also allow the sample cell to be pre-cooled at a set temperature, have a special system to quickly close the sample cell, operate at a pressure of several hundred MPa without leakage, even at sub-zero temperatures, have a glass or plastic sample cell to avoid problems of metal ion contamination, and be easily interfaced to conventional spectrophotometers.

The actual high pressure cell that we have recently constructed is similar to the device that we described 20 years ago [17]. The cell core, built by S.O.F.O.P. (Rodez, France), was made of Marval X 12 special steel allowing experiments up to 700 MPa in absorbency or in fluorescence mode. Around this core an envelope was added in which the thermostating alcohol circulated. The geometry of the assembly was miniaturized so that the cell could be placed in the sample compartment of a Cary 3 E spectrophotometer for absorbency. The windows were made of synthetic sapphire (diameter 10 mm, $h = 8$ mm) from RSA (Jarrie, France) which has a transmittance at 250 nm of 50% for a light path of 8 mm. The windows were placed directly upon the optical opturators. The tightness was assured by the reversible deformation of a polyurethane seal, which was placed between the external face of the obturator and metallic counterpart. Pressure was generated by a manual pump (700 MPa) and controlled by a metallic gauge type PR 851 C from Top Industries (Dammarie les Lys, France). The working pressure was measured by an Infinity Indicator, Model Infcp from Newport Electronique (Trappes, France). The pressure vector, water, was mediated through stainless steel capillaries (outer diameter 1/16 inch, inner diameter 0.2 mm). Protein samples were measured in a polyethylene stretch film maintained by a rubber O-ring. The stability of the temperature was ±0.2°C. At 700 MPa the stability of the pressure was ±1 MPa.

3.2. Treatment of the spectrophotometric signal: the derivative procedure

In contrast to many commercial spectrophotometers which impose fixed values for spectral band widths, we have developed in the laboratory a tunable method allowing the optimization of the necessary compromise between spectral amplitude and resolution for the aromatic amino acids. A critical analysis of the shift method applied to a mathematical model and comparison with derivatives obtained either analytically or with the method of Savitzky and Golay [18] has been made [2]. This has provided the basis for the application of the method to the analysis of the solvent and pressure effects on the fourth derivative spectra of the aromatic amino acids.

The derivation procedure was based on the method suggested by Butler [19]: the original spectrum is shifted over a small distance $\Delta\lambda$, to longer wavelengths. The calculated difference between the shifted and the original spectrum is taken as the first derivative spectrum (but shifted $\lambda/2$ to the red). Derivatives to higher order are obtained by repeating this procedure. Two types of noise suppression were used: (1) the successive derivatives were computed using slightly different $\Delta\lambda$ values, (2) the resulting derivative spectra were convoluted with a shape function with a window of 1.05 nm for phenylalanine, 0.9 nm for tryptophan and 1.1 nm for tyrosine.

The list of applications of fourth derivative spectroscopy for pressure-induced protein conformational changes is long. We will present here some typical examples (the list is not restrictive) where the method has been successfully used for different pressure-induced phenomena: reversible or irreversible folding–unfolding, conformational stability, conformational change with ligand binding, oligomeric dissociation. However, we must point out that for a complete interpretation of the data, other spectroscopic methods should be used in conjunction with derivative spectroscopy. The other techniques include fluorescence spectroscopy (mainly where tryptophan residues are present in the protein), FTIR (see the contribution of Y. Taniguchi in this volume) or spectroscopic kinetics for enzyme activity determination under high pressure.

4. Reversible folding–unfolding transitions

Protein unfolding studies have gained a lot of interest in the past few years, mainly because the pathological role of partly denatured proteins in a variety of diseases has been found. Despite numerous studies, no generally accepted mechanistic model is available. Extreme conditions of temperature were exploited in the past; more recently, the pressure parameter has been used to provide an elegant alternative, perturbing reversibly the folding–unfolding equilibrium. The derivative methods were generally used together with other spectroscopic methods such as fluorescence. In the following examples, we mainly stress on the results obtained using the fourth derivative UV spectroscopic methods.

4.1. Ribonuclease A and mutants

A good example of application of the fourth derivative method is given by the protein structural changes of bovine ribonuclease A (RNase A) in the course of its denaturation by pressure. The UV spectrum of RNase is dominated by the absorbance of tyrosine – this RNase does not contain tryptophan. As shown in Fig. 5, an increase of pressure from 1 to 500 MPa results in a blue shift of the fourth derivative maximum from 285.7 ± 0.05 to 283.5 ± 0.05 nm. This shift of 2.2 nm corresponds to an increase of the mean dielectric constant from 25 to 59. It is characteristic of the exposure to the aqueous solvent of a portion of the six tyrosines, as is expected for partial denaturation. The transition is fully reversible with clear isosbestic points. The pressure effect can therefore be described by a simple two-state model between the native ($\varepsilon_r = 25$) and the partially denatured ($\varepsilon_r = 59$) state. A simulation on the basis of this model permitted us to determine the thermodynamic parameters of this transition: $\Delta G_0 = 10.3$ kJ/mol and $\Delta V = -52$ ml/mol. A comparison with results obtained by other methods indicates that the ($\varepsilon_r = 59$) state corresponds to an intermediate in the unfolding process which has molten globule-like characteristics [3]. It thus appears that fourth derivative spectroscopy under high pressure is a suitable technique to investigate protein structural changes occurring during reversible unfolding of protein.

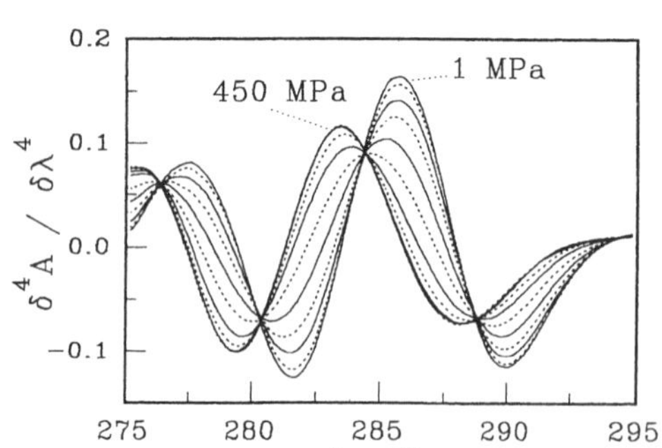

Fig. 5. Effect of pressure on the fourth derivative UV spectrum of ribonuclease A. RNase, 73 μM, at pH 2 and 25°C. (From [3], with permission.)

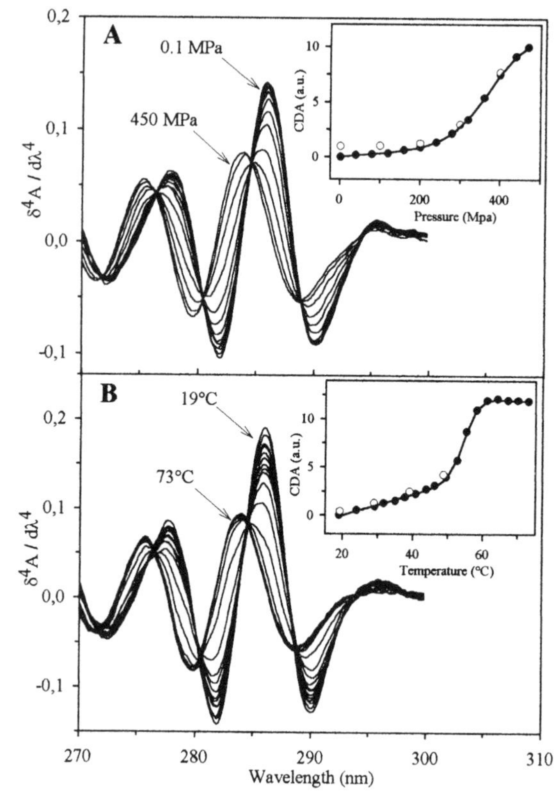

Fig. 6. Fourth derivative UV spectra of RNase variant V116A as a function of pressure (A) and temperature (B), respectively. The insets show the pressure and temperature unfolding curves. Protein concentration: 1 mg/ml in 50 mM sodium acetate buffer pH 5. ●, increasing pressure or temperature; ○, decreasing pressure or temperature. (From [20], with permission.)

To investigate the characteristics of the postulated carboxy-terminal chain folding initiation site in bovine pancreatic RNase A (residues 106–118), important in the early stages of the folding pathway, J. Torrent et al. have engineered by site-directed mutagenesis a set of 14 predominantly conservative hydrophobic variants of the protein [20]. The stability of each variant has been compared by pressure-induced unfolding monitored by fourth derivative UV absorbance spectroscopy. It was shown (as for the wild-type protein) that simple two-state, reversible unfolding transitions were observed, suggesting that the disruption of tertiary structure of each protein at high pressure (and at high temperature) is strongly cooperative (see Fig. 6). From thermodynamic analysis, the decreases in free energy of unfolding upon pressure denaturation of all variants suggest that the hydrophobic residues and side chain packing interactions of this region play an important role in maintaining the conformational stability of RNase A.

4.2. *The 33 kDa protein from spinach photosystem II particle*

Ruan et al. have recently studied the 33 kDa protein isolated from the spinach photosystem II particle which is an ideal model to explore high pressure protein unfolding [21]. The protein has a very low free energy as previously reported by chemical unfolding studies, suggesting that it must be easy to modulate its unfolding transition by rather mild pressure. The protein molecule contains only one tryptophan residue (Trp241) and eight tyrosine residues; these are convenient probes of protein conformation and structural changes under pressure using either fluorescence spectroscopy or fourth derivative UV absorbance spectroscopy. These two different experimental methods indicated that at 20°C and pH 6, the 33 kDa protein showed a reversible two-state unfolding transition from atmospheric pressure to about 180 MPa, a value much lower than those previously found for the unfolding of most proteins studied so far (see Fig. 7). The very clear isosbestic points implied that the pressure induced a cooperative two-state transition unfolding. The change in standard free energy (ΔG_0 about 14 kJ/mol) and in volume (ΔV about −130 ml/mol) for the transition (these values depend of the experimental conditions) is consistent with that obtained by chemical denaturation. Complementary experiments using different pH and different temperatures showed that physicochemical conditions can modulate unfolding reaction by influencing the value of ΔG_0.

4.3. *Pressure-induced changes in the wheat γ-46 gliadin*

The unfolding of the wheat γ-46 gliadin was inves-

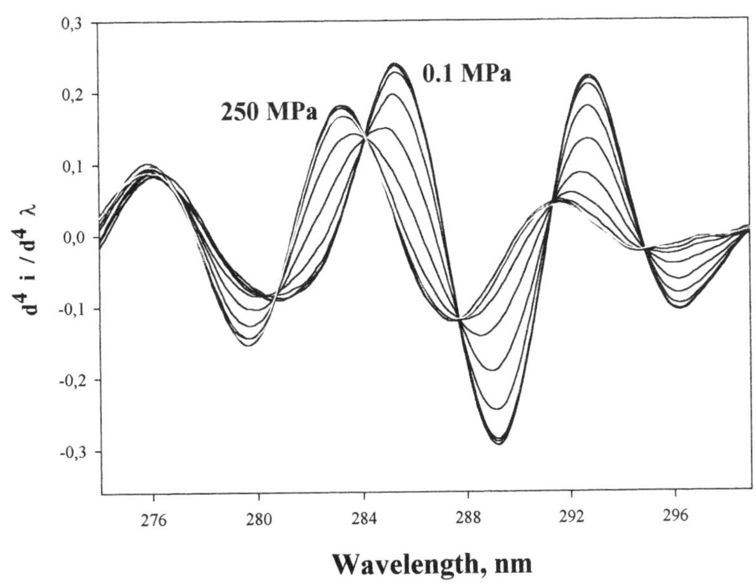

Fig. 7. The fourth derivative UV spectra of the 33 kDa protein as a function of pressure between room pressure and 250 MPa. Protein concentration: 0.5 mg/ml in 0.05 M MES buffer, pH 6.0; T = 20°C. (From [21], with permission.)

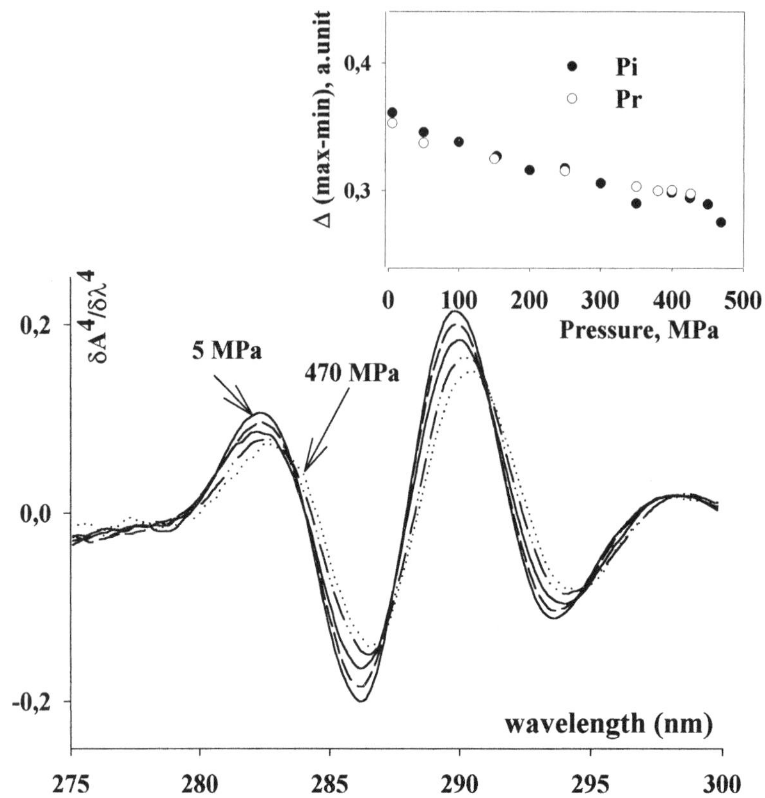

Fig. 8. Fourth derivative UV spectra of the γ-46 gliadin as a function of pressure between atmospheric pressure and 470 MPa. The protein concentration was 56 μM in 50 mM sodium citrate buffer (pH 3), the temperature was 20°C. Characteristic spectra recorded at 5, 100, 200, 400 and 470 MPa (from left to right) are shown. (Insert) Difference between maxima and minima of the fourth derivative spectra of the γ-46 gliadin as a function of pressure. Pi, compression; Pr, pressure release. (From [22], with permission.)

tigated by Lullien-Pellerin et al., in aqueous solutions, under high pressure or temperature by the use of ultraviolet absorption and fluorescence spectroscopic techniques [22]. This class of protein is involved in the extensibility of the gluten which forms a unique viscoelastic network (gluten properties are used to process food products such as bread, pasta and semolina). The rheological properties of gluten are dependent on the protein interactions that are formed in the network and these properties were found to be mainly influenced by physical parameters such as temperature and pressure [23,24]. Gliadins account for about one half of the total wheat gluten proteins.

The γ-type gliadins are single monomeric proteins with only intra-chain disulfide bonds with a N-terminal repetitive domain based on a heptapeptide repeat motif rich in proline and glutamine and a non-repetitive C-terminal domain which contains all the cysteine residues. Little information is available on the γ-gliadin group due to the difficulty in obtaining its individual members in a highly purified state.

With the aim to further understand how the extensibility character of the gliadins could be related to their conformations, high pressure has been used to investigate the structural stability of gliadins. It was found that high pressure increases the solvent accessible surface of proteins and therefore the balance between intramolecular and solvent–protein interactions [1]. The conformational changes of the γ-46 gliadin were investigated under high temperature or pressure and probed by fourth derivative and fluorescence spectroscopies in order to describe the effects of these physical parameters on the polarity of the environment of aromatic amino acids.

It was found that high pressure (above 400 MPa) induces a change in the protein conformation that results in a decrease of the polarity of the environment of aromatic amino acids. Thermodynamic parameters of this conformational change were also

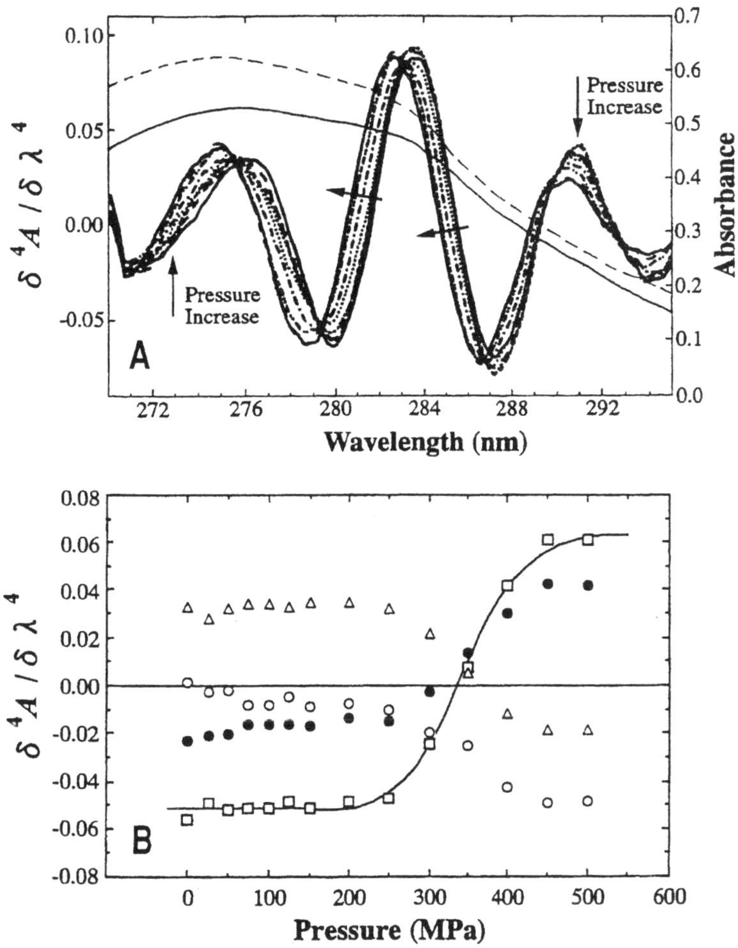

Fig. 9. The change in the fourth derivative absorption spectra of thermolysin during the pressure raising process (A) and the pressure dependence of the derivative spectra at 278 (○), 281 (●) and 285 (△) nm in 0.1 M MES/NaOH buffer at pH 6.5 (B). After releasing the pressure, a spectrum almost identical to that at 500 MPa was recorded. (From [30], with permission.)

measured ($\Delta G_0 = 38$ kJ/mol, $\Delta V = -95$ ml/mol). The high pressure (as shown in Fig. 8) and temperature stabilities of γ-46 gliadin could be related to its elastic character as observed changes were always found reversible.

5. Irreversible pressure-induced changes

A typical example is that of thermolysin, a thermostable microbial neutral protease containing zinc as a cofactor [25]. Previous studies had shown considerable activation by pressures up to 150 MPa which can be utilized for high pressure enzyme catalysis inducing peptide condensations and protein processing [26–28]. However, a higher pressure caused a time-dependent deactivation. To investigate these effects, Kunugi et al. used both the high pressure stopped-flow method [29] for the high pressure catalytic determinations and fourth derivative spectroscopy to explore the tyrosine and the tryptophan residues of the protein [30]. For enzyme activity studies under high pressure, we currently use the high pressure stopped-flow apparatus which is a mixing device working up to 200 MPa. This procedure has a dead time of some milliseconds. Our stopped-flow apparatus has two solution container syringes inside a high pressure vessel allowing the mixing of the sample solutions (and the optical detection) under indicated pressure [31].

Fig. 9 shows the change in the fourth derivative spectrum during the pressure elevation from 0.1 to 500 MPa. It is clear that the bands near 283 nm and 290 nm are mainly tyrosine and tryptophan residue contributions, respectively. On increasing the pressure, a blue shift is observed suggesting that the mean polarities of both tyrosine and tryptophan were increased, being more exposed to solvent. From these shifts, it was possible to estimate the dielectric constant variation from 55 to 70. After pressure release, the change in the spectrum did not return to the atmospheric level. This irreversible spectral change starts above 250 MPa and reaches a plateau at 450 MPa. The transition observed is an irreversible two-state transition accompanied by a large negative change in $\Delta V = -70$ ml/mol at 25°C. At the same temperature and using a dipeptide amide substrate the activation volume of the catalytic reaction was −70 ml/mol in the pressure range 0.1–200 MPa. This large activation volume is related to some structural changes up to a maximum pressure of about 200 MPa. At higher pressure the structural deformation is unfavorable to maintain the enzyme activity and the solvation shell of the protein becomes more ordered. An increase in protein–water interactions may cause such an irreversible pressure denaturation.

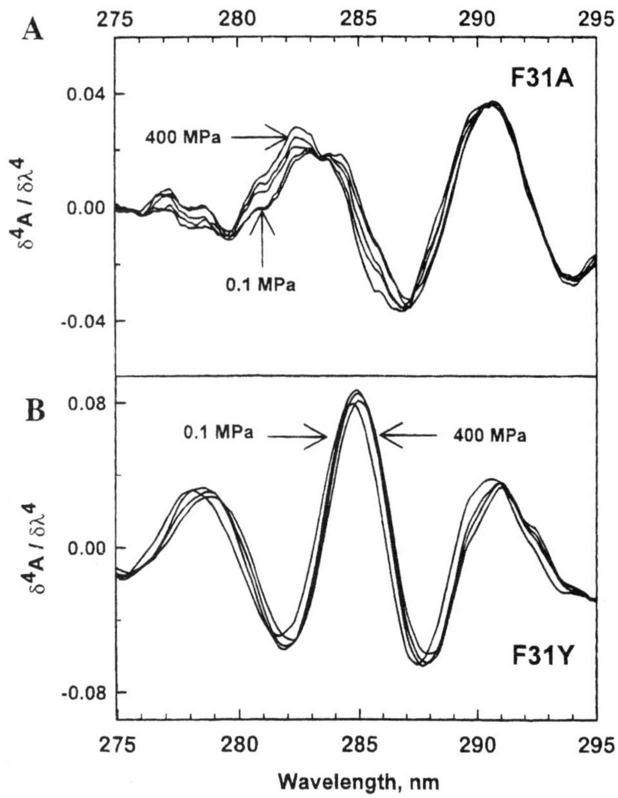

Fig. 10. Fourth derivative spectral transitions of the F31A and F31Y mutants at high pressure: MES buffer, 50 mM at pH 5, $T = 35$°C. The pressure was raised in steps of 50 MPa from room pressure to 450 MPa. Only few spectra are shown. The pressure-induced blue shift in the upper graph and a red shift in the lower graph. (From [34], with permission.)

6. Protein conformational stability

6.1. Role of a specific amino acid in the case of ribonuclease Sso 7d from Sulfolobus solfataricus

Understanding the structural changes occurring in the course of protein folding and unfolding potentially provides invaluable information about chemical interactions between particular amino acid residues and/or protein domains and the forces implied for maintaining the complex and generally unique conformation of a given protein. In that sense, thermophilic enzymes have been investigated because of their surprising resistance to denaturing agents compared to their mesophilic counterparts [32]. The ribonuclease Sso 7d (a small protein of 7 kDa molecular mass) from *S. solfataricus* [33], a thermoacidophilic archaebacterium that lives at 87°C and acidic pH in hot springs, is well suited for this type of study and has been investigated under high pressure in this laboratory [34].

As an intrinsic probe, the UV absorbance of tyrosine and tryptophan was used in the fourth derivative mode. Because, in the fourth derivative mode, the wavelength position of the maximum amplitude of tyrosine can be used to evaluate the polarity of its microenvironment, it has been possible to study the importance of interactions within the hydrophobic core for maintaining the protein stability. The structural importance of Phe31 in the hydrophobic core was studied. Heat, cold and pressure denaturation profiles of wild-type Sso 7d and its mutants F31A and F31Y were determined. The wild-type form proved to be extremely stable to temperature, pressure and pH and no variations were observed in the spectroscopic determinations up 500 MPa. However, as shown in Fig. 10, applying pressure to the F31A mutant up to 400 MPa resulted in a blue shift of the tyrosine band (282 nm) with an isosbestic point at 283.5 nm, indicating increased solvent exposure of Tyr33. The tryptophan band (near 290 nm) was not affected by pressure. From the pressure denaturation profile, the transition parameters were determined as $\Delta G_0 = 10.5$ kJ/mol and $\Delta V = -43$ ml/mol with a half-transition pressure of 256 MPa.

The introduction of a third tyrosine residue in place of Phe31 resulted in a spectral red shift of tyrosine band; this indicates that the additional tyrosine is in a strongly hydrophobic environment. An increase of pressure up to 450 MPa resulted in a red shift of the fourth derivative spectra of the tyrosine component. This indicates for this mutant that at high pressure the tyrosines are in a less polar environment: the interaction between the aromatic residues is more important at high pressure.

This stepwise change of the environment of Tyr33, a residue which participates in the maintenance of the hydrophobic core stability, indicates that pressure-induced unfolding of this protein is not a simple two-state process.

6.2. *Plasminogen conformational change*

Plasminogen, a glycoprotein present in vertebrate blood, is capable of binding to macromolecular surfaces which contain exposed carboxy-terminal lysines or some exposed amino sugars. The binding induces a conformational change and can then be activated by several proteases. Activation involves cleavage of one peptide bond; the two questions that the Kornblatt group posed were: (1) will hydrostatic pressure affect the open/closed equilibrium of plasminogen? (2) will pressure influence the interaction of plasminogen with its ligand (6-aminohexanoate)? [35]. Using fourth derivative spectroscopy (and native PAGE), it has been shown that hydrostatic pressure brings about the same conformational change as when it binds 6-aminohexanoate ($\Delta V = -33$ ml/mol). Binding and hydrostatic pressure both cause the protein to open up to expose surfaces that had previously been buried in the interior.

7. Oligomeric dissociation

Intersubunit regions of oligomeric proteins are rich in electrostatic and hydrophobic contacts, which stabilize quaternary structure; thus, it is not surprising that relatively moderate pressure may induce dissociation of oligomeric proteins by weakening these interactions. The case of yeast enolase dimer was described previously (Ch. 2.1): the pressure dissociates it into monomers. Dissociation is accompanied by a net hydration of the surface buried at the subunit interface [36,37]. Both second [13] and fourth [38] derivative spectroscopies were used. Second derivative spectroscopy has both advantages and disadvantages vis-à-vis fourth derivative spectroscopy. With second derivatives, one uses the ratio of the amplitudes, which is independent of the absolute absorbance. Therefore, one does not have to correct the spectra for compression effects and can easily compare samples studied at different times. The changes, however, are dominated by the contributions of tyrosines and provide much less information about the tryptophan residues.

To complete these structural observations and using a combination of fourth derivative spectroscopy under pressure and stopped-flow kinetics under pressure [17], Kornblatt et al. have shown that the monomers of yeast enolase produced by hydrostatic pressure are inactive. The thermodynamic parameters under different experimental conditions were determined and complementary experiments using CD spectra have allowed us to propose that optical modifications reflect changes in the conformations of the mobile loops of enolase [38].

If it has been accepted for a long time that high pressure is a rather general way to dissociate molecular assemblies, depending of the nature of the intersubunit regions, there are a few exceptions to this behavior [39]. Such is the case of an interesting example which has been found by Frank in our laboratory using methanol dehydrogenase from *Methylophaga marina* (MDH) [3]. Raising the pressure to 450 MPa results in a spectral transition detected using the fourth derivative method. Concomitant with a red shift, a high pressure decrease of the derivative amplitude is observed indicating that the tryptophans are in a less polar environment at elevated pressure (see Fig. 11). After releasing the pressure, both activity and structural modifications are reversible. The question arises: what are the origins of the structural modifications? MDH is a tetramer with two large and two small subunits. Usually, when pressure promotes oligomeric dissociation a blue shift in the UV spectrum is observed due to the hy-

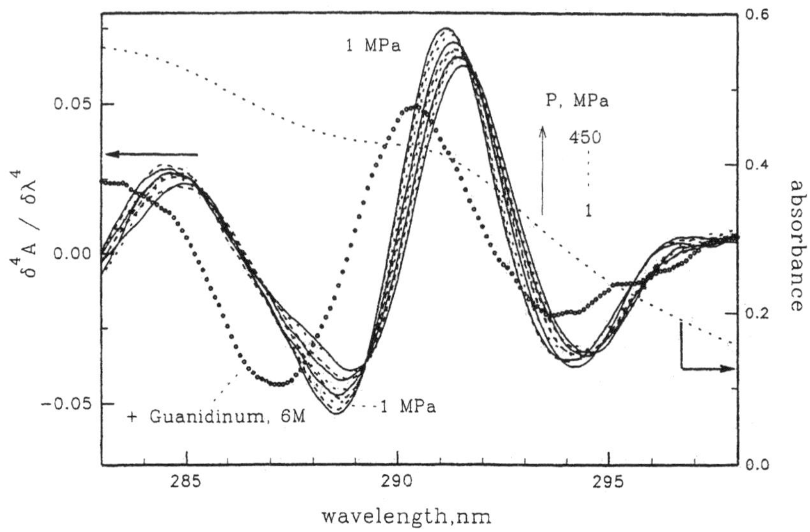

Fig. 11. Pressure effect on the fourth derivative spectrum of MDH. The original zero order spectrum is shown as a broken line, right hand scale. (From [3], with permission).

dration of hydrophobic residues upon exposure to the solvent. Here, a red shift is observed indicating that the spectral transition is, in addition to the dielectric constant, due to other parameters. On the other hand, previous electrophoretic experiments under high pressure showed that the enzyme migrated as a single peak even at 200 MPa where its activity was still measurable in the electrophoretic gel [40]. This indicates that the enzyme is maintained in its active tetrameric structure in the pressure range explored.

8. Pressure as a tool applied to some specific structural problems of cytochromes *P*-450

Cytochromes *P*-450 are enzymes of a multigenic family of heme proteins that catalyze oxidation of a variety of exogenous, as well as endogenous, substrates [41,42]. Because no liver microsomal cytochrome *P*-450 has been crystallized so far, there are numerous studies in the literature trying to get an insight into the structure of these extramembrane domain proteins by comparison with the soluble bacterial *P*-450s which were crystallized and whose three-dimensional structure is known [43,44]. These comparisons give a valuable estimate of the overall shape of the protein molecules and of the putative amino acids at their active sites. In *P*-450s, second derivative spectra in the 'aromatic' region were used to study the accessibility of *P*-450cam tyrosine side chains to solvent [45].

Recent studies monitored the changes in exposure of tyrosine and tryptophan side chains in cytochrome *P*-450 1A2 (CYP1A2). The fourth derivative was also used to examine the stability of the enzyme both in the presence and in the absence of a specific inhibitor (α-naphthoflavone), with increasing the pressure (up to 400 MPa) and temperature (up to 35°C) as perturbing factors. Anzenbacher et al. have shown that this enzyme has a stable conformation as all conformational changes were spectrally fully reversible [46].

Similar experiments were done by the same research group on the cytochromes *P*-450 3A4 (CYP3A4) and BM-3 (CYP102) also using fourth derivative spectroscopy. Conformational changes in the overall protein structures of these two enzymes due to the effects of pressure (and temperature) are reversible [47].

The central role for water in the control of spin state of cytochrome *P*-450scc has been studied by Bancel et al. This enzyme (which catalyzes pregnenolone formation in the mitochondria of steroidogenic tissues [48]) has two conformations and the conformation found at the lowest temperatures (A) displays a six-ligand high spin heme iron. As revealed by fourth derivative spectroscopy, the two conformers of the high spin state differ in the mean environment of the tryptophan residues, which was more polar in the A form. It has also been shown that the sixth ligand of the high spin form in the A conformer involves a water molecule and an unknown constraining structure [49].

9. Conclusions

The list of putative applications of the derivative methods to pressure-induced conformational studies is certainly longer. However, the interpretation of the results remains sometimes difficult – even at atmospheric pressure – as pointed out by Kornblatt et al.: 'How do the various aromatic residues of a protein contribute to the overall UV spectrum of the protein?' [50]. The overall derivative spectrum of a protein is the sum of the individual derivative spectra just as the overall UV spectrum of a protein is the sum of its component parts. In a deep study on two proteins as models: the 7 kDa Sso 7d from *S. solfataricus* (already mentioned above) and mutants from the 13 kDa iso-1-cytochrome *c* from *Saccharomyces cerevisiae*, the Kornblatt group used fourth derivative spectroscopy to evaluate the spectral contribution. The Sso 7d protein has two tyrosines and one tryptophan; the mutant forms of the 13 kDa iso-1-cytochrome *c* contain a single tryptophan and three tyrosines. They found that, if the tryptophan spectrum was not exceptional, the comparable tyrosine spectra were more complex: the derivative spectrum of individual tyrosines does not correspond to that expected on the basis of concentration. The individual tyrosines spectra of proteins may or may not differ from those in water or other solvents.

The effects of pressure on fourth derivative spectra (much as the same temperature effects) are interpreted with difficulty. In order to have as accurate as possible a view on the pressure effects on the protein structure, other techniques should be applied simultaneously. The fourth derivative method must be used jointly, when possible, with other spectroscopic methods such as fluorescence, 8-anilinonaphthalene-1-sulfonate binding, CD, Fourier transform infrared or Raman. This has been generally the case for most of the examples given above.

Acknowledgements

We wish to thank Ms. N. Bec and M.C. Valentin for their expert technical assistance with the pressure equipment and Drs. E. Anzenbacher, E. Anzenbacherova, F. Bancel, J. Frank, J.A. Kornblatt, M.J. Kornblatt, S. Kunugi, V. Lullien-Pellerin, E. Mombelli, V.V. Mozhaev, K. Ruan, and J. Torrent who spend so many hours in front of the Montpellier's high pressure apparatuses to get the nice results summarized in this paper.

References

[1] V.V. Mozhaev, K. Heremans, J. Frank, P. Masson, C. Balny, High pressure effects on protein structure and function, Proteins Struct. Funct. Genet. 24 (1996) 81–91.

[2] R. Lange, J. Frank, J.-L. Saldana, C. Balny, Fourth derivative UV-spectroscopy of proteins under high pressure. I. Factors affecting the fourth derivative spectrum of the aromatic amino acids, Eur. Biophys. J. 24 (1996) 277–283.

[3] R. Lange, N. Bec, V.V. Mozhaev, J. Frank, Fourth derivative UV-spectroscopy of proteins under high pressure. II. Application to reversible structural changes, Eur. Biophys. J. 24 (1996) 284–292.

[4] G. Talsky, Derivative Spectrophotometry; Low and High Order, VCH, Weinheim, 1994.

[5] T. Ichikawa, H. Terada, Second derivative spectrophotometry as an effective tool for examining phenylalanine residues in proteins, Biochim. Biophys. Acta 494 (1977) 267–270.

[6] T. Ichikawa, H. Terada, Estimation of state and amount of phenylalanine residues in proteins by second derivative spectrophotometry, Biochim. Biophys. Acta 580 (1979) 120–128.

[7] T. Ichikawa, H. Terada, Effect of dodecyl sulfate on the spectral properties of phenylalanine residues in serum albumin detected by second derivative spectrophotometry, Biochim. Biophys. Acta 671 (1981) 33–37.

[8] A.P. Demchenko, Ultraviolet Spectroscopy of Proteins. Springer, Berlin, 1986, pp. 121–136.

[9] R. Ragone, G. Colonna, C. Balestrieri, L. Servillo, G. Irace, Determination of tyrosine exposure in proteins by second derivative spectroscopy, Biochemistry 23 (1984) 1871–1875.

[10] H. Mach, C.R. Middaugh, Simultaneous monitoring of the environment of tryptophan, tyrosine, and phenylalanine residues in proteins by near-ultraviolet second-derivative spectroscopy, Anal. Biochem. 222 (1994) 323–331.

[11] M.T. Fisher, S.G. Sligar, Tyrosine motions in relation to the ferric spin equilibrium of cytochrome P450cam, Biochemistry 24 (1985) 6696–6701.

[12] A. Mozo-Villarias, A. Morros, J.M. Andreu, Thermal transitions in the structure of tubulin. Environments of aromatic amino acids, Eur. Biophys. J. 19 (1991) 295–300.

[13] J.A. Kornblatt, M.J. Kornblatt, G. Hui Bon Hoa, Second derivative spectroscopy of enolase at high hydrostatic pres-

sure: an approach to the study of macromolecular interactions, Biochemistry 34 (1995) 1218–1223.

[14] C. Balny, J.-L. Saldana, R. Lange, M.J. Kornblatt, J.A. Kornblatt, UV/Vis biochemical spectroscopy under high pressure, in: Ph.R. von Rohr, Ch. Trepp (Eds.), High Pressure Chemical Engineering, Elsevier Science, Amsterdam, 1996, pp. 553–558.

[15] R. Lange, N. Bec, J. Frank, C. Balny, Pressure induced protein structural changes as sensed by 4th derivative UV spectroscopy, in: R. Hayashi, C. Balny (Eds.), High Pressure Bioscience and Biotechnology, Progress in Biotechnology Series, Elsevier Science, Amsterdam, Vol. 13, 1996, pp. 135–140.

[16] C. Balny, R. Lange, Optical spectroscopic techniques in high pressure bioscience, in: R. Winter, J. Jonas (Eds.), High Pressure Molecular Science, NATO Science Series, Kluwer Academic Publishers, Dordrecht, 1999, pp. 405–422.

[17] G. Hui Bon Hoa, P. Douzou, N. Dahan, C. Balny, High pressure spectrophotometry at sub-zero temperatures, Anal. Biochem. 120 (1982) 125–135.

[18] A. Savitzky, M.J.E. Golay, Smoothing and differentiation of data by simplified least squares procedures, Anal. Chem. 36 (1964) 1627–1639.

[19] W.L. Butler, Fourth derivative spectra, Methods Enzymol. 56 (1979) 501–501.

[20] J. Torrent, J.P. Connelly, M.G. Coll, M. Robo, R. Lange, M. Vilanova, Pressure versus heat-induced unfolding of ribonuclease A: the case of hydrophobic interactions within a chain-folding initiation site, Biochemistry 38 (1999) 1592–1661.

[21] K. Ruan, C. Xu, Y. Yu, J. Li, R. Lange, N. Bec, C. Balny, Pressure-exploration of the 33-kDa protein from spinach photosystem II particle, Eur. J. Biochem. 268 (2001) 2742–2750.

[22] V. Lullien-Pellerin, Y. Popineau, F. Meersman, M.-H. Morel, K. Heremans, R. Lange, C. Balny, Reversible changes of the wheat gamma 46 gliadin conformation submitted to high pressures and temperatures, Eur. J. Biochem. 268 (2001) 5705–5712.

[23] A. Apichartsrangkoon, A.E. Bell, D.A. Ledward, J.D. Scofield, Dynamic viscoelastic behavior of high-pressure-treated wheat gluten, Cereal Chem. 76 (1999) 777–782.

[24] J. Lefebvre, Y. Popineau, G. Deshayes, L. Lavenant, Temperature-induced changes in the dynamic rheological behavior and size distribution of polymeric proteins for glutens from wheat near-isogenic lines differing in HMW-glutenin subunit composition, Cereal Chem. 77 (2000) 193–301.

[25] S. Endo, Studies on protease produced by thermophilic bacteria, J. Ferment. Technol. 40 (1962) 346–353.

[26] M. Fukuda, S. Kunugi, Pressure dependence of thermolysin catalysis, Eur. J. Biochem. 124 (1982) 157–163.

[27] S. Kunugi, K. Tanabe, K. Yamashita, Y. Morikawa, T. Ito, T. Kondoh, K. Hirata, A. Nomura, Protease-catalyzed peptide formation under high pressure, Bull. Chem. Soc. Jpn. 62 (1989) 514–518.

[28] R. Hayashi, Y. Kawamura, S. Kunugi, Introduction of high pressure to food processing: preferential proteolysis of β-lactoglobulin in milk whey, J. Food Sci. 52 (1987) 1107–1108.

[29] C. Balny, J.-L. Saldana, N. Dahan, High pressure stopped-flow spectrometry at low temperatures, Anal. Biochem. 139 (1984) 178–189.

[30] S. Kunugi, M. Kitayaki, Y. Yanagi, N. Tanaka, R. Lange, C. Balny, The effect of high pressure on thermolysin, Eur. J. Biochem. 248 (1997) 567–574.

[31] C. Balny, J.-L. Saldana, N. Dahan, High pressure stopped-flow fluorometry at subzero temperatures, Anal. Biochem. 163 (1987) 309–315.

[32] A. Fontana, Life under Extreme Conditions, Springer Verlag, Berlin, 1991.

[33] P. Fusi, G. Tedeschi, A. Aliverti, S. Rinchi, P. Tortora, A. Guerritore, Ribonuclease from extreme thermophilic archaebacterium *S. solfataricus*, Eur. J. Biochem. 211 (1993) 305–310.

[34] E. Mombelli, M. Afshar, P. Fusi, M. Mariani, P. Tortora, J.P. Connelly, R. Lange, The role of phenylalanine 31 in maintaining the conformational stability of ribonuclease P2 from *Sulfolobus solfataricus* under extreme conditions of temperature and pressure, Biochemistry 36 (1977) 8733–8742.

[35] J.A. Kornblatt, M.J. Kornblatt, C. Clery, C. Balny, The effects of pressure on the conformation of plasminogen, Eur. J. Biochem. 265 (1999) 120–126.

[36] M.J. Kornblatt, J.A. Kornblatt, G. Hui Bon Hoa, The role of water in the dissociation of enolase, a dimeric enzyme, Arch. Biochem. Biophys. 3.6 (1993) 495–500.

[37] J.A. Kornblatt, M.J. Kornblatt, G. Hui Bon Hoa, A.G. Mauk, Responses of two protein-protein complexes to solvent stress: does water play a role at the interface?, Biophys. J. 65 (1993) 1059–1065.

[38] M.J. Kornblatt, R. Lange, C. Balny, Can monomers of yeast enolase have enzymatic activity?, Eur. J. Biochem. 251 (1998) 775–780.

[39] C. Balny, Pressure effects on oligomeric dissociation, in: A. Ballesteros, F.J. Plou, J.L. Iborra, P.J. Halling (Eds.), Stability and Stabilization of Biocatalysis, Elsevier, Amsterdam, 1998, pp. 197–204.

[40] I. Heiber-Langer, C. Clery, J. Frank, P. Masson, C. Balny, Interaction of cytochrome cL with methanol dehydrogenase from *Methylophaga marina* 42. Thermodynamic arguments for conformational changes, Eur. Biophys. J. 21 (1992) 241–250.

[41] P.R. Ortiz de Montellano (Ed.), Cytochrome P450: Structure, Mechanism and Biochemistry, 2nd edn., Plenum Press, New York, 1995.

[42] D.R. Nelson, L. Koymans, T. Kamataki, J.J. Stegeman, R. Feyreisen, D.J. Waxman, M.R. Waterman, O. Gotoh, M.J. Coon, R.W. Estabrook, I.C. Gunsalus, D.W. Nebert, P450 superfamily. Update of new sequences, gene mapping, accession numbers and nomenclature, Pharmacogenetics 6 (1996) 1–42.

[43] C.A. Hasemann, R.G. Kurumbail, S.S. Boddupalli, J.A.

Peyterson, J. Deisenhofer, Structure and function of cytochromes P450: a comparative analysis of three crystal structures, Structure 3 (1995) 41–62.

[44] D.L.V. Lewis, B.G. Lake, Molecular modelling of CYP1A subfamily members based on an alignment with CYP102: rationalization of CYP1A substrate specificity in terms of active site amino acid residues, Xenobiotica 26 (1996) 723–753.

[45] M.T. Fisher, S.G. Sligar, Tyrosine motions in relation to the ferric spin equilibrium of cytochrome P-450cam, Biochemistry 24 (1985) 6696–6701.

[46] P. Anzenbacher, N. Bec, J. Hudebcek, R. Lange, E. Anzenbacherova, High conformational stability of cytochrome P450 1A2. Evidence from UV absorption spectra, Collect. Czech. Chem. Commun. 63 (1998) 441–448.

[47] E. Anzenbacherova, N. Bec, P. Anzenbacher, J. Hudecek, P. Soucek, C. Jung, A.W. Munro, R. Lange, Flexibility and stability of the structure of the cytochromes P450 3A4 and BM-3, Eur. J. Biochem. 267 (2000) 2916–2920.

[48] K. Suhara, T. Gomi, H. Sato, E. Itagaki, S. Takemori, M. Katagiri, Purification and immunochemical characterization of the two adrenal cortex mitochondrial cytochrome P450 proteins, Arch. Biochem. Biophys. 190 (1978) 290–299.

[49] F. Bancel, N. Bec, C. Ebel, R. Lange, A central role for water in the control of the spin state of cytochrome P450scc, Eur. J. Biochem. 250 (1997) 276–285.

[50] J.A. Kornblatt, M.J. Kornblatt, R. Lange, E. Mombelli, J.G. Guillemette, The individual tyrosines of proteins: their spectra may or may not differ from those in water or other solvents, Biochim. Biophys. Acta 1431 (1999) 238–248.

ELSEVIER

Biochimica et Biophysica Acta 1595 (2002) 94–102

BIOCHIMICA ET BIOPHYSICA ACTA
BBA

www.bba-direct.com

Review

High pressure static fluorescence to study macromolecular structure–function

Kangcheng Ruan [a], Claude Balny [b,*]

[a] *Institute of Biochemistry and Cell Biology, Shanghai Institutes for Biological Science, Chinese Academy of Sciences, Shanghai, PR China*
[b] *Institut National de la Santé et de la Recherche Médicale, INSERM U 128, IFR 24, CNRS, 1919, route de Mende, 34 293 Montpellier Cedex 05, France*

Received 7 September 2001; received in revised form 16 November 2001; accepted 20 November 2001

Abstract

Through some typical examples, the high pressure static fluorescence method is described. The potentiality of the intrinsic and extrinsic fluorescence probes are analyzed for structural characterizations. Special attention is given to the use of fluorescence to understand the behavior of enzymatic reactions under high pressure. The application of fluorescence polarization is also presented together with some relevant spectroscopic problems inherent in data interpretation.

Keywords: Fluorescence; Conformational change; Enzymatic activity; Fluorescence polarization; Stopped-flow; High hydrostatic pressure

1. Introduction

As a tool to explore the functionality of biomolecules, high hydrostatic pressure associated with optical detection methods has been extensively used in recent years to obtain information about molecular structure and/or molecular interactions [1–7]. In biomacromolecular high pressure studies, fluorescence spectroscopy introduced by Weber is a suitable method, in which the fluorescence spectrum, the fluorescence intensity, the fluorescence polarization and the lifetime of the intrinsic and extrinsic fluorophores can be used to detect the effect of pressure on oligomeric protein dissociation, conformational changes, folding–unfolding or on ligand affinity [8–16]. This review is restricted to the main applications of static fluorescence in high pressure study. The dynamic fluorescence method is introduced by Tauc et al. in another review in this issue.

2. Which fluorescence probes could be used: to detect what?

2.1. *Intrinsic fluorescence of proteins*

As for other spectroscopic methods, the question

Abbreviations: CSM, center of emission spectral mass; AT, *N*-acetyl-tryptophan; NATA, *N*-acetyl-tryptophanamide; TA, tryptophanamide; HPSF, high pressure stopped-flow; LADH, liver alcohol dehydrogenase; ANS, 8-anilinonaphthalinosulfonic acid; DNS, 2-(dimethylamino)naphthalene-5-sulfonic acid; FITC, fluorescein isothiocyanate; IAEDANS, iodoacetyl-amino-ethyl-aminonaphthalenesulfonic acid; IAF, iodoacetamidofluorescein

* Corresponding author. Fax: +33-4-67-52-36-81.
E-mail address: balny@falbala.crbm.cnrs-mop.fr (C. Balny).

PII: S0167-4838(01)00337-5

is: what are the most significant probes which can be used to detect molecular phenomena? There are several intrinsic fluorophores in proteins (mainly the aromatic amino acids: tryptophan, tyrosine and phenylalanine) which have been used to explore protein structure–function relationships or molecule interactions in association with other physical parameters such as temperature and pressure. Among these fluorophores, the tryptophan residue (Trp) is most often used. The characteristics of its fluorescence emission spectrum depend on its environment. The emission maximum wavelength (λ_{max}) of free Trp in aqueous solution is at about 352 nm and becomes shorter (blue shift) when the tryptophan is in an apolar solvent or in a hydrophobic environment such as inside the protein core. The more apolar the environment, the larger is the Trp λ_{max} blue shift. For instance, the λ_{max} of Azurin protein at 307 nm is the shortest λ_{max} found for tryptophan emission in proteins studied so far, indicating that the tryptophan residue is located in a very strong hydrophobic region [17]. When the protein is gradually unfolded, the tryptophan residue becomes more and more exposed to the bulk aqueous solution. Then, the λ_{max} shifts to the red and finally reaches 352 nm when Trp is totally solvated, i.e. when the protein is fully denatured. This λ_{max} at 352 nm is the value of the free Trp molecule in water. Therefore protein Trp fluorescence can be used as a probe related to the medium polarity allowing to detect and to follow protein conformational changes, from which information about protein structure–function, unfolding and molecule interaction, can be obtained. In comparison with tryptophan, tyrosine and phenylalanine residues in proteins have not been so often used for such studies mainly because their fluorescence quantum yields are lower and less sensitive to the environment, i.e. less sensitive to conformational changes. To measure the fluorescence emission coming only from the tryptophan residue, without any disturbance from the excitation of the tyrosine and phenylalanine residues, the protein sample is often excited at 295 nm or at longer wavelength, where tyrosine and phenylalanine do not absorb.

Among the parameters related to the fluorescence spectrum, the λ_{max}, the light emission intensity and the center of emission spectral mass (CSM) are most often used. For instance, in the 1970s, Weber and coworkers used the λ_{max} parameter to explore the pressure denaturation of chymotrypsinogen and lysozyme. They found plurality in the pressure-denatured forms of the proteins, an observation in good agreement with the multiple intermediate states recently reported in protein denaturation [14]. The change in λ_{max} was also used to follow the pressure-induced dissociation of protein subunits or complexes. In an F-actin study, Garcia et al. mainly used the λ_{max} to obtain the standard volume change upon association of actin (ΔV) [18]. They found a very interesting phenomenon: the presence of K^+ significantly increased the ΔV value of F-actin dissociation by a factor of about 4-fold in comparison with that measured in the presence of Mg^{2+} or Ca^{2+}.

However, it is often not very easy to determine the exact value of λ_{max} or to detect small changes in λ_{max}. In order to quantitatively analyze emission spectrum changes, Weber et al. [11,19] defined the parameter $\langle \nu \rangle$, the CSM as follows:

$$\langle \nu \rangle = \Sigma \nu_i \cdot F_i / \Sigma F_i$$

where F_i stands for the fluorescence intensity at a wave number ν_i and ΣF_i is the spectral integral in the spectrum range. Since then, $\langle \nu \rangle$ has been widely used to follow the pressure-induced dissociation of oligomeric proteins or nuclear acid–protein complexes (such as in virus), as well as the pressure-induced protein unfolding (see for example [12,13, 15,16,20,21]). The definition itself of $\langle \nu \rangle$ makes this parameter sensitive to the very small fluorescence emission spectrum changes. A good example was given by the analysis of the pressure-induced spectral red shifts of free tryptophan and its derivatives, *N*-acetyl-tryptophan (AT), *N*-acetyl-tryptophan-amide (NATA), tryptophanamide (TA) and tryptophan containing 6-polypeptides [22]. It was found by analyzing the center of spectral mass that pressure shifted the fluorescence spectra of all these species to the red direction since all the fluorescence quantum yields were reduced with pressure except free tryptophan, whose fluorescence efficiency was enhanced with pressure (see Fig. 1). These findings indicated that the intrinsic pressure effect on tryptophan must be taken into consideration to explain the phenomenon observed in pressure study of biomolecules when using the fluorescence intensities.

CSM was also recently used by Hayashi et al. to

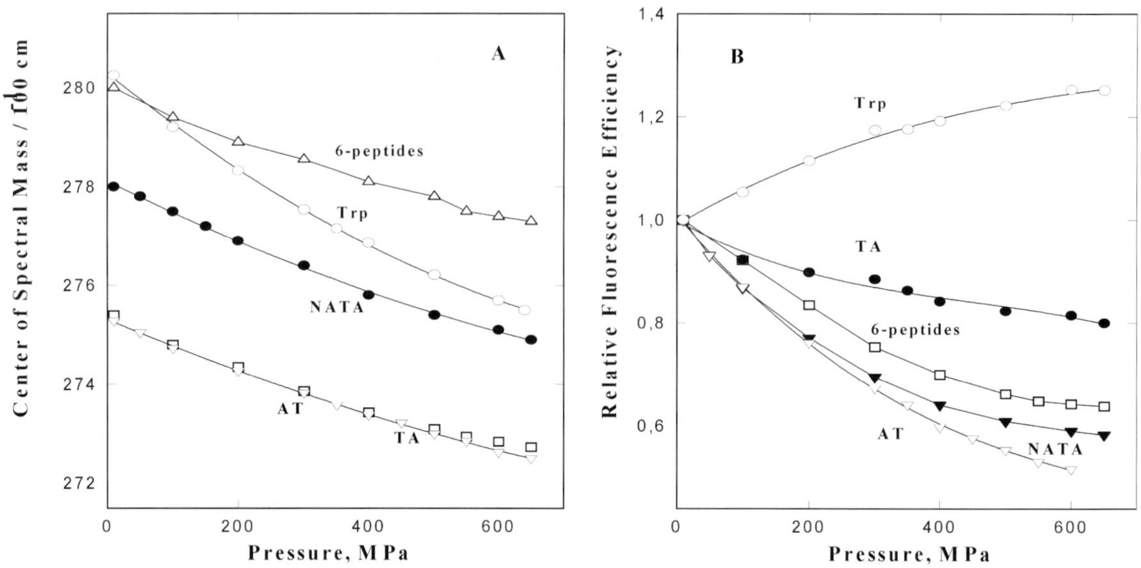

Fig. 1. Pressure effects on CMS (A) and relative fluorescence intensity (B) of free tryptophan and its derivatives. Compound concentration: 2 μM in 0.05 M Tris–HCl buffer, pH 7.3. Excitation wavelength: 280 nm, temperature, 20°C. (From [22], with permission.)

reveal at least three transition states involved in pressure-induced denaturation of either carboxypeptidase Y or unglycosylated carboxypeptidase Y [20].

Generally, the tryptophan fluorescence spectral red shift is accompanied with a decrease in quantum yield. Therefore the quantum yield, the total intensity as well as the fluorescence intensity at a given wavelength can also be used as a probe in protein study under pressure. Kunugi et al. used the tryptophan fluorescence peak intensity to explore the pH effect on the pressure-induced conformational fluctuation of apomyoglobin [23].

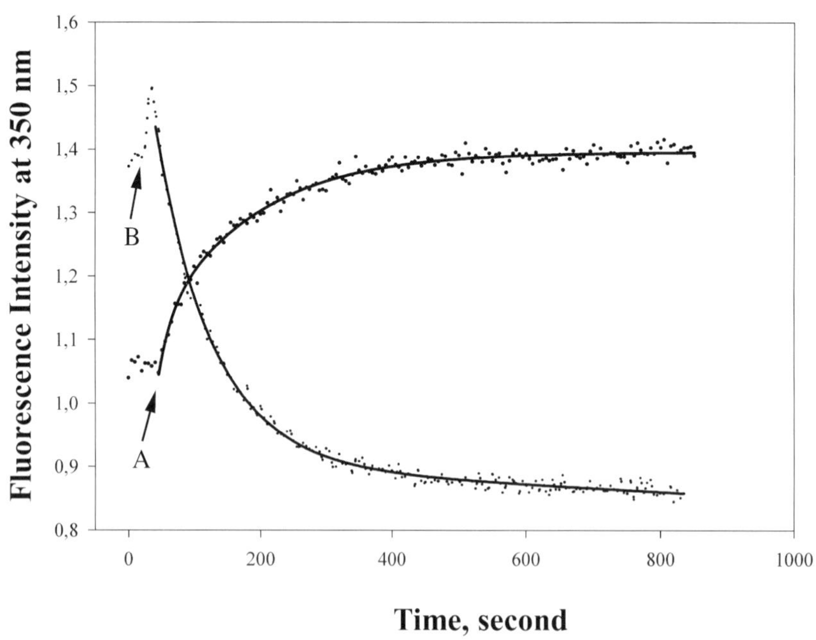

Fig. 2. Kinetics of the pressure-induced unfolding and refolding of the 33 kDa protein. Arrow A shows the start of the pressure jump from 100 MPa to 160 MPa and arrow B the start of the pressure jump from 160 MPa to 100 MPa. Protein concentration: 0.1 mg/ml in 0.05 M MES buffer pH 6 at 20°C. Excitation and emission wavelengths: 295 nm and 350 nm, respectively. (From [25], with permission.)

Depending of the system under study, and depending on the phenomenon that we would like to observe, either CSM or fluorescence intensity can be used. To have access to the reaction kinetics, the fluorescence intensity is the best parameter. A pressure relaxation study on the folding/unfolding of staphylococcal nuclease has been reported by Vidugiris et al. [24] supporting a molten globule-like model for the transition state of nuclease in which the polypeptide chain is in a collapsed, loosely packed, solvent-excluded structure. In a recent study on the pressure-induced unfolding of a 33 kDa protein isolated from spinach photosystem II, Ruan et al. [25] observed a very slow kinetics of the protein unfolding and refolding by monitoring the fluorescence intensity changes at 352 nm. Using the pressure jump method (developed at the Montpellier laboratory), the kinetics of this reaction – i.e. the relaxation time of the two-state process protein unfolding–refolding – were measured (125 and 100 s respectively) together with the thermodynamic parameters. Fig. 2 gives typical curves of the time dependence of the intrinsic fluorescence intensity upon positive and negative pressure jumps detected at 350 nm.

2.2. Fluorescence and enzymology

Fluorimetry provides a means to follow enzyme activity under high pressure when either the substrate or the product has a unique fluorescence signal. We know that in carrying out their biological functions, enzymes go through a number of conformational changes related to their dynamic structural flexibility which can be optically followed at the tryptophan level (see above). However, these conformational changes can also be related to the enzymatic activity. Then, the optical detection must be achieved either at the substrate or at the product level when they have a fluorescence signal.

We will not present here results on enzymatic steady state determinations under high pressure, but the more original approach using fluorescence fast kinetics detection. Indeed, many enzymatic processes are very rapid, and therefore difficult to record using classical detection systems. One possibility of reducing this velocity is to carry out experiments at low temperatures using cryoenzymology [26]. This artifice slows down the elementary reactions velocities which can be observed and analyzed. Using substrate (or product) fluorescence as a probe in cryoenzymic conditions, one can determine the elementary rate constants of various steps making up a reaction pathway. Under such conditions only simple rate constants are measured and precise thermodynamic data can be obtained: measurement of a composite constant (such as V_{max} or K_m) can lead to ambiguous results. This means the use of rapid reaction techniques adapted both to sub-zero temperatures and to fluorescence conditions to further increase the resolution time. Moreover, to obtain a maximum of thermodynamic information associated with the kinetic constants of interest, experiments under high pressure must be achieved. For that we have developed a stopped-flow apparatus adapted to work in a temperature range −40 to +40°C, at a pressure up to 200 MPa, operating either in absorbency or in fluorescence mode (high pressure stopped-flow (HPSF)). Its general design has been published elsewhere [27,28]. The apparatus has a dead time of 100 ms at −15°C (in fluid media composed of a mixture of water and non-denaturing organic solvent such as ethylene glycol). Under more ambient conditions (i.e. 20°C), the dead time of the system is 20 ms.

The first experiments investigated the temperature and pressure dependences of the binding kinetics of NADH to liver alcohol dehydrogenase (LADH) in the presence of DMSO (as antifreeze), a process which is accompanied by an increase in fluorescence intensity at 466 nm, with an excitation at 366 nm. Experiments were carried out under different conditions of temperature and pressure. As expected, lowering the temperature decreased the binding kinetics (reaction velocity) which are further decreased under pressure. The positive activation volume observed, increasing as the temperature increases, could be associated with the formation of a mainly hydrophobic bond between NADH and LADH [28].

Using the same approach (HPSF in fluorescence mode), in collaboration with P. Masson (La Tronche-Grenoble), we have investigated the pressure effects on the single turnover kinetics of the carbamylation of human plasma butyrylcholinesterase with *N*-methyl(7-dimethylcarbamoxy)quinolinium as substrate. The rate of carbamylation was monitored as the accumulation of the fluorescence ion, *N*-methylhydroxyquinolinium. Elevated pressure

Table 1
Comparison of the fluorescence efficiency $\Delta\Phi$ and $\Delta\nu$ of Trp, TA, NATA and AT in Tris–HCl buffer (pH 7.3) and in the presence of 100% ethanol

	Trp	TA	NATA	AT
$\Delta\Phi$ (ethanol), 0.1 MPa	−0.05	−0.08	−0.23	−0.22
$\Delta\nu$ (ethanol), 0.1 MPa	100	50	180	140
$\Delta\Phi$ (water), 650 MPa	0.22	−0.2	−0.4	−0.45
$\Delta\nu$ (water), 650 MPa	421	305	310	265

Pressures: 0.1 and 650 MPa. (From [22], with permission.)

favored formation of the enzyme–substrate complex but inhibited carbamylation of the enzyme. The large positive activation volume measured (119 ml/mol) may reflect extended structural and hydration changes [29]. Similar experiments were performed in various media: water, heavy water and water containing lithium chloride, the latter medium being known to increase water structure acting as a protein stabilizer. Kinetics and thermodynamic results from fluorescence experiments suggested that the substrate-dependent pressure effect is modulated by the solvation state of the enzyme [30].

The problem of isomerization of the actomyosin complex was known to be pressure sensitive, but the pressure sensitivity of other steps in the ATPase pathway has not been characterized. With a group of the University of Bristol, we have studied the pressure effects on the ATP actomyosin hydrolysis step of the myosin subfragment 1 ATPase, ADP binding to actomyosin subfragment 1 and the rate of ATP-induced dissociation of actomyosin subfragment 1. Kinetics were investigated by monitoring fluorescence emission at 385 nm (excitation at 366 nm) versus time at various temperatures and pressures. We were able to discuss the relationship of the changes to the observed effect of pressure on skin muscle fibers [31].

These examples showed the potential of the HPSF approach under high pressure, indicating that for the systems studied the nature of the solvent can induce variations of certain thermodynamic values, indicating that the thermodynamic macroscopic response of these systems is mainly controlled by solvent reorganization.

2.3. The problems of spectral data interpretation

As described above, tryptophan fluorescence is often used to explore the effect of hydrostatic pressure on proteins. However, some abnormal phenomena have been observed in pressure studies. If Kunugi and his coworkers [23] found that, as usual, the tryptophan fluorescence of apomyoglobin had a red shift with pressure, the fluorescence efficiency had a significant increase with pressure, an observation very different from that found for most proteins studied so far. Weber et al. also reported that in the lower pressure range, the fluorescence spectrum of chymotrypsinogen presented a red shift and an increase in intensity. A further increase in pressure made the spectral red shift accompanied with an intensity decrease [14]. Similar phenomena were also observed by ourselves in trypsin and phospholipase A_2 [16,32]. To clarify and understand these observations, as shown above, we investigated the pressure effects on fluorescence of a simple free tryptophan molecule [22]. The fluorescence spectra of free tryptophan had a red shift (3–4 nm) and an increase in fluorescence efficiency (about 1.5 times) upon pressure increasing from 0.1 to 650 MPa. However, for the tryptophan derivatives, the pressure caused a spectral red shift accompanied with a different decrease in intensity (see Fig. 1 and Table 1).

These results indicated that one should interpret carefully the pressure-induced change in tryptophan fluorescence in protein, especially in the case where the tryptophan residue is located on the surface of the protein molecules.

3. The extrinsic probes

Besides the intrinsic high pressure fluorescence of protein, some extrinsic fluorescence probes are also used. 8-Anilinonaphthalinosulfonic acid (ANS) (or bis-anilinonaphthalene sulfonate, bis-ANS) is one of the most widely used probes. This probe has a 100-fold fluorescence enhancement and a significant spectral blue shift as well as a longer emission decay upon transfer from water to a non-polar environment as may occur in binding to protein hydrophobic cores. Therefore, the binding characteristics of ANS to protein and its spectral properties deduced from the ANS fluorescence change can be used to explore protein hydrophobicity, conformational modifications, unfolding, dissociation, etc. Masson et al. found that ANS has a higher binding affinity to the pressure-induced molten globule state of protein ([33]; see [34] for a review on molten globule state). The same phenomenon was also observed by Ruan et al. in a study of the pressure denaturation of trypsin [15,32]. Using ANS, a significant difference between the native and the conformationally drifted glyceraldehyde phosphate dehydrogenase caused by dissociation–association was also observed [13].

It is worthy to be indicated that the results obtained from either intrinsic or extrinsic high pressure protein fluorescence measurements are generally consistent. A good example in a recent study was the free energy determination (ΔG) and the standard volume change (ΔV) for the unfolding transition at pH 6.0 and 20°C of the 33 kDa protein isolated from the spinach photosystem particle II obtained either from intrinsic or ANS binding fluorescence: −14.6 kJ/mol and −120 ml/mol, and −12.6 kJ/mol and 118.1 ml/mol respectively [25].

4. The application of fluorescence polarization

It is well known that the fluorescence polarization of a fluorophore excited with polarized light is related to the rotational relaxation time of the molecule containing the fluorophore. Then, the rotational relaxation time depends on its molecular weight (or molecular size). Accordingly, fluorescence polarization is widely used to detect changes related to the molecular sizes such as molecular binding interaction, dissociation or association of subunits and complexes, etc. This has been successfully used in atmospheric pressure studies. For the polarization measurement under high pressure there is an optical problem because of a polarization scrambling effect of the bomb windows with pressure. Paladini and Weber's pioneer work has allowed to solve this problem [8]. Since then, many studies have been reported on oligomeric protein, virus and complex dissociation by following the polarization changes of extrinsic or intrinsic fluorophores under pressure. For the extrinsic fluorescence polarization measurement, 2-(dimethylamino)naphthalene-5-sulfonic acid (DNS) derivatives and fluorescein 5-isothiocyanate (FITC) are widely used to label proteins. In order to obtain a suitable polarization value in the experiment, the selection of labeled probe must consider the combination of the λ_{max} fluorescence lifetime of the probe and the molecular weight of the compound to be labeled according to Perrin's equation:

$$(1/P-1/3) = (P_0/1-1/3)(1+3\tau/\rho)$$

P and P_0 are the polarization and limiting polarization respectively (or fundamental polarization), τ is the fluorescence lifetime of the fluorophore, ρ is the rotational relaxation time of the molecule to which the probe is attached. According to our experience, P_0 can be generally taken as 0.35–0.4, ρ can be estimated from the molecular weight divided by 1000. Then, the ideal polarization value of labeled protein can be obtained by considering the combination of the rotational relaxation time of the labeled molecule and extrinsic probe with a suitable lifetime τ_o. The ideal polarization value is between 0.1 and 0.3 for measurement. Ruan and Weber used polarization of a DNS–protein conjugate to follow the pressure-induced dissociation of yeast hexokinase, obtaining results consistent with those obtained from the center of spectral mass measurement [12]. Sometimes, if more than one probe molecule is covalently bound to a single protein molecule, energy transfer between the probes (homotransfer) can be observed, resulting in depolarization [17]. In this case, when the labeled probes are separated by some physical–chemical reason such as subunit dissociation, the polarization value will increase following the diminishing energy transfer upon dissociation. This phenomenon can also be used to follow subunit dissociation. A good example is that the polarization of muscle glycogen phosphorylase heavily labeled with FITC (molar ratio: FITC/phosphorylase = 4/1, FITC/subunit = 1/1) increased from 0.27 at atmospheric pressure to 0.32 when the pressure reached 200 MPa, where the protein was totally dissociated into free subunits [35]. One interesting experiment based on this phenomenon was carried out in a study of phosphorylase to follow subunit exchange during the dissociation–association equilibrium. The half-dissociation pressure – $P_{1/2}$ about 120 MPa – was applied to a mixture of protein heavily labeled with FITC and non-labeled glycogen phosphorylase (1:10), the pressure being maintained for 2 h. It was observed that the polarization increased in 100 min after a short equilibrium (see Fig. 3). This increase was attributed to the di-

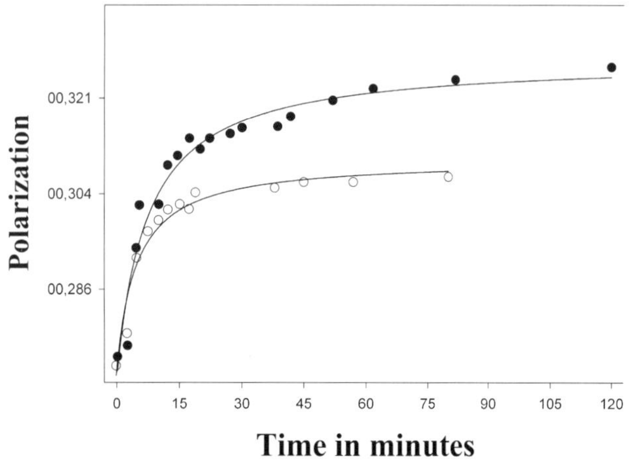

Fig. 3. Time dependence of the polarization of 0.26 μM fluorescein–glycogen phosphatase (GPA) conjugate after a rapid rise in pressure to 140 MPa at 2°C. ●, in the presence of unlabeled GPA; ○, by itself. (From [35], with permission.)

minishing energy transfer between the labeled probe caused by the continuous exchange of labeled and unlabeled subunits in the tetramer [35].

The intrinsic fluorescence polarization (mainly tryptophan fluorescence) is, when possible, also often used in pressure studies. It should be noted that the protein tryptophan fluorescence polarization is not only related to the molecular size of the protein or complex, but also to the local motion of the tryptophan residue. The consequence is that the polarization is usually smaller than that expected from molecular weight alone according to Perrin's equation. For instance, the intrinsic fluorescence polarization of the lactose repressor protein is only 0.07 for tetramer, the native form at atmospheric pressure, and 0.042 for monomer obtained at about 200 MPa [19]. Despite this observation, the related studies indicated that the intrinsic fluorescence polarization of protein can be used to follow the subunit dissociation, giving very consistent results with other methods [11–13,19,35].

In the parameters related to polarization, the intrinsic excitation polarization spectrum is sometimes used to explore the change in excitation state of intrinsic fluorophores. By comparing the excitation polarization spectra of the native glyceraldehyde phosphate dehydrogenase (GAPDH) and its conformationally drifted form obtained by pressure dissociation, Ruan and Weber found that the local mobility of the tryptophan increased and that the tryptophan transition had a blue shift in the conformationally drifted form. This conformationally drifted form was prepared by incubating a GAPDH solution at a pressure of 240 MPa for 1 h. [13]. The conformational drift has been postulated by Weber [36] to explain the loss of affinity of the protein subunits of an oligomeric aggregate when they become separated from each other. This loss of affinity is a progressive, time-dependent disorganization of the monomers occurring when, after the dissociation of the aggregate, they cease to exert influence upon each other.

Fluorescence resonant energy transfer between different fluorophores (heterotransfer) can be used to follow protein folding, protein association, etc. [37,38]. A typical example of the application of hetero-energy transfer in pressure studies can be given. The primary structures of the subunits of rabbit muscle glycogen phosphorylases A an B (GPA and GPB, respectively) are identical except that GPA is phosphorylated at the level of the Ser14. GPA is a tetramer and GPB a dimer [39] and it is not experimentally simple to demonstrate that the respective subunits from GPA and GPB can hybridize. Ruan et al. [35] used heterotransfer to obtain very strong evidence that the subunits from A and B can hybridize to each other. In this experiment, the labeled conjugate iodoacetyl-aminoethyl-amino-naphthalenesulfonic acid (IAEDANS)-GPA was used (IAEDANS being the donor), and the labeled conjugate iodoacetamidofluorescein (IAF)-GPB was used as acceptor (IAF being the acceptor). In the mixture IAEDANS-GPA and IAF-GPB (2:1) no energy transfer could be observed, IAEDANS and IAF being separately located in different subunits. The sample was subjected to a pressure of 240 MPa at which all GPA and GPB were dissociated into free subunits. After pressure release, it was found that fluorescence at 480 nm (which is the maximum emission wavelength of IAEDANS) had decreased significantly, while the fluorescence at 525 nm (the maximum wavelength of IAF) had increased, in comparison with the spectrum obtained before the pressure application. This indicated that the energy transfer from IAEDANS to IAF occurred after a pressure application and that the GPA and GPB subunits could hybridize resulting in energy transfer [38].

5. High pressure equipment in fluorescence measurement under pressure

Various vessels have being designed for high pressure fluorescence experiments. One of the first was described by Paladini and Weber [8]. Since then, different improvements have been introduced making high pressure vessels adaptable to modern fluorimeters [40]. The bodies of the vessels were constructed in steel alloy and outfitted with 3–4 removable, threaded window plugs mounted with either sapphire or quartz windows. The use of quartz windows limits the pressure to 300 MPa when sapphire windows allow a pressure up to 650 MPa. The principle of the sealing plugs is still based on Bridgman's innovation [41].

6. Conclusions

If protein fluorescence is an approach known for many years, it is under the impulsion of Weber's research group at Urbana (USA) in the 1970s and 1980s that its use for high pressure studies took a new turn. However, intrinsic protein fluorescence is limited to the tryptophan response. The natural complementary approach is the use of external fluorescence probes providing information on hydrophobic domains and hydrophobic interactions, a crucial biochemical challenge as pointed out by Kauzmann some years ago [42]. High pressure fluorescence experiments, associated with other high pressure spectroscopic methods (Fourier transform infrared spectroscopy, small-angle X-ray scattering [43], dynamic fluorescence, absorbency, etc.), yield relatively complete thermodynamic information related both to the structure and the function of the biomolecules under study.

Acknowledgements

This work was supported by a grant from INSERM/Academia China. The authors thank Dr. C. Royer for stimulating discussions and fruitful advice.

References

[1] C. Balny, R. Hayashi, K. Heremans, P. Masson (Eds.), High Pressure and Biotechnology, Colloque INSERM, Vol. 224, J. Libbey Eurotext, Montrouge, 1992.
[2] J. Markley, D.B. Northrop, C. Royer (Eds.), High-Pressure Effects in Molecular Biophysics and Enzymology, Oxford University Press, New York, 1996.
[3] V.V. Mozhaev, K. Heremans, J. Frank, P. Masson, C. Balny, Exploiting the effects of high hydrostatic pressure in biotechnological application, Trends Biotechnol. 12 (1994) 493–501.
[4] G. Weber, Protein Interactions, Chapman and Hall, London, 1992.
[5] V.V. Mozhaev, K. Heremans, J. Frank, P. Masson, C. Balny, High pressure effects on protein structure and function, Proteins Struct. Funct. Genet. 24 (1996) 81–91.
[6] C. Balny, K. Ruan, An emerging field in bioscience: high hydrostatic pressure study, Acta Biochim. Biophys. Sin. 31 (1999) 619–624.
[7] J.L. Silva, G. Weber, Pressure stability of proteins, Annu. Rev. Phys. Chem. 44 (1993) 89–113.
[8] A.A. Paladini, G. Weber, Absolute measurements of fluorescence polarization at high pressure, Rev. Sci. Instrum. 52 (1981) 419–427.
[9] A.A. Paladini, G. Weber, Pressure-induced reversible dissociation of enolase, Biochemistry 20 (1981) 2587–2593.
[10] L. King, G. Weber, Conformational drift and cryoinactivation of lactate dehydrogenase, Biochemistry 25 (1986) 3637–3640.
[11] J.L. Silva, E.W. Moles, G. Weber, Pressure dissociation and conformational drift of the β dimer of tryptophan synthase, Biochemistry 25 (1986) 5781–5786.
[12] K. Ruan, G. Weber, Dissociation of yeast hexokinase by hydrostatic pressure, Biochemistry 27 (1988) 3295–3301.
[13] K. Ruan, G. Weber, Hysteresis and conformational drift of pressure-dissociated glyceraldehyde phosphate dehydrogenase, Biochemistry 28 (1989) 2144–2153.
[14] T.M. Li, J.W. Hook, H.G. Drickamer, G. Weber, Effects of pressure upon the fluorescence of the riboflavin binding protein and its flavin mononucleotide complex, Biochemistry 15 (1976) 3205–3211.
[15] R. Ruan, R. Lange, N. Bec, C. Balny, A stable partly denatured state of trypsin induced by high pressure, Biochem. Biophys. Res. Commun. 239 (1997) 150–154.
[16] K. Ruan, R. Lange, Y. Zhou, C. Balny, Unusual effect of high hydrostatic pressure on basic phospholipase A_2 from venom of *Agkistrodon Halys Pallas*, Biochem. Biophys. Res. Commun. 249 (1998) 844–848.
[17] J.R. Lakowicz, Principle of Fluorescence Spectroscopy, 2nd edn., Kluwer Academic/Plenum Publishers, New York, 1999.
[18] C.R. Garcia, A. Amaral, P. Abrahamsohn, S. Verjovski-Almeida, Dissociation of F-actin induced by hydrostatic pressure, Eur. J. Biochem. 209 (1992) 1005–1011.
[19] C.A. Royer, G. Weber, T.J. Daly, K.S. Matthews, Dissoci-

ation of the lactose repressor protein tetramer using high hydrostatic pressure, Biochemistry 25 (1986) 8308–8315.
[20] M. Dumoulin, H. Ueno, R. Hayashi, C. Balny, Contribution of the carbohydrate moiety to conformational stability of carboxypeptidase Y, Eur. J. Biochem. 262 (1999) 475–783.
[21] S.M. Tian, K. Ruan, J.F. Qian, G.Q. Shao, C. Balny, Effects of hydrostatic pressure on the structure and biological activity of infectious bursal disease virus, Eur. J. Biochem. 267 (2000) 4486–4494.
[22] R. Ruan, S. Tian, R. Lange, C. Balny, Pressure effects on tryptophan and its derivatives, Biochem. Biophys. Res. Commun. 269 (2000) 681–686.
[23] N. Tanaka, C. Ikeda, K. Kanaori, K. Hirga, T. Konno, S. Kunugi, Pressure effect on the conformational fluctuation of apomyoglobin in the native state, Biochemistry 39 (2000) 12063–12068.
[24] G.J. Vidugiris, J.L. Markley, C.A. Royer, Evidence for a molten globule-like transition state in protein folding from determination of activation volumes, Biochemistry 34 (1995) 4909–49012.
[25] K. Ruan, C. Xu, Y. Yu, J. Li, R. Lange, N. Bec, C. Balny, Pressure-exploration of the 33-kDa protein from spinach photosystem II particle, Eur. J. Biochem. 268 (2001) 2742–2750.
[26] P. Douzou, Cryobiochemistry, an Introduction, Academic Press, London, 1977.
[27] C. Balny, J.-L. Saldana, N. Dahan, High-pressure stopped-flow spectrometry at low temperatures, Anal. Biochem. 139 (1984) 178–189.
[28] C. Balny, J.-L. Saldana, N. Dahan, High-pressure stopped-flow fluorometry at subzero temperatures: application to kinetics of the binding of NADH to liver alcohol dehydrogenase, Anal. Biochem. 163 (1987) 309–315.
[29] P. Masson, C. Balny, Effects of high pressure on the single-turnover kinetics of the carbamylation of cholinesterase, Biochim. Biophys. Acta 954 (1988) 208–215.
[30] P. Masson, C. Balny, Conformational plasticity of butyrylcholinesterase as revealed by high pressure experiments, Biochim. Biophys. Acta 1041 (1990) 223–231.
[31] D. McKillop, M.A. Geeves, C. Balny, The effect of hydrostatic pressure on the interaction of actomyosin subfragment 1 with nucleotides, Biochem. Biophys. Res. Commun. 180 (1991) 552–557.
[32] K. Ruan, R. Lange, F. Meersman, K. Heremans, C. Balny, Fluorescence and FTIR study of the pressure-induced denaturation of bovine pancreas trypsin, Eur. J. Biochem. 265 (1999) 79–85.
[33] C. Clery, F. Renault, P. Masson, Pressure-induced molten globule state of cholinesterase, FEBS Lett. 370 (1995) 212–214.
[34] P. Masson, C. Clery, Pressure-induced molten globule states of proteins, in: R. Hayashi, C. Balny (Eds.), High Pressure Bioscience and Biotechnology, Elsevier Science, Amsterdam, 1996, pp. 114–126.
[35] R. Ruan, G. Weber, Physical heterogeneity of muscle glycogen phosphorylase revealed by hydrostatic pressure dissociation, Biochemistry 32 (1993) 6295–6301.
[36] G. Weber, Phenomenological description of the association of protein subunits subjected to conformational drift. Effects of dilution and hydrostatic pressure, Biochemistry 25 (1985) 3626–3631.
[37] K. Cai, V. Schircht, Structural studies on folding intermediates of serine hydroxymethyl transferase using FRET, J. Biol. Chem. 271 (1996) 27311–27320.
[38] G.Q. Tang, T.P. Wu, K. Ruan, Hybridization of subunits of muscle glycogen phosphorylase A and B, Acta Biochim. Biophys. Sin. 27 (1995) 113–118.
[39] R.J. Fletteric, N.B. Madsen, The structures and related functions of phosphorylase A, Annu. Rev. Biochem. 49 (1980) 31–61.
[40] R. Lange, J. Frank, J.-L. Saldana, C. Balny, Fourth derivative UV-spectroscopy of proteins under high pressure. 1. Factors affecting the fourth derivative spectrum of aromatic amino acids, Eur. Biophys. J. 24 (1996) 277–283.
[41] P.W. Bridgman, The Physics of High Pressure, 2nd edn., G. Bell and Sons, London, 1949.
[42] W. Kauzmann, Thermodynamics of unfolding, Nature 325 (1987) 723–724.
[43] G. Panick, G.J. Vidugiris, R. Malessa, G. Rapp, R. Winter, C.A. Royer, Exploiting the temperature-pressure phase diagram of staphylococcal nuclease, Biochemistry 38 (1999) 4157–4164.

Biochimica et Biophysica Acta 1595 (2002) 103–115

BIOCHIMICA ET BIOPHYSICA ACTA
BBA
www.bba-direct.com

Review

Investigation of the effect of high hydrostatic pressure on proteins and lipidic membranes by dynamic fluorescence spectroscopy

Patrick Tauc [a], C. Reyes Mateo [b], Jean-Claude Brochon [a,*]

[a] *Laboratoire de Biotechnologie et de Pharmacologie Génétique Appliquée, UMR 8532 C.N.R.S., Ecole Normale Supérieure de Cachan, 61, avenue du Président Wilson, 94235 Cachan Cedex, France*

[b] *Centro de Biologia Molecular y Cellular, Universidad Miguel Hernandez, 03206 Elche-Alicante, Spain*

Received 8 October 2001; received in revised form 28 November 2001; accepted 29 November 2001

Abstract

Dynamic fluorescence spectroscopy brings new insight into the functional and structural changes of biological molecules under moderate and high hydrostatic pressure. The principles of time-resolved fluorescence methods are briefly described and the resulting type of information is summarized. A first set of selected applications of the use of dynamic fluorescence on pressure effects on proteins in terms of denaturation, ternary and quaternary structure, aggregation and also interaction with DNA are presented. A second set of applications is devoted to the effect of pressure and of cholesterol on lateral heterogeneity of lipidic membranes.

Keywords: Fluorescence lifetime; Rotational correlation time; Protein assembly; Membrane heterogeneity

1. Introduction

Fluorescence spectroscopy is widely used in the biological sciences for the study of structural organization of living systems at both molecular, supramolecular and cellular levels. Many biological molecules are intrinsically fluorescent or can be labelled by externally added fluorophores (for a general review on fluorescence see [1]). The fluorescence signal is very sensitive to the local environment of probes, i.e. within 5–10 Å distance. The physico-chemical parameters modulating fluorescence properties are mainly polarity, pH, viscosity, electric potential and ion concentration. Steady-state fluorescence, obtained upon a static excitation of the sample, measures the emission and excitation spectra, in other words the energy distribution of emitted and absorbed photons, respectively. The corresponding applications related to high pressure are treated in another chapter of this issue (K. Ruan). The fluorescence dynamics is related to the time dependence of the excited state of a fluorophore. The fluorescence lifetime is the average time between its excitation and the return to a ground state, and in addition to quantum yield and fundamental anisotropy (see below) it characterizes an individual fluo-

Abbreviations: DMPC, 1,2 dimyristoyl-3-phosphatidylcholine; DPH, 1,6-diphenyl-1,3,5-hexatriene; DPPC, dipalmitoyl-3-phosphatidylcholine; NMR, nuclear magnetic resonance; POPC, 1-palmitoyl-2-oleoylphosphatidylcholine; TMA-DPH, 1-(4-trimethylammonium-phenyl)-6-phenyl-1,3,5-hexatriene; TMAP-PDH, 3-(diphenylhexatrienyl)propyl-trimethylammonium; t-PnA, *trans*-parinaric acid

* Corresponding author. Fax: +33-1-47-40-24-79.
E-mail address: brochon@lbpa.ens-cachan.fr (J.-C. Brochon).

0167-4838/02/$ – see front matter
PII: S0167-4838(01)00338-7

rophore. Furthermore, time-resolved polarization of fluorescence provides detailed information on rotational dynamics in solution, mainly due to the Brownian motion, which is related to the hydrodynamic volume of a macromolecule as well as to any local structural flexibility. This paper reviews the applications of fluorescence dynamics in the study of the effects of hydrostatic high pressure on proteins and lipidic membranes.

2. Dynamic fluorescence principles

2.1. Theoretical background

First let us recall some fundamental considerations about time-resolved fluorescence spectroscopy. When a molecule absorbs a photon of a very short light pulse, the corresponding metastable excited state is created within a few femtoseconds and lasts from subpicoseconds to tens of nanoseconds. The deactivation of the excited state can be considered in terms of rate constants and for individual fluorophores the corresponding photon emission often follows first order kinetics. Then the fluorescence decay, $I(t)$, is a single exponential with an amplitude proportional to the absorbed light. In the case of a mixture of fluorophores or of different environments of the same molecule, the fluorescence decay is a sum of exponentials:

$$I(t) = \sum_{i=1}^{N} \alpha_i e^{-t/\tau_i} \text{ with } \alpha_i = \varepsilon_i c_i \quad (1)$$

where ε_i and c_i are the molecular absorption coefficient and concentration, respectively. When the excited pulse is linearly polarized, let us say vertically with respect to the laboratory coordinates, a chromophore having an absorption dipole parallel to that direction has the highest probability to be excited. Then an anisotropic population of fluorophores is selected through this polarized excitation at $t=0$. Between the excitation and emission times, the rotational dynamics continue and the orientation of the emission dipole of the excited molecules may change, causing a decrease in the polarization of fluorescence. The time dependence of the fluorescence polarization is more commonly expressed as the fluorescence anisotropy decay, $A(t)$, which is defined for an individual molecule as:

$$A(t) = \frac{I_{vv}(t) - I_{vh}(t)}{I_{vv}(t) + 2I_{vh}(t)} = \sum_j \beta_j e^{-t/\theta_j} \quad (2)$$

where $I_{vv}(t)$ and $I_{vh}(t)$ are the decays of fluorescence detected through a polarizer orientated vertically and horizontally, respectively (for a review see [2]). θ_j are the rotational correlation times. For a spherical molecule the rotational correlation time is related to the hydrated molecular volume V, the viscosity η, and absolute temperature T by the relation:

$$\theta = \frac{\eta V}{kT} \quad (3)$$

The Debye rotational relaxation time, ρ, is also used in some publications; $\rho = 3\theta$.

In addition, the total fluorescence intensity decay can be recovered form the sum of the two polarized components:

$$I(t) = I_{vv}(t) + 2I_{vh}(t) \quad (4)$$

2.2. Technical principles

There are mainly two techniques of time-resolved fluorescence measurements, the impulse and the harmonic excitation response techniques. In the first one, the time interval between the excited light pulse and the detection of the first fluorescence photon by a detector is repeatedly measured in order to construct the fluorescence decay profile. This time-correlated photon counting is the most efficient method to detect emission of a very diluted solution since only one photon is required to run the electronic stopwatch for time measurement. In addition, the precise determination of the Poisson statistics in photon counts is of great advantage in data analysis [3,4]. In the second technique, the exciting light is modulated sinusoidally at frequencies in the MHz range. The fluorescence emission is then delayed in phase and demodulated due to the lifetime of the excited state. The relation between the phase shift ϕ and lifetime τ is tan $\phi = \omega\tau$ where ω is the frequency of the excitation. In the same way, the phase shift between the polarized components, measured at several frequencies, leads to the rotational correlation values.

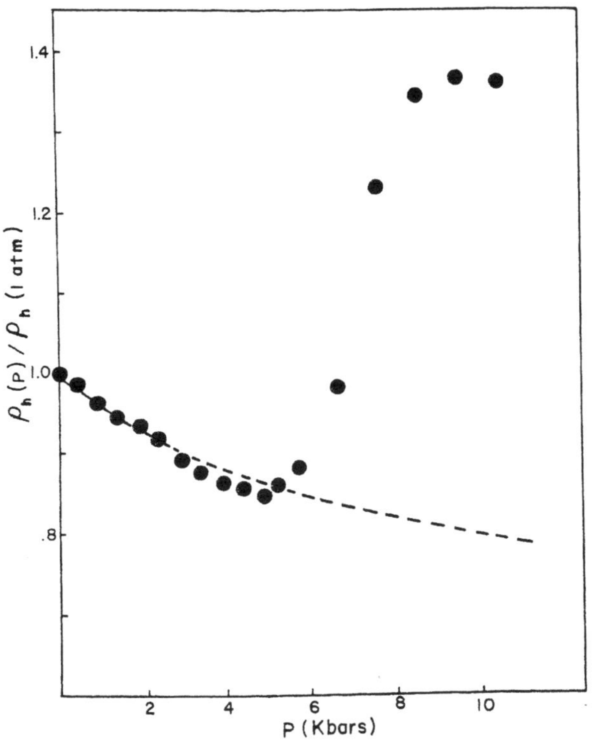

Fig. 1. Effect of high pressure on fluorescein-labelled lysozyme. $\rho_h(p)/\rho_h(\text{atm})$ is the ratio of rotational relaxation times at pressure p versus atmospheric pressure. From [18].

Descriptions of both techniques and corresponding data analysis are detailed in several reviews [1,5–7]. The construction of high pressure cells for fluorescence polarization measurements and corrections for birefringency of the windows has been described in [8,9].

3. Effects of hydrostatic pressure on proteins from time-resolved fluorescence spectroscopy

3.1. The dynamic fluorescence approach

The effects of high hydrodynamic pressure on polypeptides and proteins have been extensively studied by various physical techniques including several optical spectroscopies, nuclear magnetic resonance (NMR), and X-ray crystallography [10]. During the last three decades, steady-state intrinsic and extrinsic fluorescence of protein has been used to study the effects of high hydrostatic pressure [11,12]. In the last 10 years, the technical progress in time-resolved fluorescence measurement as well as in biochemical preparation and purification of protein allowed new experimental approaches. Since the temperature stability of protein has been extensively explored, the effect of pressure has more recently gained interest [10]. The effects of high hydrostatic pressure on proteins can be described at different levels of the molecular structure, i.e. secondary, tertiary and quaternary structural levels. In addition, the effects on higher levels of protein assembly [13] or molecular recognition such as protein–nucleic acid in virus have been also addressed [14,15]. We have to consider separately the effects on single peptide or polypeptide chains and on oligomeric proteins. In the first case, there is equilibrium between various possible conformations, leading to changes in lifetime, emission spectrum and quantum yield of the fluorescent probe; in the second case there is an additional sophistication of higher order equilibrium between possible dissociating structures.

3.2. Applications to single chain proteins

In the 70s, Weber and co-workers carried out one of the first studies on the high pressure effect on biomolecules, using dynamic fluorescence [16–18]. For example, the effect of pressure, up to 10 kbar, on the lifetime of two flavinyltryptophan peptides has been investigated [17]. The combination of lifetime and quantum yield measurements allowed the determination of the proportion of folded forms and then gave access to the compressibility of the two peptides which could be transposed into a protein denaturation study. One of the first determinations of rotational correlation times under pressure was carried out from Perrin plots and lifetime measurements of fluorescein-labelled lysozyme and chymotrypsinogen [18]. Taking into account the compressibility and the viscosity of water versus pressure there was an apparent change in the volume of lysozyme upon pressure. For the first few kilobars this change was found to be approximately parallel to that of the surrounding solvent (Fig. 1 and Table 1). Above 5–6 kbar a large increase in the rotational correlation time was observed and after 10 kbar the volume increase of 60% is complete. In contrast to lysozyme, the chymotrypsinogen volume was almost constant up to 5.5 kbar and a sharp increase by a factor of

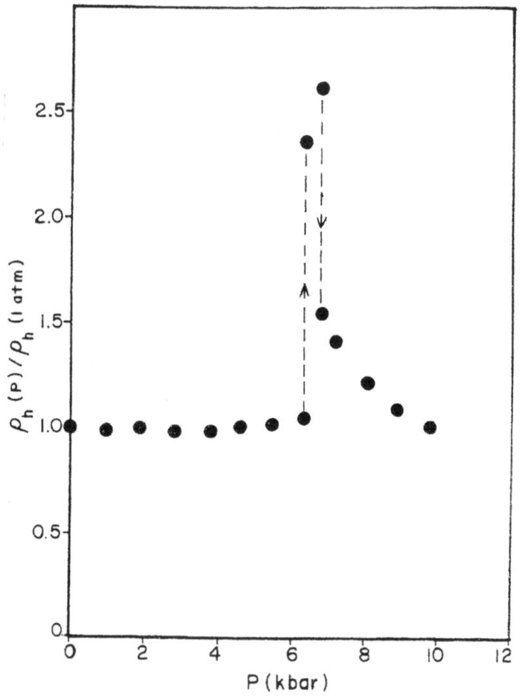

Fig. 2. Effect of high pressure on fluorescein-labelled chymotrypsinogen. From [18].

2.3 occurred at 6.3 kbar (Fig. 2). These results also provided evidence for the existence of several independent pressure denaturation steps for labelled lysozyme and an aggregation process upon pressure increase for chymotrypsinogen.

Since extrinsic fluorophores may be covalently labelled to or adsorbed into protein, tryptophan residues remain the less perturbing probes for the study of conformational changes [19–25]. Denaturation of proteins may occur over relatively variable limits of temperature, pH, chemicals and organic solvent concentrations. For example, time-resolved experiments using a tryptophan residue in one point mutation, F29W, of troponin C (TnC) were carried out at low temperature under high hydrostatic pressure [19]. An increase in pressure and a decrease in temperature drive the system to a lower entropic state, that promotes the binding of non-polar residues to water [26,12]. The lifetime distributions show that hydrostatic pressure and subzero temperatures change the TnC/N-domain to a conformation similar to that of the Ca^{2+}-bound form (Fig. 3). The decrease in temperature yields an exposure of the hydrophobic region in the N-domain in the Ca^{2+} free form, which is similar to the one obtained by binding Ca^{2+}.

3.3. Probing the stability of extremozymes

Thermophilic enzymes have been widely investigated because of their high resistance to denaturing agents compared to their mesophilic analogues. Ribonuclease Ss07d from *Sulfolobus solfataricus*, a thermoacidophilic archaebacterium living at 87°C and acidic pH in hot springs, is a model of choice for an exploration of the capability of time-resolved spectroscopy to investigate the resistance of protein structure stability to extreme conditions of denaturation. This small protein (7 kDa) exhibits a ‘melting’ temperature close to the boiling temperature of water at neutral pH [27]. The structure was determined by NMR [28] and shows a triple strand β-sheet which is orthogonal to a double-stranded β-sheet. A strongly hydrophobic core is constructed mainly around four aromatic residues, Phe5, Tyr7, Phe31 and Tyr31, and nine other residues, Val3, Val14, Ile19, Val22, Ile29, Val45, Ala50 and Leu54. This hydrophobic cluster is surrounded by a hydrophilic mantle which contains many basic residues (two arginine and 13 lysine). The substitution of Phe31 was assayed in order to investigate the role of the hydrophobic cluster in protein stability. The unique tryptophanyl residue was the intrinsic probe in use for the time-resolved fluores-

Table 1

Sample	Pressure (kbar)	Debye rotational relaxation time (ns)
Fluorescein-labelled lysozyme	0.001	13.3
	4.6	11.3
	9.8	18.1
Fluorescein-labelled chymotrypsinogen	0.001	12.1
	6.33	28.6
	6.8	18.7
	9.8	12.2

From [18].

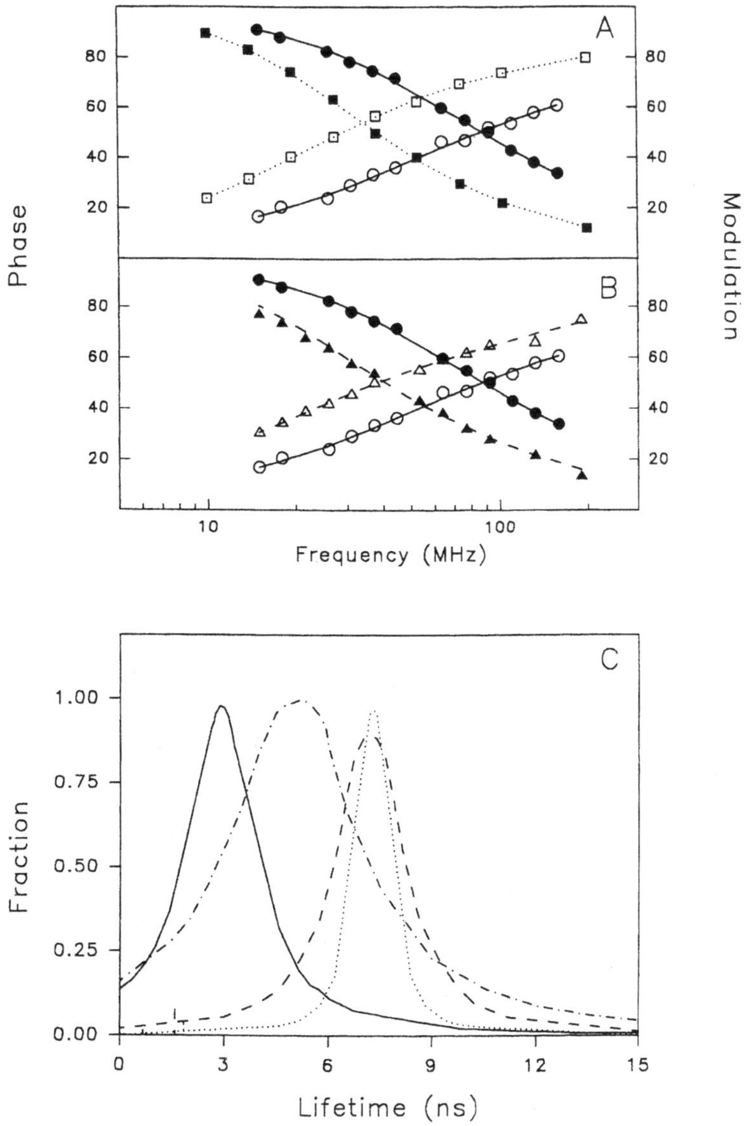

Fig. 3. From [19].

cence approach of the detection of structural changes in wild-type and mutant proteins under extreme conditions. The temperature and pressure denaturation effects on the wild-type ribonuclease and its mutant F31A have been studied by absorption spectroscopy [27].

In our laboratory a dynamic fluorescence study at low temperature was carried out without addition of cryosolvent. The sample was first pressurized to 2000 bar at 20°C and then the temperature was decreased until a strong denaturation was observed. At 1 bar the fluorescence decay of the single tryptophan of the wild-type protein is heterogeneous and corresponds to three peaks in the lifetime distribution (Fig. 5A). The simplest explanation is to assume the existence of several structural conformations for the environment of the tryptophanyl residue. The decay of the fluorescence anisotropy indicates that the indole moiety of this residue appeared to be rather immobile in the subnanosecond time scale (Fig. 5B), thus the heterogeneity is probably not due to local rapid dynamics of the indole side chain. The fluorescence decay was affected by a pressure at 2 kbar compared to 1 bar at 20°C, and the shortest component is shifted from 015–0.25 ns to 0.6 ns since a very short decay component appeared. The fluorescence decay at 10°C

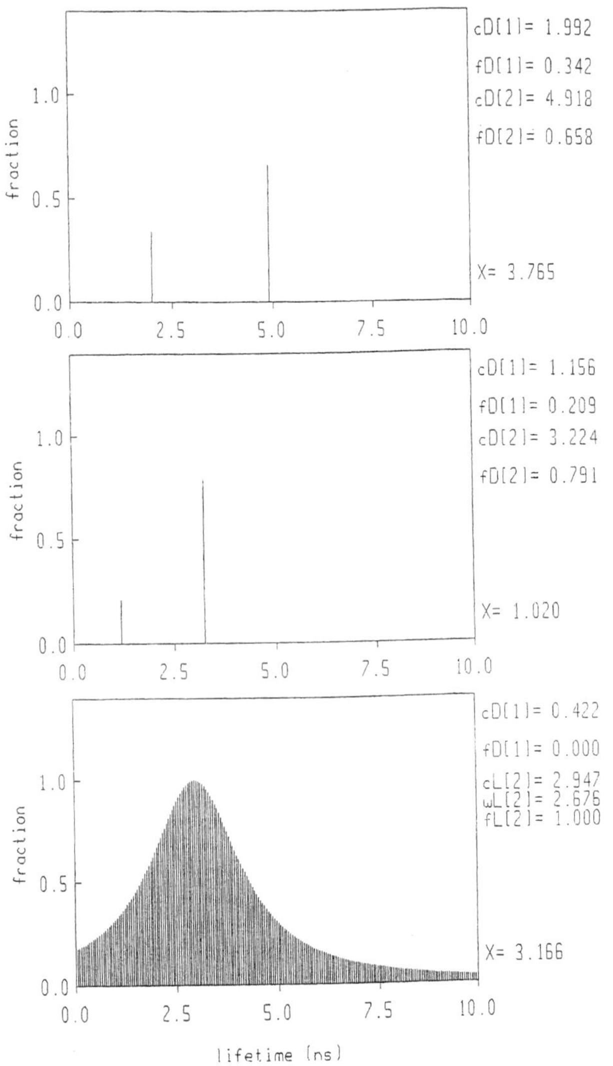

Fig. 4. Fluorescence lifetimes of the six tryptophanyl residues of bacteriophage P22 coat protein. (Top panel) Atmospheric pressure. (Middle panel) 5.5 M guanidinium chloride. The letters c and f refer to the individual lifetime value and relative pre-exponential coefficient, respectively. (Bottom panel) At 2.2 kbar the data were well fitted with a Lorentzian lifetime distribution. cL and wL refer to the centroid position and the width, respectively. Excitation was at 295 nm and emission was measured through a WG 320 filter. Protein concentration was 0.2 mg/ml. From [20].

had additional evidence for this short component that was now detectable (~60 ps). Until −14°C the distribution pattern did not change since the centre of the peak was slightly shifted to a higher value than expected for a thermal effect on the lifetime of the probe. The resistance of the wild-type enzyme to denaturation was also monitored by the rotational dynamics pattern of the molecule. From 20°C to −5°C, the protein appeared to be a rigid globular molecule. At −10°C and −14°C an additional small peak was detected in the correlation time distribution (~1 ns), suggesting an internal flexibility of small amplitude.

At normal pressure, the F31A mutant also displays three lifetime classes (Fig. 6A). The fourth lifetime class appeared with pressure at 2000 bar. At 10°C, a fifth component around 6 ns, corresponding to a tryptophan in a rather hydrophobic environment, points out the early stage of a conformational change. The relative contribution of this long lifetime component increased with cooling below 10°C. The return at 20°C and 1 bar indicates a rather limited reversibility of the strong structural changes. Mutant F31A is also a rigid globular molecule at 20°C and pressurization to 2 kbar does not affect strongly the overall motion, i.e. the hydrated volume. Nevertheless a fast internal flexibility was evidenced but the corresponding accuracy obtained by the maximum entropy method of data analysis was rather poor when linked to the low counting statistics [29,30]. The most striking effect of cooling below 0°C is the broadening of peaks in the correlation time distributions. At −14°C an internal motion of large amplitude clearly appears, indicating the beginning of an unfolding process (Fig. 6B). The peak broadness provides evidence for a dramatic structural heterogeneity and a partial state of aggregation. The immediate return to 20°C (10 min) and 1 bar reverses this conformational dynamics situation. The mutation of only one residue in the hydrophobic cluster has a dramatic effect on the temperature stability of this compact protein and the corresponding partial denaturation is made irreversible.

3.4. Applications to multimeric assemblies

An elegant work from Silva's laboratory concerns the study of the viral capsid protein P22 of bacteriophage [20]. Using fluorescence lifetime and rotational correlation time measurements, in addition to steady-state fluorescence emission and polarization as well as light scattering techniques, the pressure stability of the monomeric and polymeric forms of the protein was investigated. The monomeric form was found to be very unstable even at low pressure

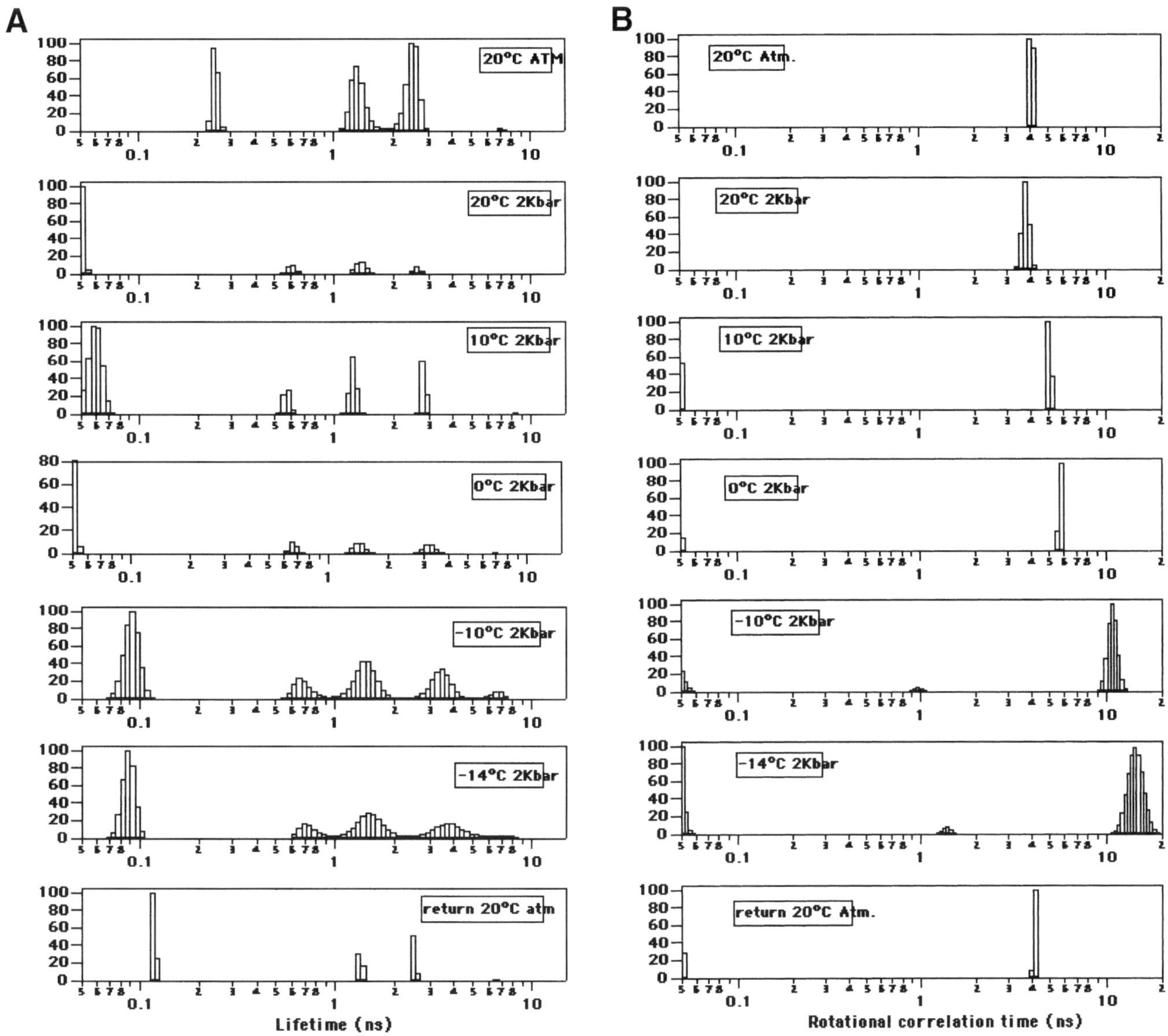

Fig. 5. (A) Fluorescence lifetime distributions of the single tryptophan residue of Ss07d protein at atmospheric pressure and at 2 kbar. (B) Structural resistance to cold denaturation of the wild-type Ss07d protein as seen by the homogeneity of the rotational correlation time distribution.

(< 0.5 kbar), which ultimately leads to denaturation. In general for monomeric globular proteins, denaturation takes place at pressure in the range of 4–10 kbar [10–12]. The best fit of the dynamic fluorescence data measured by the multifrequency technique was obtained by a sum of two exponentials (Fig. 4), which is rather unusual for a protein containing six tryptophanyl residues. The addition of 5.5 M guanidinium chloride and again a sum of only two exponentials fitted well the data but the corresponding lifetime values were then shorter. Upon 2.2 kbar pressure, a bi-exponential model cannot account for the data and only a Lorentzian distribution gave a better fit. In contrast with guanidine hydrochloride denaturation that virtually abolished any structure as indicated by a very short rotational correlation time, the pressure effect left a remaining residual structure of the monomeric form, the corresponding rotational correlation time being half that of the native form at atmospheric pressure. When polymerized into a procapsid, the protein is very stable to pressure; it does not dissociate up to 2.5 kbar but undergoes a cold

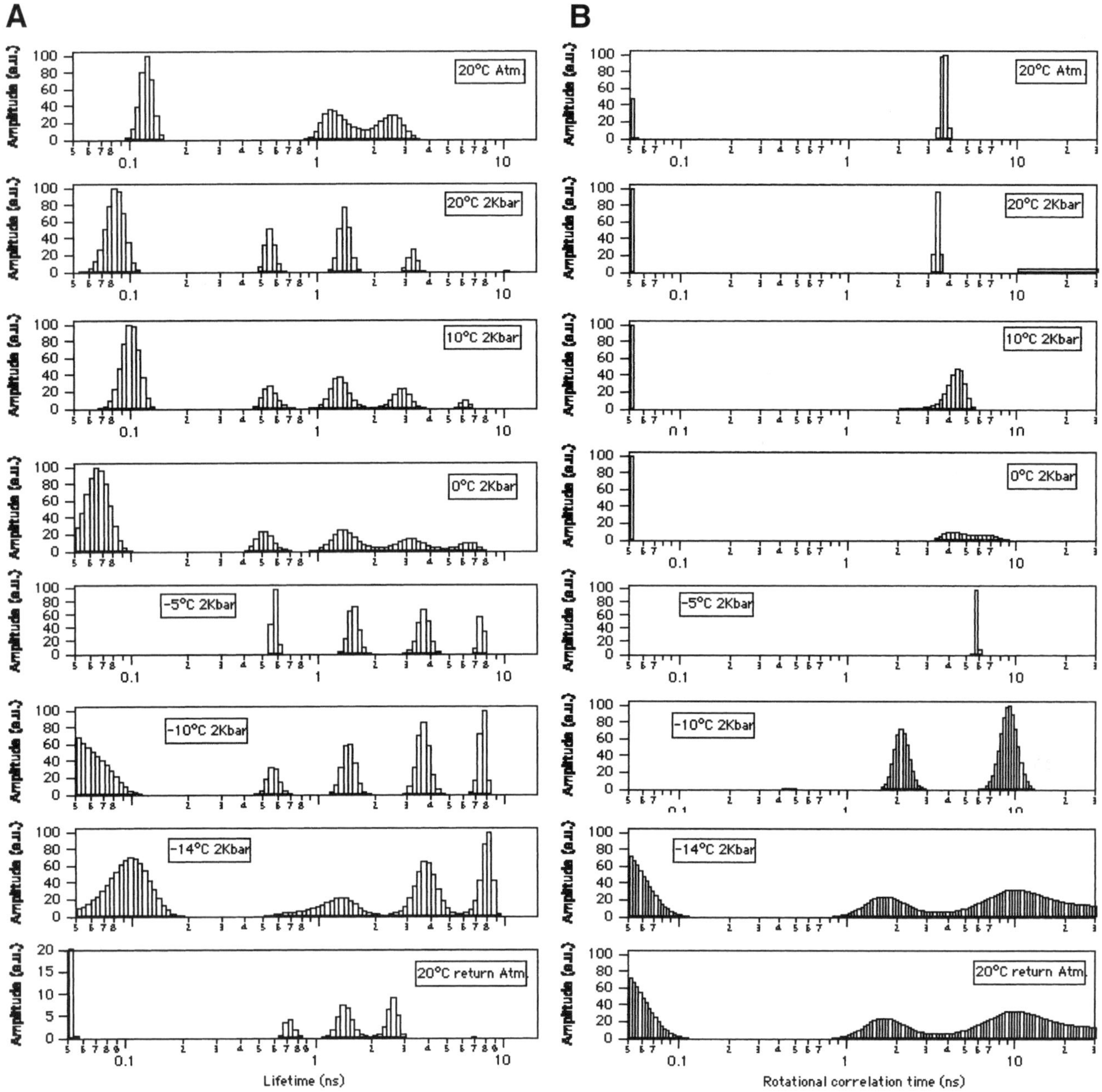

Fig. 6. (A) Fluorescence lifetime distributions of the single tryptophan residue of the F31A mutant form of Ss07d protein at atmospheric pressure and at 2 kbar. (B) Progressive cold denaturation of the F31A mutant of Ss07d protein as seen by the broadening of the rotational correlation time distribution, evidence of an irreversible process.

dissociation at 1 bar, suggesting an entropic stabilization of the assembly.

DNA-binding proteins may recognize specific sequences of DNA and the molecular mechanisms of the corresponding interactions are not yet fully understood [31]. In a recent work the dissociation of the dimeric form of LexA protein under pressure was investigated by steady-state polarization and rotational correlation time measurements. Each LexA monomer (202 residues) contains a single tryptophanyl residue at position 201. The dissociation of the dimeric form of LexA at low concentration (μM range) was monitored by pressure or urea dissociation. The highest dissociation constant was obtained at rather low temperature, 4°C (K_d = 0.74 ± 0.3 nM), where the pressure that yields 50% dissociation is 673 bar. Hydrostatic pressure generated during ultracentrifugation is usually in the range of 100–500 bar and can cause protein–protein dissociation in weakly associated oligomers [11,12]. That may explain in part why the ultracentrifugation experiment gives a higher dissociation constant. The effect of DNA binding on the dissociation of LexA dimer was also evaluated by applying pressure on the DNA–LexA complexes. Comparison of the binding stability under pressure of DNA–protein complexes with a recA operator, a half recA operator and a non-specific poly(GC) sequence, respectively, indicates that the recA operator decreases the dissociation constant by a factor 750 whereas neither the half-operator nor the non-specific oligonucleotide has any effect on the K_d. Since the in vivo protein concentration is about 1–2 μM, most of the protein would be dimeric. The authors suggest that the dimeric form may account for the ability to discriminate between non-specific and specific DNA sequences.

3.5. *The use of external fluorophore*

Several studies on the effect of high hydrostatic pressure on the state of oligomerization of proteins have been carried out by external fluorophore for protein or DNA whose intrinsic fluorescence is very weak [32–36]. A good example of the use of high pressure for validation of the native oligomeric state of a protein is the study of the porcine malate dehydrogenase [33]. Recently the aggregation state was revisited in order to clarify the disparities of the results in the literature concerning monomer–dimer equilibrium. First the authors demonstrated the stability of the dimeric form down to nanomolar concentration. The determination of the rotational correlation time from FITC-labelled enzyme allows the characterization of the dimeric form and its dissociation upon addition of guanidinium chloride at 1.5 M concentration. In contrast, pressure up to 2.5 kbar was ineffective in dissociating the dimer. Another example of the use of extrinsic fluorophores is the study of the Lac repressor tetramer and the repressor–operator complexes upon high hydrostatic pressure [34]. The protein was labelled with dansyl because of its long average lifetime which allows measurement of relatively slow rotational dynamics. Using the Perrin equation [1] and steady-state fluorescence anisotropy and lifetime data, the rotational relaxation times were determined. The application of high pressure, up to 2.2 kbar, to the tetrameric form, characterized by a rotational relaxation time of 126 ns, leads to dissociation into the respective subunits under favourable conditions of temperature and concentration. The corresponding relaxation time values of 60 ns and 30 ns were attributed to the dimeric and monomeric forms, respectively. The application of pressure results in a dissociation of the repressor–operator complex.

4. Effects of hydrostatic pressure on lipid membranes from time-resolved fluorescence spectroscopy

4.1. *Relevance of dynamic fluorescence in membrane studies*

The study of the effect of high hydrostatic pressure on natural and synthetic membranes is of fundamental importance from a wide variety of viewpoints [37]. First, the membranes of certain deep-sea organisms normally function at high pressure (up to 1 kbar) and there is considerable interest in their adaptation to this extreme condition. Second, the discovery that pressure perturbs membrane–drug interactions could provide an important tool to understand the molecular basis of anaesthesia [38,39]. Finally, experiments of high pressure applied to biomolecules (membranes, protein, nucleic acids, etc.) are also interesting because of the valuable thermodynamic, kinetic, structural and even functional properties that they can reveal.

The effects of hydrostatic pressure on lipid bilayers and cellular membranes have been studied in the last three decades by a wide variety of physical techniques, including fluorescence techniques [40–52], NMR spectroscopy [53–56], electron spin resonance [57], Fourier transform infrared [58,59], neutron diffraction [60], light scattering [61], volumetry [62,63] and electron microscopy [64]. From an experimental point of view, time-resolved fluorescence spectroscopy allows the study of pressure-induced membrane structural changes, at a molecular level, in a simple way. Since phospholipids are non-fluorescent, a suitable fluorescent probe is embedded in the lipid matrix. The success of the method depends on the interpretation of the changes observed in the photophysical parameters of this probe which are, frequently, sensitive to the lipid environment surrounding the fluorophore. Fluorescence depolarization techniques are particularly suitable to quantify the structural and dynamic properties of the lipid systems. This information (contained in r_∞, the limiting anisotropy, and θ, the rotational correlation time of the probe) is interpreted in terms of order parameter and dynamics of the lipid vesicles on the nanosecond timescale [65].

4.2. The use of fluorescence probes

The first pressure membrane experiments using fluorescent probes were carried out by Chong et al. [40], Chong and Weber [41] and Lakowicz and Thompson [42]. In these works the effect of hydrostatic pressure, ranging from 1 bar to 2 kbar, on phospholipid vesicles and cellular membranes was explored from differentially polarized phase fluorimetric studies using the well-known fluorescent probe 1,6-diphenyl-1,3,5-hexatriene (DPH). The results showed that high pressure and low temperature exert similar ordering effects upon the hydrophobic core of the lipid membranes. Isothermal pressure-induced phase transitions were observed in saturated phosphatidylcholine vesicles giving a temperature to pressure equivalence, dT/dP, between 22 and 29°C $kbar^{-1}$, depending on the length of the hydrocarbon chain. The results also demonstrated that pressure affects the location of the probe in the membrane and probably induces a change in the rotational mode [40].

Time-resolved fluorescence experiments using DPH as fluorescent probe were also carried out in brain myelin membranes from deep-sea organisms in order to investigate the evolutionary adaptation of species to temperature and pressure [45,66]. At atmospheric pressure, different membrane order parameters were determined which were correlated with the depth range inhabited by each species, whilst at their respective ambient conditions of temperature and pressure an almost identical order parameter was obtained according to the homeoviscous theory.

Although much information has been derived from these studies it is known that pressure can alter the location and/or rotation mode of free probes such as DPH, 6-propionyl-2-(dimethylamino)naphthalene or perylene [40,43,67] in the lipid bilayer and, consequently, complicate the interpretation of data. The use of cationic or anionic DPH derivatives and fluorescent fatty acids may overcome this problem to a considerable extent. In our laboratory we have used the fluorescent polyene fatty acid *trans*-parinaric acid (t-PnA) to investigate the effects of temperature and pressure on the physical properties of lipid bilayers [68,47]. The high sensitivity of the fluorescence kinetics of this probe to lipid local density [69] was exploited for this purpose. The effects of hydrostatic pressure, up to 3.2 kbar, were investigated on 1,2 dimyristoyl-3-phosphatidylcholine (DMPC) and 1-palmitoyl-2-oleoylphosphatidylcholine (POPC) lipid vesicles in the temperature range of 10–50°C [47]. Fluorescence decays were analysed by the maximum entropy method [70]. The fluorescence lifetime distributions of the probe in isotropic solvents at different pressures showed two well-resolved lifetime components; however, in pressurized lipid bilayers an abrupt change was observed in the lifetime distribution with the appearance of a long lifetime component which was associated with the isothermal pressure-induced phase transition (Fig. 7). The amplitude of this new component was found to be proportional to the fraction of probe located in the gel phase. In DMPC, the temperature to pressure equivalence, dT/dP, was found to be 24 ± 2°C $kbar^{-1}$, similar to that determined by other physical methods. This confirmed the capacity of the probe to detect, with high precision, the isotherm pressure-induced phase transition. For the unsaturated phospholipid POPC, dT/dP was found to be 14.5°C $kbar^{-1}$. This low value was attributed to the presence of the *cis* double bound on the *sn*-2 chain of POPC. The structural properties of these lipids vesicles at different pressures were also investigated from the anisotropy decays of t-PnA and DPH. The order parameter for all systems increased with pressure as was expected. In the gel phase of POPC the order parameter was smaller than that obtained in the same phase of saturated phospholipids, suggesting that due to the *cis* unsaturation an efficient packing of the POPC hydrocarbon chain is hindered.

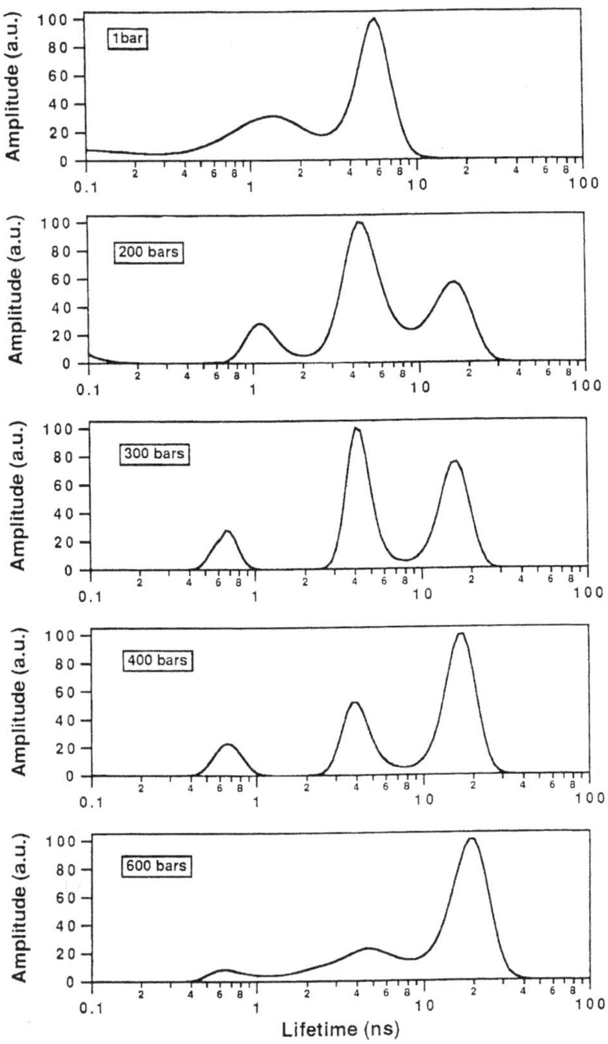

Fig. 7. Effect of hydrostatic pressure on the fluorescence lifetime distribution of *trans*-parinaric acid in DMPC large unilamellar vesicles at 30°C. Progressive increase of the relative contribution of the long lifetime component. From [47].

4.3. Study of lipid heterogeneity in membranes

The lipid component of biological membranes has usually been visualized as a physically homogeneous system. However, there is growing evidence that lipid domains of different structure and composition might exist in biological membranes, even at physiological temperatures [71–73]. Alteration of pressure will lead to changes in the structure and size of these domains and, probably, in the functional state on the membrane. Taking advantage of the fluorescent properties of t-PnA, we proposed [68] a simple and very useful method, which combines the fluorescence

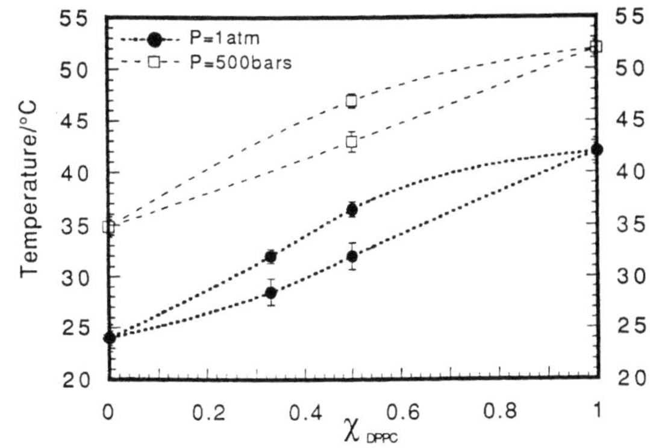

Fig. 8. Effect of pressure on the thermotropic diagram of the DMPC/DPPC lipid mixture. ●, phase diagram obtained at atmospheric pressure from the lifetime distribution of t-PnA in [66]. □, data obtained at 500 bar. At $\chi_{DPPC} = 1$ and $P = 500$ bar the temperature value is taken from [39]. From [47].

lifetime distribution with the anisotropy decay of t-PnA to detect, quantify and characterize the physical properties of lipid domains. This method was applied to the study of the pressure effects on DMPC/dipalmitoyl-3-phosphatidylcholine (DPPC) lipid mixtures, where gel and fluid phases coexist in a large temperature range of physiological interest [47]. Results showed that relatively moderate pressures of about 500 bar shifted the DMPC/DPPC phase diagram from 11.5°C to higher temperatures (Fig. 8), suggesting that relatively low pressures can modify the local fraction and composition of biological membrane lipid domains which have important biochemical implications.

Temperature and pressure show opposite effects on the order parameter of lipid bilayers. An increase in pressure or a decrease in temperature reduces the lateral intermolecular distance or volume between the components of each leaflet of the bilayer, increasing the order parameter of the lipid membrane. In a similar way, cholesterol has an ordering effect on the packing of phospholipids in their fluid phase because of the rigid chemical structure. In addition, cholesterol is not randomly distributed in lipid membranes and appears to be organized in cholesterol-rich domains (β-phase) and cholesterol-poor domains (α-phase) [74,75]. These three variables could, therefore, modulate the physical properties of lipid bilayers. For this reason, comparison of the different effects induced by temperature, pressure and cholesterol on the lipid membranes was the goal of several works in the last years [49–51,64,76].

In a recent paper [51] we use the methodology mentioned above to investigate the influence of hydrostatic pressure on cholesterol/phospholipid mixtures in a wide range of temperatures. The study showed how temperature, cholesterol and pressure modulate the formation, composition and physical properties of cholesterol domains. Pressure induces and increases the fraction of β-phase and its order parameter, but it decreases the amount of cholesterol in this phase. An increase in pressure and a decrease in temperature had similar effects on the structure and composition of phospholipid/cholesterol mixtures. The work showed that small changes in temperature/pressure and/or cholesterol concentration perturb the local fraction and composition of cholesterol domains.

Recently, Bernsdorff et al. [49,50] carried out a time-resolved fluorescence study using the probes 1-(4-trimethylammonium-phenyl)-6-phenyl-1,3,5-hexatriene (TMA-DPH) and 3-(diphenylhexatrienyl)propyl-trimethylammonium (TMAP-DPH) with a double purpose: (a) to investigate the degree of water penetration into different (saturated and unsaturated) pure phospholipid vesicles and phospholipid/cholesterol mixtures as a function of pressure, and (b) to explore the capacity of cholesterol to modulate the structural and dynamic properties of membranes. TMA-DPH and TMAP-DPH are cationic DPH derivatives which locate the fluorophore at different depths from the bilayer–water interface. Since the DPH fluorescence lifetimes are strongly influenced by the presence of water, these probes were used to detect the hydration degree of the lipid bilayer at different depths. Results showed a greater level of interfacial hydration for saturated than for unsaturated phospholipids. The addition of cholesterol had a greater effect on the increase in hydrophobicity in the interfacial region than that induced by the application of hydrostatic pressure (up to 1 kbar). From the fluorescence anisotropy decays of TMA-DPH in membranes the physical properties of these lipid systems were explored. Results showed the ability of cholesterol to adjust the structural and dynamics properties of different lipid membranes, compensating the effects induced by drastic changes in environmental conditions (high pressure or low temperature).

5. Conclusion

A better understanding of the relationships between the structural organization of a biological system and its function is linked to a better knowledge of the time-dependent mechanisms. Dynamic fluorescence spectroscopy offers both good sensitivity and high time resolution within the perturbant effect of hydrostatic pressure. In the near future, such studies extended to the single molecule level or within the confocal volume in a microscope setup will bring new insight into the dynamics and interactions of biomolecules.

References

[1] J.R. Lakowicz, in: Principles of Fluorescence Spectroscopy, 2nd edn., Kluwer Academic/Plenum Publishers, New York, 1999.
[2] P. Wahl, in: R.B. Cundall, R.E. Dale (Eds.), Time Resolved Spectroscopy in Biochemistry and Biology, Plenum Press, New York, 1980, pp. 497–521.
[3] J.M. Beechem, E. Gratton, M. Ameloot, J.R. Knutson, L. Brand, in: J.R. Lakowicz (Ed.), Topics in Fluorescence Spectroscopy, Vol. 2, Plenum, New York, 1991, pp. 241–305.
[4] P. Wahl, Biophys. Chem. 10 (1979) 91–104.
[5] D.V. O'Connor, D. Philips, Time-Correlated Single Photon Counting, Academic Press, London, 1984.
[6] R. Alcala, E. Gratton, D.M. Jameson, Anal. Inst. 14 (1985) 225–250.
[7] J.R. Lackowicz, G. Laczko, H. Cherek, E. Gratton, M. Linkeman, Biophys. J. 46 (1984) 463–478.
[8] G.S. Chryssomallis, H.G. Drickamer, G. Weber, J. Appl. Phys. 46 (1978) 3084–3087.
[9] A.A. Paladini, G. Weber, Rev. Sci. Instrum. 52 (1981) 419–427.
[10] J.L. Silva, G. Weber, Annu. Rev. Phys. Chem. 44 (1993) 89–113.
[11] K.A.H. Heremans, Annu. Rev. Biophys. Bioeng. 11 (1982) 1–21.
[12] G. Weber, H.G. Drickamer, Q. Rev. Biophys. 116 (1983) 89–112.
[13] C.A. Royer, in: Time-resolved Laser Spectroscopy in Biochemistry, SPIE Proceedings, Los Angeles, CA, 1990, p. 1024.
[14] M. Gross, R. Jaenicke, Eur. J. Biochem. 221 (1994) 617–630.
[15] J.L. Silva, D. Fogel, A.T. Da Poian, P.E. Prevelige, Curr. Opin. Struct. Biol. 6 (1996) 166–175.
[16] A.J.W.G. Visser, T.M. Li, H.G. Drickamer, G. Weber, Biochemistry 16 (1977) 4879–4882.
[17] A.J.W.G. Visser, T.M. Li, H.G. Drickamer, G. Weber, Biochemistry 16 (1977) 4883–4886.

[18] G.S. Chryssomallis, P.M. Torgerson, H.G. Drickamer, G. Weber, Biochemistry 20 (1981) 3955–3959.
[19] D. Foguel, M.C. Suarez, C. Barbosa, J.J. Rodrigues Jr., M.M. Sorenson, L.B. Smilie, J.L. Silva, Proc. Natl. Acad. Sci. USA 93 (1996) 10642–10646.
[20] P.E. Prevelige Jr., J. King, J.L. Silva, Biophys. J. 66 (1994) 1631–1641.
[21] E. Bismuto, I. Sirangelo, G. Irace, E. Gratton, Biochemistry 35 (1996) 1173–1178.
[22] R. Mohana-Borges, A.B.F. Pacheco, F.J.R. Sousa, D. Foguel, D.F. Almeida, J.L. Silva, J. Biol. Chem. 275 (2000) 4708–4712.
[23] G. Desai, G. Panick, M. Zein, R. Winter, C. Royer, J. Mol. Biol. 288 (1999) 461–475.
[24] L.M.T.R. Lima, D. Foguel, J.L. Silva, Proc. Natl. Acad. Sci. USA 97 (2000) 14289–14294.
[25] A. Di Venere, A. Rossi, F. De Matteis, N. Rosato, A. Finazzi Agro, G. Mei, J. Biol. Chem. 275 (2000) 3915–3921.
[26] A. Zipp, W. Kauzmann, Biochemistry 12 (1973) 4217–4228.
[27] E. Mombelli, M. Afshar, P. Fusi, M. Mariani, P. Tortora, J.P. Connelly, R. Lange, Biochemistry 36 (1997) 8733–8742.
[28] H. Baumann, S. Knapp, T. Lunback, R. Ladenstein, T. Hard, Nat. Struct. Biol. 1 (1994) 808–819.
[29] P. Tauc, P. Fusi, P. Tortora, R. Lange, J.C. Brochon, in: H. Ludwig (Ed.), Advances in High pressure Bioscience and Biotechnology, Springer, Berlin, 1998, pp. 191–194.
[30] E. Deprez, P. Tauc, H. Leh, J.F. Mouscadet, C. Auclair, J.C. Brochon, Biochemistry 39 (2000) 9275–9284.
[31] B. Luisi, in: D.M.J. Lilley (Ed.), DNA-protein: Structural Interactions, IRL Press, Oxford, 1995, pp. 1–48.
[32] S. Pin, C.A. Royer, E. Gratton, B. Alpert, G. Weber, Biochemistry 29 (1990) 9194–9202.
[33] S.S. Sanchez, T.L. Hazlett, J.E. Brunet, D.M. Jameson, Protein Sci. 7 (1998) 2184–2189.
[34] C.A. Royer, A.E. Chakerian, K.S. Matthews, Biochemistry 29 (1990) 4959–4966.
[35] L.P. Gaspar, J.E. Johnson, J.L. Silva, A.T. Da Poian, J. Mol. Biol. 273 (1997) 456–466.
[36] H. Stapelfeldt, C.E. Olsen, L.H. Skibsted, J. Agric. Food Chem. 47 (1999) 3986–3990.
[37] A.G. Macdonald, High Press. Biotechnol. 224 (1992) 67–75.
[38] F.H. Johnson, E.A. Flager, Science 112 (1950) 91–92.
[39] I. Ueda, H. Matsuki, H. Kamaya, P.R. Krishna, Biophys. J. (1999) 483–488.
[40] P.L. Chong, A.R. Cossins, G. Weber, Biochemistry 22 (1983) 409–415.
[41] P.L. Chong, G. Weber, Biochemistry 22 (1983) 5544–5550.
[42] J.R. Lakowicz, R.B. Thompson, Biochim. Biophys. Acta 73 (1983) 359–371.
[43] P.L. Chong, Biochemistry 12 (1988) 399–404.
[44] A.G. Macdonald, K.W. Wahle, A.R. Cossins, M.K. Behan, Biochim. Biophys. Acta 938 (1988) 231–242.
[45] M.K. Behan, A.G. Macdonald, G.R. Jones, A.R. Cossins, Biochim. Biophys. Acta 1103 (1992) 317–323.
[46] M. Sassaroli, M. Vauhkonen, P. Somerharju, S. Scarlata, Biophys. J. 64 (1993) 137–149.
[47] C.R. Mateo, P. Tauc, J.C. Brochon, Biophys. J. 65 (1993) 2248–2260.
[48] C.R. Mateo, P. Tauc, S. Tosti, G. Duportail, P. Lianos, Spectrochim. Acta A 51 (1996) 565–571.
[49] C. Bernsdorff, A. Wolf, R. Winter, Z. Phys. Chem. 193 (1996) 151–173.
[50] C. Bernsdorff, A. Wolf, R. Winter, E. Gratton, Biophys. J. 72 (1997) 1264–1277.
[51] P. Tauc, C.R. Mateo, J.C. Brochon, Biophys. J. 74 (1998) 1864–1870.
[52] L.P. Gaspar, A.F. Terezan, A.S. Pinheiro, D. Foguel, M.A. Rebello, J.L. Silva, J. Biol. Chem. 276 (2001) 7415–7421.
[53] J. Jonas, C.L. Xie, A. Jonas, P.J. Grandinetti, D. Campbell, D. Driscoll, Proc. Natl. Acad. Sci. USA 85 (1988) 4115–4117.
[54] X. Peng, A. Jonas, J. Jonas, Biophys. J. 68 (1995) 1137–1144.
[55] J. Jonas, A. Jonas, Annu. Rev. Biophys. Biomol. Struct. 23 (1994) 287–318.
[56] J. Eisenblätter, A. Zenerino, R. Winter, Magn. Reson. Chem. 38 (2000) 662–667.
[57] J.R. Trudell, D.G. Payan, J.H. Chin, E.N. Cohen, Biochim. Biophys. Acta 373 (1974) 436–443.
[58] P.L. Chong, P.T. Wong, Biochim. Biophys. Acta 1149 (1993) 260–266.
[59] O. Reis, R. Winter, T.W. Zerda, Biochim. Biophys. Acta 1279 (1996) 5–16.
[60] L.F. Braganza, D.L. Worcester, Biochemistry 25 (1986) 7484–7488.
[61] F. Ceuterick, J. Peeters, K. Heresmans, H. De Smedt, H. Olbrechts, Eur. J. Biochem. 87 (1978) 401–407.
[62] N.I. Lui, R.L. Kay, Biochemistry 16 (1977) 3484–3486.
[63] R.E. Tosh, P.J. Collings, Biochim. Biophys. Acta 859 (1986) 10–14.
[64] P. Mentre, L. Hamraoui, G. Hui Bon Hoa, P. Debey, Cell. Mol. Biol. 45 (1999) 353–362.
[65] B.R. Lentz, Chem. Phys. Lipids 50 (1989) 171–190.
[66] M.K. Behan-Martin, G.R. Jones, K. Bowler, A.R. Cossins, Biochim. Biophys. Acta 1151 (1993) 216–222.
[67] P.L. Chong, B.W. van der Meer, T.E. Thompson, Biochim. Biophys. Acta 813 (1985) 253–265.
[68] C.R. Mateo, J.C. Brochon, L.M. Pilar, U.A. Acuna, Biophys. J. 65 (1993) 2237–2247.
[69] A. Ruggiero, B.S. Hudson, Biophys. J. 55 (1989) 1111–1124.
[70] J.C. Brochon, Methods Enzymol. 240 (1994) 262–311.
[71] X. Xu, R. Bittman, G. Duportail, D. Heissler, C. Vilcheze, E.J. London, Biol. Chem. 276 (2001) 33540–33546.
[72] C. Wolf, P. Quinn, K. Koumanov, C. Chachaty, B. Tenchov, J. Soc. Biol. 193 (1999) 117–123.
[73] C.R. Mateo, M.P. Lillo, J. González-Rodriguez, A.U. Acuña, Eur. Biophys. J. 20 (1991) 53–59.
[74] M.R. Vist, J.H. Davis, Biochemistry 29 (1990) 451–464.
[75] C.R. Mateo, A.U. Acuña, J.C. Brochon, Biophys. J. 68 (1995) 978–987.
[76] A.Yu. Tyurin-Kuz'min, Membr. Cell Biol. 11 (1997) 243–251.

Biochimica et Biophysica Acta 1595 (2002) 116–130

BIOCHIMICA ET BIOPHYSICA ACTA
BBA
www.bba-direct.com

Review

Tryptophan phosphorescence and pressure effects on protein structure

Patrizia Cioni *, Giovanni B. Strambini

Istituto di Biofisica, Consiglio Nazionale delle Ricerche, Area della Ricerca di Pisa, Via Moruzzi 1, San Cataldo, 56010 Ghezzano-Pisa, Italy

Received 3 September 2001; received in revised form 5 November 2001; accepted 6 November 2001

Abstract

After a brief introduction of the potentialities of Trp phosphorescence spectroscopy for probing the conformation and flexibility of protein structure, this presentation summarizes the effects of hydrostatic pressure (up to 3 kbar) on the native fold of monomeric and oligomeric proteins as inferred from the variation of the intrinsic phosphorescence lifetime and the oxygen and acrylamide bimolecular quenching rate constants of buried Trp residues. The pressure/temperature response of the globular fold and modulation of its dynamical structure is analyzed both in terms of a reduction of internal cavities and of hydration of the polypeptide. The implications of these findings for the thermodynamic stability of proteins and for the determination of subunit dissociation equilibria under high pressure conditions are also discussed.

Keywords: High pressure; Tryptophan phosphorescence; Protein dynamics; Protein structure; Subunit dissociation; Phosphorescence quenching

1. Introduction

Oceans are populated down to the deepest levels ($\approx$ 11 800 m), which means that a significant proportion of the global biosphere is exposed to pressure of up to 1.2 kbar. Recently barophilic adaptation has been firmly established [1]. It has also been demonstrated that, in non-adapted organisms, increased pressure leads to morphological anomalies or growth inhibition and ultimately to cell death [2,3]. The detailed molecular mechanisms that account for the effect of high pressure on the structure of biological macromolecules and on life in general are still poorly understood. In vitro, pressure-induced denaturation of proteins generally occurs above 3–4 kbar [4], whereas subunit dissociation of oligomers takes place at more moderate pressures, even below the biologically relevant range (1–2 kbar) [4,5]. At subdenaturating pressures large structural changes of the polypeptide have not been detected by common biophysical techniques and therefore it is widely felt that the conformation of the globular fold is largely invariant to pressure and that inhibition of biological activity is primarily due to dissociation-induced inactivation of oligomeric enzymes [5,6]. In recent years there has been a growing realization that perhaps subtle, not yet characterized alterations of protein structure/dynamics at moderate pressure might also play an important role in determining binding/catalytic efficiencies of enzymes [1,7]. Indeed, biological activity normally requires structural flexibility, and therefore it has been proposed that inhibition

* Corresponding author. Fax: +39-50-315-2760.
E-mail address: patrizia.cioni@ib.pi.cnr.it (P. Cioni).

0167-4838/02/$ – see front matter
PII: S0167-4838(01)00339-9

of biological functions at moderate pressures could be due to the reduced flexibility of biological macromolecules, globular proteins in particular [7].

The prediction of restricted protein dynamics at high pressure is based on the direct link between segmental mobility, volume fluctuations and isothermal compressibility of the macromolecule [15]. According to protein compressibility data, a volume reduction of the protein plus solvent system is achieved by both decreasing the size of internal cavities, due to packing defects, and increasing the extent of hydration of the polypeptide [9,10]. While compaction is expected to decrease the rate and amplitude of structural fluctuations in the protein interior hydration of superficial segments is bound to increase the flexibility of the macromolecule on the periphery. Evidence of subtle effects of pressure on the native state of proteins was provided by the pressure tuning of the conformational equilibrium of myoglobin [11], horse cytochrome *c* [6] and arc repressor [12], pressure variations in the secondary structure of chymotrypsin [13] and trypsin [14] and more generally from a complex pressure dependence of catalytic rates in monomeric enzymes [7], not readily explained in terms of the activation free volumes of the reaction. Several studies have investigated the role of pressure on protein dynamics but these were generally conducted over different pressure ranges, often at a single temperature, and the results do not lead to a consistent picture of the pressure perturbation [11,15–20]. Besides, when protein dynamics are inferred from the rate of chemical reactions, such as H-exchange or CO binding, pressure may affect parameters such as reaction mechanism, rate limiting step and activity of the reagents and discordant conclusions could simply arise from the difficulty to correctly account for these concomitant changes.

In the last decade techniques based on the phosphorescence emission of tryptophan, Trp, an intrinsic probe of protein structure, have shown considerable potential for studying the flexibility/conformation of these macromolecules [21–24]. In particular, the sharp dependence of the phosphorescence lifetime, τ_0, on the fluidity of the matrix embedding the chromophore has been a useful local probe of the protein dynamical structure, capable of revealing even subtle conformational changes in enzymatic proteins as a result of their interaction with substrates [25,26], effector molecules [27], metal ions [28], stabilizing [29]/destabilizing [30] agents or change in physical state of the proteins/solvent system [31,32]. Recently, advantage has also been taken of the sensitivity of τ_0 to monitor changes in flexibility/structure of native proteins brought about by hydrostatic pressure in the 0–3 kbar range [33–37]. In these studies, in addition to τ_0, complementary information on the frequency and amplitude of structural fluctuations has been gained by determining the rate of diffusion of solutes of variable size through generally compact globular folds. The permeability of the protein matrix to oxygen and acrylamide was derived from the rate by which the phosphorescence emission of internal Trp residues is quenched by these solutes coming into their proximity [36,37].

This presentation, after a brief introduction of the main aspects of Trp phosphorescence spectroscopy, summarizes the structural perturbations induced by hydrostatic pressure (up to 3 kbar) on multiple protein systems that have been inferred from the variation of the phosphorescence lifetime and of the oxygen and acrylamide bimolecular quenching rate constants. The influence of pressure/temperature on protein flexibility, the role of internal cavities and of polypeptide hydration on the pressure response of the globular fold and their implications with regard to the stability of proteins and the determination of subunit dissociation equilibria under high pressure conditions are also discussed.

2. Tryptophan phosphorescence spectroscopy

2.1. Concepts and techniques

Relative to Trp fluorescence, whose potentialities for both structural and dynamical information on biological macromolecules are familiar to many [38], the much weaker phosphorescence emission is less known and as yet little exploited. A modified Jablonski energy diagram (Fig. 1, top) shows the relationship between the ground state (S_0) and the excited singlet (S_1) and triplet (T_1) states, which are responsible for the fluorescence and phosphorescence emission, respectively. Population of the triplet state is achieved through excitation into the singlet mani-

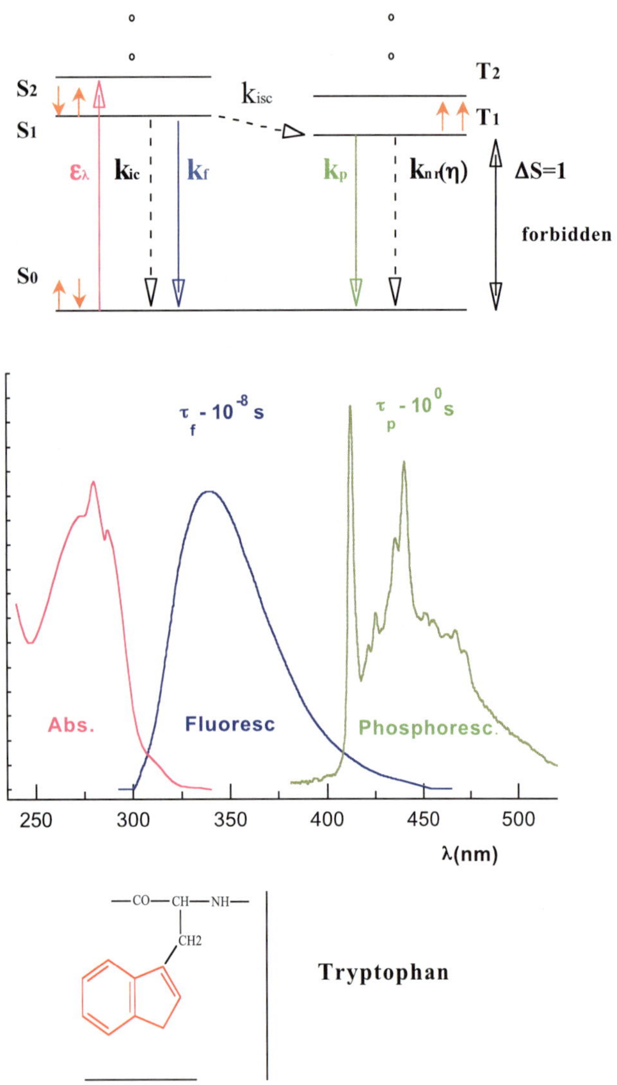

Indole

Fig. 1. Jablonski energy diagram. S_0, S_1 and T_1 refer to the ground and to lowest excited singlet and triplet state, respectively. Small red arrows indicate the spin multiplicity of the electronic state while dashed black arrows indicate radiationless transitions.

fold followed by intersystem crossing to the excited triplet state.

The phosphorescence spectrum is red shifted relative to the fluorescence spectrum and also exhibits a well resolved vibrational structure (Fig. 1, middle). Spectral energy relates to the polarity of the chromophore's surrounding and the width of the 0-0 vibrational band is therefore an indicator of the conformational homogeneity of the protein site. Because of the sharp resolution of vibronic bands it is often possible to distinguish tryptophan residues in different environments [39] and in many cases assign the peaks to specific side chains [40]. In these favorable cases the environment of individual chromophores and their interaction with other centers of the macromolecule can be studied selectively.

Phosphorescence involves a spin forbidden transition between T_1 and S_0 and, consequently, its lifetime is relatively long, 6–9 orders of magnitude longer relative to the few nanoseconds of prompt fluorescence (for Trp in a glass at 77 K $\tau_0 \approx 6.5$ s). The long lifetime has important consequences both with regard to specific radiationless relaxation pathways to S_0 and to the time scale of phenomena that can be studied by the delayed emission. Thus, the submillisecond–second phosphorescence lifetime is suitable for monitoring slow rotational motions [41] and energy [42]/electron [43,44] transfer rates in the $1–10^{-5}$ s time domain, totally complementary to fluorescence. These parameters have provided both dynamical and geometrical information on protein systems and their complexes. In fluid media, however, an undesired effect of the long-lived emission is its great susceptibility to dynamic quenching by oxygen and trace impurities, artifacts that are circumvented only after elaborate and careful sample preparation. A singular feature of the triplet state of indole is the sharp dependence of the radiationless deactivation rate (k_{nr}) on the medium viscosity (η), τ_0 decreasing from 6.5 s in a glass matrix to about 1 ms in aqueous solutions at ambient temperature (Fig. 2). The empirical correlation established between τ_0 and solvent viscosity with several Trp derivatives emphasizes that the enhancement of radiationless transitions in fluid media is largely independent of the polarity/polarizability of the solvent and therefore should apply to any protein site [45]. Fig. 2 also points out that over the same viscosity range the fluorescence lifetime drops by merely a factor of 2 and that Trp residues in rigid, well packed protein cores, such as W109 of alkaline phosphatase, exhibit a long lifetime of seconds at room temperature. Support for a correlation between τ_0 and the local viscosity has also come from a wealth of observations with Trp residues in proteins. Indeed, chromophores that are known to occupy very mobile sites, such as residues located on the surface of the macromolecule or exposed to the aqueous phase, have invariably

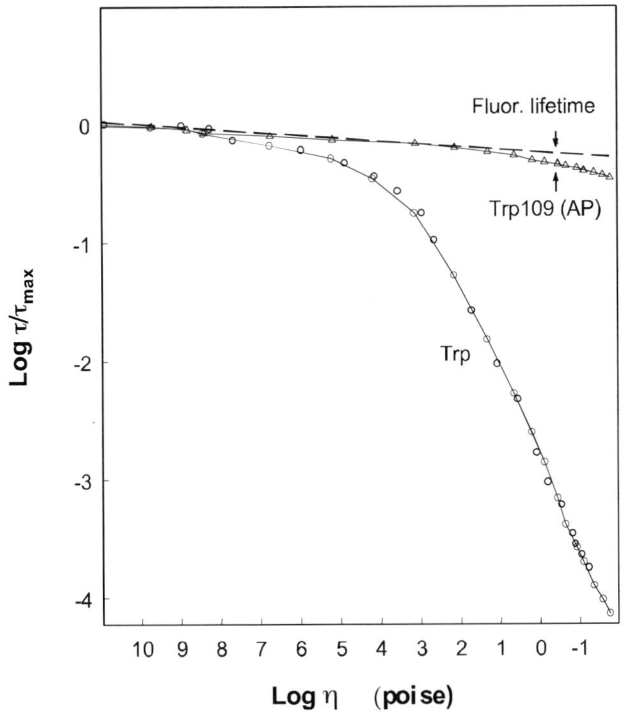

Fig. 2. Plot of the phosphorescence lifetime of tryptophan free in solution (○) and of Trp109 (△) of alkaline phosphatase as a function of the viscosity of propylene glycol/water solvent mixtures. The dependence of fluorescence lifetime of free tryptophan is also shown for comparison.

very small τ_0. Conversely, long-lived room temperature phosphorescence in proteins is a prerogative of residues located at the interior of the globular structure, in regions characterized by well ordered, rigid domains [46]. Moreover, modulation of the structural flexibility by temperature, cosolvents, pH, state of hydration, or addition of denaturants has been promptly accompanied by a corresponding change in phosphorescence lifetime [24,29,30,32,47].

The time resolution and sensitivity achieved in phosphorescence measurements with modern instrumentation allow to measure the phosphorescence from practically every protein [45]. Of course, the usefulness of τ_0 depends critically on its correct interpretation and therefore on the detailed knowledge of the factors that can influence the decay kinetics of the triplet state in the protein environment. Recent studies [46,48] have demonstrated that τ_0 can be affected by intramolecular quenching reactions with the amino acids His, Tyr, Trp and Cys lying on the immediate neighborhood of the indole ring. Thus, only when these side chains are beyond the 'solvation' shell of the indole ring does τ_0 provide a measure of the fluidity of the local structure.

In addition to τ_0, information on the protein dynamical structure can be gained by monitoring the diffusion of quencher molecules of variable size to inner regions of the globular fold. Quenching of Trp fluorescence and phosphorescence in proteins by added small solutes has been widely employed in structural studies for it can provide information both on the location of Trp residues with respect to the aqueous phase and, more importantly, on the frequency and amplitude of structural fluctuations that allow diffusion of solutes through generally well packed and extensively bonded internal regions [37,47,49,50]. Quenching studies measure the phosphorescence lifetime (τ) as a function of quencher concentration, [Q], and evaluate the bimolecular quenching rate constant, k_q, from the gradient of the Stern–Volmer plot

$$1/\tau = 1/\tau_0 + k_q[Q]$$

where τ_0 is the unperturbed lifetime. With chromophores free in solution and efficient quenching reactions k_q is practically identical to the diffusion-limited rate constant k_d, where $k_d = 4\pi r_0 D$ (r_0 is the sum of molecular radii and D is the sum of the diffusion coefficients). With Trp residues inside proteins k_q is invariably smaller even by several orders of magnitude than k_d, the slowing down of the bimolecular reaction depending on both the size of the quencher and the surface accessibility of the chromophore [47,50]. When quenching is limited by the migration of Q through the protein matrix $k_q \propto D$ and D is essentially the diffusion coefficient of Q inside the protein. Recent studies on oxygen and acrylamide quenching of Trp phosphorescence in proteins showed that in the long, millisecond–second time scale of the delayed emission both solutes have access to the innermost regions of globular proteins and that the magnitude of k_q correlates with the flexibility of the macromolecule [37,47,50].

2.2. *Instrumentation for phosphorescence measurements*

Fluorescence and phosphorescence spectra, intensities and phosphorescence decay kinetics have been obtained with a home-made apparatus [33] whose

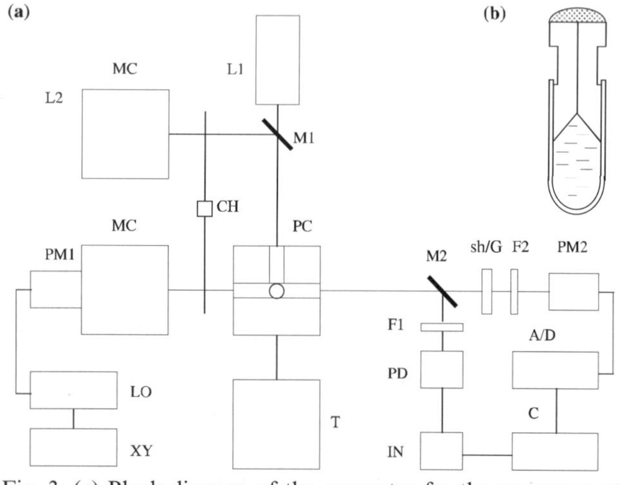

Fig. 3. (a) Block diagram of the apparatus for the measurement of protein fluorescence and phosphorescence under pressure. L1, dye laser; L2 xenon lamp; MC, grating monochromators; CH, mechanical chopper; PM1, EMI 9635SQB photomultiplier; PM2, EMI 9235QA photomultiplier; LO, lock-in amplifier; XY, XY recorder; PC, pressure cell; M1, UV mirror; M2, quartz plate; Sh/G, electronic shutter/gate; F1, 300 nm cutoff filter; F2, wide band interference filter; PD, UV-sensitive photodiode; IN, integrator circuit; A/D, computerscope; C, computer. (b) An enlargement of the sample cuvette.

layout is given in Fig. 3a. Briefly, continuous excitation for fluorescence and phosphorescence spectra is provided by a Cermax xenon lamp (LX 150 UV, ILC) and the excitation wavelength, typically 290 nm, is selected by a 0.25 m grating monochromator (model 82-410, Jarrel-Ash) with a 10 nm bandpass. The emission collected through another 0.25 m grating monochromator (Jobin-Yvon, H25) with 7 nm bandwidth is detected by an EMI9635QB photomultiplier. Its photocurrent was fed to a low noise current preamplifier (SR570, Standford) and further amplified by a lock-in detector (model 393, Ithaco). A light chopper modulates the excitation and the transition from fluorescence to phosphorescence detection is accomplished simply by shifting the chopper position such as to cut in alternative fashion both excitation and emission beams. Phosphorescence decays were obtained with pulsed excitation at 292 nm as provided by frequency-doubled dye laser pumped by either a flash lamp (UV500M, Candela) or Nd/Yag laser (Quanta Systems). The pulse duration is 1 μs/5 ns and the light energy per pulse is typically 1–10 mJ. The phosphorescence signal, collected at a right angle from the excitation beam, is filtered (410–470 nm bandpass) and detected by an EMI9235QA photomultiplier. An electronic shutter arrangement protects the photomultiplier from the intense fluorescence pulse and allows the delayed emission to be detected 2 ms after the excitation pulse. Recently, gating the photomultiplier was found to be equally effective in isolating phosphorescence from the prompt emission. The photocurrent is fed to a low noise preamplifier and digitized by a computerscope system (ISC-16, RC Electronics) capable of averaging multiple sweeps. An ultraviolet sensitive photodiode (model OSD100-7, Centronics), whose photocurrent is integrated in a charging capacitor, detects the intensity of the concomitant fluorescence signal. All phosphorescence decays are analyzed in terms of a sum of exponential components by a non-linear least-squares fitting algorithm (Global Unlimited, LFD, University of Illinois).

Luminescence measurements under pressure are carried out in a pressure cell (ISS-NOVA, ISS) provided with sapphire windows and employing highly pure ethanol as pressurizing fluid. The sample is placed in a cylindrical quartz cuvette, containing 0.12–0.15 ml of solution (Fig. 3b), tightly stoppered by a Teflon piston bored through the center by a hole 0.5 mm in diameter. The hole, which serves to expel air bubbles, is plugged on the outside by silicone grease to prevent liquid/gas exchange between sample and pressurizing fluid. With such an arrangement the phosphorescence characteristics of proteins remain constant after several compression/decompression cycles. Particular attention must be paid to assure temperature equilibration of the sample after each pressure variation, which, within the 0–60°C temperature range of the instrumentation, requires at least 5 min.

The phosphorescence lifetime of Trp decreases dramatically when the chromophore is in very flexible protein sites or is exposed to the aqueous solvent, and in practice the emission is no longer detectable. Hence, conditions that produce a drastic loosening of the protein structure about the chromophore will lead to an apparent static quenching of Trp phosphorescence. In an heterogeneous sample the fraction of chromophores still detectably phosphorescent (f_p), i.e. whose $\tau > 2$ ms, is determined by the intensity ratio $P(0)/F$, where $P(0)$ is the phosphorescence intensity extrapolated to time zero and F is the integrated fluorescence intensity obtained in the same pulse. Once duly normalized, the ratio $P(0)/F$ automatically corrects for any variability in the excitation pulse energy for changes in transmittance of the sample and pressure cell assembly as well as for any sample photobleaching. Of course, the parameter $P(0)/F$ represents an accurate measure of f_p only if the fluorescence to intersystem crossing quantum yield (Φ_F/Φ_{isc}) of Trp is pressure independent. This seems to be the case since up to 3 kbar Φ_F of Trp in aqueous solution increased by not more than 10% [52] and for deeply buried phosphorescing Trp residues in proteins Φ_F and $P(0)/F$ were found to remain constant in this pressure range [33].

2.3. Sample preparation for phosphorescence measurements

Prior to luminescence measurements all proteins are extensively dialyzed in Tris buffers whose pH is one of the least sensitive to pressure. The observation of room temperature phosphorescence of proteins in solution depends critically on the removal of the effective triplet state quencher oxygen from the solution. Deoxygenation of protein samples for high pressure experiments is obtained by adding an enzymatic system composed of 80 nM glucose oxidase, 16 nM catalase and 0.3% (w/v) glucose [51]. No emission from these proteins is detected at the amplification levels of phosphorescence measurements. The protein concentration in all phosphorescence experiments is typically between 4 and 10 μM.

In acrylamide quenching experiments, in order to avoid that oxygen radicals transiently formed during enzymatic deoxygenation trigger acrylamide polymerization, stock samples of the latter are degassed separately by repeated application of vacuum and inlet of pure N_2. Subsequently, appropriate aliquots of acrylamide stock are added to already degassed protein solutions in a nitrogen box. In the determination of O_2 phosphorescence quenching rate constants a major cause of irreproducibility is the uncertainty in $[O_2]$ in solution. Accurate control of the oxygen concentration in solution is achieved by a method that employs an internal protein reference, generally alkaline phosphatase (AP) [37]. Among proteins, AP exhibits one of the longest intrinsic lifetimes ($\tau_0 \sim 2$ s at 20°C) and smallest $k_q(O_2)$, features that allow to measure AP phosphorescence even in air saturated solutions ($\tau = 3.15$ ms at 20°C), where $[O_2]$ is known accurately. O_2 quenching experiments with AP as internal control are carried out by equilibrating the protein solution with variable O_2 partial pressures and the final $[O_2]$ in solution is determined from its effect on the phosphorescence lifetime of AP in the same solution.

3. Pressure effects on protein structure

The proteins selected for this study have well characterized phosphorescence emission properties and in each case the emission is due to a single, known chromophore, a feature that both simplifies the interpretation of the decay kinetics with respect to protein conformational homogeneity and also allows to analyze the spectroscopic changes in terms of the crystallographic structure. To test for differential effects between superficial and deeply buried sites, proteins were chosen in which the phosphorescence probe lies at variable distances (r) from the surface. Among the proteins examined are monomeric apoazurin (Az, $r = 0.8$ nm), phosphoglycerate kinase (PGK, $r = 0.35$ nm), ribonuclease T1 (RNase T1, $r = 0.2$ nm), dimeric AP ($r = 1.1$ nm), liver alcohol dehydrogenase (LADH, $r = 0.45$ nm) and tetrameric glyceraldehyde-3-phosphate dehydrogenase (GAPDH, $r = 0.55$ nm). For each protein fluorescence and phosphorescence spectra and intensities were monitored as a function of applied pressure up to 3 kbar [33–35]. Except for PGK and GAPDH, no large pressure effects were found on fluorescence under all conditions of temperature and solvent composition explored. Luminescence changes were confined to phosphorescence intensity and decay kinetics.

3.1. Pressure effects on the native structure inferred from τ_0

The effect of pressure on τ_0 (Fig. 4) is markedly similar among Az, LADH and AP, proteins in which the probe is well buried in the core of the macromolecule. Above ambient temperature, applied pressure induces a considerable lengthening of τ_0; the

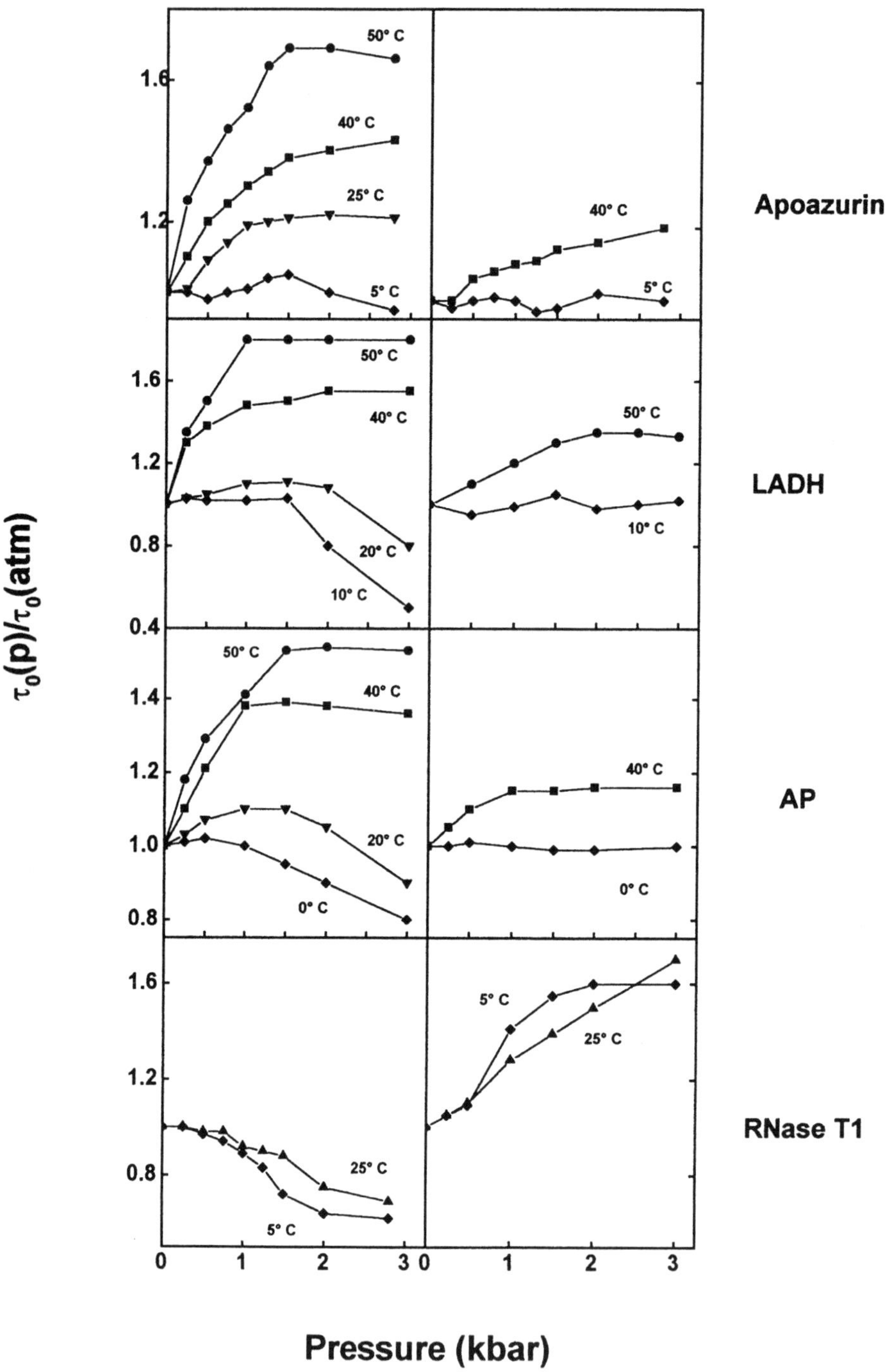

Fig. 4. Pressure dependence of the phosphorescence lifetime normalized for its value at 1 bar for apoazurin, LADH, AP and RNase T1 at selected temperatures. Comparison between buffer (left panel) and glycerol/buffer (70/30 w/w) (right panel).

change occurs entirely below 1.5 kbar and is progressively larger as the temperature increases. Above 1.5 kbar τ_0 is constant or decreases, the downward tendency becoming more pronounced at lower temperature. These responses of τ_0 to pressure appear to be completely elastic as on decompression the change is promptly reversed. From the correlation between τ_0 and the fluidity of the chromophore's environment

[45] the pressure modulation of τ_0 attests to a tightening/loosening of the protein core in the low/high pressure range, the tightening and loosening effect being enhanced at high and low temperature, respectively. The τ_0 vs. pressure profile is quite distinct for superficial W59 of RNase T1, the reduction of τ_0, which occurs mostly above 1 kbar, indicates throughout an increase in segmental flexibility of the region.

All indications are that these adaptations of protein structure to high pressure are consistent with the intrinsic compressibility of these biopolymers. The application of high pressure will promote any structural rearrangement of the macromolecule/solvent system that accommodates a reduction in its volume. According to compressibility data [9,10,53], a decrease in volume can be achieved by both the reduction of internal cavities, voids that result from imperfect packing of amino acids, and greater hydration of the polypeptide, including the penetration of water molecules into the globular fold. Reduction of cavity size and hydration exert opposing influences on protein dynamics. From the correlation between compressibility, volume fluctuations and flexibility [8], compression is expected to restrict the mobility of the polypeptide mainly of internal regions. Hydration, on the other hand, as it substitutes intrapeptide bonds with water bonds, will exert a lubricating action on segmental flexibility particularly of peripheral regions of the macromolecule that in general are already more mobile. Since cavities become larger with thermal expansion and hydration increases at low temperature, the tightening/loosening effect of pressure will be more pronounced at high/low temperature, respectively. The elastic perturbations reported by τ_0 on the above four proteins are totally consistent with these expectations. The sites of Trp-48 of Az, Trp-314 of LADH and Trp-109 of AP are quite internal and little accessible to hydration. It follows that the local mobility of side chains and backbone relies mostly on available free volume and therefore will be restricted by compression. The considerable lengthening of the phosphorescence lifetime as pressure is raised up to 1.5 kbar confirms a pronounced increase in the rigidity of the protein sites. Moreover, the tightening effect becomes amplified at higher temperature where the compressibility and the corresponding reduction in free volume are expected to be larger. On the other hand, under conditions that strongly favor protein hydration, that is very high pressure and low temperature, the lubricating action of the hydration water is expected to contrast and even dominate over any rigidity caused by the reduction in internal free volume. Indeed, above 1.5 kbar τ_0 either remains constant or decreases, the downward tendency of τ attesting to a net increase in flexibility of these internal sites that becomes more pronounced as the temperature is lowered.

Processes leading to protein hydration can be effectively contrasted by the addition of glycerol to the solution. The organic solvent stabilizes the globular state of proteins by inducing preferential hydration of the protein–water interface. As a result, the system minimizes the free energy cost of solvent segregation by adopting a compact conformation with the least surface area [54]. This can be achieved by reducing clefts as well as by assuming a smaller size and a more spherical shape. Glycerol is, therefore, expected to both compact the macromolecule [55], making it more rigid, and oppose further hydration of the polypeptide on compression. The results (Fig. 4) show, indeed, that τ_0 is larger in glycerol solutions and, more importantly, that on compression the triplet lifetime increases to a smaller extent than in buffer (Az, LADH and AP), suggesting that glycerol and pressure act similarly on decreasing the internal free volume. Likewise glycerol effectively inhibits any pressure-induced decrease in τ_0, particularly evident with superficial W59 of RNase T1 and with the other proteins at low temperature, that is attributed to polypeptide hydration.

3.2. Pressure effects on the native structure inferred from the bimolecular quenching rate constant, k_q

It is significant that such pressure modulation of protein flexibility reported by τ_0 is largely confirmed by the change in permeability of oxygen and acrylamide as determined in quenching experiments [36,37]. Representative lifetime Stern–Volmer plots for acrylamide quenching of AP at selected pressures are shown in Fig. 5. At all pressures, $1/\tau$ increases linearly with the quencher concentration as should be expected for a dynamic quenching reaction. The results show that the accessibility of acrylamide to in-

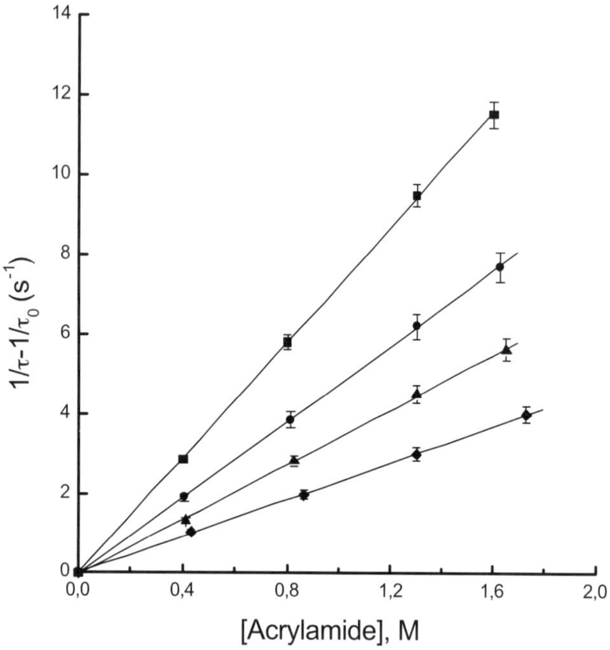

Fig. 5. Lifetime Stern–Volmer plots for the quenching of AP phosphorescence by acrylamide as a function of applied pressure at 50°C: 0.001 kbar (■), 0.5 kbar (●), 1.0 kbar (▲) and 2.5 kbar (♦). The error bars indicate the range of τ variations.

ternal W109 of AP decreases monotonically with applied pressure, the magnitude of k_q, derived from the gradient of these plots, decreasing by 3.5-fold from atmospheric pressure to 3.0 kbar. The reduction of k_q reflects a pressure-induced tightening of the globular structure that further hinders diffusion of acrylamide to its interior. The pressure dependence of the acrylamide quenching constant for four proteins, at 50°C, is reported in Fig. 6. It must be pointed out that the pressure profiles of k_q were found to be quite analogous between O_2 and acrylamide quenching [37]. In these plots we note that the trend of decreasing k_q with pressure, similar among AP, LADH and Az, is roughly a mirror image of that describing the increase in τ_0. They confirm the initial compacting effect of pressure as well as its attenuation above 1.5–2 kbar. For RNase T1 k_q is insensitive to pressurization, a result probably explained by the rather superficial location of its Trp residue. The separation of W59 from the aqueous interface is only about 2 Å, a distance over which through-space quenching by acrylamide in the solvent is competitive with diffusive pathways [47], in which case pressure effects on k_q are not contemplated.

For acrylamide quenching, the pressure dependence of k_q, at selected temperatures, is summarized in the logarithmic plots of Fig. 7. When the rate is

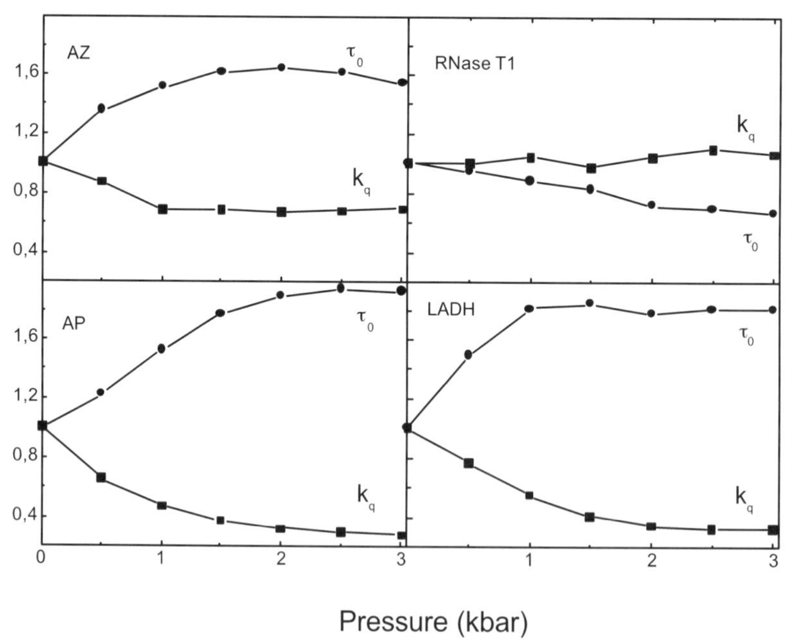

Fig. 6. Pressure dependence of k_q (■) and of intrinsic phosphorescence lifetime (●), τ_0, normalized for their values at 1 bar, for the proteins Az, AP and LADH at 50°C and for RNase T1 at 25°C.

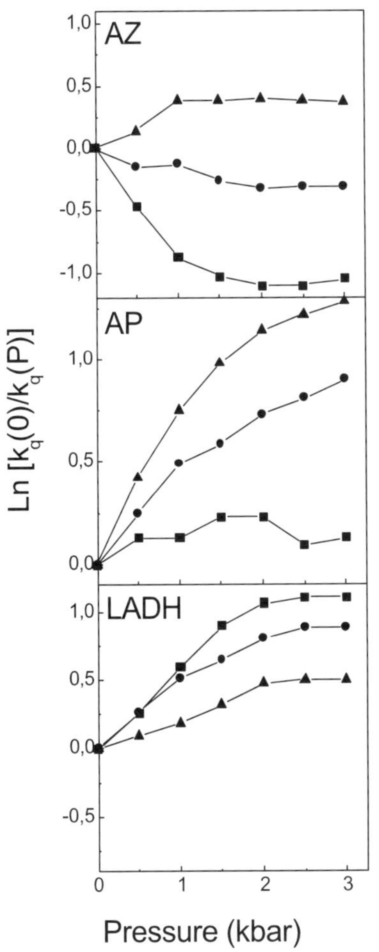

Fig. 7. Logarithmic plots of pressure dependence of $1/k_q$ at 10°C (■), 30°C (●) and 50°C (▲) for acrylamide quenching of Az, LADH and AP.

analyzed within the framework of transition state theory, $k(p) = k(0)\exp(-p\Delta V^{\ddagger}/RT)$, the slope of the log plot represents the activation volume ($\Delta V^{\ddagger}$) for acrylamide diffusion inside the protein. A prominent feature of $\ln k_q$ vs. pressure profiles of Az, LADH and AP is their non-linearity, pointing out that $\Delta V^{\ddagger}_{obs}$ generally decreases with pressure, especially above 1–1.5 kbar. Further, $\Delta V^{\ddagger}_{obs} < 20$ ml mole^{-1} is small compared to the volume of acrylamide (~65 ml mole^{-1}) and can also become negative. This suggests that diffusion jumps do not require the formation of molecular size cavities and that two or more processes with distinct activation volumes contribute to acrylamide migration inside the proteins. Considering that pressure can act on proteins through both elastic compaction of empty cavities and by promoting more extensively hydrated conformations, we propose that $\Delta V^{\ddagger}_{obs}$ be conveniently regarded as the resultant of two main contributions

$$\Delta V^{\ddagger}_{obs} = \Delta V^{*}_{cav} + \Delta V_{h}$$

where ΔV^{*}_{cav} and ΔV_h reflect the relative importance of cavity expansion (elastic) and of the degree of hydration of the polypeptide (conformational) for the structural fluctuations underlying acrylamide diffusion. Whereas ΔV^{*}_{cav} is positive and is a true activation parameter that represents an increase in barrier height due to $p\Delta V^{*}$ work, ΔV_h is negative because pressure-induced shifts to more hydrated states enhance the fluidity of the protein matrix. In fact, at sufficiently high pressure the latter process invariably leads to largely unfolded, highly flexible configurations of the polypeptide [56,57]. As the two contributions to $\Delta V^{\ddagger}_{obs}$ oppose each other, the combined effect is to attenuate pressure modulation of protein dynamics. The net result will be an inhibition or an enhancement in structural flexibility depending on which of the two factors, cavity expansion or internal hydration, will be dominant for a given protein and experimental condition. At 50°C, the acrylamide diffusion coefficient inside Az, LADH and AP shows that, below 1–1.5 kbar, $\Delta V^{\ddagger}_{obs}$ is positive (from +10 to +20 ml mole^{-1}) but becomes much smaller (0–5 ml mole^{-1}) on approaching 3 kbar. Thus, at relatively warm temperatures ΔV^{*}_{cav} exceeds ΔV_h in the lower pressure range, but the difference is progressively attenuated at higher pressure confirming an increasing role of ΔV_h. On approaching the freezing point of water, pressure-induced protein hydration is expected to be more efficacious because exposure of hydrophobic side chains to the solvent becomes less costly in free energy terms [58] and, furthermore, owing to a larger difference in volume between bulk and hydration water, the hydration drive of pressure becomes stronger. The results at 10°C show that $\Delta V^{\ddagger}_{obs}$ never exceeds +5 ml mole^{-1} for LADH and is practically zero for AP. For Az, which displays the smallest activation volume at 50°C, $\Delta V^{\ddagger}_{obs}$ turns negative already at 30°C (−3 ml mole^{-1}) and becomes more so at 10°C (−20 ml mole^{-1}). It would seem that for this protein, which lacks a structuring copper ion, pressure-induced hydration is rather effective in promoting acrylamide diffusion over and above any inhibition of migration pathways involving volume expansion.

Table 1 points out that the above general trends and also the magnitude of the activation volumes apply equally to oxygen migration and to τ_0. Because these are totally independent monitors of protein flexibility that respond to structural fluctuations of different frequency and amplitudes [37], this finding would indicate that pressure affects similarly a wide range of segmental motions and that its influence extends over large regions of the macromolecule. Such a conclusion is also supported by the relatively small difference in $\Delta V^{\ddagger}_{obs}$ among the protein examined, regardless of the depth to which the monitoring probe is buried or the great variability in the actual flexibility of the polypeptide as inferred from the magnitude of k_q and of τ_0 (Table 1).

3.3. Pressure effects on protein stability

The common observation of an inverse correlation between structural flexibility and thermal stability [16,59,60] suggests that the above pressure modulation of protein flexibility may also be important for understanding the ambivalent effect of pressure with regard to their thermal denaturation. Although high pressure invariably leads to protein denaturation, it is generally found that below 1–1.5 kbar, pressure stabilizes against thermal unfolding, a phenomenon that is depicted in the elliptical shape of temperature–pressure phase diagrams [61,62]. Naturally, the simple two-state model usually adopted to interpret pressure denaturation does not contemplate this ambivalence as the change in free energy of the reaction is simply given by $\Delta G_D = \Delta G^0_D + p\Delta V^0_D$ with the change in standard free energy and volume between denatured (D) and native (N) states, considered to be pressure independent. However, the above pressure-induced variations in the dynamical structure of the native fold emphasize that its free energy is a function of pressure and that, depending on whether $\Delta V^{\ddagger}_{obs}$ is positive or negative, the globular structure is either stabilized or destabilized on pressurization. According to $\Delta V^{\ddagger}_{obs}$ the stabilizing–destabilizing effect of pressure depends on temperature. For each protein there is a critical temperature (T_c) above which moderate pressure (below 1.5 kbar) enhances the stability of N, and the effect is amplified at higher temperature. Although the actual magnitude of the free energy change of N is not known, its stabilization in the low pressure range would offer a reasonable explanation for elliptical *P–T* phase diagrams. Vice versa, below T_c, pressure weakens intramolecular interactions and N becomes increasingly unstable. For Az, T_c is between 30 and 50°C whereas for LADH and AP it is around 10°C or below. On the application side, where pressurization of protein samples is employed to break up inter- or intramolecular interactions so as to induce subunit dissociation, solubilization of protein aggregates or unfolding of globular structures these findings point out that the effectiveness of the pressure treatment is expected to increase at lower temperature.

Among the proteins examined PGK is singular because according to its Trp luminescence the macromolecule undergoes a considerable change in structure at relatively low pressure between 1 and 2 kbar and, further, these effects are not promptly reversed on decompression. Although no details of the structural changes are available, fluorescence spectra, intensities and phosphorescence lifetimes concur that the change is best described by partial unfolding of the protein [33]. Raising the pressures above 1 kbar an increasing fraction of the sample becomes no lon-

Table 1
Comparison of the internal protein viscosity, η_{prot}, and the related activation volume, $\Delta V^{\ddagger}_{max}$, monitored by the same Trp probe according to $k_q(O_2)$, k_q(acryl) and τ_0

Protein	η^{a}_{prot} (cP)			$\Delta V^{\ddagger}_{max}$ [b] (ml)		
	$k_q(O_2)$	k_q(acryl)	τ_0	$k_q(O_2)$	k_q(acryl)	τ_0
LADH	2.0×10^2	1.2×10^5	4.9×10^4	16.0	16	20
Az	6.7×10^2	4.6×10^7	4.6×10^4	9.0	10	13
AP	4.0×10^3	1.5×10^{10}	7.5×10^5	9.8	20	18

[a]Data at 20°C. η_{prot} is obtained from either $k_{q(solv)}/k_{q(prot)}$ or from the empirical relationship between τ_0 and solvent viscosity in model studies [45].
[b]Data at 50°C.

ger phosphorescent, an unequivocal indication that the chromophore is brought into a very fluid environment where τ_0 is shorter than the detection limit. In the process its fluorescence is red shifted to an extent commensurate with large exposure of the chromophore to the polar aqueous phase. On decompression, recovery of the luminescence characteristics is delayed from minutes to several hours, the delay increasing the greater are the pressure and the length of time of incubation. This behavior is commonly observed upon pressure-induced dissociation of oligomeric proteins, a process known as conformational drift. In the case of PGK the dominance of the hydration drive is clearly manifested by the greater effectiveness of pressure at lower temperature and the ability of glycerol to abolish entirely any unfolding and hysteresis. The results with PGK represent perhaps the first example of pressure-induced conformational drift in a monomeric protein. They emphasize that pressure by itself, even in the moderate 1–2 kbar range, can cause major alterations of the native fold with a slow and continuous reorganization of the loose structure. It may be significant that this phenomenon was observed with PGK and also with staphylococcal nuclease [63], which, compared to the other proteins examined, stand out for their moderate stability.

3.4. *Pre- and post-dissociational effects of pressure on oligomeric proteins*

In the last decade the application of hydrostatic pressure has become an increasingly common tool for promoting the dissociation of oligomeric proteins, or even larger protein assemblies such as viral capsids, into individual subunits [4,64–66]. In principle, the method affords the determination of very small dissociation constants and, consequently, of the thermodynamic parameters (ΔH^0, ΔS^0 and ΔV^0) that characterize very strong subunit associations. In these studies it is assumed that pressure does not influence protein structure and some peculiarities of these equilibria, such as the decrease in subunit affinity at large degrees of dissociation (α), insensitivity of α to protein concentration and hysteresis in α/pressure diagrams are imputed to either structural heterogeneity and/or to the slow conformational drift of isolated subunits [64,67]. However, the demonstration that subdenaturing pressures affect protein dynamics and structure raises concerns that the pressure required to cause subunit dissociation may by itself influence the conformation of the oligomer and/or isolated subunit and alter the thermodynamic parameters derived from these studies.

The pressure dependence of the dissociation free energy $\Delta G(p)$ is described by

$$\Delta G(p) = \Delta G^0 + p\Delta V$$

where ΔG^0 is the dissociation free energy at atmospheric pressure, p is the applied pressure above 1 atm and ΔV is the volume change of the reaction. ΔG^0 and ΔV are estimated from the linear plot assuming that both parameters are pressure independent [67], which requires that the structure of both oligomer and isolated subunit be not affected by hydrostatic pressure and that their compressibility be negligibly small. Until now, the prevailing view is that pressure effects are negligible and that any change in the monomer structure is slow and driven primarily by the hydration of newly exposed surface areas [4,67,68]. A detailed model of oligomer dissociation was developed [68,69] according to which the progressive deterioration of the monomer structure would account for both hysteresis effects and the decrease in subunit affinity of pressurized samples. An assumption of this model is that the drift is slow relative to the time needed to establish the dissociation equilibrium (10–40 min). As such, the two processes, dissociation and drift, are separated in time and effectively do not interfere with each other.

An example of predissociational effects of high pressure is provided by dimeric LADH. The conformation of the native dimer was monitored as a function of pressure and temperature by means of Trp fluorescence and phosphorescence as well as by binding of the 1-anilinonaphthalene-8-sulfonic acid (ANS) fluorophore [34]. The phosphorescent Trp (W314) of LADH is buried at the subunit interface and any change in the phosphorescence characteristics with pressure will signal alterations of the structure at the interface. At 0°C and below 3 kbar LADH is still a dimer but both the phosphorescence intensity and τ_0 undergo considerable changes in time. The kinetics of the decrease in intensity (P/P_0) and of τ/τ_0 at 2.5, 2.75 and 3 kbar are shown in Fig. 8. The time course of these changes distinguishes two

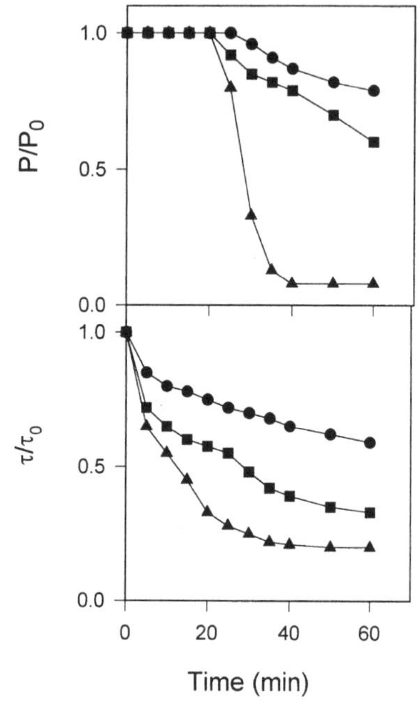

Fig. 8. Time dependence of the normalized phosphorescence intensity (P/P_0) and of the intrinsic lifetime (τ/τ_0) of LADH (5 μM) in buffer at 0°C at pressures of 2.5 kbar (●), 2.75 kbar (■) and 3 kbar (▲).

consecutive steps of the process. There is an initial stage characterized by a steady drop of τ/τ_0 (3–5-fold), a small (2–3 nm) red shift of Trp fluorescence spectrum and an increased yield of ANS fluorescence. While the drop of τ attests to a progressive loosening of the β-sheet enveloping Trp-314, the accompanying fluorescence red shift is indicative of an increasingly polar environment about the chromophore that likely reflects the penetration of water molecules into the intersubunit region. After a lag period, typically 25 min at 0°C and 3 kbar, a second phase initiates in which deeper transformations of the protein structure cause complete quenching of the phosphorescence intensity, a considerable reduction in ANS fluorescence intensity and a further spectral red shift. The loss of P/P_0 implies that τ is reduced by at least 100-fold (from ~200 ms to <2 ms) to a magnitude that is typical of fluid matrices. At the same time the fluorescence yield and λ_{max} suggest that Trp-314 has become partly exposed to the solvent.

The different nature of the structural changes during the first and the second stage is also emphasized by their opposite effect on the fluorescence yield of ANS. The initial increase in yield is consistent with the production of a molten globule-like state [70], a compact form with little tertiary structure, whereas its subsequent decrease points to the disruption of hydrophobic sites that probably follow the unfolding of large domains. An important aspect of these transformations is that they persist for relatively long times after pressure is released and are also accompanied by a slow and often incomplete recovery of the enzymatic activity. Hence, the response of the system to pressure changes will be sluggish and depend on the history of the sample; in other words, pressure cycles will necessarily exhibit pronounced hysteresis. Further, as the perturbation results in a progressive hydration/loosening of the subunit interface region it is reasonable to foresee that the subunit affinity will decrease at higher pressure, as generally observed in dissociation studies.

An example of post-dissociational effects of pressure is given by GAPDH [35]. For this tetramer, high pressure dissociation equilibria and kinetics, as well as subunit exchange rates, have been studied in great detail under a variety of conditions [68,71,72]. Also, among the tetramers examined so far, GAPDH dissociation showed the largest hysteresis effects [4]. The protein exhibits intense and long-lived phosphorescence whose interpretation in terms of changes in protein structure is facilitated by the fact that the single Trp residue (Trp-84) emitting in each subunit [40] is located in the catalytic domain at a site quite removed from the subunit interface [73]. Consequently, its environment is not altered by subunit dissociation per se and changes in τ_0 will necessarily reflect variation in the conformation of the far catalytic domain.

According to the lifetime, pressure perturbs the structure of both GAPDH tetramer and monomer [35]. On the tetramer the change is modest and elastic, involving a 40% reduction of τ_0 at 2.4 kbar, where dissociation is complete. However, the decrease in τ_0 for the isolated subunit is dramatic, over 100-fold, the increased local mobility being consistent with partial unfolding. As tightly packed cores of the globular structure, such as the one hosting Trp-84, are generally responsible for the stability of entire domains, it is likely that the structural changes observed by the phosphorescence probe extend to a large portion of the subunit. The close correspondence between the reduction of τ_0 and the loss of catalytic activity implies significant deteriorations of the native structure. Indeed, even if at atmospheric pressure monomers reassociate rapidly and the fluorescence characteristics are quickly restored, recovery of the enzymatic activity is a very slow and thermally activated process that occurs concomitantly with the recovery of the original phosphorescence intensity and lifetime. Comparisons between the kinetics of dissociation and of the lifetime change have established that pressure is an important driving force of the change of monomer structure and that its alteration normally occurs before the apparent dissociation equilibrium is reached. The kinetics of recovery at atmospheric pressure emphasize that when monomers are formed in the lower pressure range they can rapidly regain the native fold but in the higher pressure range these transformations are presumably too drastic to be readily reversed.

An inescapable consequence of these pre- and post-dissociational pressure effects on protein structure is that the free energy change describing oligomer dissociation at high pressure is a composite quantity with contributions from both the dissociation reaction and pressure dependent conformational changes. $\Delta G_{\mathrm{exp}}(p)$ may then be represented as the sum of two contributions

$$\Delta G_{\mathrm{exp}}(p) = \Delta G_{\mathrm{diss}} + \Delta G_{\mathrm{conf}}(p)$$

where $\Delta G_{\mathrm{diss}} = \Delta G^0 + p\Delta V$ represents the perturbation free dissociation free energy change while $\Delta G_{\mathrm{conf}}(p)$ takes into account the pressure dependent conformational free energy changes involved in the transformation of the oligomer and of the monomer. Unless the conformational term is negligibly small the subunit affinity will appear to be pressure dependent. Also, as in general pressure-induced structural rearrangements of the monomer are expected to be larger than those of the more stable oligomer (clearly the case of GAPDH) ΔG_{conf} will be negative and the subunit affinity will decrease with pressure as commonly observed [4]. Note that a non-negligible conformational term could also explain the sometimes large discrepancy in dissociation free energy when high pressure and dilution studies are compared [74].

In conclusion, the sensitivity of Trp phosphores cence to variations in protein conformation has demonstrated distinct pressure effects on the structure of native oligomers and on the dissociated monomers. The extent and the predicted generality of these pre- and post-dissociational changes caution of possible errors in the derivation of the thermodynamic parameters of protein dissociation, particularly for rather stable subunit assemblies, requiring relatively large dissociating pressures.

References

[1] M. Gross, R. Jaenicke, Eur. J. Biochem. 221 (1994) 617–630.

[2] R. Jaenicke, Annu. Rev. Biophys. Bioeng. 10 (1981) 1–67.

[3] R. Jaenicke, Eur. J. Biochem. 202 (1991) 715–728.

[4] G. Weber, Protein Interactions, Chapman and Hall, New York, 1992.

[5] G. Weber, in: R. Van Eldyk, J. Jonas (Eds.), High Pressure Chemistry and Biochemistry, Vol. 97, NATO ASI Ser. Math. Phys. Sci., Reidel, Dordrecht, 1987, pp. 401–420.

[6] K. Heremans, in: R. Winter, J. Jonas (Eds.), High Pressure Chemistry, Biochemistry and Material Science, Vol. 401, NATO ASI Ser. Math. Phys. Sci., Kluwer Academic Publishers, Dordrecht, 1993, pp. 443–469.

[7] M. Gross, G. Auerbach, R. Jaenicke, FEBS Lett. 321 (1993) 256–260.

[8] A. Cooper, Proc. Natl. Acad. Sci. USA 73 (1976) 2740–2741.

[9] K. Gekko, Y. Hasegawa, J. Phys. Chem. 93 (1989) 426–429.

[10] T.V. Chalikian, M. Totrov, R. Abagyan, K.J.J. Breslauer, Mol. Biol. 260 (1996) 588–603.

[11] H. Frauenfelder, N.A. Alberding, A. Ansari, D. Braunstein, B.R. Cowen, M.K. Hong, I.E.T. Iben, J.B. Johnson, S. Luck, M.C. Marde, J.R. Mourant, P. Ormos, L. Reinisch, R. Scholl, A. Schulte, E. Shymsunder, L.B. Sorensen, P.J. Steinbach, A. Xie, R. Young, K.T. Yue, Proteins Press. Phys. Chem. 94 (1990) 1024–1037.

[12] X. Peng, J. Jonas, J. Silva, Proc. Natl. Acad. Sci. USA 90 (1993) 1776–1780.

[13] L. Heremans, K. Heremans, Biochim. Biophys. Acta 999 (1989) 192–197.

[14] L. Heremans, K. Heremans, J. Mol. Struct. 214 (1989) 305–314.

[15] J.V. Carter, D.G. Knox, A. Rosenberg, J. Biol. Chem. 253 (1978) 1947–1953.

[16] G. Wagner, Q. Rev. Biophys. 16 (1983) 1–57.

[17] K.T. Hitchens, R.G. Bryant, Biochemistry 37 (1998) 5878–5887.

[18] E.J. Fuentes, J.A. Wand, Biochemistry 37 (1998) 9877–9883.

[19] R. Lange, I. Heiber-Langer, C. Bonfils, I. Fabre, M. Negishi, C. Balny, Biophys. J. 66 (1994) 89–98.

[20] K. Wuthrich, G. Wagner, R. Richarz, W. Baum, Biophys. J. 32 (1980) 549–560.

[21] M.L. Saviotti, W.C. Galley, Proc. Natl. Acad. Sci. USA 71 (1974) 4154–4158.

[22] J.M. Vanderkooi, in: J.R. Lakowicz (Ed.), Topics in Fluorescence Spectroscopy, Vol. 3, Biochemical Applications, Plenum, New York, 1991, pp. 113–136.

[23] J.A. Schauerte, D.G. Steel, A. Gafni, Methods Enzymol. 278 (1999) 49–71.
[24] G.B. Strambini, J. Mol. Liq. 42 (1989) 155–165.
[25] G.B. Strambini, M. Gonnelli, Biochemistry 29 (1990) 196–203.
[26] E. Gabellieri, S. Rahuel-Clermont, G. Branlant, G.B. Strambini, Biochemistry 35 (1996) 12549–12559.
[27] P. Cioni, G.B. Strambini, J. Mol. Biol. 207 (1989) 237–247.
[28] P. Cioni, L. Piras, G.B. Strambini, Eur. J. Biochem. 185 (1989) 573–579.
[29] M. Gonnelli, G.B. Strambini, Biophys. J. 65 (1993) 131–137.
[30] G.B. Strambini, M. Gonnelli, Biochemistry 25 (1986) 2471–2476.
[31] G.B. Strambini, E. Gabellieri, Biophys. J. 70 (1996) 971–976.
[32] G.B. Strambini, E. Gabellieri, Photochem. Photobiol. 39 (1984) 725–729.
[33] P. Cioni, G.B. Strambini, J. Mol. Biol. 242 (1994) 291–301.
[34] P. Cioni, G.B. Strambini, J. Mol. Biol. 263 (1996) 789–799.
[35] P. Cioni, G.B. Strambini, Biochemistry 36 (1997) 8586–8593.
[36] P. Cioni, G.B. Strambini, J. Mol. Biol. 291 (1999) 955–964.
[37] G.B. Strambini, P. Cioni, J. Am. Chem. Soc. 121 (1999) 8337–8344.
[38] J. Lakowicz, Principles of Fluorescence Spectroscopy, Kluwer Academic/Plenum Publishers, New York, 1999.
[39] R.M. Purkey, W.C. Galley, Biochemistry 9 (1970) 3569–3574.
[40] E. Gabellieri, G.B. Strambini, Biochemistry 28 (1989) 160–166.
[41] G.B. Strambini, W.C. Galley, Nature 260 (1976) 554–555.
[42] W.C. Galley, G.B. Strambini, Nature 261 (1976) 521–522.
[43] J.M. Vanderkooi, S.W. Englander, S. Papp, W.W. Wright, C.S. Owen, Proc. Natl. Acad. Sci. USA 87 (1990) 5099–5103.
[44] G.B. Strambini, E. Gabellieri, J. Phys. Chem. 95 (1991) 4352–4356.
[45] G.B. Strambini, M. Gonnelli, J. Am. Chem. Soc. 117 (1995) 7646–7651.
[46] M. Gonnelli, G.B. Strambini, Biochemistry 34 (1995) 13847–13857.
[47] P. Cioni, G.B. Strambini, J. Am. Chem. Soc. 120 (1998) 11749–11757.
[48] G.B. Strambini, E. Gabellieri, M. Gonnelli, S. Rahuel-Clermont, G. Branlant, Biophys. J. 74 (1998) 3165–3172.
[49] M.R. Eftink, in: J.R. Lakowicz (Ed.), Topics in Fluorescence Spectroscopy, Vol. 2, Principles, Plenum, New York, 1991, pp. 53–126.
[50] G.B. Strambini, Biophys. J. 52 (1987) 23–28.
[51] S.W. Englander, D.B. Calhoun, J.J. Englander, Anal. Biochem. 161 (1987) 300–306.
[52] T.M. Li, J.W. Hook, H.G. Drickamer, G. Weber, Biochemistry 15 (1976) 3205–3211.
[53] D.P. Kharakoz, Biophys. J. 79 (2000) 511–525.
[54] K. Gekko, S.N. Timasheff, Biochemistry 20 (1981) 4667–4676.
[55] A. Priev, A. Almagor, S. Yedgar, B. Gavish, Biochemistry 35 (1996) 2061–2066.
[56] A.C. Oliveira, L.P. Gaspar, A.T. Da Poian, J. Silva, J. Mol. Biol. 240 (1994) 184–187.
[57] G. Hummer, S. Garde, A.E. Garcia, M.E. Paulatis, L.R. Pratt, Proc. Natl. Acad. Sci. USA 95 (1998) 1552–1555.
[58] P. Privalov, G.I. Makhatadze, J. Mol. Biol. 232 (1993) 660–679.
[59] P. Zavodsky, J. Kardos, A. Svingor, G.A. Petsko, Proc. Natl. Acad. Sci. USA 95 (1998) 7406–7411.
[60] K.E.S. Tang, K.A. Dill, J. Biomol. Struct. Dyn. 16 (1998) 397–411.
[61] A. Zipp, W. Kauzmann, Biochemistry 12 (1973) 4217–4228.
[62] K. Heremans, L. Smeller, Biochim. Biophys. Acta 1386 (1998) 353–370.
[63] C.A. Royer, A.P. Hinck, S.N. Loh, K.E. Prehoda, X. Peng, J. Jonas, J.L. Markley, Biochemistry 32 (1993) 5222–5232.
[64] G. Weber, in: R. Winter, J. Jonas (Eds.), High Pressure Chemistry, Biochemistry and Material Science, Vol. 401, NATO ASI Ser. Math. Phys. Sci., Kluwer Academic Publishers, Dordrecht, 1993, pp. 471–509.
[65] J. Jonas, A. Jonas, Annu. Rev. Biophys. Biomol. Struct. 23 (1994) 287–318.
[66] J.L. Silva, D. Foguel, A.T. Da Poian, P.E. Prevelige, Curr. Opin. Struct. Biol. 6 (1996) 166–175.
[67] J.L. Silva, G. Weber, Annu. Rev. Phys. Chem. 44 (1993) 89–113.
[68] K. Ruan, G. Weber, Biochemistry 28 (1989) 2144–2153.
[69] G. Weber, Biochemistry 25 (1986) 3626–3631.
[70] O.B. Ptitsyn, in: T.E. Creighton (Ed.), Protein Folding, Freeman and Co., New York, 1992, pp. 243–300.
[71] L. Erijman, G. Weber, Biochemistry 30 (1991) 1595–1599.
[72] L. Erijman, G. Weber, Photochem. Photobiol. 57 (1993) 411–415.
[73] A.G.W. Leslie, A.J. Wonacott, J. Mol. Biol. 165 (1983) 375–391.
[74] K. Ruan, G. Weber, Biochemistry 32 (1993) 6295–6301.

ELSEVIER

Biochimica et Biophysica Acta 1595 (2002) 131–144

BIOCHIMICA ET BIOPHYSICA ACTA
BBA
www.bba-direct.com

Review

Fourier transform infrared spectroscopy in high-pressure studies on proteins

Wojciech Dzwolak [a,b], Minoru Kato [a], Yoshihiro Taniguchi [a,*]

[a] *Department of Applied Chemistry, College of Science and Engineering, Ritsumeikan University, 1-1-1 Nojihigashi, Kusatsu, Shiga 525-8577, Japan*

[b] *High Pressure Research Center, Polish Academy of Sciences, Sokolowska 29/37, 01-142 Warsaw, Poland*

Received 31 August 2001; received in revised form 2 November 2001; accepted 5 November 2001

Abstract

Several aspects of the application of Fourier transform infrared spectroscopy (FTIR) in high-pressure studies on proteins are reviewed. Basic methodological considerations regarding spectral band assignments, quantitative analysis, and choice of pressure calibrants are also placed within the scope of this paper. This work attempts to evaluate recent developments in the field of high-pressure FTIR of proteins and its prospects for future. Particular attention is paid to the phenomenon of protein aggregation. © 2002 Published by Elsevier Science B.V.

Keywords: High pressure; Fourier transform infrared spectroscopy; Protein; Amide band; Diamond anvil cell

1. Introduction

Since the early works by Suzuki [1] and Hawley [2], the high-pressure research on biochemical systems, proteins in particular, has been enjoying a steadily growing interest of the scientific community. This is largely due to the realization that the most widely studied aspect of protein stability, namely the thermal denaturation, is just a one-dimensional section of the two-dimensional pressure–temperature (*p–T*) diagram of protein stability [3]. Therefore, in order to understand thermodynamics behind the protein unfolding, besides heat, high pressure, an intensive parameter that affects exclusively the volume of a given system, has to be taken into account. The subsequent reports on the ability of the hyperbaric treatment to dissociate oligomeric proteins, viruses, immunological complexes further stimulated research in this field [4–6]. Finally, high pressure turned out to be a factor that would solidify biomembranes [7] and, which had been expected since the pioneering work of Bridgman [8], denaturate proteins of microorganism without changing the taste to food. The later observations have opened to high-pressure bright perspectives in food-processing industry [9,10].

A good overview of various methods applicable for the characterization of proteins under high pressure was done elsewhere [3]. Thus we will only briefly treat other than Fourier transform infrared spectroscopy (FTIR) techniques to indicate what kind of data they could provide and how high pressure complicates using them for protein research.

Most of pressure-induced elastic or conformational changes in proteins occur under pressure of 500–

* Corresponding author. Fax: +81-77-561-2659.
E-mail address: taniguti@se.ritsumei.ac.jp (Y. Taniguchi).

0167-4838/02/$ – see front matter © 2002 Published by Elsevier Science B.V.
PII: S0167-4838(01)00340-5

600 MPa [3]. As this range of pressure is significantly high, the number of available research tools is limited. This is best illustrated by the case of high-pressure studies on proteins. Pressure of 500 MPa is high enough to make the crystallographic methods of no avail since protein crystals would simply mechanically succumb to enormous shear force (although a crystallographic study of a lysozyme crystal at 100 MPa was carried out [11]). On the other hand, the coupling of high pressure to circular dichroism (CD) has not been successful, as this posed some unsolved technical obstacles. Yet, when pressure-induced changes in protein structure are irreversible, CD could detect those changes after, but not during, the pressure treatment [12]. Among other methods applicable for protein structure analysis under high pressure a fluorescent probe and small angle X-ray scattering – SAXS (both enable observation of the changes in protein compactness, e.g., [13,14]), UV, Raman, nuclear magnetic resonance (NMR) and IR spectroscopy remain. High-pressure fluorescence spectroscopy whether intrinsic (tryptophan residues) or extrinsic (fluorescent hydrophobic probe) is a local probe useful for monitoring changes in the protein compactness, rather than detailed characterization of its overall structure [3,15]. The method is also demanding in terms of construction of a high-pressure optical cell, which limits the range of applicable pressure. On the other hand, low sample concentration needed for a fluorescence measurement, as compared to IR or NMR spectroscopy, often allow to study a protein unfolding without causing its self-aggregation [16]. UV spectroscopy is sensitive to pressure-induced changes in dielectric constant around aromatic residues. The method, though able to follow pressure-induced denaturation, is not a structure-determining tool [17]. Structure-sensitive Raman spectroscopy is rarely used. The troublesome devising of a Raman cell appropriate for high-pressure studies adds up to the problems stemming from fluorescence of the aromatic residues and decomposition of a protein sample upon excitation with a laser beam. NMR, undeniably bearing an enormous potential for the examination of various structural features in detail, is limited by the relatively low molecular mass of suitable proteins (up to 15–20 kDa) and the technical problems with constructing a NMR probe resistant to pressure higher than 500 MPa. As opposed to that, FTIR spectroscopy, while coupled with the application of a diamond anvil cell (DAC) enables investigation of the protein secondary structure as well as changes in DNA folding even under extremely high pressure.

It is necessary to stress that the subject of this concise paper had been already attaining significant attention of many researchers, which did result in a number of review papers sharing, at least in part, the scope of this work [3,4,18]. As some excellent review works devoted to the pressure effect on proteins [3,4,19–21], as well as FTIR spectroscopy of proteins [22–26], have been already published, we chose to focus more on what is particular and unique for high-pressure FTIR studies on proteins.

High-pressure FTIR studies on nucleic acids and polysaccharides are beyond the scope of this paper and as such are not treated here. Research in this field has only recently started. In a pioneering work by Lin et al. [27] several bands arising from DNA chains were examined in the context of their possible diagnostic value for high-pressure FTIR. On the other hand, Heremans' group have started investigating gelation of starch by means of high-pressure FTIR [28].

2. Elastic and conformational pressure effects

Detailed treating of elastic and conformational (plastic) effects is beyond the scope of this article and would be repetitious with reference to an excellent review published by Heremans and Smeller [3]. Elastic effects are confined to changes in the length of chemical bonds, the hydration of a protein molecule and reductions in size of molecular cavities. Elastic effects take place within relatively low range of pressure (up to 400–500 MPa). They are always reversible and do not alter the molecular folding. In infrared spectra of proteins elastic effects appear as rather minute linear to pressure spectral shifts (see the next section).

Plastic effects consist in the cooperative rotational changes around backbone bonds of a protein, which lead to a global change in the molecule's folding. Plastic effects require higher pressure (usually above 500 MPa) and for many proteins are irreversible. The amide I band's spectral position becomes a sigmoid-shaped function of pressure. Having learnt the particular conditions of pressure and temperature, under which the plastic effect (unfolding of a given protein) occurs, one can plot an elliptic p–T phase diagram of the protein stability. The thermodynamics behind such studies are extensively treated elsewhere [3,29]. The basic theoretical approach to the thermodynamics of pressure denaturation of proteins is based on the consideration of the compressibility of molecules and changes of partial molar volumes accompanying any pressure-induced process. Since water molecules are packed more tightly around ions than around uncharged solutes, the pressure-induced rupture of the structure-stabilizing salt bridges and the following unfolding of a protein molecule under high pressure would be readily understandable. The role of hydrogen bonds and polar interactions in protein stability is known, albeit this is the formation of a hydrophobic cluster inside of a protein's molecule which is now understood as a driving force for the folding of proteins in aqueous solutions. As it was pointed out by Kauzmann [30], this assumption seemingly contradicts the finding that the partial molar volumes of the transfer of hydrocarbons into water is negative at low pressure but becomes positive at high pressure, namely when the pressure-induced denaturation of proteins occurs. However, a very recent study based on the information theory model showed that this contradiction is only apparent if the high-pressure denaturation of proteins is considered as a slow penetration of the protein's core by water rather than an extensive exposure of the hydrophobic core to the bulk solvent [31]. This remains in accordance with the experimental evidence showing that pressure-denaturated proteins, unlike thermally denaturated proteins, retain a compact structure [32].

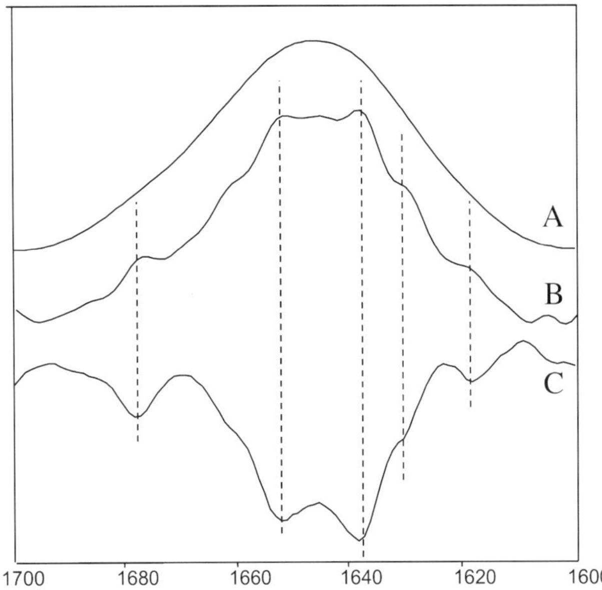

Fig. 1. Narrowing of the amide I′ band of α-lactalbumin: (A) the original absorption spectrum; (B) the spectrum after self-deconvolution; (C) the second derivative spectrum.

3. FTIR spectroscopy

The variety of chemical groups present in amino acid residues, the peptide linkage itself, and possible presence of prosthetic groups (e.g., in the case of hemoglobin) give rise to plenty of vibrational bands which can be sensitive to structural changes induced in the protein molecule upon pressure treatment. Table 1 summarizes data of vibrational bands which were investigated in high-pressure FTIR studies on proteins.

Out of seven existing infrared bands stemming from the peptide linkage only the amide I band proved to bear significant information for a detailed characterization of protein secondary structure. In terms of vibrational energy, the amide I band origins

Table 1
Vibrational bands of diagnostic value for high-pressure FTIR studies of proteins

Vibration	Diagnostic specification	Wavenumber (cm^{-1})	Ref.
Amide I band	Secondary structure	1610–1690	[23,24]
Amide II/II′ bands	Progress of H/D-exchange, protein compactness	1515–1560	[23,24]
Tyrosine ring	Tertiary structure compactness	1515	[3,33]
Antisymmetrical stretching of $-COO^-$ groups	Type of chelation of Ca^{2+} in calcium-binding proteins	1560–1590	[34]
C=O stretching in CO-bound cytochrome P-450_{cam}	Heme pocket environment	1900–1980	[35]

prevailingly (85%) from amide's C=O stretching, which is coupled with CN stretching and CCN deformation. As the energy of this complex vibration depends strongly on the presence and geometry of hydrogen bonding, the frequency of the amide I band is affected by the secondary structure, in which the particular amide bond is incorporated [24]. The bandwidths of vibrations arising from different secondary structure components are large compared to the separation between the corresponding frequencies, which results in the amide I band becoming strongly overlapped and the need for a resolution enhancement (Fig. 1). Table 2 shows typical assignments of the secondary structure components to wavenumbers obtained after resolution-enhancement of original absorption spectra. This has been, however, excellently reviewed elsewhere [23,37]. It should be also stressed that these frequencies might depend themselves on factors such as solvation of the polypeptide chain and possible distortions within helices, as studies on poly(L-lysine) show [38]. Although, likewise in the case of amide I band, secondary structure assignments of the component frequencies to the amide II band were carried out, such data have not become a reference standard for protein studies [23,24]. We are not aware of any high-pressure FTIR study on proteins that would take advantage of the amide II band other than monitoring of the pressure-enhanced H/D-exchange [39].

As the remaining three bands of those indicated in Table 1 are rather infrequently discussed in IR studies, we will focus on them only briefly when referring to the relevant works in following sections.

As high pressure causes 'compression' of chemical bonds equivalent to changes of the force constants between the involved atoms, pressure-induced frequency-shifts are frequently observed [3,40,41]. Interestingly, while this effect is usually revealed by blue-shifting IR bands, the hydrogen-bonded systems, such as proteins, show a red-shift (Fig. 2). This is due to the strengthening of the O...H hydrogen bond 'at the expense' of the C=O bond. The frequency versus pressure plots generally feature a negative slope for inter- or intra-chain hydrogen-bound structures (as α-helix and β-sheet), and a positive slope for water-bound random coil [40,41]. This phenomenon is clear evidence for the elastic pressure effect on proteins.

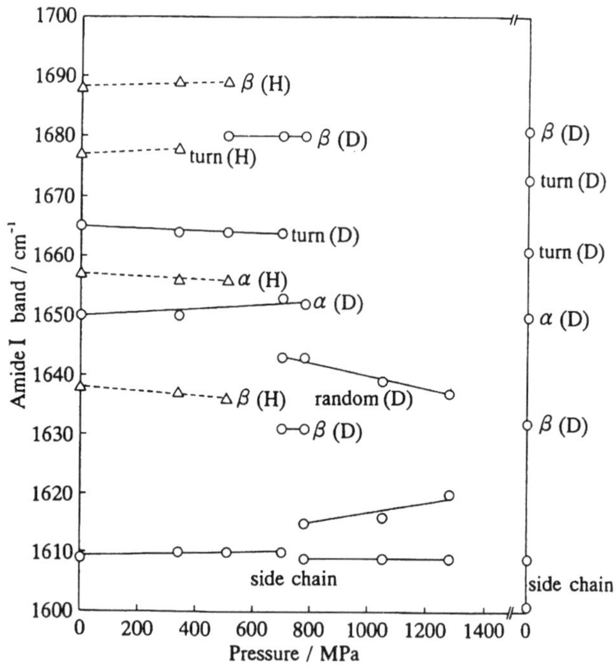

Fig. 2. Elastic shifts of the secondary structure components of ribonuclease A (from the second derivative spectra), according to [41].

Qualitative analysis of the secondary structure components can be carried out in a rather straightforward manner by conventional resolution enhancement method (such as second derivative [23,42] or Fourier self-deconvolution [23,37]), whereas quanti-

Table 2
Typical assignment of the elements of protein secondary structure after complete H/D-exchange

Structure	Deuterated form (cm^{-1})	H/D-caused red-shift (cm^{-1})	Ref.
α-Helix	~1654	~5	[23]
β-Sheet	~1637	~5	[23]
Antiparallel β-sheet (intermolecular)	1615 and 1685		[24]
Random coil	1645	~10	[23]
3_{10}-Helix	1639	~15	[36]
Turns	1682–1662	5	[23]

tative analysis is complicated by several factors. Firstly, when attempting to estimate transient contributions of particular elements of the protein secondary structure under given pressure, one has to keep in mind that the pressure-induced changes of spectral bands do not stem exclusively from the plastic (conformational) changes, but reflect the elastic effect (red shift) as well. The latter fact significantly hampers any application of factor analysis [23] by requiring a set of the reference data on pressure-induced elastic shifts of the relevant spectral components. Obviously, this problem will only be exacerbated when a rather tiny elastic shift is accompanied by a larger red shift induced by pressure-enhanced H/D-exchange in a protein dissolved in D_2O. On the contrary to that, both the second derivative method and Fourier self-deconvolution do not demand any additional set of data. Second derivative is widely deemed to be objective (no need for an arbitrary parameters setup). Yet, this will not hold true for Savitzky-Golay computational algorithm [42]. Moreover, a second derivative function gives rise to negative sidebands, enabling only semi-quantitative estimations. It has been advocated elsewhere that deconvolution coupled with multiband-fitting is the best option for quantitative analysis of protein high-pressure infrared spectra [43,44].

Upon increasing pressure, the optical pathlength and the aperture of a diamond anvil cell decrease, which causes fluctuations of the intensities of all spectral bands. This problem has to be addressed in an intelligent way equally in terms of experimental technique and data processing in studies seeking to monitor accurately pressure-induced changes of spectral intensities. That different elements of protein secondary structure apparently feature different molar absorption coefficients [38,45] only complicates handling of this problem (e.g., by undermining the assumption that an integral intensity of the amide I/I′ band will remain constant throughout pressure-induced unfolding of a protein).

4. Other experimental aspects

Of all the factors affecting stability of biopolymers, high pressure is most troublesome in handling and thus the most demanding in terms of equipment. In this regard, an obvious choice for high-pressure transmission FTIR would be a diamond anvil cell, whose various constructions and applications in physics were reviewed by Jayaraman [46]. When it comes to examining vibrational bands arising from molecules of proteins, the proper type of diamond, i.e., transmitting in the spectral range of interest, should be employed. Out of the four existing types of diamond, three contain impurities decreasing their spectral transparency in the IR range (types Ia and Ib contain nitrogen and type IIb boron) [47]. The costly type IIa is pure diamond. In the wavenumber range corresponding to the amide I range types Ia, IIa and IIb bear similar spectral characteristics, yet the significantly lower transmission of Ia diamonds rules out IR studies between 1500 and 1000 cm^{-1}. All kinds of diamonds absorb strongly around 2000 cm^{-1}, which leaves sapphire windows [48] as the only option for high-pressure studies of the ligand conformation in carbon monoxide-bound proteins [35]. Sapphire, however, has the mechanical properties inferior to diamond limiting both its durability and the range of accessible pressure. Interestingly, there was a report about construction of a high-pressure anvil cell with calcium fluoride windows for pressure not exceeding 550 MPa [49].

IR protein studies often call for D_2O as the solvent. This is due to the strong absorption of H_2O in the amide I spectral range. The isotopic substitution of solvent is followed by H/D-exchange of exchangeable protein protons, which would be revealed as a red shift of amide I and amide II bands (completely deuterated bands are marked as amide I′ and amide II′, respectively). The red-shifts are rather insignificant for the former, but large for the later band. Undeniably, H/D-exchange complicates protein spectra, but, as its rate depends strictly on the accessibility of the polypeptide chain to solvent [22], it conveys important information about the protein compactness and its changes under high pressure [21]. Although IR studies of proteins dissolved in H_2O may be useful, the requirement for D_2O becomes mandatory in high-pressure IR. The reason is twofold. Firstly, when a diamond anvil cell is used, the spacer (gasket) between the two diamonds has to be thick enough (that is usually 0.05 mm, instead of only 0.006 mm for H_2O-based protein solutions) to resist enormous shear force. This, by the elongation of the pathlength, would magnify absorption of the overlapping H_2O band. Secondly, recording an IR spectrum of a protein dissolved in H_2O requires its high concentration (at times even 10%). Should a pressure treatment be applied, it is very likely to result in irreversible aggregation.

The last important issue we would like to address in this section is choice of a pressure calibrant [50]. Although a conventional high-pressure gauge has been successfully employed in protein spectroscopic studies (e.g., [51]), the construction of a diamond anvil cell virtually rules out determination of transient values of pressure other than by an internal calibrant, whose spectral response to pressure has been empirically established. Data on some pressure calibrants suitable for high-pressure FTIR studies on biopolymers is summarized in Table 3.

The ruby method, prime in physics, has numerous disadvantages in biophysics. First of all an additional spectrometer for the analysis of ruby's R_1 fluorescence line is needed. The sample with ruby chips is irradiated with a strong laser beam, which per se destabilizes proteins and nucleic acids. Finally, the ruby method is neither sensitive nor precise below 1000 MPa, namely, in the range of pressure of the biggest interest in biophysics. In spite of this, a ruby chip becomes a calibrant of choice for the high-pressure studies with a sapphire anvil cell. Apart from this single case, α-quartz and $BaSO_4$ are the most convenient and most widely used pressure calibrant for FTIR.

5. High-pressure FTIR studies on proteins

5.1. Protein aggregation

The first attempt to inquire into pressure-induced unfolding of a protein by FTIR spectroscopy was made by Wong and Heremans [39]. The study revealed that the susceptibility of the target protein – chymotrypsinogen – to pressure unfolding depends on the rate of pressure tuning itself. Chymotrypsinogen was observed to undergo irreversible unfolding at 760 MPa upon rapid increase of pressure, or at 880 MPa when pressure was tuned at a more moderate rate. Interestingly, the work pointed out the presence of highly-protected amide protons, as the complete H/D-exchange (reflected by the vanishing of the amide II) was not achieved until the secondary structure was completely unfolded.

A following FTIR study by Ismail et al. [56] was specifically focused on the aggregation of chymotrypsinogen and the differences between structural changes induced by cooling, heating, and increasing pressure. The protein's response to cooling and pressure-treatment was similar and consisted in the reversible appearance of bands between 1627 and 1622 cm^{-1} (that the pressure approached only 420 MPa explains the reversibility). Contrary to this fact, heating chymotrypsinogen above 50°C yielded two peaks at 1618 and 1685 cm^{-1} indicative of intermolecular β-sheet, and therefore, of irreversible aggregation. This conclusion supports the NMR studies suggesting a high degree of similarity between the pressure- and cold-denaturated as opposed to the heat-denaturated protein [32].

Another work devoted to the phenomenon of protein aggregation dealt with influence of pressure on the secondary structure of poly(L-lysine) [57]. Poly-(L-lysine) may serve as a model substance for research on protein aggregation, as the polypeptide can appear (depending on the chain length, pH, temperature, and pressure) in random coil, α-helix or antiparallel β-sheet conformations. It was shown that the hyperbaric treatment at 310 MPa is antagonistic to heating as it triggers off refolding of the antiparallel β-sheet into α-helix (Fig. 3). Moreover,

Table 3
The most applicable pressure calibrants for FTIR studies on proteins

Calibrant	Monitoring of pressure	Spectral position of the pressure-sensitive band/fluorescence line*	Ref.
Ruby chip	Fluorescence	694.2 nm*	[52]
α-Quartz	IR	801/695 cm^{-1}	[53]
$BaSO_4$	IR	983 cm^{-1}	[54]
Uncoupled O–D/O–H stretches	IR	3410 cm^{-1}	[55]

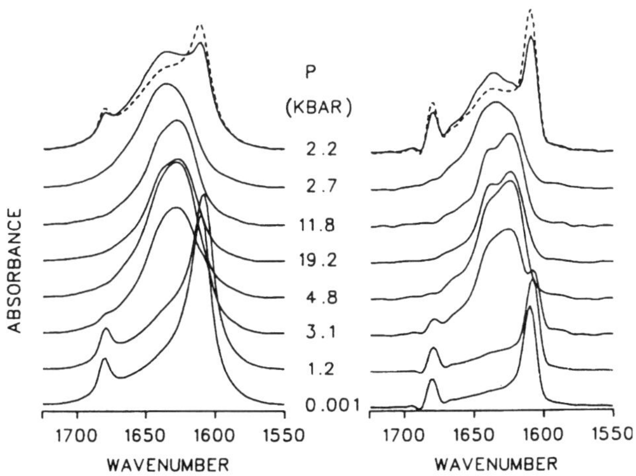

Fig. 3. Pressure-induced refolding of antiparallel β-sheet to α-helix in poly(L-lysine). The left panel shows the original spectra, whereas the right one shows the spectra after Fourier self-deconvolution; the spectra were recorded from bottom to top, according to [57].

the pressure of 900 MPa brings about the folding of unstructured poly(L-lysine) into α-helix. That the pressure-induced β-to-α refolding might simply rely on the preference for the more hydrated conformation is firmly supported by an ensuing work of the same authors [58], which proved that the hydrophobic environment of lipidic bilayers may stabilize the β-conformation of poly(L-lysine) even against pressure as high as 1.9 GPa. In unison with that, a subsequent calorimetric and infrared study by Chiou et al. [59] demonstrated that high pressure acts as an antagonist towards anesthetics with respect to the β-to-α refolding of the poly(L-lysine). Namely, high pressure rehydrates the polypeptide chain, whose hydration-shell has been destroyed by an anesthetic.

These works seem to maintain the widely accepted notion, that high pressure has a dissociative effect on non-covalent polypeptide assemblies, such as oligomeric enzymes, viruses, antigen–antibody complexes, or even aggregated proteins [60]. Paradoxically, other studies indicate that high pressure can induce conformational states which are prone to irreversible unfolding. The heat-induced aggregation of horse heart metmyoglobin was investigated by means of FTIR by Smeller et al. [61]. It was found that the high-pressure treatment of the protein produces an intermediate, which has a strong tendency to aggregate. As opposed to thermal denaturation of metmyoglobin, intermolecular β-sheet is not formed under high pressure, but only after pressure is released. After the pressure treatment, and at room temperature the aggregation process virtually does not advance, while upon modest heating (still far below the temperature, at which the thermal unfolding normally occurs) the aggregation is favored. Having made further observations that the stability of the aggregated protein against a following pressure treatment is also a function of temperature and time, which would pass between both high-pressure incubations, the authors came up with a hypothesis that the pressure-treatment of myoglobin generates folding intermediates which have the potential to unfold and to refold, but also to form a hydrogen-bonds-stabilized network. While such an intermediate is still pressure-sensitive and its immediate assemblies are prone to pressure dissociation, it slowly undergoes further conformational changes into a pressure insensitive, irreversibly aggregated protein network. These results suggested that the partial unfolding of a protein has to take place prior to aggregation, and that the mild pressurization could prevent the forming of protein aggregates under in vivo conditions. The former observation is particularly important in light of research on the molecular basis of amyloidosis and amyloidogenic proteins [62,63]. Recently it was found that native transthyretin (TTR), a thyroxine-transporting protein, whose amyloids are associated with certain neuropathological conditions, can be converted by high hydrostatic pressure into the amyloidogenic state [64].

That high pressure can push protein conformation on and off the folding pathway which leads to amyloid-formation remains an unsolved puzzle. High-pressure FTIR is apparently the best-suited method for this cause, as the formation of amyloid fibrils is usually accompanied by the formation of intermolecular antiparallel β-sheet, which is easily distinguishable from IR spectra of native proteins. Yet, to date, we are aware of only a single high-pressure FTIR work with an amyloid. The study conducted by Dubois et al. [65] examines structural changes in amyloid A induced by pressure, temperature, and basic environment. Unlike heating and alkaline solution, pressure seems to have only a limited dissociating effect on amyloid A. Whether this is related to the 'self-conflicting' effects of high pressure on amyloids and amyloidogenic proteins yet remains to be inquired.

5.2. Nucleases

A complete FTIR-monitored unfolding of a protein is expected to result in the amide I′ band becoming broad, featureless and centered around 1645 cm^{-1}. However, studies on some small proteins like insulin or bovine pancreatic trypsin inhibitor BPTI (e.g., [66]) show that even in its pressure-unfolded state the protein would not become completely structureless. It has been suggested that the pressure-unfolded state retaining more of its secondary structure that the corresponding temperature-unfolded state is a common case among monomeric proteins [67]. Another significant difference between the pressure denaturation and thermal denaturation, that can be followed by FTIR, stems from the fact that the latter is often accompanied by an irreversible formation of intermolecular β-sheet [3]. This is visible in IR spectra as the characteristic double band. Irreversible aggregation is, however, less likely to occur when a protein undergoes pressure-unfolding. An ensuing work by Takeda et al. [68] revealed that the difference between the pressure-induced and thermal unfolding could be confined to more subtle phenomena related to the dynamics of protein refolding. According to this study, despite similarity of the thermal and pressure-induced unfolding of ribonuclease A, the latter process shows a hysteresis. As pressure - release was done in a stepwise manner, the refolding process was mostly taking place under relatively high pressure (as opposed to an alternative abrupt one-step depressurization to 0.1 MPa). The results are easily understandable in light of the theoretical findings by Hummer et al. indicating that high pressure could slow down folding and unfolding kinetics [31]. The study done by Panick and Winter further evidences this claim [67].

A comparative FTIR study on the pressure unfolding of bovine pancreatic ribonuclease A and S [69] revealed that, although both the species of ribonuclease unfold reversibly under high pressure, the unfolding of ribonuclease A requires pressure about 300 MPa higher than the unfolding of proteolytically cleaved ribonuclease S. The authors also observed that the elevated pressure, even at the values not affecting the secondary structure, considerably loosens the protein compactness, which was manifested by the accelerated H/D-exchange. During the high-pressure treatment of the ribonuclease S some nonnative turns appeared, which was not the case of the A form. Finally, the pressure-induced random coil of the S form was shown not to be as featureless as that of the A form. The latter findings are particularly interesting to recall in light of a recent high-pressure FTIR study on the ribonuclease HI from *Escherichia coli* [70]. This protein was shown to be much more prone to the pressure unfolding than the proteins investigated in previous work; namely, first symptoms of unfolding could be seen at pressure as low 240 MPa. On the other hand, the minor spectral features overlapping the random-coil-assigned band did not disappear even at the highest achieved pressure of 1090 MPa. This was interpreted as the presence of some residual secondary structure within the pressure-denaturated state of the ribonuclease HI.

The peculiarity of staphylococcal nuclease consists in the lack of disulfide bonds in its structure. A high-pressure FTIR/SAXS study on staphylococcal nuclease [71] enabled drawing certain conclusions about the role played by the disulfide bonds in the protein stability to pressure. Having assumed that the pressure-induced transition of the protein is a two-state process, the authors used the FTIR pressure profiles recorded in the pressure range between 0.1 and 300 MPa for calculation of the Gibbs free energy of unfolding, 16(±4) kJ/mol, and the partial molar volume change accompanying this process, −80(±20) cm^3/mol. Very recently FTIR, in addition to fluorescence and SAXS, was used for plotting a temperature-pressure diagram of the protein [72]. FTIR proved to give results similar to those obtained by the alternative methods.

A comparative FTIR study on the influence of ten different amino acid mutations in a hydrophobic region near the C-terminal of ribonuclease A on the stability of the protein against pressure- and temperature-unfolding has been carried out recently by Torrent et al. [33] The two-state unfolding scenario was not altered by any of the mutations. The authors observed that, for each examined protein variant, the minute positive or negative changes in the stability to pressure are accompanied by similar respectively: positive or negative changes of the stability to temperature. Therefore, in terms of the *p–T* diagram, the ellipse shrinks or swells, but the original shape is retained.

5.3. Binding of ligands to proteins

IR and NMR spectroscopy were employed elsewhere [73] for a comparative examination of the pressure-enhanced H/D-exchange in two milk proteins: α-lactalbumin and β-lactoglobulin. Deconvoluted infrared spectra of both the proteins after the pressure treatment at 200 MPa showed that α-lactalbumin still retains remarkably more protected protons than β-lactoglobulin. In other words, pressure at this range can significantly reduce compactness of the latter protein only. A more detailed and strictly high-pressure FTIR study on the α-lactalbumin was published [34]. This is also the first high-pressure FTIR work with a calcium-binding protein which simultaneously takes into account spectral changes in both the calcium-ion-binding carboxyl groups' band and the amide I/I′ vibrational band. The spectral response to high pressure was compared for the calcium-binding holoform and the calcium-depleted apoform. This enabled discussion of the stabilizing role played by the calcium ion. Authors of this study found that the binding of a calcium ion remarkably stabilizes the α-lactalbumin against pressure, as it is followed approximately by a 200 MPa increase of the value of pressure at which denaturation occurs. A quantitative analysis of the band of antisymmetrical stretching vibrations of the calcium-binding carboxyl groups revealed that the pressure-induced changes in the calcium-binding loop occur in two stages. Binding of the calcium ion seemingly increases the pressure stability of the calcium-binding loop to a higher degree than the pressure stability of the secondary structure of the α-lactalbumin.

Van Riel et al. examined influence of binding of a calcium ion on the stability of Nereis Sarcoplasmatic Calcium binding Protein (NSCP) [74]. Likewise in the case of α-lactalbumin, the calcium ion contributed significantly to the overall protein stability.

Bombardier et al. employed differential scanning calorimetry along with FTIR to inquire into the stabilizing effect of various nucleotide phosphates [75]. It was demonstrated that the binding of ATP stabilizes the protein against pressure, particularly the region rich in β-sheet.

5.4. Two-dimensional FTIR of proteins

Two-dimensional (2D) correlation IR spectroscopy is a relatively new method, the principle of which lies in the correlation analysis of the external perturbation-induced variations of spectral intensities. The methodology and the possible range of applications of 2D-COS FTIR have been discussed elsewhere [76]. To put it briefly, 2D-FTIR enables tracing of minor changes in complex spectra by highlighting and separating these which are in-phase and those out-of-phase. The coupling of high pressure, as an external perturbation, with 2D-FTIR and applying it to protein research could prove to be an insightful method for the investigation of proteins. With few papers published so far in field, whether this would happen remains to be seen.

Authors of the first high-pressure 2D-FTIR works on a protein showed that the method can tell the H/D-exchange-related spectral changes from the spectral changes reflecting the protein conformation being altered under the elevating pressure [77,78].

Our previous high-pressure 2D-FTIR study [79] showed that the method could distinguish between temperature- and pressure-enhanced H/D-exchange in α-lactalbumin, when the conventional FTIR does not tell the difference immediately. The results also showed that the different components of the secondary structure undergo H/D-exchange at various rates.

5.5. Pressure-assisted cold denaturation

While cold denaturation of protein is now considered to be a very general phenomenon, its detection and researching encounters numerous obstacles, mainly because of the midpoint of a cold-transition expected for the most proteins far below the freezing point of water. Many works attempt to overcome this problem by using co-solvents or co-denaturants; however, pressure-assisted cold treatment makes an alternative. This approach takes advantage of the fact that under the hydrostatic pressure of 220 MPa the freezing point of heavy water drops to −22°C. This method has been successfully applied in the NMR studies of the cold-denaturation of proteins [32].

FTIR was employed for the examination of pressure-assisted cold-unfolding of a protein for the first time by Meersman et al. [80]. The study demonstrated that cold-denaturated state of metmyoglobin, just like pressure-denaturated state but opposite to heat-denaturated state, features some persistent secondary structure. The authors have also observed that the cold-induced spectral changes in the protein correlate with the changes in water spectral bands, hence it was inferred that the cold-denaturation of metmyoglobin could be a consequence of the dramatically changing solvating properties of water. In a study by Dzwolak et al. [81] the application of a similar approach revealed that α-lactalbumin undergoes cold-unfolding only in the calcium-depleted form, while cooling of the native calcium-bound α-lactalbumin resulted in formation of non-native α-helices. A comparative analysis of amide I/I′ band and a band assigned to the calcium-ion-binding carboxyl groups pointed out that a cold-treatment at −15°C affects the secondary structure and the calcium-binding loop in a more reversible manner than does heat treatment at 80°C.

5.6. Miscellaneous studies

An elegant high-pressure FTIR study proved that IR spectroscopy not only allows determination of pressure-induced changes in the secondary structure of a protein, but it also allows simultaneous investigation of the enzyme kinetics under high pressure [82]. In this case, enzyme – alkaline phosphatase – while converting *p*-nitrophenol phosphate into *p*-nitrophenol, was examined in the range of pressure from 0.1 up to 1400 MPa. The researchers took advantage of the fact that, during the enzymatic reaction the IR bands assigned to the substrate of this reaction would gradually vanish. Thus, calculation of the rate at which the 887.6 cm^{-1} band assigned to *p*-nitrophenol phosphate disappeared (upon the pressure-tuned enzymatic reaction), gave the possibility to determine kinetic parameters of this enzymatic reaction. Meanwhile, the amide I′ band was providing insight into how the enzyme structure itself reacts to the elevating pressure. It was concluded that the pressure increasing up to 720 MPa steadily accelerates the enzymatic reaction, and that protein structure within this range of pressure remains stable. The calculated activation volumes of this reaction were: −15.6 cm^3/mol (under pressure below 300 MPa) and −4.4 cm^3/mol (above 300 MPa). Under the pressure as high as 830 MPa first changes in the enzyme's secondary structure appeared (visible in the amide I′ band) and were immediately followed by inactivation of the enzyme, as the rate of the catalyzed hydrolysis of the substrate dropped to zero. The pressure-induced structural changes in the alkaline phosphatase were recognized as the refolding of the native α-helices into β-sheet.

Results obtained through IR and fluorescence spectroscopy were juxtaposed in an investigation of high-pressure stability of bovine pancreas trypsin by Ruan et al. [16]. The reversible pressure denaturation indicated by fluorescence was denied by the corresponding spectra. The authors stressed, however, that this discrepancy could come from relatively high, thus prone to irreversible aggregation, protein concentration used in the IR study.

Another interesting work is an example of high-pressure FTIR being applied in order to gain insight into energetic characteristics of ^{2}H-bonds placed inside the helical segments of myoglobin, rather than to investigate a pressure effect per se [83]. As we have stated before, elevating pressure leads to 'shrinking' of interatomic distances. For a protein molecule under ambient pressure, these distances are largely determined by the presence, geometry and energy of hydrogen bonding. By studying influence of high pressure on the spectral position of amide I′ band (which is determined by hydrogen bonding in polypeptide chain) under given conditions of solution, the relative strength of the ^{2}H-bonds could be estimated. The study concluded that ^{2}H-bonds strength is a rather complex function of pD and salt concentration.

A study on CO-bound cytochrome P-450$_{cam}$ took advantage of the C=O stretching in carbon monoxide-bound hemoproteins being a sensitive gauge for changes in the local environment of the heme pocket [35]. As the stretching frequency of the captured ligand is slightly above 1900 cm^{-1}, a sapphire cell had to replace a diamond anvil cell. Hydrostatic pressure and osmotic pressure (controlled by the concentration of glycerol) were shown to have contrary effects on intensities and wavenumbers corresponding to CO-stretching sub-bands. This was interpreted as the water molecules placed inside of the heme pocket and within the protein hydration shell playing an important role in topological arrangements of the Fe–CO group.

6. Conclusions

The intention of this brief paper was only to outline recent developments in the very exciting field of high-pressure infrared spectroscopy of proteins. FTIR spectroscopy is relatively free from many technical or methodological obstacles, which frequently rule out application of other techniques for examination of protein structure under high pressure. In our opinion, high-pressure FTIR spectroscopy enters a period of robust development with special respect to the research on protein aggregation and amyloid formation. The extensively reviewed studies on the nuclease family represent another important aspect of the molecular biophysics, namely protein folding thermodynamics related to single structural features. Other examples of high-pressure FTIR studies were mentioned in a somewhat superficial and random manner, yet they were intended to indicate that by an intelligent coupling to other research tools, high-pressure FTIR can launch insightful inquires into various fields of biochemistry and biophysics. How far-reaching the applications of high-pressure FTIR could be is exemplified in works by Wong et al. where the method was used for diagnostics of normal and malignant tissues [84,85]. Finally, the method not only remains a valuable tool for fundamental research on protein folding, but can also greatly contribute to the discovery of new biotechnological and food-processing solutions.

References

[1] K. Suzuki, Studies on the kinetics of protein denaturation under high pressure, Rev. Phys. Chem. Japan 29 (1960) 91–98.

[2] S.A. Hawley, Reversible pressure-temperature denaturation of chymotrypsinogen, Biochemistry 10 (1971) 2436–2442.

[3] K. Heremans, L. Smeller, Protein structure and dynamics at high pressure, Biochim. Biophys. Acta 1386 (1998) 353–370.

[4] M. Gross, R. Jaenicke, Proteins under pressure, Eur. J. Biochem. 221 (1994) 617–630.

[5] J.L. Silva, G. Weber, Pressure stability of proteins, Annu. Rev. Phys. Chem. 44 (1993) 89–113.

[6] G. Weber, Pressure dissociation of smaller oligomers: dimers and tetramers, in: R. Winter, J. Jonas (Eds.), High Pressure Chemistry, Biochemistry and Material Science, Vol. 401, NATO ASI Series. Math. Phys. Sci., Kluwer Academic Publishers, Dordrecht, 1993, pp. 471–509.

[7] P.T.T. Wong, D.J. Siminovitch, H.H. Mantsch, Structure and properties of model membranes: new knowledge from high-pressure vibrational spectroscopy, Biochim. Biophys. Acta 947 (1988) 139–171.

[8] P.W. Bridgman, The coagulation of albumin by pressure, J. Biol. Chem. 19 (1914) 511–512.

[9] B.R. Thakur, P.E. Nelson, High-pressure processing and preservation of food, Food Rev. Int. 14 (1998) 427–447.

[10] J. Yuste, M. Capellas, R. Pla, D.Y.C. Fung, M. Mor-Mur, High pressure processing for food safety and preservation: a review, J. Rap. Methods Autom. Microbiol. 9 (2001) 1–10.

[11] C.E. Kundrot, F.M. Richards, Crystal structure of hen egg-white lysozyme at a hydrostatic pressure of 1000 atmospheres, J. Mol. Biol. 193 (1987) 157–170.

[12] H. Ikeuchi, S. Kunugi, K. Oda, Activity and stability of a neutral protease from Vibrio sp. (vimelysin) in a pressure-temperature gradient, Eur. J. Biochem. 267 (2000) 979–983.

[13] M. Kato, T. Fujisawa, High-pressure solution X-ray scattering of protein using a hydrostatic cell with diamond windows, J. Synchrotron Radiat. 5 (1998) 1282–1286.

[14] T. Fujisawa, M. Kato, Y. Inoko, Structural characterization of lactate dehydrogenase dissociation under high pressure studied by synchrotron high-pressure small-angle X-ray scattering, Biochemistry 38 (1999) 6411–6418.

[15] N. Tanaka, K. Nakajima, S. Kunugi, The pressure-induced structural change of bovine α-lactalbumin as studied by a fluorescence hydrophobic probe, Int. J. Peptide Protein Res. 48 (1996) 259–264.

[16] K. Ruan, R. Lange, F. Meersman, K. Heremans, C. Balny, Fluorescence and FTIR study of the pressure-induced denaturation of bovine pancreas trypsin, Eur. J. Biochem. 265 (1999) 79–85.

[17] R. Lange, J. Frank, J.L. Saldana, C. Balny, Fourth derivative UV-spectroscopy of proteins under high pressure, Eur. Biophys. J. 24 (1996) 277–283.

[18] K. Heremans, F. Meersman, H. Pfeiffer, P. Rubens, L. Smeller, Pressure effects on biopolymer structure and dynamics, High Press. Res. 19 (2000) 623–630.

[19] V.V. Mozhaev, K. Heremans, J. Frank, P. Masson, C. Balny, High pressure effects on protein structure and function, Proteins Struct. Funct. Genet. 24 (1996) 81–91.

[20] J.V. Carter, G.D. Knox, A. Rosenberg, Pressure effects on folded proteins in solution, J. Biol. Chem. 253 (1978) 1947–1953.

[21] D.P. Kharakoz, Protein compressibility, dynamics, pressure, Biophys. J. 79 (2000) 511–525.

[22] E. Goormaghtigh, V. Cabiaux, J.M. Ruysschaert, Determination of soluble and membrane protein structure by Fourier transform infrared spectroscopy. II Experimental aspects, side chain structure, and H/D exchange, in: H.J. Hilderson, G.B. Ralston (Eds.), Subcellular Biochemistry, Vol. 23, Plenum, New York, 1994, pp. 363–403.

[23] E. Goormaghtigh, V. Cabiaux, J.M. Ruysschaert, Determination of soluble and membrane protein structure by Fourier transform infrared spectroscopy. III Secondary structures, in: H.J. Hilderson, G.B. Ralston (Eds.), Subcellular Biochemistry, Vol. 23, Plenum, New York, 1994, pp. 405–450.

[24] J. Bandekar, Amide modes and protein conformation, Biochim. Biophys. Acta 1120 (1992) 123–143.

[25] M. Jackson, H.H. Mantsch, The use and misuse of FTIR spectroscopy in the determination of protein structure, CRC Rev. Biochem. Mol. Biol. 30 (1995) 95–120.

[26] J.T. Pelton, L.R. McLean, Spectroscopic methods for analysis of protein secondary structure, Anal. Biochem. 277 (2000) 167–176.

[27] M.C. Lin, P. Eid, P.T.T. Wong, R.B. Macgregor, High pressure fourier transform infrared spectroscopy of poly(dA)poly(dT), poly(dA) and poly(dT), Biophys. Chem. 76 (1999) 87–94.

[28] P. Rubens, K. Heremans, Pressure-temperature gelatinization phase diagram of starch: an in situ Fourier transform infrared study, Biopolymers 54 (2000) 524–530.

[29] L. Smeller, K. Heremans, Some thermodynamic and kinetic consequences of the phase diagram of protein denaturation, in: K. Heremans (Ed.), High Pressure Research in the Biosciences and Biotechnology, Leuven University Press, Leuven, 1997, pp. 55–58.

[30] W. Kauzmann, Protein stabilization – thermodynamics of unfolding, Nature 325 (1987) 763–764.

[31] G. Hummer, S. Garde, A.E. Garcia, M.E. Paulaitis, L.R. Pratt, The pressure dependence of hydrophobic interactions is consistent with the observed pressure denaturation of proteins, Proc. Natl. Acad. Sci. USA 95 (1998) 1552–1555.

[32] J. Zhang, X. Peng, A. Jonas, J. Jonas, NMR study of the cold, heat, and pressure unfolding of ribonuclease A, Biochemistry 34 (1995) 8631–8641.

[33] J. Torrent, P. Rubens, M. Ribo, K. Heremans, M. Vilanova, Pressure versus temperature unfolding of ribonuclease A: an FTIR spectroscopic characterization of 10 variants at the carboxy-terminal site, Protein Sci. 10 (2001) 725–734.

[34] W. Dzwolak, M. Kato, A. Shimizu, Y. Taniguchi, Fourier-transform infrared spectroscopy study of the pressure-induced changes in the structure of the bovine α-lactalbumin: the stabilizing role of the calcium ion, Biochim. Biophys. Acta 1433 (1999) 45–55.

[35] C. Jung, O. Ristau, H. Schulze, S.G. Sligar, The CO stretching mode infrared spectrum of substrate-free cytochrome P-450cam-CO. The effect of solvent conditions, temperature, and pressure, Eur. J. Biochem. 235 (1996) 660–669.

[36] S.J. Prestrelski, D.M. Byler, M.P. Thompson, Effect of metal ion binding on the secondary structure of bovine α-lactalbumin as examined by infrared spectroscopy, Biochemistry 30 (1991) 8797–8804.

[37] M.D. Byler, H. Susi, Examination of the secondary structure of proteins by deconvolved FTIR spectra, Biopolymers 25 (1986) 469–487.

[38] M. Jackson, P.I. Haris, D. Chapman, Conformational transitions in poly(L-lysine): studies using Fourier transform infrared spectroscopy, Biochim. Biophys. Acta 998 (1989) 75–79.

[39] P.T.T. Wong, K. Heremans, Pressure effects onprotein secondary structure and hydrogen deuterium exchange in chymotrypsinogen: a Fourier transform infrared spectroscopic study, Biochim. Biophys. Acta 956 (1988) 1–9.

[40] K. Heremans, K. Goossens, L. Smeller, Pressure-tuning spectroscopy of proteins: Fourier transform infrared studies in the diamond anvil cell, in: J.L. Markley, D.B. Northtrop, C.A. Royer (Eds.), High-Pressure Effects in Molecular Biophysics and Enzymology, Oxford University Press, New York, 1996, pp. 44–61.

[41] Y. Taniguchi, N. Takeda, High-pressure FTIR studies of the secondary structure of proteins, in: J.L. Markley, D.B. Northtrop, C.A. Royer (Eds.), High-Pressure Effects in Molecular Biophysics and Enzymology, Oxford University Press, New York, 1996, pp. 87–95.

[42] A. Savitzky, J.E. Golay, Smoothing and Differentiation of data by simplified least squares procedures, Anal. Chem. 36 (1964) 1627–1639.

[43] L. Smeller, K. Heremans, In situ observation of the pressure-induced changes in proteins with infrared spectroscopy in the diamond anvil cell, in: K. Heremans (Ed.), High Pressure Research in the Biosciences and Biotechnology, Leuven University Press, Leuven, 1997, pp. 131–134.

[44] L. Smeller, K. Goossens, K. Heremans, Determination of the secondary structure of proteins at high pressure, Vib. Spectrosc. 8 (1995) 199–203.

[45] H.H.J. Jongh, E. Goormaghtigh, J.-M. Ruysschaert, The different molar absorptivities of the secondary structure types in the amide I region: an attenuated total reflection infrared study on globular proteins, Anal. Biochem. 242 (1996) 95–103.

[46] A. Jayaraman, Diamond anvil cell and high-pressure physical investigations, Rev. Mod. Phys. 55 (1983) 65–108.

[47] P.T.T. Wong, D.D. Klug, Reevaluation of type I diamonds for infrared and Raman spectroscopy in high-pressure diamond anvil cells, Appl. Spectrosc. 37 (1983) 284–286.

[48] K. Furuno, A. Onodera, S. Kume, Sapphire-anvil cell for high pressure research, Jpn. J. Appl. Phys. 25 (1986) L646–L647.

[49] P.T.T. Wong, D.J. Moffat, Calcium fluoride anvil cell for high-pressure infrared spectroscopy, Appl. Spectrosc. 38 (1984) 599–601.

[50] L. Smeller, On the precision of pressure determination by internal calibrants, Physica B 265 (1999) 268–271.

[51] H. Frauenfelder, N.A. Alberding, A. Ansari, D. Braunstein, B.R. Cowen, M.K. Hong, I.E.T. Iben, J.B. Johnson, S. Luck, M.C. Marden, J.R. Mourant, P. Ormos, L. Reinisch,

R. Scholl, A. Schulte, E. Shyamsunder, L.B. Sorensen, P.J. Steinbach, A. Xie, R.D. Young, K.T. Yue, Proteins and pressure, J. Phys. Chem. 94 (1990) 1024–1037.

[52] G.J. Piermarini, S. Block, J.D. Barnett, R.A. Forman, Calibration of the pressure dependence of the R_1 ruby fluorescence line to 195 kbar, J. Appl. Phys. 46 (1975) 2774–2780.

[53] P.T. T Wong, D.J. Moffatt, F.L. Baudais, Crystalline quartz as an internal pressure calibrant for high-pressure infrared spectroscopy, Appl. Spectrosc. 39 (1985) 733–735.

[54] P.T.T. Wong, D.J. Moffat, A new internal pressure calibrant for high-pressure infrared spectroscopy of aqueous systems, Appl. Spectrosc. 43 (1989) 1279–1281.

[55] P.T.T. Wong, D.J. Moffat, The uncoupled O–H or O–D stretch in water as an internal pressure gauge for high-pressure infrared spectroscopy of aqueous systems, Appl. Spectrosc. 41 (1987) 1070–1072.

[56] A.A. Ismail, H.H. Mantsch, P.T.T. Wong, Aggregation of chymotrypsinogen: portrait by infrared spectroscopy, Biochim. Biophys. Acta 1121 (1992) 183–188.

[57] D. Carrier, H.H. Mantsch, P.T.T. Wong, Pressure-induced reversible changes in secondary structure of poly(L-Lysine): an IR spectroscopic study, Biopolymers 29 (1990) 837–844.

[58] D. Carrier, H.H. Mantsch, P.T.T. Wong, Protective effect of lipidic surfaces against pressure-induced conformational changes of poly(L-lysine), Biochemistry 29 (1990) 254–258.

[59] J.S. Chiou, T. Tatara, S. Sawamura, Y. Kaminoh, H. Kamaya, A. Shibata, I. Ueda, The α-helix to β-sheet transition in poly(L-lysine): effects of anesthetics and high pressure, Biochim. Biophys. Acta 1119 (1992) 211–217.

[60] B.M. Gorovits, P.M. Horowitz, High hydrostatic pressure can reverse aggregation of protein folding intermediates and facilitate acquisition of native structure, Biochemistry 37 (1998) 6132–6135.

[61] L. Smeller, P. Rubens, K. Heremans, Pressure effect on the temperature-induced unfolding and tendency to aggregate of myoglobin, Biochemistry 38 (1999) 3816–3820.

[62] J.W. Kelly, The alternative conformations of amyloidogenic proteins and multi-step assembly pathways, Curr. Opin. Struct. Biol. 8 (1998) 101–106.

[63] A.L. Fink, Protein aggregation: folding aggregates, inclusion bodies and amyloid, Fold. Des. 3 (1998) R9–R23.

[64] A.D. Ferrao-Gonzales, S.O. Souto, J.L. Silva, D. Foguel, The preaggregated state of an amyloidogenic protein: hydrostatic pressure converts native transthyretin into the amyloidogenic state, Proc. Natl. Acad. Sci. USA 97 (2000) 6445–6450.

[65] J. Dubois, A.A. Ismail, S.L. Chan, Z. Alikhan, Fourier transform infrared spectroscopic investigation of temperature- and pressure-induced disaggregation of amyloid A, Scand. J. Immunol. 49 (1999) 376–380.

[66] N. Takeda, K. Nakano, M. Kato, Y. Taniguchi, Pressure-induced structural rearrangements of bovine pancreatic trypsin inhibitor studied by FTIR spectroscopy, Biospectroscopy 4 (1998) 209–216.

[67] G. Panick, R. Winter, Pressure-induced unfolding/refolding of ribonuclease A: static and kinetic Fourier transform infrared spectroscopy study, Biochemistry 39 (2000) 1862–1869.

[68] N. Takeda, M. Kato, Y. Taniguchi, Pressure- and thermally-induced reversible changes in the secondary structure of ribonuclease A studied by FT-IR spectroscopy, Biochemistry 34 (1995) 5980–5987.

[69] N. Takeda, M. Kato, Y. Taniguchi, Pressure-induced secondary structure changes of ribonuclease A and ribonuclease S studied by FTIR spectroscopy, Biospectroscopy 1 (1995) 207–216.

[70] K. Yamasaki, Y. Taniguchi, N. Takeda, K. Nakano, T. Yamasaki, S. Kanaya, M. Oobatake, Pressure-denaturated state of *Escherichia coli* ribonuclease HI as monitored by Fourier transform infrared and NMR spectroscopy, Biochemistry 37 (1998) 18001–18009.

[71] G. Panick, R. Malessa, R. Winter, G. Rapp, K.J. Frye, C.A. Royer, Structural characterization of the pressure-denatured state and unfolding/refolding kinetics of staphylococcal nuclease by small-angle X-ray scattering and Fourier-transform infrared spectroscopy, J. Mol. Biol. 275 (1998) 389–402.

[72] G. Panick, G.J.A. Vidugiris, R. Malessa, G. Rapp, R. Winter, C.A. Royer, Exploring the temperature-pressure diagram of staphylococcal nuclease, Biochemistry 38 (1999) 4157–4164.

[73] N. Tanaka, S. Kunugi, Effect of pressure on the deuterium exchange reaction of α-lactalbumin and β-lactoglobulin, Int. J. Biol. Macromol. 1 (1996) 33–39.

[74] A. van Riel, P. Rubens, K. Heremans, Y. Engelborghs, Pressure and temperature stability of holo- and apo-form of Nereis sarcoplasmic calcium-binding protein studied by Fourier-transform infrared spectroscopy, in: K. Heremans (Ed.), High Pressure Research in the Biosciences and Biotechnology, Leuven University Press, Leuven, 1997, pp. 63–66.

[75] H. Bombardier, P. Wong, C. Gicquaud, Effects of nucleotides on the denaturation of F actin a differential scanning calorimetry and FTIR spectroscopy study, Biochem. Biophys. Res. Commun. 236 (1997) 798–803.

[76] I. Noda, Two-dimensional IR spectroscopy: theory and applications, Appl. Spectrosc. 44 (1990) 550–561.

[77] L. Smeller, K. Heremans, 2D FT-IR spectroscopy analysis of the pressure-induced changes in proteins, Vib. Spectrosc. 19 (1999) 375–378.

[78] L. Smeller, P. Rubens, J. Frank, J. Fidy, K. Heremans, Two-dimensional Fourier-transform infrared correlation spectroscopy study of the high-pressure tuning of proteins, Vib. Spectrosc. 22 (2000) 119–125.

[79] W. Dzwolak, M. Kato, A. Shimizu, Y. Taniguchi, Comparative two-dimensional Fourier transform infrared correlation spectroscopic study on the spontaneous, pressure-, and temperature-enhanced H/D exchange in alpha-lactalbumin, Appl. Spectrosc. 54 (2000) 963–967.

[80] F. Meersman, L. Smeller, K. Heremans, Pressure-assisted cold unfolding of proteins and its effects on the conformational stability compared to pressure and heat unfolding, High Press. Res. 19 (2000) 653–658.

[81] W. Dzwolak, M. Kato, A. Shimizu, Y. Taniguchi, FTIR study on heat-induced and pressure-assisted cold-induced changes in structure of bovine alpha-lactalbumin: Stabilizing role of calcium ion, Biopolymers 62 (2001) 29–39.

[82] P.T.T. Wong, D.W. Armstrong, FT-IR spectroscopic kinetic analysis of alkaline phosphatase under hyperbaric manipulation, Biochim. Biophys. Acta 1159 (1992) 237–242.

[83] V. Le Tilly, O. Sire, B. Alpert, P.T.T. Wong, An infrared study of 2H-bond variation in myoglobin revealed by high pressure, Eur. J. Biochem. 205 (1992) 1061–1065.

[84] P.T.T. Wong, S. Lacelle, H.M. Yazdi, Normal and malignant human colonic tissues investigated by pressure-tuning FT-IR spectroscopy, Appl. Spectrosc. 47 (1993) 1830–1836.

[85] P.T.T. Wong, R.K. Wong, M.F.K. Fung, Pressure-tuning FT-IR study of human cervical tissues, Appl. Spectrosc. 47 (1993) 1058–1063.

ELSEVIER

Biochimica et Biophysica Acta 1595 (2002) 145–159

BIOCHIMICA ET BIOPHYSICA ACTA
BBA
www.bba-direct.com

Review

High-resolution nuclear magnetic resonance studies of proteins

Jiri Jonas *

Beckman Institute for Advanced Science and Technology, University of Illinois at Urbana, Urbana, IL 61801, USA

Received 17 September 2001; received in revised form 14 November 2001; accepted 15 November 2001

Abstract

The combination of advanced high-resolution nuclear magnetic resonance (NMR) techniques with high-pressure capability represents a powerful experimental tool in studies of protein folding. This review is organized as follows: after a general introduction of high-pressure, high-resolution NMR spectroscopy of proteins, the experimental part deals with instrumentation. The main section of the review is devoted to NMR studies of reversible pressure unfolding of proteins with special emphasis on pressure-assisted cold denaturation and the detection of folding intermediates. Recent studies investigating local perturbations in proteins and the experiments following the effects of point mutations on pressure stability of proteins are also discussed. Ribonuclease A, lysozyme, ubiquitin, apomyoglobin, α-lactalbumin and troponin C were the model proteins investigated.

Keywords: High resolution; Nuclear magnetic resonance; High pressure; Protein; Unfolding; Folding intermediate

1. Introduction

Protein folding, the relationship between the amino acid sequence and the structure and dynamic properties of the native conformation of proteins, represents one of the central problems of biochemistry and biotechnology. The stability of proteins has been extensively studied by temperature and chemical perturbations. Most studies dealing with protein denaturation have been carried out at atmospheric pressure with various physico-chemical perturbations, such as temperature, pH, or denaturants, as experimental variables. Compared to varying temperature, which produces simultaneous changes in both volume and thermal energy the use of pressure to study protein solutions perturbs the environment of the protein in a continuous controlled way by changing only intermolecular distances [1]. In addition, by taking advantage of the phase behavior of water high pressure [2] can substantially lower the freezing point of an aqueous protein solution. Therefore, by applying high pressure one can investigate in detail not only pressure-denatured proteins, but also cold-denatured proteins in aqueous solution.

It is well known that pressure affects chemical equilibria and reaction rates. The following standard equations define the reaction volume, ΔV, and the activation volume, $\Delta V^{\ddagger}$:

$$\Delta V = -\left(\frac{RT\partial \ln K}{\partial P}\right)_t ;\quad \Delta V^{\ddagger} = -\left(\frac{Rt\partial \ln k}{\partial P}\right)_t \qquad (1)$$

where K is the equilibrium constant, and k is the reaction rate. The knowledge of ΔV and $\Delta V^{\ddagger}$ allows one to draw conclusions about the nature of the reaction and its mechanism. In addition, studies of

* University of Illinois at Urbana, Champaign, 7709 Santa Margherita Way, Naples, FL 34109, USA.
Fax: +1-217-244-0987.
E-mail address: j-jonas@uiuc.edu (J. Jonas).

0167-4838/02/$ – see front matter
PII: S0167-4838(01)00341-7

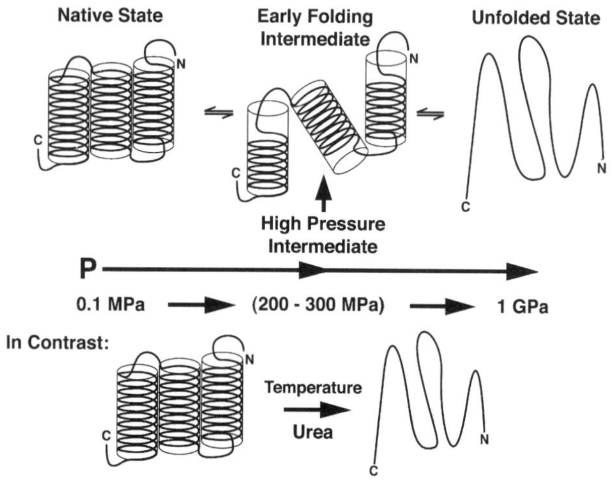

Fig. 1. Schematic diagram compares the use of pressure and temperature in protein denaturation studies.

simple liquids [3] have indicated that quite often, volume effects determine the mechanism of a specific dynamic process, whereas temperature only changes the frequency of the motions without actually affecting the mechanism.

It should be noted that the range of pressures used to investigate biochemical systems is from 0.1 MPa (1 bar) to 1 GPa (10 kbar); such pressures only change intermolecular distances and affect conformations but do not change covalent bond distances nor bond angles. In fact, pressures in excess of 3 GPa (30 kbar) are required to change the electronic structure of a molecule [4]; such pressures are not generally useful for biochemical studies since they modify covalent bonds. There are several excellent general reviews [5–7] of the wide spectrum of biochemical problems investigated by high-pressure techniques.

Increasing attention has recently been focused on denatured and partially folded states of proteins, since determination of their structure and stability may provide novel information on the mechanisms of protein folding. The native conformations of hundreds of proteins are known in great detail from structural determinations by X-ray crystallography and, more recently, nuclear magnetic resonance (NMR) spectroscopy. However, a detailed knowledge of the conformations of denatured and partially folded states is lacking, and represents a serious shortcoming in current studies of protein stability and protein folding pathways [8,9].

From the results obtained so far in our laboratory [10,11] and elsewhere it is clear that pressure denaturation is more easily controlled and represents a less drastic perturbation of protein structure than thermal or chemical denaturation (Fig. 1). Indeed, the main conclusions from our work can be summarized as follows: pressure and pressure-assisted cold denaturated states exhibit significant residual secondary structure; cold-denatured state is different from heat denatured state; pressure allows stabilization of intermediate states; pressure-assisted cold denatured states have residual secondary structure resembling early folding intermediates; mutations in the core of proteins affect differently their pressure stability than their thermal stability or resistance to chemical denaturation. The combination of advanced high-resolution NMR techniques with high-pressure capability represents a powerful new experimental tool in studies of protein folding. The main advantages of using high resolution, high-pressure NMR are as follows:

- One-dimensional (1D) NMR allows determination of structural and dynamic changes in different regions of the protein
- 2D NMR allows distance specific information to be obtained between amino acid residues in different regions of the protein
- Hydrogen exchange experiments using 2D NMR provide information about solvent (H_2O/D_2O) accessibility to specific amino aid residues in the protein

Recent advances in superconducting magnets make it possible to attain a high homogeneity of the magnetic field over the sample volume so that even without sample spinning one can achieve high resolution. It is now possible [12] to achieve an NMR linewidth of 1.5 Hz for a sample diameter of 8 mm at the proton frequency of 500 MHz at 1 GPa (10 kbar). The experimental capability of recording high-resolution NMR spectra on dilute spin systems opened new directions for high-pressure NMR spectroscopy on biochemical systems. The study of Arc repressor discussed in detail in the original publications [13,14] and a review [5] represents an excellent example of the detailed structural information obtained by high-pressure, high-resolution 2D NMR experiments.

This review is organized as follows. The experimental section deals with the instrumentation needed for high-pressure, high-resolution NMR experiments. The hydrogen exchange experiments at high pressure are also discussed as these experiments can be performed without the need for any specialized high-pressure NMR equipment.

The main section of this review is devoted to various NMR studies of reversible pressure unfolding of proteins with special emphasis on the pressure-assisted cold denaturation of proteins and the detection of folding intermediates. Recent studies investigating local perturbations in proteins and the experiments following the effects of point mutations on pressure stability of proteins are also discussed.

2. Experimental

2.1. *High-pressure NMR probe*

The autoclave design historically represents the first type of high-pressure NMR probe [15–17]. This approach places both the NMR sample and the electronic RF circuit inside a high-strength, non-magnetic vessel, with either a gas or a liquid medium used to transmit pressure externally to the glass sample cell. Schematic drawing of the high-pressure NMR probe and sample cell is given in Fig. 2. The pressure range for the autoclave is limited only by the strength of the metallic vessel, and is not limited in any way by the strength of the glass sample cell. This fact is advantageous in that it allows one to use a thin-walled glass sample cell and still achieve high pressures. The use of a high-strength metal vessel also contributes some degree of safety to this approach.

Pressure limits for the autoclave design depend both on the material used to construct the vessel and on the vessel dimensions. The highest reported pressure range for an autoclave NMR probe used in biochemical applications is 900 MPa. High spectral resolution ($<3.0\times10^{-9}$) in a high-pressure vessel is achieved through the use of thin-walled glass sample

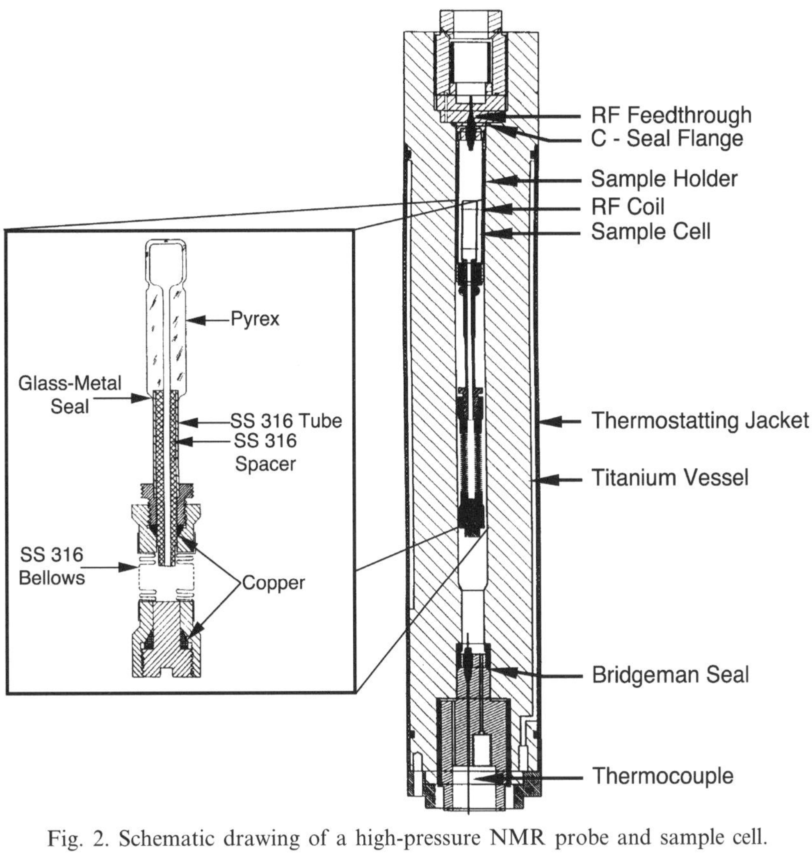

Fig. 2. Schematic drawing of a high-pressure NMR probe and sample cell.

cells with precise tolerances. These thin-walled glass sample cells allow high filling factors and large sample volumes (5–10 mm diameter sample sizes). In general, the autoclave approach is well-established with an increasing number of biochemical studies being reported.

2.2. High-pressure cell

A second method for high-pressure NMR is the high-pressure cell approach [18,19], where the sample is contained in a high strength sample cell (usually quartz, sapphire, or fused silica), with pressure applied directly (internally) to the sample. The pressure range in this case is limited by the strength of the glass walls. The obvious advantage of the high-pressure cell is that it can be used on any commercial NMR spectrometer without the need to construct special NMR probes. This advantage allows one to achieve high spectral resolution and allows the user ready access to advanced multinuclear NMR techniques on high-field superconducting magnets. The proteins such as BPTI [20] and lysozyme [21] have recently been studied up to 200 MPa on a 750 MHz instrument using the high-pressure cell approach.

Yet despite its advantages in implementation, the high-pressure cell faces serious limitations in strength. For example, thick-walled quartz pressure tubes have a working pressure limit of $\sim$200 MPa and recently LaSalle et al. [22] reported experiments at pressures up to 330 MPa. These modest pressure ranges are adequate for many types of studies, but they do not allow studies of complete protein unfolding process in many systems. In addition, the thick sample cell walls necessary for high-pressure applications significantly limit the sensitivity. This, in turn, would necessitate that one should use a higher magnetic field strength or a higher sample concentration to compensate for the loss in signal-to-noise ratio. Still, even with these limitations, the ease of implementation for high-pressure cells has contributed to a growing number of biochemical studies (e.g., [23–27]) performed with this technique.

2.3. Hydrogen exchange kinetics at high pressure

A widely used approach for studying protein structure and dynamics [28] is kinetic measurements of reactions involving the labile amide backbone protons. The amide–water hydrogen exchange rates are very sensitive to the local environment of the protons. Thus, fluctuations in the local conformations of proteins can be probed by measuring the hydrogen exchange rates of the individual backbone amide protons. The effect of the local structure on the hydrogen exchange rate constants is typically quantified in terms of the protection factor (P), which is the ratio of the observed rate constant of each amide proton in the protein to a calculated rate constant for the amide proton in an analogous unfolded structure. The calculated unfolded rate constants are determined from the rate constant for the model compound, polyalanine, by factoring in side-chain effects on the acid-, base-, and water-catalyzed exchange rate constants and by using the activation energies of exchange to correct the rate constants for temperature differences. At temperatures of −15°C and lower, the exchange rate of amide protons, even in a random coil polypeptide, is sufficiently slow ($k < 0.1$/min) that exchange rates for many of them are readily measured without any special equipment, e.g., stopped-flow apparatus. As a result the cold denaturation regime is well suited for kinetic studies. Furthermore, the exchange of protons for deuterons is conveniently measured by proton NMR, since the signal intensity is directly related to the number of protons remaining non-exchanged.

NMR work in our group has generally relied on taking spectra while the sample is under high pressure, which limited experiments to specially modified NMR spectrometers and probes [12] capable of handling the extreme pressures needed. Observing proton exchange as a result of denaturation is not subject to this restriction [29,11]. Any amide protons visible to NMR after brief exposure to D_2O are sufficiently resistant to exchange in a native protein (typically protection factor $P > 1000$) that extensive exchange can occur only when the protein is denatured. After exposing a protein to cold denaturing conditions, warming and depressurizing back to ambient conditions renatures the protein and quenches the exchange of any protons rendered exchangeable by denaturation. Consequently, the samples need not actually be under pressure while spectra are taken, and thus the spectra needed for kinetic work can be obtained on any NMR spectrometer. Additionally,

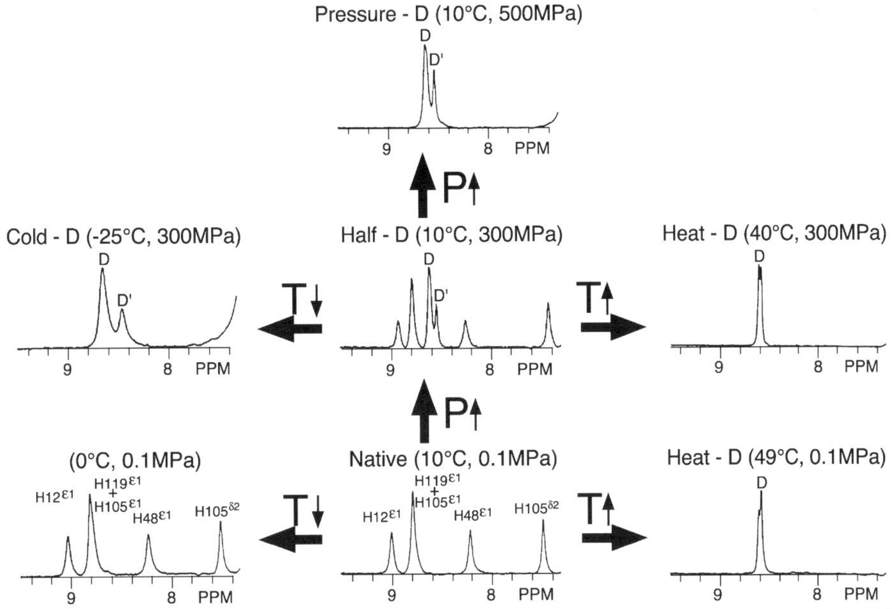

Fig. 3. Comparison of the histidine region of the 1D NMR spectra among native, pressure-denatured, cold-denatured and heat-denatured states of RNase A at pH* 2.

since pressure and temperature are the only variables used to permit or quench exchange, there is no need to change the sample's chemical environment (e.g., sample pH or denaturant concentration).

3. Results and discussion

3.1. Cold denaturation of proteins

Cold denaturation of proteins has generally been difficult to study since the denaturation transition temperature is below 0°C for most protein solutions. As a result, experimenters have had to either perturb the protein or its solvent; ways of doing this have included addition of denaturing agents [30] such as Gdn-HCl, use of cryo-solvents [31] site-directed mutagenesis [32], and emulsions of protein solution in oil [33]. By taking advantage of the anomalous phase behavior of water [2], using moderately high hydrostatic pressures (200–300 MPa) permits one to work with liquid aqueous solutions at temperatures as low as −20°C, which is below the cold denaturation temperature of many proteins.

Our group has chosen ribonuclease A (RNase A) as a system for the study of cold denaturation for a variety of reasons. RNase A is a stable, well-characterized protein. It is large enough to be of interest as a physiologically relevant polypeptide, but is small enough to have a well understood structure, including a nearly complete assignment of the solution proton NMR spectrum. Furthermore, our work [34,35] with RNase A has shown that its pressure assisted cold denaturation is completely reversible.

3.1.1. High-resolution 1D NMR study of cold denaturation of RNase A

At 300 MPa the RNase A solution can be cooled to −25°C without freezing. It should be noted here that the protein solution in the high salt buffer freezes well below the freezing point given in the phase diagram of water [2] at 300 MPa (−15°C). The phase behavior of water allows us to obtain the completely cold denatured state of RNase A with the assistance of pressure.

To better understand the states of denatured RNase A under different denaturation conditions, we have compared the NMR spectra of cold-, heat-, and pressure-denatured RNase A in the aromatic and aliphatic regions [34,35]. Fig. 3 compares only the histidine regions of the NMR spectra of native, pressure-denatured, cold-denatured, and heat-denatured RNase A. At 10°C, 0.1 MPa RNase A is in the native state. As pressure is increased the

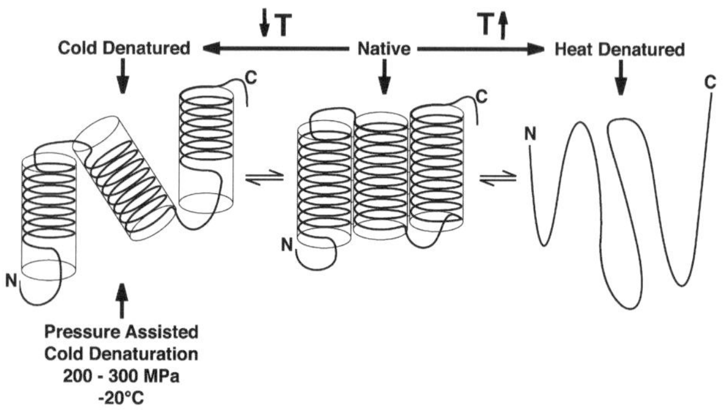

Fig. 4. Schematic comparison of heat-denatured and cold-denatured states of proteins.

protein begins to unfold. When the pressure reaches 300 MPa the protein is half-denatured, as evidenced by the appearance of denatured peaks D and D′. This half-denatured state is a pivotal state from which one can obtain the pressure-denatured state by increasing pressure to 500 MPa, the cold-denatured state by decreasing temperature to −25°C, or the heat-denatured state by increasing temperature to 40°C. The heat-denatured spectra at 300 MPa and 0.1 MPa lack resonance peak D′ while both pressure- and cold-denatured spectra show a D′ peak in addition to the D peak. By analogy with the thermal refolding intermediate reported by Blum et al. [36], the D′ peak is attributed to the His-12 residue that is located in the *N*-terminal α-helix region. Clearly, the structures of pressure- and cold-denatured states are different from the heat-denatured state in that both the pressure- and cold-denatured states have some kind of residual helical structure around the His-12 residue, while such structures are absent in the heat-denatured states at both high and atmospheric pressures. Comparison of the spectra of cold- and pressure-denatured spectrum reveals that the D′ peak is closer to the D peak in the pressure-denatured spectrum than in the cold-denatured spectrum, indicating that the cold-denatured state is more structured than the pressure-denatured state in the N-terminal region. As discussed in detail in the original studies [34,35] the percent denaturation of each histidine residue can be calculated and a pressure-temperature phase diagram of RNase can be constructed.

3.1.2. Hydrogen exchange study of pressure and cold denatured states of RNase A

The appearance of a new histidine resonance D′ in the cold denatured and pressure denatured RNase A spectra, compared to the absence of this resonance in the heat denatured state, indicated that the pressure denatured state and cold denatured state (see Fig. 4) contain partially folded structures, which may be similar to that of the early folding intermediate found in the temperature-jump experiment reported by Blum et al. [36].

Since 1D NMR spectra did not allow one to arrive at ambiguous conclusions about the presence or absence of secondary structure in the pressure and/or cold denaturated states, hydrogen-exchange experiments [34,35] were carried out to test for the presence of partially folded structures in the pressure-denatured and cold-denatured states.

Stable hydrogen-bonded structures protecting the backbone amide hydrogens from solvent exchange were observed in the pressure-denatured state. These experimental results suggested that pressure-denatured RNase A displays the characteristics of a molten globule. Fig. 5 compares the protection factors (P) as a function of residue position for pressure-assisted cold-denatured RNase A (A), and pressure-denatured RNase A (B). In contrast, no protection factor higher than $P \approx 2$ were detected for heat-denatured state.

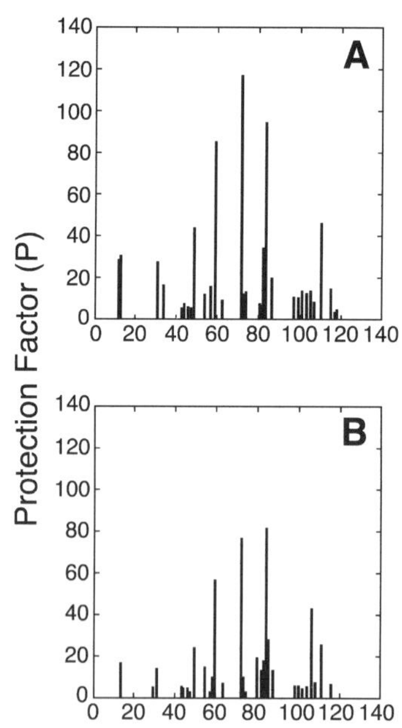

Fig. 5. Comparison of the protection factor (P) obtained as a function of residue number for (A) cold-denatured RNase A; (B) pressure-denatured RNase A.

3.2. Detection of folding intermediates

The finding that the residual structural features of pressure-assisted cold denaturation of RNase A resemble the structures of early folding intermediate provided motivation for high-pressure studies of additional proteins. The well-characterized solution structure and the presence of folding or absence of folding intermediates were the criterions for selection of the model proteins such as lysozyme [29]. Furthermore, in the case of ubiquitin, Gladwin and Evans [37] found no evidence for significant protection from exchange at early stages of folding; therefore, to test the hypothesis of this parallel, we have also investigated hydrogen exchange in the pressure-assisted, cold-denatured state of ubiquitin [38].

As an additional proof of the concept that high-pressure NMR can be used to identify and characterize early folding intermediates we studied two additional proteins, apomyoglobin [39] and α-lactalbumin (K. Thirumoorthy, J. Jonas, unpublished results), for which folding intermediates have been detected by other methods. Apomyglobin was chosen as model protein as the folding intermediate has been reported, and this protein does not contain disulfide bridges. In contrast bovine alpha-lactalbumin has four disulfide bonds and its folding intermediate has also been reported.

3.2.1. Lysozyme

Fig. 6A shows the protection factors in cold-denatured lysozyme ($T = -13°C$, $p = 375$ MPa) as a function of residue number and their correspondence with secondary structural elements in the native state [29].

From the hydrogen exchange rate data, it is clear that many regions of lysozyme are markedly protected from exchange, with P values exceeding 10, in the cold-denatured state. The range of P factors runs from 1.19 for F3 to 71.0 for R114. It should be made clear that the pressure-assisted, cold-denatured state is not a 'pure' cold-denatured state, because the pressure affects the protein directly. Nevertheless, any effect that pressure has is clearly very different from that induced by high temperature of chemical denaturants such as urea.

By using pulsed-labeling hydrogen exchange studies and confining exchange to the dead time of their instrumentation (~3.5 ms), Gladwin and Evans [37] observed relatively early stages in the folding of lysozyme. Slowing of exchange measured during this time can be quantified by comparing the measured exchange rate to that for a random coil, as in other

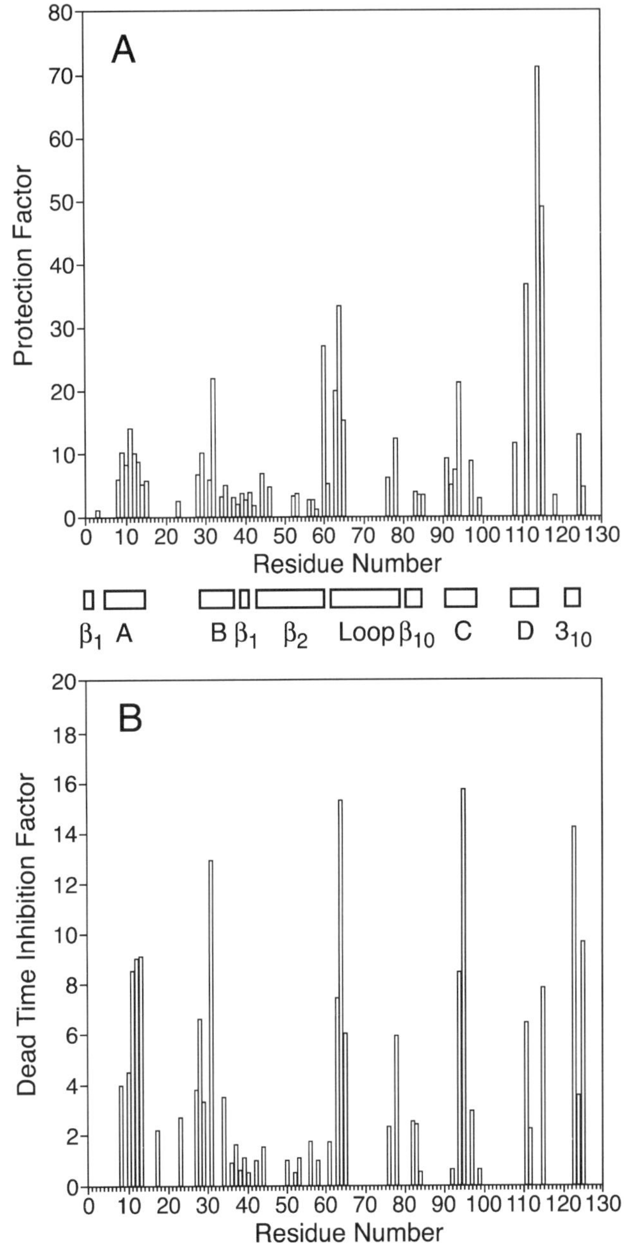

Fig. 6. (A) Protection factors for the pressure-assisted cold-denatured state ($p = 375$ MPa; $T = -13°C$) of lysozyme. Secondary structural elements are indicated below: A–D denote the four α-helices; β_1 and β denote the two- and three-stranded β sheet regions, respectively. (B) Dead-time inhibition factors obtained for the first 3.5 ms of lysozyme folding.

hydrogen exchange studies. The resulting quantity, although similar to a protection factor, is determined from a rate that changes considerably as the protein refolds; as a result, it is best referred to as a 'dead-time inhibition factor' (I_D) rather than a true protection factor. Nevertheless, I_D values provide information similar to that provided by P values in equilibrium denatured states. During the first 3.5 min of folding lysozyme shows moderate degrees of protection ($I_D > 5$), not only in the α-helices and the C-terminal 3^{10} helix, but also in the loop region from residues 60–65, as well as residue 78. We have thus observed a strong parallel between a stable, denatured form of lysozyme and the transient species observed during folding. Fig. 6B compares the inhibition of hydrogen exchange observed during the first 3.5 min of folding [37] to the protection factors [29] obtained for pressure-assisted, cold-denatured lysozyme, as shown in Fig. 6A.

In the case of lysozyme and RNase A, the patterns of protection against hydrogen exchange were similar to those observed in early (refolding time < 10 ms) folding intermediates for these proteins, leading to the idea that the cold-denatured state is structurally similar to such intermediates. To help test this idea, ubiquitin, which has folding kinetics that are markedly different from those of either RNase A or lysozyme, was investigated [38].

3.2.2. Ubiquitin

Cold-denatured ubiquitin ($p = 225$ MPa, $T = -16°C$) shows little deviation from a random coil in its hydrogen exchange kinetics, with no P values above 5 and most below 2. These values are typical of highly denatured proteins such as the urea-denatured state of lysozyme.

It is important to point out that in ubiquitin, the dead-time inhibition study [37] showed no evidence for protection from exchange, with the largest dead-time inhibition factors being ~ 2. Moreover, there was no correlation of even these modestly elevated factors within significant elements of secondary structure, as there were in lysozyme.

3.2.3. Apomyoglobin

Apomyoglobin is known to form both kinetic and equilibrium intermediates, and therefore is likely to form a partially folded structure under high pressure as well. The pH denaturation of the native state leads to the I_1 equilibrium intermediate [40], which retains structure in the A, G, and H helices although the AGH cores is not as well packed as in the native state [41]. The structure of the I_1 state in 10% ethanol has been solved by NMR [42]. The addition of trichloroacetic acid to the acid intermediate or acid denatured states results in the formation of a new intermediate, I_2, in which the B helix as well as the A, G, and H helices are stable [43,44]. A 'generalized charge effect' causes the unfolding of I_1 to U as the pH is further reduced.

High-pressure 1D ^{1}H NMR and ^{1}H–^{1}H NOESY (nuclear Oberhauser effect spectroscopy) NMR techniques [39] were used to show that apomyoglobin pressure denatures to form a pressure denatured state that still retains some structure in the AGH core. However, due to the low sensitivity of MR, higher concentrations of apomyoglobin (0.5 mM) were necessary which resulted in the aggregation of myoglobin during pressure denaturation. The use of 10 mM TCA prevented precipitation and allowed high-pressure denaturation to be completely reversible.

Pressure denaturation was monitored by measuring the decrease in peak area for the four histidine peaks H24, H36, H48, and H113 at 7.9, 8.07, 8.15, and 8.2 ppm, respectively. Therefore while the four histidines are dispersed throughout the protein, all of them report on the structure outside the interface of the A-G-H helices. Peak assignments were made by comparison to literature values.

No changes in the spectra were observed between 0.1 and 80 MPa. As pressure was further increased, the native peaks disappeared by 200 MPa and the denatured peaks at 8.5 and 8.6 ppm increased in intensity. The denatured peaks lost fine structure between 200 and 400 MPa. The total peak area between 7.8 and 8.8 ppm remained constant at all pressures. The pressure denaturation was completely reversible and the denaturation and renaturation spectra matched for each pressure. Also, the peak area measured during denaturation and renaturation were the same within error.

High-pressure 2D ^{1}H–^{1}H NOESY on wild-type apomyoglobin was investigated to further describe the structure of the apomyoglobin high pressure intermediate, and to determine if even higher pressures

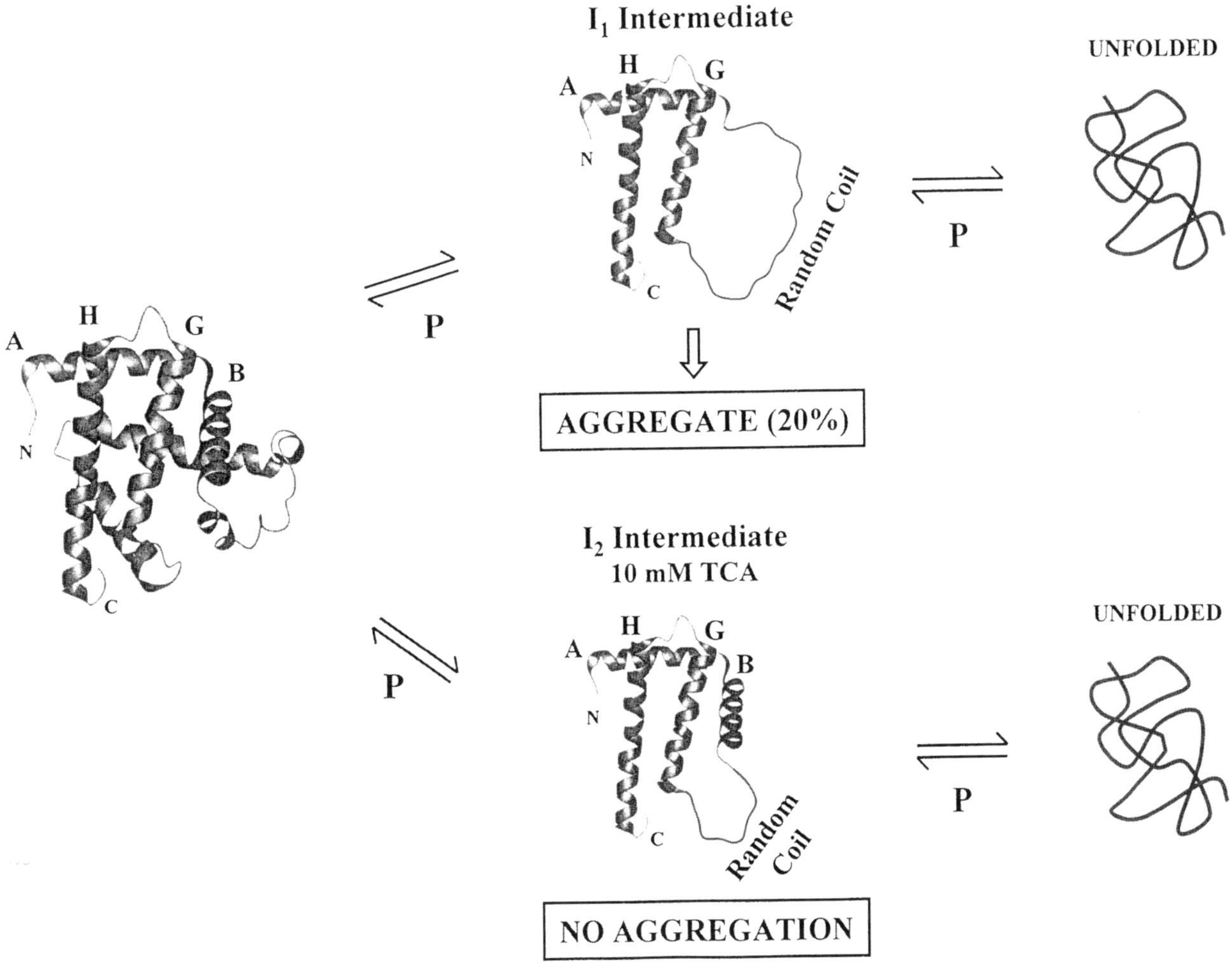

Fig. 7. Schematic diagram for pressure unfolding of apomyoglobin.

could denature the residual structure. Volume measurement of peaks from the 0.1 and 80 MPa 1D spectra show that the peak intensities of the 0.1 and 80 MPa spectra are similar. NOESY spectra at 0.1 and 80 MPa were also very similar and show no evidence for pressure denaturation at these low pressures.

The protein is in the high-pressure intermediate state Ip at 200 MPa. The high-pressure NOESY shows that the Ip state contains a cluster of aromatic residues as well as residual structure in the ABGH core. The aromatic cluster includes a strong interaction between F33 and F43. Cross-peaks were also observed between H36 and H106, and F33 and L40.

Analysis of the 1D and 2D NMR data for apomyoglobin allowed construction of a schematic diagram for pressure unfolding of apomyoglobin. Fig. 7 shows the pressure unfolding process both for apomyoglobin in the absence and presence of trichloroacetic acid. The high-pressure intermediate, Ip, is close in structure to I_2 intermediate. It is interesting to note that the existence of partial structure in the B helix prevents aggregation of apomyoglobin during depressurization. One can conclude that the I_1 intermediate with AGH helices is susceptible to aggregation.

3.2.4. α-Lactalbumin

α-Lactalbumin is structurally similar to lysozyme but has very different functions and folding behavior, α-lactalbumin is very well known to form a molten globule by acid and temperature denaturation.

Only several key results of this high-pressure experiment are presented. Both 1D and 2D NMR techniques were used to investigate the pressure unfolding of α-lactalbumin. A plateau exists in the pressure

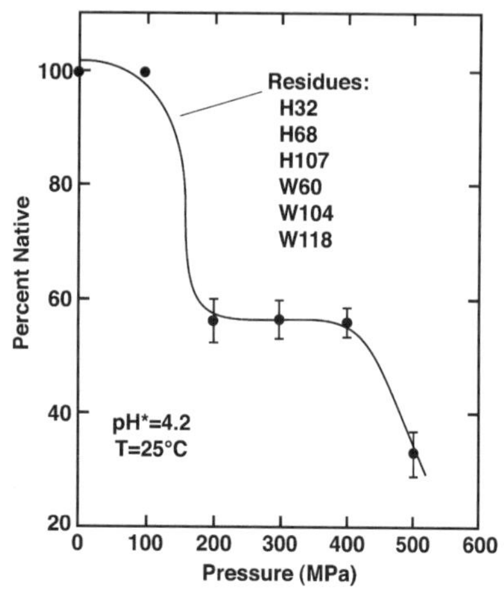

Fig. 8. Denaturation of α-lactalbumin with pressure. The experimental points represent averages from peak integration of W104, W118, W60, H32, H68 and H107 proton signals.

dependence of denaturation of α-lactalbumin in the pressure range 200–400 MPa as shown in Fig. 8.

α-Lactalbumin is found to be completely denatured by 500 MPa, and no further changes occur at higher pressures, up to 800 MPa. At 300 and 400 MPa pressures the α-lactalbumin spectrum resembles that of the molten globule spectrum from pH denaturation except for a slight change in chemical shift values of those peaks that are affected by pH. High-pressure 2D NOESY of α-lactalbumin was investigated to further describe the structure of α-lactalbumin at high pressures and to determine if any other structural details could be found. About 40 cross peaks belonging to 14 aromatic residues could be identified from the NOESY spectrum.

The experimental results show that α-lactalbumin is completely unfolded under pressure of 500 MPa at pH 4.2 and temperature of 25°C. We were also able to see the presence of α-lactalbumin in a molten globule state beyond 200 MPa pressure and up to 400 MPa pressure. The denaturation curves showed that the peak areas for residues remain constant up to 100 MPa which may be due to reorganization of the protein interior to a state of less volume by reducing the void spaces within the molecule, without changing the inter-residue interactions. It appeared that the molten globule was a stable intermediate which is plastic enough to withstand the change in pressure for about 200 MPa.

The results for pressure denaturation of apomyoglobin and α-lactalbumin provided additional evidence of the concept that high-pressure NMR experiments can be used to detect folding intermediates in proteins. Table 1 gives the summary of the recent experimental results. These results are interesting per se, but they become more important when related to modern protein folding studies. Structures that persist in a protein upon denaturation, even relatively mild denaturation, are likely to be high stable. If no appreciable barriers exist to the rapid formation of such structure, they should be among the first to appear in the refolding of a protein. Consequently, mildly denatured states may serve as useful models for species present early in refolding, with the advantage that such species could be studied for hours or even days, rather than fractions of a second. There is already considerable evidence that the equilibrium collapsed unfolded states that occur when the protein begins to fold.

3.3. Local perturbations in proteins

Recent studies [45,46] of avian troponin C (TnC) illustrate how the high-pressure, high-resolution NMR experiments allow one to investigate local perturbations in proteins. TnC is one of three members of the troponin complex. Along with TnI and TnT, TnC functions by regulating muscle contraction in response to calcium. Calcium binds to TnC and causes conformational changes. Through protein–

Table 1
High-pressure detection of folding intermediates

Protein	Disulfide bonds	Folding intermediate	Pressure intermediate	Ref.
Lysozyme	Yes	Yes	Yes	[11,29]
Ribonuclease	Yes	Yes	Yes	[34,35]
α-Lactalbumin	Yes	Yes	Yes	Unpublished results
Apomyoglobin	No	Yes	Yes	[39]
Ubiquitin	No	No	No	[38]

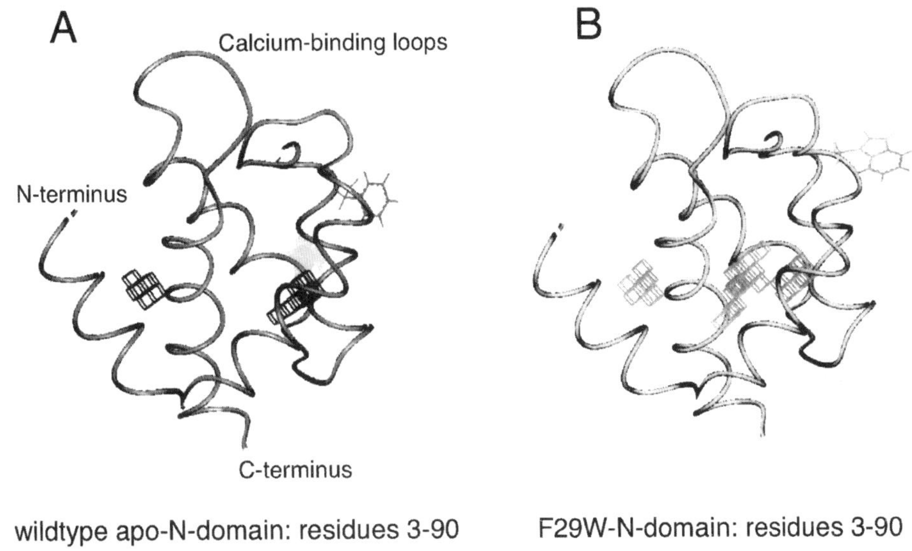

Fig. 9. (A) Simulated structure of the N-domain of TnC wild-type protein with indicated cavities. (B) Simulated structure of the F29W mutant form with cavities.

protein interactions, the conformational signal is propagated via TnI and TnT to the tropomyosin–actin filament affecting the latter's interaction with myosin heads of the thick filament and the contraction/relaxation cycle.

Because of its physiological role in muscle contraction, the conformational changes that accompany metal-binding to TnC make this protein an attractive system for study. To simplify and facilitate these studies, the isolated N- and C-domains have been prepared by recombinant methods [47] and their Ca^{2+}-binding and spectral properties shown to be identical (N-domain) or similar (C-domain) to those in intact TNC. In addition, since chicken skeletal TnC contains no naturally occurring tryptophans nor tyrosines, spectral probes have been incorporated into the protein by mutagenesis. These mutants, namely F29W in the N-domain and F105W or F154W in the C-domain, have been previously characterized by fluorescence and other spectral methods [47–50] and allowed biding events in each domain to be monitored separately.

3.3.1. N-domain of TnC

We have recently studied the N-domain of chicken TnC and its F29W mutant using high-pressure NMR approach [45] in order to address several questions: How does a conservative (F → W) point mutation of a residue partially exposed at the surface of the N-domain affect its stability, structure, and dynamics under pressure? Does pressure affect different regions of the N-domain differently? The dynamic simulation of the N-domain of TnC for the wild-type and mutant was the starting point of the study.

Despite the strong similarity between the wild-type and the mutant simulated structures, there is a significant difference in cavities predicted within their structures (see Fig. 9A,B). Using a probe of 1.25 Å in radius, which was employed in previous cavity-searching studies, we found that the wild-type contains two cavities while the F29W mutant contains three cavities. The two cavities in the wild-type are situated on opposite ends of the protein, capping the hydrophobic core of the N-domain. These two cavities are similar in dimensions, and each has a computer-generated solvent accessible volumes of approximately 18 $Å^3$. In comparison to the wild-type, two of the cavities in the mutant were virtually identical in size, shape, and location. In addition, the mutant displays a smaller third cavity between the original two cavities, with a solvent accessible volume of approximately 15 $Å^3$. This new cavity is located directly within the hydrophobic core of the protein and is bounded by the aromatic rings of Phe-75 and Phe-78. This result suggests that the core of the F29W has altered packing of hydrophobic side chains and perturbed hydrophobic interactions that may lead to a less stable protein form. In

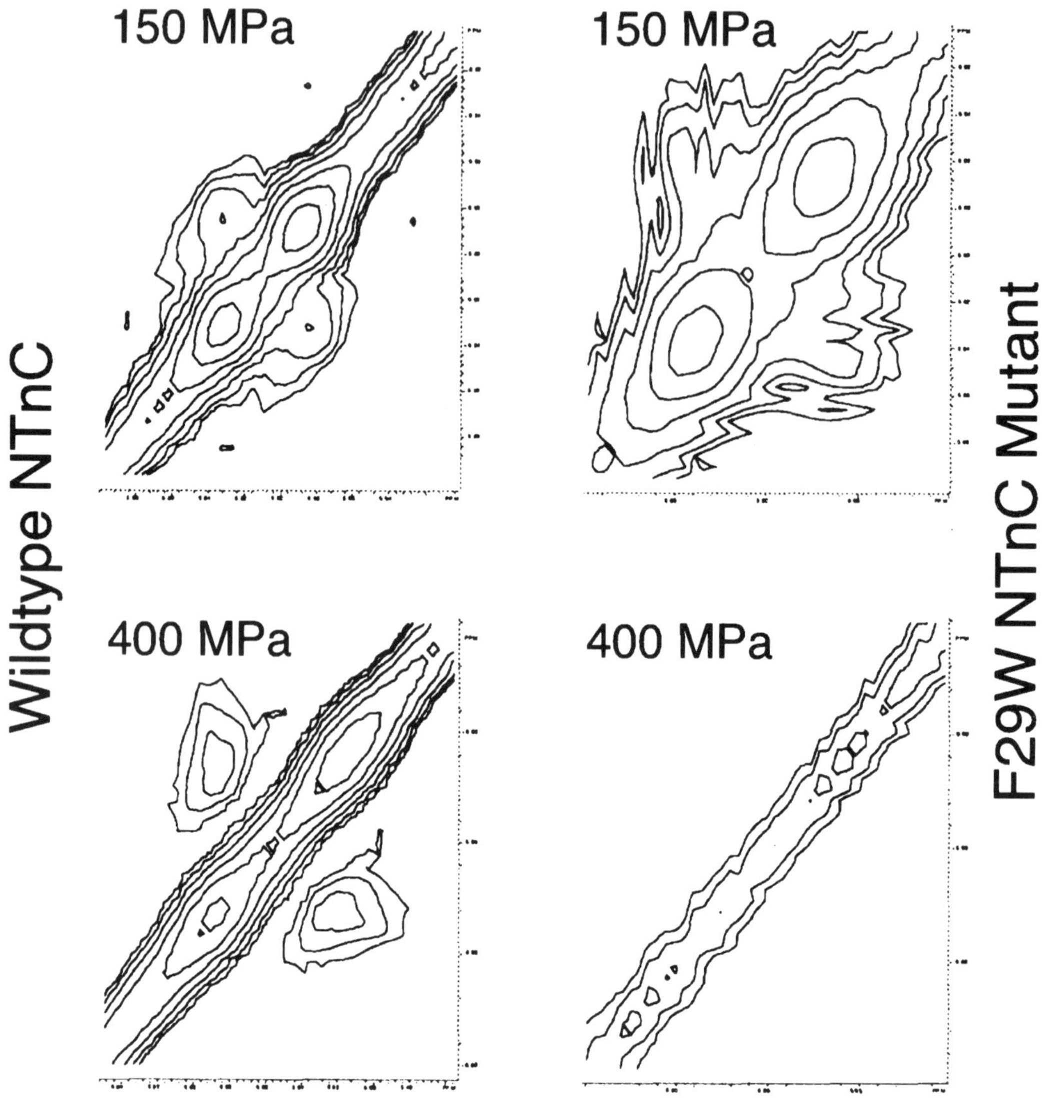

Fig. 10. Selected region of the 2D ^1H–^1H NOESY spectra at 500 MHz of the wild-type and F29W mutant TnC N-terminal domains at 150 and 400 MPa.

order to test this prediction the effects of pressure and temperature on the hydrophobic core of the wild-type and the F29W mutant of the N-domain of TnC were investigated by NMR.

The results of both the 1D and 2D NMR experiments confirmed that the presence of additional cavities in the F29W mutant will destabilize the protein. At this point, only our selected results of the 2D NMR experiments are mentioned.

The close proximity of the D36 and D74 residues of the F29W N-domain at 0.1 MPa was confirmed by a NOESY cross-peak between the D36α and D74α peaks (see Fig. 10). With increasing pressure, the intensity of the cross-peak for F29W mutant weakened significantly and essentially disappeared by 400 MPa. This indicated that the β-sheet structure was disrupted at high pressures. In contrast to the mutant the cross-peak maintains its intensity to 400 MPa for the wild-type protein. These results show that the wild-type protein is more resistant to pressure than the F29W mutant as predicted by the molecular dynamics simulations.

3.3.2. C-Domain of TnC containing Ca^{2+} or Mg^{2+}

Unique to the metal-binding sites in the carboxy-terminal domain (C-domain) is the ability to reversibly bind Mg^{2+}, although this occurs with a lower affinity than Ca^{2+}. The ability of the same sites to differentially bind two metal ions provided an opportunity to compare the changes in structure and stability within this metal–protein complex. Circular dichroism, fluorescence, 1D and 2D ^{1}H NMR were utilized to analyze the structures of isolated Ca^{2+} and Mg^{2+}-bound C-domain fragments [46]. The Mg^{2+}-bound sample was more susceptible to perturbation by the effects of high temperature, high pressure, and chemical denaturants than the Ca^{2+}-bound sample, as monitored by various spectroscopic techniques. In particular, the high-pressure 1D and 2D ^{1}H NMR experiments revealed that the substitution of Mg^{2+} in the Ca^{2+}-binding loops results in a decrease in the free energy of unfolding (ΔG_u) and an increase in the volume change (ΔV_u) associated with local pressure-induced perturbations of the metal-binding loops of the domain. This suggested that the metal-binding sites are perturbed upon coordination of the smaller Mg^{2+} ion and develop larger cavities in its presence. Our experiments demonstrated the utility of high pressure as a tool for probing the packing efficiency within specific regions of a protein.

In addition to the local structural and dynamic differences due to Mg^{2+}-binding, major changes occur in the stability of CTnC when Mg^{2+} replaced Ca^{2+}. Thermal denaturation experiments show that CTnC is markedly destabilized: the T_m shifts from 96°C for the Ca^{2+}-bound protein to 81°C for the Mg^{2+}-bound form. Similarly, urea effects on the Mg^{2+}-bound form are much more pronounced than on the Ca^{2+}-bound CTnC. The mid-point of denaturation for the Mg^{2+}-bound protein is at 5 M urea and the calculated ΔG_u in the absence of urea at 25°C and atmospheric pressure, is 2.4±0.2 kcal/mol. In contrast, the Ca^{2+}-bound form is impervious to the highest urea concentration that we were able to attain at 25°C.

From the changes in area of individual proton resonance peaks with increasing pressure, ΔV_u values can be calculated for the corresponding amino acid or group of amino acids. The significance of these parameters is clearly different from thermodynamic parameters obtained for the overall denaturation of a protein. Nevertheless, these values represent local reversible structural perturbations that can be analyzed as 2-state equilibria between a structured and an unstructured (or locally unfolded) state. As shown in Table 2, the ΔV_u values for the Mg^{2+}-bound CTnC are almost 2-fold greater than Ca^{2+}-bound protein, suggesting that cavities have been created upon binding of Mg^{2+}. Alternatively, the changes in ΔV_u could also reflect changes in the electrostriction of charged and polar groups that become exposed to solvent upon unfolding or effects of the transfer of hydrophobic groups from the protein interior to water.

3.3.3. Mutagenesis effects on protein stability under pressure

The earlier section dealing with N-domain of TnC illustrated how a single point mutation (F→W) affects its stability under pressure [45]. A better understanding of the effects of mutations on protein stability under pressure can be obtained by investigating several mutants of the same protein.

Staphylococcal nuclease does not contain sulfhydryl residues nor disulfide linkages; therefore, it is less stable under pressure and other denaturing con-

Table 2
Experimental values (ΔV_u) and the pressures at 50% denaturation (p_{50}) for selected residues in the Ca^{2+}- and Mg^{2+}-saturated wild-type CTnC samples at 25°C

	Spectral region A[a] G111NH, I1113NH, G147NH		Spectral region B[a] D114NH, I119NH, F154NH	
	ΔV_u (ml/mol)	p_{50} (MPa)[b]	ΔV_u (ml/mol)	p_{50} (MPa)[b]
Ca^{2+}-Saturated	−12±1	600	−14±2	540
Mg^{2+}-Saturated	−21±5	120	−24±1	160

[a]Region A represents positions 6 (G111 and G147) and position 8 (I113) in the 12-residue EF-hand motif. Region B represents position 8 (I149), positions 9 (D114 and D150), as well as F154 which is located at the N-terminus of helix H (immediately follows site IV in the C-domain).

[b]The pressures MPa at 50% denaturation (P_{50}) were determined by relative peak areas (see [46]).

ditions. The native state of this protein exhibits structural heterogeneity linked to the *cis-trans* isomerization of prolyl peptide bonds. Royer et al. [51] used time-resolved and steady-state fluorescence, as well as high-resolution NMR, to investigate the effects of amino acid substitutions on the high-pressure denaturation of staphylococcal nuclease. These authors studied the pressure denaturation of two wild-type forms of this protein, WT (Foggi strain nuclease) and H124L (V8 strain nuclease), and several mutants of the V8 nuclease. The NMR experiments were performed on 3 to 5 mM solutions of both wild-type forms (WT and H124L) and the G79S and P47G+P117G mutants of the V8 nuclease, at 37°C and pH 5.1. The Hisε-1 signals of His-8, His-46, and His-121 were recorded as a function of pressures up to 500 MPa. Calculated ΔV values ranged from −47 to −65 ml/mol, and the midpoints of the pressure denaturation curves were between 92 and 310 MPa, depending on the stability of the protein. In fact, the P47G+P117G mutant did not denature even at pressures as high as 500 MPa. A comparison between the time-resolved fluorescence and NMR results indicated a correlation between the ΔV values reflecting the pressure stability of a particular mutant and its dynamic properties.

In the discussion, the authors pointed out that the physical basis for the volume changes accompanying denaturation is very complex. They noted that possible contributions to the decrease in volume upon denaturation (ΔV from −10 to −100 ml/mol for diverse proteins) may result, in part, from the existence of free volume in the folded proteins, from electrostriction of water caused by the exposure of buried ion-pairs, and from decreased in volume on exposure of hydrophobic residues to water. The authors concluded that a better characterization of the various contributions to ΔV will greatly improve our understanding of the relationship between structure, packing dynamics and stability of proteins. Without doubt, this study showed the great promise of combining site-directed mutagenesis with high-pressure NMR studies of protein unfolding.

Recently, we have studied the pressure unfolding of several mutants of apomyoglobin [39] including F123W, S108L, S108K and A103K mutants. The pressure denaturation was monitored by measuring the NMR peak due to histidine 24 (for details see the original study [39]) and the results are shown in Fig. 11.

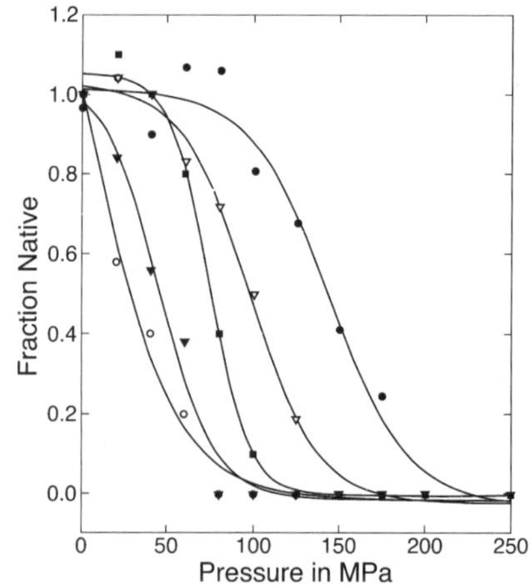

Fig. 11. 1D proton NMR denaturation curves for H24$^{\varepsilon}$H of apomyoglobin mutants. A130K (○); S108K (▼); S108L (△); F123W (■); wild-type (●).

The mutants show that high-pressure denaturation is sensitive to changes in the packing of the protein interior. The S108L and F123W mutations, which both increase the size of a well packed residue, limit the ability of apomyoglobin to repack and accommodate low pressures as well as cause native apomyoglobin to denature at pressures at least 25 MPa lower than required for wild-type. The charge burial mutants, A130K and S108K, showed an even more pronounced effect by eliminating any resistance to high pressure and denaturing at half of the pressures required for wild-type. Thus charge burial mutations are the most destabilizing to pressure stability.

The measured reaction volumes [39] for the native to pressure intermediate transition showed that the mutations do perturb significantly the pressure denaturation of apomyoglobin.

Acknowledgements

This work was supported in part by the National Science Foundation under Grant CHE 98 17558.

References

[1] G. Weber, H.G. Drickamer, Q. Rev. Biophys. 16 (1983) 89–112.

[2] J. Jonas, Science 216 (1982) 1179–1184.
[3] I. Artaki, J. Jonas, J. Chem. Phys. 82 (1985) 3360–3370.
[4] H.G. Drickamer, C.W. Frank, Electronic Transition and the High Pressure Chemistry and Physics of Solids, Chapman and Hall, London, 1973.
[5] V.V. Mozhaev, J. Heremans, J. Frank, P. Masson, C. Balny, Proteins Struct. Funct. Genet. 24 (1996) 81–91.
[6] J.L. Silva, D. Foguel, A.T. DaPoian, P.E. Prevelige, Curr. Opin. Struct. Biol. 6 (1966) 166–175.
[7] R. Winter, J. Jonas (Eds.), High Pressure Molecular Science, NATO ASI Series E 358, 1999.
[8] P.S. Kim, R.L. Baldwin, Annu. Rev. Biochem. 59 (1990) 631–660.
[9] D.R. Shortle, FASEB J. 10 (1996) 27–34.
[10] J. Jonas, A. Jonas, Annu. Revs. Biophys. Biomol. Struct. 23 (1994) 287–318.
[11] J. Jonas, L. Ballard, D. Nash, Biophys. J. 75 (1998) 445–452.
[12] L. Ballard, A. Yu, C. Reiner, J. Jonas, J. Magn. Reson. A 133 (1998) 190–193.
[13] X. Peng, J. Jonas, J. Silva, Proc. Natl. Acad. Sci. USA 90 (1993) 1776–1780.
[14] X. Peng, J. Jonas, J.L. Silva, Biochemistry 33 (1994) 8323–8329.
[15] G.B. Benedek, E.M. Purcell, J. Chem. Phys. 22 (1954) 2003–2012.
[16] J. Powles, M. Gough, Mol. Phys. 16 (1969) 349–361.
[17] J. Jonas, Rev. Sci. Instrum. 43 (1972) 643–649.
[18] H. Yamada, Rev. Sci. Instrum. 45 (1974) 640–642.
[19] H. Yamada, K. Nishikawa, M. Honda, T. Shimura, K. Akasaka, K. Tabayaski, Rev. Sci. Instrum. 72 (2001) 1463–1471.
[20] S. Sareth, H. Li, H. Yamada, G.K. Woodward, K. Akasaka, FEBS Lett. 470 (2000) 11–14.
[21] K. Akasaka, T. Tezuka, H. Yamada, J. Mol. Biol. 271 (1997) 671–678.
[22] M.W. LaSalle, H. Yamada, K. Akasaka, J. Mol. Biol. 298 (2000) 293–302.
[23] H. Li, H. Yamada, K. Akasaka, Biochemistry 37 (1998) 1167–1173.
[24] K. Akasaka, T. Tezuka, H. Yamada, J. Mol. Biol. 271 (1997) 671–678.
[25] J.L. Urbauer, M.R. Ehrhardt, R.J. Bieber, P.F. Flynn, J.A. Wand, J. Am. Chem. Soc. 118 (1996) 11329–11330.
[26] T. Yamaguchi, H. Yamada, K. Akasaka, Prog. Biotechnol. 13 (1996) 141–153.
[27] T. Yamaguchi, H. Yamada, K. Akasaka, J. Mol. Biol. 250 (1995) 689–694.
[28] S.W. Englander, L. Mayne, Annu. Rev. Biophys. Biomol. Struct. 21 (1992) 243–265.
[29] D. Nash, J. Jonas, Biochemistry 36 (1997) 14375–14383.
[30] P.L. Privalov, CRC Crit. Rev. Biochem. Mol. Biol. 25 (1993) 181–202.
[31] R.H.M. Hatley, F. Franks, Cryo-Letters 7 (1993) 226–228.
[32] L.C. Antonino, R.A. Kautz, T. Nakano, R.O. Fox, A.L. Fink, Proc. Natl. Acad. Sci. USA 88 (1993) 7715–7718.
[33] F. Franks, R.H.M. Hatley, Cryo-Letters 6 (1993) 171–174.
[34] J. Zhang, X. Peng, A. Jonas, J. Jonas, Biochemistry 34 (1995) 8631–8641.
[35] D. Nash, B.-S. Lee, J. Jonas, Biochim. Biophys. Acta 1297 (1996) 40–48.
[36] A.D. Blum, S.H. Smallcombe, R.L. Baldwin, J. Mol. Biol. 118 (1978) 305–312.
[37] S.T. Gladwin, P.A. Evans, Folding Des. 1 (1996) 407–417.
[38] D. Nash, J. Jonas, Biochem. Biophys. Res. Commun. 238 (1997) 289–291.
[39] S.E. Bondos, S. Sligar, J. Jonas, Biochim. Biophys. Acta 1480 (2000) 353–364.
[40] D. Barrick, F.M. Hughson, R.L. Baldwin, J. Mol. Biol. 237 (1994) 588–601.
[41] M.S. Kay, R.L. Baldwin, Biochemistry 327 (1998) 7859–7868.
[42] D. Eliezer, J. Yao, H.J. Dyson, P.E. Wright, Nat. Struct. Biol. 5 (1998) 148–155.
[43] S.N. Loh, M.S. Kay, R.L. Baldwin, Proc. Natl. Acad. Sci. USA 92 (1995) 5446–5485.
[44] A.L. Fink, K.A. Olberg, S. Seshardi, Fold. Des. 3 (1997) 19–25.
[45] A. Yu, L. Ballard, L. Smillie, J. Pearlstone, D. Foguel, J.L. Silva, A. Jonas, J. Jonas, Biochim. Biophys. Acta 1431 (1999) 53–63.
[46] A. Yu, L. Ballard, L. Smillie, J. Pearlstone, J.L. Silva, D. Foguel, A. Jonas, J. Jonas, Biochemistry (2002) submitted.
[47] M.X. Li, M. Chandra, J.R. Pearlstone, K.I. Racher, G. Trigo-Gonzalez, T. Borgford, C.M. Kay, L.B. Smillie, Biochemistry 33 (1994) 917–925.
[48] J. R Pearlstone, T. Borgford, M. Chandra, K. Oikawa, C.M. Kay, O. Herzberg, J. Moult, A. Herklotz, F.C. Reinach, L.B. Smillie, Biochemistry 31 (1992) 6545–6553.
[49] G. Trigo-Gonzalez, K. Racher, L. Burtnick, T. Borgford, Biochemistry 31 (1992) 7009–7015.
[50] M. Chandra, W.D. McCubbin, K. Oikawa, C.M. Kay, L.B. Smillie, Biochemistry 33 (1994) 2961–2969.
[51] C.A. Royer, A.P. Hinck, S.N. Loh, K.E. Prehoda, X. Peng, J. Jonas, J.L. Markley, Biochemistry 32 (1993) 5222–5232.

ELSEVIER

Biochimica et Biophysica Acta 1595 (2002) 160–184

BIOCHIMICA ET BIOPHYSICA ACTA

BBA

www.bba-direct.com

Review

Synchrotron X-ray and neutron small-angle scattering of lyotropic lipid mesophases, model biomembranes and proteins in solution at high pressure

Roland Winter *

Department of Chemistry, Physical Chemistry I, University of Dortmund, Otto-Hahn-Straße 6, D-44221 Dortmund, Germany

Received 6 September 2001; received in revised form 2 November 2001; accepted 5 November 2001

Abstract

In this review we discuss the use of X-ray and neutron diffraction methods for investigating the temperature- and pressure-dependent structure and phase behaviour of lipid and model biomembrane systems. Hydrostatic pressure has been used as a physical parameter for studying the stability and energetics of lipid mesophases, but also because high pressure is an important feature of certain natural membrane environments and because the high pressure phase behaviour of biomolecules is of importance for several biotechnological processes. Using the pressure jump relaxation technique in combination with time-resolved synchrotron X-ray diffraction, the kinetics of different lipid phase transformations was investigated. The techniques can also be applied to the study of other soft matter and biomolecular phase transformations, such as surfactant phase transitions and protein un/refolding reactions. Several examples are given. In particular, we present data on the pressure-induced unfolding and refolding of small proteins, such as Snase. The data are compared with the corresponding results obtained using other trigger mechanisms and are discussed in the light of recent theoretical approaches.

Keywords: Synchrotron X-ray scattering; Neutron small-angle scattering; Pressure; Pressure jump; Lipid bilayer; Protein folding

1. Introduction

The interest in pressure as a variable has been growing in physico-chemical studies of biological and other soft condensed matter systems in recent years [1–3]. Pressure affects chemical equilibria and

Abbreviations: SAXS, small-angle X-ray scattering; SANS, small-angle neutron scattering; FT-IR, Fourier transform infrared spectroscopy; PC, phosphatidylcholine; PE, phosphatidylethanolamine; PS, phosphatidylserine; FA, fatty acid; LA, lauric acid; MA, myristic acid; PA, palmitic acid; SA, stearic acid; MO, monoolein; ME, monoelaidin; GD, gramicidin D; DMPC, 1,2-dimyristoyl-*sn*-glycero-3-phosphatidylcholine (di-$C_{14:0}$); DMPS, 1,2-dimyristoyl-*sn*-glycero-3-phosphatidylserine (di-$C_{14:0}$); DTPE, 1,2-ditetradecyl-*sn*-glycero-3-phosphatidylethanolamine (di-$C_{14:0}$); DPPC, 1,2-dipalmitoyl-*sn*-glycero-3-phosphatidylcholine (di-$C_{16:0}$); DPPE, 1,2-dipalmitoyl-*sn*-glycero-3-phosphatidylethanolamine (di-$C_{16:0}$); DSPC, 1,2-distearoyl-*sn*-glycero-3-phosphatidylcholine (di-$C_{18:0}$); DOPC, 1,2-dioleoyl-*sn*-glycero-3-phosphatidylcholine (di-$C_{18:1,cis}$); DOPE, 1,2-dioleoyl-*sn*-glycero-3-phosphatidylethanolamine (di-$C_{18:1,cis}$); DEPC, 1,2-dielaidoyl-*sn*-glycero-3-phosphatidylcholine (di-$C_{18:1,trans}$); POPC, 1-palmitoyl-2-oleoyl-*sn*-glycero-3-phosphatidylcholine ($C_{16:0}$,$C_{18:1,cis}$); egg-PE, egg yolk phosphatidylethanolamine; Snase, staphylococcal nuclease; SDS, sodium dodecyl sulphate; AOT, bis[2-ethylhexyl]sulphosuccinate; TDMAO, tetradecyldimethylaminoxide; C_iE_j, *n*-alkyl polyoxyethylene ether

* Fax: +49-231-755-3901. *E-mail address:* winter@steak.chemie.uni-dortmund.de (R. Winter).

0167-4838/02/$ – see front matter
PII: S0167-4838(01)00342-9

reaction rates. The behaviour of all systems under high pressure is governed by Le Châtelier's principle, which predicts that the application of pressure shifts an equilibrium towards the state that occupies a smaller volume, and accelerates processes for which the transition state has a smaller volume than the ground state. With the knowledge of the reaction volume ΔV^0 and activation volume $\Delta V^{\neq}$, one can draw conclusions about the nature of the reaction and its mechanism. Pressures used to investigate biochemical systems range from 0.1 MPa to about 1 GPa (0.1 MPa = 1 bar, 1 GPa = 10 kbar). Such pressures only change intermolecular distances and affect conformations but do not change covalent bond distances or bond angles. The covalent structure of low molecular weight biomolecules (peptides, lipids, saccharides), as well as the primary structure of macromolecules (proteins, nucleic acids and polysaccharides), is not perturbed by pressures up to about 20 kbar. Pressure acts predominantly on the spatial (secondary, tertiary, quaternary and supramolecular) structures of macromolecules.

Besides the general physico-chemical interest in using high pressure as a tool for understanding the phase behaviour, structure and energetics of biomolecules and amphiphiles in general, high pressure is also of biotechnological (e.g. high pressure food processing) and physiological interest [1–10]. Hydrostatic pressure significantly influences the structural properties and thus functional characteristics of cell membranes. Yet this has not prevented the invasion of cold and high pressure habitats by deep-sea organisms (up to about 10 000 m depths, corresponding to 1 kbar of pressure, at 2–3°C). Without any compensatory adjustments the membranes of deep-sea organisms should be highly ordered. It is now well established that deep-sea organisms display a variety of adaptations to high hydrostatic pressure and low temperature at the molecular level of organization to keep their membranes in a fluid-like state that is a prerequisite for optimal physiological function.

In this review, we first discuss experimental techniques and results for studying the high pressure structure and phase behaviour of lyotropic lipid mesophases, surfactants and model biomembrane systems using X-ray and neutron diffraction. In addition, effects of the incorporation of additives, such as cholesterol and polypeptides, are briefly mentioned. Then we discuss the pressure jump relaxation technique for studying the kinetics of phase transformations between different lipid mesophases. Finally, we illustrate the use of synchrotron X-ray scattering for studying the effect of pressure on protein structure and un/refolding reactions of proteins.

2. Experimental techniques

2.1. *Diffraction techniques for studying soft condensed matter at high pressure*

The experimental results presented in this review are mainly based on studies using small- and wide-angle X-ray scattering (SAXS, WAXS) and small-angle neutron scattering (SANS) in combination with high pressure techniques. In this paragraph the experimental setups and the basic theory for analysing lipid– and protein–water diffraction data are briefly discussed. Small-angle neutron scattering experiments were performed at research reactors or spallation sources, at the ILL (Grenoble, France), ISIS (Didcot, UK) or the Hahn-Meitner-Institut (Berlin, Germany). The development of synchrotron radiation sources from multi-GeV electron and positron storage rings increased the flux on the sample by factors of more than 10^5 when compared to conventional laboratory X-ray sources. Use of these sources coupled with efficient electronic detectors made it possible to start collecting SAXS data on rapid time scales, which can now go down to tens of milliseconds or even below. Our time-dependent SAXS studies were performed at the synchrotron X-ray beam lines DESY (Hamburg, Germany), ESRF (Grenoble, France) and APS (Argonne, France).

We used different high pressure sample cells suitable for pressures up to 20 kbar and temperatures up to 140°C as shown in Fig. 1 [11–13]. The X-ray pressure vessels (Fig. 1a,b) are home-built and made from stainless steel or a Ni-Cr-Co alloy (NIMONIC 90) of high tensile strength. They have a high pressure connection to the pressurizing system and a bore for a thermocouple. Temperature control is achieved by circulating water from a thermostat through the outside jacket of the vessels. The sample temperature can be measured within ±0.2°C. Beryllium discs of

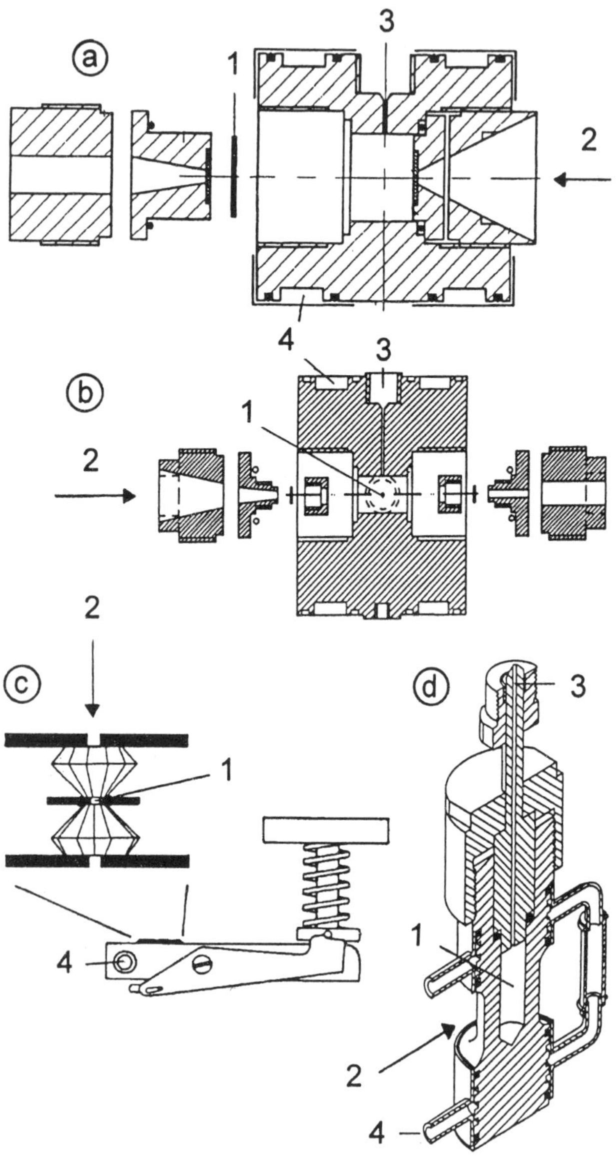

Fig. 1. High pressure sample cells. (a) Cell made from stainless steel with Be windows for X-ray scattering studies up to 2 kbar; (b) cell made from NIMONIC 90 alloy with diamond windows for X-ray scattering studies up to 8 kbar; (c) DAC for X-ray scattering studies up to 20 kbar; (d) cell made from an Al alloy for neutron scattering studies up to 2.5 kbar. 1, sample; 2, X-ray or neutron beam; 3, high pressure connection; 4, thermostatting water circuit.

1–2 mm thickness are used as X-ray windows up to pressures of 2 kbar, for higher pressures up to 8 kbar flat diamond windows of 0.5–1 mm thickness are used. The window holders are sealed with Viton-O rings and are tightened by closure nuts. The sample of 40 μl volume is held in a PTFE ring that is closed with two mylar foils glued on both sides of the ring to separate the sample from the pressurizing medium (distilled water).

The pressurizing system essentially consists of a Heise Bourdon gauge and a hand-operated pressure generator. Pressure jumps are performed by a computer-controlled opening of an air-operated valve between the high pressure cell and a liquid reservoir. With this pressure jump apparatus fast jumps (< 5 ms) are possible with variable amplitudes. To minimize adiabatic temperature changes in the course of a kinetic experiment (approx. 2 mK/bar under pure adiabatic conditions), the high pressure sample cells were constructed to hold only a very small volume of the pressurizing medium. The pressure jump technique has been shown to offer several advantages over the temperature jump approach. (1) Pressure propagates rapidly so that sample inhomogeneity is a minor problem. (2) Pressure jumps can be performed bidirectional, i.e. with increasing or decreasing pressure. (3) In the case of fully reversible structural changes of the sample, pressure jumps can be repeated with identical amplitudes to allow for an averaging of the diffraction data over several jumps and an improvement of the counting statistics.

For high pressure X-ray studies up to 20 kbar, a diamond anvil cell (DAC) from High Pressure Diamond Optics was used (Fig. 1c). Here, the sample is sealed in a metal gasket of 0.25 mm thickness with a hole of 0.37 mm diameter [11]. The pressure is calibrated by adding a small ruby crystal to the sample and by measuring the shift of the R1 ruby fluorescence line at about 694 nm. It must be noted that X-ray transmission and thus background diffraction patterns depend drastically on the density of the solvent (water) and the path length of the pressure cell, both being pressure-dependent. In order to perform an accurate background correction, one has to measure the sample and background scattering under identical conditions, which requires a high reproducibility of the path length at each pressure. Unfortunately, this cannot be achieved with a DAC and hence no accurate diffuse small-angle scattering data of low-concentrated samples (e.g. of dilute protein solutions) can be obtained. Flat diamond cells are more appropriate in these cases.

For neutron scattering experiments on lipid samples at low angles, a high pressure sample cell made from an Al alloy (AlMgSi1) was used (Fig. 1d). This alloy is characterized by a high tensile strength and allows a maximum cell pressure of about 2500 bar. The transmission of the empty cell for neutrons of 10 Å wavelength has been measured to be more than 90%. The temperature of the cell is controlled by a water circuit passing through a cell jacket above and below the neutron beam window. The cylindrical sample volume has a diameter of about 2–3 mm and is separated from the pressurizing medium (water) by a PTFE bellow. Sample cells made from stainless steel with sapphire windows of 10 mm thickness are also used for SANS studies of biomolecular systems, in particular for measurements of diffuse small-angle scattering samples, such as micellar or protein solutions.

2.2. Lipid–water diffraction patterns

Regarding the diffraction of X-rays and neutrons, the lipid–water dispersions investigated here are equivalent to powder samples that are composed of many randomly oriented microcrystals. Thus, Bragg's condition is automatically fulfilled, and all possible diffraction peaks are simultaneously recorded. While the positions of the diffraction peaks are related to periodic distances within the lyotropic lipid mesophase, their sharpness or width reflects the extent of this periodicity over large distances. The measured reciprocal spacings are given by

$$s = \frac{2}{\lambda}\sin\theta \tag{1}$$

(2θ scattering angle, λ wavelength of radiation). If a lipid–water phase is lacking any periodic structure, diffuse small-angle scattering is observed only. Lamellar lipid–water mesophases (denoted as L or P) consist of alternating layers of lipid and water molecules. This quasi-one-dimensional periodic structure exhibits diffraction patterns in the small-angle regime that are described by the equation

$$s_n = n\frac{1}{d} \tag{2}$$

where $n = 1, 2, 3, \ldots$ and d is the lamellar repeat unit, which consists of the sum of the water and lipid layer thicknesses.

The structure of the inverse hexagonal lipid–water mesophase (denoted as H_{II}) is based on cylindrical water rods, which are surrounded by lipid monolayers. The rods are packed in a two-dimensional hexagonal lattice with Bragg peaks positioned at

$$s = \frac{2}{\sqrt{3}a}\sqrt{h^2 + k^2 + hk} \tag{3}$$

where a is the distance between the centres of two neighbouring rods (lattice constant) and h, k are the Miller indices. In small-angle diffraction patterns hexagonal lipid phases are easily distinguished from lamellar phases by their ratio of Bragg peak positions, which is $1:\sqrt{3}:2:\ldots$.In a diffraction experiment Bragg peaks of cubic structures may be observed at

$$s = \frac{1}{a}\sqrt{h^2 + k^2 + l^2} \tag{4}$$

where a is the cubic lattice constant. The Miller indices h, k, l depend on the lattice type (primitive, body-centred, face-centred) and the symmetry elements of the cubic structure. In Table 1 the relative Bragg peak positions of the three cubic lipid–water phases relevant for this review are summarized.

Bragg peaks in the wide-angle region are related to the packing of the lipid hydrocarbon chains in a monolayer. Generally, this packing can be described as a centred rectangular lattice with lattice constants a_{rec} and b_{rec}, which are calculated from the d-spacings as

$$a_{rec} = 2d_{20} \tag{5}$$

$$b_{rec} = \frac{d_{11}}{\sqrt{1-[d_{11}/(2d_{20})]^2}} \tag{6}$$

When the lipid chains are rigidly packed in a hexag-

Table 1
Inverse cubic lipid–water mesophases

Lipid–water phase	Space group	Peak position ratio
Q^D_{II}	Pn3m	$\sqrt{2}:\sqrt{3}:\sqrt{4}:\ldots$
Q^P_{II}	Im3m	$\sqrt{2}:\sqrt{4}:\sqrt{6}:\ldots$
Q^G_{II}	Ia3d	$\sqrt{6}:\sqrt{8}:\sqrt{14}:\ldots$

The cubic phases are based on lipid bilayers curved in three dimensions (see Fig. 2). The midplanes of the bilayers are D, P, or G infinite periodic minimal surfaces, as denoted with the superscripts in the phase symbols.

onal lattice, only a single wide-angle Bragg peak is usually observed. Then, the lattice constant (chain–chain distance) may be obtained from Eq. 3. In the case of fluid-like disordered lipid layers, only a very broad wide-angle peak is observed.

2.3. SAXS studies of proteins in solution

Details of the experimental techniques for studying dilute solutions of proteins using synchrotron X-ray diffraction are discussed in [13–15]. The SAXS experiments were performed at synchrotron X-ray beam lines, such as ESRF or APS. The scattering data are presented as a function of momentum transfer $Q = (4\pi/\lambda)\sin\theta$ (2θ scattering angle, λ wavelength of radiation, $Q = 2\pi s$). The overall conformation of the protein can be obtained by measuring its radius of gyration R_g and pair distance distribution function $p(r)$. The pair distance distribution function $p(r)$, which depends on the molecular particle shape and on the intra-particle scattering distribution, is given by the indirect Fourier transform of the measured scattered intensity [16]. For a particle of uniform electron density, it is given by

$$p(r) = \frac{1}{2\pi^2}\int_0^\infty I(Q)Qr\,\sin(Qr)\mathrm{d}Q \quad (7)$$

The function $p(r)$ measures the distribution of pairwise distances within the volume of the scattering particle, that is, the protein molecule, with maximum dimension D_{max}. More generally, $p(r)$ is the density contrast pair correlation function of the molecule. As the use of the Guinier approximation for the determination of the radius of gyration, R_g, might lead to errors caused by concentration or aggregation effects, we used the normalized second moment of the pair distance distribution function for calculation of R_g:

$$R_g^2 = \frac{\int_0^{D_{max}} p(r)r^2\mathrm{d}r}{2\int_0^{D_{max}} p(r)\mathrm{d}r} \quad (8)$$

Additional Fourier transform infrared spectroscopy (FT-IR) experiments were performed to yield also information on the secondary structural changes of the proteins. Details of the experimental setup can be found in [14]. Besides the techniques presented here, several further techniques are used to study structural and dynamic properties of proteins under high pressure, in particular various UV/Vis, fluorescence spectroscopy and NMR techniques. They are described in detail in [17–22] and in several articles in this volume.

3. Results and discussion

3.1. Lipid mesophases and model biomembrane systems

Lyotropic lipid mesophases are organized molecular systems formed by amphiphilic molecules, such as phospholipids, in the presence of water. They exhibit a rich structural polymorphism, depending on their molecular structure and environmental conditions, such as water content, pH, ionic strength, temperature and pressure [23–25]. The basic structural element of biological membranes consists of a lamellar phospholipid bilayer matrix (see Fig. 2). In the lamellar structure the interfaces are flat and periodically stacked, forming multilamellar vesicles. Two neighbouring lipid bilayers are separated by a water layer of about 10–20 Å.

Saturated phospholipids often exhibit two thermotropic lamellar phase transitions, a gel to gel ($L_{\beta'}/P_{\beta'}$) pretransition and a gel to liquid–crystalline ($P_{\beta'}/L_\alpha$) main transition at a higher temperature T_m (for the structures see Fig. 2). In the fluid-like L_α phase, the acyl chains of the lipid bilayers are conformationally disordered, whereas in the gel phases, the chains are more extended and ordered. Because the average end-to-end distance of disordered hydrocarbon chains in the L_α phase is smaller than that of ordered (all-*trans*) chains, the bilayer becomes thinner during melting at the $P_{\beta'}/L_\alpha$ transition, even though the lipid volume increases by about 3% [26]. In addition to these thermotropic phase transitions, pressure-induced phase transformations have been observed [27–45]. Upon compression, the lipids easily adopt to volume restriction by changing their conformation and packing. Lipid systems are those biological systems which are most pressure sensitive and in general they easily undergo phase transformations under changes of environmental conditions.

It is now well known that many biological lipid

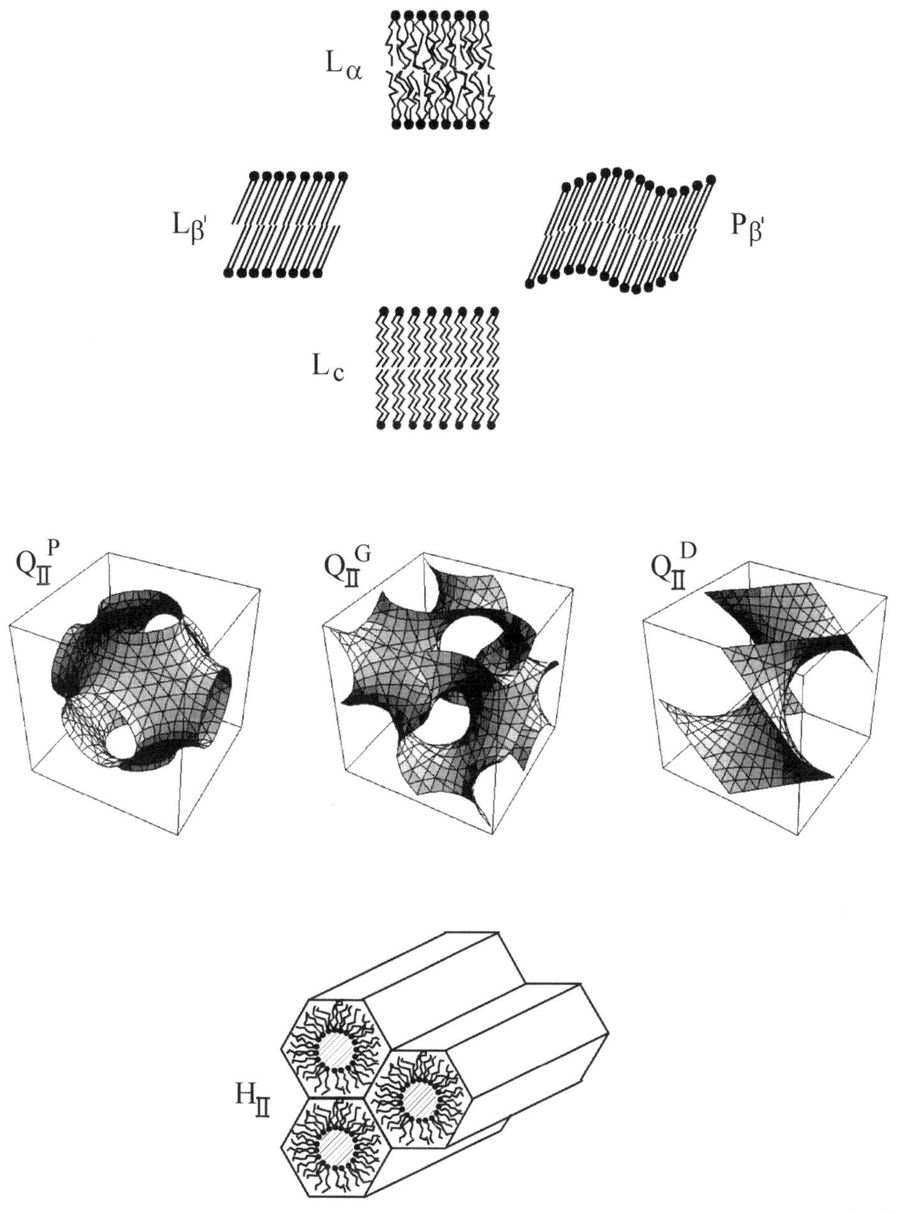

Fig. 2. Schematic drawing of lipid–water phases. L_c, lamellar crystalline; $L_{\beta'}$, $P_{\beta'}$, lamellar gel; L_α, lamellar liquid–crystalline; Q_{II}^{G}, Q_{II}^{D}, Q_{II}^{P}, inverse bicontinuous cubic; H_{II}, inverse hexagonal. The cubic phases are represented by the G, D, and P minimal surfaces, which locate the midplanes of fluid lipid bilayers.

molecules also form non-lamellar liquid–crystalline phases (see Fig. 1) [25,46–49]. Lipids, which can adopt a hexagonal phase, are present at substantial levels in biological membranes, usually with at least 30 mol% of the total lipids. Generally, micellar, hexagonal and cubic lipid dispersions are known to exist either as organizations of aggregates of amphiphiles in a continuous aqueous environment or as organizations of water aggregates in a continuous medium of the amphiphiles. In the first case they are said to be direct, or type I, structures; in the second case they are said to be inverted, or type II, structures. For the phases formed by the double-chain lipids found in membranes, the polar/apolar interface curves towards the water, hence these phases are of type II. It is generally assumed that the non-lamellar lipid structures, such as the inverted hexagonal (H_{II}) and cubic (Q_{II}) lipid phases, are also of significant

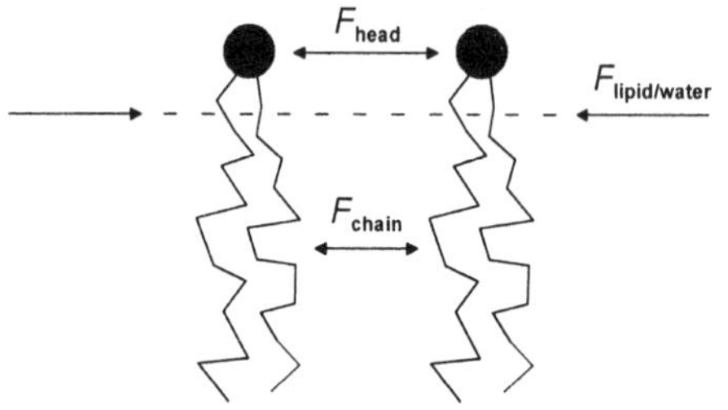

Fig. 3. Illustration of the balance of lateral forces across a lipid monolayer.

biological relevance. Fundamental cell processes, such as endo- and exocytosis, fat digestion, membrane budding and fusion, involve a rearrangement of biological membranes where non-lamellar lipid structures are probably involved. Also static cubic structures (cubic membranes) might occur in biological cells [48–50].

At a macroscopic level, lipid cubic phases are very viscous and optically isotropic. They can be sorted in two main classes: bicontinuous and micellar [25]. In the first class the amphiphiles build bicontinuous cubic phases of type II (Q_{II}), which can be visualized in terms of highly convoluted lipid bilayers, which subdivide a three-dimensional space into two disjointed polar labyrinths separated by an apolar septum. The cubic symmetries most common observed are Ia3d, Pn3m and Im3m (see Fig. 2). The structures of these phases, Q^G_{II}, Q^D_{II} and Q^P_{II}, are closely related to the Schoen gyroid (G), the Schwarz D and the Schwarz P infinite periodic minimal surfaces (IPMS). An IPMS is an intersection-free surface periodic in three dimensions with a mean curvature that is everywhere zero. The surface, that sits at the lipid bilayer midplane, thus separates the two interpenetrating but not connected water networks. In the second class of cubic structures the amphiphiles build an infinite number of finite aggregates, liquid globules or micelles, organized in regular packings of various symmetries [51].

Generally, the molecular organization within a lipid aggregate can be understood in terms of a balance of attractive and repulsive forces acting at the level of the lipid polar head groups and the non-polar acyl chains (Fig. 3). Within the head group (lipid/water) region there is an effectively attractive force $F_{l/w}$, which arises from the unfavourable contact of the hydrocarbon chains with water (the hydrophobic effect). Attractive contributions from hydrogen bonding, as in the case of phosphatidylethanolamines, may also be present. The repulsive head group pressure F_{head} is the result of hydrational, steric, and – for charged head groups – electrostatic contributions. For the acyl chain region the attractive van der Waals interactions among the CH_2 groups are compensated by the repulsive lateral pressure, F_{chain}, owing to thermally activated dihedral angle isomerizations. In a membrane bilayer at equilibrium, the various lateral pressures are balanced.

An imbalance of attractive and repulsive forces at the level of the head group and acyl chains within a given lipid monolayer yields a spontaneous curvature. Clearly the two monolayers cannot be simultaneously at a free energy minimum with regard to their intrinsic or spontaneous curvature. It follows that a symmetrical bilayer composed of non-lamellar forming lipids is under a condition of curvature elastic stress. A high curvature free energy associated with the membrane lipid–water interface then leads to the formation of non-lamellar (cubic and/or H_{II}) phases. The spontaneous curvature increases, for instance, if the lateral chain pressure increases due to an increase of *trans/gauche* isomerizations of the acyl chains at high temperatures, or if the level of head group hydration decreases (e.g. due to Ca^{2+} adsorption at the polar/apolar interface).

Until now, no full theoretical description of the lyotropic lipid phase behaviour exists, although some progress has been made in recent years [52–60]. Often a concept is used that can be explained in terms of a small set of parameters, irrespective of the precise chemical nature of the lipid molecule. Following Helfrich the surface curvature energy contribution associated with amphiphile films is described in terms of three curvature elastic parameters: the spontaneous mean curvature H_s, the mean curvature modulus κ_m, and κ_G, the Gaussian curvature modulus [23–25]. Applying differential geometry, the surface energy per unit area for small curvatures is given by

$$g_{bend} = 2\kappa_m\langle(H-H_s)^2\rangle + \kappa_G\langle K\rangle \qquad (9)$$

$H=(C_1+C_2)/2$ is the mean interfacial curvature, which is equal to half the sum of the principal curvatures $C_1=1/R_1$ and $C_2=1/R_2$ at the interface, and $K=C_1C_2$ is the Gaussian curvature at the interface,

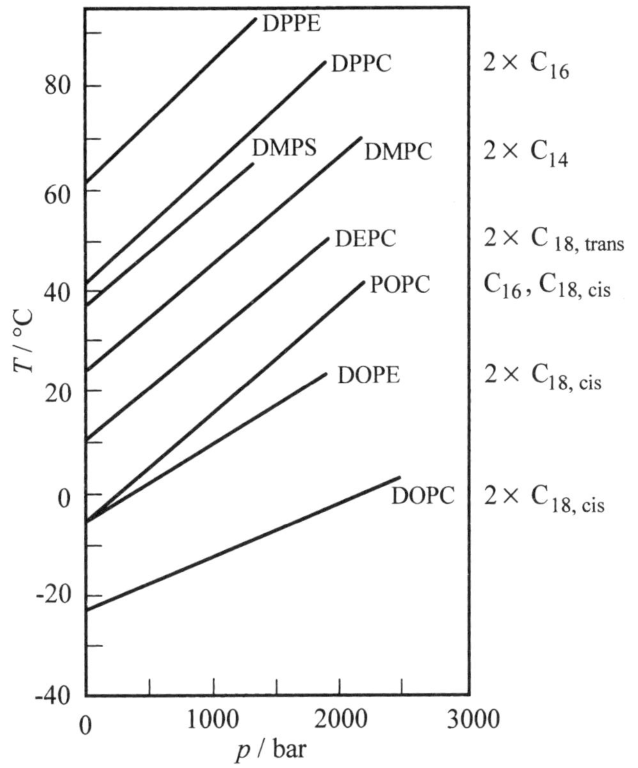

Fig. 4. *T,p* phase diagram for the main (chain melting) transition of different phospholipid bilayer systems. The fluid (liquid–crystalline) L_α phase is observed in the low pressure, high temperature region of the phase diagram.

given by the product of the principal curvatures C_1 and C_2 of the interface (e.g. $C_1 = C_2 = 0$ for a planar bilayer; $C_1 < 0$, $C_2 = -C_1 > 0$, i.e. $H = 0$, $K < 0$, for a hyperbolic saddle surface). The spontaneous mean curvature H_s is the mean curvature the lipid aggregate would wish to adopt in the absence of external constraints, and κ_m tells us what energetic cost there would be for deviations away from this. Besides the curvature energetic contribution, there are other energetic contributions. Due to the desire to fill all the hydrophobic volume by the amphiphile chains (due to the hydrophobic effect), there is a contribution quantifying an eventual packing frustration. A further contribution is due to interlamellar interactions. The curvature elastic energy is believed to be the crucial term governing the stability of non-lamellar phases and the ability of lipid membranes to bend, in particular at high levels of hydration. To probe the concept of any energetic description and the resultant set of parameters necessary to provide a general explanation of a universal lyotropic phase behaviour, one needs to scan the appropriate parameter space experimentally. To this end, also pressure-dependent studies have proven to be a very valuable tool [57–59].

3.1.1. Single-component lipid systems

3.1.1.1. Lamellar phase transitions. The increase in entropy with lipid chain rotational disorder, the increase in intermolecular entropy, and the increase in lipid head group hydration are the driving forces for the gel–fluid transition of lipid bilayer membranes (denoted as L_β–, $L_{\beta'}$– or $P_{\beta'}$–L_α transition, see Fig. 2). In a simple way, this transition is often interpreted as the melting of the lipid hydrocarbon chains. Opposing this chain melting is the increase of the internal energy due to increasing rotational isomerism, the decrease of the van der Waals attraction between the hydrocarbon lipid chains, and the increase of the polar–apolar interface resulting from the lateral bilayer expansion due to increasing chain

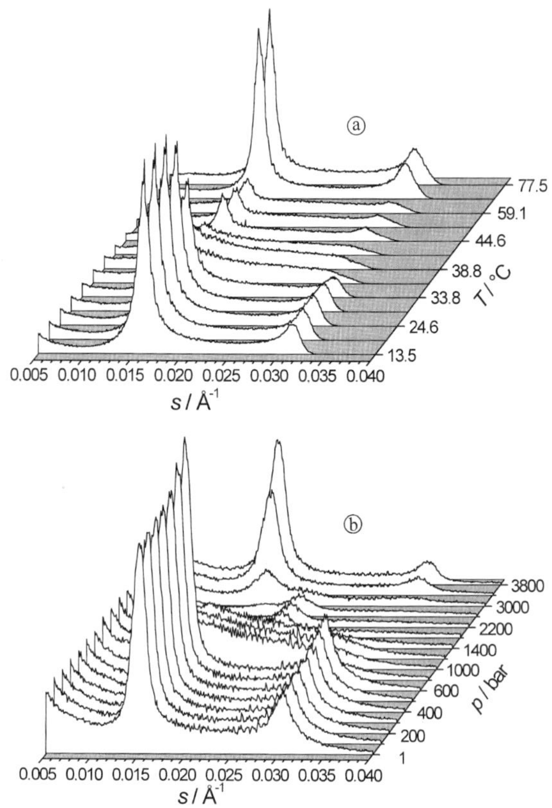

Fig. 5. Typical temperature- and pressure-dependent (at $T = 55°C$) small-angle X-ray scattering patterns of DPPC bilayers in excess water. Only one or two orders of lamellar Bragg reflections are visible.

isomerism. The balance of all these contributions to the system free energy, which depends on the lipid molecular structure, determines the main (melting) transition temperature, T_m, of a lipid bilayer. Generally, the lamellar gel phases L_β, $L_{\beta'}$, and $P_{\beta'}$ prevail at high pressures and low temperatures, whereas the lamellar fluid (liquid–crystalline) phase L_α is found as the pressure is lowered and/or the temperature is raised.

A common slope of about 22°C/kbar has been observed for the gel–fluid phase boundary of the saturated phosphatidylcholines 1,2-dimyristoyl-*sn*-glycero-3-phosphatidylcholine (DMPC), 1,2-dipalmitoyl-*sn*-glycero-3-phosphatidylcholine (DPPC) and 1,2-distearoyl-*sn*-glycero-3-phosphatidylcholine (DSPC), as shown in Fig. 4. Using the Clapeyron relation, $dT_m/dp = T_m \Delta V_m / \Delta H_m$, the positive slope can be explained by an endothermic enthalpy change, ΔH_m, and a volume increase, ΔV_m, for the gel–fluid transition, which have indeed been found in direct experimental measurements of these thermodynamic quantities (e.g. $\Delta H_m \approx 21$ kJ/mol and $\Delta V_m \approx 16$ cm^3/mol (corresponding to a ~3% volume change) for the gel–fluid transition of DMPC at pressures below 500 bar) [26,45]. Similar transition slopes have been found for the mono-*cis*-unsaturated lipid 1-palmitoyl-2-oleoyl-*sn*-glycero-3-phosphatidylcholine (POPC), the phosphatidylserine 1,2-dimyristoyl-*sn*-glycero-3-phosphatidylserine (DMPS), and the phosphatidylethanolamine 1,2-dipalmitoyl-*sn*-glycero-3-phosphatidylethanolamine (DPPE). Only the slopes of the di-*cis*-unsaturated lipids 1,2-dioleoyl-*sn*-glycero-3-phosphatidylcholine (DOPC) and 1,2-dioleoyl-*sn*-glycero-3-phosphatidylethanolamine (DOPE) have been found to be markedly smaller. While the lengths of the lipid hydrocarbon chains and the type (chemical structure) of the lipid head group do not affect the slope of the main transition significantly, these parameters determine the transition temperature. The two *cis*-double bonds of DOPC and DOPE

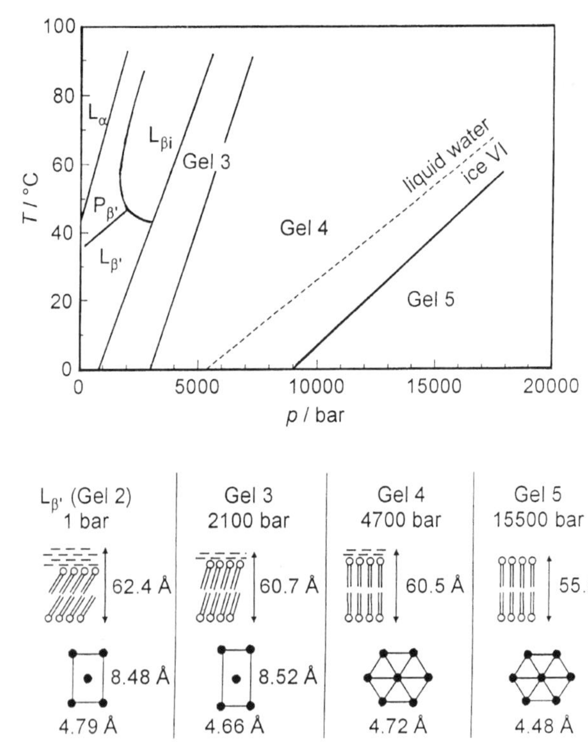

Fig. 6. T,p phase diagram of DPPC in excess water and schematic drawing of the lamellar lattice constant and lipid packing in the bilayer plane of DPPC gel phases at 23°C. It is noteworthy that an additional crystalline gel phase can be induced in the low temperature regime after prolonged cooling [49].

lead to very low transition temperatures and slopes, presumably as they impose kinks in the linear conformations of the lipid acyl chains. Thus, significant free volume is created in the bilayer and the ordering effect of high pressure is reduced.

These studies clearly demonstrate that biological organisms could modulate the physical state of their membranes in response to changes in the external environment by regulating the fractions of the various lipids in a cell membrane differing in chain length, chain unsaturation or head group structure ('homeoviscous adaptation', e.g. of deep-sea organisms under high hydrostatic pressure and at low temperature). In addition, nature has further means to regulate membrane fluidity, such as by changing the membrane concentration of cholesterol or various proteins.

As an example, Fig. 5a shows small-angle diffraction data of a DPPC bilayer in excess water as a function of temperature. Clearly the pretransition as well as the main lipid phase transition are observed as a relatively sharp shift of the Bragg peak positions at about 35 and 42°C, respectively. The lamellar lattice constant increases from ~63 Å in the $L_{\beta'}$ phase to ~72 Å in the ripple gel phase $P_{\beta'}$. Because of the highly disordered chains in the fluid L_α phase, the bilayer thickness decreases to a lattice constant of about 66 Å. Fig. 5b shows some pressure-dependent data. In DPPC dispersions at 55°C, a shift to lower scattering vectors together with a change in the line shape is observed at 800 bar which is due to the pressure-induced L_α to $P_{\beta'}$ phase transition; the corresponding lattice constant increases from 68 to 71 Å. Further increase in pressure leads to the formation of a pressure-induced interdigitated gel phase, $L_{\beta i}$ around 1400 bar, where the lipid acyl chains from opposing monolayers partially interpenetrate, which leads to a decrease of the lamellar repeat period to about 50 Å. At around 2.8 kbar the transition to the Gel 3 phase occurs at this temperature with a lattice constant that is about 10 Å larger [41].

We note that applying high pressure can lead to the formation of additional gel phases, which are not observed under ambient pressure conditions, such as the interdigitated high pressure gel phase $L_{\beta i}$ found for phospholipid bilayers with acyl chain lengths ≥C16 [28,41,42]. To illustrate this phase variety,

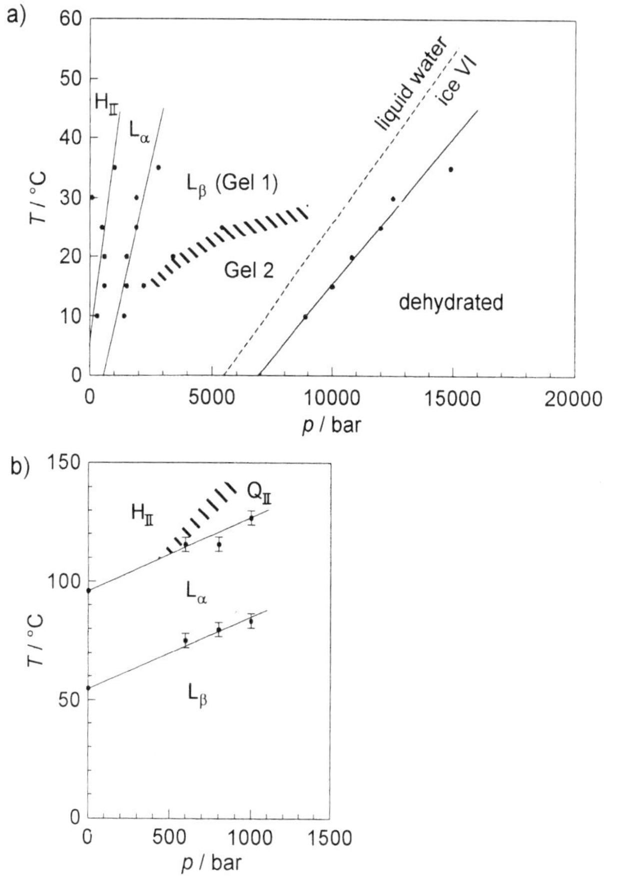

Fig. 7. *T,p* phase diagram of (a) DOPE [12] and (b) DTPE [44] in excess water (the phase boundaries of cubic phases Q_{II} are somewhat tentative).

the results of a detailed X-ray diffraction and FT-IR spectroscopy study of the *p,T* phase diagram of DPPC in excess water is shown in Fig. 6 [41]. The structures of the L_α, $P_{\beta'}$ (Gel 1), $L_{\beta'}$ (Gel 2), Gel 3, Gel 4 and Gel 5 phases are illustrated schematically in Figs. 2 and 6. In the Gel 5 phase the lipid molecules have lost essentially all the interlamellar hydrating water, which now coexists as bulk frozen water (ice VI).

3.1.1.2. Non-lamellar phase transitions. For a series of lipid molecules, also non-lamellar lyotropic phases are observed as thermodynamically stable phases or as long-lived metastable phases after special sample treatment. We will discuss lipid–water systems with the lipids taken from different types of amphiphilic molecules. Contrary to DOPC, which shows a lamellar L_β–L_α transition only (see Fig. 4),

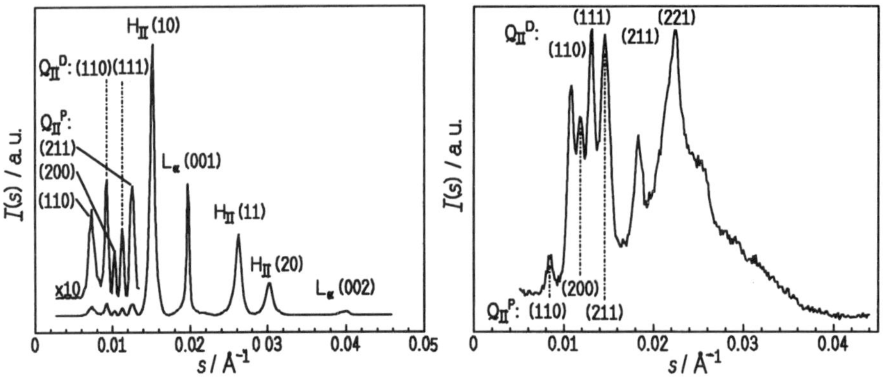

Fig. 8. SAXS patterns of DOPE in excess water, which has been pressure-cycled (40 times, 20°C, 340 bar) (left) and temperature-cycled (1400 times, 5°C, 1 bar, using a robot) across the L_α–H_{II} transition (right).

the corresponding lipid DOPE with ethanolamine as head group exhibits an additional phase transition from the lamellar L_α to the non-lamellar, inverse hexagonal H_{II} phase, when it is dispersed in water. As pressure forces a closer packing of the lipid chains, which results in a decreased number of *gauche* bonds and kinks in the chains, both transition temperatures, of the L_β–L_α and the L_α–H_{II} transition, increase with increasing pressure. In Fig. 7a, the *T,p* phase diagram of DOPE in excess water is displayed, which has been obtained by SAXS and WAXS experiments using a diamond anvil cell. The slope of the L_α–H_{II} transition is almost 3 times larger than that of the L_β–L_α transition; values of about 40 and 14°C/kbar have been found, respectively. The L_α–H_{II} transition observed in DOPE/water and also in egg yolk phosphatidylethanolamine (egg-PE)/water (egg-PE is a mixture of different phosphatidylethanolamines) is the most pressure-sensitive lyotropic lipid phase transition found to date. The reason why this transition has such a strong pressure dependence could be conjectured to be the strong pressure dependence of the chain length and volume of its unsaturated chains.

Interestingly, in the DOPE/water system inverse cubic phases, Q_{II}^{D} and Q_{II}^{P}, can be induced in the region of the L_α–H_{II} transition by subjecting the sample to extensive temperature or pressure cycles at temperature and pressure conditions close to the phase transition [61,62]. Fig. 8 (left) displays a diffraction pattern of a pressure-cycled DOPE dispersion. The Bragg peaks can be assigned as (10), (11),

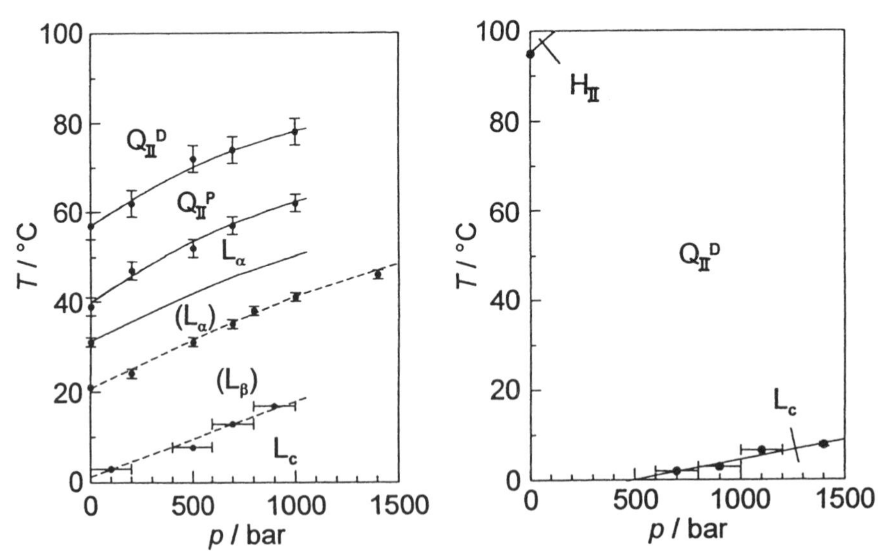

Fig. 9. *T,p* phase diagram of (a) ME and (b) MO in excess water (the L_β phase of ME is metastable).

(20) of the H_{II} phase, as (001), (002) of the L_α phase, as (110), (111) of the Q_{II}^D phase, and as (110), (200), (211) of the Q_{II}^P phase. The indexing of the cubic phases is confirmed by the observed ratio of the two cubic lattice constants, $a(Q_{II}^P)/a(Q_{II}^D) = 1.273$, which is close to the theoretically predicted value of 1.279 for these phases being in equilibrium. Fig. 8 (right) shows the diffraction pattern of a DOPE dispersion after 1400 temperature cycles between −5 and 15°C. The Bragg peaks of this X-ray diffraction pattern are also consistent with the existence of the Q_{II}^D and the Q_{II}^P phase in coexistence [12].

It has been shown that for conditions which favour a spontaneous curvature of a lipid monolayer, the topology of an inverse bicontinuous cubic phase can have a similar or even lower free energy than the lamellar or inverse hexagonal phase, as the cubic phases are characterized by a low curvature free energy and do not suffer the extreme chain packing stress predominant in the H_{II} phase. Metastable cubic phases might be formed via defect structures, which occur in passing the L_α–H_{II} transition, such as interlamellar micellar intermediates (IMI) or stalks [63,64]. Membrane fusion as a part of specific biological reactions probably also involves the formation of intermediates, such as stalks or IMIs. In the case of DOPE it is still an open question if the observed cubic phases in the region of the L_α–H_{II} phase transition, which can be induced by *T*- and *p*-cycling, represent thermodynamically stable phases with a high kinetic barrier or metastable phases. In this respect it might be helpful to inspect the *T*,*p* phase diagram of 1,2-ditetradecyl-*sn*-glycero-3-phosphatidylethanolamine (DTPE) in excess water (Fig. 7b). As found for DOPE/water, the thermotropic phase order of this system is L_β, L_α, H_{II} at atmospheric pressure. However, at pressures higher than about 500 bar an additional, intermediate region of inverse cubic phases, Q_{II}^D and Q_{II}^P, is observed [44]. Although the transition region from the cubic phases to the inverse hexagonal phase cannot clearly be resolved, this finding indicates a triple point in the *T*,*p* phase diagram and again illustrates the non-equivalence of the effects of temperature and pressure on lipid mesophases.

Finally, two further examples of single-component lyotropic lipid systems exhibiting non-lamellar phases are discussed: monoelaidin (ME) and mono-

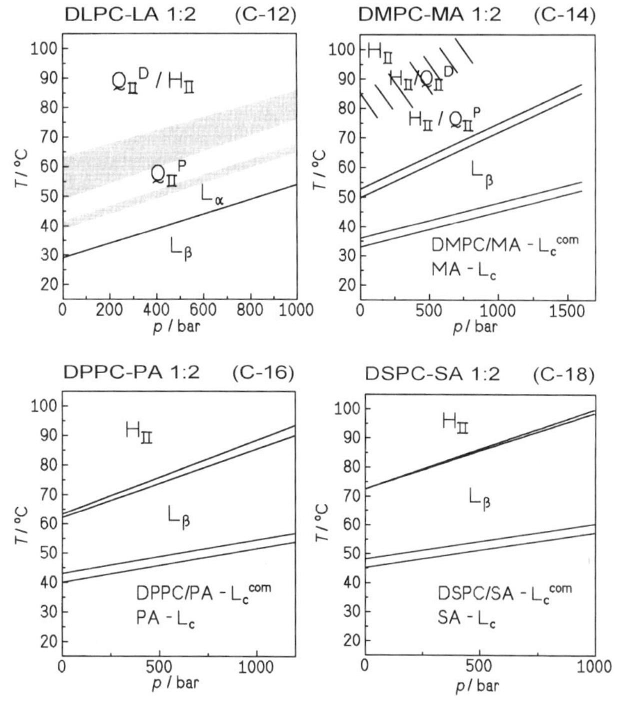

Fig. 10. *T*,*p* phase diagrams of several phosphatidylcholine/fatty acid (1:2) mixtures dispersed in excess water.

olein (MO) dispersed in excess water. ME and MO are intermediates in the fat digestion and differ only in the configuration of the double bond in their single acyl chain, which is *trans* in ME and *cis* in MO. In contrast to the preceding examples, ME and MO form spontaneously cubic phases over wide ranges of temperature and hydration [36,65–67]. The *T*,*p* phase diagrams of ME and MO in excess water are presented in Fig. 9. As can clearly be seen, the small change in the acyl chain double bond configuration, from *trans* (ME) to *cis* (MO), causes a dramatic change in the observed phase behaviour. In the system MO–water, the cubic Q_{II}^D phase is stable over wide ranges of temperature and pressure. The *cis* configuration of MO leads to a more wedge-like molecular shape and a strong tendency for a MO monolayer to curve toward the water. Hence, the formation of lamellar phases, which requires a cylindrical molecular shape, is disfavoured. Analysis of the infrared CH_2 stretching and wagging modes for evaluation of conformational states in the various disordered (L_α, Q_{II}^P, Q_{II}^D) structures of ME reveals different populations of *gauche* conformers and kinks in these fluid-like phases [67]. From the analysis of the carbonyl stretching mode vibration also small differences in the level of hydration of different bicontinuous cubic phases have been detected. Compared to the Q_{II}^{D} phase of ME, the lipid chains of the body centred cubic phase Q_{II}^{P} seem to contain a slightly higher population of *gauche* sequences, and a slightly lower level of hydration of the carbonyl group.

3.1.2. Binary lipid mixtures

3.1.2.1. Fatty acids/phosphatidylcholines. The addition of fatty acids to aqueous phosphatidylcholine dispersions changes drastically the observed *T,p* phase behaviour. Dispersions of pure phosphatidylcholines merely exhibit lamellar phases. Non-lamellar (inverse hexagonal and inverse cubic) phases can be induced by adding fatty acids, such as lauric acid (LA), myristic acid (MA), palmitic acid (PA), or stearic acid (SA). Fatty acids also influence the fusogenicity of biological membranes, because they relieve the formation of non-lamellar intermediate structures, which have to occur in the process of membrane fusion.

The gel to fluid phase transition temperature of a DMPC/MA(1:2) mixture is 25°C above the main transition temperature, T_m, of pure DMPC dispersions, which is 23.9°C. The reason for this observation is that the fatty acid molecules act as spacers between the lecithin molecules and reduce the steric repulsion between the relatively bulky lecithin head groups. Furthermore, relatively strong hydrogen bonds exist between the head groups of lecithin and fatty acid molecules, allowing the formation of relatively stable 1:2-lecithin–fatty acid complexes in the gel phase. The resulting change in the lateral pressure profile across a monolayer results in fluid non-lamellar phases being energetically favoured over the fluid lamellar L_α phase at temperatures above the chain melting temperature. However, the lamellar L_α phase has been found in the system DLPC/LA, and only under non-equilibrium conditions after pressure jumps from the gel (L_β) to the fluid (Q_{II}, H_{II}) phase region in DMPC/MA(1:2). In the liquid–crystalline phase region of DMPC/MA(1:2), DPPC/PA(1:2), and DSPC/SA(1:2) mixtures, no liquid–crystalline lamellar phase is observed under equilibrium conditions, the lamellar gel phase directly transforms to non-lamellar liquid–crystalline phases. In Fig. 10, the *T,p* phase diagrams of the 1:2-mixtures DLPC/LA, DMPC/MA, DPPC/PA, and DSPC/SA in excess water are presented [59]. In the latter three systems, a phase separation occurs at low temperatures leading to a lamellar crystalline phase L_c composed of the pure fatty acid and a lamellar crystalline phase L_c^{com} being a mixture of the fatty acid and the phosphatidylcholine lipid. At temperatures above the chain melting, the inverse hexagonal phase H_{II} is found to be the stable liquid–crystalline phase in DSPC/SA(1:2) and DPPC/PA(1:2) dispersions, whereas this phase coexists with cubic phases in the case of DMPC/MA(1:2) and is even replaced by the pure cubic Q_{II}^{P} phase over a limited temperature interval in the DLPC/LA(1:2) system. This ob-

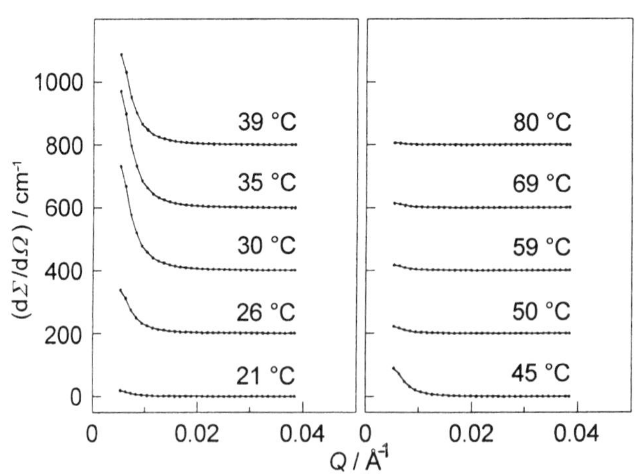

Fig. 11. SANS curves of a contrast-matched DMPC-d_{54}/DSPC(1:1) mixture dispersed in excess water at 1 bar and selected temperatures.

served trend in the phase behaviour is largely a result of the decreasing lipid chain length from DSPC/SA to DLPC/LA. The large splay of the fluid acyl chains in the DSPC/SA and DPPC/PA systems leads to a large spontaneous negative curvature of the monolayers that can only be adopted by the structure of the H_{II} phase. Contrary, the shorter lipid chains in the DMPC/MA and DLPC/LA systems increase the chain packing stress of the inverse hexagonal phase so that the formation of cubic phases is favoured.

It has been shown that the tendency for interfacial curvature can be reduced dramatically by a decreased fatty acid fraction in the lecithin/fatty acid mixtures rather than by an increase in pressure [59]. The marked differences between the effects of pressure and monolayer composition on the phase behaviour of lecithin/fatty acid mixtures reflect the fact that compositional variations cause large changes in the lateral pressure between amphiphiles, whereas hydrostatic pressure does not. Hence, pressure provides an extremely fine resolution parameter for probing the stability and geometry of lyotropic mesophases.

3.1.2.2. Phase-separated phospholipid mixtures. Phase diagrams of binary mixtures of saturated phosphatidylcholine lipids are typically characterized by a lamellar gel phase at low temperatures, a lamellar fluid phase at high temperatures, and an intermediate fluid–gel coexistence region, i.e. a phase separation. There is still a considerable controversy regarding the lateral membrane organization in the two-phase coexistence region. Recently, SANS experiments using multilamellar vesicles have shown the presence of large-scale fractal structures of coexisting gel and fluid regions in binary lipid membranes [68–71]. The mixtures investigated in these SANS studies were DMPC/DPPC(1:1), DMPC/DSPC(1:1), and DLPC/DSPC(1:1) in excess water. The fluid–gel coexistence region in the DMPC/DPPC system is rather narrow, indicating a nearly ideal mixing behaviour of the two components (isomorphous system). In comparison, the coexistence region in the DMPC/DSPC system is broader and manifests pronounced deviations from ideality. In both systems, the lipids are completely miscible in the all-gel and all-fluid phase. The strongest deviation

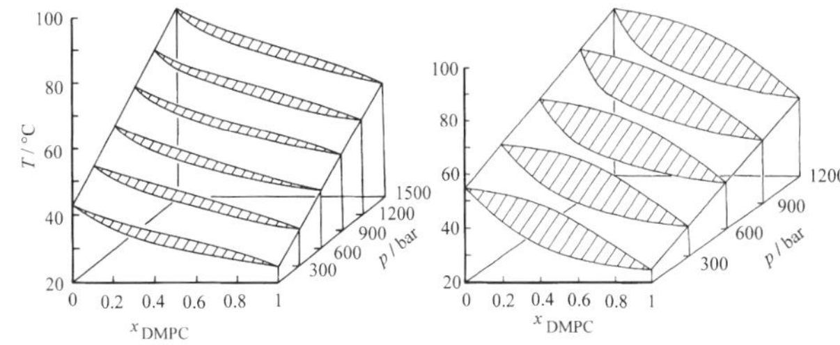

Fig. 12. *T,x* phase diagram of the equimolar binary lipid mixtures DMPC/DPPC and DMPC/DSPC in excess water as a function of pressure.

from ideality occurs in mixing DLPC and DSPC, where the acyl chain mismatch is so large that the components are essentially immiscible in the gel phase at low temperatures (peritectic system). In order to observe the concentration fluctuations caused by the gel–fluid phase coexistence in these binary phospholipid mixtures, SANS studies in combination with the H/D substitution technique were performed. One of the lipid components was deuterated in part, thus having a relatively high scattering length density (for D-atoms the scattering length is $b_D = 0.667\times 10^{-12}$ cm), whereas the scattering length density of the other lipid was significantly lower due to the smaller scattering length of H-atoms ($b_H = -0.374\times 10^{-12}$ cm). The scattering length density is given as the sum of all atom scattering lengths per volume:

$$\rho = \sum_i b_i / V \tag{10}$$

An equimolar phospholipid mixture, characterized by a mean scattering length density, was then dispersed in a H_2O–D_2O solvent mixture of the same scattering length density. Under these so-called matching conditions, no SANS signal is obtained for homogeneously mixed lipids in the all-gel or all-fluid phases, since then the scattering length density is constant over the whole sample ($\rho_{lipid} = \rho_{water}$). However, in the case of gel–fluid phase heterogeneities, SANS occurs due to the different compositions and scattering length densities of the two phases ($\rho_{gel} \neq \rho_{fluid}$). The differential coherent scattering cross-section as a function of momentum transfer Q is given as

$$\frac{d\Sigma}{d\Omega} = n\Delta\rho^2 V^2 P(Q)S(Q) \tag{11}$$

where n and V denote the number density and volume of the scattering molecular clusters, respectively. $P(Q)$, which is the square modulus of the form factor, describes the scattering contribution of individual clusters, whereas $S(Q)$, the structure factor, represents the scattering contribution due to the spatial distribution of the individual scatterers. $\Delta\rho$ is the contrast factor, i.e. the difference between the scattering length densities. As an illustrative example, several SANS curves of an equimolar DMPC-d_{54}/DSPC mixture in excess water are plotted as a function of temperature in Fig. 11. By comparison with the corresponding phase diagram (see Fig. 12), one clearly sees that significant SANS occurs within the gel–fluid coexistence region only. It is noted that the transition temperatures, T_m, of the pure lipid components at ambient pressure are 55°C for DSPC-h_{88} and 24°C for DMPC-h_{72}. Deuteration lowers T_m leading to a transition temperature of about 20°C for DMPC-d_{54} and a shift of the solidus line of the two-phase region of the mixture to slightly lower temperatures. The SANS intensity within the coexistence region may be analysed by plotting ln(dΣ/dΩ) vs. ln Q. The obtained straight lines over the whole Q-range studied are indicative of a fractal-like structure of the sample. In the case of surface fractals, which are scatterers having a fractal surface only, the scattering law is given by [72]

$$\frac{d\Sigma}{d\Omega} \propto Q^{D_s-6} \tag{12}$$

with D_s being the surface-fractal dimension ($2 < D_s < 3$). If the scatterers in the sample are mass fractals, the SANS intensity is described by

$$\frac{d\Sigma}{d\Omega} \propto Q^{-D_m} \tag{13}$$

with D_m being the mass-fractal dimension ($0 < D_m < 3$). From the double-logarithmic plots of Fig. 11 at 1 bar a slope of -3.3 ± 0.1 is obtained which yields a surface-fractal dimension of $D_s = 2.7 \pm 0.1$. Depending on the acyl chain mismatch of the lipid components, the clusters scatter like surface or mass fractals [68]. The data also imply that gel and fluid domains are correlated across many bilayers in a vesicle and that segregation into a minority and a majority phase occurs.

The results obtained imply that the real membrane

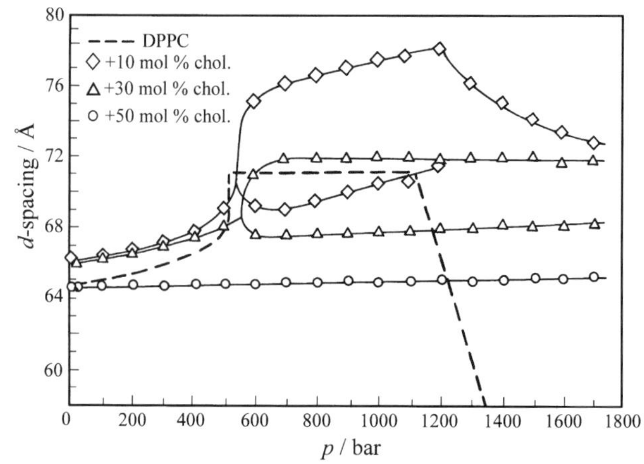

Fig. 13. Pressure dependence of the lamellar lattice constant of the system DPPC/cholesterol (0, 10, 30, 50 mol%) in excess water at 55°C.

structure in the gel–fluid coexistence region of binary phosphatidylcholine mixtures deviates strongly from the simple structure of large gel and fluid domains separated by smooth boundaries, which is expected from the equilibrium phase diagrams of the lipid mixtures and which is generally observed for macroscopically large binary fluid mixtures. The heterogeneous membrane structures observed in the two-phase coexistence regions might be due to interfacial wetting effects, such as those suggested recently in computer simulation studies [73]. In these calculations, the interface between coexisting gel and fluid phase domains is found to be enriched by one of the lipid species leading to a decrease of the interfacial tension and hence to a stabilization of non-equilibrium lipid domains. These and further results for similar binary lipid systems suggest that such strongly heterogeneous domain morphologies might be a rather common phenomenon. With increasing pressure, the gel–fluid coexistence region of a DMPC/DSPC mixture is shifted toward higher temperatures (see Fig. 12). A shift of about 22°C/kbar is observed, similar to the slope of the gel–fluid transition line of the pure lipid components (Fig. 4). We noted that the mixing behaviour of DMPC and DSPC deviates even more from ideality at 1000 bar than at 1 bar, as can be inferred from the significant small-angle scattering intensity observed at 1000 bar and 21°C, i.e. below the gel–fluid coexistence region.

3.1.3. *Effect of additives*

Not only the nature of phospholipid phase transitions, but also the way how they are affected by the incorporation of other species interacting with these membranes have attracted considerable experimental and theoretical attention. Firstly, they intimately reflect the molecular interactions of the membrane and may thus help in understanding membrane systems and function on a molecular level. Secondly, the addition of steroids, peptides, anaesthetics and drugs to membrane systems is of biological and pharmacological relevance. Generally, changes in membrane structure, phase segregation or transitions to other lamellar or to non-lamellar phases can be observed upon incorporation of additives into lipid bilayer membranes. Several additives have also been studied under high pressure conditions, such as mono- and divalent ions, anaesthetics, steroids and polypeptides [35,45,74–76]. Two examples are given here only.

3.1.3.1. Cholesterol effect. The incorporation of cholesterol into lipid membranes has a drastic effect on their structure and phase behaviour [35,45]. It is generally assumed that cholesterol causes conformational disorder in phospholipids below their gel to liquid–crystalline phase transition temperature and increases rigidity of phospholipid bilayers at temperatures above T_m. Increasing cholesterol concentration leads to a drastic reduction of the main transition enthalpy, ΔH_m, until at cholesterol contents higher than 30–50 mol% the main transition vanishes. We have examined the structure of DPPC–cholesterol mixtures in excess water. As an example, Fig. 13 shows the lamellar *d*-spacing of DPPC/cholesterol mixtures in excess water for selected cholesterol concentrations at $T = 55°C$. In pure DPPC, the *d*-spacing in the L_α phase increases slightly up to a pressure of approx. 500 bar where the transition to the rippled gel phase $P_{\beta'}$ occurs. At pressures above 1100 bar, the transition to the (partially) interdigitated $L_{\beta i}$ phase gradually takes place, which is accompanied by a drastic decrease in *d*-spacing [31]. Even up to cholesterol concentrations of 50 mol%, the lipid bilayer structure is preserved and no non-lamellar phases are formed. Addition of 10 and 20 mol% cholesterol does not lead to significant differences in the lamellar repeat unit in the disordered fluid phase. Above about 500 bar, the transition to a two-phase coexistence region is observed. In the system DPPC/10 mol% cholesterol, the lamellar stacking in one of these phases, which is probably the cholesterol-rich phase, is swelling drastically. Since the diffraction lines of each phase are relatively sharp, the phase separation is one where each phase forms separate three-dimensional domains, probably many tens of layers thick. Above 1200 bar, a phase transformation to a single high pressure gel phase region takes place. In the system DPPC/30 mol% cholesterol, the pressure-induced two-phase coexisting region is observed over the whole pressure range covered. At these higher cholesterol concentrations, the interbilayer swelling is reduced. At cholesterol concentrations of 50 mol%, the lipid main transition has disappeared and the *d*-spacing observed is almost independent of temperature and pressure. As can already be seen from this example, the incorporation of cholesterol, a major constituent of natural membranes, has a drastic effect on the lipid bilayer conformation and phase behaviour. Even low cholesterol concentrations lead to phase coexistence regions. Further fluorescence and FT-IR pressure studies [35,74] clearly demonstrate the ability of sterols to efficiently regulate the structure, dynamics and hydrophobicity of biomembranes, so that they can withstand even drastic changes in environmental conditions, such as in temperature and external pressure.

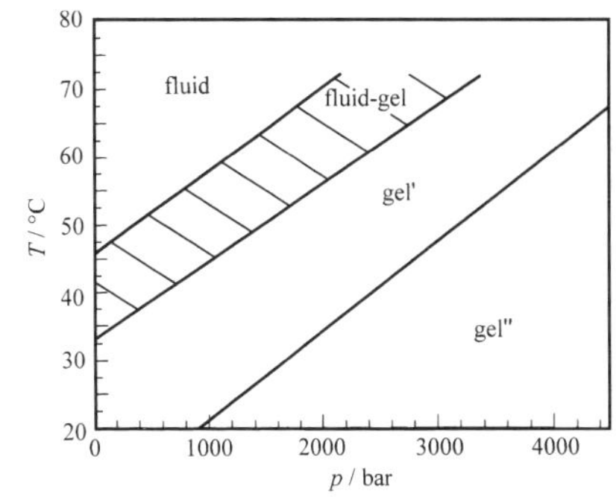

Fig. 14. Tentative phase diagram of DPPC–GD (4.7 mol%) in excess water as obtained from SAXS data and FT-IR studies.

3.1.3.2. Incorporation of polypeptides. We also investigated the effect of the incorporation of polypeptides, such as gramicidin D (GD), on the structure and phase behaviour of phospholipid bilayers of different chain lengths [75,76]. Gramicidin is a linear hydrophobic polypeptide antibiotic that forms specific channels in lipid membranes for the transport of monovalent cations [77]. Bilayers containing this channel forming polypeptide are often used as a

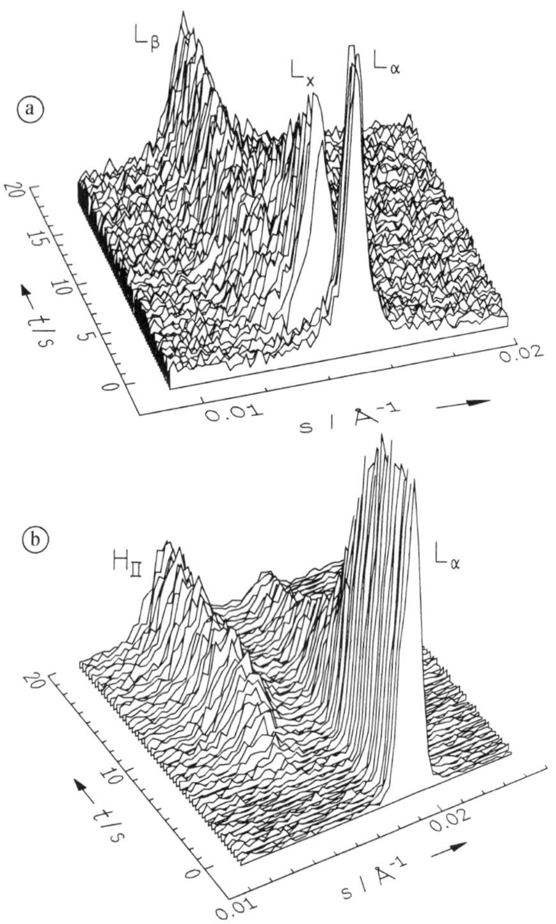

Fig. 15. Diffraction patterns of (a) DEPC in excess water after a pressure jump from 200 to 370 bar at 18°C, and (b) DOPE in excess water after a pressure jump from 300 to 110 bar at 20°C.

model for protein–lipid interaction studies. Pressure has been applied so as to be able to finely tune the lipid chain lengths and conformation and to select different lipid gel phases. Infrared spectral parameters obtained by FT-IR spectroscopy and data obtained from X-ray diffraction and ^{2}H-NMR experiments were used to detect structural changes upon incorporation of GD into DMPC, DPPC and DSPC lipid bilayers. Analysis of the infrared amide I band frequencies allowed us to determine the corresponding peptide structure adopted in the lipid environment.

Gramicidin is highly polymorphic, being able to adopt a wide range of structures with different topologies. Common forms are the dimeric single-stranded right-handed $\beta^{6.3}$-helix with a length of 24 Å, and the antiparallel double-stranded $\beta^{5.6}$-helix, being approx. 32 Å long. For comparison, the hydrophobic fluid bilayer thickness is about 30 Å for DPPC bilayers, and the hydrophobic thicknesses of the gel phases are 4–5 Å larger. Depending on the GD concentration, significant changes in the lipid bilayer structure and phase behaviour were found, such as the disappearance of some of the gel phases formed by the pure lipid bilayer systems and the formation of extensive two-phase coexistence regions at higher GD concentrations. In the liquid–crystalline phases of the phospholipid bilayers, generally more orientational order is induced in the lipid molecules by the incorporation of GD. Vice versa, also the peptide conformation is influenced by the lipid environment. Depending on the phase state and lipid acyl chain length, GD adopts at least two different types of quaternary structures in the bilayer environment, a double helical 'pore' and a helical dimer 'channel'. With regard to the changes of the bilayer thickness at the gel/fluid main phase transition of DPPC and DSPC, we showed that the conformational equilibrium of the peptide is changed. In gel-like DPPC and DSPC bilayers the equilibrium of the GD species in the lipid bilayer is shifted in favour of the double helical configuration [75]. As an example, we show the tentative p,T phase diagram for the DPPC–GD (4.7 mol%) mixture up to pressures of 4000 bar in Fig. 14. Gramicidin insertion clearly has a significant influence on the lipid topology and phase behaviour. To avoid large hydrophobic mismatch the lipid phase state is altered and broad fluid–gel coexistence regions are formed. In the phase-separated regions, interfacial adsorption and condensation phenomena may be operative. This example clearly demonstrates not only that the lipid bilayer structure and phase behaviour drastically depend on the polypeptide concentration, but also that the peptide conformation can significantly be influenced by the lipid environment. No pressure-induced unfolding of the polypeptide is observed up to 10 kbar, which might indicate that the lipid bilayer membrane has a stabilizing effect on the polypeptide conformation. For large integral and peripheral proteins, pressure-induced changes in the physical state of the membrane may lead to a weakening of protein–lipid interactions and they may be released from the membranes, however.

3.1.4. *Kinetics of phase transformations in lipid systems*

Phase transitions between lipid mesophases must be associated with deformations of the interfaces which, very often, also imply their fragmentation and fusion so that not only the symmetry changes but also the topology of the lipid organization. Depending on the topology of the structures involved, transition phenomena of different complexity are observed. In addition, the transition rates and mechanisms depend on the level of hydration of the structures involved and on the forces driving the transition. We used the synchrotron X-ray diffraction technique to record the temporal evolution of the structural changes after induction of the phase transition by a pressure jump across the phase boundary [78,79]. In the following we will discuss two representative examples.

3.1.4.1. Lamellar–lamellar transitions. First we present pressure jump experiments carried out in 1,2-dielaidoyl-*sn*-glycero-3-phosphatidylcholine (DEPC)–water dispersions to study the L_β–L_α gel to fluid main transition, which occurs at $T_m = 12°C$ at ambient pressure and which has a pressure dependence of $dT_m/dp = 20°C\ kbar^{-1}$. Selected SAXS diffraction patterns at 18°C after a pressure jump from 200 bar (L_α phase) to 370 bar (L_β phase) are depicted in Fig. 15. An intermediate structure is clearly observable here. The first order (001) Bragg reflection of the initial L_α phase ($a = 66$ Å) vanishes in the course of the pressure jump (< 5 ms). The first diffraction pattern collected after the pressure jump exhibits a Bragg reflection of a new lamellar structure L_x with a slightly larger *d*-value, which increases with time. The lattice constant of the L_β phase formed is 78 Å. The transition is complete after about 15 s. In equilibrium measurements, no such intermediate lamellar structure is detectable.

3.1.4.2. Lamellar–non-lamellar transitions. Experiments for investigating the lamellar–H_{II} transition kinetics have been performed, for example, on DOPE dispersions. The *T,p* phase diagram of DOPE in excess water is depicted in Fig. 7. Fig. 15b shows the diffraction patterns at 20°C after a pressure jump from 300 to 110 bar. Clearly, the (001) reflection of the L_α phase and the (10) reflection of the developing H_{II} phase can be identified. In this case, a two-state mechanism is observed. Interestingly, we find that successive pressure jumps lead to an acceleration of the phase transition kinetics. The half-transit time decays from 8.5 s for the first pressure jump to 5.3 s after the fourth jump. After the pressure jump, an induction period of several seconds is observed before the first Bragg reflections of the newly formed H_{II} phase appear. Upon successive pressure cycles, this induction period decreases. An explanation for this phenomenon might be the formation of defect structures, such as inverted intermicellar intermediates [63,64], which are formed during the pressure cycles and which have not healed between successive pressure cycles. This observation also shows that the history of sample preparation plays an essential role in this kind of studies. It has been found that with increasing pressure jump amplitude the induction period decreases and the rate of phase transformation increases.

Generally, as has also been found in studies of pressure and temperature jump-induced phase transitions of other systems [80,81], these results show that the relaxation behaviour and the kinetics of pressure-induced lipid phase transformations depend drastically on the topology of the lipid mesophase, and also on the temperature and the driving force, i.e. the applied pressure jump amplitude Δp. Often multicomponent kinetic behaviour is observed, with short relaxation times (probably on the nanosecond to microsecond time scale) in a burst phase referring to the relaxation of the lipid acyl chain conformation in response to the pressure change, which leads to the small changes in the observed lattice constants right after the pressure jump. The longer relaxation times measured here are due to the kinetic trapping of the system. In most cases the rate of the transition is limited by the transport and redistribution of water into and in the new lipid phase, rather than being controlled by the time required for a rearrangement of the lipid molecules. This can be inferred from the lattice relaxation experiments performed in the lipid one-phase regions and by modelling the data using simple hydrodynamics. The obstruction factor of the different structures, especially in cases where non-lamellar (hexagonal and cubic) phases are involved, controls the different kinetic components. In addition, nucleation phenomena and domain size growth

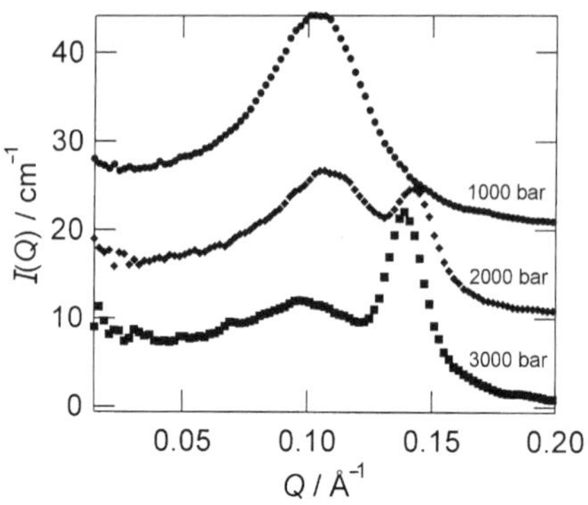

Fig. 16. SANS diffraction patterns of the ternary system AOT/*n*-octane/H_2O (45:25:30 wt%) at 25°C for selected pressures.

of the structures evolving play a role. In many cases (e.g. DEPC, DMPC–MA) a digression of the mechanism of phase transformation observed under slow-scanning equilibrium conditions appears under high free energy gradients (here large pressure jump amplitudes), and the high driving force drives the system through a correlated ordered intermediate state. In cases where the transition occurs without change in water content within the mesophase, such as in the intercubic $Q_{II}^G \rightarrow Q_{II}^D$ transition of the system DLPC–LA (1:2) at a fixed water composition of 50 wt%, the kinetics is faster [82]. As mechanism for this cubic–cubic transformation, a stretching mechanism has been proposed whereby each 4-fold junction in the Q_{II}^D phase is formed by bringing together two 3-fold junctions in the Q_{II}^G phase.

3.2. Pressure effects on surfactant phases and microemulsions

The literature which deals with the effects of high pressure on surfactant phases, microemulsions and surfactant–protein complexes is still very small. Only a few studies have been performed on the effect of pressure on the critical micelle concentration (cmc) of ionic and non-ionic surfactants (e.g. SDS, C_8E_5) in aqueous solutions [83–87]. It has been shown that the cmc vs. pressure curves have a maximum in the pressure range 0.001–4 kbar. The initial compression causes the dissociation of micelles whereas successive compression above a certain pressure causes aggregation of monomers to micelles again. The data thus imply that the partial molar volume change ΔV_{mic} on micellation is positive in the lower pressure region and negative for the higher pressures. The similar characteristic pressure dependence of the cmc for both ionic and non-ionic surfactants, as well as the existence of an alkyl chain length dependence for ionic surfactants suggest that the effect of pressure is largely independent of the hydrophilic head group of the surfactant.

Compression of concentrated micellar systems and microemulsions can lead to changes in aggregation number and will finally induce aggregate structures with smaller partial molar volumes of their amphiphiles. Only very few systems have been investigated so far. Aqueous micellar solutions of sodium and potassium decanoates were measured by FT-IR spectroscopy as a function of external pressure up to 50 kbar [88]. In these systems, phase transitions from a micellar state to a coagel state were observed. It has also been shown that the critical coagulation pressure is drastically affected by the nature of the counterion interaction. By means of SANS experiments, the dependence of pressure of the mean aggregation number of rod-like tetradecyldimethylaminoxide (TDMAO) micelles was determined, and at high pressure and low temperature the formation of a lamellar phase has been found to be very likely [89]. Furthermore, it was observed that pressurization of the bis[2-ethylhexyl]sulphosuccinate (AOT)–water–*n*-decane microemulsion system leads to a pressure-induced phase transition from a water-in-oil droplet structure to an ordered lamellar structure [90,91].

Such ternary surfactant–water–organic solvent systems were proposed as a medium for modelling the membrane environment of enzymes. Enzymes of various origins were solubilized in such systems and were found to function in different mesophases (reversed micellar, lamellar, reversed hexagonal, cubic, etc.) [92]. It has been found that phase transitions in the ternary system resulting from a change of the concentration ratio can cause dramatic changes in the enzyme catalytic activity. Interestingly, it has also been demonstrated that the application of hydrostatic pressure to enzymes placed in such nanocontainers can bring an additional advantage for both increasing the enzyme stability and modulating the enzyme activity [92,93]. In order to understand the effects of pressure on the enzyme activity, which may be caused by conformational changes of the amphiphiles or mesophase transformations, detailed studies of pressure effects on these surfactant systems are required. So far only very little data are available. As an example, Fig. 16 shows the SANS profiles of a AOT/*n*-octane/H_2O (45:25:30 wt%) mixture at 25°C for selected pressures. A broad peak at momentum transfers of about 0.1 $Å^{-1}$ is observed up to pressures of about 1000 bar. This peak is due to intermolecular correlations of the dense water-in-oil droplets (micellar phase L_2). Upon application of higher pressures, a new peak develops at about 0.14 $Å^{-1}$, which is probably due to the (001) reflection of a lamellar phase with repeat distance of ~44 Å. These data clearly indicate that already relatively low pressures can lead to drastic changes in surfactant structure and topology which might influence enzyme activity. Further studies of protein–surfactant mixtures using contrast variation techniques will lead to a more detailed understanding of enzyme–surfactant interactions and the correlation with the corresponding enzyme activity.

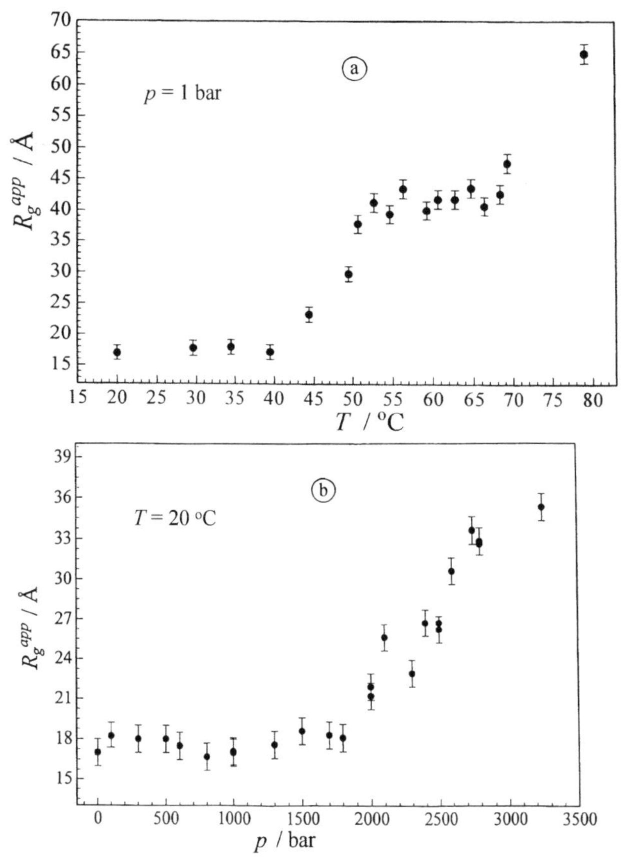

Fig. 17. Apparent radius of gyration R_g of Snase (1% (w/w), pH 5.5) as a function of temperature and pressure (at T= 20°C).

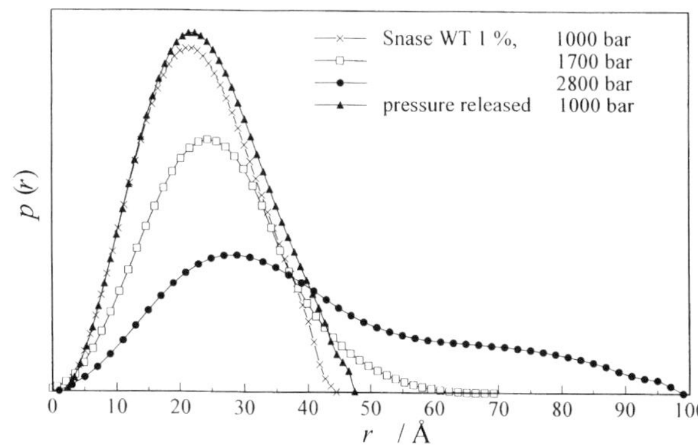

Fig. 18. Pair distance distribution function $p(r)$ of Snase (1% (w/w), pH 5.5) at selected pressures (T= 25°C).

It is clear that the application of the pressure variable in this research area has only just started and many interesting results are expected in the near future. These pressure studies may be important for the development of new technological and pharmaceutical applications, such as high pressure food processing and micellar baroenzymology.

4. Pressure effects on proteins in solution

It has long been known that the application of hydrostatic pressure results in the disruption of the native protein structure [94] due to the decrease in the volume of the protein–solvent system upon unfolding. Pressure denaturation studies thus provide a fundamental thermodynamic parameter for protein unfolding, the ΔV^0, in addition to an alternative method for perturbing the folded state, and thus extracting its stability. A number of reviews on effects of pressure on proteins discuss these volume changes in greater detail [17–21]. Denaturation is usually studied at atmospheric pressure using high temperature, guanidinium hydrochloride or urea as denaturants. However, interpretation of the results obtained using such methods is complicated by the facts that: (1) varying the temperature changes both the volume and the thermal energy of the system at the same time, and (2) the thermodynamic parameters of denaturation by guanidinium chloride or urea are influenced by the binding of these molecules to proteins. By contrast, if denaturation is induced at constant temperature in the absence of chemical denaturants, one can try to describe the change in protein Gibbs free energy as a function of the interatomic distances in the protein molecule which, in turn, are changed by variations in applied pressure. The use of pressure is also advantageous from a methodological point of view: the transition to native conditions (renaturation) is achieved simply by releasing the pressure. In general, the effects of pressure on proteins are reversible, and only seldom are they accompanied by aggregation or changes in covalent structure. The net volume change on denaturation comprises the effects of disruption of non-covalent bonds, changes in protein hydration and conformational changes. The incompressible volume of covalent bonds maintains a volume that does not change on unfolding. The reduction in the net volume seems to be predominantly the result of the disappearance of solvent-inaccessible voids inside the protein (see C. Royer in [6] and in this volume).

The fundamental understanding of the process of the folding of proteins to their native state has fascinated researchers for decades and remains one of the most interesting and challenging issues in modern biophysics. One experimental approach to understand the folding process is to characterize the nature of the barrier to folding or unfolding and the corresponding transition state. In this respect, pressure studies are also of particular use. Moreover, pressure studies present an important advantage due to the positive activation volume for folding, the result of which is to slow down folding substantially, in turn allowing for relatively straightforward measurements of structural order parameters that are difficult or even impossible to measure on much faster time scales.

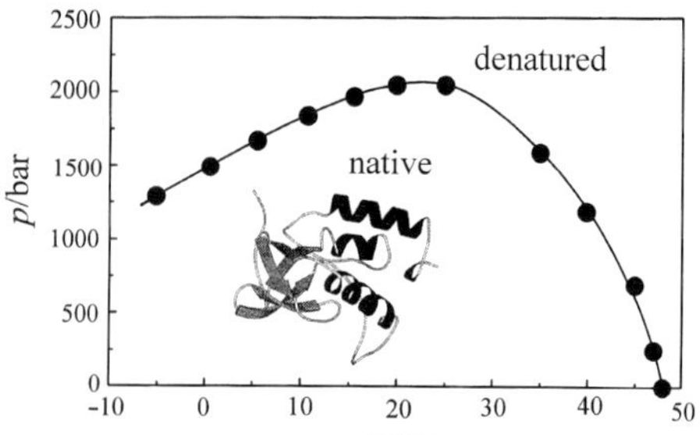

Fig. 19. T,p–stability diagram of Snase at pH 5.5.

As an example, we present data on the pressure-induced unfolding and refolding of staphylococcal nuclease (Snase). This protein has served as a model for protein folding because it is small and has a well-known native structure. These studies were performed using synchrotron small-angle X-ray scattering and Fourier transform infrared spectroscopy, which monitor changes in the tertiary and secondary structural properties of the protein upon pressurization or depressurization.

4.1. Equilibrium studies of protein denaturation

Snase is a small protein of about 17.5 kDa containing 149 amino acids and no disulphide bonds. In the crystalline state, the protein contains 26.2% helices, 24.8% β-sheets (barrel), 7.4% extended chains, 24.8% turns and loops, and 8.7% unordered chains (8.1% are uncertain). Analysis of the high pressure

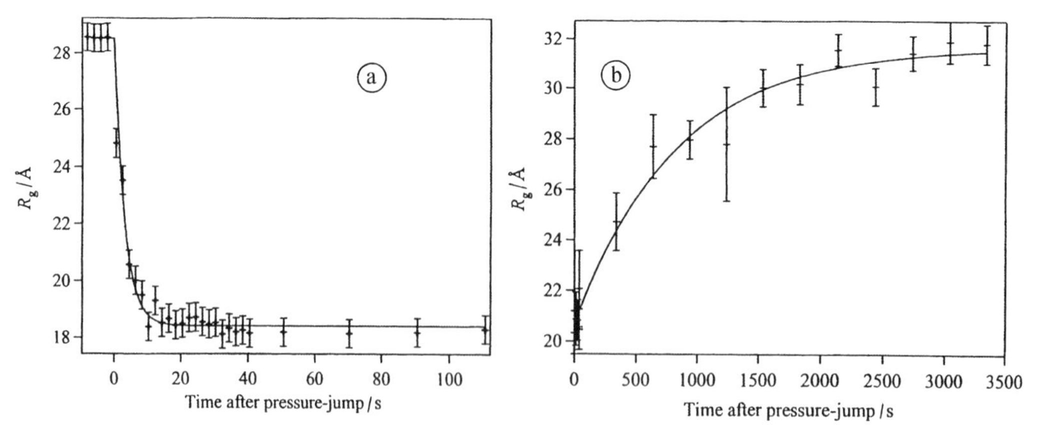

Fig. 20. (a) Time evolution of the radius of gyration R_g after a negative pressure jump from 4000 bar to 800 bar. The fit of data to a single exponential decay function (line) yields a time constant of 4.5 s. (b) Time dependence of the increase in R_g after a positive pressure jump from 1500 bar to 3000 bar. The fit of the data to a single exponential function (line) yields a time constant of 14 min.

SAXS data reveals that over a pressure range from atmospheric pressure to approx. 3 kbar, the radius of gyration R_g of the protein increases from a value near 17 Å for native Snase 2-fold to a value near 35 Å (Fig. 17). A large broadening of the pair distribution function $p(r)$ is observed over the same range, indicating a transition from a globular to an ellipsoidal or dumbbell-like structure (Fig. 18) [14,15]. Deconvolution of the FT-IR amide I′ absorption band reveals a pressure-induced denaturation process over the same pressure range as the SAXS that is evidenced by an increase in disordered and turn structures and a drastic decrease in the content of β-sheets and α-helices [14]. The pressure-induced denatured state above 3 kbar retains nonetheless some degree of β-like secondary structure and the molecules cannot be described as a fully extended random polypeptide coil, which is in accord with the SAXS results. Assuming the pressure-induced unfolding transition of Snase to occur essentially as a two-state process, analysis of the FT-IR pressure profiles yields a Gibbs free energy change for unfolding of $\Delta G_u^0 = 17 \pm 4$ kJ/mol and a volume change for unfolding of $\Delta V_u^0 = -80 \pm 20$ ml/mol at ambient temperature and pressure. For comparison, temperature-induced denaturation involves a further unfolding of the protein molecule which is indicated by a larger R_g-value of 45 Å (Fig. 17), and significantly lower fractional intensities of IR bands associated with secondary structure [14,15].

There are many indications that the conformation of a protein denatured by pressure is more compact than that of a protein denatured by temperature or chemical agents. A growing body of experimental evidence shows that, according to the characteristic structural features (the presence of secondary structure and the absence of well-organized tertiary structure), pressure-denatured proteins resemble 'molten globule'-type structures. This may not seem too surprising, as pressure is known to favour the formation of hydrogen bonds, which maintain the secondary structure network, but is unfavourable for hydrophobic interactions, which are predominantly responsible for maintaining the tertiary structure of a protein. The idea is supported by recent theoretical results which suggest water penetration into the protein interior as a likely mechanism for pressure denaturation of proteins due to a weakening of hydrophobic

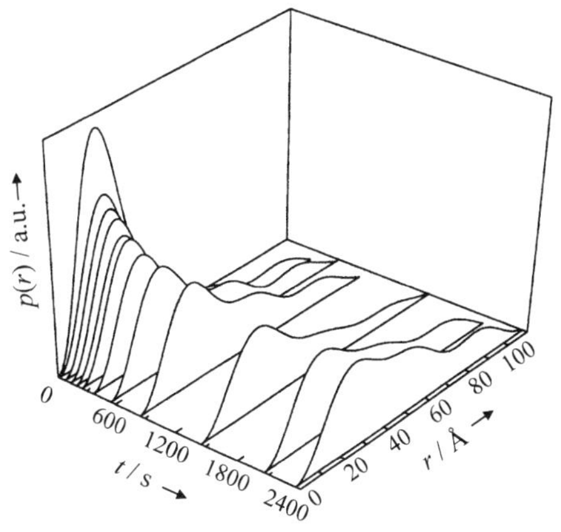

Fig. 21. Time evolution of the pair distance distribution function $p(r)$ of Snase (1% (w/w), pH 5.5) after a pressure jump from 1 to 3 kbar at $T = 25°C$.

interactions, as opposed to the temperature-induced unfolding process [95]. Formation of molten globule-like structures under pressure has been suggested for other proteins, including lysozyme, arc repressor and ribonuclease [19–21,96].

Recently we have also characterized the temperature- and pressure-induced unfolding of Snase using high precision densitometric measurements [97]. At 45°C, we calculate a decrease in apparent molar volume due to pressure-induced unfolding of -55 cm^3 mol^{-1}. The 3-fold increase in compressibility upon unfolding reflects a transition to a partially unfolded state, which is consistent with our results obtained for the radius of gyration of the pressure-denatured state of Snase. The pressure midpoints at several temperatures obtained from the FT-IR and SAXS profiles are plotted as a phase diagram in Fig. 19 [14,15]. It exhibits the curvature which is typical of heat and cold denaturation of many monomeric proteins. The phase diagram will certainly depend on the individual protein secondary structural composition and will be more complicated for larger protein systems. Also additional regions in the phase diagram may appear, such as an extended region at high temperatures where aggregation occurs. One must also be aware of the fact that the unfolded states in the p,T plane can be of considerably different structure, and that long-lived metastable phases may occur. Interestingly, pressure–temperature stability diagrams of bacteriophages, and even bacterial cells, may be described by diagrams similar to that shown for proteins in Fig. 19 [1–3,6,9].

4.2. Kinetic studies of the un/refolding reaction of proteins

By crossing the phase boundary applying a *p* jump, the folding and refolding kinetics can be studied. As can be seen in Fig. 20, the rapid decrease of pressure for a solution of Snase at 25°C from for example 4000 bar (denaturing conditions) to 800 bar (native conditions) results in a relatively rapid decrease in the value of the radius of gyration, R_g, from near 29 to 18 Å [13]. The observed pressure jump relaxation profile for the decrease in R_g fits well to a single exponential decay with a time constant τ of 4.5 s. No burst phase is apparent in the profiles of R_g, demonstrating the absence of any significantly populated collapsed intermediates in the refolding reaction of Snase under pressure. In contrast, a positive pressure jump at 25°C from 1500 bar (near-native conditions) to 3000 bar (fully denaturing conditions) results in a very slow relaxation of R_g from 20 to 31 Å. Complete relaxation is only observed after 1 h. As for the negative pressure jumps, the positive pressure jump profile is well fit by a single exponential function, with a much longer time constant of $\tau = 14$ min. Fig. 21 shows the time evolution of the pair distance distribution function $p(r)$ of Snase (1% (w/w), pH 5.5) after a pressure jump from 1 to 3 kbar at $T = 25$°C. Within the accuracy of the present time resolution and accuracy of the experiment, no evidence for a distinct intermediate form is apparent in the time evolution of the pair distance distribution function, consistent with the single exponential profiles for the evolution of R_g.

We find that the activation volume for folding is large and positive (57 ± 4 ml/mol) and that for unfolding seems to be small and negative (−23 ± 3 ml/mol) [15]. The volume of the protein solvent system in the transition state is thus significantly larger than in the unfolded state and somewhat smaller than in the folded state, so that the transition state lies closer to the folded than to the unfolded state in terms of system volumes. The present pressure jump SAXS and FT-IR data, along with previous fluorescence pressure jump results [13–15,98,99], provide strong evidence for a two-state folding/unfolding model for Snase under pressure, in that secondary structure, chain collapse and tertiary structure all exhibit relaxation profiles on similar timescales. Thus, changes in these three order parameters appear to depend upon the same rate limiting steps for unfolding and folding. The positive activation volume for the folding process, which is responsible for the large increase in the relaxation time with pressure (allowing us to observe this process without resorting to ultrafast methods), implicates dehydration in the rate limiting step for folding. Our results for this small, fast folding protein conform relatively well to the type I scenario of the funnel model [100], i.e. a relatively smooth funnel topology. In such cases, the folding rate is simply determined by the barrier height. It will be of interest to incorporate into the funnel model the observed pressure-induced decrease in the folding rate for this protein and for other protein systems studied.

This study of the time dependence of chain collapse in pressure-induced folding reactions demonstrates that novel information about the nature of protein folding transitions and transition states can be obtained from a combination of high sensitivity small-angle X-ray scattering and spectroscopic techniques with high pressure perturbation. The stage is now well set for further work addressing more complex questions, such as the study of the folding reaction of oligomers and protein complex formation as well as for studies of aggregation phenomena, such as in β-lactoglobulin [101] where denaturation results in gel-like aggregates. Only a few studies have been performed in this direction so far. At pressures of 4–8 kbar, most small monomeric proteins unfold reversibly. Oligomeric proteins and multiprotein assemblies often dissociate into individual subunits already at pressures of 1–2 kbar. The effect of pressure in promoting dissociation may be explained by the imperfect packing of atoms at the subunit interface, and negative volume changes resulting from the disruption of hydrophobic and ionic bonds in the intersubunit region. After dissociation by pressure, subunits may undergo further conformational changes. It has also been pointed our recently that pressure studies might lead to a better understanding of the interactions that lead to aggregation and will thus enhance our ability to design inhibitors and therapeutics for aggregation driven diseases, including Alzheimer's, Parkinson's and prion diseases [101–103]. These examples clearly demonstrate that pressure-dependent studies can help delineate the free energy landscape of proteins and hence help elucidate which features are essential in determining the uniqueness and stability of the native conformational state.

Acknowledgements

Financial support from the Deutsche Forschungsgemeinschaft (DFG) and the Fonds der Chemischen Industrie is gratefully acknowledged. I thank Drs. C. Royer, J. Seddon, R. Templer, J. Jonas, and K. Heremans for many valuable discussions.

References

[1] C. Balny, R. Hayashi, K. Heremans, P. Masson (Eds.), High Pressure and Biotechnology, Colloque Inseram, Vol. 224, John Libbey Eurotext, Montrouge, 1992.
[2] V.V. Mozhaev, K. Heremans, H. Frank, P. Masson, C. Balny, Trends Biotechnol. 12 (1994) 493–501.
[3] H. Ludwig (Ed.), Advances in High Pressure Bioscience and Biotechnology, Springer, Berlin, 1998.
[4] M. Kato, R. Hayashi, Biosci. Biotechnol. Biochem. 63 (1999) 1321–1328.
[5] F. Abe, C. Kato, K. Horikoshi, Trends Microbiol. 7 (1999) 447–453.
[6] R. Winter, J. Jonas (Eds.), High Pressure Molecular Science, NATO ASI E358, Kluwer Academic Publishers, Dortrecht, 1999.
[7] J.C. Rostain, E. Martinez, C. Lemaire (Eds.), High Pressure Nervous Syndrome – 20 Years Later, ARAS-SNHP Publications, Amsterdam, 1989.
[8] M. Gross, R. Jaenicke, Eur. J. Biochem. 221 (1994) 617–630.
[9] V.V. Mozhaev, K. Heremans, J. Frank, P. Masson, C. Balny, Proteins Struct. Funct. Genet. 24 (1996) 81–91.
[10] J.L. Silva, G. Weber, Annu. Rev. Phys. Chem. 44 (1993) 89–113.
[11] C. Czeslik, R. Malessa, R. Winter, G. Rapp, Nucl. Instr. Methods Phys. Res. A 368 (1996) 847–851.
[12] R. Winter, C. Czeslik, Z. Kristallogr. 215 (2000) 454–474.
[13] J. Woenckhaus, R. Köhling, R. Winter, P. Thiyagarajan, S. Finet, Rev. Sci. Instrum. 71 (2000) 3895–3899.
[14] G. Panick, R. Malessa, R. Winter, G. Rapp, K.J. Frye, C. Royer, J. Mol. Biol. 275 (1998) 389–402.
[15] G. Panick, C.J.A. Vidugiris, R. Malessa, G. Rapp, R. Winter, C. Royer, Biochemistry 38 (1999) 4157–4164.
[16] O. Glatter, O. Kratky, Small Angle X-ray Scattering, Academic Press, New York, 1982.
[17] K. Heremans, L. Smeller, Biochim. Biophys. Acta 1386 (1998) 353–370.
[18] V.V. Mozhaev, N. Bec, C. Balny, Biochem. Mol. Biol. Int. 34 (1994) 191–201.
[19] J. Zhang, X. Peng, A. Jonas, J. Jonas, Biochemistry 34 (1995) 8631–8641.
[20] X. Peng, J. Jonas, J.L. Silva, Biochemistry 33 (1994) 8323–8329.
[21] J. Jonas, L. Ballard, D. Nash, Biophys. J. 75 (1998) 445–452.
[22] M.W. Lassalle, H. Yamada, K. Akasaka, J. Mol. Biol. 298 (2000) 293–302.
[23] G. Cevc (Ed.), Phospholipids Handbook, Marcel Dekker, New York, 1993.
[24] R. Lipowski, E. Sackmann (Eds.), Structure and Dynamics of Membranes, Vols. 1A and 1B, Elsevier, Amsterdam, 1995.
[25] J.M. Seddon, Biochim. Biophys. Acta 1031 (1990) 1–69.
[26] M. Böttner, D. Ceh, U. Jakobs, R. Winter, Z. Phys. Chem. 184 (1994) 205–218.
[27] P.-L.G. Chong, G. Weber, Biochemistry 22 (1983) 5544–5550.
[28] L.F. Braganza, D.L. Worcester, Biochemistry 5 (1986) 2591–2596.
[29] P.T.T. Wong, D.J. Siminovitch, H.H. Mantsch, Biochim. Biophys. Acta 947 (1988) 139–171.
[30] J. Jonas, C.-L. Cie, A. Jonas, P.J. Grandinetti, D. Campbell, A. Driscoll, Proc. Natl. Acad. Sci. USA 85 (1988) 4115–4117.
[31] R. Winter, W.C. Pilgrim, Ber. Bunsenges. Phys. Chem. 93 (1989) 708–717.
[32] A. Landwehr, R. Winter, Ber. Bunsenges. Phys. Chem. 98 (1994) 214–218.
[33] P.T.C. So, S.M. Gruner, E. Shyamsunder, Phys. Rev. Lett. 70 (1993) 3455–3458.
[34] D.A. Driscoll, J. Jonas, A. Jonas, Chem. Phys. Lipids 58 (1991) 97–104.
[35] O. Reis, R. Winter, T.W. Zerda, Biochim. Biophys. Acta 1279 (1996) 5–16.
[36] C. Czeslik, R. Winter, G. Rapp, K. Bartels, Biophys. J. 68 (1995) 1423–1429.
[37] A. Cheng, A. Mencke, M. Caffrey, J. Phys. Chem. 100 (1996) 299–306.
[38] S. Kaneshina, H. Ichimori, T. Hata, H. Matsuki, Biochim. Biophys. Acta 1373 (1998) 1–8.
[39] H. Ichimori, H. Matsuki, T. Hata, S. Kaneshina, in: H. Ludwig (Ed.), Advances in High Pressure Bioscience and Biotechnology, Springer, Berlin, 1998, pp. 161–164.
[40] S. Krishna Prasad, R. Shashidhar, B.P. Gaber, S.C. Chandrasekhar, Chem. Phys. Lipids 43 (1987) 227–235.
[41] C. Czeslik, O. Reis, R. Winter, G. Rapp, Chem. Phys. Lipids 91 (1998) 135–144.
[42] D. Worcester, B. Hammouda, Physica B 241-243 (1998) 1175–1177.
[43] D.C. Fiech, B.B. Bonev, M.R. Morrow, Phys. Rev. E 57 (1998) 3334–3343.
[44] P.M. Duesing, J.M. Seddon, R.H. Templer, D.A. Mannock, Langmuir 13 (1997) 2655–2664.
[45] R. Winter, A. Landwehr, T. Brauns, J. Erbes, C. Czeslik, O. Reis, in: J.L. Markley, D.B. Northrop, C.A. Royer (Eds.), High Pressure Effects in Molecular Biophysics and Enzymology, Oxford University Press, Oxford, 1996, pp. 274–297.

[46] G. Lindblom, L. Rilfors, Biochim. Biophys. Acta 988 (1989) 221–256.
[47] M.W. Tate, E.F. Eikenberry, D.C. Turner, E. Shyamsunder, S.M. Gruner, Chem. Phys. Lipids 57 (1991) 147–164.
[48] S. Hyde, S. Andersson, K. Larsson, Z. Blum, T. Landh, S. Lidin, B.W. Ninham, The Language of Shape. The Role of Curvature in Condensed Matter: Physics, Chemistry and Biology, Elsevier, Amsterdam, 1997.
[49] V. Luzzati, Curr. Opin. Struct. Biol. 7 (1997) 661–668.
[50] T. Landh, FEBS Lett. 369 (1995) 13–18.
[51] H. Delacroix, T. Gulik-Krzywicki, J.M. Seddon, J. Mol. Biol. 258 (1996) 88–103.
[52] H. Chung, M. Caffrey, Nature 368 (1994) 224–226.
[53] R.H. Templer, Langmuir 11 (1995) 334–340.
[54] R.H. Templer, D.C. Turner, P. Harper, J.M. Seddon, J. Phys. II France 5 (1995) 1053–1065.
[55] P. Mariani, B. Paci, P. Bösecke, C. Ferrero, M. Lorenzen, R. Caciuffo, Phys. Rev. E 54 (1996) 5840–5843.
[56] R.H. Templer, B.J. Khoo, J.M. Seddon, Langmuir 14 (1998) 7427–7434.
[57] R.H. Templer, J.M. Seddon, N.A. Warrender, A. Syrykh, Z. Huang, R. Winter, J. Erbes, J. Phys. Chem. B 102 (1998) 7251–7261.
[58] R.H. Templer, J.M. Seddon, P.M. Duesing, R. Winter, J. Erbes, J. Phys. Chem. B 102 (1998) 7262–7271.
[59] R. Winter, J. Erbes, R.H. Templer, J.M. Seddon, A. Syrykh, N.A. Warrender, G. Rapp, Phys. Chem. Chem. Phys. 1 (1999) 887–893.
[60] U.S. Schwarz, G. Gompper, Langmuir 17 (2001) 2084–2096.
[61] E. Shyamsunder, S.M. Gruner, M.W. Tate, D.C. Turner, P.T.C. So, C.P.S. Tilcock, Biochemistry 27 (1988) 2332–2336.
[62] J. Erbes, C. Czeslik, W. Hahn, R. Winter, M. Rappolt, G. Rapp, Ber. Bunsenges. Phys. Chem. 98 (1994) 1287–1293.
[63] D.P. Siegel, W.J. Green, Y. Talmon, Biophys. J. 66 (1994) 402–414.
[64] P. Siegel, R.M. Epand, Biophys. J. 73 (1997) 3089–3111.
[65] H. Chung, M. Caffrey, Biophys. J. 69 (1995) 1951–1963.
[66] J. Briggs, H. Chung, M. Caffrey, J. Phys. II France 6 (1996) 723–751.
[67] O. Reis, R. Winter, Langmuir 14 (1998) 2903–2909.
[68] R. Winter, A. Gabke, C. Czeslik, P. Pfeifer, Phys. Rev. E 60 (1999) 7354–7359.
[69] S. Mabrey, J.M. Sturtevant, Proc. Natl. Acad. Sci. USA 73 (1976) 3862–3866.
[70] M.R. Morrow, R. Srinivasan, N. Grandal, Chem. Phys. Lipids 58 (1991) 63–72.
[71] M.B. Sankaram, T.E. Thompson, Biochemistry 31 (1992) 8258–8268.
[72] J. Teixeira, J. Appl. Crystallogr. 21 (1988) 781–785.
[73] K. Jørgensen, O.G. Mouritsen, Biophys. J. 95 (1995) 942–954.
[74] C. Bernsdorff, R. Winter, Z. Phys. Chem. 193 (1996) 151–173.
[75] M. Zein, R. Winter, Phys. Chem. Chem. Phys. 2 (2000) 4545–4551.
[76] J. Eisenblätter, R. Winter, Magn. Res. Chem. 38 (2000) 662–667.
[77] B.A. Wallace, Annu. Rev. Biophys. Chem. 19 (1990) 127–157.
[78] J. Erbes, R. Winter, G. Rapp, Ber. Bunsenges. Phys. Chem. 100 (1996) 1713–1722.
[79] J. Erbes, A. Gabke, G. Rapp, R. Winter, Phys. Chem. Chem. Phys. 2 (2000) 151–162.
[80] M. Caffrey, J. Hogan, A. Mencke, Biophys. J. 60 (1991) 456–466.
[81] M. Steinhart, M. Kriechbaum, K. Pressl, H. Amenitsch, P. Laggner, Rev. Sci. Instrum. 70 (1999) 1540–1545.
[82] A. Squires, R.H. Templer, O. Ces, A. Gabke, J. Woenckhaus, J.M. Seddon, R. Winter, Langmuir 16 (2000) 3578–3582.
[83] S. Kaneshina, M. Tanaka, T. Tomida, R. Matuura, J. Colloid Interface Sci. 48 (1974) 450–465.
[84] H.W. Offen, W.D. Turley, J. Colloid Interface Sci. 87 (1982) 442–446.
[85] M. Lesemann, K. Thirumoorthy, Y.J. Kim, J. Jonas, M.E. Paulaitis, Langmuir 14 (1998) 5339–5341.
[86] K. Hara, H. Kuwabara, O. Kajimoto, K. Bhattacharyya, J. Photochem. Photobiol. 124 (1999) 159–162.
[87] M. Kato, R. Hayashi, Biosci. Biotechnol. Biochem. 63 (1999) 1321–1328.
[88] P.T.T. Wong, H.H. Mantsch, J. Colloid Interface Sci. 129 (1989) 258–269.
[89] N. Gorski, J. Kalus, D. Schwahn, Langmuir 15 (1999) 8080–8085.
[90] M. Nagao, H. Seto, Phys. Rev. E 59 (1999) 3169–3176.
[91] H. Seto, D. Okuhara, Y. Kawabata, T. Takeda, M. Nagao, J. Suzuki, H. Kamikubo, Y. Amemiya, J. Chem. Phys. 112 (2000) 10608–10614.
[92] C. Balny, N.L. Klyachko, in: R. Winter, J. Jonas (Eds.), High Pressure Molecular Science, Kluwer Academic Publishers, Dordrecht, 1999, pp. 423–436.
[93] V.V. Mozhaev, K. Heremans, J. Frank, P. Masson, C. Balny, Proteins Struct. Funct. Genet. 24 (1996) 81–91.
[94] G. Hummer, S. Garde, A.E. Garcia, M.E. Paulaitis, L.R. Pratt, Proc. Natl. Acad. Sci. USA 95 (1998) 1552–1555.
[95] P.W. Bridgman, J. Biol. Chem. 19 (1914) 511–512.
[96] G. Desai, G. Panik, M. Zein, R. Winter, C.A. Royer, J. Mol. Biol. 288 (1999) 461–475.
[97] H. Seemannn, R. Winter, C.A. Royer, J. Mol. Biol. 307 (2001) 1091–1102.
[98] K.J. Frye, C.A. Royer, Protein Sci. 7 (1998) 2217–2222.
[99] G.J.A. Vidugiris, J.L. Markley, C.A. Royer, Biochemistry 34 (1995) 4909–4913.
[100] H.S. Cha, K.A. Dill, Proteins Struct. Funct. Genet. 30 (1998) 2–33.
[101] G. Panick, R. Malessa, R. Winter, Biochemistry 38 (1999) 6512–6519.
[102] D. Foguel, C.R. Robinson, P.C. de Sousa, J.L. Silva, A.S. Robinson, Biotechnol. Bioeng. 63 (1999) 552–572.
[103] R.J. St. John, J.F. Carpenter, T.W. Randolph, Proc. Natl. Acad. Sci. USA 96 (1999) 13029–13033.

ELSEVIER

Biochimica et Biophysica Acta 1595 (2002) 185–200

BIOCHIMICA ET BIOPHYSICA ACTA
BBA

www.bba-direct.com

Review

High pressure simulations of biomolecules

Emanuele Paci *

Laboratoire de Chimie Biophysique, ISIS, Université Louis Pasteur, 4 rue Blaise Pascal, 67000 Strasbourg, France

Received 12 September 2001; accepted 18 October 2001

Abstract

Pressure is a thermodynamic variable which is particularly suitable for exploration of the properties of biological macromolecules. For proteins, in particular, denaturation induced by pressure is different from that induced by temperature or denaturants. The response of proteins to pressure changes can provide information on properties of their native and non-native states. This review focuses on molecular dynamics studies of the effect of pressure on detailed atomic models of proteins. It also reports on other theoretical approaches, such as Monte Carlo simulations, which have been used to study simplified models. Another purpose of this review is to try to point out potential future studies that may be both interesting and feasible, with constantly increasing computing power.

Keywords: Protein; Volume; High pressure; Compressibility; Denaturation; Non-native state

1. Introduction

Exploring phase diagrams of biological systems, even outside biologically relevant regions, can provide useful information. For proteins, it is particularly interesting to explore those properties which make these molecules very peculiar heteropolymers; the most interesting property is that of possessing a unique stable and functional conformation in a specific range of temperatures and pressures and to be disordered and biologically inactive outside this range.

The present review deals primarily with theoretical approaches and molecular simulations as tools to understand how proteins react to a change in external pressure. A molecular simulation is a 'numerical experiment' in which the system being modeled is represented as a set of classical particles interacting through an empirical energy function, and the thermodynamic and kinetic properties are computed from the simulated time evolution of that system. The principal interest of molecular simulations is a very high resolution in time (1 fs) and space (0.01 nm).

Proteins are complex molecules. At or close to physiological conditions, transitions from native to non-native states and vice versa take place on the time scale of microseconds to seconds. Considering the number of degrees of freedom involved and the astronomical number of possible conformations, this time scale is quite short. Yet, it is far too large to be simulated, at least for realistic, detailed atomic models of proteins. Spontaneous folding and unfolding that take place in a test tube upon a change in the denaturant concentration, temperature, or pressure cannot be studied using molecular dynamics. Therefore, only limited regions of the conformational

* Present address: Biochemisches Institut, Universität Zürich, Winterthurerstrasse 190, 8057 Zurich, Switzerland.
Fax: +41-1-635-5712.
E-mail address: paci@bioc.unizh.ch (E. Paci).

0167-4838/02/$ – see front matter
PII: S0167-4838(01)00343-0

space close to the initial conformation (e.g. the native state) can be explored in a statistically significant way, while the whole conformational space remains too big to be explored during the simulation time. Such regions are confined by kinetic barriers which prevent the exploration of other conformations (e.g. various denatured states). Within such a limited region the system can be assumed to be approximately ergodic.

For the purpose of understanding the effect of pressure on proteins, the folding landscape of a protein can be schematically represented as in Fig. 1. The ordinate represents the free energy and the abscissa some arbitrary reaction coordinate. At physiological temperature and pressure the free energy has an absolute minimum which corresponds to the native state (N in Fig. 1).

The main effect of a moderate pressure [1] is to shift the position of the absolute minimum of the free energy (N′ in Fig. 1). The stable state of the system is mildly affected by the external pressure in such a way that the protein and the surrounding solvent occupy a smaller volume. The structural response of the protein–solvent system to moderate pressure can be called 'elastic' since it does not involve a transition from native to non-native state (there is no denaturation of the protein). From a structural point of view, elastic relaxation is characterized by a moderate change in interatomic distances, reversible when the pressure is released, but not necessarily homogeneous throughout the protein structure.

At pressure exceeding 200–500 MPa[1] [2,3], the native minimum shifts further from its original position (N″ in Fig. 1), but a more interesting effect is that N″ is no longer the absolute minimum. A different absolute minimum appears (D in Fig. 1), corresponding to a set of completely different conformations (the high pressure-denatured state). The denaturing effect of pressure is quite different from that of temperature. In some cases, pressure plays a role opposite to that of temperature: it has been observed in several

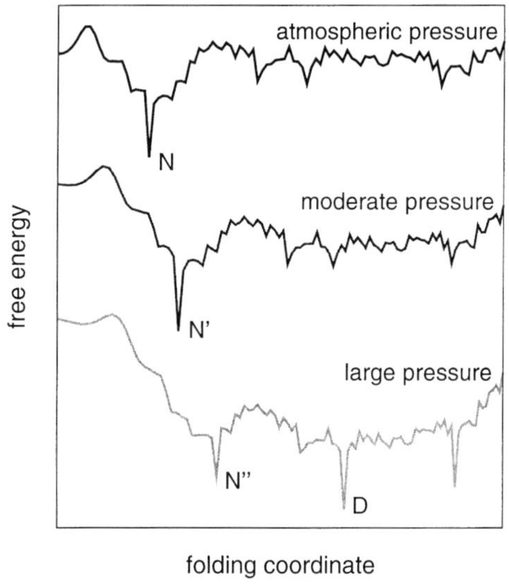

Fig. 1. Schematic folding landscape of a protein. The ordinate represents the free energy and the abscissa some arbitrary reaction coordinate. At physiological temperature and pressure the free energy has an absolute minimum which corresponds to the native state. At moderate pressure the position of the absolute minimum shifts, inducing an elastic modification of the protein structure. At larger pressure the native minimum shifts more and a new absolute minimum appears; the transition from the original to the new minimum is slow and induces a 'plastic' modification of the protein structure.

proteins that a moderate pressure stabilizes again heat denaturation [3]. Also, experimental studies [4,5] have shown that pressure-induced unfolding at room temperature is considerably slower than temperature-induced unfolding.

Simulations, even when performed at pressures large enough to destabilize the native state, usually show only elastic relaxation to a relative minimum (N″ in Fig. 1). This state (or set of states) is usually metastable. The time scales that can be currently explored by molecular dynamics simulations are too short to allow for a full relaxation from a metastable native-like minimum to the denatured state (D in Fig. 1).

To simulate transitions between states and to explore non-native states, various physical or non-physical perturbations or very high temperature were used, which either drive the system across barriers, or flatten those barriers, or both. For example, thermal unfolding in water of hen egg white lysozyme was simulated at 500 K [6], that of bovine pancreatic trypsin inhibitor (BPTI) at 423 K and

[1] Atmospheric pressure corresponds to about 0.1 MPa = 1 bar. The Pascal (Pa) is the SI unit for pressure, 1 Pa = 1 N/m^2. The pressure below 10 000 m of water (as in the deepest part of the oceans) is about 100 MPa.

498 K [7], and that of the protein barnase at 498 K and 600 K [8,9]. Other perturbations, mostly involving application of non-physical forces, have been used to unfold proteins or to study conformational preferences at non-native conditions (see e.g. [10–12]).

Since high pressure does not result in a sufficient decrease of energy barriers and, moreover, slows down the kinetics, it is clearly not a suitable perturbation to unfold proteins on the simulation time scale. What high pressure simulations allow is the study of the elastic relaxation after a pressure jump, i.e. elastic properties of both native and non-native states. Naturally, the best studied cases are native states, for which X-ray and NMR experimental techniques provide structures which are used as initial configurations in molecular dynamics simulations.

Non-native states are less well characterized, both because of their structural heterogeneity and because they have been considered of less interest from a biological perspective. However, recent experimental results have underlined the interest of some non-native states like molten globules and other folding intermediates: their elastic response to a pressure perturbation has been shown to be a good way to detect them [13] and also to understand how they differ from the native state.

2. Denaturation of proteins under pressure

According to the Le Châtelier principle [14], the effect of a large, denaturing pressure on a protein is determined by the change in solution volume before and after the unfolding; i.e. at high pressure and normal temperature a protein denatures because the denatured state occupies a smaller volume than the native one. This behavior can be contrasted with what happens at normal pressure (below 100–200 MPa), at which denatured proteins occupy a larger volume than native states.

The 'hydrophobic model' [15], i.e. the model of a folded protein stabilized by hydrophobic interactions, succeeded in explaining a large amount of experimental calorimetric data on protein denaturation. However, as Kauzmann [16] pointed out, the hydrophobic model apparently contradicts the experimentally observed high pressure protein denaturation. Indeed, liquid hydrocarbons have an opposite behavior, i.e. a transfer from non-polar to aqueous environment is accompanied by a volume decrease at atmospheric pressure and by a volume increase at higher pressures.

Theory and simulation have been used to attempt to explain the apparent failure of the hydrophobic model to explain the pressure-induced unfolding of proteins, by focusing on the pressure dependence of hydrophobic interactions [17,18]. Monte Carlo simulations [17] have been used to study the effect of high pressure on the potential of mean force between two simple hydrophobic solutes. The expression of the free energy of association in terms of partial volume and isothermal compressibility shows that the change in volume favors solute association, but the change in compressibility favors the dissociation. This means that the application of pressure can disperse the hydrophobic core of a protein in water, not because the solution containing the unfolded protein occupies a smaller volume than that containing the native state, but because it is more compressible.

Hummer and coworkers [18] suggested a solution of the pressure denaturation puzzle by considering the transfer of water into the hydrophobic core as opposed to the transfer of non-polar residues into water. Using a wide range of theoretical and simulation approaches, they computed the potential of mean force between two methane-like particles in water. Their results led to the conclusion that pressure destabilizes the configuration where two non-polar groups are in contact relative to that where a water molecule is in between the two non-polar groups (i.e. at higher pressure it is less profitable for non-polar groups to be in direct contact). This suggests that pressure denaturation is accompanied by penetration of water molecules inside the protein, rather than by a transfer of the hydrophobic interior into water, as happens in high temperature denaturation.

Using Monte Carlo simulations, Hillson et al. [19] investigated the free energy profiles and the folding kinetics of an off-lattice minimalist model of a protein. Their model included the pressure dependence of the potential of mean force between two hydrophobic solutes found by Hummer et al. [18]. It was shown that the presence of a desolvation barrier increasing with pressure results in an increase of the roughness of the energy landscape; as a consequence, folding time increases even if the intrinsic thermodynamic free energy barriers get lower. These results agree well with the experimental findings [4,5,20] that high pressure denatures proteins by slowing down the rate of folding more than that of unfolding.

Despite the extreme interest of theories and models attempting to explain the hydrophobic effect, it must be stressed that these are based on using simple non-polar spherical objects. Applying conclusions from these studies to real systems such as proteins would ignore the fact that connectivity, polar interactions, hydrogen bonds and specific packing all play an essential role in determining the stability of the native state.

3. Molecular dynamics at high pressure

3.1. Methods for high pressure simulations

Rigorous statistical mechanical studies of properties of complex molecular systems, at given pressure and temperature, have been made possible by a series of important theoretical and methodological advances that followed the seminal work of Andersen in 1980 [21]. In molecular dynamics, the equations of motion of a collection of particles in a fixed volume are integrated numerically. Energy, volume, and number of particles are conserved, and time averages equal microcanonical ensemble averages (if the system is ergodic, which can be assumed to be approximately the case at least in a limited region of the phase space). Andersen [21] introduced the so-called extended Lagrangian method where a fictitious Lagrangian with one extra degree of freedom is introduced: in the specific case addressed by Andersen, the volume in which the particles are confined becomes a dynamical degree of freedom coupled to the particles' positions. The volume varies with time so that the internal pressure equals, on the average, the external fixed pressure. The trajectory deriving from the extended Lagrangian samples an ensemble different from microcanonical: in the original case studied by Andersen, it is the iso-enthalpic–isobaric ensemble (constant enthalpy and pressure). This formalism was extended to obtain the canonical (constant temperature, or NVT) and isobaric–isothermal (constant temperature and pressure, or NPT) ensembles by Nose [22] and Hoover [23].

Other methods have also been devised to control the temperature and the pressure of the system. The most popular, and still widely used, is the *weak coupling* method of Berendsen [24], which, however, does not sample any standard statistical mechanical ensemble. The most serious consequence is that statistical mechanical relationships between fluctuations (of the energy, of the volume, etc.) and second derivatives of the thermodynamic potential (specific heat, compressibility, etc.) do not hold. The extended Lagrangian and weak coupling methods have been thoroughly compared for proteins [25]; the convergence of the total volume and of its fluctuations (related to the system compressibility) and the statistical efficiency of the two methods were studied. The most interesting finding was that observables such as volume and enthalpy obtained by extended Lagrangian and weak coupling simulations are within statistical error of each other. The same agreement was found for the compressibility and specific heat computed with finite differences by performing simulations at different pressures. Thermodynamic derivatives from fluctuation formulas in the NPT ensemble, which can only be evaluated for an extended Lagrangian simulation, agree within statistical error with results obtained from finite differences.

An improved extended Lagrangian method for performing constant pressure molecular dynamics simulations has been proposed [26]. The original equation of motion for the volume variable is replaced by a Langevin equation, thus eliminating the observed non-physical 'ringing' of the volume variable [27] which might induce problems of convergence.

A change in volume implies a change in the atomic positions; when dealing with molecules, one has the choice of coupling the volume change to each atomic degree of freedom or to each molecular center of mass [28]. In the former case, the dynamical adjustment of the volume corresponds to a space scaling that affects the position of each atom. In the latter case, the scaling occurs only for the positions of the molecular centers of mass, thus affecting only intermolecular distances. An intermediate scaling scheme has also been proposed [29] in which the centers of mass of arbitrary subsets of atoms are rescaled. This scheme has the molecular and atomic scaling as limiting cases. If the system is ergodic and the molecules do not dissociate, all the above scaling procedures have been demonstrated to be equivalent for flexible molecules [30]. In principle, different ways of coupling volume and coordinates should not affect the results of a constant pressure simulation, but rather differ in how efficiently the system relaxes and equilibrates at a certain pressure. In practice, since simulations are not exactly ergodic, the way in which the pressure is applied may affect the results. Zhu and Averbak [31] presented a modification of the Andersen method which is particularly suitable for non-homogeneous systems such as solutions containing biological macromolecules.

Molecular dynamics simulations of a large protein in explicit water with periodic boundary conditions remain extremely demanding in terms of computation time. Other approaches have been proposed where the protein is immersed in a droplet whose boundaries adjust to keep pressure constant. One such approach is the solvent boundary potential developed by Beglov and Roux [32] which accounts for the van der Waals and electrostatic interactions of an infinite bulk system surrounding simulated solute and water molecules in a sphere of variable radius; the method was improved [33] to simulate non-spherical systems by using a half-harmonic restraining potential to prevent water molecules from escaping. More recently, Lounnas et al. [34] proposed an alternative method which accounts for the dielectric properties of the bulk solvent by subjecting water molecules to forces approximating van der Waals and dipole–dipole interactions with the implicit surrounding bulk solvent. However, it has not been proved that these methods sample the NPT ensemble.

3.1.1. *Are current force fields appropriate for high pressure simulations?*

Current force fields used to simulate proteins are designed to reproduce specific properties at atmospheric pressure and room temperature. Marchi and Akasaka [35] compared two different all-atom force fields, AMBER [36] and CHARMM [37]. Structural changes occurring in hydrated BPTI at high pressure (500 MPa) were studied by performing 1 ns long constant pressure simulations. Comparing the simulation results with chemical shift measurements performed up to 200 MPa pressure [38,39] they concluded that the accuracy of available force fields is not sufficient to quantitatively study structural changes occurring in biomolecules. However, another possible reason for this disagreement is that the experimental result might contain large error bars due to a very small change in chemical shifts observed between the atmospheric and 200 MPa pressure.

Similarly, common water models were elaborated to reproduce bulk water properties at normal pressure and temperature, not at high ones. For one particular model (SPC/E [40]), a detailed study of the pressure-dependent properties [41] suggests that some experimental properties of 'real' water, like the diffusion maximum at 150 MPa (1.5 kbar), are not reproduced. For another water model (TIP3P [42]) it has been observed (Paci, unpublished results) that density and compressibility strongly depend on how interactions are truncated. Simulations of systems solvated in water would gain in reliability if the potential energy functions for water were transferable to a broader pressure and temperature range.

3.2. *Sub-nanosecond simulations at high pressure*

The first molecular dynamics simulation of a solvated protein was presented less than 10 years ago: Kitchen et al. [43] performed a simulation of BPTI in solution at a pressure of 1000 MPa, using the AMBER energy function [44] for the protein and the SPC [45] for water and the weak coupling constant pressure technique. The structural and energetic properties of the solvated protein at atmospheric and high pressure where compared along two relatively short (100 ps) simulations. No evidence of pressure-induced unfolding was observed during the high pressure simulation.

A comparison of various attempts to unfold proteins during molecular dynamics simulations using a variety of perturbations, including high pressure, has been presented by van Gunsteren and collaborators [10,46] (using the GROMOS force field [47] for the protein and the SPC model for the water, and the weak coupling method to keep pressure constant [24]). A pressure of 1000 MPa applied for a few hundred picoseconds affects the protein only marginally. Not even the onset of unfolding is detected on such a short time scale. On the other hand, the authors also observe that alternative methods, which lead to unfolding on a sub-nanosecond time scale, usually strongly bias the results; i.e. different sequences of unfolding events are observed when different perturbations are used.

Using the same force field and the same method to keep pressure constant, Wroblowski et al. [48] performed a molecular dynamics simulation of BPTI in water. Starting from atmospheric pressure, they applied pressure jumps, at 1000, 1500 and 2000 MPa, every 200 ps. The protein remained globular, but the authors noticed a conformational change that was larger than that observed in previous simulations [10,43,49]. The authors claim that this change, which mainly involves disruption of secondary structure and exposure of the hydrophobic core to water, is in good agreement with the experimental results of Goossens et al. [50] (that show denaturation of BPTI between 800 and 1400 MPa). However, other results [4] clearly show that such high pressure unfolding cannot occur on the time scale used. Another piece of evidence that simulation was too short to effectively show high pressure unfolding is that the root mean square deviation (RMSD) was never larger than 2.5 Å (i.e. the protein is very native-like even at the end of the 2000 MPa simulation); also, the RMSD was steadily increasing with time, suggesting that the equilibrium was not reached.

Paci and Marchi [51] reported a simulation of a unit cell of lysozyme containing eight proteins [52] and of superoxide dismutase (SOD) in solution at atmospheric, 1000, and 2000 MPa pressure. They used the CHARMM force field [53] for the protein and TIP3P for water [42] and a rigorous statistical mechanical method for sampling the isothermal–isobaric ensemble (see Section 3.1); this paper mostly focused on the change in the apparent and intrinsic protein volume following elastic relaxation discussed in Section 4.

Floriano et al. [54] performed a simulation of metmyoglobin up to 1200 MPa, using the AMBER force field and NPT ensemble. Instead of considering water explicitly, as in all simulations listed above in this section, they treated it as a continuum characterized by a pressure-dependent dielectric constant estimated using the Bradley and Pitzer equation [55]. They observe full unfolding passing through a molten globule-like state at about 700 MPa. It is not clear whether this result is due to the strong approximations in the solvent energy function or else to the absence of a viscous solvent. The fact that the latter increases with pressure is one of the causes of the slow kinetics of the pressure-induced unfolding both in experiments and in simulations of explicitly solvated proteins.

In the following of this section the most interesting properties observed in the high pressure MD simulations performed so far will be discussed.

3.2.1. Energetics

Simulations of solvated proteins generally show that at high pressure protein–solvent interaction is energetically more favorable than at atmospheric pressure, while the protein–protein bonded interaction energy increases (due to the moderate deformation of the protein's bond distances and angles). The more favorable interaction between protein and solvent at high pressure is accompanied by an increase in the number of protein–solvent hydrogen bonds.

3.2.2. Hydrogen bonds

At high pressure (1000 MPa), the number of protein-to-protein hydrogen bonds was found [56] to increase by 2% for SOD and by 6% for lysozyme, while those between water and protein increase more (19% and 14%, respectively). In a simulation [43] of BPTI in solution a similar trend for both protein–protein and protein–solvent hydrogen bonds after a similar pressure jump was observed. Another simulation of the same protein (using a different force field) gave a similar increase in protein–solvent hydrogen bonds but a decrease in the protein–protein hydrogen bonds. On the other hand, high pressure crystallography [52] on lysozyme shows a *decrease* of 7% in the number of water-to-protein hydrogen bonds. This result is, however, not directly comparable with the simulation since only crystallographic water molecules were included in the estimate. The number of hydrogen bonds between solvent water molecules increases by 14 and 11% in the presence of SOD and lysozyme, respectively, while in pure water the increase of hydrogen bonds is larger (19%) [56]. The considerable increase in the number of protein-to-solvent hydrogen bonds at high pressure is related not only to a higher density of donors and acceptors due to the larger compressibility of the water relative to the protein, but rather to a clear preference of water to make hydrogen bonds with the protein rather than water itself.

Despite the changes in the number of hydrogen bonds due to the increase in pressure, the average distance between acceptor and donor remains very close to 2.8–2.9 Å at all the pressures studied [35,43,51,56]. This is apparently in contradiction with the results of Akasaka and collaborators [57] which show (in BPTI) that hydrogen bonds formed by amide groups either with carbonyls or with water are shortened by pressure.

3.2.3. Atomic displacement and radius of gyration

In all the high pressure simulations reported so far the displacement of atomic positions with pressure, as measured by the RMSD from the experimental native structure, is moderate, confirming that the high pressure is not denaturing the protein over this time scale. As expected, the radius of gyration always decreases at high pressure, reflecting the increased compactness of the high pressure state. However, the variations of the radius of gyration do not provide a reliable estimate of the compressibility of the protein interior [58].

3.2.4. Atomic mobility

Most simulations of proteins at high pressure show that positional fluctuations decrease with increasing pressure, which is consistent with the intuitive argument that increasing pressure would have a dumping effect on atomic fluctuations, analogous to that produced by a decreasing temperature. Brunne and van Gunsteren [49] observed the decrease of the positional fluctuations of atoms with increasing pressure; this effect was more important for side chains than for the main chain atoms, while the mobility of the backbone ϕ- and ψ-angles was not affected. The amplitude of oscillation of side chain dihedral angles remains nearly the same for both pressures, but the rate of torsional angle transitions decreases on the average by 30% when the hydrostatic pressure increases to 5000 atm (about 500 MPa).

3.2.5. Structure modifications

Simulation allows the precise evaluation of the structural modifications occurring upon elastic relaxation after an increase of pressure. Experimentally, only the structure of lysozyme [52] has been determined at moderately high (100 MPa) pressure. When structures are available, the behavior of distinct regions within the protein under pressure changes can be singled out by computing the change in the interatomic distances d_{ij}. Simulation of the lysozyme crystal at 1000 MPa [56] led to an average estimation of $d_{ij} = -0.56$ Å as compared to -0.068 Å for the X-ray structures [52], suggesting that the deformation is very small and almost linear in pressure up to 1000 MPa. The computed and experimental d_{ij} show some common features. Lysozyme is made of two domains, one containing mostly α-helices and the other, smaller, containing a three stranded β-sheet; the latter is essentially rigid, while helices, α_1 in particular, approach the β-domain. In general, inter-domain distances decrease much more than intra-domain ones.

In one of the high pressure simulations of BPTI [48] a structural change occurred at 1000 MPa involving the central part of the β-sheet and letting few hydrophobic residues exposed to the solvent. As for the helical structure, it reacts to a change in pressure less dramatically. Another simulation of BPTI at 1000 MPa [43] also showed a very small deformation of the helices, while there was a more dramatic change in the loops.

For SOD (Paci, unpublished results), structural changes at 1000 MPa were moderate but it is interesting to note that the contact surface between the two monomers decreased by about 16%, while the total solvent-accessible surface diminished by only 2.3%. This effect seems to indicate a reduction of the dimer stability, which might prelude to a solvent-mediated dissociation prior to unfolding of the monomers.

The major structural changes observed in metmyoglobin [54] at or above 900 MPa mainly consist in the disruption of about 70% of the native secondary structure.

4. Compressibility

The apparent volume of a solute is the only directly measurable volume of a protein. It is defined as the volume change observed when a solute is added to the pure solvent:

$$V_{\mathrm{p}}^{\mathrm{a}} = (V - V_{\mathrm{s}}^{0}) \quad (1)$$

The partial volume is a well defined thermodynamic quantity which is equal to the partial derivative, at constant particle number and temperature, of the chemical potential with respect to the pressure:

$$\overline{V}_{\mathrm{p}} = \left(\frac{\partial \mu_{\mathrm{p}}}{\partial P}\right)_{\mathrm{NT}} \quad (2)$$

In the limit of infinite dilution, partial and apparent volumes are equal.

The apparent volume of proteins, as any apparent property of a molecule in solution, depends on properties of the molecule only (which are thus called *intrinsic* properties), and on the non-ideal behavior of the solvent hydrating the molecule (which can be considered a surface effect). Apparent specific volume of native globular proteins falls into a narrow range around 0.72 $\mathrm{cm^3/g}$ revealing a marginal dependence on the extent of the solvent–protein interface.

Compressibility is defined as

$$\beta \equiv -\frac{1}{V}\left(\frac{\partial V}{\partial P}\right) \quad (3)$$

where V is the volume of a microscopic sample and P is the pressure. If the derivative is measured at constant temperature or entropy (and, in all cases, constant particle number), β is isothermal (β_{T}) or adiabatic (β_{S}) compressibility, respectively.

Adiabatic compressibility of a solution, at a given concentration, can be obtained in density and sound velocity measurements [59,60]. The apparent or partial compressibility can be extracted [61,62] from the compressibilities of the solution at different concentrations which must include zero concentration (pure solvent).

The apparent and partial volumes and, consequently, apparent and partial compressibilities, can be written as the sum of an intrinsic protein term plus a solvation-related correction due to non-ideal behavior of the solvent. The intrinsic volume can, in turn, be written as the sum of the van der Waals volume plus the volume of voids due to an imperfect packing of a set of covalently connected spheres. Both decompositions can be reasonable and useful for specific purposes, but are essentially arbitrary. Several different and incompatible evaluations of the intrinsic compressibility can be found in the literature. Their difference is related to different definitions of what protein intrinsic volume is. In an important paper, Kharakoz and Sarvazyan [63] reviewed the partial compressibilities of globular proteins in water. They evaluated the contribution of hydrational and intrinsic compressibilities to experimental partial compressibility using two independent methods. The first method assumes that the non-ideal behavior of the solvent in the presence of a protein can be evaluated by assuming that each group on the surface of the protein modifies the compressibility of the solvent as it does in small molecules and that this effect is additive. The second is based on the analysis of correlation between partial compressibility and the solvent-accessible surface (which is assumed to be proportional to $M^{2/3}$, where M is the mass of the protein). They found that the isothermal intrinsic compressibility is about 14×10^{-5} $\mathrm{MPa^{-1}}$. (The volume of an object characterized by a constant compressibility of 14×10^{-5} $\mathrm{MPa^{-1}}$ decreases by 1.4% when the pressure increases by 100 MPa. In general, compressibility is not constant and decreases with increasing pressure.) This value is similar to that of solid organic polymers; it is about one third the isothermal compressibility of water at room temperature (45.2×10^{-5} $\mathrm{MPa^{-1}}$) and varies little for different proteins. The average compressibility of water in the hydration shell of proteins was found to be 20% less than that of pure water. If one takes into account the fact that proteins are usually not spherical and uses $M^{0.73}$ as a better approximation for the protein surface, the protein intrinsic compressibility becomes about one half of the compressibility of the water [64].

Using the Voronoi volume [65–67] as a definition of the intrinsic volume, Paci and Marchi [51] found very similar compressibilities for two globular proteins possessing very different secondary structures, SOD and lysozyme. On a larger set of proteins (Paci, unpublished), including barnase, lactalbumin and potato carboxypeptidase inhibitor (PCI), using pressure differences ranging from 500 to 2000 MPa, the Voronoi to pure water compressibility ratio was found to range from 0.46 to 0.54.

There are many geometrical recipes for assigning a volume to a protein. These recipes are not equivalent, and involve parameters whose definition is not always unique. The advantages of Voronoi volume is that the volume of each atom of the protein and the solvent is defined and does not depend on any parameter. It provides a plausible definition of the intrinsic volume of a protein because the compressibility estimated through it is close to the most reliable experimental estimates [61,63]; also, the compressibility of different proteins falls into a narrow range, which is expected since globular native proteins, from the point of view of condensed matter physics, are very similar objects [68].

Among other commonly used definitions of protein volume are the van der Waals volume (the volume of a collection of intersecting spheres of the atoms' van der Waals radii, therefore not taking into account cavities and voids), the *molecular volume* [69], which is the volume enclosed by the contact surface of a probe sphere rolled on the van der Waals envelope of the molecule, or the *excluded volume* [70], which is the volume excluded to the center of the probe sphere. As discussed by Paci and Velikson [58], the above definitions of protein's intrinsic volume have the contradictory feature of not reflecting the packing of solvent molecules surrounding the protein, being at the same time a function of the van der Waals radii of the protein's atoms and of the probe radius. The molecular volume of a protein, which has been used to evaluate the intrinsic protein compressibility on some occasions [43,52,58], turned out to provide an estimation of the intrinsic compressibility (about 1/10 of the water compressibility) which seems to be strongly underestimated (and comparable to the compressibility of a metal). Unlike the molecular volume, the Voronoi volume does take into account modifications occurring in the solvent due to the presence of the protein and to possible external perturbations. This is possibly a reason why it provides estimates close to the experimental ones.

Another experimental approach to estimate the compressibility of proteins consists in monitoring changes in specific protein interatomic distances under pressure and assuming that the volume behaves as the distance to the third power (the assumption would be exact in the case of homogeneous systems). This compressibility would, in principle, be an intrinsic quantity in the sense that it does not include the effect of hydration. On the other hand, simulations have revealed that, on average, short distances compress less than longer ones [51]. This non-uniform contraction induced by high pressure contrasts with the spatially uniform expansion observed at high temperature [71]. Simulation [51] has shown how, by randomly picking a pair of atoms, compressibilities ranging from very large positive to negative values can be obtained. This explained why fluorescence spectroscopy measurements [72], which probe donor–acceptor distances in heme proteins as a function of pressure, found a protein intrinsic compressibility comparable to that of bulk water. It also provides an explanation to the low compressibility (10^{-4} MPa^{-1}) obtained by the hole burning [73] technique; since average near-neighbor distances are probed, this technique is likely to lead to an underestimation of the intrinsic compressibility of proteins.

4.1. Volume fluctuations

Volume fluctuations are sometimes estimated by relating them to protein isothermal compressibility, β_T, with the equation [74]:

$$\delta V^2 = V k_B T \beta_T \tag{4}$$

where k_B is the Boltzmann constant, T the temperature, V and δV^2 are the protein's volume and the volume fluctuations squared. This relationship assumes either that the protein alone samples the NPT ensemble distribution, which is not true in general, or that the coupling between the protein and solvent volumes is small. It was shown by Paci and Marchi [51] that for Voronoi volumes the latter assumption is not even approximately true. The protein and solvent volume cross-fluctuations are of a magnitude comparable to the protein and solvent volume fluctuations alone, due to the small size of proteins and the resulting large surface-to-volume ratio. As a consequence, using Eq. 4 to derive β_T generally leads to an underestimation. Nevertheless, volume fluctuations and volume compressions of proteins and subunits of proteins are correlated. For example, the cross-correlations between the relative volume fluctuations of residue i, $\delta V_i^2/V_i$ and its compression range from +0.65 to +0.80 for a set of five proteins (Paci, unpublished data). Thus, although there is no exact relation between volume fluctuations and compressibility of residues, these quantities are strongly correlated: highly compressible residues will show larger volume fluctuations and vice versa.

The volume fluctuations of a series of proteins have been recently used by Dadarlat and Post [75] to estimate the intrinsic compressibility of a series of proteins. As protein intrinsic volume they consider the van der Waals volume where the radius of each atom is extended so that cavities and packing defects are filled (similar to the definition of *excluded volume*). The compressibility they report is about one fifth of the compressibility of water, thus below that obtained using Voronoi volumes and finite differences, but larger than the value obtained, always by finite differences, using the molecular volume [43,52]. The reason for the low compressibility obtained is probably attributable to the use of a relation (Eq. 4) between volume fluctuation and compressibility which is only approximately true for small systems. Also, the *excluded volume* as definition of intrinsic volume, similar to van der Waals and molecular volumes, tends to underestimate the compressibility as noted above.

4.2. Local compressibility

Another advantage of Voronoi volumes is that they are additive and well defined for every atom of every component of the solution. It is thus possible to attribute exact additive volumes to atoms, amino acids and substructures of a protein.

4.2.1. Substructures

For various proteins, the following results have been found ([56]; Paci, unpublished results). In lysozyme, the two domains have very similar compressibility. The less compressible structure is the β-sheet, while the three helices are more compressible than the whole protein. For SOD, the compressions of the regular and bent β-sheets do not deviate from the averaged compression. The most and least compressible structures are the two loops. The least compressible loop contains important residues for the regulation of the protein catalytic activity; its low compressibility suggests that the active site in SOD is 'screened' [1] from the effect of pressure by the deformation of its surrounding regions. In barnase, α-helices, loops, β-strands and turns show almost the same compressibility while the most and least compressible structures are the two loops. In lactalbumin, β-strands are markedly less compressible than all other structural elements.

The analysis of the Voronoi compression, on the limited set of proteins studied, does not reveal a simple relation between compressibility and structure. Different compressibility of various structural elements of a protein seems to be related to the overall protein's architecture and function rather than to the type of secondary structure elements present.

4.2.2. Residues

The compressibility of all amino acids is the same, within 20%. For a set of four proteins analyzed ([56]; Paci, unpublished results), a clear trend was found: most compressible amino acids are the hydrophobic Val, Leu and Ile, while least compressible are the charged amino acids Glu and Asp. For the four proteins, the calculated compressibilities are very similar and increase from charged to polar and to non-polar residues. These findings are in accord with the results, obtained by molecular modeling and simulation, that the free energy cost of forming atomic size cavities is significantly lower in non-polar regions of proteins than on the surface [76]. As non-polar regions contain larger portions of empty space, it is not surprising that they are more compressible. This agreement also supports the hypothesis that the free energy of cavity formation serves as a fine probe of the compressibility of the protein medium as a whole, and in particular of selected regions within the protein. Quite in disagreement with this is a result from normal mode analysis [77] (see Section 4) that the intrahelix compressibility in sperm whale deoxymyoglobin is markedly more compressible than hydrophobic clusters, even though there are large cavities in these clusters.

4.3. Normal mode calculations

Gō and collaborators studied the effects of pressure on proteins using normal modes and other techniques than molecular dynamics. The common result of the various studies is that the intrinsic compressibility of proteins (using the *excluded* volume definition as intrinsic protein volume) may be larger than was thought due to low frequency modes which are overlooked in short molecular dynamics trajectories. Studying volume fluctuation by normal mode and strain tensor analysis [77] in sperm whale deoxymyoglobin showed that interhelical regions are more compressible than the intrahelical ones. The pressure-induced deformation was shown to be heterogeneous, and, most interestingly, the contribution from low frequency normal modes was more important than that from the high frequency ones. The same group also studied the mechanical response of a TIM-barrel protein to pressure [78] by assuming the volume change to be a linear function of normal mode variables. By Delaunay tessellation [79], a procedure analogous to the Voronoi construction, the space occupied by protein atoms was partitioned and attributed to various secondary structures in the protein. The compressibility of various secondary structural elements in this protein was obtained from volume changes of respective regions in two structures with and without applied pressure; regions inside the β-barrel and between secondary structural elements appeared to be compressible, while the helix, strand, and loop segments themselves were found to be quite incompressible.

Tama et al. [80] presented an alternative approach to study volume fluctuations of proteins (they also assume that the intrinsic protein volume is the *excluded* volume). They estimated volume fluctuation of human lysozyme from a previous molecular dynamics simulation [81] using a formalism based on principal component analysis. The isothermal intrinsic compressibility was found to be larger than that reported by most experiments (from their data one can deduce the value 109.10^{-5} MPa^{-1} while the reported value, 10^3 MPa^{-1}, is probably a type). The difference between this result and others obtained by molecular dynamics (for example that of [51]) probably is due to different definitions of intrinsic protein volume, which, as we have stressed already, is a delicate issue in itself [58]. It is nevertheless interesting that the spectral analysis shows that the main contribution to fluctuation comes from low frequency modes related to large and slow motions of proteins. The authors conclude that long-time dynamics simulations are necessary to describe volume fluctuations of proteins, and the relatively short trajectories reported so far (including the 100 ps trajectory of [51]) might not be sufficient.

4.4. Compressibility of non-native states

The compact fold of native globular proteins is accompanied by a large number of non-covalent interactions which are responsible for both its structure and its physical properties. It is thus interesting to understand what discriminates, in terms of physical properties, native states from non-native ones.

Recent experimental results have shown that certain stable intermediates of various proteins (like the low pH molten globules) are characterized by an increase in the apparent compressibility as compared to the native state. This contrasts with the decrease of the apparent compressibility that occurs on protein denaturation.

Nölting and Sligar [82] first reported the measurement of the apparent adiabatic compressibility for 'molten globule' states of cytochromes *c* and b_{562}. They showed that the intrinsic compressibility in the molten globule state differs only slightly from that in the native protein.

Breslauer and coworkers [83] measured apparent volume and compressibility of native, molten globule, and unfolded states of cytochrome *c* continuously modifying the pH of the solution from 6 to 2. They found that, while the apparent volume shows only one broad transition, the apparent compressibility shows three distinguishable transitions. Based on their measurements and several assumptions, they conclude that the solvent-inaccessible core, which is 40% of the intrinsic volume of cytochrome *c*, is preserved in the molten globule and that the compressibility of this core is about 4 times larger than the intrinsic compressibility of the whole native protein. Further studies belonging to the same group of a set of globular proteins [13] led to the conclusion that changes in the apparent compressibility for single domain globular proteins correlate with the type of transition being monitored, independent of the protein. This seems to be a universal thermodynamic feature, just like the increase in the heat capacity upon unfolding [84].

Tamura and Gekko [85] also studied volume and compressibility changes upon thermal and guanidine hydrochloride-induced denaturation of ribonuclease A, concluding that apparent volume and compressibility sensitively reflect the different conformational changes induced by temperature or GuHCl. Denatured states have clearly different properties but the difficulty of splitting cavity and hydration contributions limits the possibility for a quantitative discussion of the results.

Kharakoz and Bychkova [86] measured partial specific volume and compressibility of human α-lactalbumin using standard densitometric and ultrasonic techniques. Measured variations of apparent volume and compressibility led the authors to conclude that the molten globule is liquid-like (the intrinsic compressibility is twice larger than that of the native molecule) and that it contains a large amount of water (like in the *wet molten globule* model of Finkelstein and Shakhnovich [87]).

The experimental results summarized above, although they do not always provide a clear and coherent picture, always show that compact non-native states have a larger apparent compressibility relative to the native state; this in spite of the decrease expected from the increased hydration. This increase is not negligible: it ranges from 20 to 50% of the native value. This means that the molten globule (or the ensemble of states which fall under this name) has certain physical properties which clearly distinguish it from a native or an unfolded protein. However, the transition from native to molten globule cannot be compared to melting of organic crystals, where a 3-fold increase of the compressibility is observed. Simulations might play an important role in interpreting experimental results and challenging their interpretation (often based on assumptions which are difficult to check).

Molecular dynamics simulations (Paci, manuscript in preparation) at atmospheric and high pressure have been performed on native and non-native conformations of several proteins. Non-native conformations have been generated by high temperature molecular dynamics for barnase [8] and α-lactalbumin (M. Buck, personal communication), and by reducing disulfide bonds for PCI (as described by Martí et al. [88]). In addition, polyalanines in α-helical and extended conformations have been studied. For barnase and α-lactalbumin, the intermediates considered have been selected as relatively stable conformations which, for α-lactalbumin, have several of the experimentally observed properties of the molten globular state [89].

Molecular dynamic simulation shows that both apparent and intrinsic volumes of proteins are only marginally affected by partial denaturation. In contrast with experimental results mentioned above, the apparent compressibility always decreases upon partial unfolding. The estimated increase of the intrinsic compressibility (estimated from Voronoi volume changes) is moderate (13% at most, for the barnase compact intermediate) and very far from that (3-fold) expected from a solid to liquid phase transition.

The completely denatured form of PCI, lacking any secondary structure, shows the same moderate increase in the intrinsic compressibility as the compact intermediates of barnase and lactalbumin. Also, the α-helical and extended polyalanines have identical intrinsic compressibility. Preliminary simulations do not show a strong dependence of the intrinsic compressibility on the conformation, when the Voronoi volume is used as intrinsic protein volume. A possible reason for the disagreement of the preliminary simulation results discussed here with the experiment is that the relatively stable intermediates selected at high temperature (or high temperature and low pH) molecular dynamic trajectories do not have the exact features of the molten globules studied experimentally.

5. Concluding discussion

Molecular dynamics simulations of solvated proteins show that even at pressure as high as 2000 MPa, the time scales that are computationally possible are too short to observe pressure-induced denaturation of proteins. Conformational states sampled by molecular dynamic simulations are not fully relaxed in the thermodynamic sense, but have undergone only elastic relaxation. Nonetheless, studying these pre-denatured states has provided useful information about the response of native proteins to external pressure and, as a consequence, some clues about pressure-induced denaturation.

On the other hand, studies based on simplified

models [17–19] have been shown to be very powerful tools to provide answers to questions which are important in understanding hydrophobic interactions and the role that they might play in pressure-induced unfolding of proteins. These studies show that the hydrophobic model for the protein interior explains both protein stability and the experimental evidence of protein denaturation at high pressure.

While a simple picture of a protein core as an oil drop in water can explain both stability at atmospheric pressure and denaturation at high pressure, the interior of a globular protein in the native state is characterized by a density [90,91] and a compressibility typical of a crystal [61,63]. This is so in spite of the constraints imposed by the covalent connectivity, heterogeneity and lack of long range order. Simulations, while confirming crystal-like packing and compressibility of proteins, also show that it would be simplistic to picture native globular proteins as crystal-like objects: proteins do not respond to a change in pressure in a uniform and homogeneous way. In particular, amino acids of different chemical character compress differently; polar and charged amino acids compress less than non-polar ones. The larger compressibility of hydrophobic residues is related to the much larger probability of cavity formation in hydrophobic cores than in other regions of the protein. Also, different regions or substructures within the protein compress differently; this does not seem to be correlated in a clear way to the structural patterns but depends more on the protein's specificity.

As the protein volume decreases, no systematic relative increase in the solvent-accessible surface of the non-polar residues is found during elastic relaxation, nor is there a noticeable penetration of water molecules into the hydrophobic core. However, the situation may be quite different for the high pressure-denatured state, not observable in simulation, for which it is believed that non-polar residues are transferred to the solvent from the apolar core of the protein or that water is conveyed into the protein interior.

Proteins respond to changes in pressure by increasing their affinity to form hydrogen bonds with water. The increase in the number of protein–solvent hydrogen bonds is larger than that of solvent–solvent and solvent–protein hydrogen bonds. At the same time, the protein surface accessible to the solvent decreases while protein–solvent interaction energy decreases and the solvent–solvent energy increases.

A certain amount of experimental data on high pressure properties [92] of membranes await simulations to complement the experimental picture. Some studies on volumetric properties of nucleic acids were reviewed by Chalikian and Breslauer [93]. Among the results cited, some show how compressibility data can characterize hydration properties of DNA and RNA. From such results, information on conformational and binding properties can be extracted. Simulations at various pressures of these systems would be useful to clarify the relation between measured properties, which include the effect of hydration, and intrinsic ones, which are in principle related to purely structural properties.

Much remains to be done in understanding the properties of non-native states. The increasing amount of experimental data on such states, together with new theoretical approaches to determine their structures [94], call for further simulation work aimed at understanding the physical properties of those states, and in particular their packing and compressibility.

Longer high pressure simulations (10–100 ns time scale) of solvated proteins have become possible now; performed with a correct treatment of long range electrostatics and with a rigorous statistical–mechanical implementation of constant pressure, these simulations would also constitute a reference to test the results obtained from simplified approaches. Promising techniques include solvent boundary potentials [95] and implicit solvents [96], which may allow access to time scales one or two orders of magnitude longer. These time scales may not be sufficient to directly simulate the high pressure unfolding. However, longer simulations will allow the investigation of important issues related to the convergence and to the role of transitions between different conformational substates in the estimation of compressibility and volume fluctuations.

Acknowledgements

The author thanks Annette Salmeen, Michael Gross and Boris Velikson for critically reading the manuscript, Michele Vendruscolo and Martin Karplus for useful discussions, Matthias Buck and Mare Martí for providing initial non-native conformations and Guy de Riencourt for hospitality while most of this review was written.

References

[1] H. Frauenfelder, N.A. Alberding, A. Ansari, D. Braunstein, B.R. Cowen, M.K. Hong, I.E.T. Iben, J.B. Johnson, S. Luck, M.C. Marden, J.R. Mourant, P. Ormos, L. Reinisch, R. Scholl, A. Shulte, E. Shyamsunder, L.B. Sorensen, P.J. Steinbach, A. Xie, R.D. Young, K.T. Yue, Proteins and pressure, J. Phys. Chem. 94 (1990) 1024–1037.

[2] M. Gross, R. Jaenicke, Proteins under pressure, Eur. J. Biochem. 221 (1994) 617–630.

[3] K. Heremans, L. Smeller, Protein structure and dynamics at high pressure, Biochim. Biophys. Acta 1386 (1998) 353–370.

[4] G.J.A. Vidugiris, J.L. Markley, C.A. Royer, Evidence for a molten globule transition state in protein folding from determination of activation volumes, Biochemistry 34 (1995) 4909–4912.

[5] G. Panick, R. Malessa, R. Winter, G. Rapp, K.J. Frye, C.A. Royer, Structural characterization of the pressure-denatured state and unfolding/refolding kinetics of staphylococcal nuclease by synchrotron small-angle X-ray scattering and Fourier-transform infrared spectroscopy, J. Mol. Biol. 275 (1998) 389–402.

[6] A.E. Mark, W.F. van Gunsteren, Simulation of the thermal denaturation of hen egg white lysozyme: trapping the molten globule state, Biochemistry 31 (1992) 7746–7748.

[7] V. Daggett, M. Levitt, A model of the molten globule state from molecular dynamics simulations, Proc. Natl. Acad. Sci. USA 89 (1992) 5142–5146.

[8] A. Caflisch, M. Karplus, Molecular dynamics simulation of protein denaturation: solvation of the hydrophobic cores and secondary structure of barnase, Proc. Natl. Acad. Sci. USA 91 (1994) 1746–1750.

[9] A. Caflisch, M. Karplus, Acid and thermal denaturation of barnase investigated by molecular dynamics simulations, J. Mol. Biol. 252 (1995) 672–708.

[10] P.H. Hünenberger, A.E. Mark, W.F. van Gunsteren, Computational approaches to study protein unfolding: hen egg white lysozyme as a case study, Proteins 21 (1995) 196–213.

[11] M.A. Williams, J.M. Thornton, J.M. Goodfellow, Modelling protein unfolding: hen egg white lysozyme, Protein Eng. 10 (1997) 895–903.

[12] E. Paci, L.J. Smith, C.M. Dobson, M. Karplus, Exploration of partially unfolded states of human α-lactalbumin by molecular dynamics simulation, J. Mol. Biol. 306 (2001) 329–347.

[13] T.V. Chalikian, V.S. Gindikin, K.J. Breslauer, Spectroscopic and volumetric investigation of cytochrome *c* unfolding at alkaline pH: characterization of the base-induced unfolded state at 25 degrees C, FASEB J. 10 (1996) 164–170.

[14] P.W. Atkins, Physical Chemistry, Oxford University Press, Oxford, 1978.

[15] R.L. Baldwin, Temperature dependence of the hydrophobic interaction in protein folding, Proc. Natl. Acad. Sci. USA 83 (1986) 8069–8072.

[16] W. Kauzmann, Thermodynamics of unfolding, Nature 325 (1987) 763–764.

[17] V.A. Payne, N. Matubayasi, L.R. Murphy, R.M. Levy, Monte Carlo study of the effect of pressure on hydrophobic association, J. Phys. Chem. B 101 (1997) 2054–2060.

[18] G. Hummer, S. Garde, A.E. Garcia, M.E. Paulatis, L.R. Pratt, The pressure dependence of hydrophobic interactions is consistent with the observed pressure denaturations of proteins, Proc. Natl. Acad. Sci. USA 95 (1998) 1552–1555.

[19] N. Hillson, J.N. Onuchic, A.E. García, Pressure-induced protein-folding/unfolding kinetics, Proc. Natl. Acad. Sci. USA 96 (1999) 14848–14853.

[20] G. Panick, G.J. Vidugiris, R. Malessa, G. Rapp, R. Winter, C.A. Royer, Exploring the temperature-pressure phase diagram of staphylococcal nuclease, Biochemistry 38 (1999) 4157–4164.

[21] H.C. Andersen, Molecular dynamics simulations at constant pressure and/or temperature, J. Chem. Phys. 72 (1980) 2384–2393.

[22] S. Nosé, A molecular dynamics method for simulations in the canonical ensemble, Mol. Phys. 52 (1984) 255–268.

[23] W.G. Hoover, Canonical dynamics: equilibrium phase-space distributions, Phys. Rev. A 31 (1985) 1695–1697.

[24] H.J.C. Berendsen, J.P.M. Postma, W.F. van Gunsteren, A. DiNola, J.R. Haak, Molecular dynamics with coupling to an external bath, J. Chem. Phys. 81 (1984) 3684–3690.

[25] E. Paci, M. Marchi, Constant pressure molecular dynamics techniques applied to complex molecular systems and solvated proteins, J. Phys. Chem. 100 (1996) 4314–4322.

[26] S.E. Feller, Y. Zhang, R.W. Pastor, B.R. Brooks, Constant pressure molecular dynamics simulation: the Langevin piston method, J. Chem. Phys. 103 (1995) 4613–4621.

[27] S. Nosé, M.L. Klein, Constant pressure molecular dynamics for molecular systems, Mol. Phys. 50 (1983) 1055–1076.

[28] M. Ferrario, J.-P. Ryckaert, Constant pressure-constant temperature molecular dynamics for rigid and partially rigid molecular systems, Mol. Phys. 54 (1985) 587–603.

[29] M. Marchi, P. Procacci, Coordinates scaling and multiple time step algorithms for simulation of solvated proteins in the NPT ensemble, J. Chem. Phys. 109 (1998) 5194–5202.

[30] G. Ciccotti, J.-P. Ryckaert, Molecular dynamics simulation of rigid molecules, Comput. Phys. Rep. 4 (1986) 345–392.

[31] H. Zhu, R.S. Averback, Molecular-dynamics method for simulating heterogeneous condensed matter at constant pressure, Phys. Rev. B 51 (1995) 15559–15562.

[32] D. Beglov, B. Roux, Finite representation of an infinite bulk system: solvent boundary potential for computer simulations, J. Chem. Phys. 100 (1994) 9050–9063.

[33] D. Beglov, B. Roux, Dominant solvation effects from the primary shell of hydration: approximation for molecular dynamics simulations, Biopolymers 35 (1995) 171–178.

[34] V. Lounnas, S.K. Lüdemann, R.C. Wade, Towards molecular dynamics simulation of large proteins with a hydration shell at constant pressure, Biophys. Chem. 78 (1999) 157–182.

[35] M. Marchi, K. Akasaka, Simulation of hydrated BPTI at high pressure: changes in hydrogen bonding and its relation with NMR experiments, J. Phys. Chem. B 105 (2001) 711–714.

[36] W.D. Cornell, P. Cieplak, C.I. Bayly, I.R. Gould, K.M.J. Merz, D.M. Ferguson, D.C. Spellmeyer, T. Fox, J.W. Caldwell, P. Kollman, A second generation force field for the simulation of proteins, nucleic acids, and organic molecules, J. Am. Chem. Soc. 117 (1995) 5179–5197.

[37] A.D. MacKerell Jr., D. Bashford, M. Bellot, R.L.J. Dunbrack, J.D. Evanseck, M.J. Field, S. Fischer, J. Gao, H. Guo, S. Ha, D. Joseph-McCarthy, L. Kuchnir, K. Kuczera, F.T.K. Lau, C. Mattos, S. Michnick, T. Ngo, D.T. Nguyen, B. Prodhom, W.E. Reiher, B. Roux, B. Schlenkrich, J.C. Smith, R.H. Stote, J. Straub, J. Wiórkiewicz-Kuczera, D. Yin, M. Karplus, All-atom empirical potential for molecular modeling and dynamics studies of proteins, J. Phys. Chem. B 102 (1998) 3586–3616.

[38] K. Akasaka, T. Tezuka, H. Yamada, Pressure-induced changes in the folded structure of lysozyme, J. Mol. Biol. 271 (1997) 671–678.

[39] H. Li, H. Yamada, K. Akasaka, Effect of pressure on the tertiary structure and dynamics of folded basic pancreatic trypsin inhibitor, Biophys. J. 77 (1999) 2801–2812.

[40] H.J.C. Berendsen, J.R. Grigera, T.P. Straatsma, The missing term in the effective pair potentials, J. Phys. Chem. 91 (1987) 6269–6271.

[41] K. Bagchi, S. Balasubramanian, M.L. Klein, The effects of pressure on structural and dynamical properties of associated liquids: molecular dynamics calculations for the extended simple point charge model of water, J. Chem. Phys. 107 (1997) 8561–8567.

[42] P.J. Rossky, M. Karplus, A. Rahman, A model for the simulation of an aqueous dipeptide solution, Biopolymers 18 (1979) 825–854.

[43] D.B. Kitchen, L.H. Reed, R.M. Levy, Molecular dynamics simulation of solvated protein at high pressure, Biochemistry 31 (1992) 10083–10093.

[44] S.J. Weiner, P.A. Kollman, D.T. Nguyen, D.A. Case, An all atom force-field for simulations of proteins and nucleic-acids, J. Comp. Chem. 7 (1986) 230–252.

[45] H.J.C. Berendsen, J.P.M. Postma, W.F. van Gunsteren, J. Hermans, Interaction models for water in relation to protein hydration, in: B. Pullman (Ed.), Intermolecular Forces, Reider, Dordrecht, 1981, pp. 331–342.

[46] W.F. van Gunsteren, P.H. Hünenberger, C.A. Schiffer, Investigation of protein unfolding and stability by computer simulation, Philos. Trans. R. Soc. London 348 (1995) 49–59.

[47] W.F. van Gunsteren, H.J.C. Berendsen, Groningen Molecular Simulation (GROMOS) Library Manual, Biomos, Groningen, 1987.

[48] B. Wroblowski, J.F. Diaz, K. Heremans, Y. Engelborghs, Molecular mechanisms of pressure induced conformational changes in BPTI, Proteins 25 (1996) 446–455.

[49] R.M. Brunne, W.F. van Gunsteren, Dynamical properties of bovine pancreatic trypsin inhibitor from molecular dynamics at 5000 atm, FEBS Lett. 323 (1993) 215–217.

[50] K. Goossens, L. Smeller, J. Frank, K. Heremans, Pressure-tuning the conformation of bovine pancreatic trypsin inhibitor studied by Fourier-transform infrared spectroscopy, Eur. J. Biochem. 236 (1996) 254–262.

[51] E. Paci, M. Marchi, Intrinsic compressibility and volume compression in solvated proteins by molecular dynamics simulation, Proc. Natl. Acad. Sci. USA 93 (1996) 11609–11614.

[52] C.E. Kundrot, F.M. Richards, Crystal structure of hen egg-white lysozyme at a hydrostatic pressure of 1000 atmospheres, J. Mol. Biol. 193 (1987) 157–170.

[53] E. Neria, S. Fischer, M. Karplus, Simulation of activation free energies in molecular dynamics system, J. Chem. Phys. 105 (1996) 1902–1921.

[54] W.B. Floriano, M.A. Nascimento, G.B. Domont, W.A. Goddard, Effects of pressure on the structure of metmyoglobin: molecular dynamics predictions for pressure unfolding through a molten globule intermediate, Protein Sci. 7 (1998) 2301–2313.

[55] D.J. Bradley, K.S. Pitzer, Thermodynamics of electrolytes. Dielectric properties of water and Debye-Huckel parameters to 350° and 1 kbar, J. Phys. Chem. 83 (1979) 1599–1603.

[56] E. Paci, Etude des Protéines à Haute Pression Hydrostatique par Simulation de Dynamique Moléculaire, Ph.D. thesis, Université Paris 6, 1996.

[57] H. Li, H. Yamada, K. Akasaka, Effect of pressure on individual hydrogen bonds in proteins. Basic pancreatic trypsin inhibitor, Biochemistry 37 (1998) 1167–1173.

[58] E. Paci, B. Velikson, On the volume of macromolecules, Biopolymers 41 (1997) 785–797.

[59] A.P. Sarvazyan, P. Hemmes, Relaxational contribution to protein compressibility from ultrasonic data, Biopolymers 18 (1979) 3015–3024.

[60] K. Gekko, H. Noguchi, Compressibility of globular proteins in water at 25°C, J. Phys. Chem. 83 (1979) 2706–2714.

[61] B. Gavish, E. Gratton, C.J. Hardy, Adiabatic compressibility of globular proteins, Proc. Natl. Acad. Sci. USA 80 (1983) 750–754.

[62] K. Gekko, Y. Hasegawa, Compressibility-structure relationship of globular proteins, Biochemistry 25 (1986) 6563–6571.

[63] D.P. Kharakoz, A.P. Sarvazyan, Hydrational and intrinsic compressibilities of globular proteins, Biopolymers 33 (1993) 11–25.

[64] T.V. Chalikian, K.J. Breslauer, Thermodynamic analysis of biomolecules: a volumetric approach, Curr. Opin. Struct. Biol. 8 (1998) 657–664.

[65] G.F. Voronoi, Nouvelles applications des paramètres continus à la théorie des formes quadratiques, Z. Reine Angew. Math. 134 (1908) 198–287.

[66] F.M. Richards, The interpretation of protein structures: total volume, group volume distribution and packing density, J. Mol. Biol. 82 (1974) 1–14.

[67] J.L. Finney, Volume occupation, environment and accessibility in proteins. The problem of the protein surface, J. Mol. Biol. 96 (1975) 721–732.

[68] V.N. Morozov, T.Y. Morozova, What does a protein molecule look like, Comments Mol. Cell. Biophys. 6 (1990) 249–270.

[69] F.M. Richards, Areas, volumes, packing and protein structure, Annu. Rev. Biophys. Bioeng. 6 (1977) 151–176.

[70] T.J. Richmond, Solvent accessible surface area and excluded volume in proteins, J. Mol. Biol. 178 (1984) 63–89.

[71] H. Frauenfelder, H. Hartmann, M. Karplus, I.D. Kuntz, J. Kuriyan, F. Parak, G.A. Petsko, D. Ringe, R.F. Tilton, M.L. Connolly, N. Max, Thermal expansion of a protein, Biochemistry 26 (1987) 254–261.

[72] M.C. Marden, G. Hui, F. Stetzkowsky-Marden, Heme protein fluorescence versus pressure, Biophys. J. 49 (1986) 619–627.

[73] J. Zollfrank, J. Friedrich, J. Fidy, J.M. Vanderkooi, Photochemical holes under pressure: compressibility and volume fluctuations of a protein, J. Chem. Phys. 94 (1991) 8600–8603.

[74] A. Cooper, Thermodynamic fluctuations in protein molecules, Proc. Natl. Acad. Sci. USA 73 (1976) 2740–2741.

[75] V.M. Dadarlat, C.B. Post, Insights into protein compressibility from molecular dynamics simulations, J. Phys. Chem. B 105 (2001) 715–724.

[76] J.P. Kocher, M. Prévost, S.J. Wodak, B. Lee, Properties of the protein matrix revealed by the free energy of cavity formation, Structure 4 (1996) 1517–1529.

[77] T. Yamato, J. Higo, Y. Seno, N. Gõ, Conformational deformation in deoxymyoglobin by hydrostatic pressure, Proteins 16 (1993) 327–340.

[78] N. Kobayashi, T. Yamato, N. Gõ, Mechanical property of a TIM-barrel protein, Proteins 28 (1997) 109–116.

[79] M. Tanemura, T. Ogawa, N. Ogita, A new algorithm for three-dimensional Voronoi tessellation, J. Comp. Phys. 51 (1983) 191–207.

[80] F. Tama, O. Miyashita, A. Kitao, N. Gõ, Molecular dynamics simulation shows large volume fluctuations of proteins, Eur. Biophys. J. 29 (2000) 472–480.

[81] A. Kitao, S. Hayward, N. Gõ, Energy landscape of a native protein: jumping-among-minima model, Proteins 33 (1998) 496–517.

[82] B. Nölting, S.G. Sligar, Adiabatic compressibility of molten globules, Biochemistry 32 (1993) 12319–12323.

[83] T.V. Chalikian, V.S. Gindikin, K.J. Breslauer, Volumetric characterizations of the native, molten globule and unfolded states of cytochrome *c* at acidic pH, J. Mol. Biol. 250 (1995) 291–306.

[84] P.L. Privalov, Thermodynamics problems of protein structure, Annu. Rev. Biophys. Biomol. Struct. 18 (1989) 47–69.

[85] Y. Tamura, K. Gekko, Compactness of thermally and chemically denaturated ribonuclease A as revealed by volume and compressibility, Biochemistry 34 (1995) 1878–1884.

[86] D.P. Kharakoz, V.E. Bychkova, Molten globule of human α-lactalbumin: hydration, density, and compressibility of the interior, Biochemistry 37 (1997) 1882–1890.

[87] A.V. Finkelstein, E.I. Shakhnovich, Theory of cooperative transitions in protein molecules. II. Phase diagram for a protein molecule in solution, Biopolymers 28 (1989) 1681–1694.

[88] M. Martí-Renom, R.H. Stote, E. Querol, F.X. Avilés, M. Karplus, Refolding of potato carboxypeptidase inhibitor by molecular dynamics simulations with disulfide bond constraints, J. Mol. Biol. 284 (1998) 145–172.

[89] K. Kuwajima, The molten globule state of α-lactalbumin, FASEB J. 10 (1996) 102–109.

[90] C. Chothia, Structural invariants in protein folding, Nature 254 (1975) 304–308.

[91] Y. Harpaz, M. Gerstein, C. Chothia, Volume change on protein folding, Structure 2 (1994) 641–649.

[92] J. Jonas, A. Jonas, High-pressure NMR spectroscopy of proteins and membranes, Annu. Rev. Biophys. Biomol. Struct. 23 (1994) 287–318.

[93] T.V. Chalikian, K.J. Breslauer, Volumetric properties of nucleic acids, Biopolymers 48 (1998) 264–280.

[94] M. Vendruscolo, E. Paci, C.M. Dobson, M. Karplus, Three key residues form a critical contact network in a protein folding transition state, Nature 409 (2001) 641–645.

[95] B. Roux, D. Beglov, W. Im, Generalized solvent boundary potential for computer simulations, AIP Conf. Proc. 492 (1999) 473–491.

[96] B. Roux, T. Simonson, Implicit solvent models, Biophys. Chem. 78 (1999) 1–20.

ELSEVIER

Biochimica et Biophysica Acta 1595 (2002) 201–209

BIOCHIMICA ET BIOPHYSICA ACTA
BBA
www.bba-direct.com

Review

Revisiting volume changes in pressure-induced protein unfolding

Catherine A. Royer *

Centre de Biochimie Structurale, INSERM U554, CNRS UMR 5048, 29, rue de Navacelles, 34090 Montpellier, France

Received 3 September 2001; received in revised form 16 November 2001; accepted 19 November 2001

Abstract

It has long been known that the application of hydrostatic pressure generally leads to the unfolding of proteins. Despite a relatively large number of reports in the literature over the past few decades, there has been great confusion over the sign and magnitude as well as the fundamental factors contributing to volume effects in protein conformational transitions. It is the goal of this review to present and discuss the results obtained concerning the sign and magnitude of the volume changes accompanying the unfolding of proteins. The vast majority of cases point to a significant decrease in volume upon unfolding. Nonetheless, there is evidence that, due to differences in the thermal expansivity of the folded and unfolded states of proteins reported in a half dozen manuscripts, that the sign of the volume change may become positive at higher temperatures.

Keywords: Volume change; Protein folding; Thermal expansivity; Compressibility

1. Introduction

High pressure has been used to study chemical reactions and the physical properties of liquids, solids and polymers for almost a century, and the results have provided very useful information, both fundamental and practical, which has been fully integrated by the chemical community. However, the situation is quite different in the field of protein science. Whereas studies of the effects of pressure on protein structure and function date back to 1914 [1], and have continued ever since within a small community, the interpretation of the results in the context of the larger field of protein structure and dynamics remains to be accomplished. This is in part due to the general impression that the effects of pressure on protein structure are in contradiction with the hydrophobic model for protein folding. Brandts and co-workers [2] concluded that 'the volume and compressibility changes during denaturation are not completely dominated by the hydrophobic effects which are described by the usual model compound transfer reactions', and that 'the hydrophobic model compound transfer reactions may be poor systems from which to estimate the hydrophobic effects for denaturation reactions'. Kauzmann [3] noted that 'the liquid-hydrocarbon model fails almost completely when one attempts to extend it to the effects of pressure on protein folding'. Finally, Dill [4] pointed out that 'the pressure dependence of protein stability does not resemble that of model hydrophobic compounds', and suggested that the polymeric nature of the polypeptide chain was at the basis of the discrepancy.

The effect of pressure on protein structure depends upon the changes in volume and compressibility

* Fax: +33-4-67-41-79-13.
E-mail address: royer@cbs.cnrs.fr (C.A. Royer).

0167-4838/02/$ – see front matter
PII: S0167-4838(01)00344-2

which accompany protein conformational transitions, in particular partial or complete unfolding. The change in the free energy change of unfolding, for examples, with respect to pressure between the reference pressure p_0 (atmospheric) and p is given by the following

$$\Delta G_{u(p)} = \Delta G_{u(p_0)} + \int_{p_0}^{p} \Delta V_u \mathrm{d}p$$

which can be expanded in a Taylor series about the

$$\Delta G_{u(p)} = \Delta G_{u(p_0)} + \Delta V_u^0(p-p_0) + 1/2(\mathrm{d}\Delta V_u^0/\mathrm{d}p) \times (p-p_0)^2 + ...$$

where $\mathrm{d}\Delta V_u^0/\mathrm{d}p$ is the isothermal compressibility $\Delta\beta_u$. Central to what Chalikian and Breslauer [5] have termed the *protein volume paradox* are the magnitude and sign of the change in volume upon unfolding. In fact the transfer of hydrophobic compounds to water is accompanied by a large decrease in specific volume [6] at low pressures, but an increase in specific volume at high pressures [7]. Thus, according to predictions based on the hydrophobic model compound transfer studies, the volume change upon unfolding of proteins is expected to be negative and very large in magnitude, given the large number of hydrophobic groups in the interior of proteins that become exposed to solvent upon unfolding. Moreover, still according to the model compound transfer studies, the change in volume upon unfolding should change sign, becoming positive with increasing pressure. It has been reported that the volume change for the unfolding of proteins is positive at low pressures and negative at high pressures [8] or that it is negative at low temperature, but becomes positive at high temperature [9,10]. Chalikian and Breslauer [5] make reference to small negative and *even positive* values for the volume change of unfolding. In reality, the situation is much less complicated than it may at first appear. The object of the present review is to recapitulate the accumulated data on pressure-induced unfolding with the hope of clarifying the actual experimental results and defining our present understanding of the volume change of unfolding, the underlying premise being that these volume changes are linked to chain packing and hydration, which are at the heart of the specificity of protein structure and stability.

1.1. The sign of the volume change

Not only has there been some confusion as to the magnitude and physical basis of the volume change upon unfolding, even the sign of the change in volume upon unfolding of proteins has been subject to debate. Reference has been made to positive changes in volume upon unfolding at low pressure [8] or high temperature [10]. It seems therefore worthwhile to re-examine the experimental results that have given rise to such references.

Bridgeman first observed the pressure-induced coagulation of egg white in 1914 [1]. This work was followed by studies of the pressure-induced inactivation of trypsin [11], rennin and pepsin [12], and trypsin and chymotrypsin [13]. These early studies of the effect of pressure on protein structure did not yield quantitative values for the volume changes of the transitions. Nevertheless, since pressure led to the inactivation or aggregation of these proteins, it can be concluded that pressure stabilized fully or partially unfolded forms of the proteins and that, therefore, *the volume change for the unfolding reaction was negative in sign*. Reference is made by Johnson and co-workers [14] and by Lumry and Biltonen [15] to studies showing that pressure slowed the kinetics of the heat inactivation of serum γ-globulin. Similar results were obtained for Taka amylase A [16]. This effect on the relaxation time is now understood to have arisen from a positive value of the activation volume for either folding or unfolding, since the relaxation time is the inverse sum of the rate constants for these two reactions [17]. However, the kinetic activation volumes for folding or unfolding and the overall equilibrium volume change should not be confused. A positive activation volume for the rate of protein folding has been reported recently for a number of different proteins and is entirely consistent with a negative value for the equilibrium volume change for unfolding [18–21]. Miyagawa and Suzuki [22] reported a decrease in the volume for the pressure-induced inactivation of Taka amylase, but a positive activation volume for the rate constant of inactivation, as for the γ-globulin above.

Following the technical advances made in the early 1970s in optical spectroscopy, more quantitative studies of the effect of pressure on protein structure were carried out. In 1970, Brandts and co-workers published a UV absorbance study of the unfolding of RNase A [2]. The magnitude of the volume change upon unfolding (ΔV_u) obtained from the fits of the changes in optical density as a function of pressure were dependent upon both the temperature and the pH of the sample, but in all cases the values obtained for ΔV_u were less than zero and ranged from −4.5 to −46.5 ml/mol. Gill and Glogovsky [23] also found a negative volume change for RNase A unfolding. In 1971, Hawley [10] studied the reversible unfolding of chymotrypsinogen and RNase A and proposed an elliptically shaped pressure–temperature phase diagram for these proteins. Over most temperatures, the change in volume upon unfolding would be less than zero and pressure would favor the denatured or unfolded state.

1.2. Temperature effects

In 1973 came the seminal work of Walter Kauzmann on the pressure-induced denaturation of metmyoglobin [24]. In this very complete study the effect of pressure at all pH and temperature values tested was to destabilize the folded state of the protein. Thus, under all conditions tested the volume change upon unfolding was found to be negative. Values (corrected for pressure effects on the buffers) ranged from −51 to −114 ml/mol depending upon temperature and pH. No stabilization of the protein was observed either at low pressure or at high temperature. Nonetheless, in this and all other cases in which the temperature dependence of the magnitude of the volume change upon unfolding, ΔV_u, has been examined (see Table 1), it has been observed that it decreases with increasing temperature. The decrease in the absolute value of ΔV_u with increasing temperature is equal to the overall change in the coefficient of thermal expansion between the folded and unfolded states, $\Delta\alpha_u$. Zipp and Kauzmann [24] found that the unfolded state of metmyoglobin was more susceptible to thermal expansion than the folded state by 1.8 ml/deg, and Hawley [10] reported 1.32 and 0.252 ml/deg respectively for the values of $\Delta\alpha_u$ for chymotrypsinogen and RNase A, respectively. We have found [19,25] similar values for Snase and for the *trp* repressor, near 1.0 ml/mol. A change in sign of the volume change at higher temperature is reported by Hawley [10], based on the observation that the slope of the $\Delta G_u = 0$ contour in the *P–T* phase diagram exhibits a small positive slope as it departs from the temperature axis at high temperature. The effect in this example is quite small, but nonetheless indicates that the volume change for unfolding at these high temperatures is slightly positive, at least at low pressure. Given a larger thermal expansivity for the unfolded state, it is entirely possible that at high temperatures the specific volume of the unfolded state becomes larger than that of the folded state, and thus that the change in volume for unfolding could become positive. Such behavior is necessarily associated (at some pressure and temperature) with an elliptical phase diagram. More recently, it has been reported that cytochrome *c* under relatively extreme conditions (pH 3 and >40°C) is unfolded and the application of pressure leads to a refolding of the protein [26]. This is the only direct measurement of pressure-induced protein *re*folding reported to date. More examples will need to be tested before it can be assured that the effect is general, but this observation is not inconsistent with the differences in thermal expansivity found for a number of proteins and mentioned above. However, it is of interest that the volume change associated with the thermal denaturation of RNase A is negative [27]. We note nonetheless that for most proteins and under most experimental conditions, pressure leads to a destabilization of the folded state.

1.3. Effect of pressure

Reference has been made to a (apparent) positive volume change at low pressures [28,29]. In such cases, the physical observable (i.e. tryptophan emission energy or hydrogen exchange protection factor) was observed to increase at low pressures prior to decreasing as the proteins unfold at high pressures. This *positive volume change* at low pressures should not be taken at face value and such changes in the physical observables at low pressures should not be confused with a stabilization of the folded state under these conditions. At pressures below the unfolding pressure, the folded structure responds to this perturbation by compression. Consequent to the compression of the folded state, certain parts of the protein may become more shielded from solvent (a buried tryptophan or certain amide protons) with

Table 1
Volume changes of unfolding of proteins by pressure

Protein	ΔV_u (ml/mol)	Comments	Conditions		Technique	Ref.
RNase A	−46.5	assuming non-zero $\Delta\beta$	pH 2.0	23.7°C	Absorption	[2]
	−43.0			27.5°C		
	−27.0		pH 2.6	35.5°C		
	−27.0			38.2°C		
	−20.0			41.6°C		
	−7.5		pH 4.0	47.4°C		
	−6.5			51.7°C		
	−4.5			55.1°C		
	<0	cooperative unfolding transition, no ΔV_u calculated	pH 7.0	30°C	FTIR	[30]
	−53, −61, −42, −73, −59	for different residues	pH 2.0	−10°C	NMR	[55]
	−46.5		pH 5.0	40°C	Absorption (4th der.)	[46]
		$-43.2 > \Delta V_u > -63.4$	for a series of 13 mutants			
	−59, −21	assuming a non-zero $\Delta\beta$	pH 2.0	22°C	NMR	[59]
	−35		pH 2.8		Optical rot.	[22]
RNase P2	−43	F31A mutant	pH 5.0	35°C	Absorption (4th der.)	[42]
	−23	F31Y mutant				
Chymotrypsinogen	−31.2		pH 7.6	23°C	Fluorescence	[24]
	−14	assuming non-zero $\Delta\beta$	pH 2.07		Absorption	[10]
	<0	no ΔV_u calculated, but evidence for partially unfolded intermediate	pH 6.9	∼21°C	FTIR	[29]
Lysozyme	−19.7		pH 7.6 No ligand	23°C	Fluorescence	[24]
	−25		+NAG			
	<0	no ΔV_u calculated, but evidence for partially unfolded intermediate	pH 2.0	−13°C	NMR	[57]
	∼−10	depending on the residue	pH 3.9	68.5°C	NMR	[52]
β-Lactoglobulin	<0	no ΔV_u calculated but cooperative unfolding transition with midpoint ∼200 MPa	pH 7.0	24.0°C	Fluorescence	[37]
	<0	no ΔV_u calculated but cooperative unfolding transition with midpoint ∼200 MPa	pH 7.0	20.0°C	FTIR, SAXS	[32]
	−90	core domain	pH 2.0	36.0°C	NMR	[62]
	−57	non-core domain				
Metmyoglobin	−60		pH 10.0	20°C	Absorption	[28]
	−85		pH 5.91			
	−89		pH 5.61			
	−95		pH 5.04			
	−97		pH 4.29			
	−97		pH 3.92			
	−102		pH 5.15	5°C		
	−94		pH 5.04	20°C		
	−100		pH 4.98	40°C		
	−52		pH 4.99	60°C		
	−114		pH 5.75	5°C		
	−89		pH 5.64	20°C		
	−51		pH 5.56	40°C		

Table 1 (*Continued*)

Protein	ΔV_u (ml/mol)	Comments	Conditions		Technique	Ref.
Cytochrome *c*	−15	N→I	pH 4.0	20°C	Absorption	[34]
	−15	I→U				
Cytochrome b_{562}	−102		pH 4.5	27°C	NMR, H–D exchange	[23]
Cytochrome *P*-450 CYP119	−28	*P*-450→*P*-420 inactivation transition	pH 6.5	20°C	Fluorescence, absorption	[51]
Aplysia Mb, SW Mb, *P*-450-Cam LegHb	<0	no ΔV_u calculated, but cooperative unfolding transitions apparent in all cases	pH 7.2, also pH 4.7	15°C	Fluorescence	[33]
Snase	−92	equilibration was likely incomplete	pH 5.5, 4.5, 3.5	21°C	Fluorescence	[36]
	−53		pH 5.1	37°C	NMR	
	−90		pH 5.5	2°C	Fluorescence	[31]
	−84			10		
	−70			21°C		
	−58			30°C		
	−52			40°C		
	−77		pH 5.5	21°C		[17]
	−94		pH 6.0	38°C	Fluorescence	[59]
	−74	assuming non-zero $\Delta\beta$				
	−72	assuming non-zero $\Delta\beta$	pH 5.3	1.5°C	NMR	[60]
	−83			24°C		[60]
	−42	assuming non-zero $\Delta\beta$		24°C		
	−27	assuming non-zero $\Delta\beta$		35°C		[60]
	−112	cavity mutant V66G	pH 7.0+19.1 mM xylose	21°C	Fluorescence	[44]
	−106	cavity mutant V66A	pH 7.0	21°C		
	−96	cavity mutant V66L	pH 5.5+0.5 M GuHCl	21°C		
trp repressor	−188		pH 7.0	10°C	Fluorescence	[18]
	−184			15°C		
	−162			21°C		
	−155			30°C		
	−150			40°C		
Apomyoglobin	<0	no ΔV_u calculated but cooperative N→I transition	pH 7.0	21°C	Fluorescence	[38,39]
	−70	N→I	pH 6.0	21°C	Fluorescence	[43]
	−61	I→U	pH 4.2			
	−83, −82, −70, −106	depending on the residue	pH 5.4	10°C	NMR	[47]
	−185, −126, −110, −100	for four different mutants			Fluorescence	
Azurin	−24, −55	two different mutants			Fluorescence	[48]
CI2	−35		pH 6.0	25°C	Fluorescence	[19]
Tendamistat	−41.4		pH 2.0	35°C	Fluorescence	[20]
Trypsin	−78		pH 3.0	0°C	Fluorescence	[49]
33 kDa PSII protein	−120		pH 6.0	25°C	Fluorescence	[50]
RasBD/RalGDS	−36	N→I	pH 7.3	4°C	NMR	[61]
	−78	I→U				
Thermolysin	−75				Fluorescence	[45]
$P13^{MTCP1}$	−100		pH 7.0	21°C	Fluorescence	[80]
	−100				NMR	
BSCspB	−60	N→U	pH 7.0	25°C	Fluorescence	[79]

increasing pressure at pressure below the unfolding transition. It has been shown by crystallographic studies [30] and more recently by NMR [31,32] that different regions of the proteins tested are more or less susceptible to compression. However, in addition to compression of the native state, these low pressures also shift the unfolding equilibrium towards the unfolded state [21]. However, since the destabilization at low pressures is insufficient to produce any significant population of the unfolded state, the apparent effect of pressure on the physical observable is opposite to that observed as the protein unfolds at higher pressures.

The absorption experiments of the early 1970s were followed by studies of the intrinsic fluorescence emission by Weber, Drickamer and co-workers on lysozyme and chymotrypsinogen [29]. These studies were followed in the 1980s by observations of the peptide infrared absorption of several proteins as a function of pressure [33–36]. The changes in the FTIR absorption spectra provided direct evidence that not only is protein tertiary structure disrupted by pressure, but that the secondary structure is significantly perturbed as well. Fluorescence and UV absorption spectroscopy have also provided evidence on a number of other proteins, indicating that the intrinsic aromatic residues of a number of proteins become exposed to solvent under pressure [20,37–55]. More recently, high pressure has been applied to the study of protein folding using one- and two-dimensional homonuclear and heteronuclear NMR and protein unfolding studies have been carried out [28,40,56–66]. Depending upon the protein under study, and the pressures attained in each experiment, partially or fully denatured conformations of the proteins were favored by the application of pressure. More recently, high pressure small angle X-ray scattering data have been obtained on pressure-induced unfolding of proteins, and likewise the radius of gyration, scattering intensity and pairwise distribution functions are in all cases consistent with the population of states exhibiting a substantial degree of disorder [25,35,36,67]. The volume changes obtained for the full or partial refolding of these proteins are listed in Table 1. It is not the goal of this review to discuss all of the interesting physical properties of the pressure denatured states of proteins that have resulted from the above studies. Rather, all of the studies cited provide very strong evidence that, as a general rule, excluding high temperature, the volume change upon the unfolding (partial or complete) of proteins is negative, i.e. *the specific volume of less ordered states is lower than that of the folded state.*

1.4. The magnitude of the volume change of unfolding

Also central to the *protein volume paradox* [5] is the magnitude of the volume change of unfolding. The values given in Table 1 correspond to changes in volume of less than 1% of the specific volume of the proteins, whereas one would predict values at least an order of magnitude larger in absolute value based on the model compound transfer studies [4,5,8,68–74]. The general consensus is that the model transfer studies are poor representatives of the volumetric properties of proteins because the density of liquid hydrocarbons is much smaller than that of the interior of proteins. The magnitude of the change in volume upon unfolding is so small that it may well represent the sum of large contributions of opposite sign. It is accepted widely that exposure of polar and charged groups will lead to a decrease in volume due to electrostrictive effects. Another substantial negative contribution to ΔV_u is the elimination of internal cavities and voids upon disruption of the folded structure. Experimental evidence from cavity mutants of Snase and RNase reveal generally larger negative volume changes for cavity creating mutations [48,50]. In contrast, the origin of the positive contribution to ΔV_u remains subject to debate. Chalikian and Breslauer [5] suggest that the thermal volume contributes positively to ΔV_u, while others [68,69,74,75] have suggested that the exposure of hydrophobic residues leads to an increase in volume. The observation of an increase in the specific volume of water upon transfer to organic liquids [76] supports this hypothesis. Recent theoretical calculations of the density of water in the wetting of hydrophobic surfaces lend further support to this idea [77]. A more complete understanding of the contributions of these various effects to the overall value of ΔV_u must await more systematic studies. Nonetheless, it is clear that its value is linked to changes in packing and hydration upon protein conformational transitions, and as such its detailed definition merits continued effort.

1.5. Compressibility

The change in volume upon the unfolding of proteins is thus negative and quite small for reasons that are yet to be fully determined. One last issue that must be addressed is that of compressibility changes. It should be noted that for a few of the references in Table 1, the values of the volume changes are given assuming a non-zero change in compressibility between the folded and unfolded states. The curves of the evolution of the physical observables (fluorescence, absorption NMR peak intensity, etc.) vs. pressure are typically fit in terms of the first term of the expansion. However, if there is evidence of non-linearity in the plots of ln K vs. pressure, then the second term, the change in compressibility, is non-negligible and must be taken into account. While there was no evidence for non-linear behavior in the unfolding profiles of Snase obtained by fluorescence [18], Prehoda and co-workers [63] have suggested that uncertainties in the fluorescence plateau values obscured the need to fit in terms of the $\Delta\beta$, and that fitting NMR peak intensity data revealed the importance of the $\Delta\beta$ term. However, in their high pressure H–D exchange studies on cytochrome b_{562}, Fuentes and Wand [28] noted that, 'the linearity (of the pressure response of the unfolding free energy) indicates that the associated volume change can be taken directly from the slope of the line'. Recently, direct measurements of the specific volume as a function of pressure and temperature using a high pressure vibrating tube densitometer [78] suggest that the difference in compressibility between the folded and unfolded states of Snase is negligible, although the compressibilities of the folded and unfolded states could not be measured at the same temperatures within the pressure range of the apparatus (100 MPa).

2. Conclusions

With examples of over 25 proteins, often at multiple temperatures and a variety of solution conditions, that exhibit negative volume changes of unfolding by pressure, it is clear that pressure destabilizes the folded states of proteins under almost all conditions. The volume change upon unfolding is quite small in magnitude, but nonetheless significant. This value arises most likely from the combination of contributions of relatively large magnitude (electrostriction, elimination of cavities, exposure of hydrophobic and polar moieties), but of opposite sign. The elimination of cavities and internal voids appears to represent a major negative contribution to the value of the volume change of unfolding, but more studies must be made in order to confirm this hypothesis. If indeed the magnitude of the volume change depends strongly on internal packing, then high pressure studies represent an important means of probing this fundamental property of proteins. Finally, the issue of the importance of the compressibility change between the folded and unfolded states remains open to debate. Moreover, the amplitude of $\Delta\beta$ may depend upon the identity of the protein and the conditions of the experiment. Regardless of the importance of the change in the compressibility, the effect of pressure on the native states of proteins informs on the compressibility, and thus the volume fluctuations. This information can be very important in the characterization of the dynamic properties of proteins. The contributions of high pressure to the physical characterization of proteins are already quite numerous and merit our attention. The large number of pressure studies that have appeared in the literature over the past few years attests to the growing interest in this pressure axis. Clearly, a more profound understanding of the relative importance of the different contributions to pressure effects will undoubtedly shed light on the factors controlling the structure and stability of proteins.

References

[1] P.W. Bridgman, J. Biol. Chem. 19 (1914) 511–512.
[2] J.F. Brandts, R.J. Oliveira, C. Westort, Biochemistry 9 (1970) 1038–1047.
[3] W. Kauzmann, Nature 325 (1987) 763–764.
[4] K.A. Dill, Biochemistry 29 (1990) 7133–7155.
[5] T.V. Chalikian, K.J. Breslauer, Biopolymers 39 (1996) 619–626.
[6] W.L. Masterton, J. Chem. Phys. 22 (1954) 1830–1833.
[7] S. Sawamura, K. Kitamura, Y. Taniguchi, J. Phys. Chem. 93 (1989) 4931–4935.
[8] G. Hummer, S. Garde, A.E. Garcia, M.E. Paulaitis, L.R. Pratt, Proc. Natl. Acad. Sci. USA 95 (1998) 1552–1555.
[9] W. Kauzmann, Adv. Protein Chem. 14 (1959) 1–63.
[10] S.A. Hawley, Biochemistry 10 (1971) 2436–2442.

[11] M.A. Macheboeuf, J. Basset, G. Levy, Ann. Physiol. Physiochim. Biol. 9 (1933) 713.
[12] J.E. Matthews Jr., R.B. Dow, A.K. Anderson, J. Biol. Chem. 135 (1940) 697.
[13] A.L. Curl, E.F. Jansen, J. Biol. Chem. 184 (1950) 45–54.
[14] F.H. Johnson, H. Eyring, M.J. Polissar, The Kinetic Basis of Molecular Biology, Wiley, New York, 1954.
[15] R. Lumry, R. Biltonen, in: G.D. Fasman, S.N. Timasheff (Eds.), Structure and Stability of Biological Macromolecules, Vol. 2, Marcel Dekker, New York, 1969.
[16] S. Makimoto, N. Hiroyuki, Y. Taniguchi, Biochim. Biophys. Acta 996 (1989) 233–239.
[17] S. Gladstone, K.J. Laidler, H. Eyring, The Theory of Rate Processes, McGraw-Hill Book Co., New York, 1941.
[18] G.J. Vidugiris, J.L. Markley, C.A. Royer, Biochemistry 34 (1995) 4909–4912.
[19] G. Desai, G. Panick, M. Zein, R. Winter, C.A. Royer, J. Mol. Biol. 288 (1999) 461–475.
[20] R. Mohana-Borges, J.L. Silva, J. Ruiz-Sanz, G. de Prat-Gay, Proc. Natl. Acad. Sci. USA 96 (1999) 7888–7893.
[21] G. Pappenberger, C. Saudan, M. Becker, A.E. Merbach, T. Kiefhaber, Proc. Natl. Acad. Sci. USA 97 (2000) 17–22.
[22] K. Miyagawa, K. Suzuki, Arch. Biochem. Biophys. 105 (1964) 297–302.
[23] S.J. Gill, R.L. Glogovsky, J. Phys. Chem. 69 (1965) 1515–1519.
[24] A. Zipp, W. Kauzmann, Biochemistry 12 (1973) 4217–4228.
[25] G. Panick, G.J. Vidugiris, R. Malessa, G. Rapp, R. Winter, C.A. Royer, Biochemistry 38 (1999) 4157–4164.
[26] A. Dewa, M. Tayauchi, M. Sakurai, K. Nitta, Protein Pept. Lett. 5 (1998) 265–268.
[27] Y. Suzuki, S. Imai, M. Kawakami, Y. Masuda, K. Akasaka, Bull. Environ. Contam. Toxicol. 60 (1998) 355–362.
[28] E.J. Fuentes, A.J. Wand, Biochemistry 37 (1998) 9877–9883.
[29] T.M. Li, J.W. Hook III, H.G. Drickamer, G. Weber, Biochemistry 15 (1976) 5571–5580.
[30] C.E. Kundrot, F.M. Richards, J. Mol. Biol. 193 (1987) 157–170.
[31] K. Akasaka, T. Tezuka, H. Yamada, J. Mol. Biol. 271 (1997) 671–678.
[32] K. Akasaka, H. Li, Biochemistry 40 (2001) 8665–8671.
[33] P.T. Wong, K. Heremans, Biochim. Biophys. Acta 956 (1988) 1–9.
[34] N. Takeda, M. Kato, Y. Taniguchi, Biochemistry 34 (1995) 5980–5987.
[35] G. Panick, R. Malessa, R. Winter, G. Rapp, K.J. Frye, C.A. Royer, J. Mol. Biol. 275 (1998) 389–402.
[36] G. Panick, R. Malessa, R. Winter, Biochemistry 38 (1999) 6512–6519.
[37] M.C. Marden, G. Hui Bon Hoa, F. Stetzkowski-Marden, Biophys. J. 49 (1986) 619–627.
[38] K.M. Pryse, T.G. Bruckman, B.W. Maxfield, E.L. Elson, Biochemistry 31 (1992) 5127–5136.
[39] J.L. Silva, C.F. Silveira, A. Correia Junior, L. Pontes, J. Mol. Biol. 223 (1992) 545–555.
[40] C.A. Royer, A.P. Hinck, S.N. Loh, K.E. Prehoda, X. Peng, J. Jonas, J.L. Markley, Biochemistry 32 (1993) 5222–5232.
[41] E. Dufour, G.H. Hoa, T. Haertle, Biochim. Biophys. Acta 1206 (1994) 166–172.
[42] E. Bismuto, I. Sirangelo, G. Irace, E. Gratton, Biochemistry 35 (1996) 1173–1178.
[43] E. Bismuto, G. Irace, I. Sirangelo, E. Gratton, Protein Sci. 5 (1996) 121–126.
[44] K.J. Frye, C.S. Perman, C.A. Royer, Biochemistry 35 (1996) 10234–10239.
[45] K.J. Frye, C.A. Royer, Protein Sci. 6 (1997) 789–793.
[46] E. Mombelli, M. Afshar, P. Fusi, M. Mariani, P. Tortora, J.P. Connelly, R. Lange, Biochemistry 36 (1997) 8733–8742.
[47] G.J. Vidugiris, C.A. Royer, Biophys. J. 75 (1998) 463–470.
[48] K.J. Frye, C.A. Royer, Protein Sci. 7 (1998) 2217–2222.
[49] S. Kunugi, S. Fujiwara, S. Kidokoro, K. Endo, S. Hanzawa, FEBS Lett. 462 (1999) 231–235.
[50] J. Torrent, J.P. Connelly, M.G. Coll, M. Ribo, R. Lange, M. Vilanova, Biochemistry 38 (1999) 15952–15961.
[51] S.E. Bondos, S. Sligar, J. Jonas, Biochim. Biophys. Acta 1480 (2000) 353–364.
[52] G. Mei, A. Di Venere, F.M. Campeggi, G. Gilardi, N. Rosato, F. De Matteis, A. Finazzi-Agro, Eur. J. Biochem. 265 (1999) 619–626.
[53] K. Ruan, R. Lange, F. Meersman, K. Heremans, C. Balny, Eur. J. Biochem. 265 (1999) 79–85.
[54] K. Ruan, C. Xu, Y. Yu, J. Li, R. Lange, N. Bec, C. Balny, Eur. J. Biochem. 268 (2001) 2742–2750.
[55] R.A. Tschirret-Guth, L.S. Koo, G.H. Hoa, P.R. Ortiz De Montellano, J. Am. Chem. Soc. 123 (2001) 3412–3417.
[56] S.D. Samarasinghe, D.M. Campbell, A. Jonas, J. Jonas, Biochemistry 31 (1992) 7773–7778.
[57] X. Peng, J. Jonas, J.L. Silva, Proc. Natl. Acad. Sci. USA 90 (1993) 1776–1780.
[58] X. Peng, J. Jonas, J.L. Silva, Biochemistry 33 (1994) 8323–8329.
[59] J. Zhang, X. Peng, A. Jonas, J. Jonas, Biochemistry 34 (1995) 8631–8641.
[60] D. Nash, B.S. Lee, J. Jonas, Biochim. Biophys. Acta 1297 (1996) 40–48.
[61] D.P. Nash, J. Jonas, Biochemistry 36 (1997) 14375–14383.
[62] D.P. Nash, J. Jonas, Biochem. Biophys. Res. Commun. 238 (1997) 289–291.
[63] K.E. Prehoda, E.S. Mooberry, J.L. Markley, Biochemistry 37 (1998) 5785–5790.
[64] M.W. Lassalle, H. Yamada, K. Akasaka, J. Mol. Biol. 298 (2000) 293–302.
[65] K. Inoue, H. Yamada, K. Akasaka, C. Herrmann, W. Kremer, T. Maurer, R. Doker, H.R. Kalbitzer, Nat. Struct. Biol. 7 (2000) 547–550.
[66] K. Kuwata, H. Li, H. Yamada, C.A. Batt, Y. Goto, K. Akasaka, J. Mol. Biol. 305 (2001) 1073–1083.
[67] J. Woenckhaus, R. Kohling, P. Thiyagarajan, K.C. Littrell, S. Seifert, C.A. Royer, R. Winter, Biophys. J. 80 (2001) 1518–1523.

[68] M.H. Klapper, Biochim. Biophys. Acta 229 (1971) 557–566.
[69] L. Boje, A. Hvidt, Biopolymers 11 (1972) 2357–2364.
[70] R. Jeanicke, Annu. Rev. Biophys. Bioeng. 10 (1981) 1–67.
[71] K. Heremans, Annu. Rev. Biophys. Bioeng. 11 (1982) 1–21.
[72] G. Weber, H.G. Drickamer, Q. Rev. Biophys. 16 (1983) 89–112.
[73] V.V. Mozhaev, K. Heremans, J. Frank, P. Masson, C. Balny, Proteins 24 (1996) 81–91.
[74] C.A. Royer, in: R. Winter, J. Jonas (Eds.), High Pressure Molecular Science, Kluwer Academic Publishers, Dordrecht, 1999, pp. 473–496.
[75] A. Hvidt, J. Theor. Biol. 50 (1975) 245–252.
[76] W.L. Masterton, H.K. Seiler, J. Phys. Chem. 72 (1968) 4257–4262.
[77] A. Wallqvist, E. Gallicchio, R.M. Levy, J. Phys. Chem. B 105 (2001) 6745–6753.
[78] H. Seemann, R. Winter, C.A. Royer, J. Mol. Biol. 307 (2001) 1091–1102.
[79] M. Jacob, G. Holtermann, D. Perl, T. Reinstein, M.A. Geeves, F.X. Schmid, Biochemistry 38 (1999) 2882–2891.
[80] R. Kitahara, C. Roumestad, H. Yamada, J.-L. Saldana, C.A. Royer, K. Akasaka, manuscript in preparation.

ELSEVIER

Biochimica et Biophysica Acta 1595 (2002) 210–223

BIOCHIMICA ET BIOPHYSICA ACTA
BBA

www.bba-direct.com

Review

Expanding the pressure technique: insights into protein folding from combined use of pressure and chemical denaturants

Sarah Perrett, Jun-Mei Zhou *

National Laboratory of Biomacromolecules, Institute of Biophysics, Chinese Academy of Sciences, 15 Datun Road, Chaoyang District, Beijing 100101, PR China

Received 21 September 2001; received in revised form 12 November 2001; accepted 13 November 2001

Abstract

The fundamental principles derived from in vitro protein folding experiments have practical application in understanding the pathology of diseases of protein misfolding and for the development of industrial processes to produce proteins as pharmaceuticals and biotechnological reagents. High pressure as a tool to denature or disaggregate proteins offers a number of unique advantages. The emphasis of this review is on how low concentrations of chemical denaturants can be used in combination with high pressure to extend the range and scope of this useful technique. This approach has already been used in a number of studies, which are discussed here in the context of the questions they address. These include: the origin of the volume change observed on protein unfolding, pressure-induced formation of partially structured intermediates, pressure-induced dissociation of oligomeric and aggregated proteins, and the use of volume changes to probe the structure of the transition state. Wider use of hydrostatic pressure as a denaturation tool, facilitated by combination with chemical denaturants, is likely to bring significant advances to our understanding of protein structure, stability and folding, particularly in relation to proteins associated with the amyloid and prion diseases.

Keywords: High pressure; Protein folding; Protein denaturation; Aggregation; Chemical denaturant; Transition state

1. Introduction

1.1. Overview

The ocean depths account for more than half of the earth's biosphere [1]. Deep-sea organisms must adapt not only to the effects of high pressure, but also to extremes of high or low temperature. Thus the application of potentially denaturing factors to biological systems is not simply a laboratory phenomenon. The study of protein folding is a particularly active field of research. The fundamental principles derived from in vitro folding experiments have practical application in understanding the pathology of diseases of protein misfolding [2,3]. In addition, the development of methods to produce an ever-increasing array of proteins in a pure and correctly folded form is of great importance in the biotechnology and pharmaceutical

Abbreviations: bis-ANS, 4,4′-dianilino-1,1′-binaphthyl-5,5′-disulphonic acid; CI-2, chymotrypsin inhibitor-2; GdmCl, guanidinium hydrochloride

* Corresponding author. Fax: +86-10-64-87-20-26.
E-mail address: zhoujm@sun5.ibp.ac.cn (J.-M. Zhou).

0167-4838/02/$ – see front matter © 2002 Elsevier Science B.V. All rights reserved.
PII: S0167-4838(01)00345-4

industries [4,5]. High pressure is well established as a powerful tool in the study of protein folding [1,6,7]. The combination of high pressure with chemical denaturants to modulate protein folding is showing an increasing number of applications, in both fundamental and applied research.

The purpose of this review is to highlight the advances that have been made in our understanding of protein structure, stability and folding by using hydrostatic pressure, in combination with chemical denaturants, as a denaturation tool. The topic is discussed in terms of the aspects that have been investigated by this method. Within each section, the key studies are summarised, as a means to present the circumstances under which this approach offers particular advantages and might be more widely applied.

1.2. *Hydrostatic pressure as a method of protein denaturation*

High pressure has been used extensively in the study of protein folding [1,6,7]. High pressure tends to denature proteins because the protein–solvent system for the denatured state occupies a smaller volume than that for the native state. Similarly, pressure causes oligomeric proteins to dissociate because this results in a decrease in the net volume of the system. These effects are thought to arise from a combination of factors. The presence of cavities within the folded protein structure, or at the oligomer interface, favours unfolding or dissociation [8]. Disruption of electrostatic interactions leads to a large decrease in volume caused by electrostriction of water molecules around the unpaired charged residues [9,10]. Similarly, solvation of hydrophobic and polar groups results in a decrease in volume of the water molecules. These effects compensate for the increase in volume as the crystalline-like state of the protein interior is disrupted and exposed to solvent on unfolding [11,12].

1.3. *Pressure in relation to other denaturation methods*

Pressure as a tool in the study of protein folding and stability offers a number of advantages [7]. Pressure affects only the volume of the system under study [8]. In contrast, temperature denaturation involves changes in both the volume and the thermal energy. It is found for a large number of proteins that, while denaturation by pressure or chemical denaturants is reversible, temperature denaturation results in irreversible aggregation, probably due to the increase in hydrophobic interactions with temperature. Pressure denaturation is sometimes favoured over chemical denaturation as it avoids the addition of extraneous chemical agents. A denaturation curve can be measured by gradually adjusting the pressure applied to a single sample. A jump in pressure can be used to initiate protein folding or unfolding without the difficulties inherent in achieving rapid mixing of solutions. Importantly, the measurement of activation volumes for kinetic folding or unfolding reactions is one of few methods of obtaining structural information about the transition state.

Pressures of around 3 kbar (1 bar = 0.1 MPa) are generally sufficient to cause dissociation of oligomeric proteins, whereas most monomeric proteins are stable to at least 5 kbar [6]. Some single chain proteins are observed to unfold at lower pressures at ambient temperature and close to neutral pH [13,14]. However, these are the exception rather than the rule. The pressure required to fully denature a protein will depend not only on the free energy of unfolding but also on the associated volume change [8]. Pressure-induced formation of partially folded intermediates leading to deviation from two-state denaturation behaviour may further complicate the issue.

While Bridgman succeeded in poaching an egg at 7 kbar in 1914 [15], among the pressure cells in routine use in laboratories around the world, a significant proportion have a pressure range with an upper limit of 3.5 kbar or less [14,16–19]. Some oligomeric proteins have been observed to be resistant to pressure denaturation even up to pressures of around 7 kbar [20]. These issues clearly place limitations on the protein systems that can be studied by pressure denaturation alone. However, these limitations can be circumvented by the addition of a low concentration of chemical denaturant. Subdenaturing concentrations of guanidinium hydrochloride (GdmCl) or urea can be used to poise the system towards unfolding so that pressure denaturation can be achieved within the experimentally accessible range.

The mechanism of pressure-induced unfolding is thought to involve penetration of water molecules into the protein, resulting in disruption of the structure [8]. This is not dissimilar to the action of chaotropic agents, such as GdmCl, which are thought to disrupt the protein structure by binding and stabilising the denatured state [21,22]. A linear relationship between free energy of unfolding and denaturant concentration is often observed [23,24]. When combined with pressure, the denaturant effect seems to be additive [20] with no effect on the volume change [25,26]. Thus, addition of chemical denaturants provides a logical way to extend the advantages of the pressure technique to a wider range of systems. As illustrated below, this is one of a number of reasons for using chemical denaturants in combination with high pressure.

2. Origin of the volume change observed on protein unfolding

2.1. Introduction

The small size of the volume changes observed on pressure denaturation has long puzzled scientists [27]. The changes typically account for around 1% of the total protein volume [6]. Native protein structures are relatively incompressible, due to the constraints of the covalent polypeptide backbone [8]. It has been suggested that packing defects are the principle factor contributing to the volume change observed on unfolding [8]. Experimental studies on staphylococcal nuclease indicate that neither electrostriction [13,25] nor exposure of hydrophobic surface area [28] results in a significant overall volume change in the protein–solvent system. The study of cavity mutants of staphylococcal nuclease [29] supports excluded volume effects as being the major determinant for the volume change on unfolding.

Outlined below are studies on other protein models where the addition of chemical denaturants has aided the exploration of the volume changes associated with pressure-induced structural changes.

2.2. Measuring volume changes between native, molten globule and unfolded states of apomyoglobin

The pressure-denatured state of apomyoglobin at neutral pH is a molten globule that resembles the state induced at low pH [30,31]. Pressure denaturation at low pH provides an upper limit for the volume change between molten globule and unfolded states [31]. The combination of 0.3 M GdmCl and up to 2.5 kbar of pressure allows the entire transition between the three states to be observed. The volume change between the molten globule and unfolded states is smaller than between the native and molten globule states, but is of a similar order of magnitude [31].

2.3. Volumetric equivalence of molten globule and unfolded states of α-lactalbumin

The denatured state of α-lactalbumin at 1.2 kbar of pressure corresponds to the molten globule state also observed by thermal denaturation [32]. By measuring pressure denaturation at increasing concentrations of GdmCl, the fully unfolded state is increasingly populated in preference to the molten globule state [32]. Despite the change in the structure of the denatured state, the observed volume change for the unfolding reaction shows no GdmCl dependence. This suggests that there is negligible change in the volume of the protein–solvent system between the molten globule and fully unfolded states [32].

2.4. Discussion

Pressure denaturation in the presence of low concentrations of GdmCl has allowed comparison of volume changes between native, molten globule and unfolded states [31,32]. A significant volume change was observed for the transition between molten globule and unfolded states of apomyoglobin [31] but a negligible volume change for this transition for α-lactalbumin [32]. This suggests a relatively dry molten globule state in the case of apomyoglobin consistent with the proportionally small heat capacity change between native and molten globule states for this protein [33]. In contrast, the interior of α-lactalbumin appears to be highly hydrated, as indicated by the finding that the molten globule and unfolded states of α-lactalbumin have almost the same heat capacity [34].

These studies [31,32] are consistent with the hypothesis [8] that the positive and negative contributions to the volume change upon unfolding due to hydration effectively cancel, and that the measured volume changes principally reflect elimination of void volume due to packing defects in the structured chains. This agrees with the results of other studies [8,13,25,28,29] and with estimates of the volume caused by packing defects within proteins, which are also suggested to be of the order of 2% or less of the total protein volume [35].

3. Pressure-induced formation of partially structured intermediates

3.1. *Introduction*

The pressure-denatured state and its relationship to the denatured states obtained by other methods vary from system to system. Some proteins, such as chymotrypsinogen [36,37] and RNase A [38], show loss of both secondary and tertiary structure on pressure denaturation. However, for a significant number of proteins, including apolipoprotein A-I [39], *arc* repressor [40], apomyoglobin [30,31], α-lactalbumin [32], creatine kinase [41], and a model three-helix bundle protein [19], high pressure has been found to promote formation of a molten globule-like intermediate. Lysozyme seems to deviate from two-state unfolding behaviour when pressure denatured at low temperature [42]. In some cases, the partially structured intermediates observed resemble the intermediates populated during denaturation by other methods [32,41], or under certain extremes of conditions [31,32,42]. In other cases, pressure apparently allows observation of a metastable state that is not readily trapped by other methods [19,39], but may correspond to transient intermediates formed during folding [19,42].

Below are outlined further examples of studies where the addition of chemical denaturants has aided the exploration of pressure-induced structural changes.

3.2. *Population of a metastable intermediate of human apolipoprotein A-I*

Human apolipoprotein A-I has a low free energy of folding (2.7 kcal/mol). When pressures up to 3 kbar are applied, a sigmoidal denaturation curve is observed [39]. However, the pressure-denatured state does not correspond to the 2 M GdmCl-denatured state, but appears to correspond to a metastable partially folded intermediate not observed during GdmCl denaturation of the protein. The combination of low, subdenaturing concentrations of GdmCl (0.4 M) in combination with 3 kbar of pressure allowed the fully denatured state to be attained. However, the free energy of unfolding measured by pressure denaturation is lower than that measured by GdmCl alone, consistent with the pressure-induced population of a metastable intermediate [39].

3.3. *Detection of a stable intermediate of lysozyme*

Investigation of the pressure-induced unfolding of lysozyme in the presence of various concentrations of GdmCl and at various temperatures suggests deviation from two-state unfolding at lower temperature [42]. This is consistent with the observation of a transient intermediate during folding of lysozyme and the population of a molten globule intermediate for the homologue α-lactalbumin [43]. This suggests that lysozyme has an inherent tendency to populate a partially folded state under certain extremes of conditions [42].

3.4. *The pressure-denatured state of creatine kinase is a molten globule*

For the 43 kDa dimeric protein creatine kinase, pressures of less than 1.5 kbar are sufficient to induce inactivation of the protein, which appears to precede dissociation of the dimer and is fully reversible [41]. The state induced under 6.5 kbar of pressure resembles in structure the partially folded state that is the dominant species populated at around 0.8 M GdmCl at atmospheric pressure [44]. To induce a state equivalent to the 6 M GdmCl fully unfolded state requires the presence of 1.5 M GdmCl in addition to 6.5 kbar pressure [41]. At atmospheric pressure, the native state of creatine kinase is negligibly populated in 1.5 M GdmCl and the molten globule intermediate is only approx. 50% populated relative to the fully unfolded state [44]. Thus for creatine kinase, a stepwise denaturation reaction via a molten globule intermediate is observed for both pressure and GdmCl denaturation.

3.5. Stabilising partially folded intermediates in a three-helix bundle protein

The 65 amino acid model three-helix bundle protein, α_3-1, has a relatively low free energy for complete unfolding (4.6 kcal/mol), but shows only partial unfolding under 3.5 kbar of pressure [19]. In contrast to the fully unfolded 6 M GdmCl-denatured state, the combination of subdenaturing concentrations of GdmCl (1 M) with high pressure (3.5 kbar) induces formation of a stable, partially folded intermediate which binds the fluorescent dye 4,4′-dianilino-1,1′-binaphthyl-5,5′-disulphonic acid (bis-ANS). The combination of 1 M GdmCl with cold denaturation under 3.5 kbar of pressure allowed detection of an additional, partially folded intermediate state of α_3-1 that also binds bis-ANS. The detection of intermediates for the folding of this small, single domain protein supports the results of simulation studies. It seems that the protein has an inherent tendency to form partially folded intermediates and that these intermediates can be populated under certain extreme conditions [19].

3.6. Discussion

The pressure-induced formation of molten globule or other partially structured intermediate states represents an interesting phenomenon [19,31,32, 39,40,42]. In some cases these intermediates are not observed during denaturation with chemical denaturants [19,39,42]. While agents such as GdmCl are extremely good at solvating all types of interactions that stabilise protein structure [21,22], pressure weakens electrostatic and hydrophobic interactions, but hydrogen bonds are stabilised and the volume change associated with solvation of protein–protein hydrogen bonds is small [7,45]. This may explain the more unfolded nature of GdmCl-denatured states. Having said that, pressure denaturation in the presence of GdmCl has provided evidence for residual structure even in GdmCl-denatured states of lysozyme [17] and Ure2 [20] (see Section 4.4). The apparent deviation from two-state folding behaviour at lower temperature for pressure-induced unfolding of lysozyme in the presence of GdmCl [42] may reflect destabilisation of the native state relative to a partially unfolded intermediate under these conditions. The partially structured nature of pressure-denatured states is a theme that will recur in some of the examples discussed in other contexts below.

4. Pressure-induced dissociation of oligomeric proteins

4.1. Introduction

The general finding that oligomeric proteins can usually be denatured at pressures below 3 kbar [6] possibly reflects that for many dimeric proteins, dissociation and unfolding occur in a single concerted step [46–49]. Experimental conditions where pressure will perturb the dissociation equilibrium are more readily achieved than those that allow unfolding of a single chain [8]. However, when the unfolding mechanism is more complex, so that unfolding precedes dissociation [20], or the protein forms a structured monomer [41], it is not surprising that the pressures required should be as high as or higher than those required to denature single chain proteins. In this case, combination with chemical denaturants allows the pressure effects on a wider range of systems to be studied. As illustrated below, this can provide new insight into folding mechanisms and their relationship to biological function.

4.2. Pressure unfolding of a model protein in the presence and absence of denaturant

Chymotrypsin inhibitor-2 (CI-2) is a small monomeric protein that has been used extensively as a model for protein folding studies and whose folding pathway has been elucidated in great detail [50–52]. The intact protein is unaffected by pressures up to 3.5 kbar, but can be unfolded reversibly by pressure in the presence of subdenaturing concentrations of GdmCl [53]. Proteolytic cleavage after residue 40 of the 64 amino acid protein results in two complementary fragments that form a non-covalent complex with essentially the same structure and folding mechanism as the intact protein [54]. The cleaved protein shows significant structural changes below 3.5 kbar in the absence of denaturant, consistent with its reduced stability compared to intact CI-2 [53]. The results indicate that the pressure-denatured states of both the intact and cleaved proteins contain residual structure not observed in the 6 M GdmCl-denatured state. The high pressure state is suggested to resemble a molten globule. Interestingly, pressure unfolding of the cleaved protein occurs without dissociation of the chains. This may be related to its function in vivo, which requires that the inhibitor complex remains stable after cleavage of CI-2 by the target enzyme [53].

4.3. *Enzyme inactivation precedes dimer dissociation for creatine kinase*

The investigation of the pressure-denatured state of creatine kinase [41] is outlined in Section 3.4 above. While enzyme inactivation is achieved at 1.5 kbar and is fully reversible, when pressures greater than 2 kbar are used, regain of activity becomes irreversible or only slowly reversible, for a fraction of the protein molecules. Negligible perturbation of the dimeric structure occurs at pressures up to 1.2 kbar [41]. Dimer dissociation occurs between 0.5 M and 1.0 M GdmCl at atmospheric pressure. Interestingly, in the presence of at least 1 M GdmCl the pressure-induced transition is reversible up to 6.5 kbar [41]. This suggests that the source of the irreversibility of pressure inactivation at higher pressures may be incorrect dimer formation. The reversibility in the presence of 1 M GdmCl may reflect the fact that the observed transition occurs between monomeric states.

4.4. *The yeast prion protein Ure2 dimer is resistant to dissociation*

The yeast prion Ure2 is a 40 kDa dimeric protein that has homology to glutathione *S*-transferases. The free energy of unfolding measured by GdmCl denaturation by a wide range of structural probes is 12 kcal/mol in Tris–HCl buffer pH 8.4 at 25°C [55]. This transition corresponds to unfolding of the protein, but not dissociation of the subunits ([56]; L. Zhu, J.M.Z. and S.P., unpublished results). A number of lines of experimental evidence suggest the presence of residual structure in the GdmCl-denatured state ([20,56]; L. Zhu, H. Kihara, M. Kojima, J.M.Z. and S.P., unpublished results). At pH 8.4, pressure of up to 6.5 kbar causes only minor perturbation of the Ure2 structure [20]. However, the addition of a subdenaturing concentration of GdmCl (2 M) allowed measurement of the pressure-induced unfolding transition. The free energy of unfolding measured by pressure denaturation was identical to that measured by GdmCl denaturation under the same conditions [20]. This indicates that the mechanism of pressure unfolding is the same as GdmCl denaturation, i.e. pressure induces unfolding but not dissociation of the protein. The structural basis of the extreme resistance of this protein to dissociation at pH 8.4 is still under study. However, having established that pressure can be combined with denaturants to induce unfolding opens up a number of possible lines of investigation. The unfolding mechanism is different at lower pH, conditions that also favour amyloid formation [57]. Pressure denaturation over a range of values of pH in the presence of GdmCl or urea may thus provide important clues about the molecular mechanism of Ure2 amyloid formation.

4.5. *DNA binding stabilises the dimeric E2 DNA binding protein of papillomavirus against dissociation*

The dimeric E2 DNA binding domain of HPV-16 dissociates into highly structured monomers under high hydrostatic pressure [18]. Binding of DNA stabilises the dimer and dissociation cannot be induced by pressure alone up to 3 kbar. By measuring urea denaturation curves under a range of different pressures, it was possible to assess the ability of specific and non-specific DNA sequences to stabilise E2 against dimer dissociation [18]. Binding of DNA was found to increase the free energy of dissociation. The structured monomers also appear to be competent to bind DNA. There is no volume change between specific and non-specific DNA complexes, indicating that there is no difference in surface area burial in the two modes of binding. The specific DNA sequence has a greater effect on stability than a non-specific sequence, which provides a basis for sequence discrimination in vivo [18].

4.6. *Discussion*

Pressure denaturation studies on a protein as well studied as CI-2 are particularly informative, as the wealth of information available about the structure and folding of the protein provides a context in which new experimental results can be interpreted in detail [53]. The interesting finding that the chains in the cleaved CI-2 complex denature under pressure without dissociating is consistent with the presence of residual structure in the pressure-denatured state and with its natural inhibitory activity [53]. A similar finding for the yeast prion Ure2, a very much larger, naturally occurring dimer [20], begs the question whether this too represents a phenomenon important for its biological function.

The effect of combining pressure and chemical denaturants seems to reflect the intrinsic folding properties of the protein. Ure2, which shows a two-state unfolding transition for GdmCl denaturation under the conditions used [55], showed the same unfolding mechanism when pressure was combined with a subdenaturing concentration of GdmCl [20]. Creatine kinase unfolds in a stepwise manner in GdmCl via one or more molten globule-like intermediates [44]. Similarly, high pressure could be used to step between the native and molten globule states, or between a GdmCl-induced partially unfolded state to the fully unfolded state [41]. In most cases where pressure-induced intermediates are observed, such partially structured states are thought to be pertinent to the protein folding mechanism [19,31,32,40,42]. In general it seems that the denaturant actions of pressure and GdmCl (or urea) on protein denaturation are complementary and additive.

A discussion of dissociation of oligomeric proteins would naturally include mention of viral proteins. Hydrostatic pressure is a useful tool for promoting dissociation of viral protein oligomeric structures [6]. This topic will not be discussed in detail here. However, a number of studies have been carried out using the combined effects of pressure and urea to study virus assembly, disassembly and inactivation [58–62], to which the reader may wish to refer. Notable results include the finding that the presence of subdenaturing concentrations of urea converts the pressure-induced deterministic dissociation of R17 phage into a limited stochastic equilibrium, due to a homogenising effect of urea on the virus particles [58,60], and the observation that the magnitude of the inactivating effects of urea and pressure on vesicular stomatitis virus are more than additive [62].

5. Pressure effects on protein aggregation

5.1. Introduction

It is well established that high pressure can dissociate specific oligomers [6]. It is thus a natural progression to consider whether high pressure might also cause dissociation of non-specific aggregates. Studies on protein aggregation using a combination of high pressure and chemical denaturants are outlined below.

5.2. Pressure reverses aggregation and facilitates correct folding for rhodanese

During refolding of urea-denatured rhodanese, a range of intermediates are formed which vary in size, tendency to aggregate and potential for correct folding. The application of high hydrostatic pressure is found to improve folding yields by dissociating aggregates back to folding-competent monomers [63]. Pressures of around 0.5 kbar have been shown to be effective in preventing aggregation [64], considerably lower than are generally required to dissociate specific aggregates. In addition to the important finding that pressure can assist in preventing protein aggregation [64], the use of high pressure in combination with chemical denaturant to study rhodanese folding has provided further details of the complex folding mechanism for this protein [63].

5.3. High pressure combined with low GdmCl prevents aggregation during folding

High hydrostatic pressure, combined with low, non-denaturing concentrations of GdmCl, was tested as a method of renaturing insoluble protein aggregates [4]. The model systems used were agitation-induced aggregates of human growth hormone, chaotrope-induced aggregates of lysozyme and inclusion bodies of β-lactamase. It was found that by refolding for 24 h at 2 kbar pressure in the presence of 0.7 M GdmCl, 100% recovery of soluble growth hormone was achieved, independent of protein concentration up to 9 mg/ml. Similar results were obtained for lysozyme. Pressure assisted in refolding of β-lactamase from inclusion bodies even without the addition of chemical denaturant. These results indicate that the use of high pressure may be useful in the preparation and storage of pharmaceutical agents and represents a considerably more efficient and cost-effective approach than those involving dilute protein solutions in the presence of high concentrations of chaotropic agents [4].

5.4. The mechanism of aggregation of human interferon-γ

Human interferon, which is dimeric, is observed to aggregate in the presence of 0.9 M GdmCl [65]. Sucrose binding studies indicate that aggregation proceeds through a transient expansion of the native state, N*, followed by irreversible transformation into an aggregation-competent state. This explains the observation of first order kinetics for the aggregation process. It appears that the dimeric structure remains intact before aggregation [65]. Application of pressure to this system provided information about the magnitude of the structural changes required to form the expanded native state relative to those for unfolding [66]. Pressure destabilises the native state, favouring both dissociation and aggregation. Aggregation is favoured by pressure in this case because the increase in solvation on forming N* results in a net decrease in the volume of the protein–solvent system. The results are consistent with the results of sucrose binding studies and suggest that the volume change for formation of the expanded state is independent of GdmCl concentration, pressure and temperature. The pressure-denatured state resembles the partially structured monomeric state observed at low pH [66].

5.5. Discussion

High pressure has been shown to dissociate non-specific protein aggregates formed in vitro and facilitate an increase in the final folding yield [4,64,67]. Understanding the mechanism of protein aggregation is fundamental to finding methods to control this process. Pressure was used to investigate the mechanism of GdmCl-induced aggregation of human interferon-γ [66]. In this case pressure favours aggregate formation. The results are consistent with a model for aggregation in which the transition state for aggregate formation is structurally and energetically close to the native state. This then explains the observation of aggregation over time even under strongly native conditions and has implications for the pathology of a number of diseases.

Protein misfolding or aggregation is thought to be the route to a range of human diseases, including Alzheimer's, Parkinson's and the prion diseases (for reviews see [2,3]). The proteins associated with these diseases are therefore important targets for protein folding and structural studies. Many of these proteins have a strong, inherent tendency to aggregate. The application of hydrostatic pressure may thus prove extremely useful not only as a denaturation tool, but also as an adjustable parameter when searching for the optimum solution conditions to undertake structural and mechanistic studies, whether the aim is to provoke or inhibit aggregation. This approach may be particularly suitable where there is kinetic partitioning between correct folding and aggregation [68] or where correct folding, aggregation and amyloid formation represent competing processes in at least partial equilibrium [56]. The application of high pressure to prevent aggregation has a wide range of potential applications within the biotechnology and pharmaceutical industries [4,5,67].

6. Using volume changes to probe the structure of the transition state

6.1. Introduction

The activation volumes for folding and unfolding are determined by measuring the pressure dependence of the folding and unfolding rates. The advantage of such pressure studies is that these values can then provide direct information about the structural properties of the ensemble of states referred to as the transition state for protein folding. The study of protein folding kinetics by pressure jump is limited by the small size of the pressure jumps that can be achieved experimentally, typically 100–250 bar. The system must therefore be primed to fold or unfold, which restricts conditions to within the folding transition region and generally results in small amplitudes of the relaxation kinetics. The same maximum pressures apply as for static experiments. The addition of denaturants can be used to achieve conditions under which a pressure jump in the experimentally accessible range will induce unfolding [16,25]. Alternatively, a GdmCl jump can be used to induce unfolding over a range of pressures [26], thus allowing activation volumes to be determined independent of the experimental limitations of the pressure jump technique.

6.2. Pressure studies on stabilised mutants of staphylococcal nuclease

Staphylococcal nuclease has been used extensively as a model protein for folding studies, including pressure denaturation [13,25,28,70]. While the folding kinetics is complex at atmospheric pressure [69], intermediates are not detected in kinetic experiments performed under pressure, allowing observation of the main folding transition state [70]. In addition, folding and unfolding are much slower under pressure, facilitating measurement of kinetic parameters. Pressure jump studies on staphylococcal nuclease indicate that the volume of the protein–solvent system is larger in the transition state than in either the folded or unfolded state, but the activation volume is considerably greater for folding than unfolding [70]. This then suggests a rather native-like, globular transition state, as suggested by protein engineering studies on other small proteins [51].

The addition of low concentrations of GdmCl allowed the study of the pressure denaturation of stabilised proline-to-glycine staphylococcal nuclease mutants within the same pressure range as wild-type protein [25]. Within experimental error the activation volumes and equilibrium volume changes were independent of GdmCl concentration. This suggests that the effect of exposing protein surfaces and cavities to aqueous solution of GdmCl is equivalent, in terms of system volume, to exposing them to pure water [25]. This supports the view that the pressure-denatured state is penetrated by water in a similar way to penetration of the chemically denatured state by GdmCl or urea [8]. Study of the proline mutants indicates that the kinetic complexity observed in the low pressure range is not due to proline isomerism at any of the three sites mutated [25].

6.3. Pressure jump studies on the trp repressor dimer

The presence of subdenaturing concentrations of GdmCl allows observation of dimer dissociation and unfolding of the *trp* repressor dimer at pressures below 3 kbar [16]. Equilibrium data fit to a two-state pressure unfolding transition between folded dimer and unfolded monomer. Kinetic data are also consistent with a two-state pressure-induced folding transition. This is in contrast to the complex folding behaviour observed at atmospheric pressure by chemical perturbation and suggests that formation of intermediates is no longer rate limiting in the pressure-induced reaction. The results indicate a positive activation volume for the folding/association reaction and a negative activation volume for unfolding/dissociation, indicating the rate determining transition state is intermediate in size between the native and unfolded states. No significant effect of GdmCl concentration on the value of the volume changes on folding, or of the activation volumes for folding or unfolding was observed. The temperature dependence of the volume change on unfolding/dissociation was measured in the presence of 2.5 M GdmCl and the results are consistent with burial of a significant amount of surface area in the transition state compared to the unfolded state. The results indicate that as the temperature increases, the system volume for the unfolded state increases significantly more than that for the folded state [16]. This provides an explanation for the observation that at high temperatures, the application of high pressure stabilises proteins [71].

6.4. Increasing GdmCl concentrations causes the transition state of tendamistat to become more native-like

The small all-β protein tendamistat was used as a model to investigate the GdmCl-induced folding reaction under a range of pressures, in order to obtain the volume profile for folding [26]. The observed increase in volume between the native and unfolded states is virtually independent of denaturant concentration. The activation volumes for refolding and unfolding both depend strongly on the GdmCl concentration and show that the transition state becomes significantly more native-like with increasing denaturant concentration. Above 6 M GdmCl the volume of the transition state exceeds the volume of the native state. This suggests that the transition state is largely desolvated and contains packing deficiencies at high denaturant concentrations. The results [26] are consistent with the Hammond postulate [72], which states that the transition state for a reaction becomes more product-like as the product is destabilised. Similar effects have been observed for other proteins in response to destabilising mutations, increased temperature and increased denaturant concentration [73].

6.5. Discussion

The volume changes associated with equilibrium denaturation transitions appear to be independent of GdmCl concentration [16,25,26]. The activation volumes for folding and unfolding of tendamistat show GdmCl dependence, revealing that the transition state increases linearly in size with increasing GdmCl concentration [26]. That no GdmCl dependence of the activation volumes was observed in other studies [16,25] may reflect the smaller range of GdmCl concentrations used. In those cases it is therefore difficult to extrapolate information about the relative sizes of the transition state, folded and unfolded states to outside the pressure-induced unfolding transition.

Pressure represents one of a few methods to obtain structural information about protein folding transition states. Protein engineering experiments coupled with Φ-value analysis can be used to report on the extent of formation of individual interactions within the protein structure during folding [74]. Thus the structure of the transition state (and in some cases transient intermediates) can be mapped out at the level of individual residues. Φ is defined as:

$$\Phi = \frac{\Delta\Delta G^{\ddagger}}{\Delta\Delta G_{\mathrm{U-N}}} \tag{1}$$

where $\Delta\Delta G^{\ddagger}$ is the change in energy of activation of folding on mutation (i.e. derived from kinetic folding rates) and $\Delta\Delta G_{\mathrm{U-N}}$ is the change in free energy of folding on mutation (i.e. the change in equilibrium stability) [74]. In denaturant-induced folding studies, the denaturant dependence of the kinetic rate constants can provide information about the solvent accessibility of the transition state [23]. This can be expressed in terms of the Tanford β-value:

$$\beta_{\mathrm{T}} = \frac{\left(\dfrac{\partial\Delta G^{\ddagger}}{\partial[\text{denaturant}]}\right)}{\left(\dfrac{\partial\Delta G_{\mathrm{U-N}}}{\partial[\text{denaturant}}\right)} = \frac{m^{\ddagger}}{m_{\mathrm{U-N}}} \tag{2}$$

where $m^{\ddagger}$ is the denaturant dependence of the free energy of activation (i.e. the kinetic m-value expressed in kcal mol^{-1} M^{-1}) and $m_{\mathrm{U-N}}$ is the denaturant dependence of the free energy of unfolding (i.e. the equilibrium m-value) [23,74]. β_{T} is a useful index of the compactness of the transition state and provides an approximate measure of the position of the transition state on the reaction coordinate. An analogous parameter, which we will call β_{P} (for pressure), can be derived from the volume changes associated with pressure-induced unfolding (referred to as $\Phi(V)$ in [75]):

$$\beta_{\mathrm{P}} = \frac{\left(\dfrac{\partial\Delta G^{\ddagger}}{\partial p}\right)}{\left(\dfrac{\partial\Delta G_{\mathrm{U-N}}}{\partial p}\right)} = \frac{\Delta V^{\ddagger}}{\Delta V_{\mathrm{U-N}}} \tag{3}$$

where $\Delta V^{\ddagger}$ is the activation volume and $\Delta V_{\mathrm{U-N}}$ is the volume change measured on unfolding at equilibrium. Comparison of activation and equilibrium volume changes has likewise been used to measure the extent to which the transition state resembles the native state or to estimate the position of the transition state along the reaction coordinate [16,26, 70,75]. Where β_{T} and β_{P} have been compared for the same protein, the values are similar [16,26]. The conclusions drawn from pressure studies regarding the nature of the transition state [25,26,70] are also consistent with the results of protein engineering studies [51,73].

For staphylococcal nuclease and *trp* repressor, the kinetics of pressure-induced protein folding and unfolding is slower and simplified compared to when using chemical or pH denaturation [16,70]. Pressure-induced folding and unfolding of RNase A are likewise slowed compared to other methods of denaturation, although the kinetics remains non-two-state [77]. If this slowing and simplification of the observed folding kinetics under pressure are general phenomena, this suggests that pressure could be used to help elucidate the folding pathways of proteins that have thus far been difficult to study because of very rapid or highly complex folding kinetics.

An exciting advance in pressure jump kinetic studies is the use of small-angle X-ray scattering (SAXS) as a structural probe [75,76]. SAXS provides information about changes in the shape of the protein molecule during the course of folding. Notably, information about the conformation of partially structured and denatured states can be obtained. This technique is complementary to the methods used more widely with pressure jump, such as fluorescence [14,16,25,26,70] and Fourier transform infrared spectroscopy [76,77]. Performance of kinetic studies combining SAXS and high pressure on a wider range of systems may be particularly informative.

7. Conclusions and outlook

Hydrostatic pressure provides a clean, reversible, broadly applicable method of protein denaturation that offers a number of unique advantages over other denaturation methods. It has been commented frequently in recent years that despite the advances in theoretical understanding and ease of practical application of this technique, pressure denaturation remains underused in the protein folding field [7,16]. Combination with chemical denaturants provides a logical way to extend the application of the pressure denaturation method to a wider range of systems, as demonstrated by the studies reviewed here.

This approach has allowed advances in our understanding of the origin and size of the volume changes observed on pressure denaturation [25,31,32] and formation of pressure-induced partially structured intermediates [19,31,32,39–42,53]. Pressure jump studies have provided new information about the mechanisms of protein folding and the nature of folding transition states [70,75–77]. The addition of denaturants has allowed this technique to be extended to other systems [16,25]. Denaturant-induced folding studies performed under pressure provide a powerful method to study denaturant effects on the structure of the transition state [26]. Study of protein complexes, oligomeric and aggregation prone proteins by pressure denaturation, aided by addition of subdenaturing concentrations of denaturant, has provided insight into the interplay between folding, association/dissociation and biological function [18,20,41,53,63,66,67].

An area in which the pressure technique may prove particularly fruitful in the future is in the study of the molecular basis of diseases of protein misfolding. Despite the high level of interest in this field and the efforts of many scientists around the world, the intractability of the proteins associated with diseases of protein aggregation and misfolding has severely hampered progress. The finding that in some cases pressure (with or without low concentrations of chemical denaturant) induces dissociation of non-specific aggregates [4,63,64,67] suggests that pressure may thus provide a useful, adjustable parameter in the search for solution conditions under which these proteins can be studied. The few kinetic folding studies that have been published on such proteins raise further experimental difficulties. For example, the yeast prion Ure2 shows highly complex folding and unfolding kinetics, involving a population of an unfolding intermediate and at least two folding intermediates, one of which is formed within the dead time of stopped-flow mixing [56]. The folding of the amyloidogenic protein β_2-microglobulin is also complex and involves significant structure formation on a millisecond timescale [78]. The mammalian prion protein PrP has been shown to fold too rapidly to measure by conventional stopped-flow methods [79]. Where pressure is able to slow and/or simplify the folding kinetics [16,69,77], this may provide an important means to study such proteins.

Acknowledgements

Work in this laboratory is supported by grants from the INSERM/Chinese Academy of Scieces, the National Natural Science Foundation of China (30070163) and the 973 project of the Chinese Ministry of Science and Technology (G1999075608). S.P. acknowledges support from the Chinese Academy of Sciences, the Royal Commission for the Exhibition of 1851 and the Royal Society.

References

[1] M. Gross, R. Jaenicke, Proteins under pressure, Eur. J. Biochem. 221 (1994) 617–630.

[2] J.W. Kelly, Amyloid fibril formation and protein misassembly: a structural quest for insights into amyloid and prion diseases, Structure 5 (1997) 595–600.

[3] S. Perrett, Misshapes and misfits: protein misfolding and disease, Chem. Ind. 10 (1998) 389–393.

[4] R.J. St John, J.F. Carpenter, T.W. Randolph, High pressure fosters protein refolding from aggregates at high concentrations, Proc. Natl. Acad. Sci. USA 96 (1999) 13029–13033.

[5] E. De Bernadez Clark, Protein refolding for industrial processes, Curr. Opin. Biotechnol. 12 (2001) 202–207.

[6] J.L. Silva, G. Weber, Pressure stability of proteins, Annu. Rev. Phys. Chem. 44 (1993) 89–113.

[7] V.V. Mozhaev, K. Heremans, J. Frank, P. Masson, C. Balny, High pressure effects on protein structure and function, Proteins 24 (1996) 81–91.

[8] G. Weber, H.G. Drickamer, The effect of high pressure upon proteins and other biomolecules, Q. Rev. Biophys. 16 (1983) 89–112.

[9] W. Kauzmann, Some factors in the interpretation of protein denaturation, Adv. Protein Chem. 14 (1959) 1–67.

[10] R.C. Neuman, W. Kauzman, A. Zipp, Pressure dependence of weak acid ionizations in aqueous buffers, J. Phys. Chem. 77 (1973) 2687–2691.

[11] M.H. Klapper, On the nature of the protein interior, Biochim. Biophys. Acta 229 (1971) 557–566.

[12] M. Gerstein, C. Chothia, Packing at the protein-water interface, Proc. Natl. Acad. Sci. USA 93 (1996) 10167–10172.

[13] C.A. Royer, A.P. Hinck, S.N. Loh, K.E. Prehoda, X. Peng, J. Jonas, J.L. Markley, Effects of amino acid substitutions on the pressure denaturation of staphylococcal nuclease as monitored by fluorescence and nuclear magnetic resonance spectroscopy, Biochemistry 32 (1993) 5222–5232.

[14] K. Ruan, C. Xu, Y. Yu, J. Li, R. Lange, N. Bec, C. Balny, Pressure-exploration of the 33-kDa protein from the spinach photosystem II particle, Eur. J. Biochem. 268 (2001) 2742–2750.

[15] P.W. Bridgman, The coagulation of albumin by pressure, J. Biol. Chem. 19 (1914) 511–512.

[16] G. Desai, G. Panick, M. Zein, R. Winter, C.A. Royer, Pressure-jump studies of the folding/unfolding of *trp* repressor, J. Mol. Biol. 288 (1999) 461–475.

[17] K. Sasahara, M. Sakurai, K. Nitta, The volume and compressibility changes of lysozyme associated with guanidinium chloride and pressure-assisted unfolding, J. Mol. Biol. 291 (1999) 693–701.

[18] L.M.T.R. Lima, D. Foguel, J.L. Silva, DNA tightens the dimeric DNA-binding domain of human papillomavirus E2 protein without changes in volume, Proc. Natl. Acad. Sci. USA 97 (2000) 14289–14294.

[19] A. Chapeaurouge, J.S. Johansson, S.T. Ferreira, Folding intermediates of a model three-helix bundle protein, J. Biol. Chem. 276 (2001) 14861–14866.

[20] J.M. Zhou, L. Zhu, C. Balny, S. Perrett, Pressure denaturation of the yeast prion protein Ure2, Biochem. Biophys. Res. Commun. 287 (2001) 147–152.

[21] S.N. Timasheff, The control of protein stability and association by weak interactions with water: how do solvents affect these processes?, Annu. Rev. Biophys. Biomol. Struct. 22 (1993) 67–97.

[22] J.A. Schellman, The thermodynamics of solvent exchange, Biopolymers 34 (1994) 1015–1026.

[23] C. Tanford, Protein denaturation. C. Theoretical models for the mechanism of denaturation, Adv. Protein Chem. 24 (1970) 1–95.

[24] C.N. Pace, Determination and analysis of urea and guanidine hydrochloride denaturation curves, Methods Enzymol. 131 (1986) 266–280.

[25] G.J.A. Vidugiris, D.M. Truckses, J.L. Markley, C.A. Royer, High-pressure denaturation of staphylococcal nuclease proline-to-glycine substitution mutants, Biochemistry 35 (1996) 3857–3864.

[26] G. Pappenberger, C. Saudan, M. Becker, A.E. Merback, T. Kiefhaber, Denaturant-induced movement of the transition state of protein folding revealed by high-pressure stopped-flow measurements, Proc. Natl. Acad. Sci. USA 97 (2000) 17–22.

[27] J.F. Brandts, R.J. Olivera, C. Westort, Thermodynamics of protein denaturation. Effect of pressure on the denaturation of ribonuclease, Biochemistry 9 (1970) 1038–1047.

[28] K.J. Frye, C.S. Perman, C.A. Royer, Testing the correlation between ΔA and ΔV of protein unfolding using m value mutants of staphylococcal nuclease, Biochemistry 35 (1996) 10234–10239.

[29] K.J. Frye, C.A. Royer, Probing the contribution of internal cavities to the volume change of protein unfolding under pressure, Protein Sci. 7 (1998) 2217–2222.

[30] E. Bismuto, I. Sirangelo, G. Irace, E. Gratton, Pressure-induced perturbation of apomyoglobin structure: fluorescence studies on native and acidic compact forms, Biochemistry 35 (1996) 1173–1178.

[31] G.J.A. Vidugiris, C.A. Royer, Determination of the volume changes for pressure-induced transitions of apomyoglobin between the native, molten globule and unfolded state, Biophys. J. 75 (1998) 463–470.

[32] Y. Kobashigawa, M. Sakurai, K. Nitta, Effect of hydrostatic pressure on unfolding of α-lactalbumin: volumetric equivalence of the molten globule and unfolded state, Protein Sci. 8 (1999) 2765–2772.

[33] Y.V. Griko, P.L. Privalov, Thermodynamic puzzle of apomyoglobin unfolding, J. Mol. Biol. 235 (1994) 1318–1325.

[34] Y.V. Griko, E. Freire, P.L. Privalov, Energetics of the α-lactalbumin states: a calorimetric and statistical thermodynamic study, Biochemistry 33 (1994) 1889–1899.

[35] A.A. Rashin, M. Iofin, B. Honig, Internal cavities and buried waters in globular proteins, Biochemistry 25 (1986) 3619–3625.

[36] T.M. Li, J.W. Hook, H.G. Drickamer, G. Weber, Plurality

of pressure-denatured forms in chymotrypsinogen and lysozyme, Biochemistry 15 (1976) 5571–5580.

[37] P.T. Wong, K. Heremans, Pressure effects on protein secondary structure and hydrogen deuterium exchange in chymotrypsinogen: a Fourier transform infrared spectroscopic study, Biochim. Biophys. Acta 956 (1988) 1–9.

[38] N. Takeda, M. Kato, Y. Taniguchi, Pressure- and thermally-induced reversible changes in the secondary structure of ribonuclease A studied by FT-IR spectroscopy, Biochemistry 34 (1995) 5980–5987.

[39] W.W. Mantulin, H.J. Pownall, Reversible folding reactions of human apolioprotein A-I: pressure and guanidinium chloride effects, Biochim. Biophys. Acta 836 (1985) 215–221.

[40] X. Peng, J. Jonas, J.L. Silva, Molten-globule conformation of Arc repressor monomers determined by high-pressure 1H NMR spectroscopy, Proc. Natl. Acad. Sci. USA 90 (1993) 1776–1780.

[41] J.M. Zhou, L. Zhu, C. Balny, Inactivation of creatine kinase by high pressure may precede dimer dissociation, Eur. J. Biochem. 267 (2000) 1247–1253.

[42] K. Sasahara, K. Nitta, Pressure-induced unfolding of lysozyme in aqueous guanidinium chloride solution, Protein Sci. 8 (1999) 1469–1474.

[43] K. Kuwajima, Y. Hiraoka, M. Ikeguchi, S. Sugai, Comparison of the transient folding intermediates in lysozyme and α-lactalbumin, Biochemistry 24 (1985) 874–881.

[44] Y.X. Fan, J.M. Zhou, H. Kihara, C.L. Tsou, Unfolding and refolding of dimeric creatine kinase equilibrium and kinetic studies, Protein Sci. 7 (1998) 2631–2641.

[45] R. Van Eldik, T. Asano, W.J. Le Noble, Activation and reaction volume in solution II, Chem. Rev. 89 (1989) 549–688.

[46] J.U. Bowie, R.T. Sauer, Equilibrium dissociation and unfolding of the Arc repressor dimer, Biochemistry 28 (1989) 7139–7143.

[47] H.W. Dirr, P. Reinemer, Equilibrium unfolding of class pi glutathione S-transferase, Biochem. Biophys. Res. Commun. 180 (1991) 294–300.

[48] Y.K. Mok, G. de Prat-Gay, P.J. Butler, M. Bycroft, Equilibrium dissociation and unfolding of the dimeric human papillomavirus strain-16 E2 DNA-binding domain, Protein Sci. 5 (1996) 310–319.

[49] L.A. Wallace, N. Sluis-Cremer, H.W. Dirr, Equilibrium and kinetic unfolding properties of dimeric human glutathione transferase A1-1, Biochemistry 37 (1998) 5320–5328.

[50] S.E. Jackson, A.R. Fersht, Folding of chymotrypsin inhibitor-2. 1. Evidence for a two-state transition, Biochemistry 30 (1991) 10428–10435.

[51] D.E. Otzen, L.S. Itzhaki, N.F. ElMasry, S.E. Jackson, A.R. Fersht, Structure of the transition state for the folding/unfolding of the barley chymotrypsin inhibitor 2 and its implications for mechanisms of protein folding, Proc. Natl. Acad. Sci. USA 91 (1994) 10422–10425.

[52] L.S. Itzhaki, D.E. Otzen, A.R. Fersht, Structure of the transition state for folding of chymotrypsin inhibitor-2 analysed by protein engineering methods – evidence for a nucleation-condensation mechanism for protein folding, J. Mol. Biol. 254 (1995) 260–288.

[53] R. Mohana-Borges, J.L. Silva, G. de Prat-Gay, Protein folding in the absence of chemical denaturants, J. Biol. Chem. 274 (1999) 7732–7740.

[54] G. de Prat-Gay, J. Ruiz-Sanz, B. Davis, A.R. Fersht, The structure of the transition state for the association of two fragments of the barley chymotrypsin inhibitor 2 to generate native-like protein: implications for mechanisms of protein folding, Proc. Natl. Acad. Sci. USA 91 (1994) 10943–10946.

[55] S. Perrett, S.J. Freeman, P.J.G. Butler, A.R. Fersht, Equilibrium folding properties of the yeast prion protein determinant Ure2, J. Mol. Biol. 290 (1999) 331–345.

[56] D. Galani, A.R. Fersht, S. Perrett, Folding of the yeast prion protein Ure2: kinetic evidence for folding and unfolding intermediates, J. Mol. Biol. 315 (2002) 213–227.

[57] C. Thual, L. Bousset, A. Komar, S. Walter, J. Buchner, C. Cullin, R. Melki, Stability, folding, dimerization, and assembly properties of the yeast prion Ure2p, Biochemistry 40 (2001) 1764–1773.

[58] A.T. Da Poian, A.C. Oliveira, L.P. Gaspar, J.L. Silva, G. Weber, Reversible pressure dissociation of R17 bacteriophage. The physical individuality of virus particles, J. Mol. Biol. 231 (1993) 999–1008.

[59] A.T. Da Poian, J.E. Johnson, J.L. Silva, Differences in pressure stability of the three components of cowpea mosaic virus: implications for virus assembly and disassembly, Biochemistry 33 (1994) 8339–8346.

[60] G. Weber, A.T. Da Poian, J.L. Silva, Concentration dependence of the subunit association of oligomers and viruses and the modification of the latter by urea binding, Biophys. J. 70 (1996) 167–173.

[61] C.F. Bonafe, C.M. Vital, R.C. Telles, M.C. Goncalves, M.S. Matsuura, F.B. Pessine, D.R. Freitas, J. Vega, Tobacco mosaic virus disassembly by high hydrostatic pressure in combination with urea and low temperature, Biochemistry 37 (1998) 11097–11105.

[62] C.F. Bonafe, M. Glaser, E.W. Voss, G. Weber, J.L. Silva, Virus inactivation by anilinonaphthalene sulfonate compounds and comparison with other ligands, Biochem. Biophys. Res. Commun. 275 (2000) 955–961.

[63] M. Panda, B.M. Gorovits, P.M. Horowitz, Productive and non-productive intermediates in the folding of denatured rhodanese, J. Biol. Chem. 275 (2000) 63–70.

[64] B.M. Gorovits, P.M. Horowitz, High hydrostatic pressure can reverse aggregation of protein folding intermediates and facilitate acquisition of native structure, Biochemistry 37 (1998) 6132–6135.

[65] B.S. Kendrick, J.F. Carpenter, J.L. Cleland, T.W. Randolph, A transient expansion of the native state precedes aggregation of recombinant human interferon-γ, Proc. Natl. Acad. Sci. USA 95 (1998) 14142–14146.

[66] J.N. Webb, S.D. Webb, J.L. Cleland, J.F. Carpenter, T.W. Randolph, Partial molar volume, surface area, and hydration changes for equilibrium unfolding and formation of aggregation transition state: high pressure and cosolute stud-

ies on recombinant human IFN-γ, Proc. Natl. Acad. Sci. USA 98 (2001) 7259–7264.

[67] D. Foguel, C.R. Robinson, P.C. de Sousa, J.L. Silva, A.S. Robinson, Hydrostatic pressure rescues native protein from aggregates, Biotechnol. Bioeng. 63 (1999) 552–558.

[68] T. Kiefhaber, R. Rudolph, H.H. Kohler, J. Buchner, Protein aggregation in vitro and in vivo: a quantitative model of the kinetic competition between folding and aggregation, Biotechnology 9 (1991) 825–829.

[69] H.M. Chen, J.L. You, V.S. Markin, T.Y. Tsong, Kinetic analysis of the acid and the alkaline unfolded states of staphylococcal nuclease, J. Mol. Biol. 220 (1991) 771–778.

[70] G.J.A. Vidugiris, J.L. Markley, C.A. Royer, Evidence for a molten globule-like transition state from determination of activation volumes, Biochemistry 34 (1995) 4909–4912.

[71] A. Zipp, W. Kauzmann, Pressure denaturation of metmyoglobin, Biochemistry 12 (1973) 4217–4228.

[72] G.S. Hammond, A correlation of reaction rates, J. Am. Chem. Soc. 77 (1955) 334–338.

[73] A. Matouschek, D.E. Otzen, L.S. Itzhaki, S.E. Jackson, A.R. Fersht, Movement of the position of the transition state in protein folding, Biochemistry 34 (1995) 13656–13662.

[74] A.R. Fersht, in: Structure and Mechanism in Protein Science, Freeman, San Francisco, CA, 1999.

[75] J. Woenckhaus, R. Köhling, P. Thyagarajan, K. Littrell, S. Seifert, C.A. Royer, R. Winter, Pressure-jump small-angle x-ray scattering detected kinetics of staphylococcal nuclease folding, Biophys. J. 80 (2001) 1518–1523.

[76] G. Panick, R. Malessa, R. Winter, G. Rapp, K.J. Frye, C.A. Royer, Structural characterization of the pressure-denatured state and unfolding/refolding kinetics of staphylococcal nuclease by synchrotron small-angle x-ray scattering and Fourier-transform infrared spectroscopy, J. Mol. Biol. 275 (1998) 389–402.

[77] G. Panick, R. Winter, Pressure-induced unfolding/refolding of ribonuclease A: static and kinetic fourier transform infrared spectroscopy study, Biochemistry 39 (2000) 1862–1869.

[78] F. Chiti, P. Mangione, A. Andreola, S. Giorgetti, M. Stefani, C.M. Dobson, V. Bellotti, N. Taddei, Detection of two partially structured species in the folding process of the amyloidogenic protein beta 2-microglobulin, J. Mol. Biol. 307 (2001) 379–391.

[79] G. Wildegger, S. Liemann, R. Glockshuber, Extremely rapid folding of the C-terminal domain of the prion protein without kinetic intermediates, Nat. Struct. Biol. 6 (1999) 550–553.

ELSEVIER

Biochimica et Biophysica Acta 1595 (2002) 224–234

BIOCHIMICA ET BIOPHYSICA ACTA
BBA

www.bba-direct.com

Review

High hydrostatic pressure as a tool to study protein aggregation and amyloidosis

Theodore W. Randolph [a,*], Matthew Seefeldt [a], John F. Carpenter [b]

[a] *Department of Chemical Engineering, University of Colorado, Boulder, CO 80309, USA*

[b] *Department of Pharmaceutical Sciences, School of Pharmacy, University of Colorado Health Sciences Center, Denver, CO 80262, USA*

Received 14 September 2001; received in revised form 8 November 2001; accepted 9 November 2001

Abstract

Aggregation of proteins is a serious problem, affecting both industrial production of proteins and human health. Despite recent advances in the theories and experimental techniques available to address understanding of protein aggregation processes, mechanisms of aggregate formation have proved challenging to study. This is in part because the typical irreversibility of protein aggregation processes at atmospheric conditions complicates analysis of their kinetics and thermodynamics. Because high hydrostatic pressures act to disfavor the hydrophobic and electrostatic interactions that cause protein aggregation, studies conducted under high hydrostatic pressures may allow protein aggregates to be formed reversibly, enabling thermodynamic and kinetic parameters to be measured in greater detail. Although application of high hydrostatic pressures to protein aggregation problems is rather recent, a growing literature, reviewed herein, suggests that high pressure may be a useful tool for both understanding protein aggregation and reversing it in industrial applications.

Keywords: Protein folding; Actiration volume; Fibrils

1. Introduction

1.1. Protein aggregation

Therapeutic proteins provide numerous unique and critical treatments for diseases and conditions (e.g. diabetes, cancer, hemophilia, myocardial infarction). There are already dozens of protein products on the market and hundreds more are in preclinical and clinical development [1]. Furthermore, with the recent advent of robust methods for 'humanizing' antibodies, there has been a new resurgence in biotechnology product development due to the tremendous increase in the number of antibody products being investigated for treatments of human disease [2]. With modern genomic and proteomic approaches, new safer and more effective protein therapeutics are being discovered daily. However, if a protein product cannot be stabilized adequately, its benefit to human health will never be realized. The shelf life required for economic viability of a typical protein pharmaceutical product is 18–24 months [3]. Achieving this goal is particularly difficult because of the relatively low thermodynamic stability of the protein native state [4,5].

Proteins are highly susceptible to the formation of non-native aggregates and precipitates (e.g. [3,6–8]),

* Corresponding author. Fax: +1-303-492-4341.
E-mail address: randolph@pressure.colordao.edu (T.W. Randolph).

0167-4838/02/$ – see front matter
PII: S0167-4838(01)00346-6

even under conditions that thermodynamically strongly favor the native state (e.g. neutral pH at 37°C). The biological activity of a protein in an aggregate is usually greatly reduced. More importantly, non-native protein aggregates can cause adverse reaction in patients, ranging from immune response to anaphylactic shock and even death [9–11]. The capacity of aggregates of a given protein to induce adverse responses cannot be predicted; nor can the maximal level of aggregates required for safety be determined without costly and time-consuming clinical trials [12]. Thus, a major goal of formulation engineering is to design a formulation in which aggregation is kept to an extremely low level. Generally, the goal is to have no more than 1–2% of the entire protein population form aggregates over the shelf life of the product. Achieving this goal in an aqueous formulation is a daunting challenge, which is further exacerbated by the current lacunae in the understanding of the protein aggregate state.

Furthermore, protein aggregates play a large role in human diseases. For example, both Alzheimer's disease and Parkinson's disease, the two most common neurodegenerative diseases, are associated with protein aggregation in vivo. In Alzheimer's disease there is a profuse deposition of insoluble amyloid-β fibrils in the parenchymal and vascular extracellular spaces of the cerebral cortex and leptomeningeal vessels [13]. Similarly, a pathological hallmark of Parkinson's disease is the presence of Lewy bodies, which are intracytoplasmic inclusions composed of protein and granular material [14]. Other protein amyloid diseases include pancreatic amyloid formation in non-insulin-dependent diabetes [15], Huntington's disease [16], autosomal dominant hereditary amyloidosis from lysozyme variants [17], and senile systemic amyloidosis and familial amyloid polyneuropathy putatively caused by transthyretin amyloid deposits [18]. Clearly, a better understanding of how aggregates form in vivo would be useful to guide development of new treatments for amyloidic diseases. Such advances in understanding have been slowed by challenges raised by the slow progression of the disease, and the generally irreversible nature of the amyloidic deposits.

In spite of the obvious importance of protein aggregation in both industrial applications of therapeutic proteins and in human health, relatively little is known about the thermodynamics of the aggregated protein state, and the kinetic mechanisms of its formation. Progress in understanding and preventing aggregation both in vitro and in vivo has been slow compared to advances in our understanding of other protein states (e.g. native, molten globule and denatured states). However, recent advances in conceptual understanding of the aggregation process and new experimental techniques to address aggregate structure have recently sparked a renewed interest in the aggregated state.

Understanding of protein aggregation and the aggregated state has been greatly accelerated by a number of recent conceptual advances. For example, the demonstration that aggregation proceeds through specific pathways has removed a significant conceptual bias that aggregation is a random process [19], and has led to renewed efforts to determine pathways of protein aggregation. Historically, research on protein aggregation first led to the proposal that protein aggregates formed from the fully unfolded, denatured state (reviewed in [20]). Much of this early work focused on thermally induced protein aggregation, precipitation and gelation, in which it was found that incubation of a protein at a temperature above its thermal melting point fosters aggregation. However, subsequent research documented that even under these conditions, aggregates are formed from protein molecules prior to their full unfolding. The next historical leap in insight into aggregation came when it was found that, under certain relatively destabilizing conditions (e.g. pH 3), grossly perturbed protein tertiary structure, secondary structure could be maintained [21,22]. The protein species formed was termed 'molten globule', and it is often found to be an intermediate on the protein folding pathway (e.g. [23–26]). If incubated at sufficiently high concentrations (e.g. 1 mg/ml), molten globules readily form aggregates [23,26,27]. Most recently, we have found that even under solution conditions that are not perturbing of tertiary structure and which thermodynamically greatly favor the native state, proteins can form aggregates and precipitates [28,29]. The native conformation is a dynamic structure, such that at any instant in time there exists an ensemble of species with a distribution of structural expansion/compaction. We have proposed that these aggregates are formed from species within the native state ensemble that are structurally expanded relative to the most compact conformations [28,29].

New experimental techniques to study protein aggregation and the aggregate state are also contributing to renewed interest in this difficult area. The experimental study of protein aggregation is challenging for several reasons. Among them is the generally irreversible nature of aggregate formation, which greatly complicates studies of the thermodynamics of protein aggregation. Another challenge (and opportunity, see, for example, [30–32]) is light scattering from aggregates, which makes many of the most common optical tools for studying protein structure in solution impractical. Finally, the relatively slow kinetics under physiological conditions of many protein aggregation processes in human pathologies have posed serious obstacles in the determination of causes of amyloidic diseases. Adaptation of a number of standard techniques to the study of protein aggregation, along with development of a number of new techniques has led to a recent blossoming of the field. A non-exhaustive listing of these advances includes application of surface plasmon resonance (e.g. [33]), quasielectric light scattering (e.g. [31]), flow cytometry (e.g. [34]), electron microscopy (e.g. [35]), atomic force microscopy (e.g. [36]), X-ray diffraction (e.g. [37]), and antibody mapping (e.g. [38]) to the study of aggregate structure and kinetics of formation. Another recent advance, the subject of this review, is the use of high hydrostatic pressures to modulate the kinetics and thermodynamics of protein aggregation.

1.2. Pressure effects on protein kinetics and thermodynamics

Pressure is a useful thermodynamic variable that may be manipulated to probe protein self-association, both in terms of kinetics and thermodynamic equilibrium. From transition state theory, the activation volume, ΔV^*, for a kinetic process is related to the rate constant for that process by [39]:

$$(\partial \ln k/\partial P)_{T,n} = -\Delta V^*/RT \qquad (1)$$

where k is the reaction rate constant (in concentration-independent units) for the rate limiting step, P is the pressure, R is the gas constant, and T is the temperature. The activation volume, ΔV^*, is the dif-

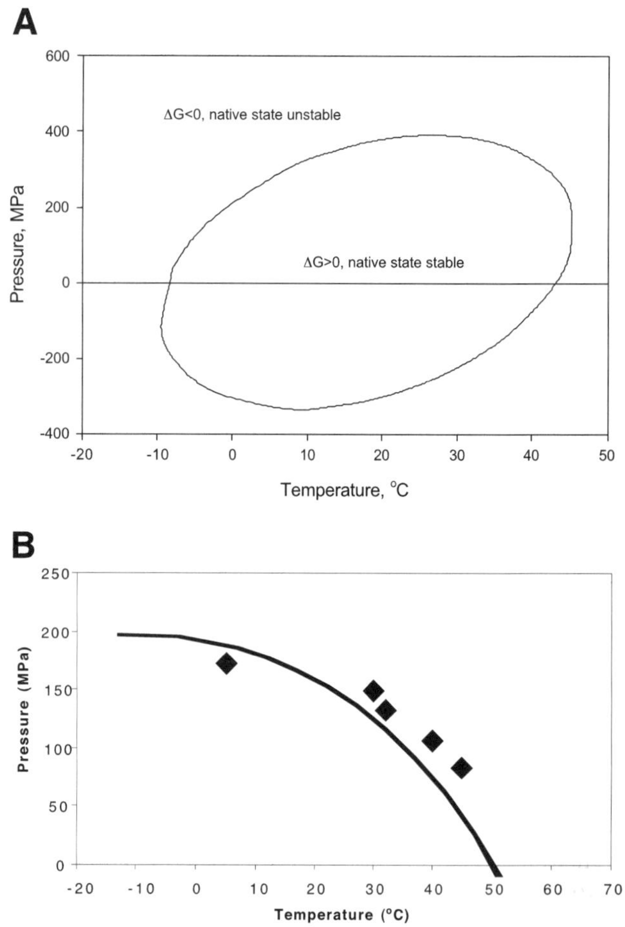

Fig. 1. (A) Temperature–pressure stability contours for chymotrypsinogen, calculated using the model and parameters of Hawley [41]. Within the contour, the native state is stable. Note that the model predicts two denaturation temperatures at ambient pressure, as well as unfolding at negative pressures, such as those that may be found at interfacial regions where protein unfolding is commonly observed. (B) Predicted and measured dissociation of recombinant human interferon-γ as a function of pressure and temperature. The solid line is a predicted contour for the free energy of dissociation, $\Delta G_{\text{dissociation}} = 0$, using values of $\Delta\alpha$ and $\Delta\beta$ for chymotrypsinogen from [41], and other thermodynamic parameters required in the model of Hawley [41] estimated from data for human interferon-γ presented in [97]. At pressure and temperature conditions to the left and below the line, the free energy of dissociation is positive, and the native dimeric structure of recombinant human interferon-γ is favored, while at those to the right and above the line the non-native monomer is favored. Data points (♦) are from experimental observations of pressure-induced unfolding of recombinant human interferon-γ ([97]; T. Randolph, 2001, unpublished), and represent temperature and pressure values where $\Delta G_{\text{dissociation}} = 0$.

ference in partial molar volumes between the transition state for the reaction and those for the reactants.

Likewise, for an equilibrium process, ΔV, the difference in partial molar volumes between the equilibrium states, is related to the equilibrium constant, K, by:

$$(\partial \ln K/\partial P)_{T,n} = -\Delta V/RT \tag{2}$$

For a protein in aqueous solution, the partial molar volume V_{protein} comprises contributions from three terms [40]:

$$V_{\text{protein}} = V_{\text{atoms}} + V_{\text{voids}} + \Delta V_{\text{hydration}} \tag{3}$$

where V_{atoms} is the sum of the volumes of the constituent atoms of the protein, V_{voids} is the volume of void space within the protein, and $\Delta V_{\text{hydration}}$ is a system volume change that results from protein–solvent interactions. V_{atoms} is a fixed quantity for a given protein. Thus, determination of ΔV or ΔV^* for an equilibrium transition or a kinetic process, respectively, provides information on how the protein structure and solvent interactions change during that transition or process.

The thermodynamics of pressure-induced protein transitions have been known for some time. Integration of the relation $d(\Delta G) = -\Delta S dT + \Delta V dP$ yields the expression for the free energy change upon protein denaturation for a two-state transition yields [41]:

$$\Delta G = \frac{\Delta\beta}{2}(P-P_0)^2 + \Delta\alpha(P-P_0)(T-T_0) -\Delta C_p\left[T\left(\ln\frac{T}{T_0}-1\right) + T_0\right] + \Delta V_0(P-P_0) -\Delta S_0(T-T_0) + \Delta G_0 \tag{4}$$

where ΔG is the free energy change of denaturation, $\Delta\alpha = (\partial\Delta V/\partial T)_P$, $\Delta\beta = (\partial\Delta V/\partial P)_T$, C_p is the heat capacity, and the subscript 0 refers to a reference temperature and pressure. This equation predicts that there is a closed stability contour in P–T space, as shown in Fig. 1. Thus, equivalent degrees of protein stabilization may be obtained by using a variety of pressure–temperature combinations. This may greatly aid in the study of protein folding, as has previously been demonstrated by manipulating other solution conditions, such as temperature and guanidine [42], to allow protein folding to be studied as a function of temperature at constant free energies of unfolding.

Processes that result in overall decreases in system volume are favored at high pressures, whereas those that result in volume increases are disfavored. For the specific case of protein structural transitions, three processes deserve particular attention. First, dissociation of chemical species into their respective ions contributes to overall reduction in system volume due to electrostriction [43]. Thus, any transition that results in deprotonation to form charged residues is favored at high pressures. Van Eldik et al. [44] report that the volume change for the reaction shown in Eq. 1a results in a change in system volume $\Delta V = -10$ ml/mol, while that of Eq. 1b results in a ΔV of -20 ml/mol.

$$\text{protein} - \text{COOH} \rightarrow \text{protein} - \text{COO}^- + \text{H}^+ \tag{1a}$$

$$\text{protein} - \text{NH}_2 + \text{H}_2\text{O} \rightarrow \text{protein} - \text{NH}_3^+ + \text{OH}^- \tag{1b}$$

Hence, structural transitions with concomitant formation of charged residues should be favored at high pressure. This may serve to foster both protein unfolding and disaggregation for at least two reasons. Non-native salt bridges may be disrupted by high pressure, and pressure-mediated increases in protein charge will serve to increase electrostatic repulsion between protein molecules, thus decreasing the rate of reassociation or reaggregation.

Second, the exposure of hydrophobic groups to water is also highly favored under pressure. The free energy of solvation of hexane in water is associated with a change in partial molar volume of -6.2 ml/mol, while the volume change of transfer for methane from hexane to water is -22.7 ml/mol [44]. In contrast, the change in volume for a third structural transition, formation of a hydrogen bond, is associated with a partial molar volume change that is approximately zero, i.e. high pressures have little effect on the tendency of a protein to form hydrogen bonds. One reference reports a slight *increase* in system volume (positive ΔV) associated with disrupting hydrogen bonds [45]. Thus, either disruption of salt bridges or hydration of hydrophobic residues is likely the dominant driving force in high pressure effects on proteins and protein aggregates.

2. High pressure studies of protein aggregation

High hydrostatic pressure has been used as a tool to study proteins and their conformational stability and aggregation behavior for nearly a century, starting with the pioneering work of Bridgman who demonstrated that high pressure could be used to denature albumin [46]. Since that time, high hydrostatic pressures have been used to explore the folding pathways of numerous proteins (for examples published within the last 3 years, see [47–62]).

2.1. High pressure effects on protein crystallization

The use of high hydrostatic pressures to study protein aggregation is relatively more recent. The area of high pressure effects on protein crystallization, which may be considered as a sub-category of protein aggregation, was sparked in 1990 by the report that glucose isomerase crystallizes an order of magnitude faster at 200 MPa than at atmospheric pressure [63]. Since that time, a number of other reports on high pressure protein crystallization have been published [64–77]. These reports have generally found, in contrast to the original report [63], that high pressure slowed the rate of protein crystallization. For example, [64–68,77] reported slowed crystallization of lysozyme at high pressures, which was at least partially explained by higher lysozyme solubility at higher pressures [64,67]. Increased lysozyme solubility at higher pressures, though not universally observed [77], was also noted in other studies [72]. Likewise, the solubility of subtilisin at 25°C increases approx. 5-fold as the pressure is increased from atmospheric pressure to 200 MPa, corresponding to a partial molar volume change of crystallization, ΔV_{cryst}, of $+21 \pm 1$ ml/mol [72]. More detailed fitting of crystallization rate constants for subtilisin crystallization found a similar result, $+30 \pm 7$ ml/mol for ΔV_{cryst}. The ability to grow or dissolve protein crystals by changing pressure may provide a valuable tool for the study of protein crystallization in general. Although the number of high pressure crystallization studies reported to date is still small, the positive values of ΔV_{cryst} suggest that pressure may also be a useful tool for inhibiting or reversing other types of protein aggregation in order to allow more detailed study of the aggregation process.

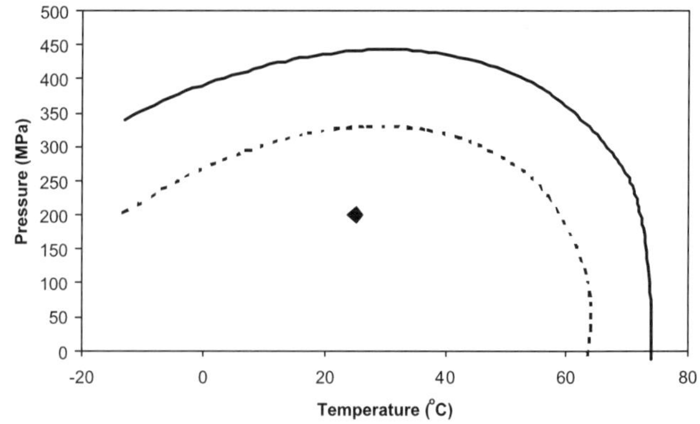

Fig. 2. Pressure–temperature stability contour for HEW lysozyme. The solid line represents conditions where $\Delta G_{unfolding} = 0$ at pH 4.0, calculated with equations presented in [41] using the data presented in [125], and estimates of $\Delta\alpha$ and $\Delta\beta$ for chymotrypsinogen from [41]. The dotted line represents a similar calculation in a solution containing 1.0 M guanidine HCl. At temperature and pressure conditions given by the diamond symbol, covalently cross-linked aggregates of lysozyme suspended in 50 mM Tris (pH 8.0, 24°C), 0.8 M guanidine HCl, 5 mM oxidized glutathione, and 2 mM dithiothreitol disaggregate and refold to active native lysozyme at approx. 70% yield [111].

2.2. High pressure effects on aggregation of food proteins

There is a substantial literature, primarily in the area of food science, that describes the use of high hydrostatic pressure as a means of inducing gelation or other aggregation in protein solutions (see, for example, recent publications [78–84]). Although a complete review of this area is beyond the scope of the current review, several pertinent observations should be noted.

Much of the high pressure work in the area of food science applications has focused on gelation of β-lactoglobulin, a protein of interest in the food industry. β-Lactoglobulin aggregates to form gels upon heating, or after application and release of high hydrostatic pressures. In the case of pressure-induced gelation of β-lactoglobulin, pressure first acts to dissociate dimer to monomer, with concomitant changes in the conformation of the polypeptide chain. Pressure induces structural changes in β-lactoglobulin at moderate pressures (below 150 MPa) [85]. β-Lactoglobulin unfolds at pressures between 150 and 300 MPa [86], and at very high pressures (1000 MPa), 90% of the native α-helix is lost [87]. Pressurization of β-lactoglobulin solutions to 150 MPa does not cause aggregation at short times, but after 30 min at this pressure, aggregation is observed after depressurization [88]. Infrared spectroscopic analysis of β-lactoglobulin A solutions under pressure showed that a band at 1614 cm^{-1} attributed to aggregates *decreases* as the pressure is increased to 730 MPa, and increases and shifts to 1612 cm^{-1} during decompression [89], indicating that aggregation occurs during and after decompression. A similar result was shown for β-lactoglobulin B during pressurization, where the characteristic aggregate band at 1614 cm^{-1} disappears completely as pressure is raised above 150 MPa. In contrast to the FTIR spectrum of β-lactoglobulin A, the spectrum for β-lactoglobulin B shows no evidence of aggregation during depressurization [89]. Increased propensity of β-lactoglobulin A to aggregate after depressurization is likely related to pressure-induced structural modifications that increase the protein's surface hydrophobicity [90]. Furthermore, many researchers who have observed aggregation after depressurization have employed relatively rapid depressurization rates (e.g. 75 MPa/s [53], 20 MPa/s [91] or 4.5 MPa/s [92]), creating the possibility of trapping non-equilibrium, aggregation-prone states. Structural changes induced at high pressures also affect the propensity of β-lactoglobulin to form non-native disulfide bonds [92,93] both during and after depressurization. β-Lactoglobulin conformational changes at high pressures allow exposure of the reactive sulfhydryl at Cys-121, which is normally buried at the monomer/monomer interface in the native dimer, resulting in enhanced rates of non-native disulfide bond formation and concomitant aggregation [94]. Blocking the free thiol moiety in each β-lactoglobulin monomer unit prevents aggregation [94].

Similar results have been described for α-lactalbumin, another protein of importance in the food industry that contains four native disulfide bonds. Changes in protein secondary structure after application of pressures up to 1000 MPa were completely reversible, and the protein remained monomeric. However, in the presence of reductive thiols (cysteine, β-mercaptoethanol, or dithiothreitol), a large fraction of the protein is found as oligomers after 30 min at high pressure [95]. Enhanced reactivity due to exposure of cysteines normally buried in the native state may be of general importance in high pressure aggregation studies, as shown by enhanced reactivity of a normally unreactive sulfhydryl at Cys-253 in rabbit muscle creatine kinase [96].

2.3. *High pressure studies of protein aggregation mechanisms*

The use of high pressure to probe the mechanisms of aggregation of therapeutically relevant proteins is more recent. The equilibrium dissociation and unfolding of recombinant human interferon-γ (rhIFN-γ) was monitored with second derivative UV spectroscopy as a function of pressure and sucrose concentration [97]. rhIFN-γ dissociates reversibly at high pressures, and the $\Delta G_{\text{dissociation}} = 0$ contour is well predicted from the thermodynamic theory shown in [41], as shown in Fig. 1B. The partial molar volume change for dissociation at 32°C was -209 ± 13 ml/mol of dimer. By determining the degree of dissociation as a function of sucrose concentration, which modulates surface tension, the specific molar surface area change for dissociation was determined to be 12.7 ± 1.6 nm^2/molecule of dimer [98,99]. The first order aggregation rate of rhIFN-γ was also studied as a function of pressure and sucrose concentration with derivative UV spectroscopy. Aggregation proceeded through a transition-state species, N*, and resulted in non-native protein precipitates containing a high level of intermolecular β-sheet. Sucrose reduced aggregation rate by shifting the equilibrium between native state (N) and N* towards the more compact N. Pressure increased aggregation rate through increased solvation of the protein, which exposes more surface area, thus shifting the equilibrium away from N towards N*. The changes in partial molar volume and specific molar surface area between the N* and N were -41 ± 9 ml/mol of dimer and 3.5 ± 0.2 nm^2/molecule, respectively. Thus, the structural change required for the formation of the transition state for aggregation is small relative to the difference between N and the dissociated state, further supporting the hypothesis that – under conditions greatly favoring the native state – aggregates form from species in the native state ensemble.

In another recent study, high hydrostatic pressures were shown to reduce aggregation of rhodanase in 3.9 M urea solutions [100]. In the same study, rhodanase was unfolded in 6 M urea, then diluted into a buffer containing 200 mM β-mercaptoethanol, 50 mM sodium thiosulfate, and 50 mM Tris–HCl, pH 7.8 and rapidly pressurized to 200 MPa. 24.4% recovery of active enzyme was reported, compared with only 5.2% when the samples were kept at atmospheric pressure after dilution. A subsequent study from the same group [58] showed that high hydrostatic pres-sures applied early in the rhodanase refolding process yielded higher levels of active enzyme than when pressure was applied after first allowing the protein to misfold during incubation. The authors suggested that the dependence of yield of native protein on the point in time during the renaturation process that high hydrostatic pressure was applied implied that pressure acts to shift equilibrium between small, aggregation-prone intermediates towards formation of monomeric folding intermediates [58].

The P22 bacteriophage tail spike protein has been extensively characterized in terms of its folding pathway and aggregation pathway at atmospheric pressure [101–104]. High hydrostatic pressures (240 MPa) applied to solutions containing aggregated P22 tail spike protein for 90 min decreased the amount of aggregates in the sample by 50%, with concomitant increases in trimeric and monomeric fractions by 25 and 38%, respectively [105]. The trimers were determined to have native-like structure and wild-type function. Because the assays for assembly state, native-like structure, and wild-type function were conducted at atmospheric conditions after depressurization, the folding pathway(s) that were affected by pressure could not be ascertained.

Human growth hormone (rhGH) is an important therapeutic protein that aggregates readily if exposed to air–water interfaces [106,107] or other hydrophobic surfaces [108]. rhGH is a 22 kDa protein formed into a four α-helical bundle [109,110], but upon aggregation the α-helix content of the protein decreases, with concomitant increases in non-native β-sheet structures [111]. Pressurization of aggregates to 200 MPa for 24 h rescues about 20% of the protein to a native, soluble monomeric form [111]. Addition of 0.75 M guanidine HCl, a concentration well below that required to unfold rhGH [112], increases the yield to essentially 100% [111]. Incubation in 0.75 M guanidine HCl at atmospheric pressure resulted in recovery of only 40% soluble monomer. An important observation was that yields of soluble, monomeric rhGH from aggregates after 24 h incubation in 0.75 M guanidine HCl at 200 MPa were essentially independent of protein concentration in the range 0.87–8.7 mg/ml [111]. This suggests that under high pressures, the intermediates in the disaggregation/refolding pathway for rhGH are not prone to aggregation, allowing high refolding yields independent of protein concentration. This is in sharp contrast to observations at atmospheric pressure, where a molten globule intermediate of rhGH formed during dilution from concentrated guanidine solutions aggregates readily [112].

2.4. *High pressure studies of proteins implicated in amyloidic disease states*

There is a growing interest in applying high pressures to the study of proteins that aggregate in vivo in amyloidic disease states. One such protein is lysozyme. In humans, an I56T mutant, which has a structure similar to wild-type protein [17,113,114], is associated with hereditary systemic amyloidosis, a fatal disease [115]. Lysozyme from hen egg white has been extensively examined as a function of pressure [60,73,116–123]. Aggregates of lysozyme can be formed by reducing the four native disulfide bonds in dithiothreitol, unfolding the reduced protein in guanidine HCl, rapidly diluting with buffer to precipitate the protein, and finally re-oxidizing the protein to form non-native disulfide bonds. The insoluble aggregate that results from this treatment has severely perturbed secondary structure, as determined by infrared spectroscopy [124]. In addition to non-native disulfide bonds, the aggregates had a large amount of non-native β-sheet structure [111,124]. After incubation for 24 h at 200 MPa in the presence of 0.8 M guanidine HCl, 5 mM oxidized glutathione and 2 mM dithiothreitol, about 70% of the native lysozyme activity could be recovered. Essentially 100% of the protein was soluble after treatment, implying that the refolding was not complete for all molecules. In contrast, controls incubated at atmospheric pressure showed less than 20% recovery of activity in most cases [111]. Recovery of active lysozyme from pressure-induced refolding displayed distinctly different kinetics at different redox conditions. Lysozyme refolded from disulfide-scrambled aggregates in solutions incubated at 200 MPa at a redox ratio (oxidized to reduced glutathione) of 4:1 regained activity with a time constant of 62 h. At a redox ratio of 1:1, refolding occurred with a time constant of 20 h, and samples pressurized with a redox ratio of 1:16 displayed a refolding time constant of 8 h [124]. It thus appears that the kinetics of lysozyme aggregate dissolution under pressure are limited by the kinetics of disulfide bond reduction. It is particularly interesting to note that the disaggregation and refolding take place under pressure, temperature and solution conditions that favor lysozyme's native state. Several studies suggest that much higher pressures are required to unfold lysozyme [117,121,122]. Based on the calorimetric data and high pressure studies in [125], and estimates of $\Delta\alpha$ and $\Delta\beta$ for chymotrypsinogen from [41], the stability contour in pressure–temperature space can be calculated for lysozyme, using Eq. 4. As shown in Fig. 2, even in the presence of 1 M guanidine HCl, the native state of lysozyme is favored thermodynamically under the disaggregation/refolding conditions of 200 MPa, 25°C, and would be even more favored in solutions containing 0.8 M guanidine HCl. This result suggests that lysozyme in the aggregated state need not completely unfold to allow the protein to refold, and furthermore suggests that refolding occurs at high pressure, rather than upon depressurization.

Another protein whose aggregates are implicated in amyloidic disease states is transthyretin [126,127]. Several studies of transthyretin under high hydrostatic pressure have been reported [128–131]. High hydrostatic pressure alters the tetrameric native form of transthyretin, which converts the native protein into a tetrameric, amyloidogenic state, T_4* [128]. The T_4* state found at 300 MPa binds bis-ANS, suggesting that new hydrophobic surfaces are exposed in the T_4–T_4* transition. Upon depressurization, transthyretin that had been pressure-treated (350 MPa, 60 min, 37°C) rapidly aggregated to form fibrils, as characterized by thioflavin T and Congo red binding. However, it is important to note that aggregation did not occur when the samples were still pressurized. Thus, high pressure provides a unique tool for studying transthyretin aggregation. Amyloidogenic intermediate states such as T_4* can be populated at high pressure and studied optically without the interference arising from aggregation-induced light scattering.

High hydrostatic pressures have also been used to study amyloid A aggregates. In reactive amyloidosis, amyloid A deposits in tissues [18]. Amyloid A is apparently derived from serum amyloid A, a protein that is found in tissues in response to injury of infection [132]. High hydrostatic pressure (1200 MPa) causes amyloidic aggregates of amyloid A isolated from murine spleens and livers to dissociate, with concomitant increases in α-helix and unordered structures and a decrease in β-sheet [133]. The protein reaggregates after pressure is released. Many amyloidic aggregates are difficult to dissociate, especially in solutions at near-physiological pH and temperature. The reversible nature of amyloid A aggregation under pressure may open new windows for the study of the kinetics and thermodynamics of its aggregation.

2.5. *Insight from the polymer literature*

Fascinating insight into the fundamentals of how protein aggregation behavior is affected by high pressure comes from recent studies of the pressure–temperature behavior of peptidyl mimic polymers and artificial proteins. The reversible aggregation of poly(*N*-vinylisobutyramide) was studied as a function of temperature and pressure [134]. At elevated pressures and/or temperatures, poly(*N*-vinylisobutyramide) undergoes a random coil to globule transition, with concomitant aggregation. The cloud-point diagrams could be fitted to within 8% using the model of Hawley [41]. Interestingly, the signs of ΔV_0 and ΔC_p relative to ΔG_0 were opposite to those observed by Hawley for chymotrypsinogen unfolding [134,135]. This might be expected because the chymotrypsi- nogen denaturation transition presumably is a globule to random coil transition, the inverse of the poly(*N*-vinylisobutyramide) transition.

The pressure–temperature behaviour of an 'artificial' protein constructed to mimic elastin, the core protein of elastic fibers within the body, has also been examined [136]. An elastin-like polypeptide, MGLDGSMG$(VPGIG)_{40}$VPLE, was synthesized by bacterial expression. A roughly ellipsoidal cloud-point diagram was observed in the experimentally accessible range of temperatures from −10 to 40°C and pressures ranging from atmospheric to 400 MPa [136], reminiscent of the denaturation contours for chymotrypsinogen [41]. At 30°C and atmospheric temperature, the artificial protein aggregates rapidly. As the pressure is raised to approx. 50 MPa, the artificial protein disaggregates. At still higher pressures (approx. 150 MPa) aggregation is once again prominent. Although this research is still in an early stage, it is apparent that the designing of proteins may greatly help us understand the mechanisms of pressure effects on protein aggregation, and in turn enable us to gain insight into protein aggregation in general.

3. Conclusions

The application of high hydrostatic pressures to the study of protein aggregation is a relatively young, but rapidly developing field. Because high hydrostatic pressure offers the potential to convert protein aggregation from an irreversible, thermodynamically intractable problem to a reversible, equilibrium process, a number of aggregation intermediate states and kinetic processes that were previously inaccessible by conventional techniques may now be amenable to spectroscopic examination. In turn, this knowledge may create avenues to new therapeutic agents to block amyloidosis, or insight into protein crystallization processes. Furthermore, high pressure dissolution and refolding from protein aggregates produced during industrial manufacturing, storage and delivery of protein therapeutics may prove to be of great value to the biotechnology industry.

References

[1] PhRMA. The Promise of Biotechnology and Genetic Research, www.phrma.org/publications/documents/back grounders//2000-12-12.191.phtml, Pharmaceutical Research and Manufacturers of America, 1100 Fifteenth Street, NW Washington, DC 20005, 2001.
[2] J.V. Gavilondo, J.W. Larrick, Biotechniques 29 (2000) 128–141.
[3] J.L. Cleland, M.F. Powell, S.J. Shire, Crit. Rev. Ther. Drug Carrier Syst. 10 (1993) 307–377.
[4] R. Jaenicke, CIBA Found. Symp. 161 (1991) 206–221.
[5] C.N. Pace, K.L. Shaw, Proteins Suppl. 4 (2000) 1–7.
[6] B.L. Chen, T. Arakawa, E. Hsu, L.O. Narhi, T.J. Tressel, S.L. Chien, J. Pharm. Sci. 83 (1994) 1657–1661.
[7] B.S. Chang, R.M. Beauvais, T. Arakawa, L.O. Narhi, A. Dong, D.I. Aparisio, J.F. Carpenter, Biophys. J. 71 (1996) 3399–3406.
[8] Y.S. Kim, J.S. Wall, J. Meyer, C. Murphy, T.W. Randolph, M.C. Manning, A. Solomon, J.F. Carpenter, J. Biol. Chem. 275 (2000) 1570–1574.
[9] W.V. Moore, P. Leppert, J. Clin. Endocrinol. Metab. 51 (1980) 691–697.
[10] R.E. Ratner, T.M. Phillips, M. Steiner, Diabetes 39 (1990) 728–733.
[11] C.A. Thorton, M. Ballow, Arch. Neurol. 50 (1993) 135.
[12] G. Murano, in: Development of Specifications For Biotechnology Pharmaceutical Products, Karger, Basel Vol. 91, 1997, pp. 3–13.
[13] A.E. Roher, Y.-M. Kuo, Methods Enzymol. 309 (1999) 58–67.
[14] M. Goedert, Nat. Rev. Neurosci. 2 (2001) 492–501.
[15] R. Kayed, J. Bernhagen, N. Greenfield, K. Sweimeh, H. Brunner, W. Voelter, A. Kapurniotu, J. Mol. Biol. 287 (1999) 781–796.
[16] E. Scherzinger, A. Sittler, K. Schweiger, V. Heiser, R. Lurz, R. Hasenbank, G.P. Bates, H. Lehrach, E.E. Wanker, Proc. Natl. Acad. Sci. USA 96 (1999) 4604–4609.
[17] D.R. Booth, M. Sunde, V. Bellotti, C.V. Robinson, W.L. Hutchinson, P.E. Fraser, P.N. Hawkins, C.M. Dobson, S.E. Radford, C.C.F. Blake, M.B. Pepys, Nature 385 (1997) 787–793.
[18] J.D. Gillmore, P.N. Hawkins, M.B. Pepys, Br. J. Haematol. 99 (1997) 245–256.
[19] M.A. Speed, D.I.C. Wang, J. King, Nat. Biotechnol. 14 (1996) 1283–1287.
[20] A. Dong, S.J. Prestrelski, S.D. Allison, J.F. Carpenter, J. Pharm. Sci. 84 (1995) 415–424.
[21] M. Ohgushi, A. Wada, FEBS Lett. 164 (1983) 21–24.
[22] D.A. Dolgikh, L.V. Abaturov, E.V. Brazhnikov, I.O. Lebedev, I.N. Chirgadze, O.B. Ptitsyn, Dokl. Akad. Nauk. SSSR 272 (1983) 1481–1484.
[23] R. Wetzel, Cell 86 (1996) 699–702.
[24] A.L. Fink, FASEB J. 11 (1997) 83.
[25] A.L. Fink, K.A. Oberg, S. Seshadri, Fold. Des. 3 (1998) 19–25.
[26] V.N. Uversky, J. Li, A.L. Fink, J. Biol. Chem. 276 (2001) 10737–10744.
[27] R. Wetzel, Adv. Protein Chem. 50 (1997) 183–242.
[28] B.S. Kendrick, J.F. Carpenter, J.L. Cleland, T.W. Randolph, Proc. Natl. Acad. Sci. USA 95 (1998) 14142–14146.
[29] Y.-S. Kim, J. Wall, J. Meyer, C. Murphy, T. Randolph, M. Manning, A. Solomon, J. Carpenter, J. Biol. Chem. 275 (2000) 1570–1574.
[30] A. Lomakin, D.B. Teplow, D.A. Kirschner, G.B. Benedek, Proc. Natl. Acad. Sci. USA 94 (1997) 7942–7947.
[31] A. Lomakin, G.B. Benedek, D.B. Teplow, Methods Enzymol. 309 (1999) 429–459.

[32] D.M. Walsh, A. Lomakin, G.B. Benedek, M.M. Condron, D.B. Teplow, J. Biol. Chem. 272 (1997) 22364–22372.
[33] D.G. Myszka, S.J. Wood, A.L. Biere, Methods Enzymol. 309 (1999) 386–402.
[34] J. Wall, A. Solomon, Methods Enzymol. 309 (1999) 460–466.
[35] S. Inoue, R. Kisilevsky, Mol. Biol. Cell 11 (2000) 240.
[36] T.T. Ding, A.C. Rochet, C. Kelly, J.H. Hafner, S.A. Lee, C.M. Lieber, P.T. Lansbury, Abstr. Pap. Am. Chem. Soc. 219 (2000) 142.
[37] L.C. Serpell, J. Berriman, R. Jakes, M. Goedert, R.A. Crowther, Proc. Natl. Acad. Sci. USA 97 (2000) 4897–4902.
[38] S. Betts, M. Speed, J. King, Methods Enzymol. 309 (1999) 333–350.
[39] K.J. Laidler, Chemical Kinetics, 2nd edn., McGraw-Hill, New York, 1965.
[40] W. Kauzmann, Adv. Protein Chem. 14 (1959) 1–63.
[41] S.A. Hawley, Biochemistry 10 (1971) 2436–2442.
[42] M.L. Scalley, D. Baker, Proc. Natl. Acad. Sci. USA 94 (1997) 10636–10640.
[43] V.V. Mozhaev, K. Heremans, J. Frank, P. Masson, C. Balny, Proteins 24 (1996) 81–91.
[44] R. Van Eldik, T. Asano, W.J. Le Noble, Chem. Rev. 89 (1989) 549–688.
[45] M. Gross, R. Jaenicke, Eur. J. Biochem. 221 (1994) 617–630.
[46] P.W. Bridgman, J. Biol. Chem. 19 (1914) 617–630.
[47] L. Smeller, P. Rubens, K. Heremans, Biochemistry 38 (1999) 3816–3820.
[48] G. Desai, G. Panick, M. Zein, R. Winter, C.A. Royer, J. Mol. Biol. 288 (1999) 461–475.
[49] G. Pappenberger, C. Saudan, M. Becker, A.E. Merbach, T. Kiefhaber, Proc. Natl. Acad. Sci. USA 97 (2000) 17–22.
[50] Y. Kobashigawa, M. Sakurai, K. Nitta, Protein Sci. 8 (1999) 2765–2772.
[51] R. Mohana-Borges, J.L. Silva, J. Ruiz-Sanz, G. de Prat-Gay, Proc. Natl. Acad. Sci. USA 96 (1999) 7888–7893.
[52] M.W. Lassalle, H. Yamada, K. Akasaka, J. Mol. Biol. 298 (2000) 293–302.
[53] P. Kolakowski, E. Dumay, J.C. Cheftel, Food Hydrocolloids 15 (2001) 215–232.
[54] K. Heremans, F. Meersman, H. Pfeiffer, P. Rubens, L. Smeller, High Press. Res. 19 (2000) 623–630.
[55] F. Meersman, L. Smeller, K. Heremans, High Press. Res. 19 (2000) 653–658.
[56] M.M. Botelho, V.L. Valente-Mesquita, K.M.G. Oliveira, I. Polikarpov, S.T. Ferreira, Eur. J. Biochem. 267 (2000) 2235–2241.
[57] H. Stapelfeldt, L.H. Skibsted, J. Dairy Res. 66 (1999) 545–558.
[58] M. Panda, B.M. Gorovits, P.M. Horowitz, J. Biol. Chem. 275 (2000) 63–70.
[59] N. Hillson, J.N. Onuchic, A.E. Garcia, Proc. Natl. Acad. Sci. USA 96 (1999) 14848–14853.
[60] K. Sasahara, K. Nitta, Protein Sci. 8 (1999) 1469–1474.
[61] A. Chapeaurouge, J.S. Johansson, S.T. Ferreira, J. Biol. Chem. 276 (2001) 14861–14866.
[62] S.T. Ferreira, A. Chapeaurouge, J.S. Johansson, Biophys. J. 80 (2001) 2488.
[63] K. Visuri, E. Kaipainen, J. Kivimaki, H. Niemi, M. Leisola, S. Palosaari, Biotechnology 8 (1990) 547–549.
[64] M. Gross, R. Jaenicke, FEBS Lett. 284 (1991) 87–90.
[65] M. Gross, R. Jaenicke, in: C. Balny, R. Hayashi, K. Heremans, P. Masson (Eds.), High Pressure and Biotechnology, Vol. 224, John Libbey Eurotext, London, 1992, pp. 83–86.
[66] M. Gross, R. Jaenicke, Biophys. Chem. 45 (1993) 245–252.
[67] B. Lorber, G. Jenner, R. Giege, J. Cryst. Growth 158 (1996) 103–117.
[68] M.V. Saikumar, C.E. Glatz, M.A. Larson, J. Cryst. Growth 151 (1995) 173–179.
[69] M.V. Saikumar, C.E. Glatz, M.A. Larson, J. Cryst. Growth 187 (1998) 277–288.
[70] R.Y. Waghmare, J.N. Webb, T.W. Randolph, M.A. Larson, C.E. Glatz, J. Cryst. Growth 208 (2000) 678–686.
[71] R.Y. Waghmare, X.J. Pan, C.E. Glatz, J. Cryst. Growth 210 (2000) 746–752.
[72] J.N. Webb, R.Y. Waghmare, J.F. Carpenter, C.E. Glatz, T.W. Randolph, J. Cryst. Growth 205 (1999) 563–574.
[73] J.N. Webb, J.F. Carpenter, T.W. Randolph, Biotechnol. Prog. 16 (2000) 630–636.
[74] K.J. Takano, H. Harigae, Y. Kawamura, M. Ataka, J. Cryst. Growth 171 (1997) 554–558.
[75] Y. Suzuki, T. Sawada, S. Miyashita, H. Komatsu, Rev. Sci. Instrum. 69 (1998) 2720–2724.
[76] Y. Suzuki, S. Miyashita, H. Komatsu, K. Sato, T. Yagi, Jpn. J. Appl. Phys. Part 2 Lett. 33 (1994) L1568–L1570.
[77] C.A. Schall, J.M. Wiencek, M. Yarmush, E. Arnold, J. Cryst. Growth 135 (1994) 548–554.
[78] L. Smeller, P. Rubens, K. Hermans, High Press. Res. 19 (2000) 713–719.
[79] E. Dumay, A. Laligant, D. Zasypkin, J.C. Cheftel, Food Hydrocolloids 13 (1999) 339–351.
[80] V.B. Galazka, E. Dickinson, D.A. Ledward, Food Hydrocolloids 13 (1999) 425–435.
[81] V.B. Galazka, D. Smith, D.A. Ledward, E. Dickinson, Food Hydrocolloids 13 (1999) 81–88.
[82] V.B. Galazka, D. Smith, D.A. Ledward, E. Dickinson, Food Chem. 64 (1999) 303–310.
[83] V.B. Galazka, E. Dickinson, D.A. Ledward, High Press. Res. 19 (2000) 515–520.
[84] V.B. Galazka, E. Dickinson, D.A. Ledward, Curr. Opin. Colloid Interface Sci. 5 (2000) 182–187.
[85] N. Tanaka, A. Koyasu, I. Kobayashi, S. Kunugi, Int. J. Biol. Macromol. 18 (1996) 275–280.
[86] E. Dufour, G.H.B. Hoa, T. Haertle, Biochim. Biophys. Acta 1206 (1994) 166–172.
[87] I. Hayakawa, Y.Y. Linko, P. Linko, Food Sci. Technol.-Lebensm.-Wiss. Technol. 29 (1996) 756–762.
[88] K. Olsen, R. Ipsen, J. Otte, L.H. Skibsted, Milchwiss.-Milk Sci. Int. 54 (1999) 543–546.
[89] T. Hosseini-nia, A.A. Ismail, S. Kubow, J. Agric. Food Chem. 47 (1999) 4537–4542.

[90] P. Pittia, P.J. Wilde, F. Husband, D.C. Clark, J. Food Sci. 61 (1996) 1123–1128.
[91] L.A. Tedford, C.J. Schaschke, Biochem. Eng. J. 5 (2000) 73–76.
[92] S. Funtenberger, E. Dumay, J.C. Cheftel, J. Agric. Food Chem. 45 (1997) 912–921.
[93] S. Funtenberger, E. Dumay, J.C. Cheftel, Food Sci. Technol.-Lebensm.-Wiss. Technol. 28 (1995) 410–418.
[94] S. Iametti, P. Transidico, F. Bonomi, G. Vecchio, P. Pittia, P. Rovere, G. DallAglio, J. Agric. Food Chem. 45 (1997) 23–29.
[95] M. Jegouic, V.Y. Grinberg, A. Guingant, T. Haertle, J. Protein Chem. 15 (1996) 501–509.
[96] N. Tanaka, T. Tonai, S. Kunugi, Biochim. Biophys. Acta 1339 (1997) 226–232.
[97] J.N. Webb, S.D. Webb, J.L. Cleland, J.F. Carpenter, T.W. Randolph, Proc. Natl. Acad. Sci. USA 98 (2001) 7259–7264.
[98] B.S. Kendrick, J.F. Carpenter, J.L. Cleland, T.W. Randolph, Proc. Natl. Acad. Sci. USA 95 (1998) 14142–14146.
[99] B.S. Kendrick, J.L. Cleland, X. Lam, T. Nguyen, T.W. Randolph, M.C. Manning, J.F. Carpenter, J. Pharm. Sci. 87 (1998) 1069–1076.
[100] B.M. Gorovits, P.M. Horowitz, Biochemistry 37 (1998) 6132–6135.
[101] C.A. Haasepettingell, J. King, J. Biol. Chem. 263 (1988) 4977–4983.
[102] A. Fuchs, C. Seiderer, R. Seckler, Biochemistry 30 (1991) 6598–6604.
[103] D.P. Goldenberg, D.H. Smith, J. King, Proc. Natl. Acad. Sci. USA 80 (1983) 7060–7064.
[104] D. Goldenberg, J. King, Proc. Natl. Acad. Sci. USA 79 (1982) 3403–3407.
[105] D. Foguel, C.R. Robinson, P.C. de Sousa, J.L. Silva, A.S. Robinson, Biotechnol. Bioeng. 63 (1999) 552–558.
[106] N.B. Bam, J.L. Cleland, J. Yang, M.C. Manning, J.F. Carpenter, R.F. Kelley, T.W. Randolph, J. Pharm. Sci. 87 (1998) 1554–1559.
[107] Y.-F. Maa, P.A. Nguyen, C.C. Hsu, J. Pharm. Sci. 87 (1998) 152–159.
[108] Y.F. Maa, C.C. Hsu, J. Pharm. Sci. 87 (1998) 808–812.
[109] A.M. de Vos, M. Ultsch, A.A. Kossiakoff, Science 255 (1992) 306–312.
[110] M.H. Ultsch, W. Somers, A.A. Kossiakoff, A.M. De Vos, J. Mol. Biol. 236 (1994) 286–299.
[111] R.J. St. John, L.F. Carpenter, T.W. Randolph, Proc. Natl. Acad. Sci. USA 96 (1999) 13029–13033.
[112] N.B. Bam, J.L. Cleland, T.W. Randolph, Biotechnol. Prog. 12 (1996) 801–809.
[113] K. Takano, J. Funahashi, K. Yutani, Eur. J. Biochem. 268 (2001) 155–159.
[114] J. Funahashi, K. Takano, K. Ogasahara, Y. Yamagata, K. Yutani, J. Biochem. 120 (1996) 1216–1223.
[115] M.B. Pepys, Philos. Trans. R. Soc. London Ser. B Biol. Sci. 356 (2001) 203–210.
[116] G.S. Chryssomallis, P.M. Torgerson, H.G. Drickamer, G. Weber, Biochemistry 20 (1981) 3955–3959.
[117] K. Akasaka, T. Tezuka, H. Yamada, J. Mol. Biol. 271 (1997) 671–678.
[118] C.E. Kundrot, F.M. Richards, J. Mol. Biol. 193 (1987) 157–170.
[119] C.E. Kundrot, F.M. Richards, Protein Eng. 1 (1987) 231–231.
[120] C.E. Kundrot, F.M. Richards, J. Mol. Biol. 200 (1988) 401–410.
[121] T.M. Li, J.W. Hook III, H.G. Drickamer, G. Weber, Biochemistry 15 (1976) 5571–5580.
[122] S.D. Samarasinghe, D.M. Campbell, A. Jonas, J. Jonas, Biochemistry 31 (1992) 7773–7778.
[123] K. Sasahara, N. Sakurai, K. Nitta, J. Mol. Biol. 291 (1999) 693–701.
[124] R.J. St. John, in: Department of Chemical Engineering, University of Colorado, Boulder, CO, 2001, p. 168.
[125] Y.F. Liu, J.M. Sturtevant, Biochemistry 35 (1996) 3059–3062.
[126] W. Colon, J.W. Kelly, Biochemistry 31 (1992) 8654–8660.
[127] S.A. Peterson, T. Klabunde, H.A. Lashuel, H. Purkey, J.C. Sacchettini, J.W. Kelly, Proc. Natl. Acad. Sci. USA 95 (1998) 12956–12960.
[128] A.D. Ferrao-Gonzales, S.O. Souto, J.L. Silva, D. Foguel, Proc. Natl. Acad. Sci. USA 97 (2000) 6445–6450.
[129] A.D. Ferrao-Gonzales, D. Foguel, J.L. Silva, Biophys. J. 78 (2000) 2480Pos.
[130] D. Foguel, A.D. Ferrao-Gonzales, M. Valory, L. Palmieri, T.C. Walkup, J.W. Kelly, J.L. Silva, Biophys. J. 80 (2001) 1731.
[131] A.D.F. Gonzales, S. Oquioni, J.L. Silva, D. Foguel, Biophys. J. 76 (1999) A418–A418.
[132] S. Chronopoulos, S.L. Chan, M.J.H. Ratcliffe, Z. Alikhan, Amyloid Int. J. Exp. Clin. Invest. 2 (1995) 191–194.
[133] J. Dubois, A.A. Ismail, S.L. Chan, Z. Ali-Khan, Scand. J. Immunol. 49 (1999) 376–380.
[134] S. Kunugi, K. Takano, N. Tanaka, K. Suwa, M. Akashi, Macromolecules 30 (1997) 4499–4501.
[135] S. Kunugi, K. Takano, N. Tanaka, K. Suwa, M. Akashi, Macromolecules 31 (1998) 2394–2394.
[136] T. Tamura, T. Yamaoka, S. Kunugi, A. Panitch, D.A. Tirrell, Biomacromolecules 1 (2000) 552–555.

ELSEVIER

Biochimica et Biophysica Acta 1595 (2002) 235–249

BIOCHIMICA ET BIOPHYSICA ACTA
BBA
www.bba-direct.com

Review

Pressure effects on intra- and intermolecular interactions within proteins

Boonchai B. Boonyaratanakornkit, Chan Beum Park, Douglas S. Clark *

Department of Chemical Engineering, University of California, 201 Gilman Hall, Berkeley, CA 94720-1462, USA

Received 3 October 2001; received in revised form 21 November 2001; accepted 22 November 2001

Abstract

The effects of pressure on protein structure and function can vary dramatically depending on the magnitude of the pressure, the reaction mechanism (in the case of enzymes), and the overall balance of forces responsible for maintaining the protein's structure. Interactions between the protein and solvent are also critical in determining the response of a protein to pressure. Pressure has long been recognized as a potential denaturant of proteins, often promoting the disruption of multimeric proteins, but recently examples of pressure-induced stabilization have also been reported. These global effects can be explained in terms of pressure effects on individual molecular interactions within proteins, including hydrophobic, electrostatic, and van der Waals interactions, which can now be studied in greater detail than ever before. However, many uncertainties remain, and thorough descriptions of how proteins respond to pressure remain elusive. This review summarizes basic concepts and new findings related to pressure effects on intra- and intermolecular interactions within proteins and protein complexes, and discusses their implications for protein structure–function relationships under pressure.

Keywords: High pressure; Molecular interaction; Protein function; Protein stability

1. Introduction

The Earth is predominantly a high-pressure environment, and 62% of the total biosphere is characterized by pressures greater than 100 bar [1]. Nonetheless, information on the effects of pressure on biomolecules has been relatively scarce until recently. Recent advances in analytical techniques for studying proteins under pressure [2] have facilitated the investigation of molecular interactions within proteins and protein complexes, and have provided new insights into the structure and function of biomolecules at high pressures. Moreover, because pressure effects on proteins often differ from those of temperature, pH, and chemical denaturants, pressure is now regarded as an important variable for examining protein structure–function relationships [3]. High pressure has also proven to be useful for refolding recombinant, oligomeric proteins as an alternative to concentrated chaotropes such as guanidine [4,5].

Pressure has long been recognized as a potential denaturant of proteins; however, the pressure range for protein denaturation differs depending on the structural characteristics of protein. Whereas primary and secondary structures of proteins are not affected by pressures as high as 10 kbar, oligomeric proteins can readily dissociate at pressures below 3 kbar [2,3]. On the other hand, hydrostatic pressure has been used to increase the activity and/or stability of several enzymes, including thermolysin [6], hy-

* Corresponding author. Fax: +1-510-643-1228.
E-mail address: clark@cchem.berkeley.edu (D.S. Clark).

0167-4838/02/$ – see front matter
PII: S0167-4838(01)00347-8

drogenase from *Methanococcus jannaschii* [7], and a putative protease from *M. jannaschii* [8] (since determined to be a proteasome).

In general, the effect of pressure on a physicochemical process at equilibrium is governed by the volume change of the process, ΔV. The relationship between ΔV and K, the equilibrium constant for the process, is given by Eq. 1:

$$\left(\frac{\partial \ln K}{\partial P}\right)_T = -\left(\frac{\Delta V}{RT}\right) \quad (1)$$

Thus the application of pressure will cause the equilibrium to shift in favor of the state with the lowest overall volume. The magnitude and sign of ΔV for changes in protein structure will depend on the specific molecular interactions [3,9], but the precise details of ΔV for different interactions in proteins are often difficult to define (a case in point is the controversy that has surrounded the role of hydrophobic interactions in pressure-induced denaturation [10,11]). Whereas the activity and stability of proteins are influenced by myriad interactions, the major factors that contribute to ΔV and hence govern pressure effects include the electrostriction of charged and polar groups, elimination of packing defects, and the solvation of hydrophobic groups [11]. This review summarizes key concepts and recent advances toward the elucidation of pressure effects on molecular interactions within proteins and protein complexes, and discusses important implications of these effects for protein activity and stability. Finally, it should be noted that the molecular forces involved in intramolecular interactions are the same as those that effect intermolecular interactions; thus, the distinction between the two types of interactions is a somewhat artificial construct intended to simplify organization of this review.

2. Intramolecular interactions

2.1. Ionic interactions

Ion pairs in proteins are attractive, short-range interactions that occur between negatively and positively-charged amino-acid side chains over a distance of 4 Å or less. Sometimes ion pairs are also termed salt bridges, but this usually refers only to ion-pair interactions that are also hydrogen bonded [12]. The role of ion pairs in proteins is to stabilize the tertiary [13] and quaternary structure. For example, interprotein salt bridges provide a mechanism for solvent exclusion from the interfacial domain of the cytochrome b_5–cytochrome *c* complex, in addition to maintaining the stability and specificity of the complex formed [14].

Dissociation of ion pairs leads to electrostriction, which is the contraction of solvents such as water and alcohols [15] due to alignment of dipolar solvent molecules in the electric field of an exposed charge. Because contraction of solvents leads to a volume decrease, pressure is expected to favor ion-pair disruption [16]. However, as discussed by Michels et al. [9], the volume reduction associated with electrostriction should be counteracted by high temperature due to disruption of the highly ordered structure of electrostricted water. Indeed, dissociation of ion pairs upon application of pressure has been shown to produce an overall volume decrease in the range of 10–30 ml/mol [17]. There is a limit to the solvent's compressibility, however, due to repulsive interactions caused by van der Waals repulsion at close distances [18].

Because pressure favors electrostriction [19], pressure can promote disruption of protein salt bridges and induce protein denaturation. In the case of chymotrypsin, disruption of a salt bridge in the proflavin binding site produced a volume change of −30 ml/mol [20]. Furthermore, when a pressure of 500 bar was applied to *Pyrococcus furiosus* rubredoxin, which contains multiple electrostatic interactions [21–23], the protein was destabilized at 110°C [7], possibly as a result of pressure-induced ion-pair dissociation.

On the other hand, pressure may bring about protein stabilization if there are weak intersubunit ion-pair interactions not exposed to solvent, which may be strengthened by pressure. In studies of glutamate dehydrogenase (GDH) from *P. furiosus* by Sun and colleagues, a tetrapeptide extension (Gly–Ser–Gly–Cys) was added to the C-terminal end of the protein [24]. The dangling tail was believed to transmit increased thermal fluctuations throughout the protein, leading to weakened intersubunit ion-pair interactions and destabilization relative to the native form. When 500 bar was applied to the protein at 103°C, the recombinant GDH was stabilized by 24-fold, whereas the native GDH was stabilized only 3-fold. The enhanced stabilization was attributed to stronger intersubunit ion-pair interactions accompanying decreased thermal fluctuations.

2.2. Hydration

Water molecules exert a profound influence on protein conformation through interactions with amino acids [25]. In proteins, water fills crevices between amino acids, which serves to screen repulsive forces [3] and facilitates motion of the side chains and polypeptide backbone [26]. Upon application of relatively low pressures (<2 kbar), increased conformational fluctuations provide pathways for water to penetrate into the interior of the native protein [27]. These fluctuations are enhanced by pressure due to increased water exchange between the protein interior and bulk solvent. On the other hand, hydration is opposed by pressure-induced reduction of cavity sizes, rendering pressure-enhanced fluctuation inapplicable to some proteins such as apoazurin and basic pancreatic trypsin inhibitor (BPTI) at pressures below 1.5 kbar and 1.2 kbar, respectively [28–30]. This is consistent with the apparent reduction of thermal fluctuations in glutamate dehydrogenases observed by Sun et al. [24] with pressures up to 750 bar. It should be noted that in some cases amino acids can act as two-way valves controlling the exchange of water between the bulk solvent and the protein, whereas in other cases amino acid residues can act as 'one-way check valves' that enable water molecules to be released from the protein interior while impeding penetration into the core [31]. This effect may be responsible for the increased conformational fluctuations with increasing pressure observed for some proteins but not others. The current level of understanding, however, does not allow us to draw general conclusions about pressure effects on protein hydration, or about water penetration in particular.

As a result of penetration of water penetration into the protein interior, pressure likely leads to conformational transitions resulting in unfolding [32]. Water penetration under pressure can induce the protein to adopt the conformation of a molten globule [32], a compact, partially folded conformation without specific tertiary structure [33]. This molten globule form has been observed in the pressure dissociated-denatured Arc repressor, and is thought to be the result of water penetration and non-polar side-chain exposure to solvent [33]. However, water penetration can also lead to increased reactivity as in the case of β-amylase [28], where enhanced activity upon an increase in pressure from 1 to 4 kbar may have been due to decreased rigidity [34] resulting from hydration of the β-amylase binding site.

At very high pressures (>5 kbar), protein aggregation and loss of secondary structure can occur due to hydration. Protein unfolding becomes irreversible at pressures above 5 kbar [35], and aggregation occurs because of increased protein exposure to solvent [36]. Wroblowski et al. [37] proposed that loss of secondary structure above 10 kbar results from the different compressibilities of water and protein, which is thought to disturb the balance of hydrophobic/hydrophilic protein interactions by allowing the solvent to interpenetrate the protein.

Besides protein unfolding and aggregation, increased hydration of the polypeptide can cause changes in protein compressibility and flexibility. Pressure can induce increased hydration of the polypeptide [29,38] leading to decreases in compressibility due to electrostriction of solvent around charged and polar groups and loss of void volume [39]. The decrease in compressibility is expected to decrease mobility of internal regions [29]. On the other hand, pressure-induced hydration can also reduce the number of intramolecular hydrogen bonds, due to the formation of intermolecular hydrogen bonds with water, resulting in increased conformational fluctuations of peripheral protein segments that are generally more mobile [29].

One way to counteract protein unfolding due to hydration under hydrostatic pressure is through application of osmotic pressure, e.g., by addition of glycerol. Addition of glycerol raises the chemical potential of the protein due to preferential exclusion of the glycerol from the protein surface, which is thermodynamically unfavorable [40]. However, the increase in chemical potential is counteracted by a contraction of the protein-solvent interface, decreasing the extent of preferential hydration and rendering the system less unfavorable thermodynamically. Compression of the protein, and reduction of the protein–solvent interface, is accomplished by the release of water from voids, which has the effect of increasing the core density and decreasing amino acid motion [41].

Addition of 40% (v/v) glycerol to α-chymotrypsin at 1.8 kbar reduced protein denaturation at 50°C [42], probably via decreased hydration as a result of the osmotic pressure produced by glycerol [43]. Supporting the beneficial effects of osmotic pressure, sodium chloride was also found to stabilize the dimeric state of the Arc repressor by decreasing the available water concentration [18]. Providing further evidence for the potentially disruptive role of water in protein–protein interactions, Kornblatt and Hui Bon Hoa showed that glycerol clearly increased the affinity of pulsed cytochrome *c* oxidase for reduced cytochrome *a* [44]. Recently, Sun et al. [24] found that adding 25% (v/v) glycerol increased the half-life of a recombinant GDH mutant from *P. furiosus* 14-fold at 105°C and 5 atm as shown in Table 1. It was proposed that pressure and glycerol affect proteins through similar mechanisms, whereby the enzyme adopts a more compact and rigid structure in some cases [29], diminishing thermal fluctuations [24]. Trehalose was also found to stabilize GDH from *Thermococcus litoralis* against thermoinactivation [45]. Oliveira and co-workers even went so far as to suggest that pressure denaturation would not occur in the absence of water by extrapolating ΔV values for dissociation of Arc repressor at glycerol concentrations of 0–50% up to 100% glycerol [33] (see Fig. 1). However, Sun et al. [24] and Mozhaev et al. [42] showed that glycerol stabilizes proteins maximally at 25% and 40%, respectively, when the addition of 0–80% (v/v) glycerol was tested for protein stabilization. These data illustrate that a single water exclusion model is insufficient to fully explain the stabilizing effects of cryosolvents for all proteins.

Table 1
Glycerol stabilization of a recombinant GDH mutant (containing an extra tetrapeptide at the C-terminus) from *Pyrococcus furiosus* at 105°C and 5 atm. Reproduced from Sun et al. [24]

Glycerol (% v/v)	Half-life[a,b] (min)	$t^a_{1/2}$ (glycerol)/$t_{1/2}$
0	13 ± 0.5	1
25	175 ± 7	14
50	120 ± 3	9.2
75	80 ± 5	6.3

[a]Half-lives are averages of two trials.
[b]Errors represent mean deviations.

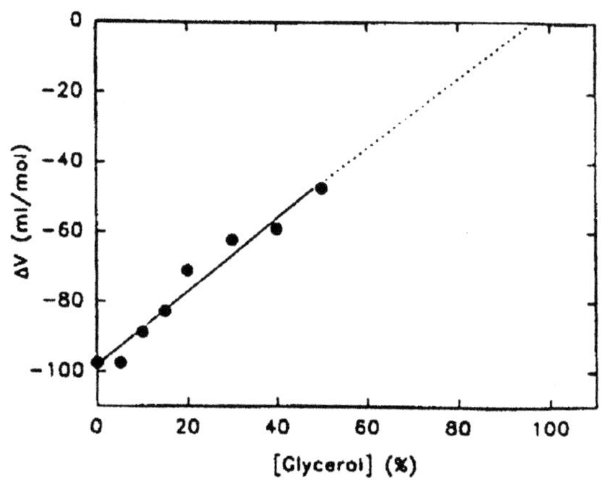

Fig. 1. Effect of glycerol on the volume change for dissociation of the Arc repressor. Extrapolating to 100% (v/v) glycerol suggests that, without the presence of water, the Arc repressor will not dissociate under pressure. Reproduced from Oliveira et al. [33].

2.3. Hydrophobic interactions

Hydrophobic interactions direct non-polar side chains to cluster inside proteins and are a major driving force for proper folding [10]. Hydrophobic interactions are expected to be stabilized by elevated pressure (although this conclusion is not universally accepted in the literature). Stabilization of these interactions by pressure has been attributed to the unfavorable volume increase that results from the solvation of apolar surfaces during protein unfolding [7,9]. This volume increase results from clathrate formation around the apolar groups when they are exposed to water.

Hydrostatic pressure has been utilized to probe the role of hydrophobic interactions in protein stability. Hei and Clark [7] investigated the stability of GAPDHs isolated from several sources in relation to the hydrophobicity of the so-called s-loop region (residues 178–201) of each subunit. A positive correlation between the hydrophobic transfer free energy (a measure of the unfavorable energy difference associated with transfer of a hydrophobic side chain from a hydrophobic to an aqueous phase) and pressure-induced stabilization was observed, as shown in Fig. 2. Thus, the observed correlation indicated that hydrophobic interactions between the s-loops, which are an important determinant of GAPDH thermal stability, were favored by increased pressure (0.5 kbar).

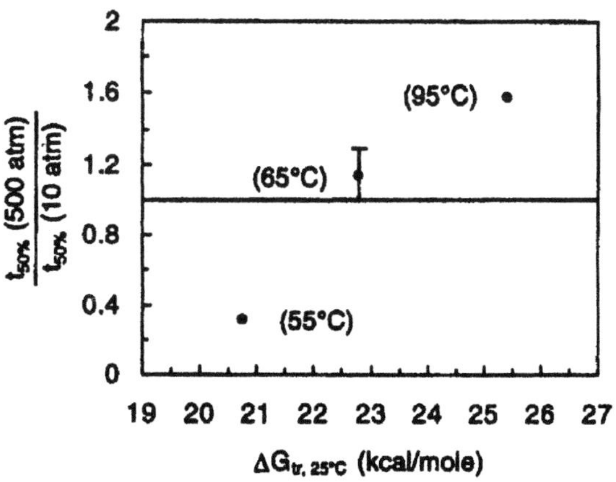

Fig. 2. Inactivation ratios for GAPDHs versus transfer free energies (ΔG_{tr}) of the s-loops. The source of GAPDH and temperatures at which the inactivation experiments were performed are, from left to right, pig muscle (55°C), *B. stearothermophilus* (65°C), and *T. maritime* (95°C). Data shown are averages of duplicate trials (in some cases, the error bars are smaller than the symbols). This graph indicates that hydrophobic interactions between s-loops (an important determinant of GAPDH thermal stability) are favored by increasing pressure. Reproduced from Hei and Clark [7].

Hydrostatic pressure was also used to study the role of hydrophobic residues in the stability of azurin, a single-chain blue-copper protein [46]. Stability of azurin is practically unaffected by hydrostatic pressure up to 3 kbar. Through site-directed mutagenesis, Mei et al. [46] substituted a hydrophobic residue, either Ile7 or Phe110, in both holo- and apo-azurin with serine, and compared the stabilities of the mutants with their wild-type counterparts under hydrostatic pressure (~3 kbar). As shown in Fig. 3, each mutation caused destabilization at high hydrostatic pressure. The positive correlation between pressure-induced stabilization and hydrophobicity of the azurin core is consistent with the earlier results obtained for GAPDH [7].

Although hydrostatic pressure is an important tool for investigating the role of hydrophobic residues in the stability of proteins, the actual behavior of hydrophobic residues within proteins under pressure remains unclear. An important and relevant question concerns the nature of the hydrophobic core within proteins. The original 'oil-drop' or 'liquid-hydrocarbon' model for the hydrophobic core of proteins, suggested by Kauzman several decades ago [47], was challenged for failing to explain the effect of pressure on protein unfolding (the infamous 'fly in the ointment') [10,11,48]. Kauzman wrote later [10] that the reason for the apparent inapplicability of the model arose from contradictory values of ΔV: ΔV upon protein unfolding was positive at moderate pressures and became negative at pressures higher than 1–2 kbar; however, ΔV for the transfer of hydrocarbons into water (a model process for unfolding of the liquid-like protein) showed the opposite signs of ΔV at both pressure ranges for protein unfolding. However, considering it is still arguable whether hydrocarbons such as low molecular mass alkanes accurately represent hydrophobic side chains in the core of a protein [49], Kauzman's ointment may have been free of flies after all.

In some respects the interiors of proteins more closely resemble a solid than a liquid [50–53]. A semi solid-like protein interior was first proposed on the basis that molecular densities of several proteins were estimated to be much higher than those of

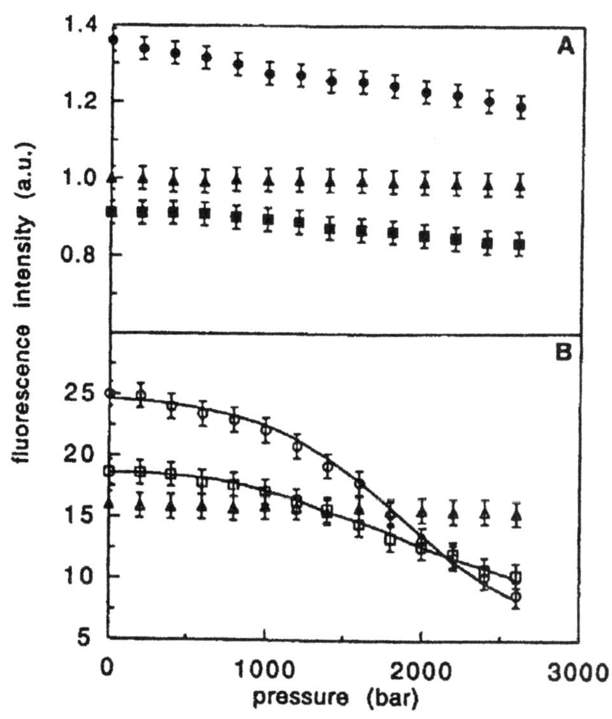

Fig. 3. Effects of pressure on azurin as indicated by relative fluorescence intensity as a function of hydrostatic pressure for the holo- (A) and apo-proteins (B). Symbols represent WT azurin (triangles), Ile7Ser mutated azurin (squares), and Phe110Ser mutated azurin (circles). The fluorescence intensity of holo-wt azurin at 1 bar has been normalized to 1. In each case, the mutations resulted in destabilization at high pressures. Reproduced from Mei et al. [46].

organic liquids, and proteins are relatively less compressible [9,50]. Supporting this model, the first reported measurement of the isothermal compressibility of an enzyme's interior showed that mesoporphyrin-substituted horseradish peroxidase behaved as a fully elastic solid at pressures up to 20 bar [51]. However, binding of a small aromatic substrate molecule in the heme pocket produced a significant change in protein compressibility, effecting a transition from a solid-like to a liquid-like structure [52]. An efficiently packed, solid-like interior was also assumed by Hummer et al. [48,53], who used information theory to predict that water molecules penetrate into the tightly packed protein interior at hydrostatic pressures up to 7 kbar.

2.4. Hydrogen bonds

Based on negative volume changes associated with hydrogen bond formation in organic compounds such as formic acid, ε-caprolactam, phenol, and poly-L-lysine [20,54], hydrostatic pressure has been proposed to promote hydrogen bond formation within proteins [9,55]. As noted in the hydration section, pressure may promote intermolecular hydrogen bonds at the expense of intramolecular hydrogen bonds thus causing increased conformational fluctuations [29].

The lengths of existing hydrogen bonds within proteins have been observed to shorten under hydrostatic pressure. Shortening of hydrogen bonds at high pressure was first detected with ^{1}H nuclear magnetic resonance (NMR) for liquid water and ethanol [56,57]. Furthermore, low-frequency Raman spectra of liquid amides (formamide, *N*-methylformamide, *N*-ethylformamide, and *N*-methylpropionamide) reflected the formation and compression of hydrogen bonds under hydrostatic pressure up to 4 kbar, due to hindered rotational motion as evidenced by reduction in the pressure dependence of low frequency bands [58]. Recently, Li et al. [59] observed compression of hydrogen bonds within BPTI at 2 kbar using two-dimensional ^{1}H NMR. However, compression was not uniform, with estimated reductions in distance varying considerably in the range of 0–0.11 Å. Thus, the shortening of hydrogen bonds, in addition to the collapse of internal cavities, can contribute to the compression of proteins under pressure [60,61].

2.5. Van der Waals forces

Van der Waals (VDW) forces are weak attractions produced by all atoms and molecules as a result of interactions related to induced polarization effects at close distances [12]. These forces can be represented by a Lennard-Jones potential. Although generally considered as a distinct force among the various intramolecular interactions that govern protein folding and stability, as well as protein–protein interactions [14], VDW forces also contribute enthalpic and entropic components to the hydrophobic effect [12,62]. Sites where local VDW interactions are weakened may lead to imperfect packing and cavity formation, which could in turn serve to undermine protein stability [63].

VDW forces are likely to be favored by pressures since they tend to maximize the packing density of proteins, thus reducing the volume. Packing density is maximized by VDW interactions because the dipole moment attractions they produce improve packing of the hydrophobic core [34,64]. This conclusion is supported by the work of Lange and co-workers, who replaced the phenylalanine residue of ribonuclease from *Sulfolobus solfataricus* with alanine, which probably weakened the former aromatic network, destabilizing the protein at 2.5 kbar [65]. This result points to the importance of the aromatic network for ribonuclease stability, although the effect may also have resulted from creation of a non-polar cavity that was filled with water under pressure.

On the other hand, VDW forces can also contribute to protein destabilization under pressure. Pressure below 3 kbar promotes dissociation of oligomers resulting in replacement of the weakest non-covalent interactions between amino acid residues with amino acid–water interactions and changes in peptide chain conformations [18]. These protein–water interactions are promoted by pressure because they produce stronger and shorter bonds that decrease the overall volume [2,18].

2.6. Covalent bonds

Even though covalent bonds are affected by temperature, these bonds are negligibly compressible under pressure [66]. This statement is consistent with the fact that covalent bonds define the primary

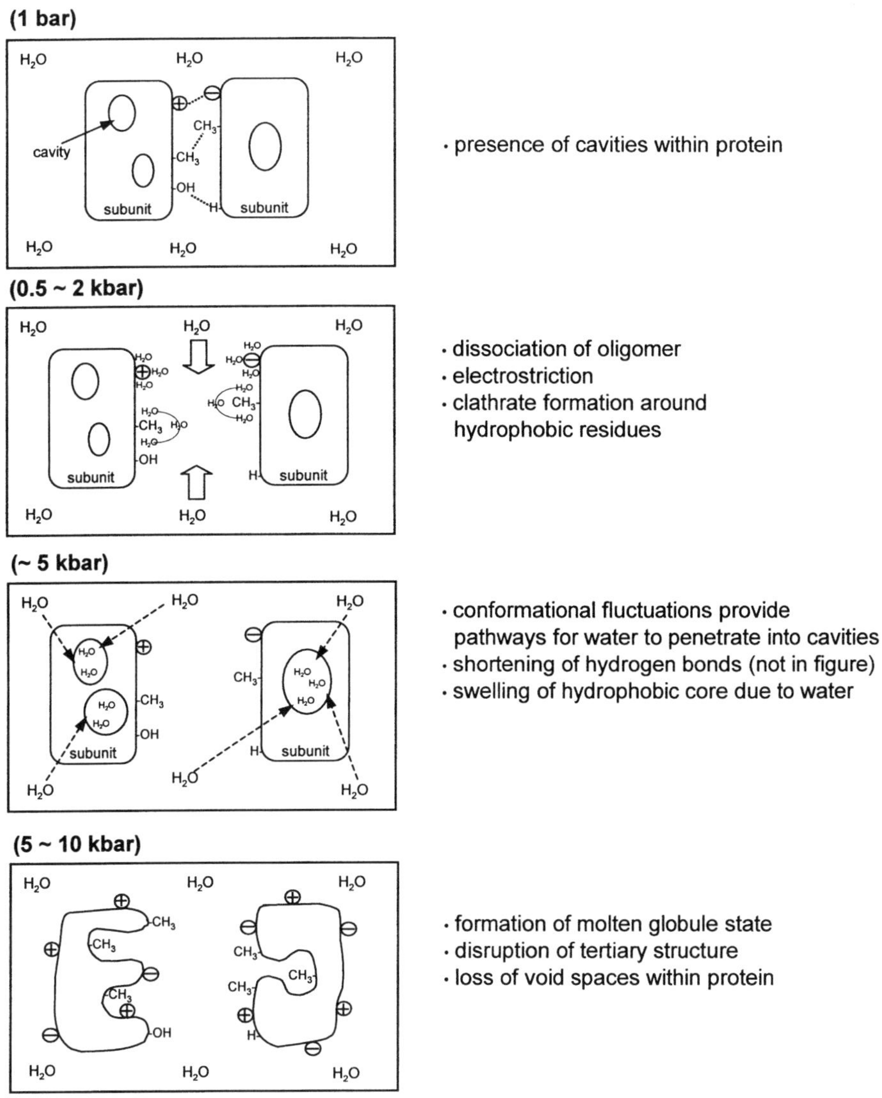

Fig. 4. Major effects of pressure on molecular interactions causing dissociation of oligomeric proteins and subunit denaturation. Note that water molecules in the 5–10 kbar range have been omitted for clarity.

structure of proteins, and the primary structure of hen egg-white lysozyme was unperturbed by pressures greater than 10 kbar [34,67]. It is also worth noting that when the active conformation of chymotrypsin was stabilized by covalent binding of a diisopropylphosphoryl residue to the active site, serine, no change in structure was observed at pressures of up to 3 kbar [68]. Because covalent bonds are highly incompressible, denatured proteins might be expected to have a low compressibility [69]. However, compressibility of the denatured protein is also influenced by solvation. As a result of solvation, the denatured form can be more compressible than the native form. Pressure can thus result in unfolded proteins with greater compressibility than the native form [69] due to increased solvation, but can also result in unfolded proteins with decreased compressibility due to loss of void volumes and solvent exclusion [38,70].

3. Intermolecular interactions

3.1. *Protein–protein*

In many cases relatively high pressure (1–2 kbar)

has been found to promote dissociation of oligomeric proteins. The dissociation of oligomeric proteins is typically accompanied by negative and relatively large ΔV values (−50 to −200 ml/mol) [18]. Oligomeric proteins such as glyceraldehyde-3-phosphate dehydrogenase (GAPDH) from yeast, lactate dehydrogenase (LDH), and malate dehydrogenase have been found to dissociate and reassociate reversibly upon application and release of hydrostatic pressure up to 2 kbar [34]. Pressure-dissociated monomers can also incur pressure-induced conformational changes, as was reported for dimeric Arc repressor [71,72]. Fig. 4 illustrates mechanisms believed to contribute to the pressure-induced dissociation and denaturation of multimeric proteins.

Hysteresis or conformational drift has been observed during pressure-mediated dissociation and association of many oligomeric proteins [2,34,73]. For example, in studies of the tetrameric proteins LDH [74] and GAPDH [75], reassociation of monomers to tetramer following pressure-induced dissociation was slow and sometimes incomplete. In a study of GAPDH from yeast by fluorescence spectroscopy, various conformations of oligomeric aggregates were observed depending on the protein concentration, temperature, pressure, and the presence of specific ligands [75]. The extent of hysteresis appears to increase as the number of protein subunit increases. According to a recent study by Panda et al. [76], pressure-dissociated subunits of tetradecameric GroEL did not readily reassociate or failed to refold at all.

Inactivation of oligomeric proteins by elevated pressure is not a universal phenomenon, especially under relatively moderate pressures (< 1 kbar). There are many examples of pressure stabilizing thermophilic and hyperthermophilic proteins against thermal inactivation [7,9,24,25]. In particular, GAPDHs from thermophiles showed enhanced thermostability at 500 bar relative to 10 bar, in contrast to GAPDHs from mesophiles, which were destabilized by the elevated pressure [7]. Among the forces helping to maintain the structures of oligomeric proteins, electrostatic repulsion between subunits was found to play an important role in the pressure stabilization of glutamate dehydrogenase (GDH) [24,45]. Sun and co-workers proposed that destabilizing electrostatic repulsion in GDH could trigger large conformational fluctuations of the entire hexamer. Studies of the wild-type and two single mutants of GDH from *T. litoralis* (D167T and T138E) indicated that the degree of pressure stabilization correlated with the magnitude of electrostatic repulsion created by charged residues at positions 138 and 167.

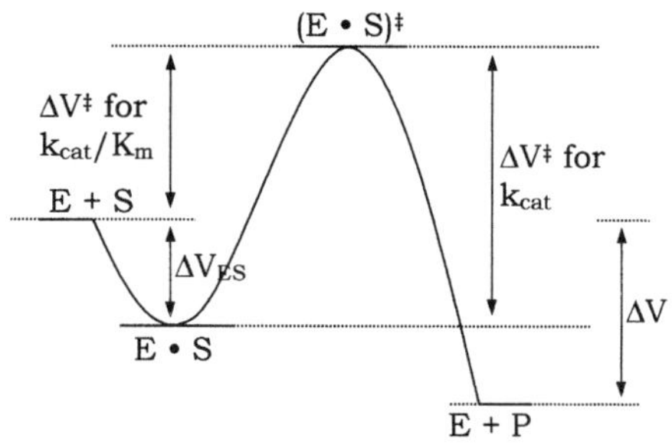

Fig. 5. Activation volumes and volume changes for an enzymatic reaction.

3.2. Enzyme–substrate

Binding of a substrate to the active site of an enzyme is typically accompanied by a small positive or negative volume change (ΔV_{ES} in Fig. 5) at atmospheric pressure [9]. Substrate binding stabilizes enzymes by shifting the unfolding equilibrium toward the native state [9]. Morita and co-workers showed that the activity of aspartase from *Escherichia coli* increased with pressure in the range of 1–1000 bar at temperatures above 45°C [77]. Above 53°C, the optimal activity occurred at 700 bar with a drop in activity at higher pressures. The increased activity was attributed to increased stability and was shown to require the presence of substrate to prevent initial heat inactivation before application of pressure. Pressure may also be a cause of pressure sensitivity for some reactions due to effects on substrate binding [9]. For example, studies of camphor analogues binding to cytochrome $P450_{camphor}$ under pressures up to 2.5 kbar showed that small substrates bind less tightly and with less efficient packing, increasing water accessibility and leading to increased likelihood of enzyme inactivation [3,78].

Due to the importance of ion pairs in substrate binding, electrostriction is one of the factors that may lead to weakened substrate binding under pressure. When water penetrates the binding site under pressure, electrostriction can occur, resulting in disruption of ion pairs between the substrate and enzyme and inactivation of the enzyme. However, electrostriction of ion pairs in the binding site may also decrease the volume of the transition state, thus leading to a more negative activation volume and a pressure-activated reaction [79,80]. The activation volume, defined here as the volume difference between the transition state and ground state of the enzyme–substrate complex, will be discussed in a following section.

3.3. Protein–non-protein

Several molecular forces act in concert to effect the specific binding between proteins and DNA [81]. For example, the specificity of protein binding to DNA is known to change depending on the salt concentration. DNA is a polyelectrolyte with a high axial charge density, and the electrostatic potential is a function of the ionic strength due to counterions close to the poly-ion surface. Furthermore, water is commonly identified at specific binding sites in crystal structures of protein–DNA complexes, mediating hydrogen bonding between the protein and DNA. Hydrophobic interactions are also known to play an important role in sequence-specific protein–DNA interactions [82].

Hydrostatic pressure has been used to probe the role of water in interactions between DNA and restriction enzymes [2,83]. For example, the binding specificity of the restriction enzymes *Eco*RI, *Bam*HI, *Pvu*II, and *Eco*RV were investigated under hydrostatic and osmotic pressure [83,84]. Osmotic pressure was found to release water molecules from the protein–DNA complex, increasing the star activity of *Eco*RI ('star activity' refers to relaxed specificity of the restriction enzyme resulting in cleavage at sites different from the normal recognition sequences). In contrast to osmotic pressure, hydrostatic pressure rehydrated the complex, restoring the natural selectivity of *Eco*RI. From these experiments, it became evident that water molecules assist in the correct recognition of the DNA restriction site by *Eco*RI.

Binding to DNA has been reported to stabilize the Arc repressor dimer against pressure-induced denaturation. Arc repressor, a small DNA-binding dimeric protein (M_r 13 000), dissociates completely under a hydrostatic pressure of about 2 kbar [71]. However, upon binding to operator DNA, the Arc repressor became resistant to pressure denaturation at pressures up to 2 kbar. Formation of the Arc-operator DNA complex was presumably accompanied by burial of nonpolar side chains and weak hydrophobic interactions between the protein and bases on the DNA [82]. The pressure-stabilization effect of DNA is highly dependent on the specificity of binding for the DNA–Arc complex [85].

Molecular interactions between antigens and antibodies have also been investigated under hydrostatic pressure. Sundaram et al. [86] observed that the sensitivity of antigen–antibody complexes to pressure differed significantly depending on the nature of the complex, although the underlying mechanism for such differences was unclear. Surprisingly, studies of hen egg white lysozyme binding to two monoclonal antibodies showed that the complex with greater electrostatic interactions (three salt bridges at the interface) did not exhibit pressure sensitivity, in contrast to what would be expected from electrostriction. Moreover, the complex with greater hydrophobic interactions (six out of 13 contacts on the antibody are tyrosines) exhibited a strong sensitivity to pressure. In the case of digoxigenin and a monoclonal antibody, their association was promoted by hydrostatic pressure (2 kbar) with a ΔV of -30 ml/mol [87]. However, when several aromatic residues of the monoclonal antibody that contact digoxigenin were replaced with histidine, a significant decrease in the pressure-induced association was observed.

4. Implications for protein function and stability

4.1. Pressure-induced thermostabilization

Generally, pressure-induced stabilization has been discussed from the viewpoint of resisting thermal unfolding at elevated temperatures [7,9,42]. A major contributor to thermal unfolding are fluctuations that arise from higher temperatures [24,42]. Pressure-induced stabilization is not applicable to all proteins, in part because pressure can increase and not decrease fluctuations [29], as observed for proteins

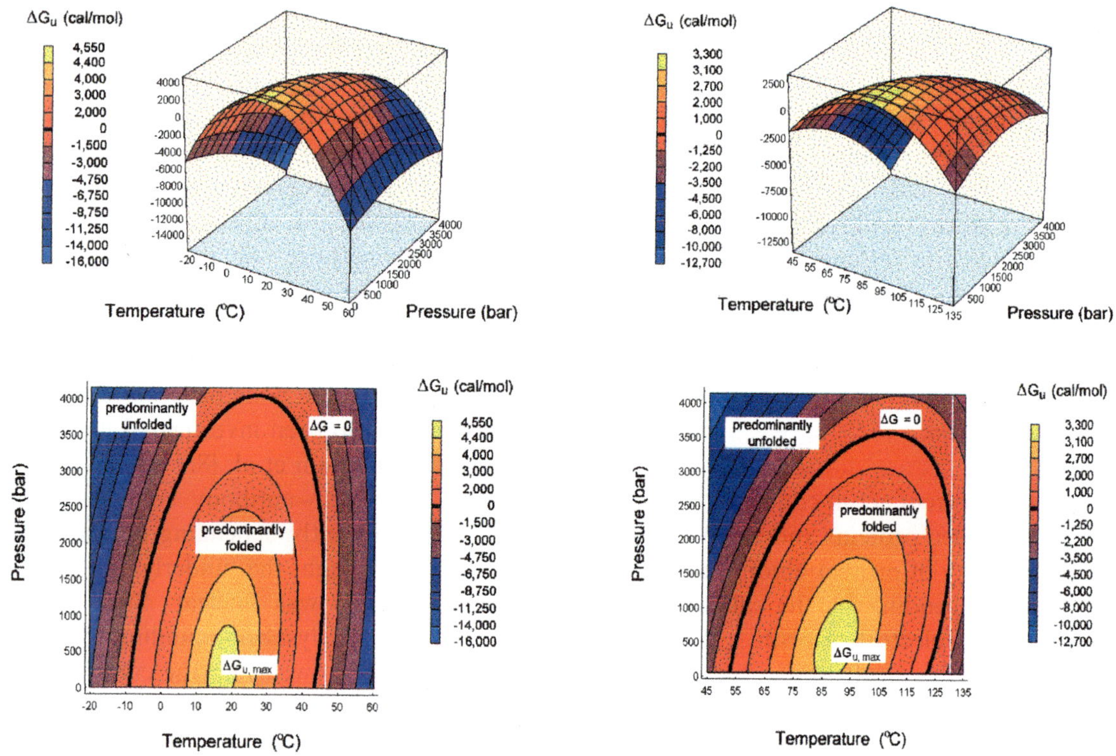

Fig. 6. Three- and two-dimensional free-energy diagrams for protein folding in P-T space: chymotrypsinogen (left) and *M. jannaschii* proteasome (right).

such as apomyoglobin, RNase T1, and phosphoglycerate kinase [28].

Some proteins are nonetheless stabilized against thermal inactivation within a limited pressure range (as illustrated by some of the examples already discussed in this review). When the protein-stability diagram was plotted for chymotrypsinogen in temperature–pressure space, it was found to have a curved boundary [9,88]. A curved protein-stability diagram has also been generated by Clark and co-workers for a putative protease from *M. jannaschii* (Fig. 6). As shown by these contour plots, there are temperatures at which each protein is stabilized by pressure, and a distinct pressure at which the concentration of folded protein is at a maximum. A limited range of pressure-induced stabilization was also shown by Rubens and co-workers, who found that β-galactosidases from *E. coli* and *Aspergillus oryzae* were stabilized maximally in the range of 2–3 kbar at 50°C and 65°C, respectively [89].

Notably, Hei and Clark reported the first examples of pressure-induced thermostabilization of proteins from extreme thermophiles [7]. Hydrogenases from *M. jannaschii* and *Methanococcus igneus* (both extreme thermophiles) were substantially stabilized by pressures of 500 bar, whereas hydrogenases from *Methanococcus thermolithotrophicus* (a moderate thermophile) and *Methanococcus maripaludis* (a mesophile) were destabilized by the same pressure at the selected inactivation temperatures. As mentioned previously, by comparing the hydrophobic s-loops of GAPDHs from various mesophiles and thermophiles, Hei and Clark found that increased thermostability correlated with increased hydrophobicity and postulated that pressure stabilization was also related to increased hydrophobicity at the intersubunit interface.

Several other proteins have also been shown tobe stabilized against thermal inactivation by pressure. The protein, α-chymotrypsin, was stabilized by pressures of 1.2 kbar against thermal inactivation at 50°C [42]. Clark and colleagues also showed that, at 125°C, a putative protease from *M. jannaschii* was stabilized by pressure. Raising the pressure from 1 bar to 500 bar increased the reaction rate 3.4-fold and the thermostability 2.7-fold [8]. Sun and co-

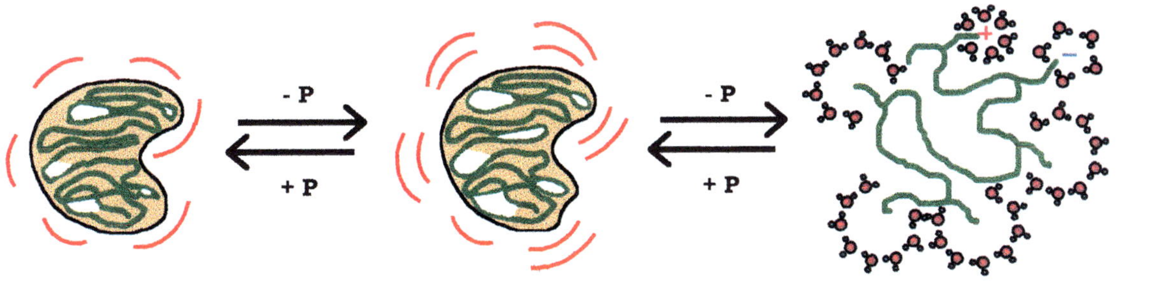

Pressure-stabilized state

- Reduced protein fluctuations
- Compression of hydrophobic cavities resulting in increased VDW forces (assume no water intrusion)

Native state

- Protein fluctuates
- Multiple hydrophobic cavities
- Multiple hydrophobic groups that can be exposed to water

Unfolded state

- Unfolding is not favored by pressure assuming ΔV>0 due to the predominant effect of hydrophobic solvation

Fig. 7. Proposed mechanisms for pressure stabilization of enzymes. Interactions between the unfolded protein and water molecules are included for the unfolded protein.

workers have recently shown that glutamate dehydrogenases (GDHs) from *T. litoralis* and *P. furiosus* are stabilized against thermal inactivation in the range of 250–500 bar [24,45]. The half-life of the *P. furiosus* GDH as measured by remaining activity at 105°C increased from 13 min at 5 bar to 360 min at 500 bar [24]. Values of $\Delta\Delta G_p^{\neq}$ (the difference in the free energy of activation for inactivation) for the GDH from *T. litoralis* verified that the degree of pressure stabilization increased with temperature [45] (see Table 2).

Table 2
Effects of temperature on thermal half-lives of wild-type (WT) and two single mutants, D167T and T138E, of the *T. litoralis* GDH. Positive values of $\Delta\Delta G_p^{\neq}$ indicate stabilization by pressure ($\Delta\Delta G_p^{\neq} = \Delta G_{500atm}^{\neq} - \Delta G_{5atm}^{\neq}$). In general, the stabilizing effect of 500 atm increased with increasing temperature. Reproduced from Sun et al. [45]

Enzyme	Temp. (°C)	$t_{1/2}$ (5 atm) (min)	$\Delta\Delta G_p^{\neq}$ (kJ/mol)
D167T	101.5	80	3.4
D167T	102.9	67	3.4
D167T	104	32	3.9
WT	98.8	97	3.8
WT	100	79	4.9
WT	101.4	29	6.3
T138E	86	68	8.0
T138E	87	60	7.6
T138E	90	22	8.6
T138E	92	10	9.6

One reason for pressure-induced thermostabilization is that temperature and pressure are opposing forces in terms of entropy. Increasing temperature leads to disorder, while increasing pressure leads to order [42,90]. This is not always true, however, as increasing pressures may lead to hydration of protein interiors resulting in increased conformational fluctuations [27] and disorder. One example of pressure rigidification is the case of apoazurin. Below 1.5 kbar, pressure reduces the flexibility of apoazurin, leading to a more closely-packed, rigid structure and reducing thermal fluctuations caused by high temperatures [29,91]. For other proteins, i.e., lysozyme and ribonuclease A, pressure has been found (according to the H–^{2}H exchange for amide protons) to initially tighten the protein structure with an opposite effect at above 2.5 kbar [92,93]. Mozhaev et al. has proposed that pressure stabilization of chymotrypsin against heat inactivation may arise from protein interactions with water [42]. Hydration of charged groups by water molecules is favored by pressure due to solvent electrostriction. Temperature disrupts the highly ordered structure of electrostricted water and reduces the strongly destabilizing effect of pressure on ion pairs [9].

Sun and co-workers invoked the concept of different conformational substates (CS) to explain the pressure thermostabilization of the glutamate dehydrogenase (GDH) hexamer [45]. Pressure favors a compact, rigid structure where thermal fluctuations are diminished, thus shifting the equilibrium toward the CS with the greatest reduction in volume. In the studies by Sun et al. [45], increased pressure stabilization was observed for a site-mutated GDH containing increased electrostatic repulsion. The apparent first-order inactivation of GDH was modeled by the following scheme:

$H_{CS0} \rightarrow H_{CS1} \rightarrow H_U \rightarrow A$

where H_{CS0} and H_{CS1} are CS that comprise the native-state ensemble of the GDH hexamer, and H_U is a partially-unfolded hexamer that is then able to aggregate to form the aggregated species A. Presumably, increased electrostatic repulsion in the protein leads to increased volume fluctuations, enabling pressure to stabilize GDH by shifting the equilibrium between H_{CS0} and H_{CS1} (the two CS that comprise the native-state ensemble of the GDH hexamer) towards H_{CS0}, the CS lower in energy and volume. Furthermore, amplification of pressure-induced stabilization at higher temperatures suggested the emergence of a new CS possessing even higher energies and volumes:

$H_{CS0} \rightarrow H_{CS1} \rightarrow H_{CS2} \rightarrow H_U \rightarrow A$

where H_{CS2} is a new CS that is populated to a significant extent only at higher temperatures. A general model for pressure-induced stabilization of proteins, based on evidence described in the aforementioned studies, is presented in Fig. 7.

4.2. Pressure-induced activation

The activation volume ($\Delta V^{\neq}$) for an enzymatic reaction is often defined as the overall molar volume difference between the transition state and ground state of the enzyme–substrate complex [9,94] (see Fig. 5). Volume changes include multiple contributions, notably volume changes in the reacting molecules themselves (including the enzyme), and volume changes arising from interactions between reacting molecules and the surrounding solvent [15].

Values of $\Delta V^{\neq}$ can be obtained by plotting ln(k) versus P where k is the enzymatic rate constant for formation of the transition state (either k_{cat} or k_{cat}/K_m depending on the reaction conditions and the particular definition of $\Delta V^{\neq}$) [66]. Such plots allow determination of the activation volume from the general relationship given by Eq. 2:

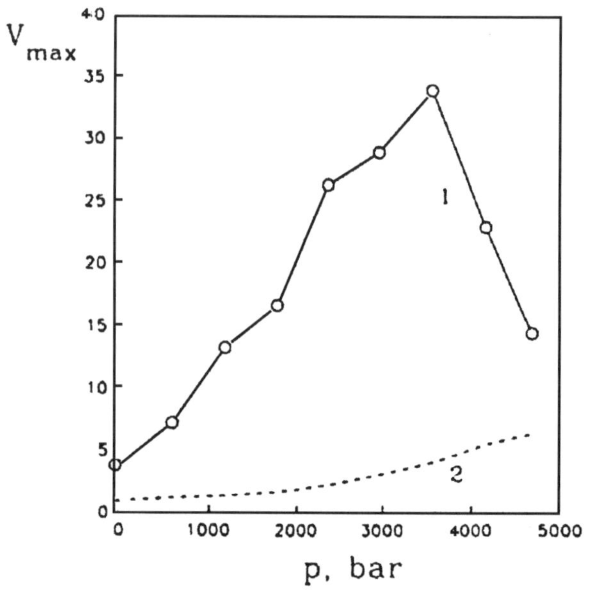

Fig. 8. Rate of hydrolysis of N-succinyl-L-phenylalanine-p-nitroanilide catalyzed by α-chymotrypsin in 50 mM Tris–HCl buffer (pH 8.0) at 50°C (curve 1) and 20°C (curve 2) as a function of pressure. V_{max} values are given in relation to rate at 1 atm and 20°C. Note the effect that temperature has on increasing enzyme activity and the decrease in activity at 50°C and pressures greater than about 3.5 kbar. Reproduced from Mozhaev et al. [100].

$$\Delta V^{\neq} = -RT\left(\frac{\partial \ln k}{\partial P}\right)_T \quad (2)$$

[9]. The plot may not be linear, however, if $\Delta V^{\neq}$ is pressure dependent (for example if the reactants or their activated states have different compressibilities [2]).

Even though activation volumes can be approximated using Eq. 2, prediction of activation volumes is quite complicated. In general, $\Delta V^{\neq}$ values for biological systems are difficult to interpret at the molecular level because of the many interactions involved [67]. Moreover, activation volumes for reactions catalyzed by enzymes with similar primary sequences or folding structures may be considerably different [95].

Eq. 2 illustrates that pressure accelerates reactions having negative activation volumes, and, the more negative the activation volume, the greater the rate acceleration with pressure. The pressure effect is especially strong for reactions accompanied by a significant change in the components' hydration state between the ground and transition state [42,96]. Pressure acceleration of reactions with negative activation volumes is supported by evidence that, in the hydrolysis of an anilide substrate by chymotrypsin at 20°C, there was a 6.5-fold increase in activity at 4.7 kbar compared to 1 bar [42]. The activity increase was attributed to the negative activation volume for this reaction [97], which may result from electrostriction in the polar transition state of peptide-bond hydrolysis [98]. Another protease, thermolysin, also exhibited enhanced activity at pressures up to 1 kbar, with activation volumes for hydrolysis of 3-(2-furyl)acryloyl-blocked substrates ranging from −25 to −35 ml/mol [6].

Activation volumes can also vary with temperature. For example, Mozhaev et al. [42] (see Fig. 8) found that increasing the pressure from 1 bar to 3 kbar increased the rate of *N*-succinyl-L-phenylalanine-*p*-nitroanilide hydrolysis by α-chymotrypsin about 7-fold at 50°C, compared to about a 4-fold increase at 20°C. These results correspond to a $\Delta V^{\neq}$ of −25 ml/mol at 50°C, which is 2.5-fold more negative than at 20°C [42]. On the other hand, a further pressure increase to 4 kbar at 50°C led to a decrease in catalytic activity, which may have resulted from weakened enzyme–substrate interactions and/or enzyme denaturation. A temperature-dependent activation volume was also observed by Kunugi and co-workers, who showed that $\Delta V^{\neq}$ for the hydrolysis of a dipeptide amide substrate by thermolysin was −71 ml/mol at 25°C and −95 ml/mol at 45°C [99].

The dependence of activation volume on pressure and hydration creates the opportunity to manipulate enzymatic reactions. For example, enzymatic activity under pressure might be regulated by varying the water content of the system since the activation volume has been proposed to be dependent on water content [90,97]. In particular, the enzyme-catalyzed transformation of two substrates can be regulated if the activation volumes differ in sign [100]. This effect was demonstrated for the trypsin-catalyzed hydrolyses of *N*-benzoyl-L-arginine ethyl ester ($\Delta V^{\neq} = -2.4$ ml/mol) and *N*-benzoyl-L-arginine-*p*-nitroanilide ($\Delta V^{\neq} = 3$–6 ml/mol) [101]. This approach was also applied to chymotrypsin catalysis, whereby pressure was used to activate hydrolysis of *N*-succinyl-L-phenylalanine-*p*-nitroanilide ($-\Delta V^{\neq}$) while inhibiting hydrolysis of *N*-carbobenzoxy-L-tyrosine-*p*-nitrophenyl ester ($+\Delta V^{\neq}$) [97].

Acknowledgements

B.B.B. and C.B.P. were supported by the National Science Foundation (the ERC Program under Award Number EEC-9731725 and the LExEn Program under Award Number BES-9816490, respectively).

References

[1] D. Prieur, V.T. Marteinsson, Adv. Biochem. Eng. 61 (1997) 23–35.
[2] V.V. Mozhaev, K. Heremans, J. Frank, P. Masson, C. Balny, Proteins Struct. Funct. Genet. 24 (1996) 81–91.
[3] C.R. Robinson, S.G. Sligar, Methods Enzymol. 259 (1995) 395–426.
[4] B.M. Gorovits, P.M. Horowitz, Biochemistry 37 (1998) 6132–6135.
[5] R.J. St. John, J.F. Carpenter, T.W. Randolph, Proc. Natl. Acad. Sci. USA 96 (1999) 13029–13033.
[6] M. Fukuda, S. Kunugi, Eur. J. Biochem. 142 (1984) 565–570.
[7] D.J. Hei, D.S. Clark, Appl. Environ. Microbiol. 60 (1994) 932–939.
[8] P.C. Michels, D.S. Clark, Appl. Environ. Microbiol. 63 (1997) 3985–3991.
[9] P.C. Michels, D. Hei, D.S. Clark, Adv. Protein Chem. 48 (1996) 341–376.
[10] W. Kauzmann, Nature 325 (1987) 763–764.
[11] K.J. Frye, C.A. Royer, Protein Sci. 7 (1998) 2217–2222.
[12] T.E. Creighton, Proteins: Structural and Molecular Properties, Freeman, New York, 1993, pp. 142–143.
[13] L. Heremans, K. Heremans, J. Mol. Struct. 214 (1989) 305–314.
[14] K.K. Rodgers, S.G. Sligar, J. Mol. Biol. 221 (1991) 1453–1460.
[15] S.D. Hamann, Rev. Phys. Chem. Jpn. 50 (1980) 147–168.
[16] T.V. Chalikian, A.P. Sarvazyan, K.J. Breslauer, J. Phys. Chem. 97 (1993) 13017–13026.
[17] F. Abe, C. Kato, K. Horikoshi, Trends Microbiol. 7 (1999) 447–453.
[18] J.L. Silva, G. Weber, Annu. Rev. Phys. Chem. 44 (1993) 90–113.
[19] A. Disteche, Symp. Soc. Exp. Biol. 26 (1972) 27–60.
[20] K. Heremans, Rev. Phys. Chem. Jpn. 50 (1980) 259–273.
[21] P.R. Blake, J.B. Park, F.O. Bryant, S. Aono, J.K. Magnuson, E. Eccleston, J.B. Howard, M.F. Summers, M.W.W. Adams, Biochemistry 30 (1991) 10885–10895.
[22] M.W. Day, B.T. Hsu, L. Joshuator, J.B. Park, Z.H. Zhou,

M.W.W. Adams, D.C. Rees, Protein Sci. 1 (1992) 1494–1507.
[23] S. Cavagnero, Z.H. Zhou, M.W.W. Adams, S.I. Chan, Biochemistry 34 (1995) 9865–9873.
[24] M.M.C. Sun, N. Tolliday, C. Vetriani, F.T. Robb, D.S. Clark, Protein Sci. 8 (1999) 1056–1063.
[25] T. Ooi, in: H. Ludwig (Ed.), Advances in High Pressure Bioscience and Biotechnology, Springer, Berlin, 1999, pp. 255–258.
[26] J. Barciszewski, J. Jurczak, S. Porowski, T. Specht, V.A. Erdmann, Eur. J. Biochem. 260 (1999) 293–307.
[27] N. Tanaka, C. Ikeda, K. Kanaori, K. Hiraga, T. Konno, S. Kunugi, Biochemistry 39 (2000) 12063–12068.
[28] N. Tanaka, D. Mitani, S. Kunugi, Biochemistry 40 (2001) 5914–5920.
[29] P. Cioni, G.B. Strambini, J. Mol. Biol. 242 (1994) 291–301.
[30] G. Wagner, FEBS Lett. 112 (1980) 280–284.
[31] P. Masson, C. Clery, P. Guerra, A. Redslob, C. Albaret, P.-L. Fortier, Biochem. J. 343 (1999) 361–369.
[32] J.S. Nehme, J.L. Silva, J.R. Meyer-Fernandes, Biochim. Biophys. Acta 1546 (2001) 164–170.
[33] A.C. Oliveira, L.P. Gaspar, A.T. Da Poian, J.L. Silva, J. Mol. Biol. 240 (1994) 184–187.
[34] M. Gross, R. Jaenicke, Eur. J. Biochem. 221 (1994) 617–630.
[35] V.A. Payne, N. Matubayasi, L.R. Murphy, R.M. Levy, J. Phys. Chem. B 101 (1997) 2054–2060.
[36] J.N. Webb, S.D. Webb, J.L. Cleland, J.F. Carpenter, T.W. Randolph, Proc. Natl. Acad. Sci. USA 98 (2001) 7259–7264.
[37] B. Wroblowski, J.F. Diaz, K. Heremans, Y. Engelborghs, in: R. Hayashi, C. Balny (Eds.), High Pressure Bioscience and Biotechnology, Elsevier Science, Amsterdam, 1996, pp. 167–170.
[38] G.J.A. Vidugiris, J.L. Markley, C.A. Royer, Biochemistry 34 (1995) 4909–4912.
[39] K. Sasahara, M. Sakurai, K. Nitta, J. Mol. Biol. 291 (1999) 693–701.
[40] K. Gekko, S.N. Timasheff, Biochemistry 20 (1981) 4667–4676.
[41] A. Priev, A. Almagor, S. Yedgar, B. Gavish, Biochemistry 35 (1996) 2061–2066.
[42] V.V. Mozhaev, R. Lange, E.V. Kudryashova, C. Balny, Biotechnol. Bioeng. 52 (1996) 320–331.
[43] M.J. Kornblatt, J.A. Kornblatt, G. Hui Bon Hoa, Arch. Biochem. Biophys. 306 (1993) 495–500.
[44] J.A. Kornblatt, G. Hui Bon Hoa, Biochemistry 29 (1990) 9370–9376.
[45] M.M.C. Sun, R. Caillot, G. Mak, F.T. Robb, D.S. Clark, Protein Sci. 10 (2001) 1750–1757.
[46] G. Mei, A.D. Venere, F.M. Campeggi, G. Gilardi, N. Rosato, F.D. Matteis, A. Finazzi-Agro, Eur. J. Biochem. 265 (1999) 619–626.
[47] W. Kauzmann, Adv. Protein Chem. 14 (1959) 1–63.
[48] G. Hummer, S. Garde, A.E. Garcia, M.E. Paulaitis, L.R. Pratt, Proc. Natl. Acad. Sci. USA 95 (1998) 1552–1555.
[49] A. Hvidt, J. Theor. Biol. 50 (1975) 245–252.
[50] M.H. Klapper, Biochim. Biophys. Acta 229 (1971) 557–566.
[51] J. Zollfrank, J. Friedrich, J. Fidy, J.M. Vanderkooi, J. Chem. Phys. 94 (1991) 8600–8603.
[52] J. Fidy, J.M. Vanderkooi, J. Zollfrank, J. Friedrich, Biophys. J. 63 (1992) 1605–1612.
[53] G. Hummer, S. Garde, A.E. Garcia, M.E. Paulaitis, L.R. Pratt, J. Phys. Chem. B 102 (1998) 10469–10482.
[54] K. Suzuki, Y. Taniguchi, Symp. Soc. Exp. Biol. 26 (1972) 103–124.
[55] R. Jaenicke, Annu. Rev. Biophys. Bioeng. 10 (1981) 1–67.
[56] J.W. Linowski, N.-I. Liu, J. Jonas, J. Chem. Phys. 65 (1976) 3383–3384.
[57] J.W. Linowski, N.-I. Liu, J. Jonas, J. Magn. Reson. 23 (1976) 455–460.
[58] K. Goossens, L. Smeller, K. Heremans, J. Chem. Phys. 99 (1993) 5736–5741.
[59] H. Li, H. Yamada, K. Akasaka, Biochemistry 37 (1998) 1167–1173.
[60] K. Gekko, H. Noguchi, J. Phys. Chem. 83 (1979) 2706–2714.
[61] K. Gekko, Y. Hasegawa, Biochemistry 25 (1986) 6563–6571.
[62] R. Jaenicke, J. Biotechnol. 79 (2000) 193–203.
[63] M.A. Williams, J.M. Goodfellow, J.M. Thornton, Protein Sci. 3 (1994) 1224–1235.
[64] R. Jaenicke, G. Bohm, Curr. Opin. Struct. Biol. 8 (1998) 738–748.
[65] E. Mombelli, M. Afshar, P. Fusi, M. Mariani, P. Tortora, J.P. Connelly, R. Lange, Biochemistry 36 (1997) 8733–8742.
[66] C. Balny, P. Masson, Food Rev. Int. 9 (1993) 611–628.
[67] V.V. Mozhaev, K. Heremans, J. Frank, P. Masson, C. Balny, Trends Biotechnol. 12 (1994) 493–501.
[68] L. Heremans, K. Heremans, Biochim. Biophys. Acta 999 (1989) 192–197.
[69] K.E. Prehoda, E.S. Mooberry, J.L. Markley, Biochemistry 37 (1998) 5785–5790.
[70] G.J.A. Vidugiris, C.A. Royer, Biophys. J. 75 (1998) 463–470.
[71] J.L. Silva, C.F. Silveira, A. Correira, L. Pontes, J. Mol. Biol. 223 (1992) 545–555.
[72] X. Peng, J. Jonas, J.L. Silva, Biochemistry 33 (1994) 8323–8329.
[73] J.L. Silva, D. Foguel, A.T. Da Poian, P.E. Prevelige, Curr. Opin. Struct. Biol. 6 (1996) 166–175.
[74] J. King, G. Weber, Biochemistry 25 (1986) 3632–3637.
[75] K. Ruan, G. Weber, Biochemistry 28 (1989) 2144–2153.
[76] M. Panda, J. Ybarra, P.M. Horowitz, J. Biol. Chem. 276 (2001) 6253–6259.
[77] R.D. Haight, R.Y. Morita, J. Bacteriol. 83 (1962) 112–120.
[78] C. Di Primo, G. Hui Bon Hoa, P. Douzou, S.G. Sligar, Eur. J. Biochem. 209 (1992) 583–588.
[79] G.D. Lockyer Jr., D. Owen, D. Crew, R.C. Neuman Jr., J. Am. Chem. Soc. 96 (1974) 7303–7307.
[80] M. Fukuda, S. Kunugi, Eur. J. Biochem. 149 (1985) 657–662.
[81] T. Härd, T. Lundbäck, Biophys. Chem. 62 (1996) 121–139.
[82] D. Foguel, J.L. Silva, Proc. Natl. Acad. Sci. USA 91 (1994) 8244–8247.
[83] C.R. Robinson, S.G. Sligar, Proc. Natl. Acad. Sci. USA 92 (1995) 3444–3448.

[84] C.R. Robinson, S.G. Sligar, Biochemistry 33 (1994) 3787–3793.
[85] J.L. Silva, C.F. Silveira, Protein Sci. 2 (1993) 945–950.
[86] S. Sundaram, C.M. Roth, M.L. Yarmush, Biotechnol. Prog. 14 (1998) 773–781.
[87] P. Roy, C.M. Roth, M.N. Margolies, M.L. Yarmush, Biophys. Chem. 83 (2000) 171–177.
[88] S.A. Hawley, Biochemistry 10 (1971) 2436–2442.
[89] P. Rubens, P. Degraeve, P. Lemay, K. Heremans, in: H. Ludwig (Ed.), Advances in High Pressure Bioscience and Biotechnology, Springer, Berlin, 1999, pp. 227–230.
[90] C. Balny, V.V. Mozhaev, R. Lange, Comp. Biochem. Physiol. 116A (1997) 299–304.
[91] P. Cioni, G.B. Strambini, J. Mol. Biol. 263 (1996) 789–799.
[92] J.V. Carter, D.G. Knox, A. Rosenberg, J. Biol. Chem. 253 (1978) 1947–1953.
[93] P. Cioni, G.B. Strambini, J. Mol. Biol. 291 (1999) 955–964.
[94] C. Balny, in: H. Ludwig (Ed.), Advances in High Pressure Bioscience and Biotechnology, Springer, Berlin, 1999, pp. 261–266.
[95] T.K. Hitchens, R.G. Bryant, Biochemistry 37 (1998) 5878–5887.
[96] S. Kunugi, S. Fujiwara, S. Kidokoro, K. Endo, S. Hanzawa, FEBS Lett. 462 (1999) 231–235.
[97] V.V. Mozhaev, N. Bec, C. Balny, Biochem. Mol. Biol. Int. 34 (1994) 191–199.
[98] K.J. Laidler, D. Chen, Trans. Faraday Soc. 54 (1958) 1026–1033.
[99] S. Kunugi, M. Kitayaki, Y. Yanagi, N. Tanaka, R. Lange, C. Balny, Eur. J. Biochem. 248 (1997) 567–574.
[100] V.V. Mozhaev, E.V. Kudryashova, N. Bec, in: R. Hayashi, C. Balny (Eds.), High Pressure Bioscience and Biotechnology, Elsevier Science, Amsterdam, 1996, pp. 221–226.
[101] S. Kunugi, M. Fukuda, N. Ise, Biochim. Biophys. Acta 704 (1982) 107–113.

ELSEVIER

Biochimica et Biophysica Acta 1595 (2002) 250–265

BIOCHIMICA ET BIOPHYSICA ACTA
BBA
www.bba-direct.com

Review

Pressure induces folding intermediates that are crucial for protein–DNA recognition and virus assembly

Jerson L. Silva [a,*], Andréa C. Oliveira [a], Andre M.O. Gomes [a], Luís Maurício T.R. Lima [b], Ronaldo Mohana-Borges [a], Ana B.F. Pacheco [c], Débora Foguel [a]

[a] *Programa de Biologia Estrutural, Departamento de Bioquímica Médica – ICB, Centro Nacional de Ressonância Magnética Nuclear de Macromoléculas, Universidade Federal do Rio de Janeiro, 21941-590 Rio de Janeiro, RJ, Brazil*

[b] *Departamento de Medicamentos, Faculdade de Farmácia, Universidade Federal do Rio de Janeiro, 21941-590 Rio de Janeiro, RJ, Brazil*

[c] *Instituto de Biofísica Carlos Chagas Filho, Universidade Federal do Rio de Janeiro, 21941-590 Rio de Janeiro, RJ, Brazil*

Received 16 October 2001; received in revised form 14 November 2001; accepted 15 November 2001

Abstract

Protein–nucleic acid interactions are crucial for a variety of fundamental biological processes such as replication, transcription, restriction, translation and virus assembly. The molecular basis of protein–DNA and protein–RNA recognition is deeply related to the thermodynamics of the systems. We review here how protein–nucleic acid interactions can be approached in the same way as protein–protein interactions involved in protein folding and protein assembly, using hydrostatic pressure as the primary tool and employing several spectroscopic techniques, especially fluorescence, circular dichroism and high-resolution nuclear magnetic resonance. High pressure has the unique property of stabilizing partially folded states or molten-globule states of a protein. The competition between correct folding and misfolding, which in many proteins leads to formation of insoluble aggregates is an important problem in the biotechnology industry and in human diseases such as amyloidosis, Alzheimer's, prion and tumor diseases. The pressure studies reveal that a gradient of partially folded (molten globule) conformations is present between the unfolded and fully folded structure of several bacteria, plant and mammalian viruses. Using pressure, we have detected the presence of a ribonucleoprotein intermediate, where the coat protein is partially unfolded but bound to RNA. These intermediates are potential targets for antiviral compounds. Pressure studies on viruses have direct biotechnological applications. The ability of pressure to inactivate viruses has been evaluated with a view toward the applications of vaccine development and virus sterilization. Recent studies demonstrate that pressure causes virus inactivation while preserving the immunogenic properties. There is substantial evidence that a high-pressure cycle traps a virus in the 'fusion intermediate state', not infectious but highly immunogenic.

Keywords: Hydrostatic pressure; Protein–nucleic acid interaction; Protein misfolding; Amyloidogenesis; Human papillomavirus E2 protein; Virus fusion; Picornavirus

1. Introduction

One should recall that a protein in water solution is already a chemically metastable system as the ad-

* Corresponding author. Fax: +55-21-2270-8647.
E-mail address: jerson@bioqmed.ufrj.br (J.L. Silva).

0167-4838/02/$ – see front matter
PII: S0167-4838(01)00348-X

dition of an appropriate protease brings about its hydrolysis to the component amino acids. Gregorio Weber (Protein Interactions, 1992).

Protein–nucleic acid complexes play crucial functions in cells and the formation of specific complexes is dictated by rules that are not completely understood [1,2]. The elusive character of protein–nucleic acid interactions may derive from the large number of interactions among its components that involve relatively weak energies (less than 10 kcal/mol). The energetic coupling among these interactions is very likely the basis of the high specificity of the recognition process. The mechanism by which a nucleic acid sequence is chosen to interact with a protein among millions of competing sequences bears analogy to the mechanism by which unstructured proteins spontaneously fold to their native functional form. To this end, it is necessary to isolate and describe the intermediate structures that occur during folding, oligomerization and interaction with the nucleic acid. Here, it is reviewed that in several cases these intermediates have been trapped under pressure and their dynamics and structure have been characterized by fluorescence, light scattering, NMR and hydrodynamic methods.

As pointed out by Weber in his book [3], the changes in the standard free energy of reaction with pressure, at constant temperature, have a simplicity that they should be examined previous to any consideration of the effects of temperature at constant pressure. Pressure affects the equilibrium between denatured or dissociated and native forms in the direction of the form that occupies a smaller volume [4–8]. The structural region of the protein that is most sensitive to pressure is the hydrophobic core, especially when cavities are present [9]. It is described below the unique characteristic of pressure to isolate molten globules or to favor the state that appears to have a segment of the protein in a partially folded state ('partial' molten globule). The study of effects of pressure on protein and protein–nucleic acid complexes has biotechnological applications. The interactions of drugs with protein–nucleic acid complexes and viruses have been tackled by pressure studies. Viruses inactivated by pressure may prove to be useful in vaccine development and virus sterilization [10–14].

Fig. 1 schematizes the possible conformational pathways of a polypeptide after its synthesis in the ribosomes. It can follow the folding pathway resulting in a functional complex, e.g., a protein that binds DNA or that assembles into a macromolecular structure, such as a virus. If instead, the protein takes an off-pathway, it gives rise to a misfolded conformation, usually forming aggregates. High pressure can explore each of these reactions as it will be discussed below.

2. The volume cavity model

The mechanisms by which a linear sequence of amino acid residues is translated into a three dimensional structure evolving in time remains one of the major challenges in Biology. Although most of the thermodynamic studies have been done with conventional perturbing agents, such as high temperature, urea, guanidine, etc, they offer limited information, especially because they cause drastic changes in protein structure, not likely to occur at physiological conditions. Folding of a protein, or association of one protein with another, is accompanied by an increase in volume because of the additive effects of the formation of solvent-excluding cavities and the release of bound solvent. Water is released as nonpolar amino acid residues are buried, as well as when salt linkages are formed [5–8]. In the former case, hydrophobic interactions have a paramount contribution. The Gibbs free energy (and the equilibrium constant) for an interprotein or intraprotein interaction will depend on the standard volume change (ΔV) of the reaction:

$$\Delta G_p = \Delta G_o + p\Delta V \quad (1)$$

$$\ln(\alpha_p^n/(1-\alpha_p)) = p(\Delta V/RT) + \ln(K_{do}/n^n C^{(n-1)}) \quad (2)$$

where ΔG_p and ΔG_o are the free energies for association/folding at pressure p and at atmospheric pressure, respectively; K_{do} is the equilibrium constant for dissociation or denaturation at atmospheric pressure; ΔV is the volume change, α_p is the extent of reaction at pressure p, C is the protein concentration and n is the number of dissociating subunits.

The packing among the different components play

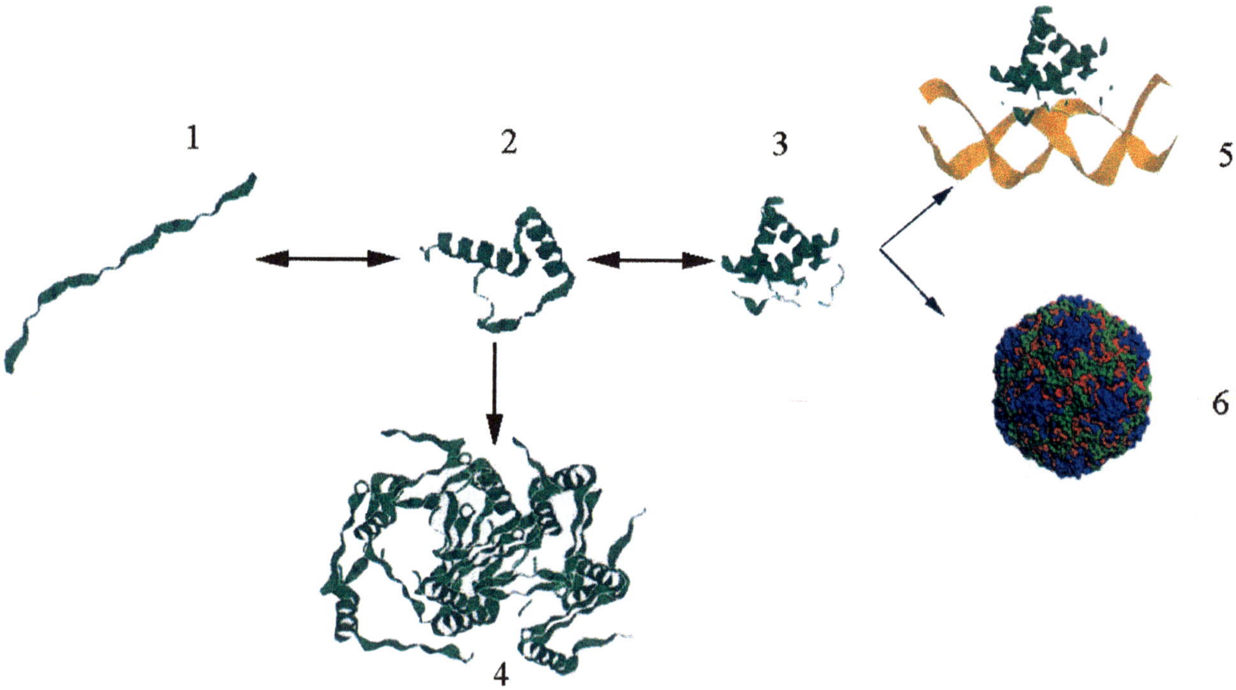

Fig. 1. Schematic representation of the folding and assembly of protein and multimolecular complexes. (1) Unfolded protein; (2) folding intermediate; (3) tertiary structure; (4) protein misfolding – aggregate; (5) protein–DNA interaction; (6) virus.

a major role in stabilizing proteins. Because of the huge number of atoms involved, packing defects cannot be avoided, which lead to formation of cavities [15]. The incompressibility of proteins shows that the atomic packing densities in globular protein are limited to the invariance of the covalent bond angles and distances up to 10 kbar [4,5]. Earlier studies revealed the anisotropic distribution of cavities and compressibilities inside the protein structure [16,17] Cavities are also related to the metastability of some proteins and macromolecular assemblages [18–21]. Recent studies show that hydrostatic pressure is a useful tool to investigate packing and cavities [9,19–21]. Royer and coworkers have obtained strong evidence [9] for the hypothesis that the internal cavities in the protein's structure contribute to the magnitude of the observed volume change. Experimental and theoretical approaches indicate that the underlying mechanism of pressure unfolding is the penetration of water into the protein matrix [22,23].

3. Protein folding intermediates, molten globules and amyloidogenic conformations isolated under pressure

It has been generally accepted that to unveil protein folding it is necessary to isolate and describe the intermediate structures. Some of these intermediates have been trapped under pressure and their dynamic and structural properties have been scrutinized.

Several model proteins have been used for studies of protein folding intermediates, including Arc repressor [22,24–26], lysozyme [27], the E2 DNA-binding domain (E2-DBD) from human papillomavirus [28,29], ribonuclease A [30], trypsin [31], apomyoglobin [32], the Ras binding domain of RalGDS [33] and dihydrofolate reductase [34]. Arc repressor dimer reversibly dissociates into partially folded subunits under pressure [24] and the structure of the pressure-denatured monomer was partially determined by two-dimensional ^{1}H nuclear magnetic resonance

(NMR) [25,26]. The partially folded conformation of pressure-denatured proteins binds a substantial amount of water [22] and pressure denaturation does not progress when water is replaced with an organic solvent. This latter result has been corroborated by theoretical approaches that show the infiltration of water inside the protein [23,35]. The protection against pressure denaturation by osmolytes has also been found for staphylococcal nuclease [36], F_0F_1-ATPase [37] and the recombinant human interferon-γ [38].

Small monomeric proteins are the best models for studying protein folding but they are often too stable for the pressures normally attainable in standard equipment. Recently, we made use of the lower stability of a noncovalent complex formed by two complementary fragments of the chymotrypsin inhibitor-2 to study the folding equilibrium [39] and kinetics [40]. The pressure-denaturation of the complex was completely independent of protein concentration, indicating the formation of a denatured heterodimer. Pressure also leads to the formation of a denatured dimer in the case of the coiled-coil dimer of tropomyosin [41]. In the case of triose phosphate isomerase, the lack of concentration dependence for the pressure-dissociation curves is due to the subunit heterogeneity [42]. The structural characteristics of the pressure denatured state of the CI-2 complex can be explained by a molten-globule conformation, similar to those obtained for E2-DBD [28,29] and Arc repressor [25,26]. Further kinetic studies under pressure confirmed a denatured CI-2 complex with residual persistent structure [40].

3.1. Misfolding intermediates

Protein folding has recently elicited a great deal of interest in Medicine and Biotechnology. The isolation of folding intermediates is crucial to the understanding of protein misfolding and protein aggregation. In the last years, several diseases, such as Alzheimer's disease, Parkinson's disease, spongiform encephalopathies (caused by prions), inherited emphysema, cystic fibrosis and likely many cancers are caused by protein misfolding [43,44]. Biotechnology companies also face many problems with proteins aggregating into inclusion bodies when they are expressed in heterologous cells. Fig. 1 shows how functional proteins and macromolecular assemblies may divert from the folding pathway to a misfolding and aggregation dead end.

Pressure allows one to explore the two pathways (aggregation/misfolding and correct folding) as recently demonstrated in rhodanese [45], in the tail spike protein of bacteriophage P22 [46], in myoglobin [47], and in the amyloidogenic protein transthyretin [48]. In the case of P22 tailspike aggregates, high pressure induces dissociation and results in formation of monomers and native folded trimers [46]. Pressure has also been utilized to dissociate large aggregates of proteins in the presence of low concentrations of guanidine hydrochloride [49]. These studies have opened new avenues to use high pressure for combating aggregation, with applications in Medicine and Biotechnology.

Results on the amyloidogenic protein transthyretin [48], and more recently on tumor suppressor p53 protein, suggest that stable intermediates can be achieved by carefully tuning high pressure and temperature. Transthyretin (TTR) is a tetrameric human plasma protein responsible for familial amyloid polyneuropathies (FAP) and senile systemic amyloidosis (SSA) [50]. Among other properties, partially folded TTR induced by pressure binds bis-ANS, similar to the acid-denatured protein that undergoes aggregation into amyloid fibrils [48]. A less stable tetrameric TTR (a 'pre-aggregated' state with loosened subunits) is formed after decompression, and aggregation can be prevented by keeping the sample at low temperature [48]. The isolation of these intermediates provides unprecedented targets for the development of antagonists capable of blocking protein misfolding and aggregation, potential drugs against amyloidogenic and tumor diseases.

More recently, we have investigated the misfolding conformation induced by pressure of the DNA-binding domain of the tumor suppressor protein p53. Most human cancers ($\approx 50\%$) result from mutations in the p53 protein, mainly at its DNA binding domain (p53/DBD), affecting the DNA binding ability and/or the protein stability [51]. Our results demonstrate that p53/DBD undergoes denaturation when submitted to hydrostatic pressures at 37°C, leading to a misfolded and aggregated state. However, when we investigated the pressure effects at 4°C, the protein denaturation was displaced to higher pressures and aggregation was noncooperative with denaturation (unpublished results). These studies show that pressure can be used to explore intermediates and determine their structures, which may be valuable to the rational design of drugs.

4. Protein–DNA interactions

Genes are switched on and off by the binding of proteins to specific DNA sequences [1,2]. Protein–nucleic acid interactions constitute the framework for the regulation of key biological functions such as transcription, translation, replication, gene regulation, virus assembly and recombination. Structural data derived from X-ray crystallography and NMR have demonstrated that the reading of a DNA sequence is achieved by a combination of van der Waals, electrostatic and hydrogen bonds [1,2,52,53]. Structural studies alone cannot separate the contribution of each of these factors. Cooperative interactions among protein subunits play an important role in the binding to the specific DNA sites [52–54]. Deciphering the interactions among the players (different domains, DNA, and binding proteins) requires biophysical studies in addition to structure. Noncovalent interactions can be perturbed reversibly using high hydrostatic pressure, which allows protein folding, protein–protein and protein–ligand interactions to be characterized thermodynamically.

Studies on the combined effects of osmotic and hydrostatic pressure on the specificity of protein–DNA complexes have been very valuable [55,56]. The results obtained in Sligar's group show that water solvation plays an important role in the site-specific recognition of DNA. Application of hydrostatic pressure counteracts the effects of osmotic pressure by restoring the hydration shell of the enzyme, and restores the correct recognition pattern [55].

In one of the first studies on pressure effects on protein–nucleic acid interactions, Royer et al. [57] found that tetrameric lac repressor, a helix-turn-helix DNA-binding protein, was stabilized by the inducer and destabilized by DNA, indicating significant coupling between subunit interactions, operator binding and inducer affinity. In several cases, the formation of a specific protein–DNA complex is followed by an increase in nonpolar interactions and release of solvent, which explain the entropy-driven character of the specific binding, while solvent is not displaced in nonspecific complexes [58,59]. On the other hand, eukaryotic transcription factors seem to be less sensitive to hydration effects and cooperative interactions among the proteins (homo- or heterointeractions) play a dominant role in the binding to the specific DNA sites.

4.1. Role of water in protein–DNA interaction

The use of high pressure has several advantages over other methods because it affects the structure of the protein but not that of the DNA [60]. With the Arc repressor system, the nucleic acid–protein recognition process is coupled to folding and oligomerization of the protein [61]. The stabilization of the native dimer is determined by the specificity of the protein–DNA interaction. This free-energy linkage is absent in a mutant Arc protein (P8L), which binds to operator and to nonspecific DNA sequences with equal, low affinity. The coupling between DNA binding and the conversion of partially folded monomers to native dimer seem to account for the ability to recognize operator DNA. By cold-denaturing the protein repressor–operator complex at sub-zero temperatures under pressure, we have shown that the entropy increases greatly when Arc repressor binds tightly to its operator sequence but not when it binds loosely to a nonspecific sequence [58]. Increase in entropy enables the protein to distinguish between a specific and a nonspecific DNA sequence. We postulate that the formation of a protein–operator DNA complex is followed by an increase in nonpolar interactions and release of solvent, which would explain its entropy-driven character, whereas solvent would not be displaced in nonspecific complexes [58]. In other words, specific protein–DNA interactions are 'dry' whereas nonspecific interactions are 'wet'. This hypothesis is represented in Fig. 2, which extends the concept to protein–nucleic acid interactions in viruses.

In general, both the hydrostatic and osmotic pressure studies have addressed several questions posed by von Hippel [2] to try to elucidate how a protein discriminates between a unique sequence of DNA and the millions of competing nonspecific sites in

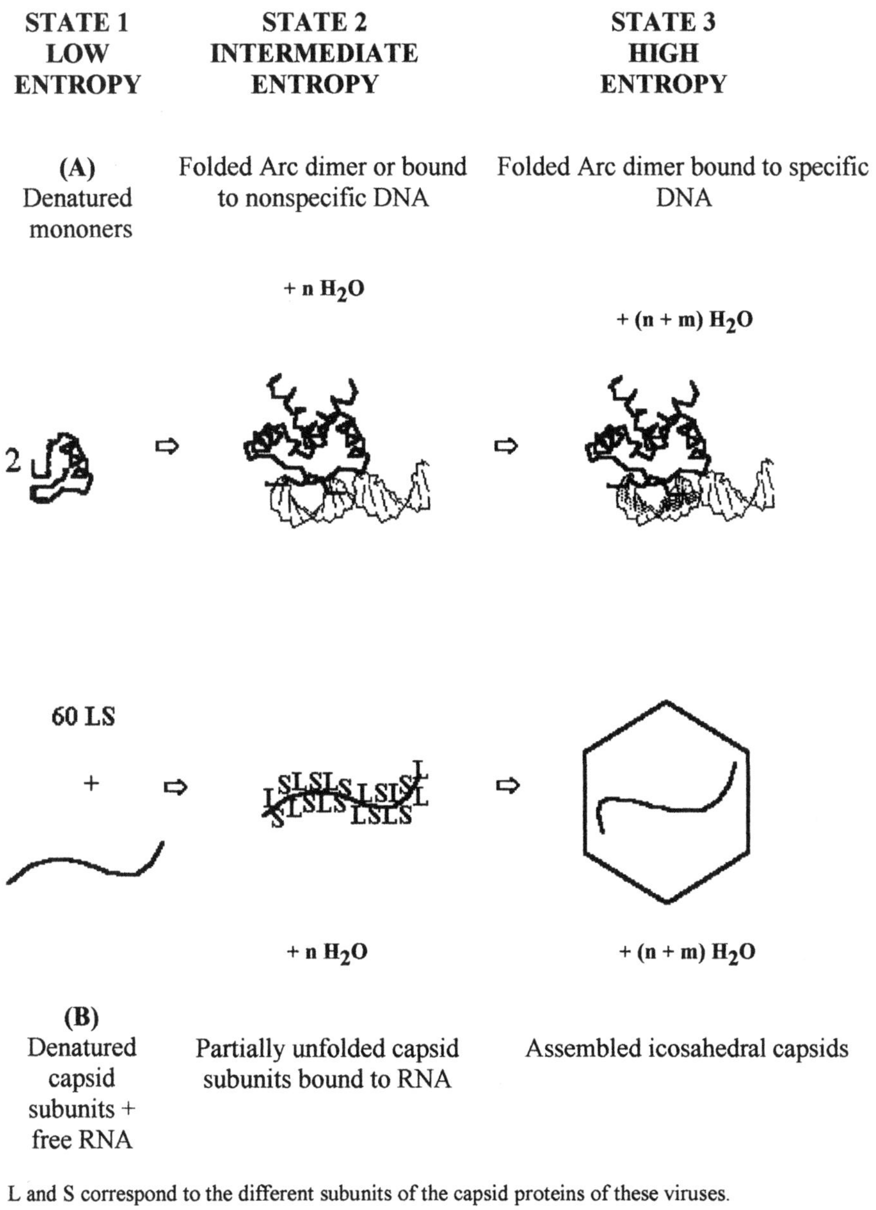

Fig. 2. Schematic representation of Arc–DNA recognition (A) and a $T = 3$ RNA virus assembly (B).

the genome. As pointed out by von Hippel, the understanding of protein–DNA recognition involves: (1) Defining the hydrophobic content of a nucleic acid surface; (2) evaluating the differences in hydration and conformation that distinguish nonspecific vs. specific protein–nucleic acid complexes. The pressure studies have partially answered these questions. Because nonpolar interactions are entropic and compressible, they are more sensitive to pressure and low temperature. Squeezing and cooling biological complexes can thus be useful thermodynamic tools for studying hydration and conformation. As a general answer to the questions raised by von Hippel, as shown in Fig. 2, specific protein–nucleic acid complexes (both in protein–DNA complexes and virus particles) are more sensitive to cold than the protein–protein interactions that stabilize a folded protein. These results demonstrate the crucial role of

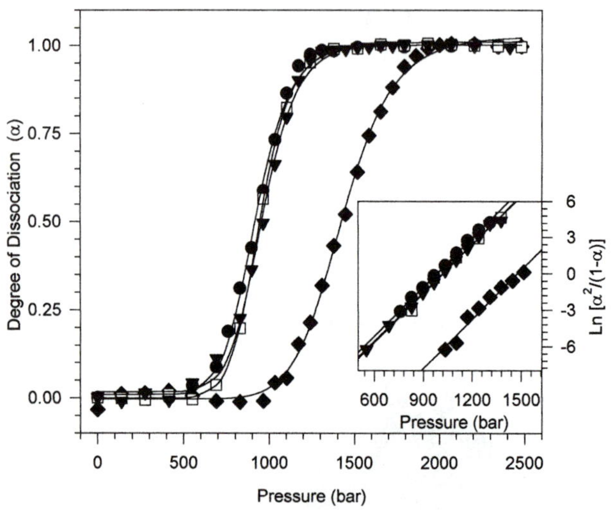

Fig. 3. Effects of DNA binding on stabilization of LexA dimers. LexA preparations (2 μM) were incubated with one of the following 34-base pair oligonucleotides: (i) *recA* operator (♦); (ii) *recA* half-operator (▼); (iii) *polyGC* (□), in which the entire SOS box is replaced by a $(GC)_{10}$ tract. Control without DNA is indicated by (●).

dehydration in the formation of the specific interactions. Especially with our studies with Arc–DNA interactions, it appears that the specific protein-operator DNA complex is more dehydrated than the nonspecific complexes.

4.2. The role of specific DNA in tightening protein–protein interactions in LexA repressor and E2 DNA-binding domain (E2c) protein of human papillomavirus 16

LexA repressor is a 202-residue protein that regulates transcription in *Escherichia coli* of the SOS system, about 17 genes that encode proteins involved in DNA repair [62]. LexA has two domains and it undergoes autocleavage at residues 84–85 in a flexible hinge between the two domains [62]. The solution structure of the DNA-binding domain was determined by high-resolution nuclear magnetic resonance [63]. With LexA repressor, we found that the protein is a dimer at nanomolar concentrations and that the dissociation constant lies in the picomolar range [64]. Whereas nonspecific DNA has no stabilizing effects, specific DNA induces tightening of the dimer and a 750-fold decrease in the K_d (Fig. 3). Accordingly, the LexA dimer only loses its ability to recognize a specific DNA sequence by RecA-induced autoproteolysis. Our findings indicate that LexA exists within the cell as a dimer at physiological conditions [64]. The dimeric structure of LexA would be crucial for the interaction with RecA/ssDNA filament as pointed out by Walker [65] in the light of the recent crystal structures of LexA mutants [66].

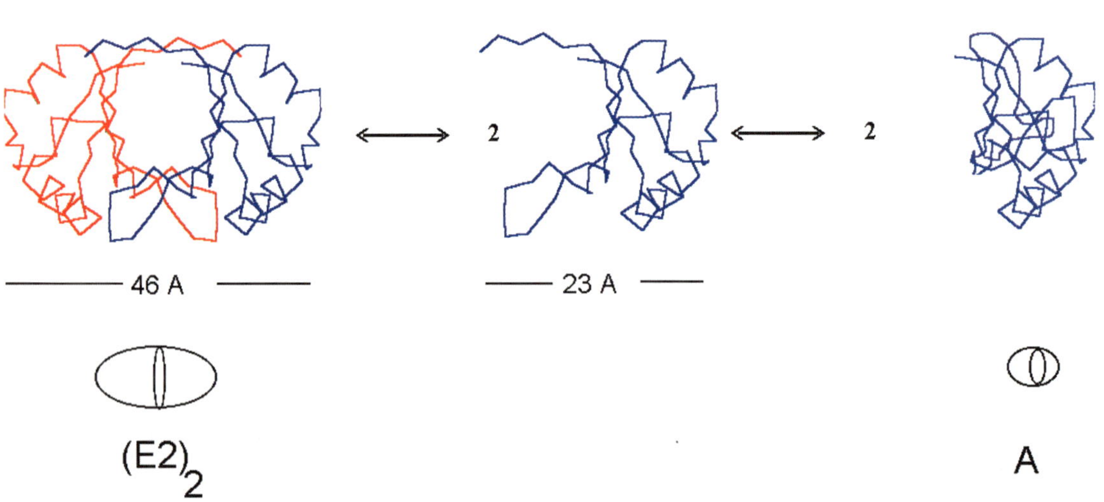

Fig. 4. Schematic representation of the equilibrium attained under high pressure by E2 DNA-binding domain (E2c). The ellipsoid dimeric structure, $(E2)_2$, has a longitudinal diameter of 46 Å as measured from its crystal structure [21] while each monomer has about 23 Å. Under high pressure the monomers is partially denatured to a state A which retains part of its tertiary structure and appears to be more compact than a separated monomer.

E2 proteins from papillomaviruses are involved in viral DNA replication and regulation of transcription [67]. The papillomavirus E2 protein is comprised of an N-terminal transactivation domain (E2n) separated by a flexible hinge region from the C-terminal DNA binding and dimerization domain (E2c). This dimeric domain, from a high-risk human papillomavirus (HPV-16), can be pressure-dissociated to a monomeric state that presents substantial residual structure, through a highly reversible transition [28]. Further unfolding of this high-pressure state can be achieved by low concentrations of urea. NMR studies have confirmed that the pressure-dissociated monomer of E2c is highly folded [29]. Fig. 4 shows the conversion of dimeric E2 into the compact, folded monomer.

The dissociation of E2c-DNA complexes (with specific and nonspecific sequences) cannot be obtained using pressure alone [29]. Taking into account that pressure and urea studies on the stability of E2c provide equivalent thermodynamic parameters, we obtained pressure isotherms in different urea concentration for both specific and nonspecific DNAs. A plot of ΔG_o versus pressure is linear for both the E2c complexes of E2c dimer [29]. In Fig. 5, pressure dissociation data derived from the pressure/urea studies show that both specific and nonspecific DNA sequences promote a large stabilization of E2c dimer when compared to the protein alone. A higher change in free energy is observed for the specific DNA sequence ($\Delta\Delta G_o = 6.7$ kcal/mol) than for the nonspecific DNA ($\Delta\Delta G_o = 4.5$ kcal/mol). The difference between the effects produced by specific and nonspecific DNA is likely the basis for the sequence discrimination.

DNA shifts the denaturation curves to higher pressures (Fig. 5A) with no changes in volume, which indicates that no extra surface area is involved when E2c dissociates on the DNA. These results indicate that E2c dimers dissociate into monomers that are still bound to DNA (Fig. 5B). The retention of a large amount of tertiary structure by the dissociated monomer as evidenced by fluorescence [28] and NMR [29] data corroborate the hypothesis of a functional monomeric unit of E2c. These studies open new ways to the development of drugs targeted to the monomer that will make it possible to prevent or treat human papillomavirus infection.

A

B

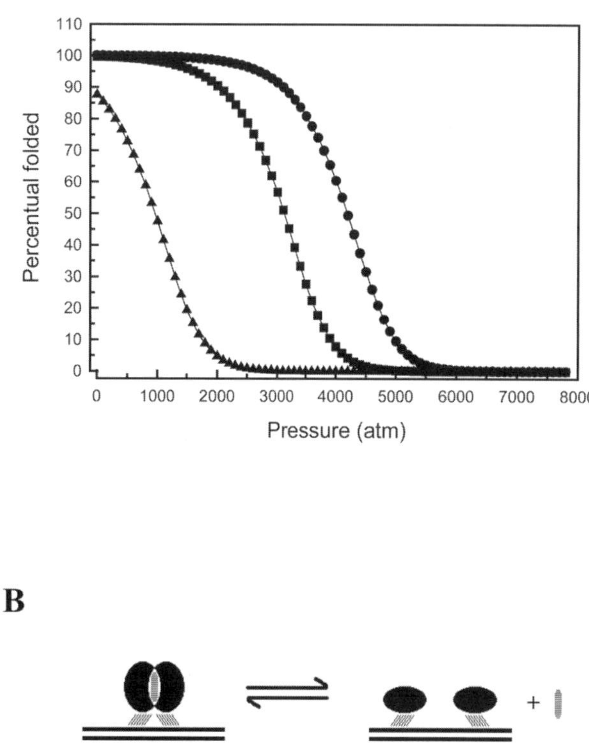

Fig. 5. (A) Pressure isotherms of E2c dissociation in the presence of DNA. Changes in association degree with pressure for 250 nM E2c dimer (●), showing the stabilization of the dimer in the presence of 250 nM of (▲) nonspecific and (■) specific DNA. Data and equation used to generate this plot are from [29]. (B) Diagram of the monomer–dimer equilibrium of E2c bound to DNA. Volume change upon protein–protein and protein–DNA interaction corresponds to hatched symbols.

5. Virus assembly

Virions have evolved to move their genome between cells of a susceptible host and between hosts. The virus particle is composed of either a membrane enveloped or non enveloped protein shell and nucleic acid [68]. Evolution has routed the protein shell to adopt multiple functions: shielding of the nucleic acid, participation in chemical reactions for particle maturation and ability to penetrate into the cell and undergo disassembly. The coat proteins are usually arranged in a shell with icosahedral shape. To pack an infectious genome, integral multiples of 60 subunits are required to form the shell, resulting in nonidentical contacts between subunits [68].

Pressure addresses how the plasticity required for successful assembly of a virus particle is coded into the native conformation of a capsid protein subunit [59]. This combined thermodynamic and structural approach has been used to try to identify the general rules that govern virus assembly. In general, it has been found that the capsid coat proteins (monomers or dimers) are much less stable to pressure than the assembled icosahedral particles [59]. The isolated capsid and the assembly intermediates assume different partially folded states in the assembly pathway [12,69–72]. Single amino acid substitutions in the hydrophobic core of the coat protein of icosahedral viruses produce large decreases in stability against pressure and chemical denaturants. High pressure has permitted us to trap a ribonucleoprotein intermediate (a potential target for antiviral drugs), where the coat protein is partially unfolded but bound to RNA [12,69–72]. Intermediate states also appear in the dissociation of empty capsids, such as P22 procapsids [71,73].

The contribution of protein folding and protein–nucleic acid interactions to virus assembly has been measured in several bacterial, plant and animal viruses [12,14,69–74]. In a typical example, the cowpea mosaic virus (CPMV), the free-energy linkage was evaluated by comparing pressure stability of empty capsids and ribonucleoprotein particles. Empty capsids were much less stable than the ribonucleoprotein particles [69]. For several viruses, we find that the isolated capsid proteins and the assembly intermediates are not fully folded, and that association of 60 or more subunits into an icosahedral particle is coupled to progressive folding of the coat protein and also to changes in interactions with the nucleic acid [59]. Similar to the protein–nucleic acid complexes of Arc repressor, several viruses undergo cold denaturation under pressure, suggesting the entropic stabilization of the particles [12,70–72].

5.1. Free-energy stabilization in the formation of icosahedral shells

The pressure-induced dissociation of tobacco mosaic virus (TMV), a helical structure, was first reported by Lauffer and Dow [75] and more recently by Bonafe et al. [76], where cold-denaturation studies under pressure were also performed. Pressure studies have revealed differences between the stability of the whole virus particle and their capsid subunits. Monomeric capsid protein of P22 bacteriophage was very sensible to pressure and underwent denaturation at pressures below 2.0 kbar [70], whereas the $T=7$ procapsid shell did not dissociate with pressures up to 2.5 kbar. P22 procapsid shells dissociated only at high pressures and low temperatures, indicating that they are stabilized by entropy [70]. Because the assembly of $T=7$ dsDNA viruses requires the scaffolding protein, the dissociation was irreversible.

Several single amino acid substitution mutants have been utilized to dissect the factors that determine the free-energy stability of the icosahedral lattice of bacteriophage P22 [73,77]. The W48Q mutant shells could be easily dissociated by pressure at room temperature with little dependence on decreasing temperature, suggesting a smaller entropic contribution. In contrast, the lower stabilities found for the G232D and T294I mutants were associated with defective protein cavities [73,77].

5.2. Isolation of a ribonucleoprotein complex and protein-folding gradient in the assembly of icosahedral viruses

A ribonucleoprotein intermediate has been detected in the pressure dissociation of several viruses [12,69–72], as shown for the case of picornaviruses (Fig. 6). We have postulated that the ribonucleoprotein intermediate would serve as a core for ready regeneration of the particle when the pressure is reduced. An unusual feature of this ribonucleoprotein intermediate is the presence of partially unfolded coat proteins bound to the RNA. RNA apparently plays a chaperone-like role during assembly of the capsid. In the absence of RNA the subunits drift to a disorganized structure and cannot renature when the perturbation is withdrawn. Ribonucleoprotein intermediates are potential target for antiviral drugs. The ribonucleoprotein intermediate seems to have a condensed structure [72], which demonstrates the high degree of plasticity of the coat-protein–RNA complex.

Another common feature found in the pressure studies of different viruses is a progressive decrease in folding structure in moving from assembled capsids to ribonucleoprotein intermediates (in the case of RNA viruses), free dissociated units (dimers or monomers), and finally unfolded monomers. We

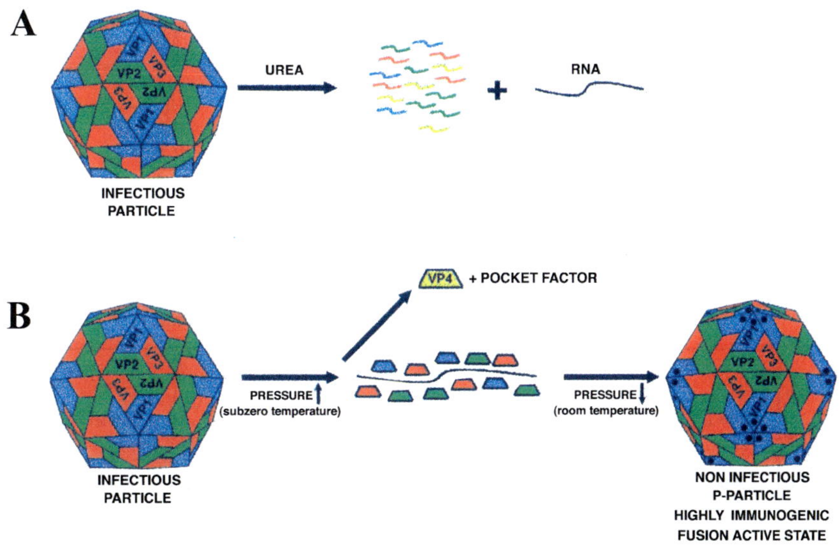

Fig. 6. Proposed model for the disassembly of picornaviruses by denaturing high concentrations of urea (A) or by pressure and low temperature (B). In A, a high urea concentration elicits complete dissociation and denaturation, with separation of the RNA from the coat protein (urea-unfolded subunits are represented as elongated random coils). In B, on the contrary, pressure plus low temperature disrupt the icosahedral structure, but the capsid proteins (VP1, VP2 and VP3) still remain bound to the RNA. This particle loses infection on return to atmospheric pressure and room temperature. This infectivity loss may be due to release of VP4 and the 'pocket factor'.

have proposed a gradient of molten-globule states between the fully structured coat protein in the capsid and the unfolded monomers [58,68–72]. High pressure would affect primarily the quaternary and tertiary structure of the capsid protein, leading to partially unfolded, molten-globule conformations. In contrast, high urea concentrations would primarily disrupt the secondary structure.

5.3. Structural transition in insect and mammalian viruses induced by pressure

Viruses are macromolecular assemblies designed to exert their biological role in a single sequential cycle: (1) assembly inside the cells; (2) release to the environment; (3) attachment to new host cells; (4) disassembly and delivery of genome; (5) replication of the genome and transcription of new viral proteins. Among these five steps, disassembly of the capsid and unpacking of the nucleic acid are the least understood [68]. To be effective, disassembly has to occur fast and at the correct time after endocytosis. The switch for this process is usually attributed to the acidic pH inside the endocytic vesicles, but in vitro many viruses are not uncoated by low pH, or the uncoating occurs slowly, not consistent with the requirement for rapid replication. High pressure provides a powerful tool to explore the uncoating of animal viruses.

The family *Picornaviridae* includes several viruses of great economic and medical importance [78]. These viruses have in common a capsid structure composed of sixty copies of four different proteins, VP1 to VP4, and their 3D structures show similar general features [78]. Oliveira et al. [12] have described the differences in stability against high pressure and cold denaturation of these viruses. Both poliovirus and rhinovirus are stable to high pressure at room temperature; pressures up to 2.4 kbar are not enough to promote viral disassembly and inactivation. Within the same pressure range, foot and mouth disease virus (FMDV) particles are drastically affected by pressure, with a loss of infectivity of more than 4 log units. The dissociation of polio and rhino viruses can be observed only under high pressure at low temperatures in the presence of low concentrations of urea (1–2 M). The pressure and low-temperature data reveal clear differences in stability among the three picornaviruses, FMDV being the most sensitive, polio the most resistant, and rhino having intermediate stability.

The changes produced by pressure and low tem-

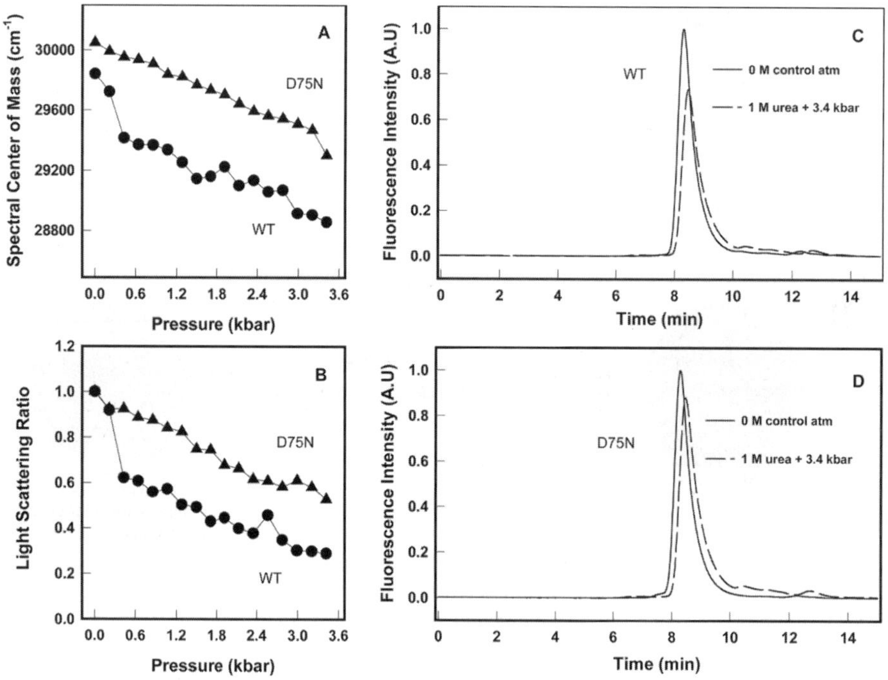

Fig. 7. Pressure sensitivity of wild-type and cleavage-defective FHV VLPs. The pressure stability of Flock House virus (FHV): wild-type (circles) and D75N (triangles) mutant virus-like particles was analyzed by the shift of intrinsic fluorescence spectra (the spectral center of mass, A) and particles average size analysis (light scattering, B). To better appreciate the effect of pressure, the reaction was poised by adding 1 M urea, a sub-dissociating concentration. Reassembly was evaluated by size-exclusion high-performance liquid chromatography for wild-type FHV (C) and mutant after decompression (D).

perature on picornaviruses is schematized in Fig. 6. The most important feature is that after a pressure cycle there is reassociation to a noninfectious particle (named P-particle). A more drastic denaturation treatment, such as with high concentrations of urea, results in irreversible unfolding. The scheme also presents a possible explanation for loss of infection that has previously been proposed by other authors to account for heat treatment of some picornaviruses [79], where the defective particle would lose VP4 and/or small molecules (pocket factors) bound to the canyon. The pressure-inactivated picornavirus may resemble the A-particle detected in poliovirus and rhinovirus upon interaction with the host cell, which also lack the internal capsid protein VP4 [78,79]. The A-particle is substantially less infectious than natural virions and has been considered an intermediate in uncoating. This state is similar to the 'fusion intermediate state' that has been found in enveloped viruses [80–82]. During the fusion process, the conformations of the coat proteins and envelope glycoproteins change, which on one hand lead to noninfectious particles and on the other may lead to the exposure of previously occult epitopes, important for vaccine development. These irreversible conformational changes evoked by high pressure that resemble the changes that occur in vivo are discussed below for most of the viruses we have studied.

We have also used high pressure to evaluate the role of maturation cleavage on flock house virus (FHV) by comparing wild-type (wt) and cleavage-defective mutant (D75N) FHV virus-like particles (VLPs) [83]. We found that maturation cleavage favors a metastable state, in which dissociation of the coat protein is coupled to unfolding. Our results demonstrated that the maturation process targets the particle for an 'off-pathway' disassembly, since dissociation is coupled to unfolding. These maturation events are crucial to make the particle infectious. The mature particles were more unstable to pressure than the cleavage-defective particles (Fig. 7), which can be explained by the metastability elicited upon maturation. We also have evidence that the γ subunit is released after a cycle of compression and decompression. A similar situation would occur with picor-

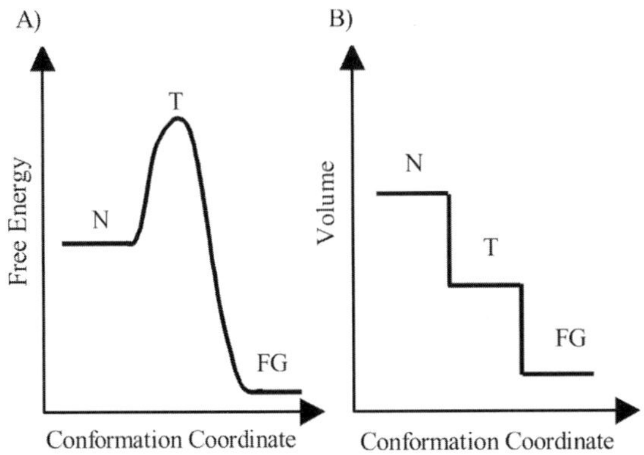

Fig. 8. Gibbs free-energy (A) and volume (B) diagrams for enveloped viruses, showing the conversion between native nonfusogenic state (N) and fusogenic state (FG). The transition intermediate state is represented by (T).

naviruses, with pressure-induced release of VP4 (Fig. 7). In both cases, pressure induces the formation of a noninfectious particle, apparently a fusion-active state of these nonenveloped viruses.

Fig. 7 also shows that the reaction was completely reversible for wild-type and mutant capsids. The greater sensitivity to pressure of cleaved wt FHV is consistent with the higher dynamics revealed by fluorescence polarization and NMR measurements [81], in the light of the well-established relation between protein flexibility (due to volume fluctuations) and its isothermal compressibility [84].

6. Fusion-active states of viruses induced by high pressure as a means to produce vaccines

Enveloped animal viruses undergo membrane fusion in order to deliver their genome into the host cytoplasm [80–82]. We have recently found that high pressure inactivates the enveloped viruses influenza and Sindbis by trapping the particles in the 'fusion intermediate state' [85]. Altogether we show that hydrostatic pressure triggers a conformational change in the glycoprotein of influenza and Sindbis viruses at neutral pHs that is very similar to the change triggered by the low pHs at the endosomes. Our results indicate that pressure rescues the conformation of the metastable, native state (larger volume) into the fusogenic state (smaller volume) by populating a transition-activated state, which although of high energy presents a smaller volume (Fig. 8). The use of pressure to populate fusion-active states can be utilized in the development of new antiviral vaccines and drugs.

The formation of noninfective particles after a cycle of compression and decompression has been demonstrated in many viruses like rotaviruses [86], infectious bursal disease virus (IBDV) [14], vesicular stomatitis virus (VSV) [10], simian immunodeficiency virus [11], HIV [87], influenza [85], lambda phage [88], picornaviruses [12] and alphavirus [85,89]. For IBDV, high pressure caused elimination of infectivity, which retained the original immunogenic properties and could elicit high titers of virus neutralizing antibodies [14]. In the case of VSV, which is a membrane enveloped virus, application of a pressure of 2.5 kbar for sufficient time abolishes the infectivity [10]. Electron microscopy of the compressed samples shows no detectable dissociation but the subunits appear displaced from their normal positions by the pressure, as shown by bulges under the virus membrane. With the membrane-enveloped viruses, the membrane glycoprotein undergoes a conformational change induced by pressure similar to the receptor-activated conformational change [85]. The finding that influenza and HIV envelope glycoproteins undergo pH- or receptor-activated conformational changes indicate that their native states are metastable. For influenza virus, this change is usually referred as the spring-loaded model, in which the fusion-peptide region is thought to insert into the target membrane at an early step of the fusion process [80–82]. For influenza, low pH elicits the spring-loaded mechanism.

We find that influenza virus subjected to pressure exposes hydrophobic domains as determined by tryptophan fluorescence and by the binding of bis-8-anilino-1-naphthalenesulfonate (bis-ANS), a well-established marker of the fusogenic state in influenza virus [85]. Pressure also produced an increase in the fusion activity at neutral pH as monitored by fluorescence resonance energy transfer using lipid vesicles labeled with fluorescence probes [85]. These data indicate the change of the labile native state of the envelope complex to a more stable, mimicking the fusion-active conformation.

7. Pressure inactivation of viruses and other microorganisms

Virus inactivation by hydrostatic pressure has been evaluated with a view toward two potential applications: vaccine development and virus sterilization [10–14,85–89]. Concerning viral vaccines, there are basically three types of immunization strategies: use of live (attenuated) particles; use of killed (inactivated) whole-virion particles; and use of subunit vaccines (hepatitis B). Immunization is the most efficient way of preventing infectious diseases in animals and humans [90]. The antibodies against pressurized virus particles were as effective as those against the intact viruses when measured by their neutralization titer in plaque reduction assay [10,12,14,86]. Pressure induced the formation of inactive rotavirus [86], the most important viral agent that causes gastroenteritis responsible for millions of fatal diarrhea in children from developing countries. The new conformation of the pressurized particle did not result in loss of immunogenicity. Pressure alters the receptor-binding protein VP4 by triggering changes similar to those produced when the virus interacts with target cells [86]. As the changes in VP4 conformation caused by pressure occur prior to virus exposure to target cells, it leads to noninfectious particles and may lead to the exposure of previously occult epitopes, important for vaccine development.

Recently, high pressure has also been used to inactivate the bacteria leptospiras [91], the causative agent of important zoonotic and human diseases. *Leptospira interrogans* serovar *hardjo* was subjected to different hydrostatic pressures. Electron microscopy showed dislocation of the outer membrane, partial loss of the helical shape and extrusion of the axial filament from the cytoplasmic cylinder of the pressurized leptospiras. When the pressure-treated leptospiras were inoculated into rabbits they were highly immunogenic and are now being evaluated to formulate a vaccine.

The high titers of the neutralizing antibodies elicited by pressure-inactivated viruses indicate that hydrostatic pressure can be used to prepare whole-virus immunogens. Effective immunization against viruses requires presentation of the whole virus particle to the immune system. The employment of high pressure to prepare antiviral vaccines may have important advantages over other methods such as attenuation, chemical inactivation or isolated subunit vaccines. Attenuated live viruses can revert after a certain time and may cause the disease that they are intended to prevent, or worsen the real disease. Immunization with isolated subunits has several problems, especially because the immune system recognizes the isolated antigen less effectively than the whole virus. A reliable physical method, such as high pressure, to prepare killed vaccines should not have the same problems.

The reason why the pressurized viruses maintain the immunogenic potential probably resides in the fact that the structural changes are very subtle. As we discussed above pressure treatment seems to mimic the changes that are produced when viruses bind to cellular receptors. We found that pressure-inactivated VSV attaches to the cellular membrane, but it is not internalized by endocytosis [92]. The most reasonable explanation is that high pressure leads the envelope protein to the fusion conformation. This would be similar to the inhibition of an enzyme by a transition-state analog; the difference here is that a physical tool, pressure, freezes the viral protein in a conformation that cannot bind to the receptor preventing progression to endocytosis. Therefore, the other steps of the infection cycle are compromised. However, the attachment of inactivated virus, such as VSV, to the host cell [92] in a nonproductive way may be crucial to evoke immune response. First, because new epitope might be exposed which leads to efficient production of neutralizing antibodies; second, because the particle, although not infectious, can be processed by the cell resulting in a cellular response ($CD8^+$ cytotoxic T lymphocytes; CTLs).

The schematic representation for picornaviruses in Fig. 6 can be extended to enveloped viruses, the difference is that the main change after pressure release occurs in the membrane protein. The black circles represent the epitopes that seem to appear in the pressure-induced fusion-active state.

The potential of using fusion complexes for HIV vaccine development has been evaluated [93]. Using the idea that epitopes with superior immunogenicity might be exposed or created as HIV-1 begins to fuse with cell membranes, they tested fusion complexes for their ability to induce neutralizing antibodies in mice. Fusion-competent HIV vaccine immunogens were generated by fixing with formaldehyde the virus–cell complex. This preparation of fixed virus and whole cells elicited antibodies capable of neutralizing infectivity of 23 of 24 primary HIV isolates. The authors suggest that these fusion-dependent immunogens may lead to a broadly effective HIV vaccine. However, the major problem of their approach is the unclean preparation of a mixture of viruses and cells. Since high pressure is able to produce the same fusion-intermediate state, at least in some of the virus studied, it is a cleaner and more controllable way to produce antiviral vaccines.

8. Conclusions and perspectives

The increasing use of structural methods, such as NMR, to characterize the pressure-denatured state has placed pressure as one of the more promising techniques to unravel the missing links in the 'protein folding problem'. Pressure has permitted to characterize the high degree of plasticity of proteins. Recent detailed studies on the kinetics of pressure unfolding and folding [94,95] together with several theoretical approaches have increased our understanding of the pressure effects. Furthermore, the isolation of folding intermediates is crucial to the understanding of protein misfolding and protein aggregation, which have been recently tackled by pressure. The population of an amyloidogenic intermediate without proceeding to aggregation is a unique property of pressure, which opens the prospect to characterize the structure of the amyloidogenic form. The combination of several pressure approaches allows dissecting the role of packing and cavities in folding and in the assembly of multimolecular structures, such as protein–DNA complexes and viruses.

At the molecular level, the pressure/cold denaturation of proteins, protein–DNA complexes and viruses indicates that these processes are accompanied by exposure to the solvent of a substantial proportion of nonpolar side chains. Against expectation, most of the interactions that lead to protein folding, DNA recognition and virus assembly are entropy driven, which means that they are dominated by weak interactions (nonpolar) and less contribution of polar or H-bonding interactions which would normally be expected to be driven by changes in ΔH.

Our model supports the view that DNA recognition and virus assembly involve large conformational changes characterized mainly by decrease in the contacts of nonpolar side chains with water – increase in 'hydrophobic bonds'. Weber [96] formulated an elegant theory that attributes the entropy-driven character of protein associations to the large number of weak protein bonds rather than to solvent–solvent interactions. Weber's proposal takes into account the distribution of protein conformations that result from the dynamic nature of these macromolecules. The classical theory of hydrophobic bond [97] and Weber's theory can equally explain the cold denaturation of proteins, protein–DNA complexes and viruses. In both theories, the net result in the production of entropy-driven complexes is solvent displacement and formation of nonpolar interactions (London dispersion forces). However, Weber's theory assigns a more active participation of the weak interactions for the generation of the increase in entropy indicating a higher degree of specificity for these interactions.

The reduced stability of virus particles at high pressure and low temperature suggests the potential application to inactivate viruses of medical importance. We have found that several viruses are inactivated by pressure at room temperature. The substantial evidence that high pressure traps viruses in the 'fusion intermediate states' (found with alphaviruses, influenza, retroviruses, etc.), not infectious but highly immunogenic, is very promising for vaccine development. Pressure-induced population of fusion-active states can be utilized in the development of new antiviral vaccines and drugs. The viruses stable against pressure at room temperature are inactivated when temperature is decreased under pressure. These studies are important not only as a potential approach to produce antiviral vaccines as well as to sterilize biological products, such as blood, plasma and derivatives.

Acknowledgements

This work was supported in part by an international grant from the Howard Hughes Medical Institute to J.L.S., and by grants from Conselho Nacional de Desenvolvimento Científico e Tecnológico (CNPq), Programa de Núcleos de Excelência (PRONEX), Fundação de Amparo à Pesquisa no Estado do Rio de Janeiro (FAPERJ), Financiadora de Estudos e Projetos (FINEP) of Brazil to J.L.S and D.F.

References

[1] C.O. Pabo, R.T. Sauer, Annu. Rev. Biochem. 61 (1992) 1053–1095.
[2] P.H. von Hippel, Science 263 (1994) 769–770.
[3] G. Weber, Protein Interactions, Chapman and Hall, New York, 1992.
[4] G. Weber, H.G. Drickamer, Q. Rev. Biophys. 16 (1983) 89–112.
[5] J.L. Silva, G. Weber, Annu. Rev. Phys. Chem. 44 (1993) 89–113.
[6] V.V. Mozhaev, K. Heremans, J. Frank, P. Masson, C. Balny, Proteins 24 (1996) 81–91.
[7] J. Jonas, A. Jonas, Annu. Rev. Biophys. Biomol. Struct. 23 (1994) 287–318.
[8] J.L. Silva, D. Foguel, C.A. Royer, Trends Biochem. Sci. 26 (2001) 612–618.
[9] K.J. Frye, C.A. Royer, Protein Sci. 10 (1998) 2217–2222.
[10] J.L. Silva, P. Luan, M. Glaser, E.W. Voss, G. Weber, J. Virol. 66 (1992) 2111–2117.
[11] E. Jurkiewicz, M. Villas-Boas, J.L. Silva, G. Weber, G. Hunsmann, R.M. Clegg, Proc. Natl. Acad. Sci. USA 92 (1995) 6935–6937.
[12] A.C. Oliveira, D. Ishimaru, R.B. Goncalves, T.J. Smith, P. Mason, D. Sa-Carvalho, J.L. Silva, Biophys J. 76 (1999) 1270–1279.
[13] A.C. Oliveira et al., in: R. Winter, J. Jonas (Eds.), High Pressure Molecular Science, Vol. 358, NATO ASI series E, Kluwer Academic Publishers, Dordrecht, 1999, pp. 497–513.
[14] S.M. Tian, K.C. Ruan, J.F. Qian, G.Q. Shao, C. Balny, Eur. J. Biochem. 267 (2000) 4486–4494.
[15] A.E. Eriksson, W.A. Baase, X.-J. Zhang, D.W. Heinz, M. Blaber, E.P. Baldwin, B.W. Matthews, Science 255 (1992) 178–183.
[16] R.F. Tilton Jr., G.A. Petsko, Biochemistry 27 (1988) 6574–6582.
[17] C.E. Kundrot, F.M. Richards, J. Mol. Biol. 193 (1987) 157–170.
[18] C. Lee, S.H. Park, M.Y. Lee, M.H. Yu, Proc. Natl. Acad. Sci. USA 97 (2000) 7727–7731.
[19] D. Foguel, C.M. Teschke, P.E. Prevelige, J.L. Silva, Biochemistry 34 (1995) 1120–1126.
[20] P.C. De Souza, R. Tuma, P.E. Prevelige, J.L. Silva, D. Foguel, J. Mol. Biol. 287 (1999) 527–538.
[21] L.P. Gaspar, A.F. Terezan, A.S. Pinheiro, D. Foguel, M.A. Rebello, J.L. Silva, J. Biol. Chem. 276 (2001) 7415–7421.
[22] A.C. Oliveira, L.P. Gaspar, A.T. Da Poian, J.L. Silva, J. Mol. Biol. 240 (1994) 184–187.
[23] G. Hummer, S. Garde, A.E. Garcia, M.E. Paulaitis, L.R. Pratt, Proc. Natl. Acad. Sci. USA 95 (1998) 1552–1555.
[24] J.L. Silva, C.F. Silveira, A. Correia, L. Pontes, J. Mol. Biol. 223 (1992) 545–555.
[25] X.D. Peng, J. Jonas, J.L. Silva, Proc. Natl. Acad. Sci. USA 90 (1993) 1776–1780.
[26] X.D. Peng, J. Jonas, J.L. Silva, Biochemistry 33 (1994) 8323–8329.
[27] D.P. Nash, J. Jonas, Biochemistry 36 (1997) 14375–14383.
[28] D. Foguel, J.L. Silva, G. Prat-Gay, J. Biol. Chem. 273 (1998) 9050–9057.
[29] L.M.T.R. Lima, D. Foguel, J.L. Silva, Proc. Natl. Acad. Sci. USA 97 (2000) 14289–14294.
[30] J. Zhang, X. Peng, A. Jonas, J. Jonas, Biochemistry 34 (1995) 8631–8641.
[31] K. Ruan, L. Lange, F. Meersman, K. Heremans, C. Balny, Eur. J. Biochem. 265 (1999) 79–85.
[32] S.E. Bondos, S. Sligar, J. Jonas, Biochim. Biophys. Acta 1480 (2000) 353–364.
[33] K. Inoue, H. Yamada, K. Akasaka, C. Herrmann, W. Kremer, T. Maurer, R. Doker, H.R. Kalbitzer, Nat. Struct. Biol. 7 (2000) 547–550.
[34] R. Kitahara, S. Sareth, H. Yamada, E. Ohmae, K. Gekko, K. Akasaka, Biochemistry 39 (2000) 12789–12795.
[35] N. Hillson, J.N. Onuchic, A.E. Garcia, Proc. Natl. Acad. Sci. USA 96 (1999) 14848–14853.
[36] K.J. Frye, C.A. Royer, Protein Sci. 6 (1997) 789–793.
[37] J. Saad-Nehme, J.L. Silva, J.R. Meyer-Fernandes, Biochim. Biophys. Acta 1546 (2001) 164–170.
[38] J.N. Webb, S.D. Webb, J.L. Cleland, J.F. Carpenter, T.W. Randolph, Proc. Natl. Acad. Sci. USA 98 (2001) 7259–7264.
[39] R. Mohana-Borges, J.L. Silva, G. de Prat-Gay, J. Biol. Chem. 274 (1999) 7732–7740.
[40] R. Mohana-Borges, J.L. Silva, J. Ruiz-Sanz, G. de Prat-Gay, Proc. Natl. Acad. Sci. USA 96 (1999) 7888–7893.
[41] M.C. Suarez, S.S. Lehrer, J.L. Silva, Biochemistry 40 (2001) 1300–1307.
[42] A.W. Rietveld, S.T. Ferreira, Biochemistry 35 (1996) 7743–7751.
[43] E.H. Koo, P.T. Lansbury Jr., J.W. Kelly, Proc. Natl. Acad. Sci. USA 96 (1999) 9989–9990.
[44] S.E. Radford, C.M. Dobson, Cell 97 (1999) 291–298.
[45] B.M. Gorovits, P.M. Horowitz, Biochemistry 37 (1998) 6132–6135.
[46] D. Foguel, C.R. Robinson, P.C. de Sousa Jr., J.L. Silva, A.S. Robinson, Biotechnol. Bioeng. 63 (1999) 552–558.
[47] L. Smeller, P. Rubens, K. Heremans, Biochemistry 38 (1999) 3816–3820.
[48] A.D. Ferrao-Gonzales, S.O. Souto, J.L. Silva, D. Foguel, Proc. Natl. Acad. Sci. USA 97 (2000) 6445–6450.
[49] R.J. St. John, J.F. Carpenter, T.W. Randolph, Proc. Natl. Acad. Sci. USA 96 (1999) 13029–13033.
[50] J.W. Kelly, Curr. Opin. Struct. Biol. 8 (1998) 101–106.
[51] P.A. Hall, D.P. Lane, Curr. Biol. 7 (1997) R144–R147.

[52] B. Luisi, in: D.M.J. Lilley (Ed.), DNA–Protein: Structural Interactions, IRL Press, 1995, pp. 1–48.
[53] R.S. Spolar, M.T. Record Jr., Science 263 (1994) 777–784.
[54] S.M.A. Holmbeck, H.J. Dyson, P.E. Wright, J. Mol. Biol. 284 (1998) 533–539.
[55] C.R. Robinson, S.G. Sligar, Proc. Natl. Acad. Sci. USA 92 (1995) 3444–3448.
[56] T.W. Lynch, S.G. Sligar, J. Biol. Chem. 275 (2000) 30561–30565.
[57] C.A. Royer, A.E. Chakerian, K.S. Matthews, Biochemistry 29 (1990) 4959–4966.
[58] D. Foguel, J.L. Silva, Proc. Natl. Acad. Sci. USA 91 (1994) 8244–8247.
[59] J.L. Silva, D. Foguel, A.T. Da Poian, P.E. Prevelige, Curr. Opin. Struct. Biol. 6 (1996) 166–175.
[60] R.B. Macgregor Jr., Biopolymers 48 (1998) 253–263.
[61] J.L. Silva, C.F. Silveira, Protein Sci. 2 (1993) 945–950.
[62] J.W. Little, B. Kim, K.L. Roland, M.H. Smith, L.L. Lin, S.N. Slilaty, Methods Enzymol. 244 (1994) 266–284.
[63] R.H. Fogh, G. Ottleben, H. Rüterjans, M. Schnarr, R. Boelens, R. Kaptein, EMBO J. 13 (1994) 3936–3944.
[64] R. Mohana-Borges, A.B. Pacheco, F.J. Sousa, D. Foguel, D.F. Almeida, J.L. Silva, J. Biol. Chem. 275 (2000) 4708–4712.
[65] G.C. Walker, Mol. Cell 8 (2001) 486–487.
[66] Y. Luo, R.A. Pfuetzner, S. Mossimann, M. Paetzel, E.A. Frey, M. Cherney, B. Kim, J.W. Little, N.C.J. Strynadka, Cell 106 (2001) 585–594.
[67] R.S. Hegde, E.J. Androphy, J. Mol. Biol. 284 (1998) 1479–1489.
[68] J.E. Johnson, Proc. Natl. Acad. Sci. USA 93 (1996) 27–33.
[69] A.T. Da Poian, J.E. Johnson, J.L. Silva, Biochemistry 33 (1994) 8339–8346.
[70] A.T. Da Poian, A.C. Oliveira, J.L. Silva, Biochemistry 34 (1995) 2672–2677.
[71] P.E. Prevelige, J. King, J.L. Silva, Biophys. J. 66 (1994) 1631–1641.
[72] L.P. Gaspar, J.E. Johnson, J.L. Silva, A.T. Da Poian, J. Mol. Biol. 273 (1997) 456–466.
[73] P.C. De Souza, R. Tuma, P.E. Prevelige, J.L. Silva, D. Foguel, J. Mol. Biol. 287 (1999) 527–538.
[74] M. Leimkuhler, A. Goldbeck, M.D. Lechner, J. Witz, J. Mol. Biol. 296 (2000) 1295–1305.
[75] M.A. Lauffer, R.B. Dow, J. Biol. Chem. 140 (1941) 509–518.
[76] C.F. Bonafe, C.M. Vital, R.C. Telles, M.C. Goncalves, M.S. Matsuura, F.B. Pessine, D.R. Freitas, J. Vega, Biochemistry 37 (1998) 11097–11105.
[77] D. Foguel, C.M. Teschke, P.E. Prevelige, J. L Silva, Biochemistry 34 (1995) 1120–1126.
[78] R.R. Rueckert, in: B.N. Fields, D.M. Knipe, P.M. Howley et al. (Eds.), Fields Virology, 3rd Edition, Lippincott-Raven, Philadelphia, pp. 609–645.
[79] M.G. Rossmann, Protein Sci. 3 (1994) 1712–1725.
[80] C.M. Carr, P.S. Kim, Cell 73 (1993) 823–832.
[81] P.A. Bullough, F.M. Hughson, J.J. Skehel, D.C. Wiley, Nature 371 (1994) 37–43.
[82] J.J. Skehel, D.C. Wiley, Annu. Rev. Biochem. 69 (2000) 531–569.
[83] A.C. Oliveira, A.M.O. Gomes, F.C.L. Almeida, R. Mohana-Borges, A.P. Valente, V.S. Reddy, J.E. Johnson, J.L. Silva, J. Biol. Chem. 275 (2000) 16037–16043.
[84] A. Cooper, Proc. Natl. Acad. Sci. USA 73 (1976) 2740–2741.
[85] L.P. Gaspar, A.M.O. Gomes, A.C.B. Silva, A.P.D. Ano-Bom, W.D. Schwarcz, D. Foguel, J.L. Silva, (2001) in press.
[86] L. Pontes, Y. Cordeiro, V. Giongo, M. Villas-Boas, A. Barreto, J.R. Araujo, J.L. Silva, J. Mol. Biol. 307 (2001) 1171–1179.
[87] T. Nakagami, H. Ohno, T. Shigehisa, T. Otake, H. Mori, T. Kawahata, M. Morimoto, N. Ueba, Transfusion 36 (1996) 475–476.
[88] D.W. Bradley, R.A. Hess, F. Tao, L. Sciaba-Lentz, A.T. Remaley, J. Laugharn Jr., M. Manak, Transfusion 40 (2000) 193–200.
[89] L.P. Gaspar, A.F. Terezan, A.S. Pinheiro, D. Foguel, M.A. Rebello, J.L. Silva, J. Biol. Chem. 276 (2001) 7415–7421.
[90] B.R. Bloom, Science 272 (1996) 1888–1890.
[91] C.C. Silva, V. Giongo, A.J. Simpson, E.R. Camargos, J.L. Silva, M.C. Koury, Vaccine 19 (2001) 1511–1514.
[92] A.T. Da Poian, A.M.O. Gomes, R.J.N. Oliveira, J.L. Silva, Proc. Natl. Acad. Sci. USA 93 (1996) 8268–8273.
[93] R.A. La Casse, K.E. Follis, M. Trahey, J.D. Scarborough, D.R. Littman, J.H. Nunberg, Science 283 (1999) 357–362.
[94] G.J. Vidugiris, J.L. Markley, C.A. Royer, Biochemistry 34 (1995) 4909–4912.
[95] J. Woenckhaus, R. Kohling, P. Thiyagarajan, K.C. Littrell, S. Seifert, C.A. Royer, R. Winter, Biophys. J. 80 (2001) 1518–1523.
[96] G. Weber, J. Phys. Chem. 27 (1993) 7108–7115.
[97] W. Kauzmann, Adv. Protein. Chem. 13 (1959) 1–63.

ELSEVIER

Biochimica et Biophysica Acta 1595 (2002) 266–276

BIOCHIMICA ET BIOPHYSICA ACTA
BBA
www.bba-direct.com

Review

The interactions of nucleic acids at elevated hydrostatic pressure

Robert B. Macgregor Jr. *

Department of Pharmaceutical Sciences, Faculty of Pharmacy, University of Toronto, 19 Russell Street, Toronto, ON, Canada M5S 2S2

Received 3 September 2001; received in revised form 5 November 2001; accepted 6 November 2001

Abstract

The application of elevated hydrostatic pressure on the order of a few thousand bar provides insight into the molecular forces responsible for stabilizing the conformations and non-covalent interactions of biological molecules in aqueous solution. In particular, the parameters derived from these studies have enabled researchers to glean information regarding the importance of hydration in the energetics and kinetics of these systems. This review presents data concerned with the application of hydrostatic pressure to study the thermodynamics, kinetics, and structure of nucleic acids and the interactions between nucleic acids and proteins, enzymes, and drugs. These complexes often form extremely stable non-covalent complexes in which electrostatic interactions play an important role. The sensitivity of these interactions to pressure offers a valuable experimental tool for investigating the energetics of the complexes.

Keywords: Hydrostatic pressure; DNA–protein interaction; DNA; DNA–drug interaction; Hydration; DNA stability

1. Introduction and scope

The effect of hydrostatic pressure on a system can be summed up by paraphrasing Le Chatelier, 'squeeze it and it gets smaller'. Because of the very low compressibility of systems composed of biological molecules in water, the mechanism by which the system reduces its volume (i.e., 'gets smaller') offers interesting insight in to the forces stabilizing the structures and interactions of these molecules. This review presents data concerning the influence of hydrostatic pressure on nucleic acids and non-covalent complexes they form with drugs and proteins. The focus will be on the complexes formed with DNA although those formed with RNA will be discussed. Other volumetric data on DNA and DNA–ligand complexes will be referred to in those cases where they can be compared with high-pressure experiments. The majority of the research articles on the effect of high pressure on nucleic acids and nucleic acid-ligand complexes have focused on either the thermodynamics of the complex or the effect of pressure on the conformational stability of nucleic acid polymer or oligomer. There has been very little work on the effect of pressure on the structure of these systems. The data concerning the effect of pressure on the conformational stability of nucleic acids have been recently reviewed [1,2]. Although new results not published at the time of the two previous reviews will be presented, only an overview of the effect of pressure on the conformational stability of nucleic acids will be discussed here.

The emphasis of this review is the effect of hydro-

* Fax: +1-416-978-8511.
E-mail address: rob.macgregor@utoronto.ca
(R.B. Macgregor Jr.).

0167-4838/02/$ – see front matter
PII: S0167-4838(01)00349-1

static pressure on thermodynamics, kinetics, and structure of non-covalent complexes with nucleic acids. The effect of osmotic pressure will be discussed only in those cases where it aids in understanding the data acquired under hydrostatic pressure. This review will not discuss the effect of pressure on molecules that bind to nucleic acids if the nucleic acid was not included in the experiment.

Interest in the effect of hydrostatic pressure on biological molecules began in the late nineteenth century with the discovery of living organisms 6000 m below the surface of the ocean at a pressure of approximately 600 bar[1] [3,4]. This discovery led to the development of qualitative and quantitative experiments to study the effect of pressure on biological and biochemical systems. The comparative biochemistry of barophilic organisms has motivated a great deal of pressure research [5,6]. Although the theoretical framework for interpreting the effect of pressure on chemical equilibria existed, the application of this description to extracting thermodynamic data regarding biological equilibria required the relatively recent development of appropriate instrumental techniques.

In comparison to other techniques used to study the physical chemistry of nucleic acids, the effect of hydrostatic pressure has received scant attention. Of the work that has been reported, the overwhelming majority of studies have been directed at investigating the conformational stability of higher order nucleic acid structures such as double and triple helices. Furthermore, most of these studies were carried out at pressures less than 4,000 bar although under conditions of static pressure, the covalent structure of nucleic acids is stable up to at least 12 kbar [7]. Rapid pressure changes do disrupt the chemical structure of polymeric nucleic acids. The large, shearing forces developed by the rapid pressure drop achieved by a French press is a useful method to break long pieces of genomic DNA into shorter strands [8]. Given the behavior of other molecules in water, it is likely that the covalent structure of DNA and RNA are stable at static pressures as great as 40 or 50 kbar; however, the pressure domain from ~4000 bar to the pressure-induced freezing of water at ~15 kbar remains essentially uninvestigated.

[1] 1 bar = 0.1 MPa = 0.9678 atm, 1 kbar = 1000 bar.

2. The conformational stability of DNA at high pressure

The effect of hydrostatic pressure on the stability of helical structure of double- and triple-stranded DNA has been recently reviewed [1,2]. In all studies, a spectroscopic property of DNA dissolved in a buffered aqueous solution is measured while the sample is subjected to pressure. Although the change in light absorption at 260 nm accompanying the helix-to-coil transitions of nucleic acids is the parameter most often monitored, other spectroscopic variables have been used, e.g., infrared absorption, and optical rotation. Generally, the spectroscopic changes are used to obtain the temperature at which the base pairing interactions of the nucleic acid are disrupted and the strands separate. The spectroscopic change and the physical event underlying it occur over a narrow temperature range and can be treated as a phase transition. The midpoint of the transition from the helical form to the coil form is referred to as the *transition temperature* or *melting temperature*, T_m. The T_m is usually measured as a function of hydrostatic pressure, in the present instance, or as a function of other experimental variables. The observed changes in T_m with hydrostatic pressure can be related to thermodynamic parameters by the Clapeyron equation, $dT_m/dP = T_m\Delta V/\Delta H$, where ΔV and ΔH are the molar volume and enthalpy change of the transition, respectively. However, it is also possible to analyze the data as an equilibrium process by calculating the equilibrium constant for the helix–coil transition at each temperature throughout the transition. From such an analysis, one can extract the equilibrium constant describing the process at each temperature.

The value of dT_m/dP has been studied for several different types of DNA, double-stranded DNA with natural sequences, non-random sequence synthetic polymers, and oligonucleotides of different lengths. These molecules have been investigated under different salt concentrations and in solution with different types of cations. Under standard experimental conditions, the molar volume change (ΔV) for transition from double-helical DNA to the complementary single strands is small and positive, typical values range from +1 to +3 ml/mol. Positive values of ΔV imply that the volume of the system that includes the double-stranded polymer, water, and cations is less than the volume of the system containing the single-stranded species. The positive value of ΔV also means that the double-stranded structure is stabilized by elevated hydrostatic pressure. Enhanced stacking of the hydrophobic bases with increasing pressure is responsible for the stabilization. The effect of pressure on the helix-to-coil transition temperature increases with salt concentration, that is, dT_m/dP becomes larger.

Single-stranded nucleic acid polymers do not exist as random coils in solution. They exhibit transient structure due to the stacking of adjacent bases. Suzuki et al. [9] reported the effect of pressure on the absorption spectra of heat denatured salmon sperm DNA in 150 mM NaCl, pH 7.8. They observed increasing hypochromicity with increasing pressure throughout the wavelength range they measured. Their results are consistent with the formation of increased stacking of the unpaired bases at higher pressure.

3. Phase diagram of nucleic acid stability

The most important new result concerning the effect of pressure on the stability of double-stranded nucleic acid polymers to arise since the recent previous reviews is the finding that the value of dT_m/dP can be positive or negative [10]. This experimental observation arose as a consequence of theoretical and experimental analyses. A theoretical pressure–temperature phase diagram describing the stability of double-stranded nucleic acids was constructed based on the observation that the there is a small but non-negligible change in the heat capacity that accompanies the transition. Thus, the enthalpy change of the helix-to-coil transition is temperature dependent. Incorporation of the temperature dependence of the enthalpy and the compressibility and volume changes that accompany the transition into an equation that describes the general stability of the polymer led to the development of the phase diagram for the stability of polymeric nucleic acids. The phase diagram describes the helical state (single- or double-stranded) of the nucleic acid to hydrostatic pressure, at different temperatures for several values of the transition temperature, T_m. The T_m is a function of the sequence of the nucleic acid polymer and the salt concentration; as such, it is an experimental variable. The stability of the double helix falls into several clearly defined regions in pressure–temperature space (Fig. 1). The model predicted that at moderate pressures, the value of dT_m/dP should be a monotonic function of the helix-to-coil transition temperature that changes from negative to positive values with the cross-over, i.e., the temperature at which $dT_m/dP = 0$ at approximately 50°C. Polymers with T_m values below 50°C should be destabilized by hydrostatic pressure. All previous studies had explored the region where dT_m/dP is greater than zero; however, the region where this parameter is negative had not been observed. The prediction of the analysis was borne out by studying the effect of pressure on the helix–coil transition of polymers with T_m values below ~50°C. By changing the T_m of poly(A)poly(U) and poly(dA)poly(U) through changes in the salt concentration, it was shown that value of dT_m/dP could be positive or negative.

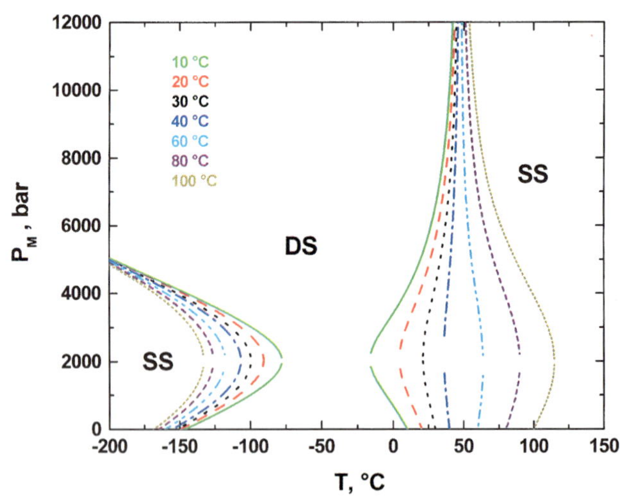

Fig. 1. A phase diagram for the stability of double-stranded nucleic acid polymers. The regions marked DS and SS indicate temperature–pressure regions where the double-stranded polymer and single-stranded polymer are stable, respectively. The different lines show the dependence of the phase boundaries on the T_m of the polymer at atmospheric pressure. From Fig. 2 of [10].

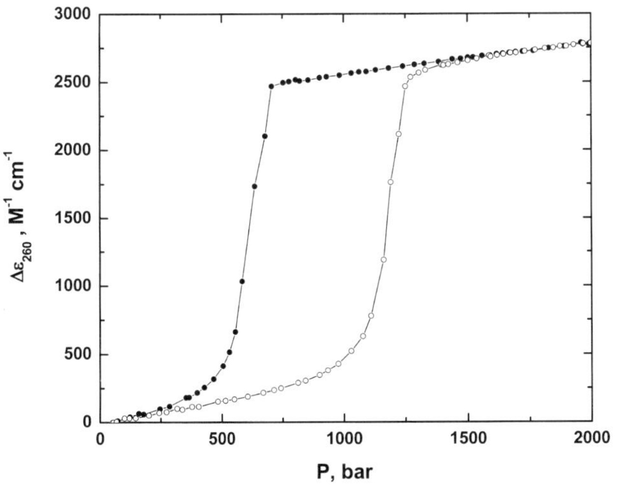

Fig. 2. A pressure-induced helix-to-coil transition. The molar extinction coefficient at 260 nm of poly(dA)poly(U) was measured as a function of pressure at 20 (○) and 25°C (●). From Fig. 5 of [10].

The value of dT_m/dP for the hybrid duplex polymer poly(dA)poly(U) was sufficiently negative that the helical structure could be disrupted by pressures less than 2000 bar. By increasing the hydrostatic pressure at constant temperature, it is possible to induce a helix-to-coil transition in this polymer, Fig. 2. This is the first experimental observation of this phenomenon. The P_m, the pressure at the midpoint of the pressure-induced transition, increased as the T_m increased, as would be expected based on the monotonic nature of the change of dT_m/dP with T_m.

Despite the success of the phase diagram to predict the effect of pressure on the stability of double-stranded polymeric nucleic acids, it clear that the model does not take into account all of the thermodynamic factors important in the stability of nucleic acids. The most obvious difficulty is the effect of pressure on the T_m of oligonucleotides. The value of dT_m/dP for a series of oligonucleotides, $dA_n{\bullet}dT_n$ is positive for $n = 11$, 15, and 19, although the T_m of these oligonucleotides were between 24 and 35°C [11]. There was a decrease in dT_m/dP with decreasing chain length attributed to the increasing contribution of the ends of the oligonucleotides; however, dT_m/dP remained positive under all conditions. Further studies will be necessary to understand whether these oligonucleotides are anomalies, if end effects are indeed important, or if some other effect is responsible for the observations.

4. Non-covalent complexes with nucleic acids

Biological regulation occurs through non-covalent interactions between DNA and proteins. On the level of the organism, the genomic sequence is not random and protein-DNA interactions may occur in a sequence specific and sequence independent manner. The physical processes and molecular mechanisms underlying the ability of specific proteins to recognize specific DNA sequences has been and remains a subject of intense research interest.

The fact that DNA is a polyanion places structural requirements on the molecules that interact with it. Thus, independent of whether the interaction is sequence specific or not, the negative charge at each phosphate group on DNA must be accommodated by the protein. Although the interface of the protein that interacts with DNA is not necessarily as highly positively charged as DNA there are a large fraction of charged and polar amino acids at the interface. In the non-complexed state, these amino acids and the DNA are hydrated. Upon formation of a non-covalent complex, the majority, and in some cases all, of the water molecules are lost into the solvent.

Many small molecules, certain drugs and dyes, also form non-covalent complexes with DNA. Small molecules exhibit two principle modes of interaction with DNA, namely intercalation and groove binding. Intercalating molecules have a multi-ring aromatic structural moiety. This aromatic moiety inserts itself between the bases of the double-stranded structure. The groove-binding motif involves the formation of hydrogen bonds between the drug and the chemical groups of the exposed to the solvent in the minor groove of the nucleic acid double helix. In both of these modes of binding, the drug must undergo dehydration prior to formation of the complex. In either mode of interaction, the hydration of the DNA also changes upon binding. To a first approximation, intercalation is not sequence specific whereas groove binding is sequence specific.

Investigations of the thermodynamics and kinetics of the non-covalent complexes formed by proteins and dyes with DNA are considered very important for understanding the mechanisms by which these molecules recognize their binding sites in vivo. There have been many studies on the thermodynamics and kinetics of the interactions of small molecules (dyes and drugs) over the past 30 years (for recent reviews see [12,13]). One of the original motivations for studying the interactions of small molecules with nucleic acids was to gain understanding of the factors important in the stabilizing nucleic acid–protein complexes. The recognition of the importance of water in the overall thermodynamics of these complexes has increased the interest in studying the pressure dependence of the interactions.

5. Small molecules

The first publication on the effect of pressure on the interaction between a dye and DNA was reported by Suzuki et al. in 1972 [9]. They studied the absorption spectrum of acridine orange bound to native and denatured salmon sperm DNA at pressures up to 4650 bar and found only minimal effects. Under the conditions of their experiments, the dye apparently remained bound at high pressure.

Since this initial publication, studies of the effect of hydrostatic pressure on the complexes formed between small molecules and nucleic acids have focused on the equilibrium and kinetics properties of the intercalator, ethidium bromide. The kinetics of the interaction of ethidium bromide with poly[d(A-T)] and poly[d(G-C)] were studied using a repetitive pressure-jump technique [14]. These were not high-pressure studies per se, in that the pressures of the jumps were small ($\sim$20 bar) and the static pressure was low ($\sim$100 bar). However, the kinetic amplitudes were used to calculate the molar volume change of intercalation. The value found using this kinetic technique agreed well with the value found in a later study that employed equilibrium high-pressure measurements [15]. These are summarized in Table 1. The results show that the volume change of intercalation is a function of the structure of the intercalator, the sequence of the DNA binding site, and the ionic strength. The state of hydration of the DNA and the dye change upon formation of the complex. The results also show that equilibrium and kinetics techniques yield similar results, which is encouraging.

A further advantage of using pressure techniques to probe biochemical equilibrium processes lies in the fact that it opens up experimental domains that cannot be studied using other methods. For example,

Table 1
Molar volume change of intercalation into dAdT polymers[a,b]

	[NaCl] (mM)	ΔV (ml/mol)	
		Ethidium	Propidium
Poly[d(A-T)]	0	−7.2	
	50	−13.0	−8.3
	100	−15.0[c]	−8.9
Poly(dA)poly(dT)	0	+4.0	+5.6
	50	+4.5	+6.8

[a]From [15].
[b]In 20 mM Tris–HCl, pH 7.2, 21°C.
[c]From [14].

pressure jump was used to study the binding of ethidium to poly[d(A-T)] in solutions containing organic solvents [16]. The kinetics of binding in the mixed solvents showed that the kinetics were consistent with the formation of an open state in the DNA structure prior to drug binding. The fraction of base pairs in an open conformation that would allow intercalation agreed well with other results on the extent of 'breathing' of DNA polymers. The second example of a novel use of pressure in the investigation of non-covalent drug–DNA complexes is a pressure-jump to study the kinetics of ethidium binding to poly[d(G-C)] in an agarose gel [17]. The concentration of agarose was sufficiently high to preclude diffusive motion of the polymer; however, the diffusion of the much smaller ethidium molecule was unaffected, to a first approximation. The results showed that, as expected, the presence of the gel had no effect on the kinetics of intercalation. The diffusion of the ethidium molecule determines the rate of molecular encounters between the dye and the DNA.

6. Proteins and enzymes

Although it is still not widely used, hydrostatic pressure is an important variable in the investigation of protein–DNA interactions. Firstly, protein–DNA complexes involve the formation of ionic interactions. Charge neutralization in aqueous solution generally proceeds with a positive volume change [18]. The sign of the volume change arises from the loss of water in the hydration shell of the interacting molecules. These hydrating waters are generally of higher density than bulk water due to electrostriction, the electrostatic interaction between the water dipole and the charge on the protein interface. Other types of interactions are also involved in the stabilization of protein–DNA complexes, specifically, hydrophobic interactions, play an important role; however, the extensive charge neutralization appears to make DNA protein complexes dissociate under hydrostatic pressure. In addition to the direct ionic contacts made by the protein and DNA, many DNA-binding proteins and most enzymes that catalyze reactions with DNA require a divalent metal ion in solution. This metal ion is often intimately involved in the mechanism of the site recognition or enzymatic catalysis. The ionic interactions between the protein–DNA complex and the divalent metal ion are also disrupted by hydrostatic pressure.

Many sequence specific DNA–protein complexes have extremely large binding constants; dissociation of these complexes with hydrostatic pressure provides one of the few experimental methods to reversibly disrupt the complex and thereby study the thermodynamics of these interactions. The generally observed dissociation of DNA-binding proteins at elevated pressures permits the dissociation constant of the complex to be shifted to a value at which analytical techniques, such as fluorescence spectroscopy, are useful.

The most informative measurements concerning the behavior of protein–DNA complexes under pressure are those in which the state of the complex is probed while the protein–nucleic acid system is under pressure. This can be achieved either by monitoring spectroscopic changes, measuring an enzymatic activity, or other parameter. By exploiting these techniques, the interaction of DNA with several enzymes and proteins has been studied at high pressure. Nevertheless, the investigation of these complexes by hydrostatic pressure has not yet found broad application.

The first study of a protein-nucleic acid system under high pressure appeared in 1972 [19]; these authors reported the effect of pressure on the activity of ribonuclease. They observed a reversible increase in the enzymatic activity with pressure. The enhanced activity at high pressure was attributed to the formation of a charged intermediate in the catalytic pathway.

Since the publication of this initial work, several laboratories have employed hydrostatic pressure as a method for investigating proteins and enzymes that interact with DNA. The first investigation into the effect of hydrostatic pressure on a restriction endonuclease demonstrated that the activity of *Eco*RI decreased with pressure and enzymatic activity was absent at 2000 bar [20]. This inhibition was reversible upon return to atmospheric pressure. The sequence specificity of the enzyme for its canonical recognition sequence, GAATTC, was maintained at high pressure and for the enzyme after return to atmospheric pressure.

Since these initial investigations, there have been several other reports on the effect of hydrostatic pressure on the interactions and specificity of DNA-binding enzymes. Sligar's group in particular has presented interesting and important data comparing the effect of osmotic pressure and hydrostatic pressure on the enzymatic activity and specificity of the restriction endonucleases *Eco*RI, *Eco*RV, *Bam*-HI, and *Pvu*II.

The effect of changes in the osmotic pressure of the solution have been exploited for the purposes of investigating changes in molecular hydration arising from the formation of a non-covalent complex (for a recent review see [21]). Significant osmotic pressures are created through the addition of uncharged osmolytes. Although large uncharged polymers, such as polyethylene glycol or dextran, may be used to increase the osmotic pressure of the solution, their molecular mass precludes achieving very high osmotic pressures. However, by using small, uncharged organic molecules, osmotic pressures greater than 100 bar may be achieved. Although the presence of an osmolyte in the solution changes the binding constant of the interaction, the osmolytes do not interact physically with the DNA or the protein. To rule out the possibility of weak interactions, several structurally unrelated osmolytes are used in each study. The presence of the osmolytes decreases the water activity in the solution and thereby provides a physical method for observing the role of water in the conformations, binding, and catalytic activity of DNA, enzymes, proteins and their ligands.

The use of small organic molecules, however, has an additional effect on the activity of restriction endonucleases. Although these enzymes cleave double-stranded DNA with extreme specificity under defined conditions of salt concentration, the type of salt, pH, etc., the specificity of many of these enzymes is relaxed if the solution contains organic solvents or certain metal ions. This is referred to as 'star' activity. For example, the canonical recognition site of *Pvu*II is CAGCTG; it will cleave both strands of DNA between the G and C residues. In the presence of organic co-solvents, the relaxed specificity leads to cleavage within the following sequences, CNGCTG, CANCTG, CAGNTG, CAGCNG, and in sequences having A or G in the 5′ position or T or C in the 3′ position [22]. *Bam*HI and *Pvu*II display a linear increase in star activity with osmotic pressure for low molecular mass osmolytes.

In their studies in which osmotic stress and hydrostatic pressure have been studied, Sligar's group has found that hydrostatic pressure reverses the effect of osmotic pressure. Presumably, hydrostatic pressure leads to a re-hydration of the molecular surfaces that are dehydrated by the presence of the osmolyte. They also report a general decrease in the activity of these enzymes with increasing hydrostatic pressure.

Robinson and Sligar have studied the balance between hydrostatic pressure and osmotic pressure for *Bam*HI, *Pvu*II, *Eco*RI, and *Eco*RV [23,24]. Hydrostatic pressure caused a moderate decrease in enzymatic activity of these enzymes, Table 2. The authors attribute the general decrease in enzymatic activity at the canonical sites to dissociation of the protein–DNA complex and the dissociation of the two subunits of the enzyme. (The four restriction endonucleases studied are dimeric proteins with identical subunits.) Robinson and Sligar report that the star activity of *Bam*HI and *Pvu*II, similar to that of *Eco*RI, correlates directly with osmotic pressure and that hydrostatic pressure reverses this star activity. Endonuclease star activity can also be induced by

Table 2
Inactivation of restriction endonucleases by hydrostatic pressure[a]

Enzyme	Percent of enzymatic activity at atmospheric pressure		
	775 bar	1030 bar	1650 bar
*Pvu*II	100	98	90
*Bam*HI	100	96	79
*Eco*RV	94	82	41

[a]From [23].

changes in the ionic strength of the solution, pH, and the type of ions in the solution. It is interesting that hydrostatic pressure only reversed the star activities induced by small, uncharged, organic molecules implying that different mechanisms lead to the same star activities. No star activity was observed for *Eco*RV upon the addition of osmolytes. Furthermore, the *Eco*RV star activity caused by the pH and ionic strength was unaffected by increasing hydrostatic pressure.

The magnitude of the changes in the kinetics parameters reported by Robinson and Sligar is not large enough to account for the loss of activity at 2000 bar observed in the study by Macgregor [20]. In the work by Robinson and Sligar, the highest hydrostatic pressure used was 500 bar whereas the earlier report employed higher pressure [23]. It is possible that different molecular events occur in the two pressure ranges. Regardless of the origin of the apparent discrepancy in the data, it is evident that hydrostatic pressure causes a general decrease in the enzymatic activity of restriction endonucleases. Finally, although there is an increase in star activity with increasing osmolyte concentration that is reversed by the application of hydrostatic pressure, it is interesting that the total specificity of the enzymes (i.e., standard and star) is maintained. That is, no new star activity is found under any of the conditions.

Kabata et al. have carried out quantitative studies of the effect of pressure on *Eco*RI and *Bam*HI [25]. Their results are apparently at odds with some of the data from other laboratories. They report finding that hydrostatic pressure had a limited effect upon the overall reaction of *Eco*RI. These authors report that the V_{max} changes from $6.2 \pm 0.2 \times 10^{-2}$ nM/min at 1 bar to $7.0 \pm 0.2 \times 10^{-2}$ nM/min at 2000 bar, while the K_m changes from 19 ± 2 nM to 28 ± 2 nM over the same pressure range. They also observed that hydrostatic pressure decreased the extent of star activity. The effect of pressure on *Eco*RI was found to be reversible, that is, the activity returned to upon return to atmospheric pressure. However, a substantial irreversible inactivation of *Bam*HI was found.

The only report on the effect of hydrostatic pressure on the structure of a DNA–protein complex used high resolution DNA footprinting [26]. The complex formed between *Eco*RI and its canonical binding site was studied at pressures up to 1400 bar. In the absence of magnesium ions, *Eco*RI will recognize and bind to its canonical site; however, it will not cleave the DNA. At atmospheric pressure the enzyme protects 10 bases from cleavage by hydroxyl radicals. These bases are centered on the recognition sequence, GAATTC. With increasing pressure, the bases peripheral to the binding site displayed an increased susceptibility to hydroxyl radical cleavage, i.e., they were more exposed to the solvent at high pressure than at atmospheric pressure. The two bases at the point of enzymatic cleavage showed a large increase in the rate of cleavage by hydroxyl radicals. (The enzyme cleaves both strands of DNA between the G and A residues of the canonical sequence.) The bases at the center of the recognition sequence showed no change in their solvent accessibility. These results imply that the contacts between the DNA binding site and *Eco*RI do not respond uniformly to changes in hydrostatic pressure. Some contacts are more sensitive than others, suggesting that the overall effect of hydrostatic pressure can be divided into 'local' pressure effects of each protein–DNA contact.

Erijman and Clegg have investigated the effect of hydrostatic pressure on the activity of RNA polymerase, an enzyme that forms a ternary DNA/RNA/enzyme complex [27,28]. They have quantitatively studied the activity of the enzyme at high pressure and they have investigated the effect of pressure on the interaction of the subunits of the enzyme using gel electrophoresis at high pressure.

Erijman and Clegg found that the extent of inactivation of the enzyme caused by pressure correlates with the extent to which the subunits are dissociated by pressure [27]. Pressure halted chain elongation; however, the activity of the enzyme was completely restored upon return to atmospheric pressure. They found that their sample, which was homogeneous by other measures, acted heterogeneously towards pressure. At a given pressure, one fraction of the enzyme lost its activity while the remainder retained its atmospheric pressure activity. They have interpreted these results as arising from heterogeneity in the subunit interactions of the enzyme. The heterogeneous subunit interactions respond differentially to hydrostatic pressure.

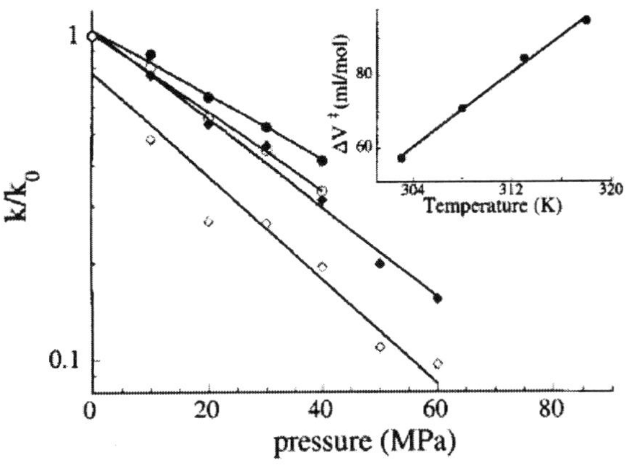

Fig. 3. Temperature dependence of the activation volume in the synthesis of RNA catalyzed by *E. coli* RNA polymerase (from Fig. 5 of [28]). The reaction was carried out at 30°C (●), 35°C (○), 40°C (■), and 45°C (□) and monitored by the increase in fluorescence at 500 nm upon incorporation of a fluorescent base analog into the nascent RNA. The template was poly[d(A-T)]. The inset shows the activation volume as a function of temperature.

In their second paper, Erijman and Clegg report that actively transcribing elongation complexes do not dissociate from the DNA template at pressures as high as 1800 bar [28]. This allowed them to determine a lower limit for the ternary complex, DNA, RNA, RNA polymerase, of 20 kcal/mol. The decreased rate of chain elongation with pressure was consistent with an activation volume of 55–60 ml/mol at 35°C. The effect of hydrostatic pressure on the rate of synthesis of RNA is shown in Fig. 3.

The response of lambda exonuclease to elevated hydrostatic pressure has reported by Rudd [29]. This is a third class of enzymes that uses DNA as a substrate. Lambda nuclease is a monomeric enzyme that, like the restriction endonucleases, cleaves DNA and, like RNA polymerase, is processive enzyme, acting on successive bases before releasing the substrate. The data show that lambda exonuclease is greater than 99% inhibited at 3360 bar, 20°C; the inhibition is reversible upon return to atmospheric pressure. The inhibition by pressure is greater at 15°C than at 20°C.

7. DNA-binding proteins

Silva and coworkers have presented several papers that deal with the effect of hydrostatic pressure on the thermodynamics of DNA-binding proteins [30–33]. This group has also studied other protein systems as a function of hydrostatic pressure; however, these investigations are outside of the scope of this review.

In their studies of the complexes formed between DNA and the two repressors Arc and LexA, they have employed hydrostatic pressure in order to probe details of the thermodynamics of the protein–DNA complexes that are not accessible at atmospheric pressure. These proteins interact with DNA in a sequence-specific manner and display large favorable entropy changes upon binding. Both LexA and Arc are dimeric proteins.

Pressure causes a transition of Arc from the native state to a molten globule state [30]. Folding and dimerization of the protein occur with a large, positive change in entropy. The stability of the protein tertiary and quaternary structure is greatly enhanced by binding to DNA. The authors found that the protein exists as denatured monomers at 22°C in the absence of the operator sequence DNA. The complex formed between the Arc repressor and a non-specific DNA sequence was 93% dissociated at 22°C and completely dissociated as the temperature was decreased. In the presence of the operator sequence the complex was stable at 22°C but it also denatured with decreasing temperatures. The authors observed the denaturation of the Arc–operator complex in liquid water at −15°C, exploiting the fact that high pressure lowers the freezing temperature of water. These results underscore the importance of entropy ($T\Delta S = 33$ kcal/mol at 22°C) in this sequence-specific DNA–protein interaction.

Mohana-Borges et al. showed that the LexA repressor binds to its operator site as a dimer, contradicting earlier results that implied that the two halves of the protein bound to DNA independently [33]. They measured the monomer–dimer dissociation constant for LexA by monitoring the fluorescence polarization as a function of hydrostatic pressure and urea concentration, Table 3. The dissociation constant is in the picomolar range and the volume change for the interaction of the two monomers is 340 ml/mol. This is a large volume change for the association of two 202-residue proteins. The authors interpret this value as arising from the burial of a large non-polar surface. Interaction with DNA leads to a large, additional increase in the stability of the LexA monomer–dimer complex, the dissociation constant decreased by a factor of 750. However, the increased stabilization arises only upon binding to the operator sequence.

Silva and coworkers have also examined the energetics of nucleic acid-protein interactions of the Cowpea Severe Mosaic Virus capsid protein [31]. They have studied the disassembly of the virus with hydrostatic pressure and found that protein–RNA interactions are important for establishing a protein conformation that is amenable to assembly of the virus. The different pathways leading to disassembly of the virus were evident in the susceptibility of the viral RNA to enzymatic digestion. Exposing the virus to hydrostatic pressure in the presence of ribonuclease did not lead to any digestion of RNA. However, at −10°C under pressure, the viral RNA was digested by ribonuclease. Using fluorescence spectroscopy and trypsin digestion, the results imply that the viral proteins have a different conformation if they are denatured by low temperature at high pressure, or if they are denatured by urea at atmospheric pressure.

8. Conclusions and future work

Pressure is an important variable for the study of biochemical equilibria, and increasingly, it is being used to study the non-covalent complexes with nu-

Table 3
Effect of specific and non-specific DNA binding upon the pressure-induced dissociation of dimers of LexA repressor[a]

DNA	ΔV (ml/mol)	K_d (pM)[b]	$\Delta p_{1/2}$ (bar)
None	334 ± 5	15 ± 3	
RecA operator	337 ± 25	0.02 ± 0.01	510 ± 11
Non-specific DNA	334 ± 22	8 ± 3	24 ± 7

[a]From [33].
[b]pM = 10^{-12} M.

cleic acids. The recent results of Dubins et al. [10] underscore that there is still much work to be done to understand the pressure–temperature behavior of nucleic acids. The use of pressure to study the interactions of nucleic acids and proteins and drugs is still very limited; however, the results outline above make a strong case for its continued application. This is perhaps especially true for nucleic acid–protein interactions, where entropy plays such an important role in the energetics of binding.

Although the results obtained so far are interesting and significant and often unattainable by alternative approaches, the widespread use of hydrostatic pressure for the study of biochemical equilibria will likely not occur until there are a greater variety of experimental techniques available. The overwhelming majority of reports using hydrostatic pressure rely on the measurement of either the absorption or fluorescence of the sample, which, in general, are broad, featureless signals. It is true, that in the case of fluorescence, one can measure several possible variables, e.g., the peak position (or center of spectra mass), polarization, or fluorescence lifetime; however, the application of different techniques will greatly enhance the utility of pressure studies. Nuclear magnetic resonance and electrophoresis are the two most obvious methods that could greatly enhance pressure studies. High-pressure versions of both of these techniques have been developed; however, they have not yet been exploited for nucleic acid studies. The report by Erijman and Clegg [27] is the only one to employ high-pressure electrophoresis to study a DNA–protein complex despite the nearly universal application of electrophoresis to study these complexes at atmospheric pressure.

It would also appear that the field of high-pressure biophysical chemistry would benefit by expanding the range of pressures used in the investigations. Very few studies are carried out at pressures above 2500 bar and at this pressure it is not always obvious that the parameter being measured has achieved its limiting value. Because a two-state model is used to interpret the data in most studies, failure to establish a true high-pressure limit can result in erroneous values of the thermodynamic parameters the investigators seek to measure. In the temperature range of interest to biophysical studies, water solidifies above 12 000 bar; thus, there is a wide range of pressures available. The conformational changes caused by using higher pressures will complicate matters somewhat but this argues for the development of alternatives to UV-visible optical spectroscopy for monitoring the effects of pressure.

References

[1] R.B. Macgregor Jr., Effect of hydrostatic pressure on nucleic acids, Biopolymers 48 (1998) 253–263.

[2] J. Barciszewski, J. Jurczak, S. Porowski, T. Specht, V.A. Erdmann, The role of water structure in conformational changes of nucleic acids in ambient and high-pressure conditions, Eur. J. Biochem. 260 (1999) 293–307.

[3] A. Certes, Note relative à l'action des hautes pressions sur la vitalité des microorganismes d'eau douce et d'eau de mer, Comp. Rend. Acad. Sci. 36 (1884) 220–222.

[4] P. Regnard, Note relative à l'action des hautes pressions sur phénomènes vitaux (movement des cils vibratiles, fermentation), Comp. Rend. Acad. Sci. 36 (1884) 187–188.

[5] G.N. Somero, Biochemical ecology of deep-sea animals, Experientia 48 (1992) 534–543.

[6] H. Ludwig, W. Scigalla, B. Sojka, Pressure and temperature-induced inactivation of microorganisms, in: J.A. Markley, D.B. Northrup, C.A. Royer (Eds.), High-Pressure Effects in Molecular Biophysics and Enzymology, Oxford University Press, New York, 1996.

[7] M.-C. Lin, P. Eid, P.T.T. Wong, R.B. Macgregor Jr., High pressure Fourier transform infrared spectroscopy of poly(dA)poly(dT), poly(dA) and poly(dT), Biophys. Chem. 76 (1999) 87–94.

[8] C.P. Ordahl, T.R. Johnson, A.I. Caplan, Sheared DNA fragment sizing: comparison of techniques, Nucleic Acids Res. 3 (1976) 2985–2999.

[9] K. Suzuki, Y. Taniguchi, Y. Miyosawa, The effect of pressure on the absorption spectra of DNA and DNA–dye Complex, J. Biochem. 72 (1972) 1087–1091.

[10] D.N. Dubins, A. Lee, R.B. Macgregor, Jr., T.V. Chalikian, On the stability of double-stranded nucleic acids, J. Am. Chem. Soc. 123 (2001) 9254–9259.

[11] R.B. Macgregor Jr., Chain length and oligonucleotide stability at high pressure, Biopolymers 38 (1996) 321–327.

[12] J.G. Chaires, Drug–DNA interactions, Curr. Opin. Struct. Biol. 8 (1998) 314–320.

[13] B.S. Reddy, S.K. Sharma, J.W. Lown, Recent developments in sequence selective minor groove DNA effectors, Curr. Med. Chem. 8 (2001) 475–508.

[14] R.B. Macgregor Jr., R.M. Clegg, T.M. Jovin, Pressure-jump study of the kinetics of ethidium bromide binding to DNA, Biochemistry 24 (1985) 5503–5510.

[15] L.A. Marky, R.B. Macgregor Jr., Hydration of dAdT polymers: role of water in the thermodynamics of ethidium and propidium intercalation, Biochemistry 29 (1990) 4805–4811.

[16] R.B. Macgregor Jr., R.M. Clegg, T.M. Jovin, Viscosity de-

pendence of ethidium–DNA intercalation kinetics, Biochemistry 26 (1987) 4008–4016.

[17] R.B. Macgregor Jr., R.M. Clegg, Ethidium–DNA binding kinetics in an agarose gel, Biopolymers 26 (1987) 2103–2106.

[18] N.S. Isaacs, in: Liquid Phase High Pressure Chemistry, Wiley, New York, 1981, pp. 165–166.

[19] R.K. Williams, C. Shen, The effect of pressure on the activity of ribonuclease, Arch. Biochem. Biophys. 152 (1972) 606–612.

[20] R.B. Macgregor Jr., Reversible inhibition of *Eco*RI with elevated pressure, Biochem. Biophys. Res. Commun. 170 (1990) 775–778.

[21] R.P. Rand, V.A. Parsegian, D.C. Rau, Intracellular osmotic action, Cell. Mol. Life Sci. 57 (2000) 1018–1032.

[22] M. Nasri, D. Thomas, Alteration of the specificity of *Pvu*II restriction endonuclease, Nucleic Acids Res. 15 (1987) 7677–7687.

[23] C.R. Robinson, S.G. Sligar, Hydrostatic pressure reverses osmotic pressure effects on the specificity of *Eco*RI–DNA interactions, Biochemistry 33 (1994) 3787–3793.

[24] C.R. Robinson, S.G. Sligar, Heterogeneity in molecular recognition by restriction endonucleases: osmotic and hydrostatic pressure effects on *Bam*HI, *Pvu* II, and *Eco*RV specificity, Proc. Natl. Acad. Sci. USA 92 (1995) 3444–3448.

[25] H. Kabata, A. Nomura, N. Shimamoto, S. Kunugi, Effect of high pressure on *Eco*RI reactions, J. Mol. Recognit. 7 (1994) 25–30.

[26] R.B. Macgregor Jr., Footprinting of *Eco*RI endonuclease at high pressure, Biochim. Biophys. Acta 1129 (1992) 303–308.

[27] L. Erijman, R.M. Clegg, Heterogeneity of *E. coli* RNA polymerase revealed by high pressure, J. Mol. Biol. 253 (1995) 259–265.

[28] L. Erijman, R.M. Clegg, Reversible stalling of transcription elongation complexes by high pressure, Biophys. J. 75 (1998) 453–462.

[29] E.A. Rudd, Reversible inhibition of lambda exonuclease with high pressure, Biochem. Biophys. Res. Commun. 230 (1997) 140–142.

[30] D. Foguel, J.L. Silva, Cold denaturation of a repressor–operator complex: the role of entropy in protein–DNA recognition, Proc. Natl. Acad. Sci. USA 91 (1994) 8244–8247.

[31] L.P. Gaspar, J.E. Johnson, J.L. Silva, A.T. Da Poian, Partially folded states of the capsid protein of cowpea severe mosaic virus in the disassembly pathway, J. Mol. Biol. 273 (1997) 456–466.

[32] L.M.T.R. Lima, D. Foguel, J.L. Silva, DNA tightens the dimeric DNA-binding domain of human papillomavirus E2 protein without changes in volume, Proc. Natl. Acad. Sci. USA 97 (2000) 14289–14294.

[33] R. Mohana-Borges, A.B.F. Pacheco, F.J.R. Sousa, D. Foguel, D.F. Almeida, J.L. Silva, LexA repressor forms stable dimers in solution, J. Biol. Chem. 275 (2000) 4708–4712.

ELSEVIER

Biochimica et Biophysica Acta 1595 (2002) 277–282

BIOCHIMICA ET BIOPHYSICA ACTA
BBA

www.bba-direct.com

Review

Experimental and theoretical high pressure strategies for investigating protein–nucleic acid assemblies

T.W. Lynch [1,a], S.G. Sligar [a,b,*]

[a] *Department of Biochemistry, University of Illinois, 505 South Goodwin, Urbana, IL 61801, USA*
[b] *Department of Chemistry, University of Illinois, 505 South Goodwin, Urbana, IL 61801, USA*

Received 12 September 2001; received in revised form 28 November 2001; accepted 29 November 2001

Abstract

A method was developed to investigate the stability of protein–nucleic acid complexes using hydrostatic pressure during electrophoretic gel mobility shift analysis. The initial system probed by this technique was the well-characterized cognate *Bam*HI–DNA complex. Band shift analysis at several elevated pressures found the equilibrium dissociation (K_d) constant to be dependent on pressure, which allowed the volume change of dissociation (ΔV) to be calculated. In order to describe the effects of pressure on the specific *Bam*HI–DNA complex at the molecular level, molecular dynamics simulations at both ambient and elevated pressure was performed. Comparison of the simulation trajectories identified several individual *Bam*HI–DNA contacts that are disrupted due to pressure. The disruption of these contacts can be attributed to an observed pressure-induced increase in hydration at the protein–DNA interface during the elevated pressure simulation.

Keywords: Gel shift assay; Protein–DNA recognition; Molecular dynamics; High pressure electrophoresis; Hydration

1. Introduction

Numerous important cellular processes are dependent on site-specific protein–nucleic acid recognition including replication, recombination, transcription, and restriction/modification. A major objective of biophysical studies on protein–nucleic acid interactions is to understand the molecular basis for these processes and identify the key factors involved in site-specific binding. The large number of X-ray and nuclear magnetic resonance structures of protein–nucleic acid complexes have been useful in the identification of elements necessary for sequence specificity and complex stability [1,2]. Together with structural analysis, detailed biochemical and thermodynamic investigations offer a complimentary understanding of DNA binding and recognition by proteins.

For several years, it has been appreciated that water molecules play a significant thermodynamic role [3–7] and contribute importantly to the specificity in protein-DNA complexes [8–10]. Thermodynamic investigations have illustrated that specific protein–DNA binding is generally accompanied with a large and negative change in heat capacity (ΔC_p) [11–13]. This usually indicates that solvent-accessible surface has been removed from bulk water upon protein binding to DNA [11,12]. A number of

* Corresponding author. Fax: +1-217-244-7100.
E-mail address: s-sligar@uiuc.edu (S.G. Sligar).

[1] Present address: Department of Biochemistry and Molecular Biology, University of Chicago, 920 East 58th Street, Chicago, IL 60637, USA.

0167-4838/02/$ – see front matter
PII: S0167-4838(01)00350-8

investigations have focused on the change in macromolecular solvation (ΔV) during specific and nonspecific DNA binding by proteins [10,14–22]. Structurally, perhaps the best case of using water molecules during recognition is found in the *trp* repressor–operator system. As shown in the x-ray structure, each contact between protein and the three bases of the operator half site are water-mediated [23,24]. Interestingly, in the unbound DNA structure, the DNA bases of the half site are occupied and the water molecules are regarded as recognition elements used as a target by the repressor [25].

This short review describes a new method for investigating the role of water molecules and dissecting the energetic and structural contributions toward protein–DNA complex stability. The system used to examine the effects of hydrostatic pressure during gel mobility shift assay and MD simulations in this review, is the specific *Bam*HI–DNA complex. This assembly is an ideal candidate for this study due to its remarkable specificity, stability, and the availability of detailed structural information.

2. Protein–DNA recognition: the restriction endonucleases

The *Bam*HI endonuclease is a member of the type II restriction/modification class of enzymes [26]. The endonuclease recognizes the DNA sequence 5′-G′GATCC-3′ and cleaves after the first guanine base on each strand in the presence of Mg^{2+} [27]. *Bam*HI binds DNA as a homodimer (213 amino acids/monomer) with a dissociation constant $\sim 10^{-9}$ M [19]. The main feature of *Bam*HI structure is a large six-stranded mixed β-sheet surrounded by α-helices [28]. Numerous investigations have focused on the structure of *Bam*HI bound to DNA containing the cognate recognition sequence [29–31]. The most interesting structural feature in the specifically bound structure is the unfolded C-terminal helices (α7) of each monomer. The disordered helix from one monomer folds back towards the protein core and makes contacts with the DNA phosphate backbone, while the helix from the opposite monomer inserts into the DNA minor groove [30]. The disordered helix that inserts into the DNA minor groove has been termed the 'recognition arm'. In the non-specific *Bam*HI–DNA complex, the two C-terminal helices remain intact, indicating that the arms play a key role during site-specific DNA binding and catalysis [32].

3. Gel shift analysis at elevated hydrostatic pressure

A principal method of investigating macromolecular hydration during protein–DNA association events is through the application of hydrostatic pressure. The goal of our work has been to develop a method to incorporate hydrostatic pressure into the electrophoretic mobility shift assay (EMSA) [33–35]. The gel shift assay consists of adding a DNA-binding protein to a solution with a DNA oligomer containing the specific recognition sequence. Following equilibration, the protein–DNA complexes are separated from the free components (DNA, protein) by electrophoresis. Using a combination of pressure and electrophoresis, the stoichiometry of association and thermodynamic parameters can be determined with the added advantage of the direct visualization of the pressure effects on protein–DNA complexes. It has been suggested that methods, which directly show the state of association, are more suitable for thermodynamic investigations using hydrostatic pressure [36].

High-pressure electrophoresis has long been a method for separating biomolecules according to stoichiometry and conformation. Earlier investigations have employed this technique to understand the properties of oligomeric proteins [36–38], protein conformations [39,40], and macromolecular association [41] at elevated pressure. While this technique is very effective for characterizing these uni- and bimolecular interactions, the method of electrophoretic mobility shift assay to study protein–nucleic acid complex stability at elevated hydrostatic pressure has heretofore not been utilized. To accomplish this goal, the system described by Paladini et al. [37] was used as a basis for our flatbed gel high-pressure electrophoresis apparatus design. An electrophoresis box was also constructed and developed for routine high-pressure binding analysis and accommodates eight 10-μl samples.

Standard gel shift assays were performed, in the same manner previously described [19], at ambient and elevated pressures using the aforementioned apparatus. The applied hydrostatic pressure during gel shift analysis of the specific *Bam*HI–DNA complex causes a shift in the equilibrium dissociation constant towards reduced affinity [42]. The effect is clearly illustrated in Fig. 1, whereby the band corresponding to *Bam*HI bound to a DNA 26-mer containing the recognition sequence has reduced intensity over an identical protein titration at an elevated pressure of 300 bar versus that of an ambient pressure assay. The dissociation constants at several pressures were determined (0.7 ± 0.1 nM (ambient), 1.5 ± 0.1 nM (150 bar), 2.5 ± 0.3 nM (300 bar), 4.6 ± 0.4 nM (500 bar)), and the volume of specific *Bam*HI dissociation was calculated ($\Delta V = -92 \pm 8$ ml/mol).

The application of high-pressure gel mobility shift analysis allows the determination of the pressure dependence on protein–DNA binding affinity. These measurements also enabled the calculation of the volume change of pressure-induced dissociation. This method can be applied to numerous protein–DNA systems and has proven to be a useful tool in thermodynamic investigations.

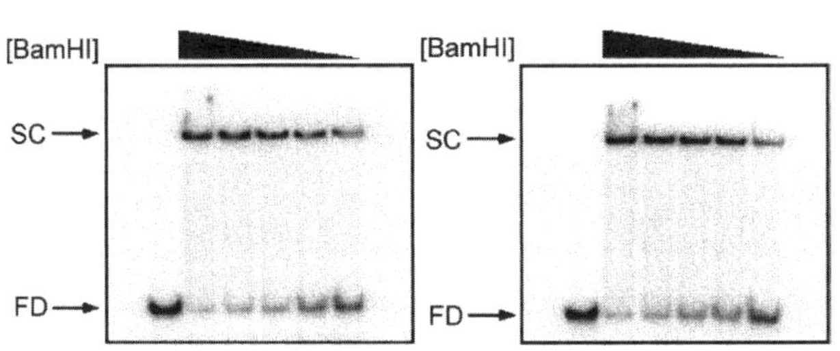

Fig. 1. The observed effect of hydrostatic pressure during the gel shift analysis of *Bam*HI bound to a DNA 26mer containing the cognate recognition sequence. In the gel shift experiments, the ^{32}P-labeled DNA, at a concentration of 1 pM, was incubated with various concentrations of *Bam*HI at 25°C in 10 mM Tris–HCl, 150 mM NaCl, 100 μg/ml bovine serum albumin, 0.1 mM EDTA (pH 8.0), 1 mM dithiothreitol, pH 7.9 for at least 30 min. The samples were incubated in the absence of divalent cations (Mg^{2+}) so DNA strand hydrolysis would not commence. The samples were loaded onto 15% polyacrylamide gels (37.5:1) in 0.5× TBE (45 mM Tris borate, 1 mM EDTA) buffer and electrophoresed at 9 V/cm for 1.5 h at both ambient pressure (left) and 300 bar (right). Lanes for each gel correspond to samples with equal DNA concentration, and protein concentrations (from left): no protein added, 6.0 nM, 3.0 nM, 1.5 nM, 0.75 nM, 0.38 nM. SC, specific *Bam*HI–DNA complex; FD, free DNA.

4. The importance of theory in understanding the origin of pressure effects

Molecular dynamics (MD) has a long history and has become a key theoretical tool for investigating biomolecular systems (for reviews see [43,44]). MD has routinely been used to monitor the structural and energetic contributions of specific contacts and the effects of DNA base substitutions in several protein–DNA systems [45–49]. Theoretical simulations have also been applied to DNA recognition sequences of well-characterized DNA-binding proteins, such as the *Bam*HI endonuclease, to investigate the structure and hydration of the DNA oligomer alone [50]. However, simulations of protein–DNA systems have not incorporated the variable of pressure into the calculations to better understand the recognition process.

High-pressure MD simulation analysis has, however, been used to monitor the pressure induced structural perturbations in unimolecular systems. Extensive work on bovine pancreatic trypsin inhibitor (BPTI) using molecular dynamics has been performed in order to better understand protein folding and compressibility [51–54].

To explore the effects of pressure on the specific *Bam*HI–DNA complex at the molecular level, a set of MD simulations was performed [42]. The protein–DNA-solvent system was constructed using the specific *Bam*HI–DNA structure (Protein Data Bank entry: 1 bhm) with a solvent of cube with the dimensions 88 Å^3. The first one nanosecond (1 ns) simulation was performed on the system at ambient pressure, while the second incorporated a pressure gradient of 50 bar/100 picosecond leading to a final pressure of 400 bar. The two simulations were compared and the differences in the trajectories were documented using interaction energy analysis of the *Bam*HI–DNA contacts observed in the X-ray structure.

The ambient pressure trajectory reveals that the specific contacts between *Bam*HI and DNA are maintained throughout the simulation. The results of elevated pressure simulation, however, illustrate that several protein contacts (Lys-193, Asp-196, Gly-197, and Met-198) with DNA are disrupted at elevated pressure. Most notably, the recognition arm is forced out of the DNA minor groove as shown in

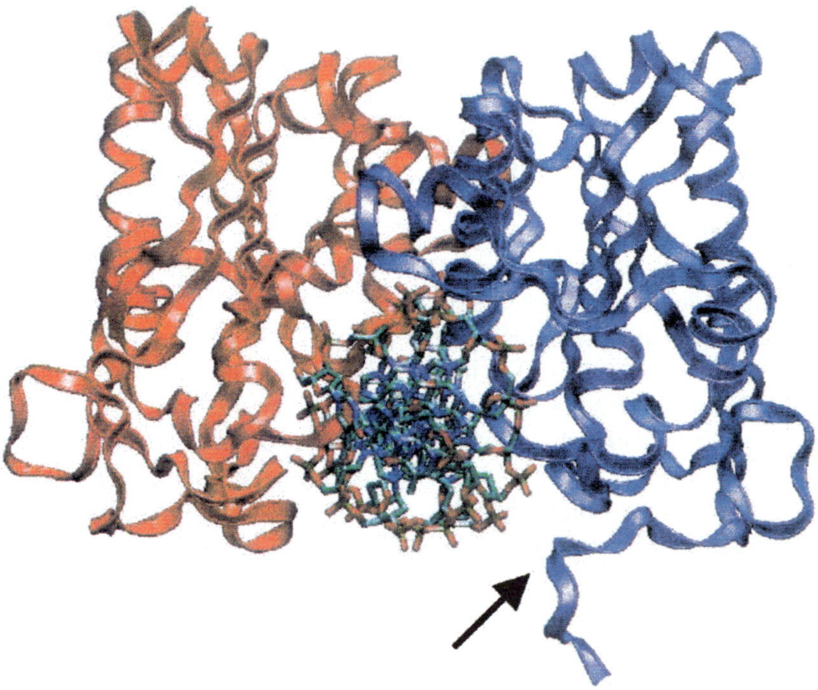

Fig. 2. The effect of pressure on the *Bam*HI–DNA system during the elevated pressure trajectory. The C-terminal arm loses contact with the DNA minor groove as hydrostatic pressure is applied.

Fig. 2. Fig. 3 displays the loss of interaction between Lys-193 and the DNA phosphate backbone. This contact is broken following the removal of the arm from the minor groove.

Investigation of the hydration at the protein–DNA interface was monitored to determine the role of water molecules, those 3 Å from both *Bam*HI and DNA, in disrupting the specific complex. In comparing the two simulations, pressure acts to force water molecules into the DNA minor groove and are sequestered at the *Bam*HI–DNA interface (Fig. 4). The molecular dynamic simulations have offered an explanation, at the molecular level, of the effects of pressure on the *Bam*HI–DNA binding equilibrium during elevated pressure gel mobility shift analysis.

5. Summary

This review describes a new method for investigat-

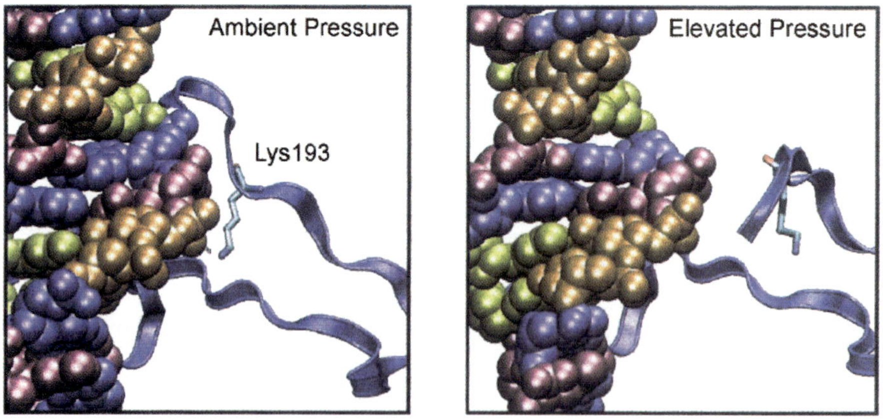

Fig. 3. The pressure-induced disruption of the contact between Lys-193 and the DNA phosphate backbone. Simulation snapshots at ambient pressure (left) and elevated pressure (∼400 bar) (right).

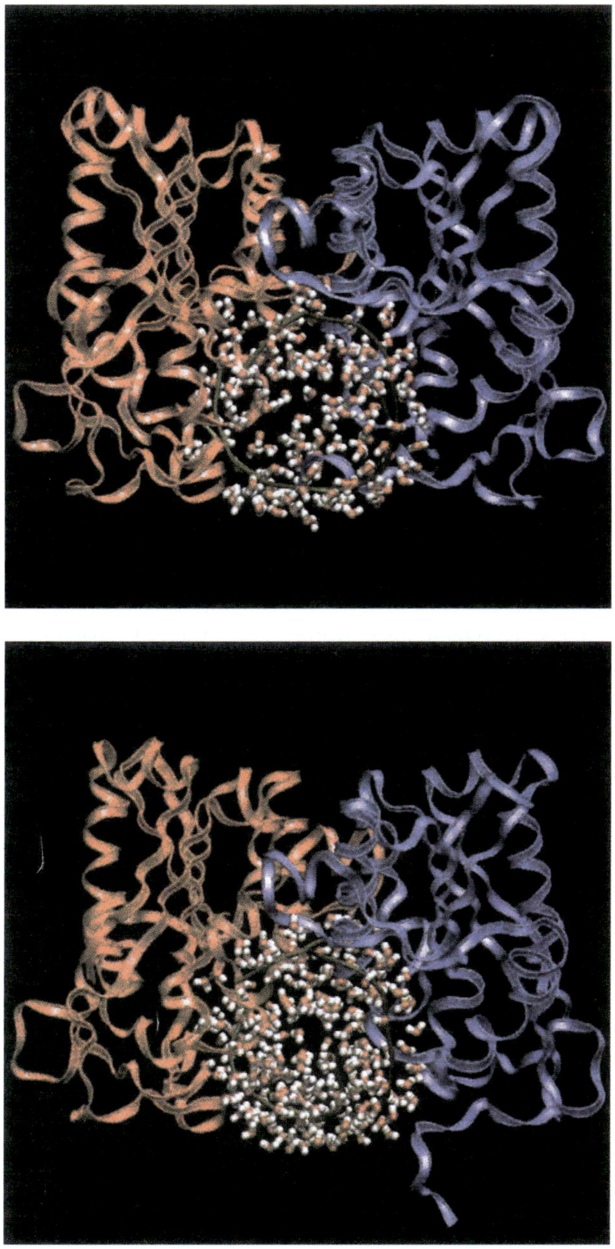

Fig. 4. The *Bam*HI–DNA interfacial hydration at both ambient pressure (top) and elevated pressure (∼400 bar) (bottom). Interfacial water molecules (colored in red and white) are those 3 Å from both the DNA and protein (blue and red).

ing the stability of protein–DNA complexes and dissecting the structural and energetic contributions in these systems. Several factors, however, must be considered before applying the high-pressure gel mobility shift technique. First, complexes should be stable. The specific *Bam*HI–DNA complex was chosen for the initial study because the complex runs as a tight and discrete band during electrophoresis. Structures with gel lifetimes that exceed the duration of the experiment migrate as discrete bands, whereas short-lived complexes that dissociate during electrophoresis yield diffuse zones of product [55]. Protein–nucleic acid complexes that run as diffuse bands indicate a loose complex, an occurrence that complicates data analysis. Electrophoretic dead time also plays a significant role when combining high-pressure electrophoresis and the gel mobility shift assay [35]. High-pressure electrophoresis requires the loading of an equilibrium mixture into the sample wells and subsequently pressurizing the polyacrylamide gel before electrophoresis. This problem can be assisted through the application of high voltages at the beginning of electrophoresis and minimization of sample volumes.

Molecular dynamic simulations of protein–nucleic acid systems at elevated pressure add an important component to the understanding of protein–nucleic acid recognition. They link a bulk thermodynamic quantity to specific and local structural interactions. The recent advances in MD have allowed expedient calculations to be performed on macromolecular systems, including those that contain large polyanionic structures such as DNA. Ideally, investigators would like to use this tool to identify the location of the experimentally determined volume change; however this was not a part of our analysis.

The investigation on the specific *Bam*HI–DNA complex demonstrates the importance of combining experimental data with theoretical simulations. The MD analyses provided insight into the mechanism of pressure dissociation and the structural elements that provide stability in the specific complex. All of the results presented here are an indication of the feasibility in using this combination of strategies as a tool for dissecting the stability of protein–DNA complexes.

Acknowledgements

We owe a great deal of gratitude to both Dr. Dorina Kosztin and Dr. Klaus Schulten at the University of Illinois Urbana-Champaign for their assistance in performing the theoretical analysis. This work was supported by grants GM33775 and GM31756 from the National Institutes of Health.

References

[1] A.A. Travers, in: DNA–Protein Interactions, Chapman and Hall, London, 1990.

[2] S.C. Harrison, Nature 353 (1991) 715–719.

[3] D.R. Lesser, M.R. Kurpiewski, L. Jen-Jacobsen, Science 250 (1990) 776–785.

[4] L. Jen-Jacobsen, Biopolymers 44 (1997) 153–180.

[5] J.W.R. Schwabe, Curr. Opin. Struct. Biol. 7 (1997) 126–134.

[6] M.P. Brown, A.O. Grillo, M. Boyer, C.A. Royer, Protein Sci. 8 (1999) 1276–1285.

[7] J. Janin, Struct. Fold. Des. 7 (1999) R277–R279.

[8] C.R. Robinson, S.G. Sligar, J. Mol. Biol. 234 (1993) 302–306.

[9] C.R. Robinson, S.G. Sligar, Proc. Natl. Acad. Sci. USA 92 (1995) 3444–3448.

[10] C.R. Robinson, S.G. Sligar, Proc. Natl. Acad. Sci. USA 95 (1998) 2186–2191.

[11] J.-H. Ha, R.S. Spolar, M.T. Record Jr., J. Mol. Biol. 209 (1989) 801–816.

[12] R.S. Spolar, M.T. Record Jr., Science 263 (1994) 777–784.

[13] J.E. Ladbury, J.G. Wright, J.M. Sturtevant, P.B. Sigler, J. Mol. Biol. 238 (1994) 669–681.

[14] P. Douzou, Proc. Natl. Acad. Sci. USA 91 (1994) 1657–1661.

[15] D. Foguel, J.L. Silva, Proc. Natl. Acad. Sci. USA 91 (1994) 8244–8247.

[16] M.M. Garner, D.C. Rau, EMBO J. 14 (1995) 1257–1263.

[17] N.Y. Sidorova, D.C. Rau, Proc. Natl. Acad. Sci. USA 93 (1996) 12272–12277.

[18] N.Y. Sidorova, D.C. Rau, J. Biomol. Struct. Dyn. 17 (1999) 19–31.

[19] T.W. Lynch, S.G. Sligar, J. Biol. Chem. 275 (2000) 30561–30565.

[20] V.A. Parsegian, R.P. Rand, D.C. Rau, Methods Enzymol. 259 (1995) 43–94.

[21] C.R. Robinson, S.G. Sligar, Methods Enzymol. 259 (1995) 395–427.

[22] V.J. LiCata, N.M. Allewell, Methods Enzymol. 295 (1998) 42–62.

[23] Z. Otwinowski, R.W. Schevitz, R.G. Zhang, C.L. Lawson, A. Joachimiak, R.Q. Marmorstein, B.F. Luisi, P.B. Sigler, Nature 335 (1988) 321–329.

[24] C.L. Lawson, J. Carey, Nature 366 (1993) 178–182.

[25] Z. Shakked, G. Guzikevich-Guerstein, P.B. Sigler, Nature 368 (1994) 469–473.

[26] G.A. Wilson, F.E. Young, J. Mol. Biol. 97 (1975) 123–125.

[27] P. Hensley, G. Nardone, J.G. Chirikjian, M.E. Wastney, J. Biol. Chem. 265 (1990) 15300–15307.

[28] M. Newman, T. Strzelecka, L.F. Dorner, I. Schildkraut, A.K. Aggarwal, Structure 2 (1994) 439–452.

[29] T. Strzelecka, M. Newman, L.F. Dorner, R. Knott, I. Schildkraut, A.K. Aggarwal, J. Mol. Biol. 239 (1994) 430–432.

[30] M. Newman, T. Strzelecka, L.F. Dorner, I. Schildkraut, A.K. Aggarwal, Science 269 (1995) 656–663.

[31] H. Viadiu, A.K. Aggarwal, Nat. Struct. Biol. 5 (1998) 910–916.

[32] H. Viadiu, A.K. Aggarwal, Mol. Cell 5 (2000) 889–895.

[33] M.G. Fried, Electrophoresis 10 (1989) 366–376.

[34] J. Carey, Methods Enzymol. 208 (1991) 103–117.

[35] M.G. Fried, J.L. Bromberg, Electrophoresis 18 (1997) 6–11.

[36] A.A. Paladini, J.L. Silva, G. Weber, Anal. Biochem. 161 (1987) 358–364.

[37] A.A. Paladini, G. Weber, L. Erijman, Anal. Biochem. 218 (1994) 364–369.

[38] L. Erijman, R.M. Clegg, Rev. Sci. Instrum. 67 (1996) 813–817.

[39] L. Erijman, A.A. Paladini, G.H. Lorimer, G. Weber, J. Biol. Chem. 268 (1993) 25914–25919.

[40] J.A. Kornblatt, M.J. Kornblatt, C. Clery, C. Balny, Eur. J. Biochem. 265 (1999) 120–126.

[41] P. Masson, J. Reybaud, Electrophoresis 9 (1988) 157–161.

[42] T.W. Lynch, D. Kosztin, M.A. McLean, K. Schulten, S.G. Sligar, Biophys. J. 82 (2002) 93–98.

[43] M. Karplus, G.A. Petsko, Nature 347 (1990) 631–639.

[44] M.E. Tuckerman, G.J. Martyna, J. Phys. Chem. B 104 (2000) 159–178.

[45] T.C. Bishop, K. Schulten, Proteins 24 (1996) 115–133.

[46] Y. Duan, P. Wilkosz, J.M. Rosenberg, J. Mol. Biol. 264 (1996) 546–555.

[47] D. Kosztin, T.C. Bishop, K. Schulten, Biophys. J. 73 (1997) 557–570.

[48] S. Sen, L. Nilsson, Biophys. J. 77 (1999) 1801–1810.

[49] A. Suenaga, C. Yatsu, Y. Komeiji, M. Uebayasi, T. Meguro, I. Yamato, J. Mol. Struct. 526 (2000) 209–218.

[50] T. Castrignano, G. Chillemi, A. Desideri, Biophys. J. 79 (2000) 1263–1272.

[51] D.B. Kitchen, L.H. Reed, R.M. Levy, Biochemistry 31 (1992) 10083–10093.

[52] B. Wroblowski, J.F. Diaz, K. Heremans, Y. Engelborghs, Proteins 25 (1996) 446–455.

[53] M. Marchi, K. Akasaka, J. Phys. Chem. B 105 (2001) 711–714.

[54] V.M. Dardarlat, C.B. Post, J. Phys. Chem. B 105 (2001) 715–724.

[55] A. Revzin, J.A. Ceglarek, M.M. Garner, Anal. Biochem. 153 (1986) 172–177.

Biochimica et Biophysica Acta 1595 (2002) 283–296

www.bba-direct.com

Review

Effect of pressure on electron transfer reactions in inorganic and bioinorganic chemistry

Joanna Macyk, Rudi van Eldik *

Institute for Inorganic Chemistry, University of Erlangen-Nürnberg, Egerlandstrasse 1, 91058 Erlangen, Germany

Received 3 September 2001; received in revised form 20 November 2001; accepted 21 November 2001

Abstract

Kinetic and thermodynamic studies involving the application of different high-pressure techniques, are very useful in gaining mechanistic information on the basis of volume changes that occur during inorganic and bioinorganic electron transfer reactions. The most fundamental type of electron transfer reaction concerns self-exchange reactions, for which the overall reaction volume is zero, and activation volumes can be measured and discussed. In the case of non-symmetrical electron transfer reactions, intra- and intermolecular processes can be studied and volume profiles can be constructed. Precursor complex formation can in some cases be recognized kinetically in such systems. Typical values of activation and reaction volumes are reviewed for various reversible and irreversible electron transfer reactions. Mechanistic conclusions reached on the basis of these parameters are presented. Volume profiles for electron transfer reactions enable a simplistic presentation of the reaction mechanism on the basis of intrinsic and solvational volume changes along the reaction coordinate.

Keywords: Pressure effect; Electron transfer reaction; Activation volume; Reaction volume; Volume profile

1. Introduction

The value of mechanistic information that emerges from kinetics measurements over a series of elevated pressures for solution reactions in inorganic chemistry has been realized for some time [1]. However, many inorganic reactions are too fast to follow using conventional instrumentation. Hence the momentum regarding investigations at high pressures vis-a-vis organic reactions was delayed somewhat until adaptation of rapid reaction techniques for operation at high pressures had been achieved, mostly in the period from 1975 to 1985. This fertile period has been recorded in reviews, in conference proceedings, and in monographs, and readers may obtain a thorough background and sense of historical development by consulting this literature [2–13]. Even until quite recently, suitable instrumentation was not widely available.

The purpose of this article is to familiarize the reader with the current status of activities in the application of hydrostatic pressure to mechanistic studies on electron transfer reactions in inorganic chemistry, as well as in the blossoming field of bioinorganic chemistry. Since this is not intended to be a comprehensive review, only a selection of representative reactions and typical examples, in

* Corresponding author.

0167-4838/02/$ – see front matter
PII: S0167-4838(01)00351-X

many cases from our own work in this area, will be described. The basic principles involved in high-pressure kinetics for reactions in general have been the subject of many reports [14–16].

The parameter that is derived from high-pressure kinetic experiments in solution when the transition state theory applies, is the difference in partial molar volume between the activated complex of transition state theory and the reactant state, and is known as the volume of activation, $\Delta V^{\#}$. If the particular reaction is reversible and the system experimentally accessible, $\Delta V^{\#}$ for the reverse reaction can also be obtained and thus, except for solvent exchange and self-exchange electron transfer reactions in which there is no net volume change, the difference between these two quantities results in the reaction volume, ΔV. The latter quantity, in some cases, may also be determined by measuring the equilibrium constant (K) for the reaction as a function of pressure or from the partial molar volumes of the reactants and products, derived from solution density measurements. The volume of activation itself is determined from measurements of the reaction rate constant k at different hydrostatic pressures p at a given absolute temperature T, since $(\partial \ln k/\partial p)_T = -\Delta V^{\#}/RT$ (R is the ideal gas constant), an equation developed within transition state theory based upon the analogous equilibrium constant relationship $(\partial \ln K/\partial p)_T = -\Delta V/RT$. The former equation, upon integration, can be employed to determine $\Delta V^{\#}$ from a plot of ln$k$ versus p. Providing the pressure is no higher than 200 MPa, in the vast majority of cases $\Delta V^{\#}$ is pressure independent and the plot is linear. A non-linear behavior is usually encountered when dealing with a compressible solvent where both the reaction and activation volume become pressure sensitive. In this article the focus will be on reactions in which there is a negligible or absence of pressure dependence of the volume of activation. In general, volume of activation data quoted in this report will refer to ambient conditions, i.e., close to room temperature, and readers are advised to consult the cited literature for more detailed information on the exact experimental conditions employed.

Equilibrium and kinetic parameters obtained as a function of temperature permit the drawing of diagrams illustrating the Gibbs free energy (G), enthalpy (H) and entropy (S) changes in proceeding in the sequence reactant state/transition state/product state, and including intermediates when they are formed. Correspondingly, a volume diagram or volume profile can chart the respective volume changes along the reaction coordinate, and when appropriate actual partial molar volumes are known, on an absolute rather than a relative basis, something realizable for S, but not for H and G. Hence if reactants A and B form a product AB and no intermediates are formed, i.e., there is a single-step reaction, a volume profile in which the reaction volume is, for example, negative and the volume of activation is such that the transition state is almost halfway between reactant and product states, is depicted in Fig. 1.

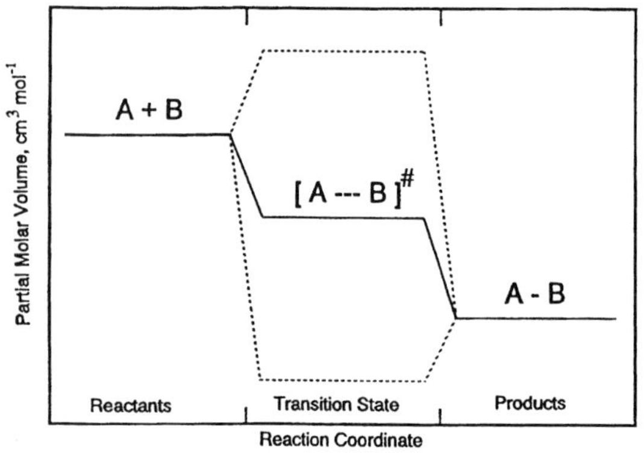

Fig. 1. Volume profile for the overall reaction A+B→AB. The activated complex is [A- - -B]$^{\#}$.

As shown in Fig. 1, other forms of the volume profile are possible depending on the particular character of the system. Thus, in principle, a volume profile represents a simple and lucid way of describing a reaction and diagnosing the mechanism, but with the caveat that mechanistic diagnosis is uncomplicated when only intrinsic changes (changes in bond lengths, bond angles for example) occur. In many actual reactions, when charged species are produced or neutralized during the reaction, or increases or decreases in polarity occur, then there is also a change in the volume occupied by the solvent molecules surrounding the system by virtue of an increase or decrease in (at least) the first solvation layer. Volume reduction of solvent from this source is known as electrostriction. Thus the facile interpretation of measured values of ΔV or $\Delta V^{\#}$ can be compromised by the existence of the two and difficulty quantifiable contributions.

2. Electron-transfer reactions

A fundamental understanding of oxidation–reduction reactions is vital to the inorganic chemist in the context of energy transduction, corrosion processes, metallurgy, redox processes in environmental chemistry, and metalloenzymes and metalloproteins involved in electron transfer. Electron-transfer reactions of transition metal complexes are accompanied by a change in the oxidation state of the metal atom and the overall charge on the complex ion. This can cause both intrinsic and solvational volume changes, such that it is reasonable to expect that electron-transfer reactions should exhibit characteristic volumes of activation.

In the following sections the effect of pressure on different types of electron-transfer processes will be discussed systematically. Some of our work in this area was reviewed as part of a special symposium devoted to the complementarity of various experimental techniques in the study of electron transfer reactions [17]. Swaddle and Tregloan recently reviewed electrode reactions of metal complexes in solution at high pressure [12]. The main emphasis in this section will fall on some of the most recent work that we have been involved in, dealing with 'long-distance' electron-transfer processes involving cytochrome *c*. However, by way of introduction, a short discussion on the effect of pressure on self-exchange (symmetrical) and non-symmetrical electron-transfer reactions between transition metal complexes that have been reported in the literature, will be presented.

3. Self-exchange reactions

Self-exchange reactions are the most simple type of electron-transfer reactions, they are symmetrical processes for which both the reaction free energy and reaction volume are zero, and ideal for theoretical modeling. They form the basis for much of the discussion and interpretation of non-symmetrical reactions in a similar way as solvent exchange reactions form a basis for the understanding of ligand substitution reactions. Swaddle and co-workers have made a significant contribution in this area. They have studied the effect of pressure on the self-exchange reactions (volumes of activation are quoted in brackets in $cm^3\ mol^{-1}$) $[Fe(H_2O)_6]^{3+/2+}$ (−11.1) [18], $[Fe(phen)_3]^{3+/2+}$ (−2.2) [18], $[Fe(CN)_6]^{3-/4-}$ (+22) [19], $[MnO_4]^{2-/-}$ (−23) [20], $[Co(sep)]^{3+/2+}$ (−6.4) [21], $[Co([9]aneS_3)]^{3+/2+}$ (−4.8) [21], $[Co(diamsarH_2)]^{5+/4+}$ (−9.6) [22], $[Co(diamsar)]^{3+/2+}$ (−10.4) [22], $[Co(en)_3]^{3+/2+}$ (−15.5) [23], and $[Co(phen)_3]^{3+/2+}$ (−17.6) [24]. In the majority of cases the self-exchange reaction is significantly accelerated by pressure, with the exception of the $[Fe(CN)_6]^{3-/4-}$ system which goes in exactly the opposite way, and the observed volume of activation is in agreement with the significantly negative entropy of activation reported for such systems. In addition, they have also gone through impressive efforts to calculate the volumes of activation theoretically based on the Marcus–Hush–Stranks treatment and their own modifications and additions [25,26]. For a large number of systems a good agreement between the experimental and theoretically calculated volumes of activation were found as shown in Fig. 2.

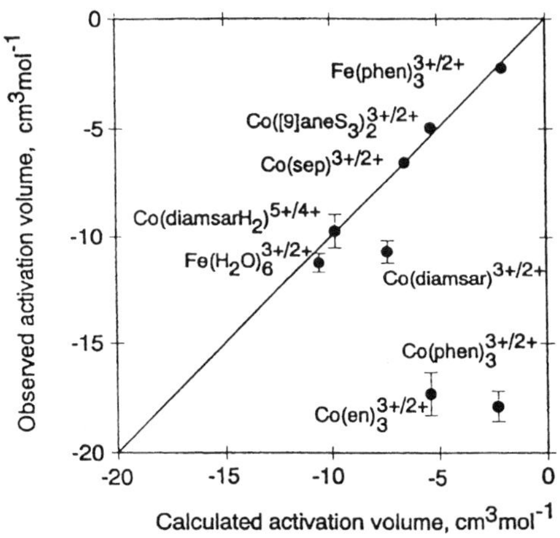

Fig. 2. Plot of $\Delta V^{\#}$(observed) versus $\Delta V^{\#}$(calculated) for a series of self-exchange electron-transfer reactions.

In most cases solvent reorganization accounts for the largest contribution towards the observed volume of activation. Large deviations were found only for the $[Co(en)_3]^{3+/2+}$ and $[Co(phen)_3]^{3+/2+}$ systems where the theoretical volume of activation is between 10 and 15 $cm^3\ mol^{-1}$ more positive than the experimental value [23,24]. This deviation is most probably related to the participation of a high-spin to low-spin changeover associated with the electron-transfer process, which can account for an additional volume collapse of ca. 10 to 15 $cm^3\ mol^{-1}$ [22].

$$MnO_4^- + MnO_4^{2-} \rightarrow MnO_4^{2-} + MnO_4^- \qquad (1)$$

A notable feature of anion-anion electron transfer reactions such as that of Reaction 1 and also in non-symmetrical reactions described below, is the fact that some alkali metal ions can catalyze the electron transfer process [27–31]. Thus far in most cases no detailed explanation from the point of view of volume changes is available since extraction of the volumes of activation for the catalyzed and uncatalyzed pathways from the high-pressure kinetics results is often not possible. Lithium ions yield no catalytic influence, and the effect is only modest or very small when sodium ions are present in the case of Reaction 1. However, lithium ions do have a significant catalytic effect, and sodium ions quite a substantial one, on cyanometalate self-exchange reactions in water [30]. Reaction acceleration is quite pronounced in the presence of potassium ions, even more so when Rb^+ or NH_4^+ are present in examples where they have been employed. The catalysis no doubt arises from the cation shielding the repulsion between the anionic partners. However, it is not yet established which specific property such as the radius or the charge density results in efficient catalysis by some ions and not by others. In the case of Reaction 1, where the observed $\Delta V^{\#}$ for the uncatalyzed path equals $-21\ cm^3\ mol^{-1}$ and the observed rate constant has been dissected into the values for the catalyzed and uncatalyzed pathways, it is clear that the catalyzed path involves a dramatically lower volume change in reaching the transition state ($\Delta V^{\#} = +3\ cm^3\ mol^{-1}$ (Na^+) and $-1\ cm^3\ mol^{-1}$ (K^+)) [20]. The catalytic efficiency may be correlated with the polarizability of M^+ or with the inverse of the hydrated ion ($M(aq)^+$) radius. In the uncatalyzed path the volume reduction may be regarded as arising from the compression of the ellipsoidal solvent cavity in which the two ions are enclosed and the closer approach of the two Mn centers. The solvational effects are much less in the cation catalyzed pathway.

More recently, Swaddle and coworkers [32–34] have demonstrated that for aqueous systems a good correlation exists between the volume of activation for a homogeneous self-exchange reaction and the activation volume for the heterogeneous self-exchange reaction at an electrode. The slope of the line in Fig. 3 for 10 sets of data is 0.50 and in precise agreement with an extension of the Marcus theory. This led the authors to suggest a 'fifty percent rule' with which volume of activation data can be predicted for self-exchange reactions, which cannot be measured directly for technical reasons, on the basis of electrochemical data recorded as a function of pressure.

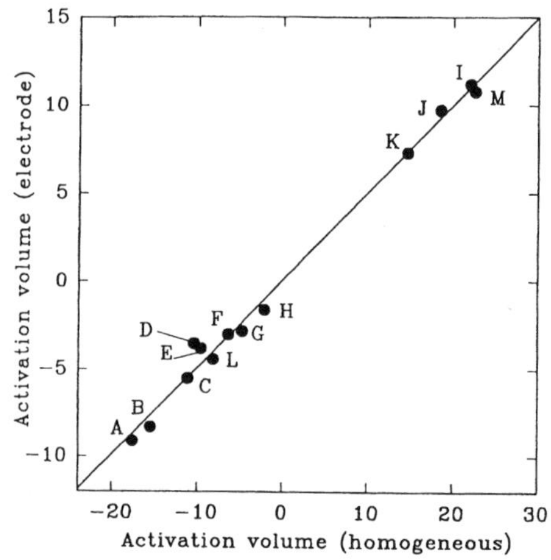

Fig. 3. Plot of $\Delta V^{\#}$(electrode) versus $\Delta V^{\#}$(homogeneous) for a series of self-exchange couplets: A, $Co(phen)_3^{3+/2+}$; B, $Co(en)_3^{3+/2+}$; C, $Fe(H_2O)_6^{3+/2+}$; D, $Co(diamsar)^{3+/2+}$; E, $Co(diamsarH_2)^{5+/4+}$; F, $Co(sep)^{3+/2+}$; G, $Co(ttcn)_2^{3+/2+}$; H, $Fe(phen)_3^{3+/2+}$; I, $Fe(CN)_6^{3-/4-}$; J, $Os(CN)_6^{3-/4-}$; K, $Mo(CN)_8^{3-/4-}$; L, $Mo(CN)_8^{3-/4-}$; M, $W(CN)_8^{3-/4-}$.

The above quoted reactions all proceed via an outer-sphere electron-transfer mechanism. By way of comparison, the volume of activation for the exchange reaction in $[Fe(H_2O)_5OH]^{2+}/[Fe(H_2O)_6]^{2+}$ was reported to be $+0.8\ cm^3\ mol^{-1}$, i.e., significantly more positive than found for $[Fe(H_2O)_6]^{3+/2+}$ [18]. This difference was ascribed to the release of a solvent molecule associated with the formation of an hydroxo-bridged intermediate in terms of an inner-sphere mechanism.

4. Non-symmetrical reactions

For the mechanistic interpretation of activation volume data for non-symmetrical electron-transfer reactions, it is essential to have information on the overall volume change that can occur during such a process. This then enables the construction of a volume profile for the electron transfer process and allows a detailed interpretation of the activation volume in terms of the overall volume associated with the reaction. The reaction volume can be calculated from the partial molar volumes of reactant and product species, when these are available, or can be determined from density measurements. Efforts have in recent years focused on the electrochemical determination of reaction volume data from the pressure dependence of the redox potential. Tregloan and coworkers [35,36] have demonstrated how such techniques can reveal information on the magnitude of intrinsic and solvational volume changes associated with electron-transfer reactions of transition metal complexes. The measured reaction volumes have to be corrected for the effect of pressure on the reference electrode, for which they conducted a well-designed set of experiments on a series of $Fe^{II/III}$ couples in which the ligands were systematically varied in order to adjust the overall charge of the complex, and therefore the solvational contribution towards the overall reaction volume. A good correlation was found between the reaction volume and the difference in the square of the charge on the oxidized and reduced forms of the complex as shown in Fig. 4 [35]. By interpolation to $\Delta z^2 = 0$, the reaction volume for the Ag/Ag^+ reference electrode was determined to be -11.9 ± 0.5 cm^3 mol^{-1}. Measurements on a series of Cr, Co and Ru complexes enabled a systematic differentiation to be made between intrinsic and solvational volume contributions associated with the redox process [36].

It has in general been the objective of many mechanistic studies dealing with inorganic electron-transfer reactions to distinguish between outer-sphere (intermolecular) and inner-sphere (intramolecular) mechanisms, since these describe the way the reactants interact with each other prior to electron transfer. In the case of electron-transfer reactions between proteins, the location of the protein at a specific site usually leads to an intramolecular electron-transfer process. Along these lines high-pressure kinetic methods and the construction of reaction volume profiles have also been employed to contribute towards a better understanding of the intimate mechanisms involved in such processes. The differentiation between outer-sphere and inner-sphere mechanisms depends on the nature of the precursor species, Ox//Red in

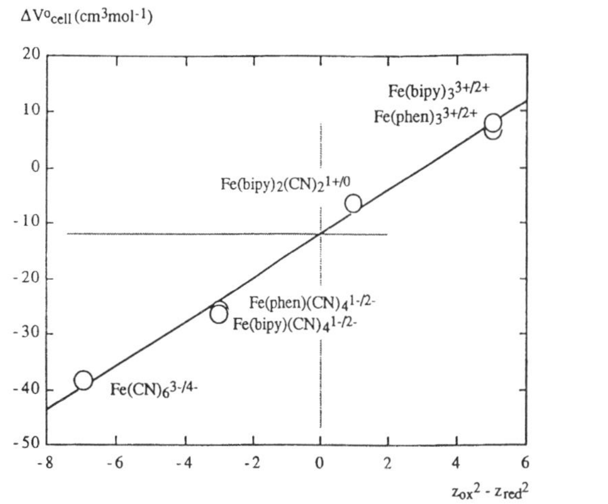

Fig. 4. Plot of ΔV°_{cell} versus $Z^2_{ox}-Z^2_{red}$ for a series of Fe(II)/Fe(III) redox systems.

the following scheme, which can either be an ion pair or encounter complex, or a bridged intermediate, respectively.

$$\mathrm{Ox + Red \rightleftharpoons Ox//Red},\ K \tag{2}$$

$$\mathrm{Ox//Red \rightarrow Ox^-//Red^+},\ k_{ET} \tag{3}$$

$$\mathrm{Ox^-//Red^+ \rightleftharpoons Ox^- + Red^+} \tag{4}$$

The coordination sphere of the reactants remains intact in the former case and is modified by ligand substitution in the latter, which will naturally affect the associated volume changes.

A general difficulty encountered in kinetic studies of outer-sphere electron-transfer processes concerns the separation of the precursor formation constant (K) and the electron-transfer rate constant (k_{ET}) in the reactions outlined above. In the majority of cases, precursor formation is a diffusion controlled step, followed by rate-determining electron transfer. In the presence of an excess of Red, the rate expression is given by:

$$k_{obs} = k_{ET}K[\mathrm{Red}]/(1 + K[\mathrm{Red}]) \tag{5}$$

In many cases K is small, such that this equation simplifies to $k_{obs} = k_{ET}K[\mathrm{Red}]$, which means that the observed second-order rate constant and the associated activation parameters are composite quantities, viz. $\Delta V^{\#} = \Delta V^{\#}(k_{ET}) + \Delta V(K)$. When K is large enough such that $1+K[\mathrm{Red}] > 1$, it is possible to separate k_{ET} and K kinetically and also the associated activation parameters, viz. $\Delta V^{\#}(k_{ET})$ and $\Delta V(K)$ [37]. A series of reactions were studied where it was possible to resolve K and k_{ET}, i.e., $\Delta V(K)$ and $\Delta V^{\#}(k_{ET})$. In this case oppositely charged reaction partners were selected as indicated in the following reactions [38–40]:

$$[Co(NH_3)_5X]^{(3-n)+} + [Fe(CN)_6]^{4-} \rightleftharpoons \{Co(NH_3)_5X^{(3-n)+}, Fe(CN)_6^{4-}\},\ K \quad (6)$$

$$\{Co(NH_3)_5X^{(3-n)+}, Fe(CN)_6^{4-}\} \rightarrow Co^{2+} + 5NH_3 + X^{n-} + [Fe(CN)_6]^{3-},\ k_{ET} \quad (7)$$

$$X^{n-} = H_2O,\ Me_2SO,\ py,\ Cl^-, N_3^-$$

Throughout the series, ion-pair formation is accompanied by significantly negative $\Delta S°$ values and close to zero ΔV values. The latter is rather surprising, since it is generally accepted that ion-pair formation should involve considerable charge neutralization accompanied by strong desolvation due to a decrease in electrostriction. The values of ΔV therefore indicate that the reaction partners most probably exist as solvent-separated ion pairs, i.e., with no significant charge neutralization accompanied by desolvation. The activation parameters demonstrate that the electron-transfer steps exhibit a strong pressure deceleration, most systems have a $\Delta V^{\#}$ value of between +25 and +34 $cm^3\ mol^{-1}$. These values indicate that electron transfer is accompanied by extensive desolvation, most probably related to charge neutralization associated with the electron-transfer process [40]. A simplified model based on partial molar volume data, in which electron transfer occurs from the precursor ion pair $\{Co(NH_3)_5X^{(3-n)+}, Fe(CN)_6^{4-}\}$ to the successor ion pair $\{Co(NH_3)_5X^{(2-n)+}, Fe(CN)_6^{3-}\}$, predicts an overall volume increase of ca. 65 $cm^3\ mol^{-1}$. This means that according to the reported $\Delta V^{\#}$ values the transition state for the electron-transfer process lies approximately halfway between the reactant and product states on a volume basis for the precursor and successor ion pairs. The largest volume contribution arises from the oxidation of $[Fe(CN)_6]^{4-}$ to $[Fe(CN)_6]^{3-}$, which is accompanied by a large decrease in electrostriction and an increase in partial molar volume. Theoretical calculations also confirmed that the transition state for these reactions lies approximately halfway along the reaction coordinate on a volume basis [40]. This first information on the nature of the volume profile for an outer-sphere electron-transfer reaction proved to be in good agreement with subsequently reported results for systems with low driving forces in which it was possible to construct a complete volume profile by studying the effect of pressure on both the forward and reverse reactions, as well as on the overall equilibrium constant (see further discussion). Data have also been reported for a series of related complexes containing phosphor-oxo ligands [41–43], and the results can be interpreted in terms of major solvational changes associated with the oxidation of $[Fe(CN)_6]^{4-}$.

5. Bioinorganic reactions

A number of electron transfer reactions of biological interest have been studied using high-pressure techniques [2,3]. These include the oxidation of L-ascorbic acid by $[Fe(CN)_6]^{3-}$ [44], $[Fe(CN)_5NO_2]^{3-}$ [45], and $Fe(phen)_2(CN)_2]^-$ [46]. The first two reactions are characterized by volumes of activation of −16 and −10 $cm^3\ mol^{-1}$, respectively, which indicate that solvent rearrangement as a result of an increase in electrostriction must account for the volume collapse on going to the transition state. In comparison, the third reaction exhibited almost no pressure dependence, in line with significantly less charge formation than in the case of the 3-charged complexes. The oxidation of deoxy- and oxymyoglobin by $[Fe(CN)_6]^{3-}$ was also studied as a function of pressure [47]. The oxidation of deoxymyoglobin is ca. seven times faster than the oxidation of oxymyoglobin at 298 K, and the corresponding volume of activation, determined under limiting kinetic conditions, is −20 $cm^3\ mol^{-1}$. This value is in good agreement with that reported above for the reduction of $[Fe(CN)_6]^{3-}$ to $[Fe(CN)_6]^{4-}$. The oxidation of oxymyoglobin is characterized by an activation volume of −3 $cm^3\ mol^{-1}$, which must be corrected for the reaction volume of −19 $cm^3\ mol^{-1}$ associated with the binding of dioxygen to myoglobin [48]. The resulting −22 $cm^3\ mol^{-1}$ is once again in good agreement with that expected for the reduction of $[Fe(CN)_6]^{3-}$.

A challenging question concerns the feasibility of

the application of high-pressure kinetic and thermodynamic techniques in the study of such reactions. Do 'long-distance' electron transfer processes exhibit a characteristic pressure dependence, and to what extent can a volume profile analysis reveal information on the intimate mechanism?

The systems that we investigated in collaboration with others involved intermolecular and intramolecular electron-transfer reactions between ruthenium complexes and cytochrome *c*. We also studied a series of intermolecular reactions between chelated cobalt complexes and cytochrome *c*. A variety of high-pressure experimental techniques, including stopped-flow, flash-photolysis, pulse-radiolysis and voltammetry, were employed in these investigations. As the following presentation will show, a remarkably good agreement was found between the volume data obtained with the aid of these different techniques, which clearly demonstrates the complementarity of these methods for the study of electron-transfer processes.

Application of pulse-radiolysis techniques revealed that the following intramolecular and intermolecular electron-transfer reactions all exhibit a significant acceleration with increasing pressure. The reported volumes of activation are -17.7 ± 0.9, -18.3 ± 0.7, and -15.6 ± 0.6 cm^3 mol^{-1}, respectively, and clearly demonstrate a significant volume collapse on going from the reactant to the transition state [49].

$$(NH_3)_5Ru^{II}-(His33)cyt\ c^{III} \rightarrow (NH_3)_5Ru^{III}-(His33)cyt\ c^{II} \quad (8)$$

$$(NH_3)_5Ru^{II}-(His39)cyt\ c^{III} \rightarrow (NH_3)_5Ru^{III}-(His39)cyt\ c^{II} \quad (9)$$

$$[Ru^{II}(NH_3)_6]^{2+} + cyt\ c^{III} \rightarrow [Ru^{III}(NH_3)_6]^{3+} + cyt\ c^{II} \quad (10)$$

At this stage it was uncertain what the negative volumes of activation really meant since overall reaction volumes were not available. There were, however, data, now in the literature [36], that suggested that the oxidation of $[Ru(NH_3)_6]^{2+}$ to $[Ru(NH_3)_6]^{3+}$ is accompanied by a volume decrease of ca. 30 cm^3 mol^{-1}, which would mean that the activation volumes quoted above could mainly arise from volume changes associated with the oxidation of the ruthenium redox partner.

In order to obtain further information on the magnitude of the overall reaction volume and the location of the transition state along the reaction coordinate, a series of intermolecular electron-transfer reactions of cytochrome *c* with pentaammineruthenium complexes were studied, where the sixth ligand on the ruthenium complex was selected in such a way that the overall driving force was low enough so that the reaction kinetics could be studied in both directions [50,51]. The selected substituents were isonicotinamide (isn), 4-ethylpyridine (etpy), pyridine (py), and 3,5-lutidine (lut). The overall reaction can be formulated as:

$$[Ru^{III}(NH_3)_5L]^{3+} + cyt\ c^{II} \rightleftharpoons [Ru^{II}(NH_3)_5L]^{2+} + cyt\ c^{III} \quad (11)$$

For all the investigated systems, the forward reaction was significantly decelerated by pressure, whereas the reverse reaction was significantly accelerated by pressure. The absolute values of the volumes of activation for the forward and reverse processes were indeed very similar, demonstrating that a similar rearrangement occurs in order to reach the transition state. In addition, the overall reaction volume for these systems could be determined spectrophotometrically by recording the spectrum of an equilibrium mixture as a function of pressure, and electrochemically by recording cyclic and differential pulse voltammograms as a function of pressure [52]. A comparison of the ΔV data demonstrates the generally good agreement between the values obtained from the difference in the volumes of activation for the forward and reverse reactions, and those obtained thermodynamically. Furthermore, the values also clearly demonstrate that $|\Delta V^{\#}| \approx 0.5|\Delta V|$, i.e., the transition state lies approximately halfway between the reactant and product states on a volume basis independent of the direction of electron transfer. The typical volume profile in Fig. 5 presents the overall picture, from which the location of the transition state can clearly be seen.

Similar results were obtained for the redox reactions of a series of cobalt diimine complexes with cytochrome *c* [53,54]. In general, a good agreement exists between the kinetically and thermodynamically determined parameters, and the typical volume profile in Fig. 6 once again demonstrates the symmetrical location of the transition state with respect to the reactant and product states.

At this point it is important to ask the question where these volume changes really come from. We have always argued that the major volume change arises from changes on the redox partner and not on cytochrome *c* itself. This was suggested by the fact that the change in partial molar volume associated with the oxidation of the investigated Ru(II) and Co(II) complexes as obtained from electrochemical and density measurements, almost fully accounted for the observed overall reaction volume. Thus the reduction of cytochrome *c* can only make a minor contribution towards the overall volume change.

These arguments were apparently in contradiction with electrochemical results reported by Cruanes et al. [55], according to which the reduction of cytochrome *c* is accompanied by a volume collapse of 24 $cm^3\ mol^{-1}$. This value is so large that it almost represents all of the reaction volume found for the investigated reactions discussed above. A reinvestigation of the electrochemistry of cytochrome *c* as a function of pressure, using cyclic and differential pulse voltammetric techniques [52], revealed a reaction volume of $-14.0 \pm 0.5\ cm^3\ mol^{-1}$ for the reaction

$$\text{cyt}\ c^{\text{III}} + \text{Ag(s)} + \text{Cl}^- \rightarrow \text{cyt}\ c^{\text{II}} + \text{AgCl(s)} \qquad (12)$$

A correction for the contribution from the reference electrode can be made on the basis of the data published by Tregloan et al. [35], and a series of measurements of the potential of the Ag/AgCl(KCl sat'd) electrode relative to the Ag/Ag^+ electrode as a function of pressure. The contribution of the reference electrode turned out to be $-9.0 \pm 0.6\ cm^3\ mol^{-1}$, from which it then followed that the reduction of cytochrome c^{III} is accompanied by a volume decrease of $5.0 \pm 0.8\ cm^3\ mol^{-1}$. This contribution is significantly smaller than concluded by Cruanes et al. [55], and is also in line with the other arguments referred to above. Thus we conclude that the observed activation and reaction volumes mainly arise from volume changes on the Ru and Co complexes, which in turn will largely be associated with changes in electrostriction in the case of the ammine complexes. The oxidation of the Ru(II) ammine complexes will be accompanied by a large increase in electrostriction and almost no change in the metal–ligand bond length, whereas in the case of the Co complexes a significant contribution from intrinsic volume changes associated with the oxidation of Co(II) will partially account for the observed effects [36].

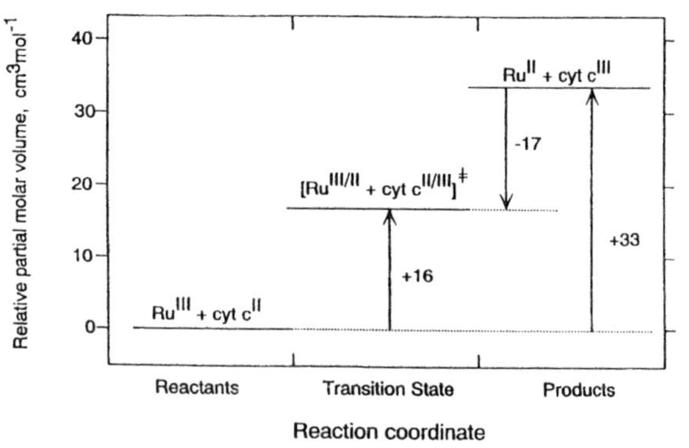

Fig. 5. Volume profile for the reaction $Ru(NH_3)_5isn^{3+}$+cyt $c^{\text{II}} \rightleftharpoons Ru(NH_3)_5isn^{2+}$+cyt c^{III}.

The available results nicely demonstrate the complementarity of the kinetic and thermodynamic data obtained from stopped-flow, UV-vis, electrochemical and density measurements. The resulting picture is very consistent and allows a further detailed analysis of the data. The overall reaction volumes determined in four different ways are surprisingly similar and underline the validity of the different methods employed. The volume profiles in Figs. 5 and 6 demonstrate the symmetric nature of the intrinsic and solvational reorganization in order to reach the

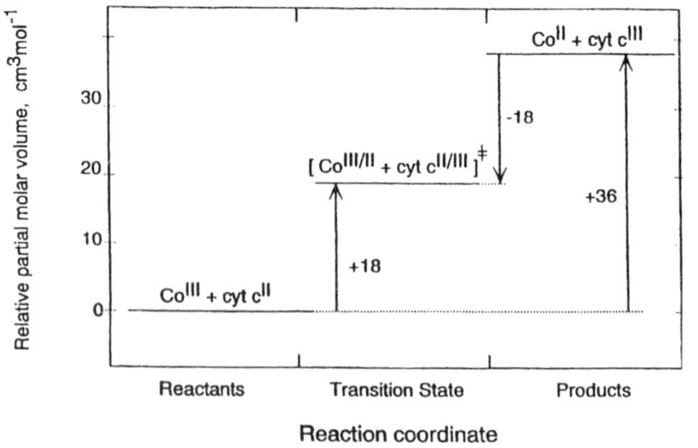

Fig. 6. Volume profile for the reaction $Co(terpy)_2^{3+}$+cyt $c^{\text{II}} \rightleftharpoons Co(terpy)_2^{2+}$+cyt c^{III}.

transition state of the electron-transfer process. In these systems the volume profile is controlled by effects on the redox partner of cytochrome *c*, but this does not necessarily always have to be the case. The location of the transition state on a volume basis will reveal information concerning the 'early' or 'late' nature of the transition state and reveal details of the actual electron-transfer route followed.

In the above quoted examples involving cytochrome *c*, precursor formation between the reactants is too weak to enable the kinetic separation of the precursor formation constant K and the electron transfer rate constant k_{ET}, with the result that the observed second-order rate constant is a composite quantity and represents $k_{ET}K$. For this reason much attention was paid in subsequent studies to reactions in which the separation of these constants should in principle be possible. Under conditions of effective precursor formation (for instance in the case of reaction partners of opposite charge), saturation kinetics should be observed for which the electron transfer reaction then represents the limiting rate constant.

The first example is the irreversible reaction between the anionic *trans*-bis(2-ethyl-2-hydroxybutanoato(2-))oxochromate(V) complex and cytochrome c^{II} as given in Reaction 13 [56].

$$[Cr^{V}(L)_2]^{1-} + \text{cyt } c^{II} \rightarrow [Cr^{IV}(L)_2]^{2-} + \text{cyt } c^{III} \quad (13)$$

$$L = 2\text{-ethyl-}2\text{-hydroxybutanoato}(2-)$$

The low stability of the chromium complex complicated the experimental measurements at high complex concentrations. Nevertheless, the concentration dependence of the observed pseudo-first order rate constant showed clear saturation kinetics, for which $K = 37 \pm 5$ M^{-1} and $k_{ET} = 1510 \pm 170$ s^{-1} at pH 4.8 and 15°C. The thermal activation parameters determined in the low concentration range (i.e., where the second-order rate constant equals Kk_{ET}), show a low value for the activation enthalpy, $\Delta H^{\neq} = 20.9 \pm 0.6$ kJ mol^{-1}, and a high negative value for the activation entropy, $\Delta S^{\neq} = -80 \pm 2$ J K^{-1} mol^{-1}. The significantly negative activation entropy suggests a highly-structured transition state. What can we say about the location of the transition state in terms of volume changes along the reaction coordinate?

The activation volumes were found to be -9.2 ± 0.6 and -11.1 ± 0.8 cm^3 mol^{-1} at pH 5.0 and 4.8, respectively. From electrochemical measurements on the $Cr^{V/IV}$ couple, the dependence of the redox potential on pressure was determined, from which reaction volumes of -11.8 ± 0.5 (pH 5.0) and -14.7 ± 0.7 cm^3 mol^{-1} (pH 4.8) for this redox couple could be calculated. In this case the overall volume changes result from the reduction of Cr^{V} to Cr^{IV}, and the change in the overall charge from -1 to -2. From earlier electrochemical studies it is known that the oxidation of cytochrome c^{II} to cytochrome c^{III} is accompanied by a small volume increase of 5.0 ± 0.8 cm^3 mol^{-1} [52]. It follows that the overall oxidation of cytochrome c^{II} by the Cr^{V} oxo species is associated with an overall volume decrease of 7 and 10 cm^3 mol^{-1}, respectively. The solvational effects arising from the increase in electrostriction account for the overall negative reaction volumes. The volume profile for electron transfer in this system exhibits a non-symmetrical nature as shown in Fig. 7. The transition state lies on the product side and can be characterized as 'late' [56].

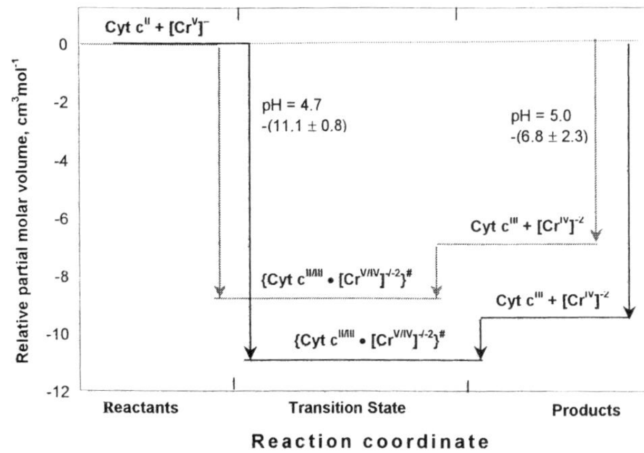

Fig. 7. Volume profile for the reaction of cyt c^{II} with *trans*-bis(2-ethyl-2-hydroxybutanoato(2-))oxochromate(V).

Saturation kinetics were also found for the irreversible reaction between cytochrome c^{II} and trisoxalatocobalt(III) as shown in Reaction 14 [57].

$$[Co^{III}(ox)_3]^{3-} + \text{cyt } c^{II} \rightarrow [Co^{II}(ox)_3]^{4-} + \text{cyt } c^{III} \quad (14)$$

The activation volume for the combined formation of the ion-pair and subsequent electron transfer reaction was obtain for low concentrations of the cobalt complex and found to be $\Delta V^{\neq} = \Delta V(K) + \Delta V^{\neq}(k_{ET}) = -11 \pm 1$ cm^3 mol^{-1}, since $k_{obs} = Kk_{ET}$-[Co^{III}] under these conditions. At high Co(III) concentrations, $\Delta V^{\neq}(k_{ET})$ was found to be -5.5 ± 0.6 cm^3 mol^{-1}, since $k_{obs} = k_{ET}$ under such conditions. The $\Delta H^{\neq}$ and $\Delta S^{\neq}$ values are larger and $\Delta V^{\neq}$ is more negative for lower concentrations of $[Co(ox)_3]^{3-}$, i.e., under conditions where the two steps of the reaction contribute to these values. At the high concentration limit only the electron transfer process determines the activation parameters.

The difference between the $\Delta V^{\neq}$ data reported for the reaction at high and low trisoxalatocobalt(III) concentrations, represents the value for $\Delta V(K)$ and equals -5.5 ± 1.0 cm^3 mol^{-1}. Thus ion-pair formation between cytochrome c^{II} and $[Co(ox)_3]^{3-}$ must involve a significant overlap of the van der Waals radii of the reactants without significant charge neutralization which would induce desolvation. The ion-pair can then be depicted as solvent separated. The $\Delta V^{\#}(k_{ET})$ value of -5.5 ± 1.0 cm^3 mol^{-1} indicates that the transition state for the electron transfer reaction within the precursor (ion-pair) complex is more compact, which is most probably related to a volume decrease on the $[Co^{III}(ox)_3]^{3-}$ partner. As we showed earlier, the oxidation of cytochrome c^{II} to c^{III} is accompanied by a volume increase of 5 cm^3 mol^{-1} [52]. When $[Co^{III}(ox)_3]^{3-}$ is reduced to $[Co^{II}(ox)_3]^{4-}$, the increase in overall negative charge will cause an increase in electrostriction which could account for the observed negative volume of activation.

Recent investigations on a series of intramolecular electron transfer reactions, closely related to the series of intermolecular reactions described above, revealed non-symmetrical volume profiles [58]. Reactions of the type shown in Reaction 15

$$(NH_3)_4(L)Ru^{III}-(His33)-cyt\ c^{II} \rightleftharpoons (NH_3)_4(L)Ru^{II}-(His33)-cyt\ c^{III} \quad (15)$$

where L = isonicotinamide, 4-ethylpyridine, 3,5-lutidine, and pyridine, all exhibited volumes of activation for the forward reaction of between +3 and +7 cm^3 mol^{-1}, compared to overall reaction volumes of between +19 and +26 cm^3 mol^{-1}. This indicates that electron transfer from Fe to Ru is characterized by an 'early' transition state in terms of volume changes

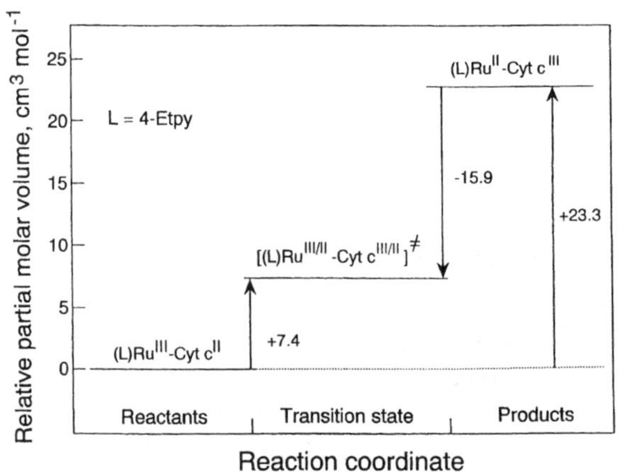

Fig. 8. Volume profile for the reaction $(NH_3)_4Ru^{III}$(4-Etpy)-cyt $c^{II} \rightleftharpoons (NH_3)_4Ru^{II}$(4-Etpy)-cyt c^{III}.

along the reaction coordinate (see Fig. 8). The overall volume changes could be accounted for in terms of electrostriction effects centered around the ammine ligands on the ruthenium center. A number of possible explanations in terms of the effect of pressure on electronic and nuclear factors were offered to account for the asymmetrical nature of the volume profile [58].

One system has been investigated where the effect of pressure on the electron-transfer rate constant revealed information on the actual electron-transfer route. In this study the effect of pressure on distant electronic coupling in $Ru(bpy)_2(im)$-modified His33 and His72 cytochrome c derivatives, for which the electron transfer from Fe(II) to Ru(III) is activationless [59]. In the case of the His33-modified system the electron-transfer rate constant exhibited no dependence on pressure within experimental error limits. However, the rate constant for the His72-modified protein increased significantly with increasing pressure, corresponding to a $\Delta V^{\#}$ value of -6 ± 2 cm^3 mol^{-1}. Since this value is exactly opposite to that expected for the reduction of Ru(III), the result was interpreted as an increase in electronic coupling at elevated pressure. The application of moderate pressures will cause a slight compression of the protein that in turn shrinks the through-space gaps that are key units in the electron-tunneling pathway between the heme and His72. A decrease of 0.46 A in the tunneling path length at a pressure of 150 MPa can account for the observed increase in rate constant. This in turn means that there is an average

Table 1
Activation and reaction volumes for typical outer-sphere electron transfer reactions

Reaction	$\Delta V^{\#}$, $cm^3\ mol^{-1}$	ΔV, $cm^3\ mol^{-1}$	Ref.
$[Co(NH_3)_5OH_2]^{3+}$+$[Fe(CN)_6]^{4}$	+26.5 ± 2.4	−15 ± 8	[38]
$[Co(NH_3)_5py]^{3+}$+$[Fe(CN)_6]^{4-}$	+29.8 ± 1.4	+23 ± 3	[38]
$[Co(NH_3)_5Me_2SO]^{3+}$+$[Fe(CN)_6]^{4-}$	+6 ± 2	−11 ± 3	[38]
$[Co(NH_3)_5(HP_2O_7)]$+$[Fe(CN)_6]^{4-}$	+36 ± 3 (308 K)	–	[41]
$[Co(NH_3)_5(P_2O_7)]^-$+$[Fe(CN)_6]^{4-}$	+13 ± 1 (314 K)	–	[41]
β-$[Co(NH_3)_5(P_3O_{10})]^{2-}$+$[Fe(CN)_6]^{4-}$	+13 ± 2 (308 K)	–	[41]
γ-$[Co(NH_3)_5(P_3O_{10})]^{2-}$+$[Fe(CN)_6]^{4-}$	+32 ± 2 (317 K)	–	[41]
$[Co(NH_3)_5(HPO_3)]^+$+$[Fe(CN)_6]^{4-}$	+22 ± 2 (314 K)	–	[41]
$[Co(NH_3)_5(H_2PO_2)]^{2+}$+$[Fe(CN)_6]^{4-}$	+30 ± 1 (314 K)	–	[41]
$[Co(NH_3)_5(HPO_4)]^+$+$[Fe(CN)_6]^{4-}$	+37 ± 4	–	[42]
$[Co(NH_3)_5(PO_4)]$+$[Fe(CN)_6]^{4-}$	+17 ± 1	–	[42]
$[Co(NMeH_2)_5(HPO_4)]^+$+$[Fe(CN)_6]^{4-}$	+44 ± 5	–	[42]
$[Co(NMeH_2)_5(PO_4)]$+$[Fe(CN)_6]^{4-}$	+32 ± 1	–	[42]
$[CoL(HPO_4)]^{+a}$+$[Fe(CN)_6]^{4-}$	+36 ± 2	–	[42]
L-Ascorbic acid+$[Fe(CN)_6]^{3-}$	−16.6 ± 0.5 (pH 0.3)	–	[44]
	−15 ± 1 (pH 5)	–	[44]
	−16.3 ± 0.4 (pH 5.25)	–	[44]
L-Ascorbic acid+$[Fe(CN)_5(NO_2)]^{3-}$	−10.0 ± 0.5	–	[45]
$[Ru(NH_3)_6]^{2+}$+cyt c^{III}	−15.6 ± 0.6	–	[49]
$[Ru(NH_3)_5isn]^{3+}$+cyt c^{II}	+15.9 ± 0.7	–	[50]
$[Ru(NH_3)_5isn]^{2+}$+cyt c^{III}	−17.2 ± 1.5	–	[50]
$[Ru(NH_3)_5lut]^{3+}$+cyt c^{II}	+16.9 ± 1.4	33.6 ± 1.7	[51]
$[Ru(NH_3)_5lut]^{2+}$+cyt c^{III}	−17.8 ± 1.6	34.7 ± 2.1	[51]
$[Ru(NH_3)_5etpy]^{3+}$+cyt c^{II}	+14.7 ± 1.8	26.9 ± 1.8	[51]
$[Ru(NH_3)_5etpy]^{2+}$+cyt c^{III}	−14.9 ± 1.1	29.6 ± 1.4	[51]
$[Ru(NH_3)_5py]^{3+}$+cyt c^{II}	+17.4 ± 1.5	33.4 ± 1.9	[51]
$[Ru(NH_3)_5py]^{2+}$+cyt c^{III}	−17.7 ± 0.8	35.1 ± 1.7	[51]
$[Co(bpy)_3]^{3+}$+cyt c^{II}	+12.5 ± 0.9	21.8 ± 0.7	[54]
$[Co(bpy)_3]^{2+}$+cyt c^{III}	−12.6 ± 1.5	25.1 ± 1.7	[54]
$[Co(phen)_3]^{3+}$+cyt c^{II}	+17.0 ± 0.9	37.9 ± 2.0	[54]
$[Co(phen)_3]^{2+}$+cyt c^{III}	−16.2 ± 1.0	34.2 ± 1.7	[54]
$[Co(terpy)_3]^{3+}$+cyt c^{II}	+18.4 ± 1.2	33 ± 3	[54]
$[Co(terpy)_3]^{2+}$+cyt c^{III}	−18.0 ± 1.4	36 ± 2	[54]
$[Cr^V]^-$+cyt c^{II}	−11.1 ± 0.8 (pH 4.7)	−9.7 ± 2.5 (pH 4.7)	[56]
	−9.2 ± 0.6 (pH 5)	−6.8 ± 2.3 (pH 5)	[56]
$[Co(ox)_3]^{3-}$+cyt c^{II}	−5.5 ± 1.0 (for k_{et})	−5.4 ± 1.6 (for K)	[57]
HAO+cyt c^{III}	−27.7 (I = 20 mM, T = 4°C)	–	[62]
	−24.3 (I = 20 mM, T = 20°C)	–	[62]
	−5.6 (I = 420 mM, T = 20°C)	–	[62]
MDH+cyt c_L^{III}	+122 (I = 20 mM, T = 4°C)	–	[62]
	+54 (I = 10 mM, T = 25°C)	–	[62]
	+70 (I = 110 mM, T = 25°C)	–	[62]

[a] L = *cis*-10-Amino-10-methyl-1,4,8,12-tetraazacyclopentadecane.

decrease in the space-gap of 0.1 A. The absence of an effect for the His33-modified species is understandable since electronic coupling through covalent and hydrogen bonds will be less pressure sensitive than coupling via van der Waals gaps [59].

Morishima and co-workers [60] investigated the effect of pressure on electron transfer rates in zinc/ruthenium modified myoglobins. The rate constant for electron transfer from photoexcited $^3ZnP^*$ to a covalently attached $[Ru(NH_3)_5]^{3+}$ moiety on the surface of the protein decreased from 5×10^7 to 55 s^{-1} upon increasing the distance from 9.5 to 19.3 Å when the Ru complex is attached to His70 and His83, respectively. This decrease in the rate constant was accompanied by an increase in $\Delta V^{\#}$ from +4 to +17 cm^3 mol^{-1}. Within the context of the results reported above and the volume changes associated with the reduction of the Ru(III) ammine complexes, the gradual increase in $\Delta V^{\#}$ with increasing donor-acceptor distance and with decreasing rate constant could be a clear demonstration of 'early' (for the fast) and 'late' (for the slow reactions) transition states. Volume changes mainly associated with changes in electrostriction on the Ru ammine center will control the solvent reorganization and so account for the 'early' (reactant-like) and 'late' (product-like) transition states.

Recently, Morishima and co-workers [61] re-examined their earlier results for electron transfer reactions within myoglobin derivatives [60]. They found that the differences in the activation volumes determined before, are wrong due to significant contributions from dissolved oxygen. The dissolved oxygen in the sample enhanced the decay of the triplet exited state of ZnMb. The results under completely anaerobic conditions show a negative activation volume for the forward reaction in the case of His83-Mb. The pathways of electron transfer in His83-Mb and His81-Mb are very similar constructed by many hydrogen bonds. Both histidines are exposed to the solvent, however hydrogen bonds of His81-Mb restrain the thermal fluctuations near His81, which is reflected in the reduced pressure perturbation on the D-A distance of His81-Mb. The pressure dependence of the electron transfer reaction is sensitive to slight structural deviations of the redox center. The D-A distance is shorter for His48-Mb, for which $\Delta V_{\mathrm{f}}^{\#}$ was found to be positive. The $\Delta V_{\mathrm{b}}^{\#}$ values for the backward reactions in all three derivatives were found to be negative.

Balny and co-workers have studied the effect of pressure on many protein-protein electron transfer processes, for which they also studied the effect of temperature and ionic strength. For the reduction of cytochrome *c* by hydroxylamine oxidoreductase, the activation volume is independent of temperature, but strongly dependent on the ionic strength. With decreasing ionic strength, the activation volume becomes more negative. At higher ionic strength, solvational effects play a more important role and the activation volume is more positive. Another behavior was observed for the reaction of methanol dehydrogenase with cytochrome c_{L} (both isolated from the *Methylophaga marina*). Values of the activation vol-

Table 2
Activation and reaction volumes for typical inner-sphere electron transfer reactions

Reaction	$\Delta V^{\#}$, cm^3 mol^{-1}	ΔV, cm^3 mol^{-1}	Ref.
(hh)[a]cyt c^{III}–$(NH_3)_5Ru^{II}$–His33	−17.7 ± 0.9	–	[49]
(Ck)[b]cyt c^{III}–$(NH_3)_5Ru^{II}$–His39	−18.3 ± 0.7	–	[49]
(hh)[a]cyt c^{III}–$(NH_3)_4$(isn)Ru^{II}	+3.2 ± 0.3	+21.1 ± 1.0	[58]
(hh)[a]cyt c^{III}–$(NH_3)_4$(Etpy)Ru^{II}	+7.4 ± 0.6	+24.9 ± 1.1	[58]
(hh)[a]cyt c^{III}–$(NH_3)_4$(lut)Ru^{II}	+5.8 ± 0.4	+23.3 ± 0.9	[58]
(hh)[a]cyt c^{III}–$(NH_3)_4$(py)Ru^{II}	+7.2 ± 0.2	+23.3 ± 0.6	[58]
MbZnP–$(NH_3)_5Ru^{III}$(His48)	+6	–	[60]
MbZnP–$(NH_3)_5Ru^{III}$(His70)	+4	–	[60]
MbZnP–$(NH_3)_5Ru^{III}$(His81)	+17	–	[60]
MbZnP–$(NH_3)_5Ru^{III}$(His83)	+9	–	[60]
MbZnP–$(NH_3)_5Ru^{III}$(His48)	+6.5 ± 1.0	–	[61]
MbZnP–$(NH_3)_5Ru^{II}$(His48)	−6.2 ± 0.5	–	[61]
MbZnP–$(NH_3)_5Ru^{III}$(His81)	+3.7 ± 0.1	–	[61]
MbZnP–$(NH_3)_5Ru^{II}$(His81)	−5.3 ± 0.5	–	[61]
MbZnP–$(NH_3)_5Ru^{III}$(His83)	−1.6 ± 0.2	–	[61]
MbZnP–$(NH_3)_5Ru^{II}$(His83)	−11.0 ± 0.5	–	[61]
cyt a_3–cyt *a*	+41 ± 5		[63]
cyt *a*–Cu_A	+28 ± 4		[63]

[a](hh), horse heart.
[b](Ck), *Candida krusei*.

ume depend on the temperature. In the Arrhenius plot a break is observed, which was interpreted as resulting from the existence of two different conformations of cytochrome c_L [62].

Larsen [63] used a pressure-dependent transient absorption spectroscopy technique for studying the intramolecular electron transfer reaction in CO-mixed derivative of bovine heart cytochrome *c* oxidase (CcO). Two phases of electron transfer were distinguished. The first one is the fast intramolecular ET between cytochrome *a* and cytochrome a_3. The rate constant $k_{et} = (6.7 \pm 0.9) \times 10^5\ s^{-1}$ and activation volume $\Delta V^{\#} = +45 \pm 5\ cm^3\ mol^{-1}$ were determined. The value of $\Delta V^{\#}$ for this process arises from the structural changes localized at cytochrome a_3 upon heme reduction. The second phase is a slow electron transfer process between cytochrome *a* and Cu_A. The rate constant k_{et} was found to be $(5.9 \pm 1.7) \times 10^4\ s^{-1}$ at ambient pressure while $\Delta V^{\#} = +28 \pm 4\ cm^3\ mol^{-1}$. The activation volume results from structural changes at Cu_A during the reduction. The conformation regulation during electron transfer reaction was not observed on the surface of cytochrome *a*. Information about conformational changes upon reduction resulted from the crystal structures of reduced and oxidized forms of these cytochromes. Analysis of similar examples from the literature suggested that in the presented system the activation volume for the intramolecular electron transfer process depends on two contributions, viz. reorganization ($\Delta V^{\#}_{FC}$) and electronic coupling ($\Delta V^{\#}_{EC}$) terms.

6. Conclusions

Studies on the effect of pressure on electron transfer reactions can assist the elucidation of the intimate reaction mechanism of such processes. The experimental data for symmetrical (self exchange) electron transfer reactions only allow the interpretation of activation volumes, which turn out to be in good agreement with theoretical predictions. In the case of non-symmetrical electron transfer reactions, the first objective is to distinguish between intramolecular and intermolecular processes. A summary of the available activation and reaction volumes for typical outer-sphere and inner-sphere electron transfer reactions is given in Tables 1 and 2, respectively. When the electron transfer reaction is reversible, it is in some cases possible to study the reaction in both the forward and backward directions and to construct a volume profile. For reactions that occur irreversibly, it is more difficult to determine all the parameters that are necessary to construct a volume profile. Nevertheless, these problems can be overcome in some cases by using alternative techniques such as electrochemical measurements at elevated pressure. The application of high pressure does not only contribute to an improved mechanistic understanding of electron transfer reactions in model inorganic systems, but can also contribute to the understanding of more complex processes as found in long distance electron transfer reactions in biological systems.

Acknowledgements

The authors gratefully acknowledge financial support from the Deutsche Forschungsgemeinschaft and the Volkswagen Foundation.

References

[1] D.R. Stranks, Pure Appl. Chem. 38 (1974) 303–323.
[2] H. Kelm, High Pressure Chemistry, Reidel, Dordrecht, 1978.
[3] R. van Eldik, Inorganic High Pressure Chemistry, Kinetics and Mechanism, Elsevier, Amsterdam, 1986.
[4] R. van Eldik, J. Jonas, High Pressure Chemistry and Biochemistry, Reidel, Dordrecht, 1987.
[5] M. Kotowski, R. van Eldik, Coord. Chem. Rev. 93 (1989) 19–57.
[6] R. van Eldik, T. Asano, W.J. le Noble, Chem. Rev. 89 (1989) 549–688.
[7] J.W. Akitt, A.E. Merbach, NMR Basic Princ. Prog. 24 (1990) 189.
[8] R. van Eldik, A.E. Merbach, Comments Inorg. Chem. 12 (1992) 341.
[9] S.F. Lincoln, A.E. Merbach, Adv. Inorg. Chem. 42 (1995) 1.
[10] L. Helm, A.E. Merbach, Coord. Chem. Rev. 187 (1999) 151–181.
[11] G. Stochel, R. van Eldik, Coord. Chem. Rev. 187 (1999) 329–374.
[12] T.W. Swaddle, P.A. Tregloan, Coord. Chem. Rev. 187 (1999) 255–289.
[13] R. Winter, J. Jonas, High Pressure Chemistry, Biochemistry and Materials Science, Kluwer, Dordrecht, 1993.
[14] C.D. Hubbard, R. van Eldik, Instrum. Sci. Technol. 23 (1995) 1–41.

[15] R. van Eldik, C.D. Hubbard, Chemistry under Extreme or Non-Classical Conditions, Wiley/Spektrum, New York/Heidelberg, 1997.
[16] W.R. Holzapfel, N.S. Isaacs, High Pressure Techniques in Chemistry and Physics, Oxford University Press, Oxford, 1997.
[17] R. van Eldik, in: D. Nocera, J.F. Wishart (Eds.), Photochemistry and Radiation Chemistry: Complementary Methods in the Study of Electron Transfer, Adv. Chem. Series 254, Washington, DC, 1998.
[18] W.H. Jolley, D.R. Stranks, T.W. Swaddle, Inorg. Chem. 29 (1990) 1948–1951.
[19] H. Takagi, T.W. Swaddle, Inorg. Chem. 31 (1992) 4669–4673.
[20] L. Spiccia, T.W. Swaddle, Inorg. Chem. 26 (1987) 2265–2271.
[21] H. Doine, T.W. Swaddle, Inorg. Chem. 30 (1991) 1858–1862.
[22] R.D. Shalders, T.W. Swaddle, Inorg. Chem. 34 (1995) 4815–4820.
[23] W.H. Jolley, D.R. Stranks, T.W. Swaddle, Inorg. Chem. 29 (1990) 385–389.
[24] M.R. Grace, T.W. Swaddle, Inorg. Chem. 32 (1993) 5597–5602.
[25] T.W. Swaddle, Inorg. Chem. 29 (1990) 5017–5025.
[26] M.R. Grace, H. Takagi, T.W. Swaddle, Inorg. Chem. 33 (1994) 1915–1920.
[27] T.W. Swaddle, High Pressure Inorganic Chemistry: Kinetics and Mechanisms, Elsevier, Amsterdam, 1987.
[28] B. Goyal, S. Solanki, S. Arora, A. Prakash, R.N. Mehotra, J. Chem. Soc. Dalton Trans. (1995) 3109.
[29] J.M. Leal, P.L. Domingo, B. Garcia, S. Ibeas, J. Chem. Soc. Faraday Trans. 89 (1995) 3571.
[30] P.D. Metelski, T.W. Swaddle, Inorg. Chem. 38 (1999) 301–307.
[31] C.D. Hubbard, A. Gerhard, R. van Eldik, unpublished observation.
[32] Y. Fu, T.W. Swaddle, Chem. Commun. (1996) 1171-1172.
[33] T.W. Swaddle, Can. J. Chem. 74 (1996) 631–638.
[34] Y. Fu, T.W. Swaddle, J. Am. Chem. Soc. 119 (1997) 7137–7144.
[35] J.I. Sachinidis, R.D. Shalders, P.A. Tregloan, Inorg. Chem. 33 (1994) 6180–6186.
[36] J.I. Sachinidis, R.D. Shalders, P.A. Tregloan, Inorg. Chem. 35 (1996) 2497–2503.
[37] R. van Eldik, High Press. Res. 6 (1991) 251.
[38] I. Krack, R. van Eldik, Inorg. Chem. 25 (1986) 1743–1747.
[39] I. Krack, R. van Eldik, Inorg. Chem. 29 (1989) 851–855.
[40] I. Krack, R. van Eldik, Inorg. Chem. 29 (1990) 1700–1704.
[41] M. Martinez, M.-A. Pitarque, R. van Eldik, J. Chem. Soc. Dalton Trans. (1996) 2665-2671.
[42] M. Martinez, M.-A. Pitarque, R. van Eldik, J. Chem. Soc. Dalton Trans. (1995) 4107.
[43] M. Martinez, M.-A. Pitarque, R. van Eldik, Inorg. Chim. Acta 256 (1997) 51–59.
[44] B. Bänsch, P. Martinez, J. Zuluaga, D. Uribe, R. van Eldik, Z. Phys. Chem. 170 (1991) 59–71.
[45] A. Wanat, R. van Eldik, G. Stochel, J. Chem. Soc. Dalton Trans. (1998) 2497-2501.
[46] M. Matsumoto, T. Tarumi, K.-I. Sugimoto, N. Kagayama, S. Funahashi, H.D. Takagi, Inorg. Chim. Acta 255 (1997) 81.
[47] E. Ilkowska, R. van Eldik, G. Stochel, J. Biol. Inorg. Chem. 2 (1992) 603.
[48] H.-D. Projahn, C. Dreher, R. van Eldik, J. Am. Chem. Soc. 112 (1990) 17–22.
[49] J.F. Wishart, R. van Eldik, J. Sun, C. Su, S.S. Isied, Inorg. Chem. 31 (1992) 3986–3989.
[50] B. Bänsch, M. Meier, P. Martinez, R. van Eldik, C. Su, J. Sun, S.S. Isied, J.F. Wishart, Inorg. Chem. 33 (1994) 4744–4749.
[51] M. Meier, J. Sun, R. van Eldik, J.F. Wishart, Inorg. Chem. 35 (1996) 1564–1570.
[52] J. Sun, J.F. Wishart, R. van Eldik, R.D. Shalders, T.W. Swaddle, J. Am. Chem. Soc. 117 (1995) 2600–2605.
[53] M. Meier, R. van Eldik, Inorg. Chim. Acta 225 (1994) 95.
[54] M. Meier, R. van Eldik, Chem. Eur. J. 3 (1997) 39–46.
[55] M.T. Cruanes, K.K. Rodgers, S.G. Sligar, J. Am. Chem. Soc. 114 (1992) 9660–9661.
[56] M. Körner, R. van Eldik, Eur. J. Inorg. Chem. (1999) 1805-1812.
[57] J. Macyk, R. van Eldik, J. Chem. Soc. Dalton Trans. (2001) 2288-2292.
[58] J. Sun, C. Su, M. Meier, S.S. Isied, J.F. Wishart, R. van Eldik, Inorg. Chem. 37 (1998) 6129–6135.
[59] M. Meier, R. van Eldik, I.J. Chang, G.A. Mines, D.S. Wuttke, J.R. Winkler, H.B. Gray, J. Am. Chem. Soc. 116 (1994) 1577–1578.
[60] Y. Sugiyama, S. Takahashi, K. Ishimori, I. Morishima, J. Am. Chem. Soc. 119 (1997) 9582–9583.
[61] Y. Furukawa, K. Ishimori, I. Morishima, J. Phys. Chem. 104 (2000) 1817.
[62] C. Balny, High Pressure and Biotechnology., Vol. 224, Editions John Libbey Eurotext, Montrouge, 1992, pp. 155–157.
[63] R.W. Larsen, FEBS Lett. 462 (1999) 75–78.

ELSEVIER

Biochimica et Biophysica Acta 1595 (2002) 297–308

BIOCHIMICA ET BIOPHYSICA ACTA
BBA
www.bba-direct.com

Review

High pressure, a tool for exploring heme protein active sites

Gaston Hui Bon Hoa [a,*], Mark A. McLean [b], Stephen G. Sligar [b]

[a] *INSERM U473, 84 rue du Général Leclerc, 94276 Le Kremlin Bicêtre cedex, France*
[b] *Department of Biochemistry, School of Molecular and Cellular Biology, University of Illinois, Urbana, IL 61801, USA*

Received 12 September 2001; received in revised form 23 November 2001; accepted 28 November 2001

Abstract

High pressure is an interesting and suitable parameter in the study of the dynamics and stability of proteins. The effects of pressure on proteins delineates its volumic ($\Delta V°$) and energetic ($\Delta G°$) parameters. An enormous amount of effort has been invested by several laboratories in developing basic theory and high pressure techniques that allow the determination of barotropic parameters. Cytochrome P450s, one of the largest super families of heme proteins, are good models for high pressure studies. Two distinct pressure-induced spin transitions of the heme iron in the active site and a P450 to P420 inactivation process have been characterized. The obtained reaction volumes of these two processes for a series of analog-bound cytochrome P450s are compared. We have shown that both the spin volume and the inactivation volume are dependent on the substrate analogs which are known to modulate the polarity and hydration of the heme pocket. Several linear correlations were found between these reaction volumes and the physico-chemical properties of the heme protein such as the polarity-induced exposure of tyrosines, the hydration of the cytochrome CYP101 heme pocket, and the mobility and binding of the substrates indicate that they constitute the main contribution to the complex thermodynamic reaction volume parameters. This interpretation allows us to conclude that cytochrome CYP101, CYP2B4 and CYP102 possess a similar mechanism of substrate binding. Interestingly the barotropic behaviors of monomeric cytochrome P450s are quite different from those of oligomeric and hetorooligomeric cytochrome P450s. The interactions of heterooligomeric subunits influence the stability of individual cytochrome P450s and the asymmetric organization of subunits which can control and modulate the activity and the recognition with NADPH–cytochrome P450 reductase.

Keywords: High pressure effect; Stability of P450; P450 dynamics; Pressure-induced P420; Active site hydration; Oligomerization dynamics

1. Introduction

It is surprising to discover that cells living deep below the ocean surface can withstand pressures as high as 60 MPa. The reason for cells extraordinary adaptation to hydrostatic pressure lies in the properties of cellular water and in particular in the water associated with macromolecules. These associated waters, or structured waters, form a hydration shell around macromolecules. The densities of the structured waters are heterogeneous and reflect the densities of water molecules which interact with different side-chain and main chain functionalities within the protein structure. The variations in the water density and volume can be large which makes the system quite sensitive to pressure. The reorganization of

* Corresponding author.
E-mail address: gasthuiz@netscape.net (G. Hui Bon Hoa).

0167-4838/02/$ – see front matter
PII: S0167-4838(01)00352-1

water structures under pressure necessitates that a change in the macromolecular conformation is occurring. Heme-containing enzymes, particularly the cytochrome P450s, are an interesting class of proteins which not only undergo changes in their hydration shell under pressure but display changes in the active site hydration and structure. For example, moderate pressures can inactivate a cytochrome P450 without any easily detectable change in secondary structure. Due to the presence of the heme chromophore, these changes can be detected through various spectroscopic techniques. It is thus important to use high pressure as a tool to explore subtle changes in the active site of these important enzymes of central metabolism. The purpose of this review is to present to the general reader the potential of high pressure as a tool to study the hydration, stability and dynamics of the active site, and to also give insight into both the substrate recognition and structural reorganization of subunit interactions of the cytochrome P450 super family. The barotropic behaviors of cytochrome CYP101, CYP102, CYP2B4 and CYP1A2 are presented.

2. High-pressure effects on proteins

High pressure is a thermodynamic parameter that permits to isolate the effects that depend exclusively upon volume changes. On the contrary, temperature changes result in simultaneously changes in volume (through the coefficient of thermal expansion) and thermal energy (through the increase of temperature) in a way which makes it difficult to separate their effects. In fact to obtain the temperature coefficient at constant volume, the use of pressure is indispensable. Following Le Chatelier's principle, the application of pressure to any system in equilibrium forces the system to favor the equilibrium state which has the smaller volume. The amplitude and the direction of the equilibrium shift depend on the value of the reaction volume and its sign:

$$\delta(\ln K_{eq})/\delta p = -(\Delta V^\circ)/RT \qquad (1)$$

where K_{eq} is the equilibrium constant of the system, p is the pressure (usually reported in MPa, 1 MPa = 10 bar), R is the universal gas constant, T is the absolute temperature and ΔV° is the reaction volume change of the system in cm^3/mol. The reaction volume is equal to the difference between the partial molar volumes of the products and the reactants:

$$\Delta V^\circ = V^\circ_{prod} - V^\circ_{react} \qquad (2)$$

A positive reaction volume implies a shift of the equilibrium towards the reactants upon increase of pressure. The opposite effect will occur when the reaction volume is negative. For kinetic analysis in the simplest Van't Hoff formalism:

$$\delta(\ln k)/\delta p = -\Delta V^{\neq}/RT \qquad (3)$$

where k is a rate constant and $\Delta V^{\neq}$ is the activation volume.

The origins of a volume change in a protein are numerous. The contributions to the apparent partial molar volume of a protein include several terms:

$$V^\circ_{app} = V^\circ_{intr} + V^\circ_{cav} + V^\circ_{solv} \qquad (4)$$

Where V°_{intr} is the intrinsic volumes of all the atoms, V°_{cav} is the void volumes within the protein due to the imperfect packing of the polypeptide chain, the void volume is big enough for the binding of one molecule of water or xenon, and V°_{solv} is the solvation volume due to the interactions of chemical groups with the solvent. For proteins, only the contributions of the last two terms are usually significant. For cytochromes P450, the heme pocket is a large cavity (79 Å^3 for CYP101 and 343 Å^3 for CYP102, results from X-ray crystallographic structures, Dr. Shelley Maves, personal communication) that can have a varying degree of compressibility. We shall see that the solvation of the heme pocket of cytochrome P450 plays an important role in controlling substrate binding, activity and enzyme stability. The sensitivity of such a reaction will be described in the present review.

Pressure acts not only on the differences between equilibrium states of proteins but also on the way solvent interacts with chemical groups on the protein surfaces and can also perturb the active site, depending on it's compressibility and hydration state. The breaking of an ionic salt linkage, substrate dissociation [1,2] and the interaction of water molecules with the apolar domains [3] give rise to negative volume changes. The density of these reorganized water are greater than 1 [4,5]. Charge-transfer interactions and stacking of aromatic residues also exhibits negative volume changes [6].

When studying biological systems that are sensitive to pH, the effect of pressure on the buffering system must be taken into account due to the fact that acid-base reactions also obey Le Chatelier's Principle. Anionic buffers such as phosphate are highly pressure sensitive ($\Delta V^{\circ} \sim -10$ cm^3/mol) while amine buffers such as Tris and MES buffers are rather insensitive to pressure ($\Delta V^{\circ} \sim -1$ cm^3/mol) [7].

The effects of pressure on enzyme kinetics depend on the conformation of the transition states involved in the reaction mechanism. The theory of such an effect was first formulated by Laidler in 1950 [8], followed recently by an overview given by Morild, [9]. In direct comparison to the equilibrium reaction, enzyme kinetics obey the same Le Chatelier principle and are controlled by the magnitude of the activation volume changes ($\Delta V^{\neq}$ which is the excess of the partial molar volume of the transition state over the sum of the partial molar volumes of the reactants). If the activation volume changes are negative (contraction of atomic bonds or rearrangement of surrounding solvent molecules during reaction) then the rates are accelerated. On the contrary, formation of single bonds are usually accompanied by an expansion in interatomic distances leading to a slowing down of the rates.

3. High-pressure methodologies

To analyze the effects of pressure on proteins, several laboratories have developed specific techniques which utilize hydrostatic pressure to yield information about the volume profile of a reaction. Since cytochromes P450 display distinct absorbance as well as fluorescence properties in the visible and UV regions due to their aromatic amino acid side chains on the heme prosthetic group, we focus only on the description of high-pressure optical spectroscopy.

A high-pressure system incorporating an optical cell that is suitable to follow UV/visible pressure-induced spectra changes has been described [10]. The major components of this system are a 700 MPa high-pressure chamber equipped with sapphire windows (10 mm thick), several Bridgman's high-pressure seals, a metallic transducer for monitoring the pressure and a 700 MPa high-pressure generator. A quartz optical cell containing the enzyme solution is sealed by a flexible Teflon membrane and placed inside the high-pressure chamber which is filled with the compression fluid (*n*-pentane, silicon oil or ethanol). Pressure can then be applied inside the bomb through a high-pressure capillary connected to the generator. The high-pressure chamber can reach, without any leakage, 700 MPa and −40°C using a thermocryostat and is also designed to fit inside the sample compartment of a Cary 3E spectrophotometer or an SLM 4800 spectrofluorometer. A popular method to investigate the effect of pressure on reaction kinetics is to apply a pressure jump to a system which is at equilibrium and then monitor the rate at which the system reaches a new equilibrium. We have constructed a system capable of producing a pressure jump of 20 MPa in both directions with a dead time of less than 3 ms. The originality of such a setting lies in the addition of a reservoir isolated from the high-pressure cell and the generator by high-pressure valves. The reservoir is brought to a pressure higher or lower than that of the optical chamber in order to create an excess of pressure which is liberated when the pneumatic valve is opened suddenly. Positive as well as negative P-jumps can be achieved [12].

4. Cytochrome P450 monooxygenases

Cytochrome P450s belong to a super family of monooxygenases which metabolize xenobiotics and participate in a variety of biosynthetic pathways such as steroid hormone biosynthesis. They share a common proposed reaction cycle which involves substrate binding, reduction of the heme iron followed by dioxygen binding to yield a reduced oxygenated protein. This oxygenated intermediate then accepts a second electron followed by reductive scission of the dioxygen bond and hydrogen abstraction from the substrate. Finally, in a process introduced by Groves [13] termed 'oxygen rebound', oxygen is incorporated into the substrate, and the product is released, leaving the enzyme in its ferric resting state. Since P450s are involved in the metabolism of many pharmaceuticals, studies of these enzymes and their reaction mechanisms have occupied a central place in the interests of pharmacologists and toxicologists.

4.1. CYP101 ($P450_{cam}$)

CYP101 is a 45 kDa cytosolic bacterial heme protein from *Pseudomonas putida* which hydroxylates its natural substrate, camphor, both stereo- and regiospecifically. It was the first P450 enzyme to have the crystallographic structure determined [14,15]. The crystal structure showed that the heme is held by a loop on the end of the L-helix which provides the cysteine thiolate ligand to the iron atom. Binding of the camphor substrate above pyrrole rings C and D is promoted by a hydrogen bond between the camphor oxygen and Tyr-96 and also by hydrophobic contacts between the substrate and Phe-87, Leu-244 and Val-295. In the absence of camphor, a network of six water molecules occupies the active site while one water molecule acts as the sixth ligand of the heme iron [14]. Displacement of these water molecules upon substrate binding would therefore give rise to a gain in entropy. This entropic effect serves as molecular model for the mechanism of high spin to low spin transition observed for many cytochrome P450s.

4.2. CYP2B4 ($P450_{2B4}$)

CYP2B4 is a microsomal membrane heme protein from rabbit liver which hydroxylates benzphetamine and testosterone at position 16α. It is an oligomeric enzyme in solution and thought to be so within the microsomal membrane. Electron microscopy studies of solubilized CYP2B4 to be hexameric with a combined molecular mass of 360 kDa. Since there is no crystal structure available for CYP2B4, the structure of CYP102, which has the highest homology with microsomal P450s, serves as model for these membrane-bound hemeproteins [16].

4.3. CYP102 ($P450_{BM3}$)

CYP102 is a bacterial cytochrome P450 from *Bacillus megaterium*, which metabolizes fatty acids such as arachidonic and palmitic acids [17]. CYP102 differs from other P450s due to the fact that the reductase and heme enzyme reside on the same polypeptide chain. The holoenzyme has a molecular mass of 119 kDa while the heme domain has a weight of 55 kDa. The crystal structure is known [18] and shows that the active site is accessible through a long hydrophobic channel formed by the β domain and the B′ and F helices of the α domain. There are large structural differences between CYP102 and CYP101 near the substrate binding pocket and the region important for redox partner interactions [19]. As mentioned earlier, CYP102 possesses a high sequence homology with microsomal P450s and has served as model for the studies of mammalian cytochrome P450s [16].

5. Ferric spin-state transition in the cytochromes P450

Many cytochrome P450s display a conversion of low spin ($S = 1/2$) ferric heme to a high spin ($S = 5/2$) configuration when substrates bind at the active site and displace a water ligand to the heme iron. This is most easily observed as a blue shift of the wavelength maximum of the Soret band from roughly 417 nm to 392 nm [20]. EPR spectroscopy of the iron indicates that the transition is due directly to a change in the iron spin state distribution from a low-spin d-electron configuration ($S = 1/2$) to a high-spin configuration ($S = 5/2$). The spectral shifts are accompanied by a change in the redox potential from −300 mV to −170 mV [21]. This shift in potential may have a physiological rate, as in the case of CYP101 in *Pseudomonas* since the normal redox partner, putidaredoxin, has a redox potential of −196 mV allowing the reduction of CYP101 to be much more favorable in the substrate bound form [21]. Similar spin state changes in some hepatic microsomal cytochromes P450 have also been shown to modulate the redox potential [22,23]. In order to probe localized protein structural changes in the active site, second-derivative UV-visible spectroscopy as outlined by Ragone et al. [24] has been performed to determine changes in the degree of exposure of tyrosine residues were linked to ferric spin equilibrium. As shown by the three-dimensional X-ray structure [14,15], tyrosine-96, which is situated at the active site close to the sixth ligand of the heme iron, is a potential candidate for the local conformational change. By using various substrate analogs which modulate the spin state

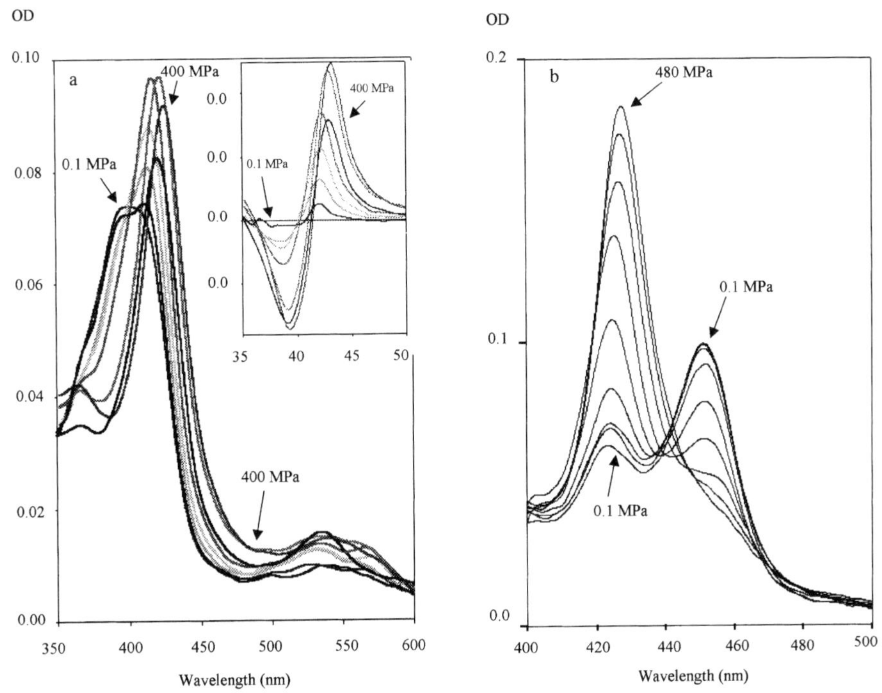

Fig. 1. Pressure-induced changes of absorbance spectra of cytochrome P450s. (a) Pressure-induced spin transition followed by P450 to P420 transition of the oxidized $P450_{BMP}$, 2 μM in the presence of 8 μM palmitic acid. Absolute spectra were measured every 30 MPa from 0.1 to 400 MPa. Buffer conditions: 100 mM Na–HEPES (pH 7.4), 1 mM EDTA, 1 mM dithiothreitol, 25°C. The inset shows the results as the difference between the spectra at a particular pressure and the initial spectrum at 0.1 MPa. (b) Pressure-induced P450 to P420 inactivation of the ferrous carbonyl cytochrome $P450_{2B4}$, 4 mM in 100 mM Na–HEPES buffer (pH 7.4), 1 mM EDTA, 1 mM dithiothreitol, 25°C. Absolute spectra were measured every 30 MPa from 0.1 to 480 MPa.

Table 1
Degree of tyrosine exposure versus high spin fraction in CYP101

Substrate	% High spin	No. of tyrosines exposed[a]
None	4	4.20
TMCH	14	4.06
Norcamphor	45	3.92
Fenchone	46	3.95
Camphenilone	55	3.92
Thiocamphor	61	3.83
5-Ketocamphor	64	3.83
3-*endo*-Bromocamphor	67	3.83
Adamantane	79	3.76
5-*exo*-Bromocamphor	81	3.78
Camphor	95	3.58
Adamantanone	99	3.66

[a]Calculated from the average of 30 spectra.

of cytochrome CYP101, tyrosine motions are demonstrated to be coupled to the binding of these substrate analogs [25].

Table 1 indicates that the degree of exposure of tyrosine-96 is observed to decrease as the percentage of high spin CYP101 is increased, and is consistent with one tyrosine being embedded in the interior of the protein matrix as the protein shifts to the high spin form. There is a linear relationship between the degree of tyrosine exposure and the free energy of spin change. Such behavior is compatible with the mechanism of a spin transition controlled by changes of the polarity of the heme pocket due to the presence of water molecules [14,15]. The initial spin state equilibrium (or initial free energy of spin state change at 0.1 MPa and room temperature, $\Delta G°_i = -RT\ln K_{spin}$) can be used as a standard to characterize the hydration state of the heme pocket and will also be used to probe the relationship between spin state changes and spin volume changes as well as inactivation volume changes.

6. Pressure effects on heme optical spectra changes of cytochrome P450s

The application of high pressure perturbs the spectra of cytochromes P450 in a complex way. Using a factor analysis method [11] to analyze the spectral changes in the Soret region versus pressure, we can resolve two particular pressure-induced processes and determine their barotropic parameters ($\Delta V°$ and $P_{1/2}$). The first process is a reversible pressure-induced high-spin to low-spin shift which is related to conformational changes of the substrate free cytochrome P450 as well as the substrate-bound form. For the intermediate substrate concentrations, the pressure-induced spin transition is associated to pressure-induced substrate dissociation/association processes. The second process is the pressure-induced inactivation of oxidized cytochrome P450 to cytochrome P420, as well as the pressure-induced reduced cytochrome P450–CO complex to reduced cytochrome P420–CO form. This process is quasi-reversible for CYP101 in the absence of excess cysteine and reversible in its presence, the pressure profiles in both conditions are the same. For the other cytochromes P450 CYP2B4, CYP102 and CYP1A2, the pressure-induced inactivation process is reversible. The pressure-induced inactive states have a Soret maximum near 420 nm in both the oxidized and the reduced carbonyl forms, albeit with different molar extinctions (Fig. 1a,b).

6.1. *Volume changes during spin state transitions, substrate binding and protein inactivation*

Pressure-induced spectral changes accompanying

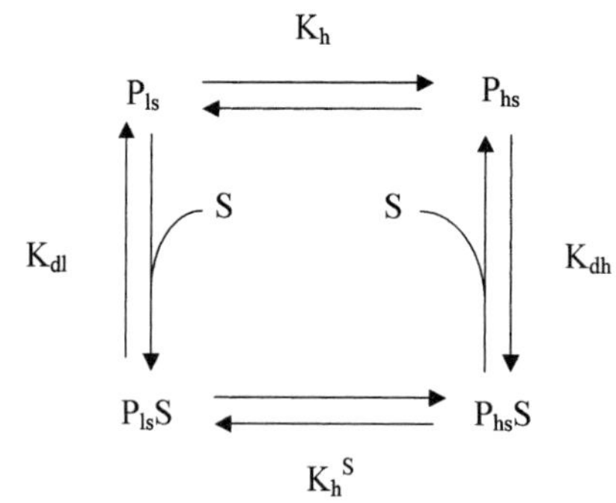

$$K_h = [P_{hs}]/[P_{ls}]\ ;\ K_h^S = [P_{hs}S]/[P_{ls}S]$$

$$K_{dl} = [P_{ls}][S]/[P_{ls}S]\ ;\ K_{dh} = [P_{hs}][S]/[P_{hs}S]$$

Scheme 1.

the high spin to low spin transition for camphor-bound cytochrome CYP101 at saturation of substrate or at its intermediate concentration, as well as the pressure-induced P450 to P420 transition are proportional to the concentration changes for both forms of P450. A plot of the natural logarithm of their concentrations versus pressure deduced from Eq. 1 allows the determination of the spin reaction volume ($\Delta V°_{spin}$), substrate dissociation volume ($\Delta V°_{diss}$) and the inactivation reaction volume ($\Delta V°_{inact}$). It is shown that the volume changes were negative indicating that pressure shifted the equilibrium to the more compact low spin and P420 states. For spin equilibria and substrate binding studies, a four-state model of the transitions was used (Scheme 1) where 'P_{ls}' and '$P_{ls}S$' correspond to substrate-free P450 and its complex with the substrate 'S', respectively. The subscripts 'ls' and 'hs' designate the low-

Table 2
Spin reaction volume and equilibrium constant, substrate dissociation constant and volume, and inactivation reaction volume of cytochrome P450s, at 277 K

P450 $\Delta V°_{spin}$ (cm^3/mol)	CYP101	CYP102	CYP2B4
Substrate-free	−20	−23	−21
Substrate-bound	−91	−16	−49
Spin equilibrium constant K_{hs}^S at 0.1 MPa	62	1.2	8.3
Substrate dissociation constant (K_{dh}) (μM) at 0.1 MPa	0.047	0.61	36
Substrate dissociation volume $\Delta V°_{dh}$ (cm^3/mol)	−29	−13	−23
$\Delta V°_{inact}$ (cm^3/mol)			
Substrate-free $P420^{3+}$	−80	−99	−32
Substrate-bound $P420^{3+}$	−160	−42	−32
$P420^{2+}$–CO	−107	−99	−50

and high-spin states of the proteins.The spin reaction volumes (ΔV°_{spin}) for substrate free and bound cytochrome P450s, the substrate dissociation volumes (ΔV°_{dh}) for the high-spin forms of the cytochromes and the inactivation volumes ($\Delta V^{\circ}_{inact.}$) are given in Table 2. The spin state volume changes for all of the substrate free P450s are quite similar, yet conversely much different in their substrate-bound forms, ranging from −16 cm^3/mol for CYP102–palmitic acid, −49 cm^3/mol for CYP2B4–benzphetamine and −91 cm^3/mol for CYP101–camphor. It was shown that for camphor-bound cytochrome CYP101, the extent of the spin volume changes depends on the physicochemical conditions of the medium (potassium ion and proton concentration) [26]. Inactivation volumes of cytochrome CYP101 seem to depend on the hydration state of the heme pocket (see below). Binding of camphor increased the difference between native and inactive partial molar volumes ($\Delta V^{\circ}_{inact} = -160$ cm^3/mol) and stabilized the active site structure by an increase of the half pressure of inactivation of the substrate-bound complex ($P_{1/2\,inact} = 170$ MPa) compared to the substrate-free form ($P_{1/2\,inact} = 110$ MPa). Such a dependency of the inactivation volume on substrate concentration was also observed in the case of cytochrome CYP102-palmitic acid complexes. In the absence of substrate the inactivation volume was −99 cm^3/mol, while it decreased to −42 cm^3/mol by the addition of 32 μM palmitic acid with an increase of P_{inact} from 210 to 310 MPa. However these values should be considered as 'apparent' because of some pressure-induced dissociation of the substrates [27]. Table 2 shows that the dissociation reaction volumes of high spin CYP101, CYP102 and CYP2B4 are negative. For cytochrome CYP2B4 the pressure-induced P420 conversion showed no dependency on the substrate concentration ($\Delta V^{\circ}_{inact} = -32$ cm^3/mol) with a half pressure of inactivation equal to $P_{1/2} = 156$ MPa. The inactivation volumes of the reduced carbonyl complexes were negative in the order CYP101 < 0.

6.2. *Active site hydration and correlation to spin state and inactivation volume changes*

In order to avoid complex contributions to the volume changes due to drastic changes in the solvent conditions, we have fixed a typical buffer condition

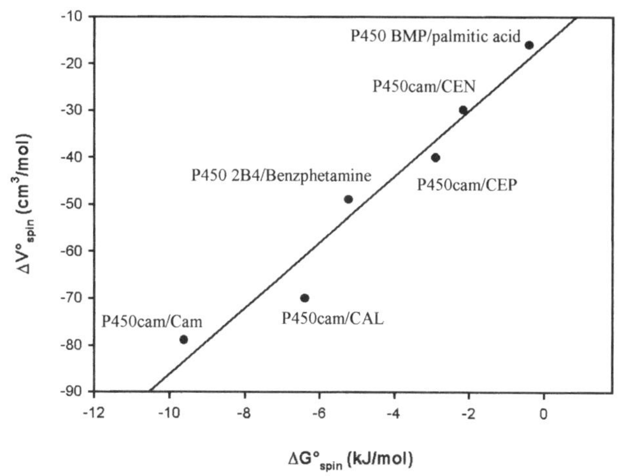

Fig. 2. Linear correlation between spin volume changes derived from pressure-dependence spin transition studies and the initial free energy of spin states ($\Delta G^{\circ}_{i} = -RT\ln K_{spin}$ at 0.1 MPa and room temperature) for substrate-bound cytochrome P450$_{cam}$ complexes, P450$_{2B4}$–benzphetamine and P450$_{BMP}$–palmitic acid complexes. The new analogs of camphor are: CAL, *endo*-borneol *O*-allyl ether of camphor; CEP, *endo*-borneol *O*-propyl ether of camphor; CEN, *endo*-borneol *O*-methyl ether of camphor. Buffer conditions were 50 mM Tris–HCl (pH 7.4) in the presence of 400 μM of analogs of camphor and 240 mM KCl for cytochrome P450$_{cam}$, and 100 mM Na–HEPES buffer (pH 7.4) in the presence of 1 mM EDTA, 1 mM dithiothreitol for cytochrome P450$_{2B4}$ and P450$_{BMP}$, 25°C.

(50 mM Tris–HCl (pH 7.2), 240 mM KCl, 400 μM camphor) and varied the ferric spin equilibrium by using substrate analogs in order to determine whether there is a relationship between the magnitude of the observed volume changes and the extent of the changes in spin state [26]. To further gain insight into the role of active site water molecules in the control of spin volume changes of substrate-bound complexes, we have synthesized a series of compounds (endo-borneol methyl ether, endo-borneol propyl ether, endo-borneol allyl ether and endo-borneol dimethyl allyl ether) with substituents at the camphor carbonyl oxygen to displace the cluster of water molecules in the CYP101 active site and to occupy an empty region of the substrate binding pocket or any solvent occupying site [28,29]. Coumpounds with a 3-carbon chain substituent were predicted to match the size of the unoccupied region optimally and thus have the tightest association. The spin state volumes of these new analog of substrate–CYP101 complexes were then measured. The results shown in Fig. 2 indicated that the whole series

of $\Delta V^{\circ}_{spin}/\Delta G^{\circ}_{spin}$ data pairs for the spin state transition of the substrate-bound P450s obey the same linear dependence [27]. Camphor-bound CYP101 exhibits a higher spin state ($K_{HS}S = 62$, where $K_{HS}S$ is the spin equilibrium constant for substrate-bound enzyme, and $\Delta G^{\circ}_{spin} = -RT\ln K_{spin}$) and a lower exposure of tyrosine residues. Its heme pocket is less polar and thus it is less hydrated (Table 1). Increasing pressure will shift the spin state to low spin and push water molecules into the pocket; the volume changes for such a process has a higher value ($\Delta V^{\circ}_{spin} = -91$ cm^3/mol). CYP102–palmitic acid complex exhibits a more low spin state due to the fact that the initial water content of its heme pocket is higher than in the case of CYP101, thus the difference of the hydration state between low spin and high spin forms will be smaller leading to smaller spin volume changes ($\Delta V^{\circ}_{spin} = -16$ cm^3/mol). The case of CYP2B4–benzphetamine complex is intermediate between the previous two cytochromes. The linear relationship between ΔV°_{spin} and ΔG°_{spin} appears to be a general feature of P450 heme proteins and suggests a common mechanism of the shift of spin equilibrium by substrate binding for the three heme proteins investigated. The spin transition of the substrate-bound P450 form exhibits a flux of water molecules that is dependent on the initial hydration state of the heme pocket (11 for CYP101, 17 for CYP102 and 16 for CYP2B4) [27]. Unlike the substrate bound enzyme where there are significant changes in the active site hydration, the substrate-free heme protein exhibits a much smaller effect where only one estimated water molecule enters the active site during the spin transition.

The same linear relationship has been found between the free energies of spin state transition (ΔG°_{spin}) which measures the initial degree of hydration of the heme pocket, and the inactivation volumes (ΔV°_{inact}) [30]. The different free energies of

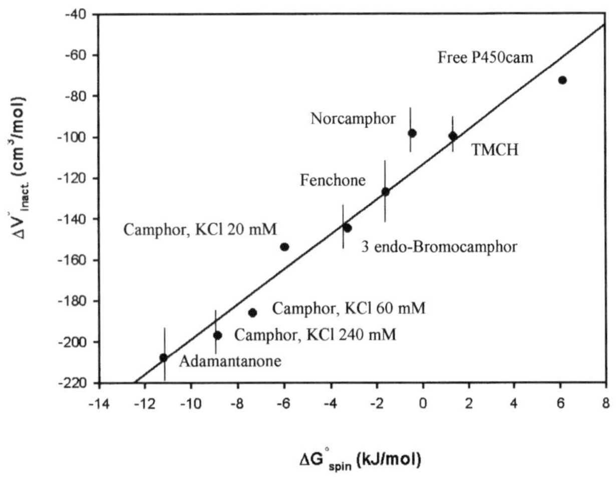

Fig. 3. Linear correlation between inactivation volume derived from pressure-dependence P450 to P420 inactivation studies and the initial free energy of spin states ($\Delta G^{\circ}_{i} = -RT\ln K_{spin}$ at 0.1 MPa and room temperature) of substrate-bound cytochrome P450$_{cam}$. Buffer conditions were the same than in Fig. 2.

spin state shown in Fig. 3 were obtained by modulating the spin state by the addition of substrate analogs as indicated in Table 1. As for the oxidized substrate-bound complexes, the larger value obtained for the inactivation volume is due to a hydration change within the heme pocket resulting from the displacement or mobility of the substrate during compression and subsequent entrance of water molecules. In addition the stability of the protein against compression is correlated with water accessibility to the active site [30]. Camphor and adamantanone, which strongly interact with the protein and exclude water from the active site, enhances the stability by 30 kJ/mol at 20°C as compared to the free protein. Norcamphor, a loosely bound substrate that is more mobile because all of the methyl groups are missing and therefore does not completely exclude water, increases the stability of the protein by only 7 kJ/mol. Finally, these results suggest that pressure-induced cytochrome P420 is a unique form where the active

Table 3
Thermodynamic parameters of spin equilibrium, substrate binding of (1*R*)-camphor and (1*S*)-camphor-bound cytochrome CYP101 complexes

Substrate	K_{spin} (297 K)	K_d (μM) (297 K)	$P_{1/2(spin)}$ (MPa)	ΔV°_{spin} (cm^3/mol)	$\upsilon_{CO(I)}$ (20% GlOH) (cm^{-1}) (290 K)	$\upsilon_{CO(II)}$ (20% GlOH) (cm^{-1}) (290 K)
(1*R*)-Camphor	251	0.8	175	−52	1939.7	1930.0
(1*S*)-Camphor	87	1.2	78	−36	1940.2	1946.3

CO-stretching frequencies of the reduced-CO cytochrome P450$_{cam}$ heme–iron environment are also shown.

site is hydrated and camphor is displaced from its binding site.

6.3. Mobility and binding of substrate in the heme pocket and spin volume changes

The correlations between spin volume, spin state and accessibility of water molecules in the heme pocket could be the result of several factors including, substrate affinity or mobility, and conformational flexibility. A comparative study of enantiomeric substrates such as (1*R*)-camphor and (1*S*)-camphor suggests that non specific interactions and mobilities of the substrates plays an important role in the accessibility of water into the active site of cytochrome CYP101 [31]. (1*S*)-camphor is the optical antipode of (1*R*)-camphor. The mirror-symmetry of the enantiomeric structure is obtained by changing the position of one methyl group of the camphor molecule without changes of the physical and chemical properties. Such a small perturbation of the camphor structure gives rise to many changes in the thermodynamic and physico-chemical parameters between substrate and cytochrome CYP101 interactions (see Table 3) [31]. The enantiomer (1*S*)-camphor is loosely bound in the heme pocket compared to the natural substrate (1*R*)-camphor. The low value of the spin state ($K_{spin} = 87$) as well as the smaller volume ($\Delta V°_{spin} = -36$ cm^3/mol) indicates that it is more hydrated.

The half-pressure for the spin state transition is lower for (1*S*)-camphor ($P_{1/2(spin)} = 78$ MPa) than that of (1*R*)-camphor ($P_{1/2(spin)} = 175$ MPa) bound CYP101. In addition, the CO stretching frequencies of (1*S*)-camphor are slightly higher (1940.2 cm^{-1} for the major band and 1946.3 cm^{-1} for the minor band), suggesting a loosening of the contact between the iron-bound CO ligand and some amino acids of the I-helix. This weak contact of the iron ligand may explain the higher amount of uncoupling of the cytochrome P450 reaction cycle (11%) as compared to the reaction cycle in the presence of camphor (3%) [32].

To further probe the role of methyl groups within the active site in the control of the mobilities of the substrates and the water accessibility in the active site, a series of norbornane molecules, which differ in the number and position of the methyl- and keto groups, have been used as substrates [33]. The mobility of the substrate inside the heme pocket was characterized experimentally by cryotrapping the spin state during its relaxation to a new spin state equilibrium [30]. After rapidly cooling the protein samples to 77 K, the new spin equilibrium state was followed kinetically in the Soret region. Depending on the strength of substrate binding and its re-

Table 4
P450→P420 transitions for oligo- and heterooligomers

Cytochrome P450s		$P_{1/2\,inact}$ (MPa)	$\Delta V°_{inact}$ (cm^3/mol)	Amplitude (%)
CYP101 Fe^{3+}		110	−80	100
CYP101 Fe^{3+}–S		170	−160	100
CYP2B4 Fe^{3+}		156	−32	67
CYP2B4 Fe^{2+}–CO		300	−50	65–70
CYP2B4 Fe^{3+}+0.2% Triton N-101		120	−86	99.9
CYP2B4 Fe^{2+}–CO+0.2% Triton N-101		130	−104	95
Molar content %				
CYP2B4 Fe^{2+}–CO	CYP1A2 Fe^{2+}–CO			
0	100	420	−28	8.3
33.3	66.6	292	−29	42
50	50	298	−26	25
66.6	33.3	264	−25	31
100	0	297	−61	68
2B4 Fe^{2+}–CO and 1A2 Fe^{2+}–CO	Mixture w/o pre-incubation			
50	50	183	−44	38

If not indicated, the mixtures of the hemeproteins were pre-incubated overnight at 4°C prior to the experiments.

laxation rates, an intermediate extent of spin state was trapped by the low temperature before reaching its new equilibrium. It was shown previously [25] that the rates of spin dissociation compete with the rates of water rebinding to the sixth ligand of the heme. Different amounts of low spin iron was detected in the first cooling period followed by a stabilization of the low spin state and its freezing around the glass transition temperature (180 K for glycerol cryocosolvent). The same low spin content was found at 77 K, indicating that the spin was trapped by such a temperature jump. The percent low spin was calculated and its value depended on the mobility and binding constant of the substrate analog. Mobile substrates result in a higher content of low spin iron after a negative temperature jump. This result confirms the finding of Fisher and Sligar [25] and shows that the detected low-spin content in the heme pocket is correlated with the higher substrate mobility. To quantitate experimentally such a behavior, the slope of the fast temperature-induced kinetic profiles was determined and used as a useful parameter which indicates the extent of the substrate mobility. For practical reasons, an inverse slope of the observed temperature-induced spin transition profile was used and defined as the substrate rigidity factor (SRF):

$$\mathrm{SRF} = \text{Substrate rigidity factor} = 1/\text{slope of T-jump profile} = (\mathrm{T_i}-\mathrm{T_f})/(\mathrm{HS_i}-\mathrm{HS_f}) \quad (5)$$

Where T_i and T_f refer to 297 K and the glass transition temperature of the protein/glycerol solution (180 K), respectively. HS_i is the initial high-spin content at 297 K and HS_f the high-spin content of the cryogenically trapped sample.

A plot of the substrate rigidity factor versus free energy of the initial spin (ΔG^i_{spin} at 297 K) gives a straight line showing empirically that a correlation exists between mobilities of the substrate and the initial spin state content, which is the measure of the hydration (or water content) of the heme pocket. Mobile substrates allow for more hydration of the active site therefore facilitating water access to the heme pocket. The same plot is valid for the rigidity factor and the natural logarithm of the substrate dissociation constant ($\ln K_d^{(297\,K)}$) [33]. Finally because of the correlation described above, a linear relationship was found for the spin volumes and substrate rigidity factors as shown in Fig. 4.

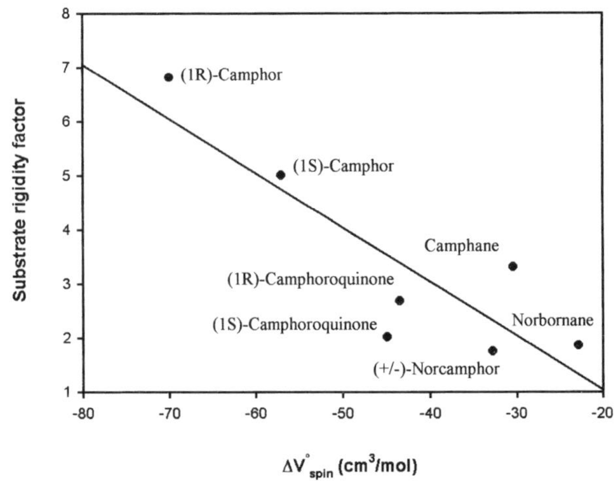

Fig. 4. Linear correlation between substrate rigidity factor derived from negative temperature jump studies and the spin volume changes derived from pressure-induced spin transition studies for analog of substrate-bound cytochrome $P450_{cam}$ complexes. The experiments were carried out using a cryostat DN 704 (Oxford) and a spectrophotometer UV 2101 PC (Shimadzu) up to 77 K. To obtain a glassy sample and hence to avoid light scattering at low temperature, these experiments were performed with 0.1 M potassium phosphate–glycerol 60% (w/w) buffer, pa_H 7.15. The glass transition temperature was around 180 K for glycerol buffer studies.

7. Implications of homo- and heterooligomerization on the barotropic behaviors of cytochrome P450s

Analysis of the pressure-induced spectra transitions (spin state interconversion or the P450 to P420 transitions) for monomeric cytochrome P450 imply that during the transformation from one state to another, the sum of the different states is unity. For an oligomeric protein such as cytochrome CYP2B4, remarkable barotropic behaviors were obtained. As shown in Table 4, 100% of the CYP101 was converted to CYP101 under pressure while only a portion (67% ± 5%) of the total ferric cytochrome CYP2B4 was transformed into P420 species with an inactivation volume of −32 cm^3/mol and a half pressure of inactivation of 156 MPa [11]. The remaining fraction (31% ± 5%) was exposed to pressure and remarkably the spin equilibrium was not affected by the pressure-induced inactivation. Such a peculiarity was encountered in the ferrous carbonyl form of cytochrome CYP2B4 for which 65–70% of the total Fe^{2+}–CO hemeprotein was converted into P420 ($\Delta V°_{inact} = -50$ cm^3/mol, $P_{1/2\,inact} = 300$ MPa). The remaining portion was stable at pressures up to 600 MPa. Upon dissociation of the oligomers in the presence of 0.20% Triton N-101, the inhomogeneity behaviors vanished as indicated in Table 4.

Oligomerization results in a considerable stabilization of the hemeprotein which is less flexible (smaller inactivation volume). We interpret the pressure behaviors of the CYP2B4 oligomers as evidence for the presence of two conformers of CYP2B4 which have different pressure stabilities. This special feature is due presumably to the architecture of the subunit assemblies leading to different conformation and/or orientation of the subunits. Heterooligomeric subunit interactions will also lead to similar rearrangements of the subunit's architecture. It is known that in the endoplasmic reticulum membrane, different P450 isoforms coexist and compete for reductase and cytochrome b_5. We have shown that in solution hetero-oligomeric-subunit exchanges occur in a reversible way as probed by fluorescence resonance energy transfer techniques [34]. The effect of such intermolecular interactions between cytochrome CYP2B4 and cytochrome CYP1A2, for example, will change the stability and the accessibility of cytochrome CYP2B4 towards small molecules. When taken separately cytochrome $P450^{2+}$ CYP1A2–CO complex was extremely stable towards pressure inactivation ($P_{1/2\,inact} = 420$ MPa) with only 8.3% of the hemeprotein being affected by pressure and a very small inactivation volume change ($\Delta V°_{inact} = -28$ cm^3/mol). While cytochrome $P450^{2+}$ CYP2B4–CO complex is very sensitive to pressure effects, 68% of the total CYP2B4 protein was susceptible to pressure-induced inactivation leading to a low inactivation pressure ($P_{1/2\,inact} = 297$ MPa) and a higher negative inactivation volume ($\Delta V°_{inact} = -61$ cm^3/mol). When oligomers of CYP2B4 and CYP1A2 at different molar concentrations were pre-incubated overnight at 4°C, the formation of heterooligomers has a dramatic effect on the barotropic behavior of $CYP2B4^{2+}$–CO complex which absorbs at 451.4 nm while $CYP1A2^{2+}$–CO complex has its maximum Soret band centered at 448.3 nm. In pre-incubated 1:1 oligomer mixtures, no more than 25% of the total CYP2B4 content undergoes P450→P420 inactivation, with a partial molar inactivation volume of -26 cm^3/mol, similar to those found for pure CYP1A2. When data is taken immediately after mixing, the amplitude of inactivation was 38%. In the heterooligomers of CYP1A2 and CYP2B4, the CYP2B4 heme protein appears to be protected from the pressure-induced inactivation. In addition to this effect, interactions between CYP1A2 and CYP2B4 resulted in a displacement of the Soret band of the ferrous carbonyl complex to shorter wavelength (from 451.3 to 448.4 nm) and largely strengthens the dependence of the Soret band wavenumber on hydrostatic pressure below 200 MPa. This effect is related to the pressure-induced compressibility of the heme pocket (for the theory see the review of Jung in this issue), and important hydration of the cytochrome CYP2B4 heme moiety in response to the interactions with cytochrome CYP1A2. The formation of hetero-oligomers causes a conformational change in CYP2B4 causing the stabilization of some CYP2B4 subunits and the additional hydration of their heme pockets. In addition, hetero-subunit interactions in the heterooligomers affect the activity and the interactions with NADPH–cytochrome P450 reductase [35,36].

Acknowledgements

The research discussed in this review was supported in part by grants from the Institut National de la Santé et de la Recherche Médicale, France and the National Institutes of Health. The authors are very grateful to C. Di Primo, E. Deprez, C. Jung, H. Schulze and D.R. Davydov for their participation and advice in this work.

References

[1] G. Weber, H.G. Drickamer, Q. Rev. Biophys. 16 (1983) 89–112.
[2] D.R. Stranks, Pure Appl. Chem. 38 (1974) 303–323.
[3] F.H. Stillinger, Science 209 (1980) 451–457.
[4] W. Kauzmann, Adv. Protein Chem. 14 (1959) 1–63.
[5] D.P. Kharakov, V.E. Bychkova, Biochemistry 36 (1997) 1882–1890.

[6] A.J.W.G. Wisser, T.M. Li, H.G. Drickamer, G. Weber, Biochemistry 16 (1977) 4883–4886.

[7] R.C. Neuman, W. Kauzman, A. Zipp, J. Phys. Chem. 77 (1973) 2687–2691.

[8] K.J. Laidler, Arch. Biochem. Biophys. 30 (1950) 226.

[9] E. Morild, Adv. Protein Chem. 34 (1981) 93–167.

[10] G. Hui Bon Hoa, P. Douzou, N. Dahan, C. Balny, Anal. Biochem. 120 (1982) 125–135.

[11] D.R. Davydov, T.V. Knyshko, G. Hui Bon Hoa, Arch. Biochem. Biophys. 320 (1992) 330–344.

[12] G. Hui Bon Hoa, C. Di Primo, A. Des Alvarez, T. Cao, in: 1er Forum de Technologie des Hautes Pressions, La Londe les Maures, France, 1997.

[13] J.T. Groves, G.A. McClusky, J. Am. Chem. Soc. 98 (1976) 859–861.

[14] T.L. Poulos, B.C. Finzel, A.J. Howard, Biochemistry 25 (1986) 5314–5322.

[15] T.L. Poulos, B.C. Finzel, A.J. Howard, J. Mol. Biol. 195 (1987) 687–700.

[16] D.F. Lewis, B.G. Lake, Xenobiotica 27 (1997) 443–478.

[17] L.O. Narhi, A.J. Fulco, J. Biol. Chem. 261 (1986) 7160–7169.

[18] S.S. Boddupalli, C.A. Hasemann, K.G. Ravichandran, J.-Y. Lu, E.J. Goldsmith, J. Deisenhofer, J.A. Peterson, Proc. Natl. Acad. Sci. USA 89 (1992) 5567–5571.

[19] K.G. Ravichandran, S.S. Boddupalli, C.A. Hasemann, J.A. Peterson, J. Deisenhofer, Science 261 (1993) 731–736.

[20] T. Tsai, C.A. Yu, I.C. Gunsalus, J. Peisach, W. Blumberg, W.H. Orme-Johnson, H. Beinart, Proc. Natl. Acad. Sci. USA 66 (1970) 1157–1163.

[21] S.G. Sligar, I.C. Gunsalus, Proc. Natl. Acad. Sci. USA 73 (1976) 1078–1082.

[22] S.G. Sligar, D.L. Cini, G.G. Gibson, J.B. Schenkman, Biochem. Biophys. Res. Commun. 90 (1979) 925.

[23] H. Rein, O. Ristau, R. Misselwitz, E. Buder, K. Ruckpaul, Acta Biol. Med. Ger. 38 (1979) 187–200.

[24] R. Ragone, G. Colonna, C. Balestrieri, L. Servillo, G. Irace, Biochemistry 23 (1984) 1871–1875.

[25] M.T. Fisher, S.G. Sligar, Biochemistry 24 (1985) 6696–6701.

[26] G. Hui Bon Hoa, M.C. Marden, Eur. J. Biochem. 12 (1982) 311–315.

[27] D.R. Davydov, G. Hui Bon Hoa, J.A. Peterson, Biochemistry 38 (1999) 751–761.

[28] V. Helms, E. Deprez, E. Gill, C. Barret, G. Hui Bon Hoa, R.C. Wade, Biochemistry 35 (1996) 1485–1499.

[29] V. Helms, R.G. Wade, Proteins 32 (1998) 381–396.

[30] C. Di Primo, G. Hui Bon Hoa, P. Douzou, S.G. Sligar, Eur. J. Biochem. 209 (1992) 583–588.

[31] H. Schulze, G. Hui Bon Hoa, V. Helms, R.C. Wade, C. Jung, Biochemistry 35 (1996) 14127–14138.

[32] S. Kadkhodayan, E.D. Coulter, D.M. Mayniak, T.A. Bryson, J.H. Dawson, J. Biol. Chem. 270 (1995) 28042–28048.

[33] H. Schulze, G. Hui Bon Hoa, C. Jung, Biochim. Biophys. Acta 1338 (1997) 77–92.

[34] D.R. Davydov, N.A. Petushkova, A.I. Archakov, G. Hui Bon Hoa, Biochem. Biophys. Res. Commun. 276 (2000) 1005–1012.

[35] D.R. Davydov, N.A. Petushkova, E.V. Bobroynikova, T.V. Knyushko, P. Dansette, in: R. Snyder et al. (Eds.), Biologically Active Intermediates VI: Chemical and Biological Mechanisms in Susceptibility and Prevention of Environment Disease, Kluwer Academic/Plenum, Dordrecht/New York, 2000.

[36] W.L. Blakes, C.J. Batie, G.F. Cawley, Biochemistry 37 (1998) 12852–12859.

ELSEVIER

Biochimica et Biophysica Acta 1595 (2002) 309–328

BIOCHIMICA ET BIOPHYSICA ACTA
BBA
www.bba-direct.com

Review

Cytochrome *P*-450-CO and substrates: lessons from ligand binding under high pressure

Christiane Jung *

Max-Delbrück-Center for Molecular Medicine, Protein Dynamics Laboratory, Robert-Rössle-Strasse 10, 13125 Berlin, Germany

Received 26 July 2001; received in revised form 22 October 2001; accepted 23 October 2001

Abstract

An overview of the application of high-pressure studies on the carbon monoxide complex of cytochrome *P*-450 is given. Different approaches to characterize ligand binding steps, the conformational states and substates and the compressibility of the ligand-bound complex are reviewed. A particular focus is the effect of substrates on these properties. It is shown that substrate mobility, compressibility and water accessibility are interrelated and may have functional meaning.

Keywords: High pressure; Cytochrome *P*-450; Ligand binding; Fourier transform infrared spectroscopy; UV-visible spectroscopy; Compressibility

1. Introduction

Binding of small molecules to proteins is a fundamental process and essential for organisms to live. From the large number of protein–ligand pairs, the binding of diatomic molecules such as CO, O_2 or NO to the heme iron in myoglobins and hemoglobins is the classical example which has been studied extensively over several decades. Since the discovery that small diatomic ligands such as NO and probably also CO function as signal molecules, studies of ligand binding in various heme proteins became even more important and of increased interest for medicine and pharmacology. The study of ligand binding in heme proteins is therefore a broad field and high-pressure approaches have been widely applied.

This review will focus on cytochrome *P*-450 (*P*-450) as the primary example and demonstrate how studies of the effect of high hydrostatic pressure help in understanding structural and functional properties. In particular, studies on the carbon monoxide complex of *P*-450 in the absence and in the presence of substrates turned out to be very informative and will be emphasized in this review. Studies on other heme proteins will be discussed if relevant comparisons can be made. Understanding pressure effects on the protein structure also requires the understanding of the temperature effect. Therefore, the effect of temperature will also be discussed for selected parameters.

2. Ligand binding steps in heme proteins

Ligand binding will be separated into two main steps: P+L → {P:L} → P-L, (i) the association be-

* Fax: +49-30-94-06-33-29.
E-mail address: cjung@mdc-berlin.de (C. Jung).

0167-4838/02/$ – see front matter
PII: S0167-4838(01)00353-3

tween ligand L and protein P with entry of the ligand into the heme pocket and (ii) the coordination of the ligand to the heme iron. Subsequently, there are three main forms which are sensitive to pressure: (i) the ligand unbound protein P; (ii) the transient ligand–protein association complex {P:L}; and (iii) the ligand-bound protein P-L.

When using a spectroscopic probe to monitor ligand binding and to study the effect of hydrostatic pressure, one has to differentiate between two qualitatively different responses. The spectral signals of the ligand-unbound and ligand-bound protein usually differ significantly because they are chemically different species. This spectral change may be used to study the thermodynamics and kinetics of the ligand binding step itself. However, the spectral signals of each of these chemically different species are slightly changed by pressure, and also by temperature. These spectral changes reflect the pressure-induced changes of the intramolecular interactions near the spectroscopic probe and may be used to characterize the conformation and subconformations of the protein. Both spectral changes overlap when ligand binding is monitored. The latter spectral change is usually ignored because it seems to be marginal compared to the changes induced by ligand coordination. However, even these small effects give access to important quantities such as compressibilities and thermal expansion coefficients which are related to protein dynamics.

3. Cytochrome *P*-450 proteins and relevant spectroscopic probes

Cytochrome *P*-450s are thiolate heme proteins and represent the terminal oxidase in monooxygenase systems found in almost all organisms [1]. These monooxygenase systems are responsible for the oxidative degradation as well as for the biosynthesis of many compounds. From the large number of known *P*-450 sequences the crystal structures of only few *P*-450s have been resolved so far. They show the same overall fold and tertiary structure. However, some structural differences exist which are responsible for the different substrate specificities [2]. As seen in the reaction cycle of cytochrome *P*-450 (Fig. 1), binding of the dioxygen ligand to the heme iron is the essen-

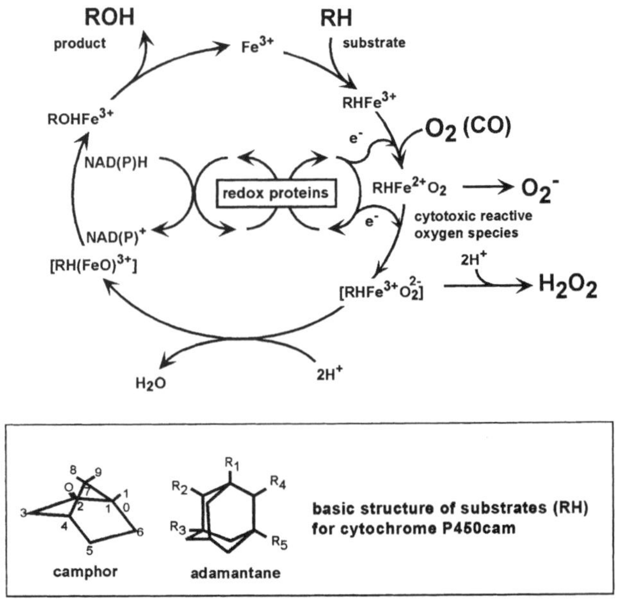

Fig. 1. Reaction cycle of cytochrome *P*-450. Fe^{3+} and Fe^{2+} stand for the heme protein in the ferric and ferrous form, respectively. RH and ROH are the substrate and product, respectively. Two side reactions, leading to the formation of reactive oxygen species, are indicated. Carbon monoxide (CO) binds at the same iron coordination position as dioxygen (O_2). On the bottom the basic chemical structure of two important classes of substrates for cytochrome *P*-450cam from *P. putida* are shown.

tial prerequisite for function [1]. Besides dioxygen, other small ligands can bind. Carbon monoxide plays a specific role because its binding causes the characteristic shift of the intense electronic absorption band (Soret band) to approx. 450 nm from which the name *P*-450 is derived.

CO binds only to the iron(II) form. Because *P*-450s in the resting state are in the iron(III) form (oxidized *P*-450), electron transfer from NAD(P)H via redox proteins is required for binding of these ligands. The oxidized *P*-450 exists in two states which differ in the total spin of the five electrons in the *d*-orbitals of the heme iron. The low-spin state with a total spin of $S = 1/2$ is six-coordinated with a water cluster as distal ligand *trans*-standing to the cysteine axial ligand in the proximal position (Fig. 2) [3]. In the high-spin state the total spin is $S = 5/2$ and the distal coordination position is not occupied [4]. When a substrate binds to *P*-450 the low-spin state is completely or partially shifted to the high-spin state. All the mentioned forms and states of *P*-450 differ significantly in the electron absorption

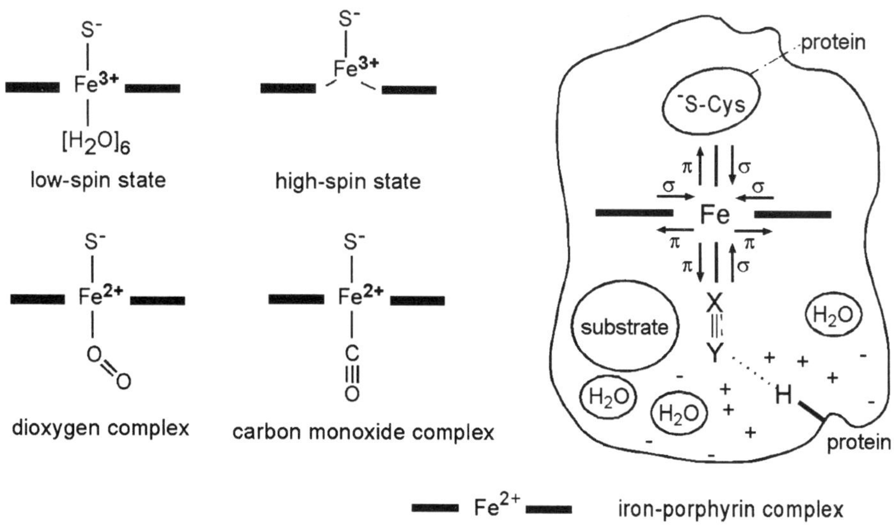

Fig. 2. Sketch of the coordination sphere of the cytochrome *P*-450 heme. The main parameters which influence the heme and coordinated ligand X-Y are sketched on the right side. These are the relations of ligand σ-donation and iron π-backdonation which are strongly modulated by electrostatic effects, solvent, substrate and proximal iron ligand [11]. In cytochrome *P*-450 the proximal ligand is the negatively charged sulfur of a cysteine.

spectrum [5] and can therefore be used to monitor the effect of pressure on the protein structure and ligand binding. Besides the electronic absorption spectra, the infrared active stretch vibration mode of the carbon monoxide ligand gives a further important spectroscopic probe to study the effect of hydrostatic pressure on the protein structure. These two spectral signals will be used in the following chapters to discuss the effect of hydrostatic pressure.

4. Effect of pressure on the ligand-unbound protein

When pressure is applied to the ferric form of cytochrome *P*-450, the conformational change of the protein is accompanied by a change in the total spin of the heme iron. As learned from crystal structure analysis [3] and various other experiments [6] this spin change results from the binding of a water ligand to the heme iron. The main response in the electronic absorption spectrum is therefore dominated by this chemical reaction step and reflects the high-spin/low-spin state equilibrium, mentioned above. Over the years, extensive studies have been done on the pressure effect on this equilibrium in various cytochromes *P*-450 with and without substrate and under various solvent conditions [7,8]. Increasing pressure shifts the equilibrium to the low-spin state. The most important finding, however, is that the reaction volume for the pressure-induced low-spin state formation is negative and the absolute value increases with increasing high-spin state content existing initially at 0.1 MPa before the pressure is applied [9,10]. This observation is explained by a hydration phenomenon and can be explained with a simple model. If *P*-450 exists in the high-spin state, most of the active site water molecules present in the heme pocket in the absence of a substrate are replaced by the substrate. When pressure is applied, water molecules are pressed into the protein which increases the probability that water binds to the heme iron and produces the low-spin state. The more water molecules exist initially in the protein, the fewer water molecules have to be pressed into the protein to get the low-spin state. This fundamental observation is reviewed in more detail by G. Hui Bon Hoa and S.G. Sligar in this issue. For the carbon monoxide complex discussed here this fundamental finding on the ferric *P*-450 is very important as seen below.

5. Effect of pressure on the ligand binding step

Small external ligands may bind to the ferric and/or ferrous form of the heme protein [11]. The experimental methods used to study this process are stopped-flow and flash photolysis under pressure. The spectroscopic probe is the intense Soret electronic absorption band of the heme. The exploration of the ligand binding and the effect of the substrate in cytochrome *P*-450 benefits significantly from the observations obtained on other heme proteins, in particular on myoglobin, which will briefly be sketched before discussing *P*-450.

The relevant parameters obtained from pressure studies are the reaction volume ΔV and the activation volume $\Delta V^{\#}$ determined from the pressure dependence of the equilibrium constant K and rate constant k according to Eq. 1.

$$\left|\frac{\partial \ln K}{\partial P}\right|_T = -\frac{\Delta V}{R \cdot T} \quad ; \quad \left|\frac{\partial \ln k}{\partial P}\right|_T = -\frac{\Delta V^{\#}}{R \cdot T} \qquad (1)$$

In addition, activation volumes of intermediate steps have also been determined from flash photolysis studies using laser-induced transient grating and photoacoustic calorimetric techniques [12]. Although these methods do not use pressure as variable parameter, such studies provide complementary data for volume changes of intermediates. Volume profile analysis has been shown to be very helpful in understanding the ligand binding mechanism [13].

5.1. Ligand binding steps in myoglobin

Flash photolysis studies carried out over more than three decades revealed that ligand binding in heme proteins is very complex [14–18] (Scheme 1).

The sequence of events starts from the ligand-bound state ('A state' [19]). The iron–ligand bond is broken due to light absorption by the heme. Within few picoseconds the dissociated CO ligand docks

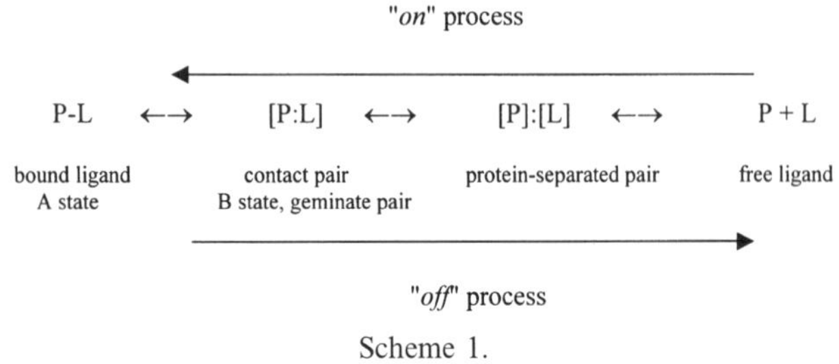

Scheme 1.

to sites near the heme and forms the so-called 'contact pair' [20], 'B state' [19] or 'geminate pair' [21] before it travels along various pathways (via several docking sites further from the heme [22]) through the protein.

The B state has been detected by flash photolysis FTIR studies on the CO stretch mode of the dissociated CO ligand whose frequency is observed between 2100 cm^{-1} and 2150 cm^{-1} [23–25] close to the stretch mode frequency of the free CO molecule.

Using laser-induced transient grating and photoacoustic calorimetric techniques an intermediate in horse muscle myoglobin with a lifetime of ~180 ns has been detected which shows a negative activation volume ($\Delta V^{\#} = -5.2$ cm^3/mol) (contraction) and likely corresponds to an intermediate within the migration step from the contact pair to another docking site [12]. CO exits the protein and enters the solvent where it becomes solvated. Escape of CO to the solvent has a decay time of ~700 ns and is accompanied by a positive activation volume change of ~14.7 cm^3/mol (expansion). This value is in the range of 11.7 cm^3/mol observed from the pressure dependence study of the CO flash photolysis on sperm whale myoglobin [20]. Because the partial molar volume of CO is about 36–37 cm^3/mol, one may conclude that the myoglobin protein itself contracts by about 20 cm^3/mol during the release of CO to the solvent.

Rebinding from the solvent starts by forming a so-called 'protein-separated pair' where CO must be desolvated again before it can enter the protein. The activation volume for entry of the ligand into the protein should have a positive activation volume because of the ligand desolvation [20] although its value has not been determined. The ligand entry rate depends on the ligand size which has been explained by a protein 'gate'. After CO has migrated again through the protein and formed the contact pair, the iron–CO bond is formed ('geminate rebinding') which is accompanied by significant volume contraction.

A comparative flash photolysis study on ferrous heme model complexes and myoglobin show [20] that the sign of the activation volume for the overall *on* reaction has been found to be dependent on the nature of the ligand indicating two main processes, the iron–ligand bond formation (negative $\Delta V^{\#}$) and the entry of the ligand into the protein (positive $\Delta V^{\#}$), which can be rate-limiting. The overall activation volume $\Delta V^{\#}$ results from compensation of these two contributions and may have a negative or positive net value. It has been shown that for the CO ligand the overall activation volume is negative (~ -10 cm^3/mol for myoglobin), indicating that the bond formation is the rate-limiting process. As learned from flash photolysis studies at low temperatures, the Fe–CO bond formation step itself is very fast and independent of the CO concentration once the CO molecule has found the optimal place close to the iron (B state). It is the probability to find this optimal place which contributes to the rate limitation.

5.2. Ligand binding steps in cytochrome P-450

Considering the results on myoglobin-CO and various other heme proteins with histidine as proximal ligand, it was surprising that cytochromes *P*-450 do not seem to show the same behavior. Lange et al. [26] have determined the activation volumes for CO binding in several cytochromes *P*-450 in the absence of a substrate using the stopped-flow technique under pressure. It turned out that all the proteins which have a thiolate as proximal ligand, including chloroperoxidase, have a small positive activation volume in the order from +1 cm^3/mol to 6 cm^3/mol (Table 1). It was concluded that the transition state in the sulfur ligand class proteins is structurally very close to the ground state and that the negatively charged sulfur from the cysteine ligand produces specific electronic properties which may be the origin [26]. However, flash photolysis studies under pressure for *P*-450cam (CYP101) in the presence of various substrate analogues indicated that substrate binding regulates whether ligand entry or the bond formation process is rate-limiting for the overall binding [27]. This has also been confirmed recently by stopped-flow studies under pressure for *P*-450cam bound with various camphor analogues (C. Jung, N. Bec and R. Lange, unpublished data, Table 1). For example, 1*R*-camphor-bound *P*-450cam shows a strongly negative activation volume of approx. -20 cm^3/mol while norbornane, which has nor bulky methyl groups, reveals a positive activation volume of approx. 8 cm^3/mol (Table 1). In the latter case, the binding is very fast (Table 1).

Considering the large number of studies with different methods on *P*-450cam bound with various substrate analogues [10,28–35] one can conclude that a high substrate mobility in the heme pocket and an increased access of the protein structure for water molecules favors the geminate rebinding and increases the overall rebinding rate. In these cases the CO binding is driven by entropic effects although the activation enthalpy is positive in all cases. As discussed in [21,28] the activation entropy for the *on* process is markedly positive in substrate-free *P*-450cam ($\sim$+99 J/kmol [28]), similar to the value of +145 J/kmol for a porphyrin model complex which is strongly exposed to the solvent [36]. It is clear that in such cases the activation volume should be positive because of many transient desolvation/solvation steps during the entry of CO into the protein and its migration to the heme iron. This reflects the general relation between entropy and volume fluctuation [37]. In contrast, camphor-bound *P*-450 shows a negative entropy of ~ -44 J/kmol and the activation volume is negative. So, the transition state is highly ordered in the camphor-bound protein and the energetic contribution of the entropic part ($-T{\cdot}\Delta S^{\#}$) leads to an increase of the activation free enthalpy and subsequently to a significant decrease of the binding rate compared to the substrate-free protein.

The *off* process – the dissociation of CO and its escape to the solvent – has also been studied using photoacoustic calorimetry [38] for the time window of 1 ns–1 μs. CO dissociation in substrate-free *P*-450cam has a transient shorter than 10 ns with a volume change of 1.8 cm^3/mol. However, when 1*R*-camphor binds, an intermediate with a lifetime of 130 ns and a volume change of +4.4 cm^3/mol is detected. A further expansion by 5.9 cm^3/mol is found when CO escapes from the protein. So, the CO dissociation process in *P*-450cam matches qualitatively the behavior observed in myoglobin. Based on a comparative photoacoustic calorimetry study on 1*R*-camphor-bound wild type and D251N mutant *P*-450cam, Di Primo et al. [39] concluded that the bifurcated salt bridge Arg186-Asp251-Lys178 may transiently solvate and desolvate when CO escapes

Table 1
Activation volume for CO binding in cytochromes *P*-450 and other heme proteins obtained from stopped-flow (SF) and flash photolysis (F) studies under pressure monitored at the Soret band

Protein	Method	Proximal ligand	$\Delta V^{\#}$ (cm^3/mol)	k_{on} (10^4 M^{-1} s^{-1})	Ref.
Chloroperoxidase	SF, 25°C	S^- (cys)	1 ± 2	14	[26]
P-450scc, free	SF, 25°C	S^- (cys)	2 ± 2	23	[26]
P-450LM2, free	SF, 25°C	S^- (cys)	3 ± 2	310	[26]
P-450LM3, free	SF, 25°C	S^- (cys)	6 ± 5	220	[26]
P-4507α	SF, 25°C	S^- (cys)	2 ± 2	2.8	[26]
P-450cam, free	SF, 5°C	S^- (cys)	1: 4.6 ± 2.0 (< 114 MPa)	29.5	u.p.
			2: 10.2 ± 2.1 (< 114 MPa)	297	
P-450cam, free	F, 20°C	S^- (cys)	4 ± 1	850	[27]
P-450cam, norcamphor	SF, 5°C	S^- (cys)	7.6 ± 1.1	381	u.p.
P-450cam, norcamphor	F, 20°C	S^- (cys)	3 ± 2	1000	[27]
P-450cam, norbornane	SF, 5°C	S^- (cys)	8.4 ± 0.9	332	u.p.
P-450cam, adamantane	F, 20°C	S^- (cys)	7 ± 3	1300	[27]
P-450cam, 1*R*-camphor	SF, 5°C	S^- (cys)	-19.6 ± 0.9	3	u.p.
P-450cam, 1*R*-camphor	F, 20°C	S^- (cys)	-31 ± 2	1	[27]
P-450cam, fenchone	F, 20°C	S^- (cys)	-20 ± 1	150	[27]
P-450cam, TMCH	F, 20°C	S^- (cys)	-14 ± 1	75	[27]
P-450cam, bromocamphor	F, 20°C	S^- (cys)	-32 ± 2	55	[27]
P-450cam, camphane	SF, 5°C	S^- (cys)	1: -10.6 ± 1.1, < 115 MPa	1.6	u.p.
			2: -18.2 ± 3.8, < 115 MPa	7.8	
			1: -53.1 ± 21.9, > 115 MPa	2.5	
			2: -138.8 ± 27.8, > 115 MPa	24.2	
P-420	SF, 25°C	N (his)?	-25 ± 5	0.66	[26]
			-11 ± 6	1.2	
			-42 ± 8	0.096	
Lactoperoxidase		N (his)	-10 ± 3	0.08	[26]
Horseradish peroxidase		N (his)	−24	0.28	[71]
Sperm whale myoglobin		N (his)	−9	38	[72]
			−10	52	[20]
Hemoglobin		N (his);R	-9 ± 0.7		[27]
		N (his);T	-31.7 ± 2.4		
Hemoglobin	SF, 25°C	N (his)	−3	280	[71]
Hemoglobin		N (his)	−21	0.6	[26]

u.p., C. Jung, N. Bec and R. Lange, unpublished data. Studies on *P*-450cam are performed in 100 mM potassium phosphate buffer, pH 7.3, 10% (w/w) glycerol, 0.4 mM 1*R*-camphor, 0.4 mM camphane, 4 mM norcamphor, 2 mM norbornane or no substrate, respectively. Fit of the time curves was performed with a bimolecular kinetics except for camphane and substrate-free complex where two bimolecular processes were required. *P*-450cam-camphane shows two pressure ranges with different parameters. The contribution of k_{on1} is independent of the pressure below 115 MPa and has an average value of 52%. This contribution increases up to approx. 70% above 115 MPa. The k_{on1} and k_{on2} of the camphane complex for the higher pressure range are given at 115 MPa. For substrate-free *P*-450cam the contribution of the faster reaction increases from 54% to 65% when pressure increases up to 114 MPa. TMCH, tetramethylcyclohexanone.

to the solvent and this may explain the positive activation volume for the *off* process.

5.3. Structural assignment for ligand binding steps in P-450

Although these thermodynamic considerations are straightforward, many efforts have also been made to assign the structural changes accompanied with CO photodissociation and rebinding. Jung et al. [40] have measured by FTIR at 20 K the stretch mode of the photodissociated CO ligand in the B state. These studies indicated that there is an ensemble of several B states in substrate-free *P*-450cam which is in line with the existence of an ensemble of A states (iron-bound CO) (Table 2). In the presence of 1*R*-cam-

Table 2
Stretch vibration frequencies of the photodissociated CO ligand ('B state', 'contact pair', 'geminate pair') at 15–20 K for cytochromes *P*-450 and other heme proteins

Protein	A state (cm^{-1})	B state (cm^{-1})	Ref.
P-450cam, 1*R*-camphor[a]	1933.3, 1941.7	2121, 2131.5	[40]
P-450cam, no substrate[a]	1911.4, 1927.6, 1948.9, 1953.1	2115, 2117, 2128, 2131, 2139, 2148	[40]
P-450lin, linalool[a]	1949.9, 1952.7	2118	[40]
P-450lin, no substrate[a]	1944.3, 1946.9	2113, 2124, 2136	[40]
$iNOS_{ox}$, no substrate, $-H_4B$	1935.4, 1944.0, 1954.7	2123.2, 2127.8, 2133.6	[73]
$iNOS_{ox}$, no substrate, $+H_4B$	1945.6	2126	[73]
$iNOS_{ox}$, arginine, $-H_4B$	1903.5	2120.4, 2130.0, 2144.5	[73]
$iNOS_{ox}$, arginine, $+H_4B$	1901.8, 1922.3	2118.7, 2130.7, 2150,2	[73]
Sperm whale myoglobin	1931.9, 1945.5, 1966.4	2118.8, 2130.6. 2145.5	[23,24]
Leghemoglobin	1948	2134	[25]

[a]Sample and solvent conditions are described in [40]. Frequencies of the A states (iron-bound CO) obtained from flash photolysis experiments may not exactly match the values in static low-temperature studies because of different fast rebinding rates within the inhomogeneous substates. The sequence of the listed A state and B state frequencies is not an assignment of the B state to a particular A state (see [73]). H_4B, tetrahydrobiopterin.

phor, only two B states corresponding to the two A states are observed. Using step-scan time-resolved FTIR in the time window of 5 μs–20 ms at 300 K it was found that in the 1*R*-camphor and in the camphane complex the dissociation of CO is accompanied with structural changes in a salt link (presumably between Asp297, Arg299 and the heme propionate group) and in the secondary structure [29]. Rebinding of CO and protein relaxation monitored on the CO stretch mode band at about 1940 cm^{-1} and on the amide I band (1700–1600 cm^{-1}) and salt link signal (~1720 cm^{-1}), respectively, are tightly coupled as seen from the same k_{on} rate constant (Table 3). For complexes with substrates missing the methyl groups (norcamphor, norbornane) or in the absence of a substrate, the protein tends to relax faster than CO rebinds to the iron although both parts show rate constants larger by 10–40 times com-

Table 3
Bimolecular rate constants (k_{on}) for the CO ligand binding in cytochrome *P*-450cam obtained from flash photolysis step-scan time-resolved FTIR studies [29]

Substrates	k_{on} (10^4 M^{-1} s^{-1})				
	CO ligand	Salt link	Turns	α-Helix	β-Sheet
1*R*-Camphor, wt, D_2O	9.8 ± 0.1	8.6 ± 0.6	11.9 ± 1.0	7.4 ± 1.5	14.5 ± 1.5
1*R*-Camphor, wt, H_2O	21.3 ± 0.3	25.5 ± 2.2	34.2 ± 6.8	19.1 ± 6.1	–
1*R*-Camphor, D251N, D_2O	10.3 ± 0.1	8.7 ± 0.4	12.6 ± 1.6	5.1 ± 1.5	10.4 ± 1.6
Camphane[a], wt, D_2O	10.4 ± 0.3 (0.61, 1939.4)	5.8 ± 0.4	6.3 ± 0.5	9.7 ± 1.2	
	49.7 ± 1.8 (0.25, 1949.2)				34.5 ± 3.0
Norcamphor, wt, D_2O	340.8 ± 7	406.3 ± 55	474.4 ± 76		
Norbornane, wt, D_2O	343.5 ± 6	109.0 ± 41	562.5 ± 85		648.8 ± 135
Substrate-free[a], wt, D_2O	158.1 ± 5 (0.54, 1941.2)	235.9 ± 86	271.6 ± 55		
	132.5 ± 13 (0.08, 1951.9)				
	381.3 ± 29 (0.31, 1960.1)				

[a]The different values of k_{on} for the CO ligand correspond to the different substates. Values in parentheses are the population of the substate and the frequency in cm^{-1}.
wt, wild type; D251N, mutant. The rate constants were determined from the time dependence of the signals corresponding to the different structural elements (CO ligand: 1938–1960 cm^{-1}; salt link: ~1720 cm^{-1}; turns: ~1667–1676 cm^{-1}; α-helix: 1638–1655 cm^{-1}; β-sheet: 1620–1637 cm^{-1}).

pared to the rate constants for the 1*R*-camphor and camphane complex. It is a general trend that within an ensemble of substates (A states) always those corresponding to a higher CO stretch mode frequency rebind or relax faster than the others with a lower frequency (Table 3). This can also be explained by the entropy effect because higher CO stretch mode frequencies are connected with a larger solvent accessibility in this substate (see below). Although the solvation/desolvation model may explain many features of CO binding as a function of the substrate in cytochrome *P*-450, several other factors such as heme and protein deformation and relaxation due to substrate movement and flipping between different substrate binding orientations must also be considered, as discussed recently, to explain the bimodal enthalpy distribution for the CO rebinding kinetics [18,41].

There are a number of flash photolysis [42–47] and stopped-flow [48,49] studies on various other cytochromes *P*-450 under ambient pressure. All these studies show an effect of the substrate on the CO binding kinetics whereby the efficiency of this influence and whether the CO binding is accelerated or slowed down vary in the different *P*-450s. Only few studies under high hydrostatic pressure have been reported so far, which are discussed above. However, already these few examples show that high-pressure studies can provide further details to uncover enzyme–substrate interaction in cytochromes *P*-450.

6. Effect of pressure on the ligand-bound protein

As mentioned above, there are two important spectroscopic signals which can be used to monitor pressure effects on the cytochrome *P*-450 structure. These are the Soret band in the electronic absorption spectrum and the stretch vibration mode of the CO ligand measured by FTIR, which both probe structural properties near the heme and the substrate binding site. The Soret band, usually regarded as less informative in the CO complex, reveals to be a good probe to determine isothermal compressibilities for the different substrate complexes. The CO stretch mode allows analysis of subconformational behavior, not seen in the Soret band. Therefore the two methods are complementary.

6.1. FTIR CO stretch mode vibration

6.1.1. Substate equilibrium

It is a fundamental observation from many experimental and theoretical approaches that protein structures exist in an equilibrium of various conformational substates [14,15,19]. There are, however, only few methods which can resolve directly these substates and quantitate their thermodynamic and structural parameters. Infrared spectroscopic studies of the CO ligand stretch mode contributed significantly to uncover this behavior of proteins. Extensive studies of the combined effect of pressure and temperature on the CO stretch mode infrared spectra of myoglobin, horseradish peroxidase and chloroperoxidase have been performed [50,51]. However, in particular for cytochrome *P*-450, such studies brought a deeper insight into the enzyme–substrate interaction [31,51,52]. Most of such studies have been performed on *P*-450cam from *Pseudomonas putida* [11]. In the absence of a substrate a broad structured infrared band is observed which results from several overlapping CO stretch mode signals at about 1915 cm^{-1} (band I), 1930 cm^{-1} (band II), 1946 cm^{-1} (band IVa), 1953 cm^{-1} (band IVb) and 1965 cm^{-1} (band V) [30]. These different signals represent different conformational substates classified as 'taxonomic substates' or 'CS^0' substates according to Frauenfelder's model of a hierarchy of substates [14,19]. Their populations change strongly with temperature, pressure and solvent conditions. A complete set of spectra for the temperature and pressure dependence has been published in several papers [11,30,53] and will not be shown here again. Fig. 3B gives a general impression of the spectra. Decreasing temperature shifts the substate equilibrium to the lower-frequency mode at ~1930 cm^{-1} (band II) while increasing hydrostatic pressure favors the substates with the higher-frequency mode at ~1946 cm^{-1} (band IVa) and ~1953 cm^{-1} (band IVb) which often merge to one band at about 1950 cm^{-1} (band IV) [30,51]. When *P*-450cam binds the substrate 1*R*-camphor, the number of substates is significantly restricted (Fig. 3C). However, low temperature and high-pressure studies reveal that at least two substates are present at about 1935 cm^{-1} (band 1) and 1942 cm^{-1} (band 2) [51–53]. Jung et al. [30] have determined the enthalpy ΔH_{ir} and entropy ΔS_{ir} as well as volume ΔV_{ir} differences between substate i and the reference substate r, the thermal expansion coefficient α_i and isothermal compressibility β_i (Table 4) by a global fit analysis of the populations A_i (A_r) of the substates i (reference r) using the fundamental relations given in Eqs. 2–7.

$$A_i/A_r = \exp[-\Delta G_{ir}/RT];$$

$$\Delta G_{ir} = \Delta E_{ir} + P\cdot\Delta V_{ir} - T\cdot\Delta S_{ir} \tag{2}$$

$$\Delta S_{ir}(T,P) = \Delta S_{ir}(0,0) + \Delta s_i\cdot T + \Delta \upsilon_i\cdot P \tag{3}$$

$$\Delta V_{ir}(T,P) = \Delta V_{ir}(0,0) - \Delta \upsilon_i\cdot T + \Delta q_i\cdot P \tag{4}$$

$$\Delta H_{ir}(T,P) = \Delta E_{ir}(0,0) + P\cdot\Delta V_{ir}(0,0) + \frac{1}{2}\Delta s_i\cdot T^2 + \frac{1}{2}\Delta q_i\cdot P^2 \tag{5}$$

where (0,0) refers to $T = 0$ K and $P = 0$ MPa. Δs_i, $\Delta \upsilon_i$ and Δq_i are fit parameters. The latter two parameters are related to the isobar thermal expansion coefficient α_i and the isothermal compressibility β_i with V being the partial molar volume of the protein.

$$\beta_i = -\frac{1}{V}\left[\frac{\partial V_{ir}}{\partial P}\right]_T = -\frac{1}{V}\cdot\Delta q_i \tag{6}$$

$$\alpha_i = \frac{1}{V}\left[\frac{\partial V_{ir}}{\partial T}\right]_P = \frac{1}{V}\cdot\Delta \upsilon_i \tag{7}$$

Three important results were obtained from these studies. (i) The volume differences between the substates are rather small for substrate-free P-450cam (5.6–9.0 cm^3/mol, at 298 K). In contrast, in the presence of 1R-camphor, a large volume difference of 65.1 cm^3/mol at 298 K was obtained. Thus different large volumes may reflect a similar hydration phenomenon as observed for the reaction volume of the high-spin/low-spin state equilibrium discussed above. (ii) Substates differ in isothermal compressibility and thermal expansion coefficient. (iii) In the substrate-free protein, an inverse effect of hydrostatic pressure and osmotic pressure (induced by glycerol) on the population of the substate, which are characterized by a high CO stretch mode frequency, was found. For example the volume difference of substate V (at about 1965 cm^{-1}) minus IV (at about 1951 cm^{-1}) is negative ($\Delta V = -5.8$ cm^3/mol) when hydrostatic pressure is applied. However, it is +93 cm^3/mol for the osmotic pressure. The volume differences induced by hydrostatic pressure are smaller in absolute values than by osmotic pressure, which has also been found for the high-spin/low-spin state equilibrium [34]. One may conclude that these 'high-frequency' substates are characterized by a higher access of the heme pocket for water molecules.

6.1.2. CO stretch mode frequency shifts

6.1.2.1. Elastic and conformational frequency shifts. Pressure affects the frequency by two effects: the elastic and the conformational effect. In a uni-

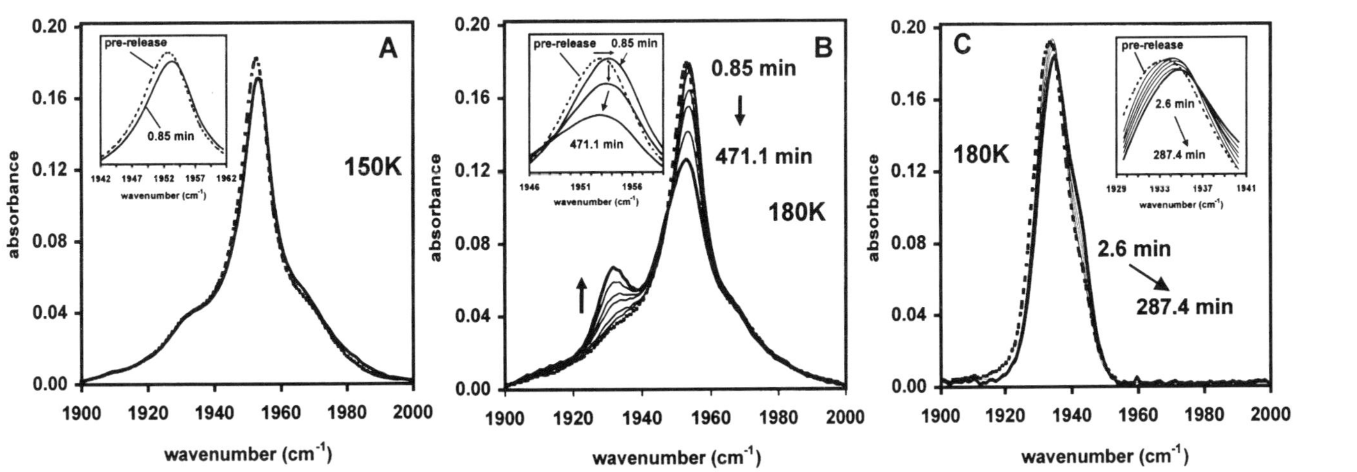

Fig. 3. CO stretch mode infrared spectra of cytochrome P-450cam-CO, in the absence (A,B) and in the presence of 1R-camphor (C), monitored in pressure 'fast-release' experiments (200–40 MPa) [40,51]. 100 mM potassium phosphate buffer pH 7, 50% glycerol (1.5 mM 1R-camphor for substrate bound). Experimental setup for the pressure experiments and the FTIR measurements are described in [50,51]. The insets give an expanded plot of the bands around the maximum.

Table 4
Thermodynamic parameters, thermal expansion coefficients and isothermal compressibilities for the taxonomic substates (CS^0) of cytochrome *P*-450cam-CO [30] and myoglobin-CO [50] at 298 K

Protein	ΔH (kJ/mol)	ΔS (J/kmol)	ΔV (cm^3/mol)	ΔG (kJ/mol)	α (10^{-6} K^{-1})	β (GPa^{-1})
P-450cam, substrate-free						
I	15.3	66.5	7.6	−4.5	5.6	0.042
II	16.3	79.9	9.0	−7.5	0.3	0.003
IV	25.5	122.4	5.8	−10.9	2.9	0.017
P-450cam, camphor-bound						
I	44.9	215.0	65.1	−19.1	12.8	−0.048
Sperm whale myoglobin						
pH 6.6	7.5	−3.4	−12.2	8.5	8.3	–
pH 5.5	2.5	−18.9	−3.3	8.1	9.8	–

Reference states: *P*-450cam substrate-free: V (~1963 cm^{-1}); *P*-450cam camphor-bound: 2 (~1942 cm^{-1}); myoglobin: A_1 (~1945 cm^{-1}); CS^0 means conformational substate on level 0 (taxonomic substate) according to [19].

form solid material, as for example a crystal, pressure decreases the distance between the atoms whereby the relative arrangement of the atoms is not changed. This is the elastic effect and leads to a shift of the frequency to lower values ($\delta\nu$) which is linear in pressure and strictly reversible. It should respond instantaneously to pressure change and occur at all temperatures [50]. The value of the shift depends on the dielectric constant ε (Eq. 8) of the medium surrounding the spectroscopic probe [54] which changes with pressure. C and ν_0 are a constant and the frequency at a reference state, e.g. 0.1 MPa.

$$\frac{\delta\nu}{\nu_0} = C\frac{\varepsilon - 1}{2\cdot\varepsilon + 1} \tag{8}$$

In soft materials such as proteins the pressure effect is not so simply understood and the conformational effect becomes very important. The sign of the frequency shift induced by a conformational change can be positive or negative depending on the kind and the relative position of the spectroscopic probe in the protein and to internal solvent molecules. If a change in the equilibrium of the conformational substates is induced by pressure, as discussed above, apparent frequency shifts may be observed. Curve fitting helps to get a more realistic insight into the real shift of each band corresponding to a particular taxonomic substate. Population exchange and frequency shifts reflect conformational changes on a different level in the hierarchy of substates [19] which can be differentiated by pressure jump (fast-release) experiments at different temperatures.

6.1.2.2. Pressure jump (fast-release) effects. Pressure jump (fast-release) experiments have been done on myoglobin, horseradish peroxidase and chloroperoxidase, and on cytochrome *P*-450cam in the absence and in the presence of the substrate 1*R*-camphor [40,50,51]. The protein is pressurized to 200 MPa above the glass transition temperature T_g, and then cooled down to different target temperatures at which pressure is suddenly released from 200 MPa to 40 MPa. One observes two different effects when monitoring the time course of the frequency change at different temperatures. Fig. 3A shows the CO stretch mode spectrum of substrate-free *P*-450cam at 150 K. Immediately after the pressure release the whole spectrum shifts up by approx. 0.6 cm^{-1} without significant further shift with time. This has been classified as an elastic shift [50] which is negative at pressure increase and positive at pressure release. However, with passing the glass transition temperature T_g (~180 K) a population exchange between the taxonomic substates (CS^0 level [19]) is observed (Fig. 3B) and the frequency for the single band IV (after curve fitting) shifts up first, followed by a decrease in the frequency (Fig. 4A,B). The latter change (down-shift for pressure release) corresponds to a conformational change of the statistical substates (CS^1 level [19]) and matches that observed when pressure is increased at the target temperature (up-shift). Obviously there are two conformational transitions corresponding to changes within the taxonomic substate CS^0 (FIM^1) and to a transition between CS^0 substates (FIM^0) in accordance with the model of hierarchy of substates [14,19]. FIM stands for 'functional important motions' [19]. Interestingly, substrate-bound *P*-450cam does not show an elastic frequency shift but only a conformational shift (Fig. 4C). This might be explained by a lower local dielectric constant near the CO ligand compared to the substrate-free protein where more water molecules have access to the heme pocket. Such interpretation is in agreement with observations in myoglobin where substate A_0 at about 1966 cm^{-1} shows an elastic shift of about 0.6 cm^{-1} while for substate A_1 at about 1945 cm^{-1} the elastic shift is zero [50,51]. This seems to be reasonable. It is suggested from different experiments that A_0 corresponds to a more 'open' pocket with the swung-out conformation of the distal histidine [55], which may facilitate the access for water molecules, while substate A_1 should reflect a 'closed' form.

The kinetics of the frequency shifts has been analyzed in more detail. Below the glass transition the relaxation function $\Phi(T,t)$ (Eq. 9) follows the stretched exponential [50] and a modified Arrhenius equation for the rate constant $k(T)$ (Eq. 10), both being typical glass-like behaviors. A second process above the glass transition is clearly described by a single exponential and the Arrhenius equation (Eq. 10).

$$\theta(T,t) = \frac{|O(P_f,T,t) - O(P_f,T,t\to\infty)|}{|O^+(P_f,T,t=0) - P(P_f,T,t\to\infty)|} \quad (9)$$

$$\Theta(T,t) = \exp[-(k(T)\cdot t)^{\beta'}]$$

$$k(T) = A\cdot\exp\left[-\left(\frac{E_a}{k_B\cdot T}\right)^x\right] \quad (10)$$

For $T < T_g$: $\beta' < 1$ and $x = 2$; for $T > T_g$: $\beta' = 1$ and $x = 1$. $O(P_f,T,t)$, $O(P_f,T,t\to\infty)$ and $O^+(P_f,T,t=0)$ are the experimental values for the selected parameter (here the frequency) at the final pressure P_f (40 MPa), the temperature T and at time (t, $t\to\infty$ and $t=0$, respectively). $O^+(P_f,T,t=0)$ corresponds to the frequency in the equilibrium minus the elastic frequency shift.

Table 5 compares the relaxation parameters for the process below the glass transition (FIM^1 process [19,50]) for *P*-450cam with the values for myoglobin and chloroperoxidase. The proteins have similar parameters except the very low stretch parameter β' for camphor-bound *P*-450, which is a factor of 10 smaller than in the absence of the substrate. This is not yet understood but it indicates a broader or asymmetric distribution of the statistical substates.

6.1.2.3. Compressibility. Pressure-induced frequency shifts are related to the isothermal compressibility β generally defined by Eq. 6. Assuming that

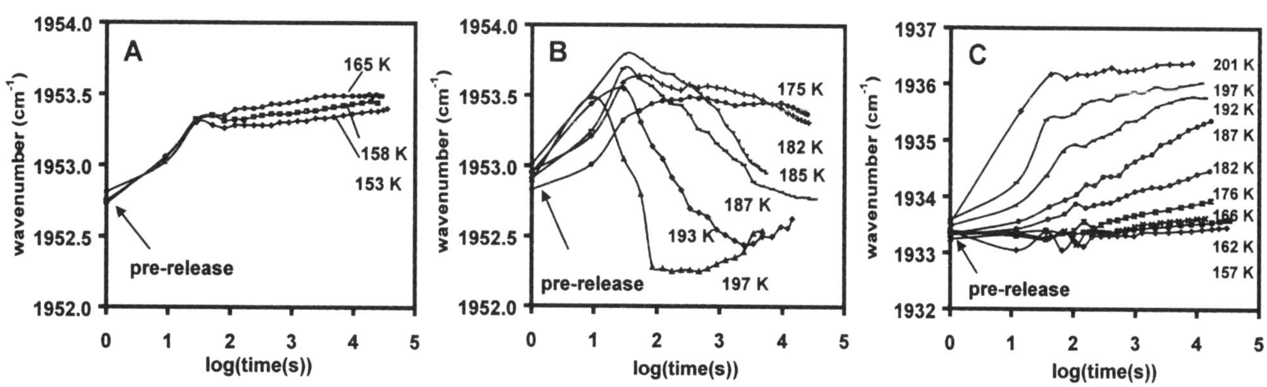

Fig. 4. CO stretch mode frequency of cytochrome *P*-450cam-CO, in the absence (A,B) and in the presence of 1*R*-camphor (C), as a function of the decadic logarithm of the time (in s) in pressure 'fast-release' experiments (200–40 MPa) at different temperatures [40,51]. A and B show the frequency of the main band of substrate-free cytochrome *P*-450cam after curve fitting (see Fig. 3A,B, band IV according to [30]). C represents the frequency of the main band in 1*R*-camphor-bound cytochrome *P*-450cam after curve fitting. Experimental conditions are described in Fig. 3.

Table 5
Elastic frequency shift, relaxation parameters and activation energy for conformational substates (CS[1]) for cytochrome *P*-450cam and other heme proteins obtained from pressure fast-release experiments using FTIR [40,50,51]

Protein/substate	Elastic shift		Relaxation parameters for conformational shift (FIM[1])			
	$\lvert\Delta\nu\rvert$ (cm^{-1})	ΔP (MPa)	β'	log A (s^{-1})	E_a (kJ/mol)	fit range (K)
P-450cam, substrate-free ν ($\sim$1952 cm^{-1})	0.6	160	0.40	9	7–8	153–173
P-450cam, camphor ν ($\sim$1935 cm^{-1})	0.0	160	0.05	12	8	163–176
Chloroperoxidase pH 6, ν ($\sim$1935 cm^{-1})	0.5–0.6	160	0.50	12	9	160–173
Horseradish peroxidase pH 6, ν ($\sim$1905 cm^{-1})	0.2–0.3	115	–	–	–	–
Myoglobin ν ($\sim$1966 cm^{-1})						
pH 5.5	0.7	175	0.30	7	7	155–170
pH 6.6	0.4	175	0.20	8	7	150–165
pH 6.9	0.5	160	0.50	7	7	155–170

CS[1] and FIM[1] mean the conformational substate on level 1 ('statistical substate') and 'functional important motion' within level 1 according to [19]. The parameters are defined in Eqs. 9 and 10.

volume and frequency are related via the Grüneisen parameter [37,56] one may define the compressibility as given in Eq. 11.

$$\beta = const\cdot\frac{1}{\nu_0}\left[\frac{\partial\nu}{\partial P}\right]_T \tag{11}$$

Eq. 11 actually holds only for phonons in crystals [37,56]. When applying it to the stretch mode of the iron-bound CO ligand one implicitly assumes that the CO vibration is influenced by phonons or low-frequency modes of the protein. Proteins show a density-of-states spectrum below 300 cm^{-1} [57,58]. These low-frequency modes may couple with the CO stretch mode or other modes which affect the CO ligand. Such a 'coupling' model has been used previously to explain the temperature dependence of the CO stretch mode width in cytochrome *P*-450lin [53]. The coupling energy (E/h, h Planck constant) has been found to be in the order of 0.06 cm^{-1}.

The quantity *const* in Eq. 11 depends on the nature of the vibrating group and the type of vibration [37]. However, regarding the same vibration – here the stretch mode of the CO ligand – a comparison between different heme proteins or substrate-bound and substrate-free *P*-450 may be justified. Because in most cases the CO stretch mode is shifted to lower frequencies by pressure we will define the negative slope of the shift to produce a positive compressibility. Therefore, *const* should be negative here by definition.

Although the compressibility is pressure dependent in a broad pressure range [56], in the narrow range one might expect a frequency shift linear in pressure and the slope should give β as the linear compressibility. The intercept of the linear pressure-dependent frequency at 0 MPa is taken for ν_0 in Eq. 11.

Fig. 5 shows that indeed up to 200 MPa the shift is almost linear. The shift is positive for substrate-free *P*-450cam and negative for the camphor-bound protein. Within the experimental error the compressibility is the same at 290 K, 200 K and 180 K for camphor-bound *P*-450cam but becomes lower or even changes the sign for substrate-free *P*-450cam going to 180 K (Table 6), again indicating a complex conformational behavior in the absence of a substrate.

Regarding only the absolute values of the compressibility one concludes at first glance that substrate-free *P*-450cam is less compressible than in the presence of camphor, which would be in contrast to conclusions drawn from a number of different experiments (see below). The compressibility determined from the frequency shift is approx. 5 times smaller than the compressibility differences between the taxonomic substates obtained from the global fit analysis of the substate populations (Table 4). This result seems to be reasonable because volume differences between statistical substates within the same taxonomic substate should be smaller than between the different taxonomic substates.

6.1.2.4. Thermal expansion. Frequency shifts of

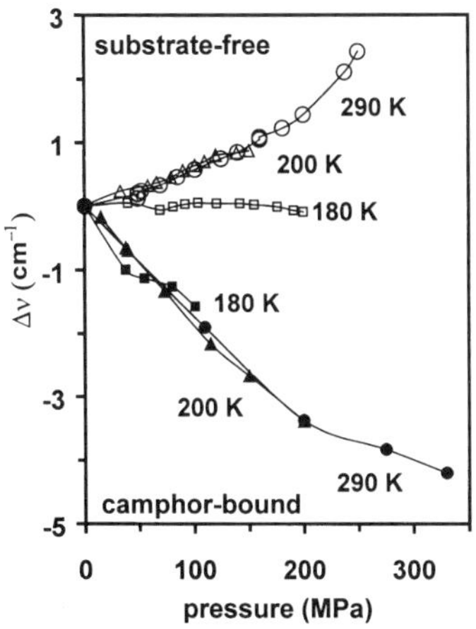

Fig. 5. Pressure-induced shift of the CO stretch mode frequency of cytochrome *P*-450cam-CO. Open symbols are in the absence and filled symbols in the presence of 1*R*-camphor. Squares, 180 K; triangles, 200 K; circles, 290 K. The shift refers to the frequency at 0.1 MPa: substrate-free *P*-450cam: 1951.93 cm^{-1} (180 K), 1951.29 cm^{-1} (200 K) and 1949.70 cm^{-1} (290 K); camphor-bound: 1936.63 cm^{-1} (180 K), 1937.28 cm^{-1} (200 K) and 1939.12 cm^{-1} (290 K). Experimental conditions are described in [30,52].

the CO stretch mode are not only induced by pressure but also by temperature which leads to contraction or expansion of the sample. The temperature effect has been studied on a number of substrate complexes of *P*-450cam [31,53,59,60]. Fig. 6 shows a plot of the CO stretch mode frequency versus the temperature for selected substrate complexes of *P*-450cam. Linear thermal expansion coefficients can be estimated from the frequency shifts (Eq. 12) for the range below and above the glass transition region between 150 K and 230 K assuming the same considerations as discussed above for Eq. 11.

$$\alpha = const\cdot\frac{1}{\nu_0}\left[\frac{\partial\nu}{\partial T}\right]_P \qquad (12)$$

In general, the thermal expansion coefficients above 230 K are larger than below 150 K (Table 7). In a few cases even slightly negative expansion coefficients are obtained in the low temperature range. There seems to be no correlation between the value of thermal expansion coefficient and any structural property of the substrate (e.g. volume) or a substrate-induced property of the protein (e.g. high-spin state content). The thermal expansion coefficient of 1*R*-camphor *P*-450cam determined from the frequency shift (3.3×10^{-6} K^{-1}) is approx. 4 times smaller than obtained from the global fit analysis for the substate populations (12.8×10^{-6} K^{-1}, Table 4), indicating that thermal expansions on the CS^0 and CS^1 levels differ, as discussed above for the compressibilities.

The thermal expansion coefficients obtained for *P*-450 from the frequency shift are smaller by a factor of 10–20 compared to thermal expansion coefficients reported for other proteins [37], indicating that one has to think more about the meaning of the quantity *const* in Eq. 12.

The pressure dependence of the CO stretch mode spectrum for the complexes with various substrates other than 1*R*-camphor has not been performed until now. Therefore, the comparison between pressure and temperature effects on the frequency shifts can

Table 6
Isothermal compressibilities for cytochrome *P*-450cam determined from the frequency shift of the CO stretch modes corresponding to different substates

Protein	*T* (K)	Slope (cm^{-1}/MPa)	ν_0 (P=0 MPa) (cm^{-1})	β (GPa^{-1})
Substrate-free				
II (~1931 cm^{-1})	180	−0.0032	1931.18	0.0017
	200	0.0034	1931.56	−0.0018
	290	–	–	–
IV (~1951 cm^{-1})	180	0.0000	1951.29	0.0000
	200	0.0062	1951.29	−0.0032
	290	0.0076	1949.54	−0.0039
Camphor-bound				
1 (~1937 cm^{-1})	180	−0.0145	1936.44	0.0075
	200	−0.0175	1937.28	0.0090
	290	−0.0169	1939.11	0.0087

Determined from the data published in [30,51,52] using Eq. 11 with *const* = 1.

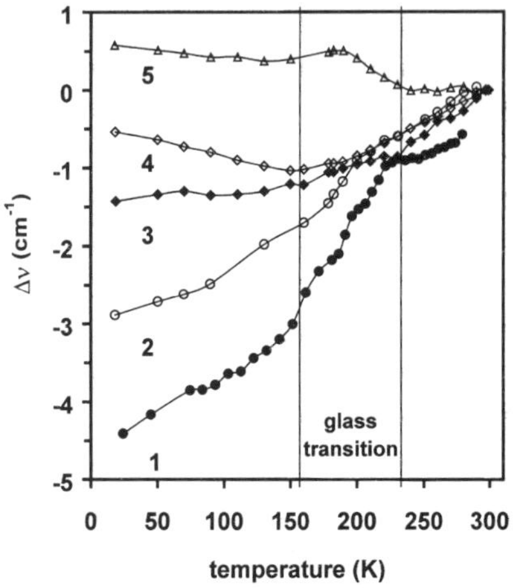

Fig. 6. Temperature-induced shift of the CO stretch mode frequency of cytochrome *P*-450cam-CO in the presence of different substrates. The shift refers to the frequency at 298 K: 1, 1*R*-camphor (1939.14 cm^{-1}); 2, norbornane (1952.80 cm^{-1}); 3, adamantanone (1941.50 cm^{-1}); 4, tetramethylcyclohexanone (1933.46 cm^{-1}); 5, norcamphor (1946.15 cm^{-1}). 100 mM potassium phosphate buffer, pH 7, 60% (by mass) glycerol and substrate as described in [60].

only be made for the substrate-free and 1*R*-camphor complex. Increasing pressure produces qualitatively the same changes as decreasing temperature, pointing to the conclusion that volume changes are most relevant. In addition, for substrate-free *P*-450cam, osmotic pressure (glycerol) influences the CO stretch mode frequency in an inverse manner to the hydrostatic pressure, as was also observed for the substate populations discussed above. Although the compressibility as well as the thermal expansion coefficient are appropriate quantities to compare the different substrate complexes of *P*-450, they are still not well understood in terms of the atomic interactions behind the volume change.

6.2. Pressure-induced Soret band shifts

It is generally observed that the Soret band in carbon monoxide complexes of heme proteins shifts to lower wave numbers with increasing pressure [32,61] (Fig. 7). In contrast to the intense Soret band, the weak Q bands in the visible region are not affected by pressure in *P*-450 or other heme proteins, such as e.g. myoglobin. Two different formalisms have been proposed to explain this observation [62]. The first approach attributes the red shift to a solute–solvent interaction [63] which increases with density. In the other approach the red shift is accounted for by means of a negative volume change associated with a more compressible excited state relative to the ground state. However, Jung et al. [32] found for cytochrome *P*-450cam bound with various substrates that the slope of the Soret band red shift is related to the high-spin state content of the substrate complex in the ferric form of *P*-450. Because the high-spin/low-spin state equilibrium is strongly connected with solvent access of the protein, the first approach seems to describe better the effect, at least for *P*-450.

The microscopic theory for solute–solvent interaction and for reversible pressure effects on electronic absorption bands developed by Laid and Skinner [64] have been used and adapted for cytochrome *P*-450 to calculate isothermal compressibilities.

According to this theory the absorption frequency $\nu_i(R,P)$ consists of different contributions:

$$\nu_i(R,P) = \nu_{vac} + \nu_{si}(R_0, P=0) + \alpha_i'\cdot P \quad (13)$$

R is the distance between the solvent molecule and the solute molecule (chromophore) at pressure *P*. ν_{vac} is the absorption wave number in the absence of solvent, that means in 'vacuum'. $\nu_{si}(R_0,P=0)$ is the change of the absorption wave number caused by the interaction with the solvent molecules in a

Table 7
Thermal expansion coefficients for cytochrome *P*-450cam-CO bound with various substrates, obtained from temperature-induced shift of the CO stretch mode at 0.1 MPa

Substrate	α (<150 K) (10^{-6} K^{-1})	α (>230 K) (10^{-6} K^{-1})
1*R*-Camphor	5.28	3.30
1*R*-Camphor-quinone	0.00	0.00
1*S*-Camphor	4.62	10.67
1*S*-Camphor-quinone	4.33	12.46
Norcamphor	−0.75	0.00
Norbornane	4.08	3.48
Tetramethyl cyclohexanone	2.01	4.49
Adamantanone	0.63	6.16
1-Cl-adamantane	−0.34	7.16
1-Br-adamantane	0.50	1.42
1-I-adamantane	1.63	28.11
1-Cl-dimethyl-adamantane	1.53	20.66
1-Br-dimethyl-adamantane	1.73	14.97
Various proteins [33]		40–110

Determined with Eq. 12 using *const* = 1 from experimental spectra collected in [59] (Fig. 6).

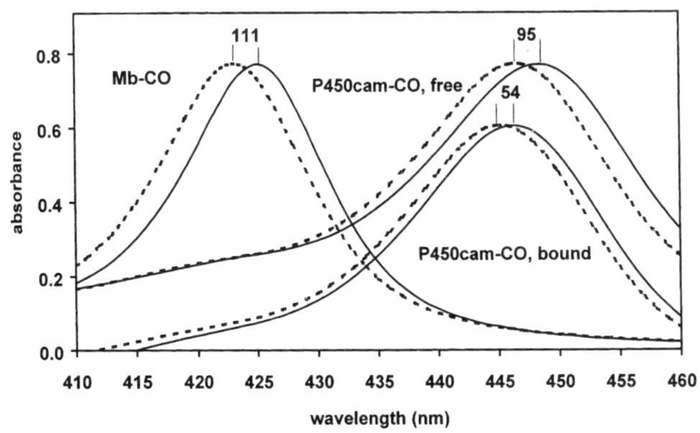

Fig. 7. Soret band of cytochrome *P*-450cam-CO and myoglobin-CO at different pressure. Spectra for *P*-450 in the absence (free) and in the presence of 1*R*-camphor (bound) and of sperm whale myoglobin at 0.1 MPa (dotted line) and at 360 MPa (full line). 100 mM potassium phosphate buffer, pH 7, 60% (by mass) glycerol. The numbers at the band maxima indicate the pressure-induced shift of the Soret band given in cm^{-1}. The intensity of the Soret band shown in this spectrum does not correspond to the real molar extinction coefficient. The spectra are scaled in such a way to get the best view of the band shift. All experimental conditions are given in [32].

distance R_0 at the pressure $P = 0$ MPa ('solvent shift'). Increasing pressure shortens the distance R and changes the solvent shift. α_i' is the coefficient of the pressure dependence of the absorption frequency. α_i' is related to the compressibility β_i of the chromophore environment (Eq. 14). Because the heme as chromophore is deeply buried in the protein, the protein itself including internal water molecules has to be regarded here as 'solvent'. β_i corresponds therefore to the compressibility of the protein in the substrate analogue complex *i*.

$$\alpha_i' = \left[\frac{\partial \nu_{si}(R,P)}{\partial R}\right]\cdot\frac{\partial R}{\partial P} \quad ; \quad \frac{\partial \mathrm{R}}{\partial \mathrm{P}} = -\frac{\mathrm{R}}{3}\cdot\beta_i \qquad (14)$$

The change of the transition energy with decreasing the solvent–solute distance is described by an inverse distance power dependence (Eq. 15) where n is 6 for long-range dispersive interactions.

$$\nu_{si}(R) = \nu_{si}(R_0)\cdot\left[\frac{R_0}{R}\right]^n \qquad (15)$$

It follows:

$$\nu_i(R,P) = \nu_{vac} + \nu_{si}(R_0, P=0) + \frac{n}{3}\cdot\beta_i\cdot\nu_{si}(P)\cdot P \qquad (16)$$

It was found that α'_i tends to decrease in absolute values with increasing high-spin state content HS_i (in %): $\alpha'_i = 0.00159 \cdot HS_i - 0.0331$ ($r^2 = 0.561$).

To calculate the compressibilities, $\nu_{si}(R)$ was approximated by $\nu_{si}(R) = c_i \nu_s'$. ν_s' is the solvent shift corresponding to one water molecule and c_i is the number of the water molecules. ν_s' was estimated from $\nu_s/[c_{ls} - c_{hs}] = -70.7/[c_{ls} - c_{hs}]$ cm^{-1}. c_{ls} and c_{hs} are the number of water molecules in the protein for 100% low-spin state and 100% high-spin state, respectively, which influence the heme. The value -70.7 cm^{-1} corresponds to the maximal solvent shift derived from the linear relation between the high-spin state content and the intercept at $P = 0$ MPa of the pressure-induced shift for each substrate complex. This corresponds to ($\nu_{vac} + \nu_{si}(R_0, P = 0)$) in Eq. 16. Using the experimentally observed relation between α_i and HS_i and $a_i = c_i/[c_{ls} - c_{hs}]$ one obtains:

$$a_i \cdot \beta_i = -0.0112 \cdot HS_i + 2.34[GPa^{-1}] \qquad (17)$$

It turned out that a_i should be about 20 to get values for the compressibility which are in the range observed for other proteins (0.020–0.15 GPa^{-1} [37]). Using this value and crystal structure data on the distribution of water molecules in substrate-free and camphor-bound *P*-450cam, it was concluded that 40–60 water molecules may have a long-range dispersive interaction with the heme which is relevant for the pressure-induced red shift of the Soret band [65]. 2–3 water molecules more may stay near the heme in the CO complex for a low-spin state inducing substrate compared to a high-spin state inducing substrate. The fractional low-spin state content observed for the different substrate complexes may then be interpreted dynamically by different residence times of these 2–3 water molecules. This dynamical behavior should also be valid for the CO complex.

The compressibility may also be determined by Eq. 11 as described above for the pressure-induced shift of the CO stretch mode frequency. The relation between the compressibility and the high-spin state content in various substrate-bound *P*-450cam complexes holds even better than discussed before.

$$\beta_i = -0.000154 \cdot HS_i + 0.02223; \; r^2 = 0.722 \qquad (18)$$

So, the latter method is very simple and might be suitable to also apply to other cytochromes *P*-450 or heme proteins (Table 8). Interestingly, for 1*R*-camphor-bound *P*-450cam-CO the compressibility obtained in this way is the same as that determined from the pressure-induced CO stretch mode frequency shift assuming *const* = 1 (Soret: 0.00713 GPa^{-1}; IR-CO: 0.00870 GPa^{-1}). In contrast, in substrate-free *P*-450cam the compressibilities obtained from the Soret band and from the CO stretch mode differ in sign and value (Soret: 0.01228 GPa^{-1}; IR-CO: −0.00389 GPa^{-1}) which might indicate a specific effect caused by a water cluster. The experimentally determined compressibility for substrate-free *P*-450 obtained from the Soret band shift is smaller by a factor of two than one would expect from the correlation with the high-spin state content obtained for the different substrate analogues (Eq. 18). One explanation might be that in contrast to a substrate which itself is regarded to be incompressible and might produce voids because of sterical reasons, a water cluster could adjust better to the environmental protein and heme group by making hydrogen bonds. Such a possibility has already been discussed previously to explain why substrate-free *P*-450cam-CO does not fit in the relation between the CO ligand stretch mode frequency ν(CO) and the percentage content of high-spin state (HS) in the oxidized form (Eq. 19) which has been found for a number of different substrate *P*-450cam complexes [60].

$$\nu(\mathrm{CO}) = -0.21 \cdot \mathrm{HS} + 1962; \quad r^2 = 0.89 \tag{19}$$

Studies on a large number of different *P*-450s bound with a broad variety of substrates will show in the future whether high-pressure studies of the Soret band red shift and the determination of the compressibility can be implemented as a standard approach to characterize *P*-450–substrate interactions. The first few examples other that of *P*-450cam have been studied [66,67] (Table 8).

Table 8
Isothermal compressibilities for cytochromes *P*-450 determined from the Soret band red shift

Protein	Maximal *P* (MPa)	High spin (%)	$\beta_{i,2}$ (GPa^{-1})	$a \times \beta_{i,1}$ (GPa^{-1})	Ref.
CYP101 (*P*-450cam)					
Substrate-free	305	4.7	0.01228	2.283	[32]
Norcamphor	400	35.4	0.01445	1.944	[32]
Dimethylallyl borneol ether	400	39.4	0.01975	1.899	[32]
Adamantane	400	58.0	0.01130	1.690	[32]
Fenchone	400	82.0	0.01271	1.422	[32]
Camphane	400	86.2	0.00638	1.375	[32]
Bromocamphor	400	88.9	0.00981	1.344	[32]
Allylborneol ether	400	93.1	0.00499	1.297	[32]
1*R*-Camphor	400	97.3	0.00713	1.250	[32]
1*S*-Camphor	400	100.0	0.00753	1.220	[32]
Adamantone	400	100.0	0.00786	1.220	[32]
CYP102 (*P*-450 BM3)					
Substrate-free	300		0.01686		[66]
Palmitate	300		0.01729		[66]
CYP3A4					
Substrate-free	300		0.01099		[66]
Triacetyloleandomycin	300		0.01524		[66]
CYP1A2					
Substrate-free	200		0.01131		
CYP2B4					
Substrate-free	100		0.00406		[67]
Myoglobin, sperm whale	400		0.01244		[32]
Proteins			0.02–0.15		[37]

$a \times \beta_{i,1}$ according to Eq. 17; $\beta_{i,2}$ determined from Eq. 11 using the slope and $\nu_{i,0}$ given in [32,66,67] and *const* = 1. Regression analysis for $\beta_{i,2}$ gives Eq. 18. *P*-450cam, CYP2B4 at 4°C and *P*-450BM3 at 25°C.

7. Outlook

Studies of the effect of high hydrostatic pressure on ligand binding reveal interesting insights into the structural behavior of proteins which might be relevant for their function. The CO ligand is an appropriate probe. Its use for heme proteins, in particular for *P*-450, is widely motivated by assuming that CO models the physiologically relevant dioxygen. This is certainly true if only the binding locus is regarded. However, it is clear that CO differs from O_2 in many aspects which are relevant for function [11,68]. It has recently been found by cryocrystallography [68] that two water molecules bridge the terminal oxygen atom of the bound dioxygen ligand with Thr252 in the I-helix. These water molecules are not present in the CO complex. So, CO might not always be a good model for O_2. However, with respect to the analysis of enzyme–substrate interaction it reveals that the CO complex of *P*-450 is particularly useful. It has two spectroscopic probes – the Soret band and the CO stretch mode vibration – which both give complementary data for compressibility values which, however, have still to be assigned to relevant interatomic interactions. Flash photolysis and stopped-flow experiments performed under high pressure are particularly informative to complementarily characterize intermediate ligand binding steps. It would be a great challenge to combine the flash photolysis technique and step-scan time-resolved FTIR with studies under pressure allowing the study of changes in the heme pocket as well as in the protein.

Summarizing the outcome of all studies on *P*-450cam under high pressure it seems that the accessibility of the protein for water molecules is a relevant property which is modulated by substrate binding. However, the term 'water accessibility' must still be clarified in more detail because the spectroscopic probes cannot directly prove the existence of water molecules but reflect the influence of an electrostatic field in the microenvironment of the probe [69,70] which may be changed by water.

How would this finding help understand important aspects of the reaction mechanism of this enzyme? One characteristic feature of the reaction cycle of *P*-450 shown in Fig. 1 is the so-called 'uncoupling', that means the formation of the cytotoxic hydrogen peroxide as a side reaction which competes with substrate hydroxylation. It is widely observed for various *P*-450 systems that in the presence of some substrates large amounts of hydrogen peroxide are produced while other substrates almost suppress this side reaction. Studies of ligand binding under pressure provided an important clue to uncover this puzzling mechanism. At least for the camphor-hydroxylating cytochrome *P*-450cam system one may conclude that a not optimal fit of the substrate in the heme pocket increases the mobility of the substrate, facilitates the access for water molecules and makes the heme pocket more compressible. Under these conditions the tight structural coupling for the specific proton transfer is disturbed, favoring the formation of hydrogen peroxide over substrate hydroxylation [11,60].

Acknowledgements

The author thanks Hans Frauenfelder, Gaston Hui Bon Hoa and Reinhard Lange for giving the possibility to perform high-pressure studies in their laboratories. The former PhD students Reinhard Scholl, Corinne Mouro, Heike Schulze and Jörg Contzen are acknowledged for their work on cytochrome *P*-450 which is partially reviewed. Financial support came from the Deutsche Forschungsgemeinschaft (Ju229/1-1, Ju229/3-1, Sk35/3-1,2), INSERM and the Deutscher Akademischer Austauschdienst.

References

[1] D.F.V. Lewis, Cytochromes P450 – Structure, Function and Mechanism, Taylor and Francis, London, 1996.

[2] P.A. Williams, J. Cosme, V. Sridhar, E.F. Johnson, D.E. McRee, Mammalian microsomal cytochrome P450 monooxygenase: structural adaptations for membrane binding and functional diversity, Mol. Cell 5 (2000) 121–131.

[3] T.L. Poulos, B.C. Finzel, A.J. Howard, Crystal structure of substrate-free *Pseudomonas putida* cytochrome P-450, Biochemistry 25 (1986) 5314–5322.

[4] T.L. Poulos, B.C. Finzel, A.J. Howard, High-resolution crystal structure of cytochrome $P450_{cam}$, J. Mol. Biol. 195 (1987) 687–700.

[5] C. Jung, O. Ristau, H. Rein, The high-spin/low-spin equilibrium in cytochrome P-450 – a new method for determination of the high-spin content, Biochim. Biophys. Acta 1076 (1991) 130–136.

[6] S.B. Philson, P.G. Debrunner, P.G. Schmidt, I.C. Gunsalus, The effect of cytochrome P-450cam on the NMR relaxation rate of water protons, J. Biol. Chem. 254 (1979) 10173–10179.

[7] F. Bancel, N. Bec, C. Ebel, R. Lange, A central role for water in the control of the spin state of cytochrome P-450scc, Eur. J. Biochem. 250 (1997) 276–285.

[8] G. Hui Bon Hoa, C. Di Primo, I. Dondaine, S.G. Sligar, I.C. Gunsalus, P. Douzou, Conformational changes of cytochrome P-450cam and P-450lin induced by high pressure, Biochemistry 28 (1989) 651–656.

[9] G. Hui Bon Hoa, C. Di Primo, P. Douzou, The role of electrostriction in the control of reaction volume of biochemical processes, in: C. Balny, R. Hayashi, K. Heremans, P. Masson (Eds.), High Pressure and Biotechnology, Vol. 224, Colloque INSERM/John Libbey Eurotext, London, 1992, pp. 139–145.

[10] H. Schulze, G. Hui Bon Hoa, C. Jung, Mobility of norbornane-type substrates and water accessibility in cytochrome P-450cam, Biochim. Biophys. Acta 1338 (1997) 77–92.

[11] C. Jung, Insight into protein structure and protein-ligand recognition by Fourier transform infrared spectroscopy, J. Mol. Recognit. 13 (2000) 325–351.

[12] M. Sakakura, S. Yamaguchi, N. Hirota, M. Terazima, Dynamics of structure and energy of horse carboxymyoglobin after photodissociation of carbon monoxide, J. Am. Chem. Soc. 123 (2001) 4286–4294.

[13] G. Stochel, R. van Eldik, Elucidation of inorganic reaction mechanisms through volume profile analysis, Coord. Chem. Rev. 187 (1999) 329–374.

[14] H. Frauenfelder, The complexity of proteins, in: H. Flyvbjerg, J. Hertz, M.H. Jensen, O.G. Mouritsen, K. Sneppen (Eds.), Physics of Biological Systems – from Molecules to Species, Springer-Verlag, Berlin, 1997, pp. 29–60.

[15] A. Ansari, C.M. Jones, E.R. Henry, J. Hofrichter, W.A. Eaton, Conformational relaxation and ligand binding in myoglobin, Biochemistry 33 (1994) 5128–5145.

[16] T.G. Traylor, D. Magde, D.J. Taube, K.A. Jongeward, D. Bandyopadhyay, J. Luo, K.N. Walda, Geminate recombination of carbon monoxide complexes of heme and heme proteins, J. Am. Chem. Soc. 114 (1992) 417–429.

[17] J.S. Olson, G.N. Phillips Jr., Kinetic pathways and barriers for ligand binding to myoglobin, J. Biol. Chem. 271 (1996) 17593–17596.

[18] C. Tétreau, C. Di Primo, R. Lange, H. Tourbez, D. Lavalette, Dynamics of carbon monoxide binding with cytochrome P-450, Biochemistry 36 (1997) 10262–10275.

[19] A. Ansari, J. Berendzen, S.F. Bowne, H. Frauenfelder, I.E.T. Iben, T.B. Sauke, E. Shyamsunder, R.D. Young, Protein states and protein quakes, Proc. Natl. Acad. Sci. USA 82 (1985) 5000–5004.

[20] D.J. Taube, H.-D. Projahn, R. van Eldik, D. Magde, T.G. Traylor, Mechanism of ligand binding to heme and heme-proteins. High-pressure study, J. Am. Chem. Soc. 112 (1990) 6880–6886.

[21] M. Kato, R. Makino, T. Iizuka, Thermodynamic aspects of the CO-binding reaction to cytochrome P-450cam. Relevance with their biological significance and structure, Biochim. Biophys. Acta 1246 (1995) 179–184.

[22] I. Schlichting, K. Chu, Trapping intermediates in the crystal: ligand binding to myoglobin, Curr. Opin. Struct. Biol. 10 (2000) 744–752.

[23] D.P. Braunstein, K. Chu, K.D. Egeberg, H. Frauenfelder, J.R. Mourant, G.U. Nienhaus, P. Ormos, S.G. Sligar, B.A. Springer, R.D. Young, Ligand binding to heme proteins. III. FTIR studies of His-E-E7 and Val-E11 mutants of carbonmonoxymyoglobin, Biophys. J. 65 (1993) 2447–2454.

[24] M. Lim, T.A. Jackson, P.A. Anfinrud, Mid-infrared vibrational spectrum of CO after photodissociation from heme: evidence for a ligand docking site in the heme pocket of hemoglobin and myoglobin, J. Chem. Phys. 102 (1995) 4355–4366.

[25] F. Stetzkowski, R. Banerjee, M.C. Marden, D.K. Beece, S.F. Bowne, W. Doster, L. Eisenstein, H. Frauenfelder, C. Jung, Dynamics of dioxygen and carbon monoxide binding to soybean leghemoglobin, J. Biol. Chem. 260 (1985) 8803–8809.

[26] R. Lange, I. Heiber-Langer, C. Bonfils, I. Fabre, M. Negishi, C. Balny, Activation volume and energetic properties of the binding of CO to hemeproteins, Biophys. J. 66 (1994) 89–98.

[27] M. Unno, K. Ishimori, Y. Ishimura, I. Morishima, High-pressure flash photolysis study of heme proteins: effect of substrate analogues on the recombination of carbon monoxide to cytochrome $P450_{CAM}$, Biochemistry 33 (1994) 9762–9768.

[28] J. Contzen, O. Ristau, C. Jung, Time-resolved Fourier-transform infrared studies of the cytochrome P-450cam carbonmonoxide complex bound with (1R)-camphor and (1S)-camphor, FEBS Lett. 383 (1996) 13–17.

[29] J. Contzen, C. Jung, Step-scan time-resolved FTIR spectroscopy of cytochrome P-450cam carbon monoxide complex: a salt link involved in the ligand rebinding process, Biochemistry 37 (1998) 4317–4324.

[30] C. Jung, O. Ristau, H. Schulze, S.G. Sligar, The CO stretching mode infrared spectrum of substrate-free cytochrome P-450cam-CO: effect of solvent conditions, temperature, and pressure, Eur. J. Biochem. 235 (1996) 660–669.

[31] H. Schulze, G. Hui Bon Hoa, H. Helms, R. Wade, C. Jung, Structural changes in cytochrome P-450cam effected by the binding of the enantiomers (1R)-camphor and (1S)-camphor, Biochemistry 35 (1996) 14127–14138.

[32] C. Jung, G. Hui Bon Hoa, D. Davydov, E. Gill, K. Heremans, Compressibility of the heme pocket of substrate analogue complexes of cytochrome P-450cam-CO: the effect of hydrostatic pressure on the Soret band, Eur. J. Biochem. 233 (1995) 600–606.

[33] C. Di Primo, G. Hui Bon Hoa, S.G. Sligar, Heme-pocket-hydration change during the inactivation of cytochrome P-450camp by hydrostatic pressure, Eur. J. Biochem. 209 (1992) 583–588.

[34] C. Di Primo, E. Deprez, G. Hui Bon Hoa, P. Douzou, Antagonistic effects of hydrostatic pressure and osmotic

pressure on cytochrome P-450cam spin transition, Biophys. J. 68 (1995) 2056–2061.

[35] M.T. Fisher, S.F. Scarlata, S.G. Sligar, High-pressure investigations of the cytochrome P-450 spin and substrate binding equilibria, Arch. Biochem. Biophys. 240 (1985) 456–463.

[36] E.F. Caldin, B.B. Hasinoff, Diffusion-controlled kinetics in the reaction of ferroprotoporphyrin IX with carbon monoxide, J. Chem. Soc. Faraday Trans. I 71 (1975) 515–527.

[37] K. Heremans, L. Smeller, Protein structure and dynamics at high pressure, Biochim. Biophys. Acta 1386 (1998) 353–370.

[38] C. Di Primo, G. Hui Bon Hoa, E. Deprez, P. Douzou, S.G. Sligar, Conformational dynamics of cytochrome P-450cam as monitored by photoacoustic calorimetry, Biochemistry 32 (1993) 3671–3676.

[39] C. Di Primo, E. Deprez, S.G. Sligar, G. Hui Bon Hoa, Origin of the photoacoustic signal in cytochrome P-450cam: role of the Arg186-Asp251-Lys178 bifurcated salt bridge, Biochemistry 36 (1997) 112–118.

[40] C. Jung, R. Scholl, H. Frauenfelder, G. Hui Bon Hoa, Structural multiplicity in the active center of cytochrome P-450, in: A.I. Archakov, G.I. Bachmanova (Eds.), Cytochrome P450, INCO-TNC Joint Stock Company, Moscow, 1992, pp. 33–38.

[41] C. Tétreau, M. Tourbez, D. Lavalette, Conformational relaxation in heme proteins: the cytochrome P-450cam case, Biochemistry 39 (2000) 14219–14231.

[42] S.V. Smith, A.P. Koley, R. Dai, R.C. Robinson, H. Leong, A. Markowitz, F.K. Friedman, Conformational modulation of human cytochrome P450 2E1 by ethanol and other substrates: a CO flash photolysis study, Biochemistry 39 (2000) 5731–5737.

[43] M.A. McLean, H. Yeom, S.G. Sligar, Carbon monoxide binding to cytochrome P450BM-3: evidence for a substrate dependent conformational change, Biochimie 78 (1996) 700–705.

[44] A.P. Koley, R.C. Robinson, A. Markowitz, F.K. Friedman, Interaction of polycyclic aromatic hydrocarbons and flavones with cytochromes P450 in the endoplasmic reticulum: effect on CO binding kinetics, Biochemistry 34 (1995) 1942–1947.

[45] A.P. Koley, J.T.M. Buters, R.C. Robinson, A. Markowitz, F.K. Friedman, CO binding kinetics of human cytochrome P450 3A4: specific interaction of substrates with kinetically distinguishable conformers, J. Biol. Chem. 270 (1995) 5014–5018.

[46] T. Shimizu, O. Ito, M. Hatano, Y. Fujii-Kuriayama, CO binding studies of engineered cytochrome P-450ds: effect of mutations at putative distal sites in the presence of polycyclic hydrocarbons, Biochemistry 30 (1991) 4659–4662.

[47] F. Mitani, T. Iizuka, H. Shimada, R. Ueno, Y. Ishimura, Flashphotolysis studies on the CO complexes of ferrous cytochrome P-450scc and cytochrome P-45011β: effects of steroid binding on the photochemical and ligand binding properties, J. Biol. Chem. 260 (1985) 12042–12048.

[48] R.D. Gray, Kinetics and mechanism of carbon monoxide binding to purified liver microsomal cytochrome P-450 isozymes, J. Biol. Chem. 257 (1982) 1086–1094.

[49] I.F. Sevrioukova, J.A. Peterson, Reaction of carbon monoxide and molecular oxygen with P450terp (CYP108) and P450BM-3 (CYP102), Arch. Biochem. Biophys. 317 (1995) 397–404.

[50] H. Frauenfelder, N.A. Alberding, A. Ansari, D. Braunstein, B.R. Cowen, M.K. Hong, I.E.T. Iben, J.B. Johnson, S. Luck, M.C. Marden, J.R. Mourant, P. Ormos, L. Reinisch, R. Scholl, A. Schulte, E. Shyamsunder, L.B. Sorensen, P.J. Steinbach, A. Xie, R.D. Young, K.T. Yue, Proteins and pressure, J. Phys. Chem. 94 (1990) 1024–1037.

[51] R. Scholl, Relaxation Dynamics in Heme Proteins, PhD Thesis, University of Illinois at Urbana-Champaign, IL, 1991.

[52] H. Schulze, O. Ristau, C. Jung, The carbon monoxide stretching modes in camphor-bound cytochrome P450cam – the effect of solvent conditions, temperature, and pressure, Eur. J. Biochem. 224 (1994) 1047–1055.

[53] C. Jung, F. Marlow, Dynamic behavior of the active site structure in bacterial cytochrome P-450, Stud. Biophys. 120 (1987) 241–251.

[54] A.D. Buckingham, Solvent effects in infra-red spectroscopy, Trans. Faraday Soc. 56 (1960) 169–182.

[55] J. Vojtěchovský, K. Chu, J. Berendzen, R.M. Sweet, I. Schlichting, Crystal structures of myoglobin-ligand complexes at near-atomic resolution, Biophys. J. 77 (1999) 2153–2174.

[56] D.P. Kharakoz, Protein compressibility, dynamics, and pressure, Biophys. J. 79 (2000) 511–525.

[57] D.A. Case, Normal mode analysis of protein dynamics, Curr. Opin. Struct. Biol. 4 (1994) 285–290.

[58] B. Melchers, E.W. Knapp, F. Parak, L. Cordone, A. Cupane, M. Leone, Structural fluctuations of myoglobin from normal-modes, Mössbauer, Raman, and absorption spectroscopy, Biophys. J. 70 (1996) 2092–2099.

[59] H. Schulze, Strukturmultiplizität im Cytochrom P-450cam, PhD Thesis, Humboldt-University, Berlin, 1997.

[60] C. Jung, H. Schulze, E. Deprez, The role of the polarity of the heme environment for the CO stretch modes in cytochrome P-450cam-CO, Biochemistry 35 (1996) 15088–15094.

[61] Q.H. Gibson, F.G. Carey, Effect of hydrostatic pressure on spectra of heme compounds, J. Biol. Chem. 252 (1977) 4098–4101.

[62] R.G. Alden, J.D. Satterlee, J. Mintorovitch, I. Constantinidis, M.R. Ondrias, B.I. Swanson, The effects of high pressure upon ligated and deoxyhemoglobin and myoglobin, J. Biol. Chem. 264 (1989) 1933–1940.

[63] E.G. McRae, Theory of solvent effects on molecular electronic spectra frequency shifts, J. Phys. Chem. 61 (1957) 562–572.

[64] B.B. Laird, J.L. Skinner, Microscopic theory of reversible pressure broadening in hole-burning spectra of impurities in glasses, J. Chem. Phys. 90 (1989) 3274–3281.

[65] C. Jung, Compressibility of the heme pocket in cytochrome P-450cam, in: K. Heremans (Ed.), High Pressure Research

in the Biosciences and Biotechnology, Leuven University Press, Louvain, 1997, pp. 35–38.

[66] E. Anzenbacherová, N. Bec, P. Anzenbacher, J. Hudeèek, P. Souèek, C. Jung, A.W. Munro, R. Lange, Flexibility and stability of the structure of cytochromes P450 3A4 and BM-3, Eur. J. Biochem. 267 (2000) 2916–2920.

[67] D.R. Davydov, N.A. Petushkova, A.I. Archakov, G. Hui Bon Hoa, Stabilization of P450 2B4 by its association with P450 1A2 revealed by high-pressure spectroscopy, Biochem. Biophys. Res. Commun. 276 (2000) 1005–1012.

[68] I. Schlichting, J. Berendzen, K. Chu, A.M. Stock, A.S. Maves, D.E. Benson, R.M. Sweet, D. Ringe, G.A. Petzko, S.G. Sligar, The catalytic pathway of cytochrome P450cam at atomic resolution, Science 287 (2000) 1615–1622.

[69] M. Laberge, Intrinsic protein electric fields: basic non-covalent interactions and relationships to protein-induced Stark effects, Biochim. Biophys. Acta 1386 (1998) 305–330.

[70] E.S. Park, S.S. Andrews, R.B. Hu, S.G. Boxer, Vibrational Stark spectroscopy in proteins: a probe and calibration for electrostatic fields, J. Phys. Chem. B 103 (1999) 9813–9817.

[71] C. Balny, F. Travers, Activation thermodynamics of the binding of horseradish peroxidase. Role of pressure, temperature, and solvent, Biophys. Chem. 33 (1989) 237–244.

[72] B.B. Hasinoff, Kinetic activation volumes of the binding of oxygen and carbon monoxide to hemoglobin and myoglobin studied on a high-pressure flash photolysis apparatus, Biochemistry 13 (1974) 3111–3117.

[73] C. Jung, D. Stuehr, D.K. Ghosh, FT-infrared spectroscopic studies of the iron ligand stretch mode of iNOS oxygenase domain: effect of arginine and tetrahydrobiopterin, Biochemistry 39 (2000) 10163–10171.

ELSEVIER

Biochimica et Biophysica Acta 1595 (2002) 329–344

BIOCHIMICA ET BIOPHYSICA ACTA
BBA

www.bba-direct.com

Review

Cold denaturation of proteins under high pressure

Shigeru Kunugi *, Naoki Tanaka

Laboratory for Biopolymer Chemistry, Department of Polymer Science and Engineering, Kyoto Institute of Technology, Matsugasaki, Sakyo, Kyoto 606 8585, Japan

Received 30 August 2001; received in revised form 25 October 2001; accepted 26 October 2001

Abstract

The advantageous usage of the high pressure technique in studies of cold denaturation of proteins is reviewed, with a brief explanation of the theoretical background of this universal phenomenon. Various experimental results are presented and discussed, explaining the plausible image of the cold denatured state of proteins. In order to understand more clearly this phenomenon and protein structure transition in general, several studies on model polymer systems are also reviewed.

Keywords: High pressure; Cold denaturation; Globular protein; Model polymer; Heat capacity change; Maximum stability

1. Introduction

It has been known for many years that some enzymes or proteins lose, reversibly or irreversibly, their activities while they are stored at low temperatures, even at refrigerator temperature [1–4]. Such kind of cold inactivation or cryoinactivation has, in principle, two reasons. One is cold dissociation, which is often observed for oligomeric (usually tetrameric or higher) proteins. Lactate dehydrogenase and glyceraldehyde phosphate dehydrogenase are well-known examples. At low temperatures, these enzymes dissociate into a lower association level (monomer) and the activities, which require the native oligomeric structures, are lost. These changes are almost reversible and thus incubation at, e.g., room temperature for several hours or days will restore the activity by re-association into the native oligomeric structures. The characteristic feature of this cold inactivation is the concentration dependence of the process, for both disactivation and reactivation. In some cases, however, dissociative cold inactivation shows irreversibility, in which irreversible aggregations occur as the following step, after the initial cold induced dissociation. King and Weber [5,6] proposed that cryoinactivation is brought about by a process of dissociation, conformational drift of the separated subunits, and reassociation into inactive aggregates, similar to those observed after high pressure dissociation followed by decompression. Recently increasing numbers of results on this type of cold inactivation have been reported for proteins of much higher order association, such as virus proteins, including the effects of pressure on these processes. These will be reviewed by Silva in this issue [7].

There is another type of cold inactivation of enzymes and proteins [8–11]. This is a process on indi-

* Corresponding author. Fax: +81-75-724-7800.
E-mail address: kunugi@ipc.kit.ac.jp (S. Kunugi).

0167-4838/02/$ – see front matter
PII: S0167-4838(01)00354-5

vidual monomeric or dimeric globular proteins and observed in situ, which does not involve a dissociation process as the primary step. In principle, such a process is caused by the loss, more or less, of the original structure of the protein (denaturation) at low temperatures and directly related with the stability, unfolding, and refolding of the native protein structures in general. Thus, this type of inactivation is often called cold denaturation, which is the subject of this review.

Cold denaturation of proteins has been observed for a variety of globular proteins. One of the earliest reports on this phenomenon has been given by Brandts [12,13], in which he showed that chymotrypsinogen exhibited a parabolic curve of free energy of transition and that the most stable temperature (vertex) was at around 10°C. This result predicted the existence of unstable regions (free energy change of transition becomes negative) at subzero temperatures, besides, as well known, at higher temperatures (> 54°C). Since the obtainable data at low temperatures in aqueous solutions are limited by the freezing point of water, various kinds of techniques, such as shift in pH and addition of chemical denaturants, were used in order to perturb the protein structures and to lower the freezing temperature of the solvent. Recently, high pressure has been utilized as an effective and useful tool for this purpose and here some of the results, as well as their theoretical background, are reviewed.

2. Theoretical background

Thermodynamic explanations of the cold denaturation of proteins have been given either in terms of temperature as a sole variable or, in a more general manner, within a temperature and pressure plane. The latter strategy will be reviewed by Smeller in this issue [14]. With a number of calorimetric data, Privalov gave a clear-cut thermodynamic explanation [8]. Since then his explanation has been presented in many reviews, and therefore some important parts are briefly traced here.

When the transition of protein structure is a reversible process and expressed in a single step process between a native state (N) and a denatured state (D) with an equilibrium constant K as

$$\mathrm{N} \overset{K}{\Leftrightarrow} \mathrm{D} \tag{1}$$

then the Gibbs energy of transition (ΔG) is given by

$$\Delta G = G^{\mathrm{D}} - G^{\mathrm{N}} = -RT \ln K \tag{2}$$

and the enthalpy (ΔH) and entropy (ΔS) of transition are given by

$$\Delta H = H^{\mathrm{D}} - H^{\mathrm{N}} = R(\partial \ln K / \partial(1/T)) \tag{3a}$$

$$\Delta S = S^{\mathrm{D}} - S^{\mathrm{N}} = RT(\partial \ln K / \partial H) + R \ln K \tag{3b}$$

(Note that, as far as the transition is expressed by equilibrium steps, the following treatment can also be applied to a multi-step process, whatever species we are looking at.)

At the midpoint of transition (transition equilibrium) at $T = T_{\mathrm{m}}$,

$$\Delta G(T_{\mathrm{m}}) = \Delta H(T_{\mathrm{m}}) - T_{\mathrm{m}} \Delta S(T_{\mathrm{m}}) = 0 \tag{4a}$$

$$\Delta S(T_{\mathrm{m}}) = \Delta H(T_{\mathrm{m}}) / T_{\mathrm{m}} \tag{4b}$$

From the definition of the heat capacity change of the transition under constant pressure ($\Delta C_P = (\partial \Delta H(T) / \partial T)_P$),

$$\Delta H(T) = \Delta H(T_{\mathrm{m}}) + \int_{T_{\mathrm{m}}}^{T} \Delta C_P(T) \mathrm{d}T \tag{5}$$

When $\Delta C_P(T)$ can be assumed to be constant (ΔC_P) in the experimental range of temperature,

$$\Delta H(T) = \Delta H(T_{\mathrm{m}}) + (T - T_{\mathrm{m}}) \Delta C_P \tag{6}$$

Since $(\partial S/\partial T)_P = (\partial S/\partial H)_P (\partial H/\partial T)_P = T^{-1} C_P$, $\mathrm{d}S = \Delta C_P / T \,\mathrm{d}T$. Then

$$\Delta S(T) = \Delta S(T_{\mathrm{m}}) + \int_{T_{\mathrm{m}}}^{T} (\partial \Delta S(T) / \partial T) \mathrm{d}T =$$

$$\Delta H(T_{\mathrm{m}}) / T_{\mathrm{m}} + \int_{T_{\mathrm{m}}}^{T} (\Delta C_P(T)) \mathrm{d} \ln T$$

$$= \Delta H(T_{\mathrm{m}}) / T_{\mathrm{m}} + \Delta C_P \ln(T / T_{\mathrm{m}}) \tag{7}$$

Thus, from Eqs. 6 and 7,

$$\Delta G(T) = c\Delta H(T) - T \Delta S(T) =$$

$$\{(T_{\mathrm{m}} - T)/T_{\mathrm{m}}\} \Delta H(T_{\mathrm{m}}) +$$

$$(T_{\mathrm{m}} - T) \Delta C_P - T \Delta C_P \ln(T / T_{\mathrm{m}}) \tag{8}$$

From Eq. 8, $\partial^2\Delta G(T)/\partial T^2 = \partial(\Delta H(T_m)/T_m - \Delta C_P \ln(T/T_m))/\partial T = -\Delta C_P/T$. If $\Delta C_P > 0$, then $\partial^2\Delta G(T)/\partial T^2 < 0$ and a convex curve is given with a maximum stability at $T = T_S$, where $\partial\Delta G(T)/\partial T = 0$. This means that one more temperature (T_m') other than T_m exists, where $\Delta G(T)$ becomes zero.

T_S is estimated by solving $\partial\Delta G(T)/\partial T = \Delta S(T) = 0$ with an approximation that $(T_m - T)/T \ll 1$ and $\ln(T_m/T) \approx (T_m - T)/T$.

$$T_S \approx T_m/\{(\Delta H(T_m)/(T_m\Delta C_P) + 1\} \quad (9)$$

Since protein denaturation is an endothermic process ($\Delta H(T_m) > 0$), the denominator of Eq. 9 is larger than 1 with $\Delta C_P > 0$ and then $T_S < T_m$. Thus T_m' is lower than T_m, the cold denaturation temperature.

The approximated value of T_m' is obtained by solving Eq. 8 = 0 by considering up to the second order term of Taylor's expansion of $\ln(T_m/T)$ ($\approx (T_m - T)/T - (T_m - T)^2/T^2/2$) as below.

$$\{(T_m - T)/T_m\}\Delta H(T_m) + (T - T_m)\Delta C_P - T\Delta C_P \ln(T/T_m) = \{(T_m - T)/T_m\}\Delta H(T_m) + (T - T_m)\Delta C_P + T\Delta C_P[(T_m - T)/T - (T_m - T)^2/T^2/2] = \Delta H(T_m)/T_m + \Delta C_P\{-(T_m - T)/T/2] = 0$$

Then

$$T_m' \approx T_m^2/\{2\,\Delta H(T_m)/\Delta C_P + T_m\} \quad (10)$$

Assume that a protein has, e.g., $T_m = 60°C$, $\Delta H(T_m) = 500$ kJ mol^{-1}, and $\Delta C_P = 10$ kJ K^{-1} mol^{-1}, then this protein shows its maximum stability at 17°C (T_S; Eq. 9) and denatures at −17°C (T_m'; Eq. 10). Fig. 1 shows the temperature dependence of thermodynamic parameters for such a protein. $\Delta G(T)$ calculated from Eq. 8 gives a convex hyperbolic-like curve ($\Delta G(1)$) and the vertex (maximum stability, where $\Delta S(T) = 0$) is at around 287 K (14°C), slightly lower than the above approximated temperature ($\Delta G(2)$). The transition point of low temperature denaturation is calculated as 243 K (−30°C), also lower than the approximate value given above. It is to be noted that this convex shape is performed by a subtle balance of the ΔH term (dotted curve) and the ΔS term (broken curve).

In the above discussion, as easily understood, the positive value of ΔC_P is critical for the observation of a denaturation at lower temperature. The heat capacity changes of transition have been measured for several simple globular proteins [8], such as myoglobin (holo and apo) and microbial nuclease. In these examples, peaks of C_P were observed at each end (high temperature side and low temperature side) of the native state and the ΔC_P was positive; C_P showed a larger value in the denatured states. Fig. 2 shows the observable C_P values calculated for the above hypothetical protein. At T_m and T_m', two peaks are observed and the differences in C_P of both ends from the central part (ΔC_P) are naturally positive.

Molecular simulations [15] and theoretical studies [16] for collapse transition of polymers in solution have been performed in order to solve the protein folding problem. In the recent trials, the contribution of solvent (water) to the transition has been considered in the model in order to properly involve the hydrophobic effects. A widely used class of models in these theoretical studies is the coarse grained self-avoiding chain on a lattice (two- or three-dimensional lattice) representation. Dinner et al. used the random energy model (REM) accounting the solvent effects by using a parameter independent of the temperature, and computed the phase diagrams of protein-like chains [17]. They found three different regions: a native state, a globular state and a random coil denaturation, whereas no characteristics of a cold denaturation were exhibited. Collet proposed the REM model based on a study of the thermodynamic and structural properties of the aqueous environment of a protein, and found both cold and heat denaturations [18]. The key issue here is again whether the low temperature peak and the positive difference in ΔC_P can be reproduced.

The HP model, introduced by Dill and Chan, is a sophisticated method in which the protein conformation is coarse grained as hetero-polymer from binary monomer on the lattice [15]. The main force stabilizing the native conformation is assumed to be the hydrophobicity of non-polar amino acids, while the chemical and electrostatic property of the chain was simplified. The elements of this chain are just polar (P) and non-polar (H) units, and the hydrophobicity is introduced as an effective attraction between H units. De los Rios and Caldarelli pointed out that the degrees of freedom of the solvent is neglected in the HP model, and they introduced a modified model accounting for the interaction between water and the chain [19]. In their model, the HP polymer was put back into water; every site of the lattice unoccupied by the polymer is occupied by water. The energy of each H unit in contact with water was simplified by a bimodal effective model. They calculated the free energy, enthalpy, and entropy differences between the denatured and the native conformations at different temperatures, and reproduced the cold and heat denaturation of protein. Trovato et al. also obtained similar results using a lattice model for a hetero-polymer with random hydrophilic–hydrophobic charges interacting with the solvent [20].

The zipper model is another simple method for the energetic calculation of the cooperative transition in protein conformation. This model was originally introduced to describe the helix coil transition, in which the relevant degrees of freedom (conformational angles) are modeled through a binary variable as in the case of the 1D Ising model. Each variable is either matching the ordered structure (helix), or in a coiled state. Zwanzig applied this model to the 3D folding transition, describing it in terms of variables of 'true ' (fold) and 'false' (unfold) [21]. Such simple binary variables enabled the calculation of the Hamiltonian of the chain. Hansen et al. modified this method by taking into account the energy of the interaction between protein and water [22]. They calculated the order parameters of the two critical points of the model protein, which correspond to the cold and heat denaturation transitions of the real protein, and could reproduce two peaks in the calculated C_p as a function of T.

3. Cold denaturation and high pressure

As seen in Eq. 10 in Section 2, although it only gives a rough estimation, $T_m{}'$ increases with T_m, almost linearly, and decreases with $\Delta H(T_m)$. A larger ΔC_P gives larger $T_m{}'$. As for the above discussed hypothetical protein, T_m values higher than 80°C (when other parameters are fixed), $\Delta H(T_m)$ values below 350 kJ mol^{-1}, or ΔC_P larger than 14 kJ K^{-1} mol^{-1} will give a $T_m{}'$ higher than 0°C. For ordinary

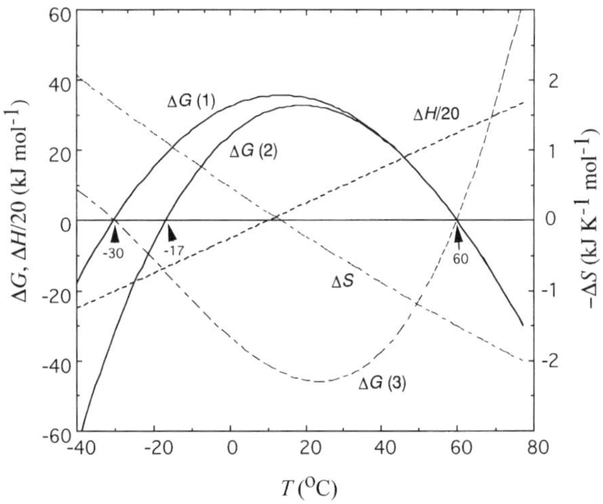

Fig. 1. Temperature dependence of ΔG, ΔH, and ΔS for a hypothetical protein which has a plausible set of T_m, $\Delta H(T_m)$, and ΔC_P as 60°C, 500 kJ mol^{-1}, and 10 kJ K^{-1} mol^{-1}, respectively. $\Delta G(1)$ was calculated by Eq. 8 and $\Delta G(2)$ with an approximation of $\ln(T_m/T) \approx (T_m - T)/T$. $\Delta G(3)$ is for a polymer with $\Delta C_P = 10$ kJ K^{-1} mol^{-1} at −30°C and $\partial \Delta C_P/\partial T = -0.25$ kJ K^{-2} mol^{-1}. $\Delta G(3)$ values are halved.

proteins, however, it is difficult to experimentally observe the cold denaturation phenomenon above the freezing temperature of aqueous solvent under usual conditions. The attempts to introduce thermoresistant proteins, because of their higher T_m values, have not been successful, probably due to their small ΔC_P.

Hence, the commonly used key technique in the studies of cold denaturation of proteins is to reduce the stability of the protein by several methods [8]. These include shifts of pH by using, e.g., acetate buffer, or addition of chemical denaturants, such as urea, guanidium chloride (GdnHCl) and methanol. In other words, these techniques have been used to shift the convex curve, as in Fig. 1, downward, or, namely, to shift the $\Delta G = 0$ level upward. By doing so, T_m becomes smaller and $T_m{}'$ shifts to a higher temperature, even above 0°C. Another strategy is to lower the freezing point of water by adding some antifreeze reagent, such as glycerol (and also methanol), or by studying in supercooled water.

High pressure can be used in studying this interesting behavior of protein in both senses of these two strategies. By increasing the pressure to an extreme value, protein can be denatured, though the transition pressure varies from a very low value (< 200

MPa) to a very high value (> 700 MPa), depending on the structure and the nature of each protein, as reviewed throughout this issue. Besides, it is well known that the molar volume change of melting water is positive and that the freezing point of water decreases as pressure increases up to 200 MPa, where water freezes at about −22°C. Thus pressure can be utilized as a destabilizer of structure and, at the same time, as antifreeze.

As reviewed by Smeller in this issue [14], a general thermodynamic treatment of two state transition in the temperature–pressure plane, as described by Hawley [23,24] or by Zipp and Kauzmann [25], gives the free energy of transition as Eq. 11:

$$\Delta G = \Delta\beta(P-P_0)^2/2 + \Delta\alpha(P-P_0)(T-T_0) - \Delta C_P[T(\ln(T/T_0)-1) + T_0] + \Delta V_0(P-P_0) - \Delta S_0(T-T_0) + \Delta G_0 \quad (11)$$

where $\Delta\beta$ and $\Delta\alpha$ are the changes of (absolute) compressibility ($=(\partial\Delta V/\partial P)_T$) and (absolute) expansivity ($=(\partial\Delta V/\partial T)_P = -(\partial\Delta S/\partial P)_T$), respectively. Note that here the standard state is denoted by the suffix '0' and usually taken at 0°C and 0.1 MPa. This equation is approximated by assuming $(T-T_0)/T_0 \ll 1$ as Eq. 12.

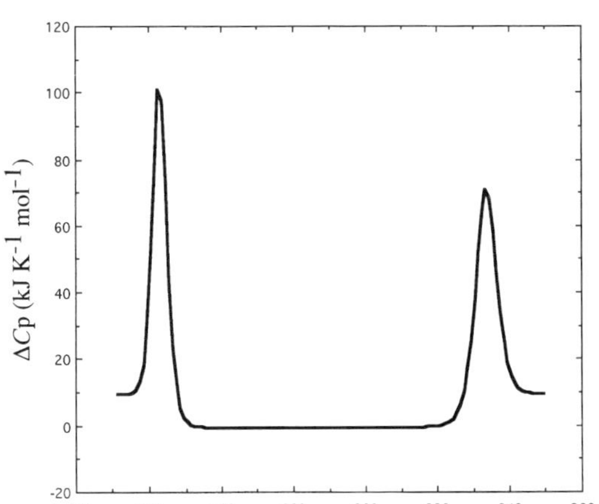

Fig. 2. The observable C_P calculated for the hypothetical protein from Fig. 1 by accounting for the ratio of native and denatured protein with K. At T_m and T_m', two peaks are observed and the differences in C_P of both ends from the central part (ΔC_P) are positive.

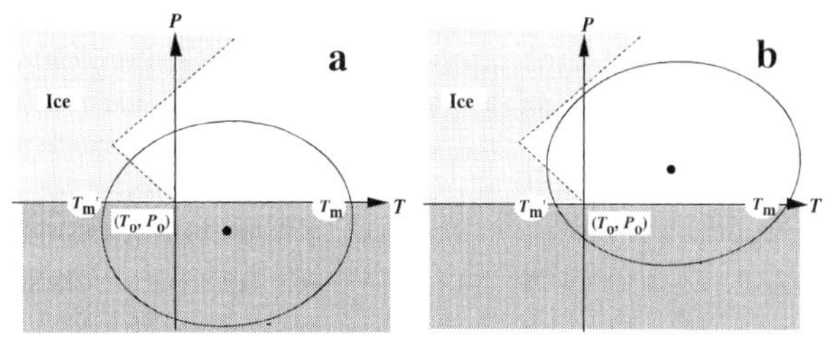

Fig. 3. Schematic presentation of the two possible origins of the observable pressure assisted cold denaturation. The ellipse shows the transition point (T, P), where the ΔG becomes zero. T_m' and T_m indicate the cold denaturation and heat denaturation temperature under atmospheric pressure, respectively. (T_0, P_0) indicates the reference state in Eqs. 11 and 12. The lightly dotted area is the ice phase of water, and the heavily dotted area is the negative pressure zone. (a) When the center of the ellipse (●) is below the T-axis, the cold denaturation temperature increases with an increase in pressure, up to the liquid phase of water and becomes observable. (b) Even when the center of the ellipse (●) is above the T-axis, the liquid phase of water is realized under a certain range of high pressure, and then the cold denaturation can be observed in some cases.

$$\Delta G = \Delta\beta(P-P_0)^2/2 + \Delta\alpha(P-P_0)(T-T_0) - \Delta C_P(T-T_0)^2/2/T_0 + \Delta V_0(P-P_0) - \Delta S_0(T-T_0) + \Delta G_0 \quad (12)$$

and the latter equation is a quadratic equation to give an ellipse when $-\Delta\beta\Delta C_P/T_0 - \Delta\alpha^2 > 0$.

Two representative cases regarding the position of the ellipse origin can be considered, although, in some cases, the inclination of the ellipse axes becomes important [26]. When the center of the ellipse is in the negative (non-real) pressure region ($\Delta V_0\Delta C_P/T_0 - \Delta S_0\Delta\alpha < 0$; Fig. 3a), it is clear that pressure can be simply a destabilizing agent of the protein structure to increase the lower transition temperature, since the most stable pressure is in the negative value region, and T_m' can enter the liquid phase of water at a certain pressure. The other case is somehow sophisticated. When $\Delta V_0\Delta C_P/T_0 - \Delta S_0\Delta\alpha > 0$ (Fig. 3b), the maximum stability of the protein is attained at a certain elevated pressure and the possibility of the observation of cold denaturation is determined by the position of the curve within the expanded liquid phase of water. These two cases are related to the observation of the antagonism of temperature and pressure effects at the heat denaturation point (T_m). In the former case, no such antagonism would be observed, as was the case for ribonuclease A [23] and myoglobin [25], and, in the latter case, pressure can depress the heat denaturation at T_m, as was the case for chymotrypsinogen [23].

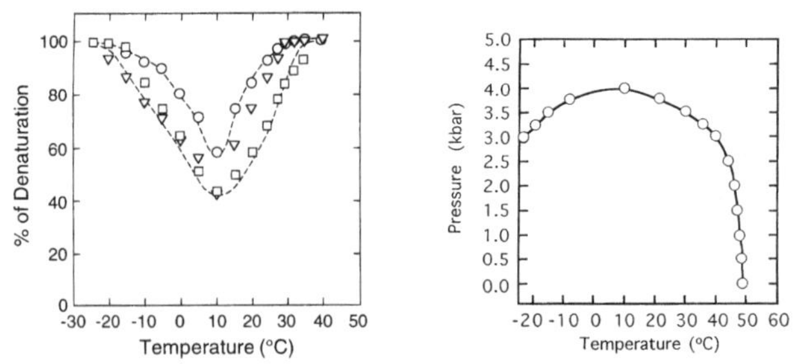

Fig. 4. Temperature (high and low) denaturation curves detected by histidine shift in NMR spectra for RNase A at 300 MPa (left) and the obtained phase diagram (right) at pH* 2.0. (Left) ○, His 12$^{\varepsilon 1}$; □, His 48 $^{\varepsilon 1}$; ▽, His 119$^{\varepsilon 1}$+His 105$^{\varepsilon 1}$. Taken from [28].

4. Experimental results for several proteins

4.1. Single domain protein

The best-studied protein with regard to cold denaturation might be ribonuclease A (RNase A) from bovine pancreas. Hawley [23] showed the possible existence of T_m' at subzero temperature. This reversible cold unfolding of RNase A (and RNase A-inhibitor complex as well) was studied by 1D and 2D ^{1}H NMR spectroscopy by Jonas' group [27–29]. The cold denaturation was carried out at 300 MPa, where the protein solution can be practically cooled down to −25°C without freezing (Fig. 4, left), and a temperature–pressure phase diagram of RNase A was constructed (Fig. 4, right). The appearance of a new histidine resonance in the cold denatured RNase A spectra (a pressure denatured one at 25°C as well) was in contrast to the absence of this resonance in the heat denatured state, which indicated that cold denatured states may contain partially folded structures. This structure was similar to that of the early folding intermediate found in the temperature jump experiment [30].

To probe the extent of a residual secondary structure in the denatured state, hydrogen exchange experiments were performed at low temperature and high pressure [31]. The protection factors obtained from the kinetic data [32,33] were similar to those obtained for RNase A denatured by exposure to high pressure near its temperature of maximum stability, where stable hydrogen bonded structures protecting the backbone amide hydrogens from solvent exchange were observed to confirm the presence of partially folded structure. Both results differed markedly from those in heat denatured RNase A.

Some other nucleic acid hydrolyzing enzymes have been studied as for their protein folding and unfolding processes. Staphylococcal nuclease (Snase) was one of the best. The temperature dependence of the pressure induced equilibrium unfolding of this enzyme was determined by fluorescence of its single tryptophan residue, FTIR absorption for the amide I′ and tyrosine OH bands, and small angle X-ray scattering (SAXS) [34]. These three techniques gave similar results, but the stability measured by fluorescence was slightly lower than that measured by the others (Fig. 5). The free energy change of unfolding showed a typical parabolic-like shape and the obtained phase diagram showed a convex curvature. The volume change for unfolding became more negative with decreasing temperature, consistent with a larger thermal expansivity ($\Delta\alpha$) for the unfolded state than for the folded state. A less stable mutant of this enzyme (V66A) showed lower pressure stability throughout the measured temperature range and seemed to give a higher cold denaturation temperature.

The thermodynamics of the temperature and pressure induced conformational changes of ribonuclease P2 (RNase P2) from a thermophilic archaebacterium *Sulfolobus solfataricus* was studied for the wild type (WT) and the single point mutants of the hydrophobic core phenylalanine (F31A and F31Y), by analyzing the fourth derivative UV absorbance [35] (Fig. 6). This enzyme is a small protein (7 kDa) and the wild type protein is extremely stable under all conditions of temperature and pressure, due to a strong van der Waals interaction energy in the hydrophobic core. The heat and cold denaturation of both mutants, as well as denaturation by pressure of the F31A mutant, led to significant blue shifts of the derivative spectrum, indicating increased solvent exposure of

Tyr33. For the F31Y mutant, high pressure (400 MPa) protected the protein against thermal denaturation (*T–P* antagonism). This study, probing the properties of the hydrophobic aromatic core, complemented a thermal unfolding study that probed the overall structural changes of RNase P2. The differences observed in response to extremes of temperature and pressure were rationalized by an unfolding mechanism involving larger parts of the peripheral protein, while the integrity of the hydrophobic core is maintained.

At high (> 350 MPa) pressures and low (< −10°C) temperatures, hen egg white lysozyme denatures readily and reversibly. The structure of the pressure assisted cold denatured state of lysozyme was studied by the amide hydrogen exchange method [36]. Protection factors were obtained for 52 backbone amide protons, and the extent of protection of many of these protons was markedly different from that in lysozyme denatured by high temperature, high urea concentration, or chemical modification. Specifically, the protection factors are higher and are strongly correlated with elements of secondary structure present in the native state. Furthermore, the pattern of protection factors is similar to that observed in lysozyme during refolding from highly denatured states, particularly during the early stages (< 3.5 ms) of refolding (Fig. 7) [37].

For this enzyme, the effects of temperature on the pressure induced unfolding in aqueous GdnHCl solutions were also investigated by means of ultraviolet spectroscopy [38]. Assuming a two state transition

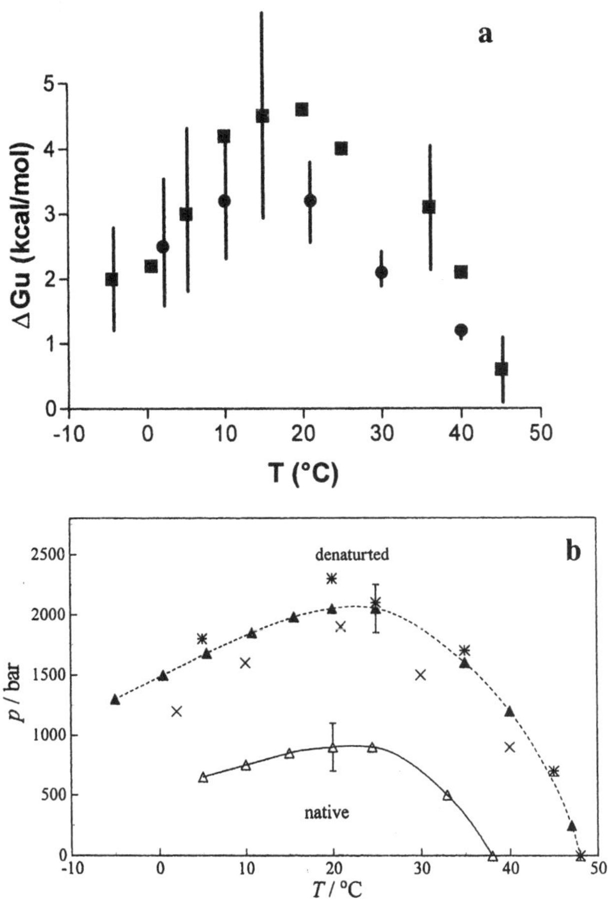

Fig. 5. Temperature dependence of ΔG of unfolding for Snase (a) as determined from fluorescence (●) and FTIR (■), and the *T–P* phase diagram (b) as detected by (*) SAXS with WT enzyme, (▲) FTIR with WT enzyme, (×) fluorescence with WT enzyme, and (△) FTIR with Snase V66A. Taken from [34].

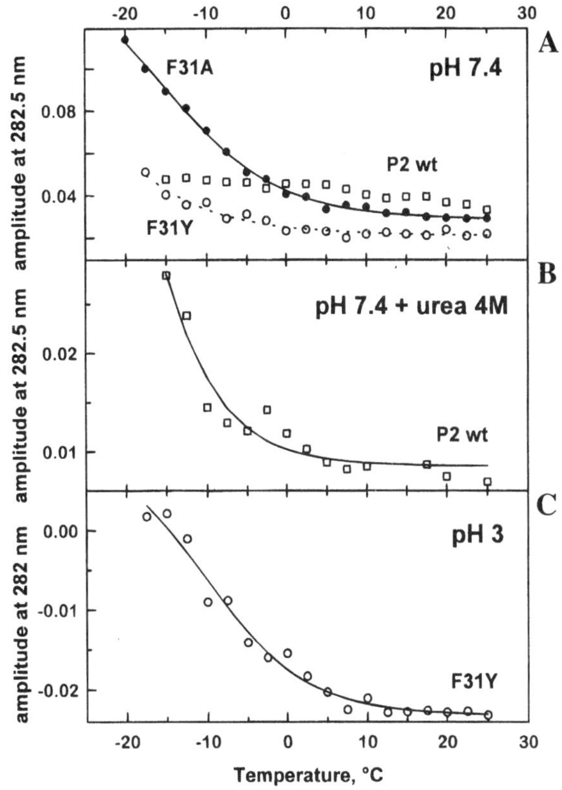

Fig. 6. Cold denaturation profiles of WT P2, F31A P2 and F31Y P2, as detected by fourth derivative amplitude in the tyrosine region. pH 7.4 (50 mM phosphate). F31Y data were upshifted for a better comparison. In the presence of 4 M urea, pH 7.4 (as in A). pH 3 (50 mM pyrophosphate). Taken from [35].

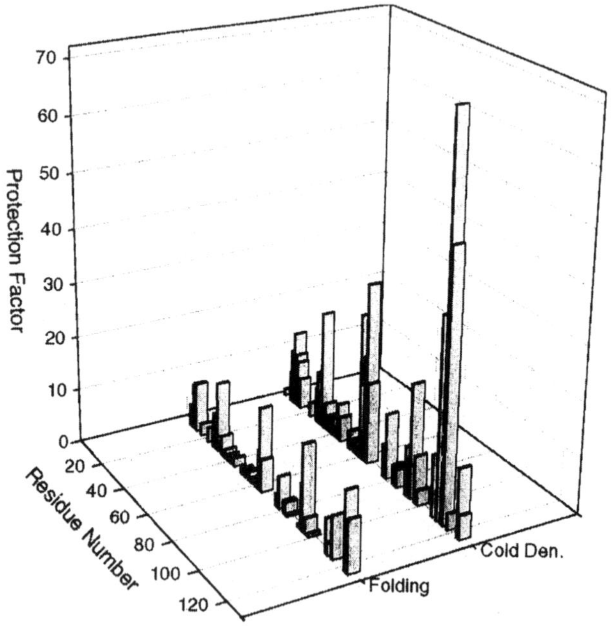

Fig. 7. Comparison of protection factors observed in cold denatured lysozyme to the dead time inhibition factors obtained for the first 3.5 ms of lysozyme folding. Taken from [36]. Folding kinetics data were from [37].

model, volume changes were calculated over a temperature range of 10–60°C (Fig. 8). At higher temperatures (between 25 and 60°C), almost constant volume changes were observed in the transition region. On the other hand, different volume changes were observed at temperatures lower than 25°C. These data suggested that a two state model is not appropriate at low temperature, and one or more intermediates are present under low temperature conditions.

The pressure assisted cold denatured state of ubiquitin in aqueous solution was studied by high resolution NMR [39]. Ubiquitin is a small 76 residue protein found ubiquitously in cells and it covalently modifies other proteins for degradation by the 26S proteasome. This protein has been known to show no evidence for significant protection in hydrogen exchange kinetics at early stages of folding [37]. In contrast to cold denatured ribonuclease A and lysozyme, as explained above, the cold denatured ubiquitin showed little persistent secondary structure.

From all these results, the residual structures of pressure assisted cold denatured states are strongly related with the structure of early folding intermediates provided that they exist.

α-Lactalbumin (α-LA), a milk protein, has a strong structural homology with lysozyme. Pressure assisted cold induced changes in the secondary structure of this protein (bovine) were carried out using the second derivative FTIR method for the native holoprotein and the calcium ion depleted apoprotein, in a temperature range from −15 to 80°C, during pressure assisted cold treatment and heat treatment [40]. Under the solution conditions applied (close to neutral pD and low ionic strength), the apoprotein remained in the molten globule state and the holoprotein maintained its native tertiary structure. The compactness and secondary structure of the protein were considerably stabilized against temperature and pressure perturbation by the presence of a calcium ion. Therefore, under the experimental conditions, only the apoprotein was susceptible to cold denaturation. In contrast, an unexpected linear increase of the α-helical content was observed upon cooling of the holoprotein under high pressure.

Another milk protein, β-lactoglobulin (β-LG), was also investigated for the effects of hydrostatic pressure on its dimeric structure at low temperatures [41]. Application of pressures of up to 350 MPa induced a significant red shift (similar to 11 nm) and a 60% increase in intrinsic fluorescence emission. These changes were very similar to those induced by GdnHCl, which caused subunit dissociation and un-

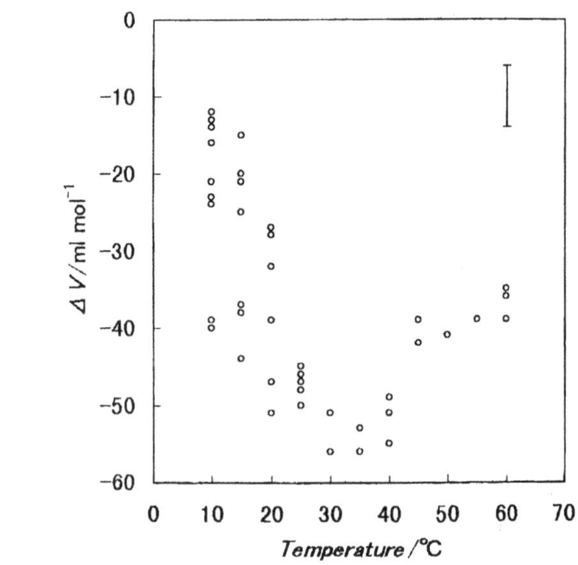

Fig. 8. Relation between temperature and volume change of lysozyme unfolding in aqueous GdnHCl solution. Taken from [38].

folding. A large hysteresis in the recovery of fluorescence parameters was observed upon decompression. Pressure induced dissociation and unfolding were not fully reversible, because of the formation of a non-native intersubunit disulfide bond that hampered correct refolding of the dimer. A marked destabilization of β-LG was observed at low temperature. Interestingly, pressure induced dissociation/unfolding was completely independent of β-LG concentration, in apparent violation of the law of mass action.

Conformation changes in pure β-LG induced at 200 MPa by low or subzero temperatures in liquid phase were also monitored by UV spectroscopy at pH 7.0 [42]. Decreasing the temperature under 200 MPa from +20 to +4°C, then to −15°C, led to marked 'blue shifts' in the fourth derivative UV spectra, revealing that tyrosine and tryptophan residues became significantly more exposed to an aqueous microenvironment at low temperatures. These shifts may be interpreted as different β-LG conformations existing under various *P*/*T* conditions: the pressure denatured state at 200 MPa/+20°C and the fully hydrated, cold denatured state at 200 MPa/−15°C. In this case β-LG aggregates (36–500 kDa) were detected by gel permeation chromatography when pressurized at +25°C but not at lower temperatures.

4.2. Multi-domain proteins

All of the examples explained above are single peptide and single domain proteins of relatively small molecular masses (ranging from 7 kDa (RNase P2) to 18.4 kDa (β-LG)). The pressure assisted cold denaturation phenomenon can be observed even for a larger size protein of multi-domain structure, although such a domain structure might show a multi-step transition in the cold denaturation process, as was observed for phosphoglycerate kinase [43], and some transitions might not be truly reversible.

Carboxypeptidase Y (CPDY) (EC 3.4.16.1) is a serine-containing carboxypeptidase from yeast with molecular mass ~66 kDa composed of two domains. We have examined the changes of its intrinsic fluorescence spectrum, as well as its catalytic activity, for a range of temperatures and pressures (Fig. 9) [26]. This protein showed at least three states of pressure induced transition and the third stage seemed to be irreversible. This multi-stage denaturation process can be explained by assuming the existence of independent structural domains in the protein [44,45]. The last stage was also observed in the fluorescence change corresponding to the increased binding of a fluorescent hydrophobic probe molecule (ANS) to this protein, implicating the occurrence of a molten globule-like state. However, the CD spectrum of the protein after incubation under high pressure showed practically no evidence of irreversible changes in the protein's secondary structure. The highest fluorescent intensity was recorded at ambient pressure and the lowest temperature, while the intensity decreased

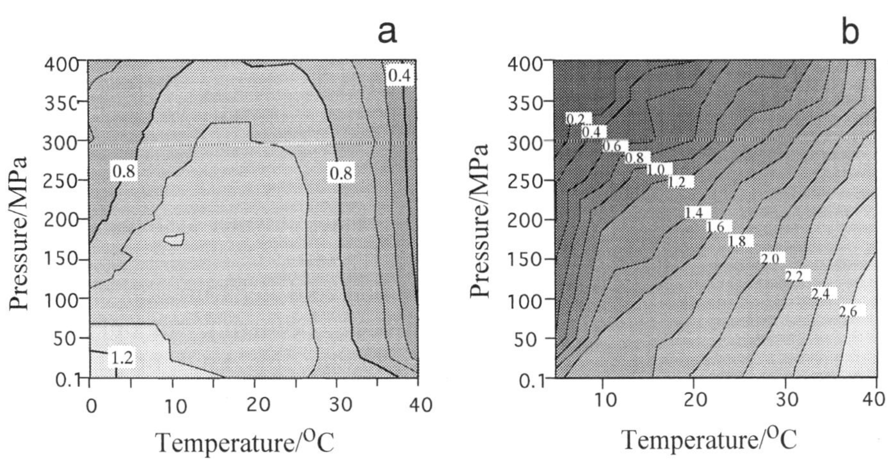

Fig. 9. Temperature and pressure dependence of the intrinsic fluorescence intensity (a) and the apparent second order rate constant (b) of CPDY at pH 7 (0.1 M HEPES/NaOH containing 2% Me_2SO). Activity was measured for Suc-Ala-Ala-Pro-Phe-MCA at [S] = 3 mM and [E] = 87–440 nM (depending on the activity). Numbers on the contours denote the intensity relative to that at 25°C and 0.1 MPa in a and the $\log(k_{cat}/K_{m(app)})$ (M^{-1} s^{-1}) in b. Taken from [26].

with increasing temperature or pressure. The intensity also decreased with decreasing temperature under elevated pressure. The resulting diagram showed a characteristic feature of pressure assisted cold denaturation. The catalytic activity became very small at a higher pressure and lower temperature region. In this example also, pressure behaved as a perturbation agent to increase the cold transition temperature.

Calorimetric measurements of CPDY by the DSC technique revealed that the ΔH_m and ΔC_P of the heat denaturation of CPDY were 1370 kJ mol^{-1} and 25.9 kJ mol^{-1} K^{-1}, respectively, and these values, coupled with the peak temperature of the endothermic thermogram ($T_p = 335.4$ K or 62.2°C) predicted that the approximated T_m' of the cold denaturation of this protein is at around 255 K or −18°C (Eq. 10), with an approximation that T_p is practically equivalent to the heat transition temperature (T_m).

CPDY is a glycoprotein and Dumoulin et al. examined the role of the carbohydrate moiety in its response to pressure and low temperature, by measuring in situ fluorescence (Fig. 10) [45]. The unglycosylated CPDY showed a similar three transition process, but the first two transitions clearly occurred at lower pressures than those for glycosylated CPDY, which indicated that the carbohydrate moiety protects CPDY against pressure induced denaturation.

Thermolysin is a zinc containing protease from thermophilic bacteria (*Bacillus thermoproteolyticus*) of 35 kDa composed of two distinct domains. We have studied the pressure and temperature dependence of its spectroscopic properties as well as of its catalytic properties (Figs. 11 and 12) [46–48]. Metalloproteases of this kind were found to exhibit fairly large pressure activation, at least up to approx. 200 MPa, but a prolonged incubation under high pressure caused a time-dependent partially irreversible disactivation. In spite of this irreversibility and its multi-domain structure, the changes detected through the fourth derivative absorbance spectrum and the intrinsic fluorescence were explained by a simple two state transition model, accompanied by a large and negative change in the volume of transition. The temperature and pressure diagram for this enzyme, drawn as an iso-intensity contour in the *P–T* plane (Fig. 11a), is compared with that for another metalloprotease, vimelysin from *Vibrio* sp. T1800, a psychrophilic enzyme (Fig. 11b). Thermolysin showed a larger fluorescence decrease at high pressure and high temperature, although this enzyme is thermoresistant at ambient pressure. Positive tangential lines are observed at the low temperature and high pressure region for thermolysin, whereas the tangential lines at high temperatures and low pressures seem to have a slightly positive slope for vimelysin. The former results mean that thermolysin will show a kind of cold denaturation upon decreasing temperature, e.g. from 10°C to −5°C at constant pressure above 350 MPa, whereas the latter result means the pressure induced stabilization of vimelysin towards heat denaturation. This difference might be related to that in their optimum temperatures or adapted environments.

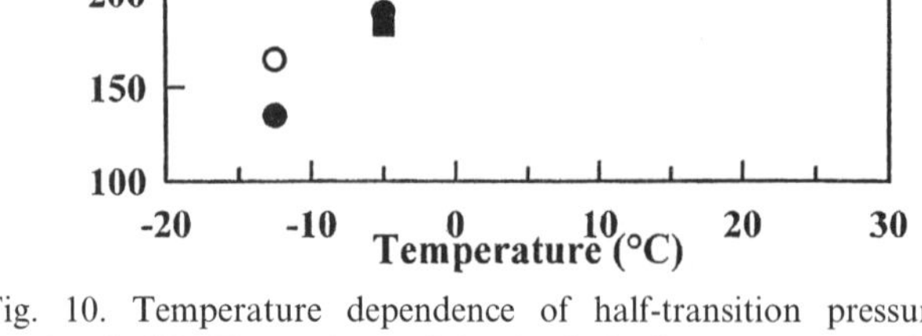

Fig. 10. Temperature dependence of half-transition pressure ($P_{1/2}$) of CPDY. ○, detected by intrinsic fluorescence; □, by ANS binding fluorescence. Open symbols, WT CPDY; closed symbols, unglycosylated CPDY. [E] = 0.1 mg ml^{-1} in 50 mM Tris–HCl (pH 7). Taken from [45].

5. Model polymer systems

The cold denaturation phenomenon is a universal process, occurring not only in protein systems [49]. Including this kind of denaturation, experimental studies of folding and unfolding processes from a more general point of view by using very simple model polymers have been successfully performed. As explained in the former section, the statistical and numerical approaches to the cold denaturation phenomenon always consider protein as a kind of

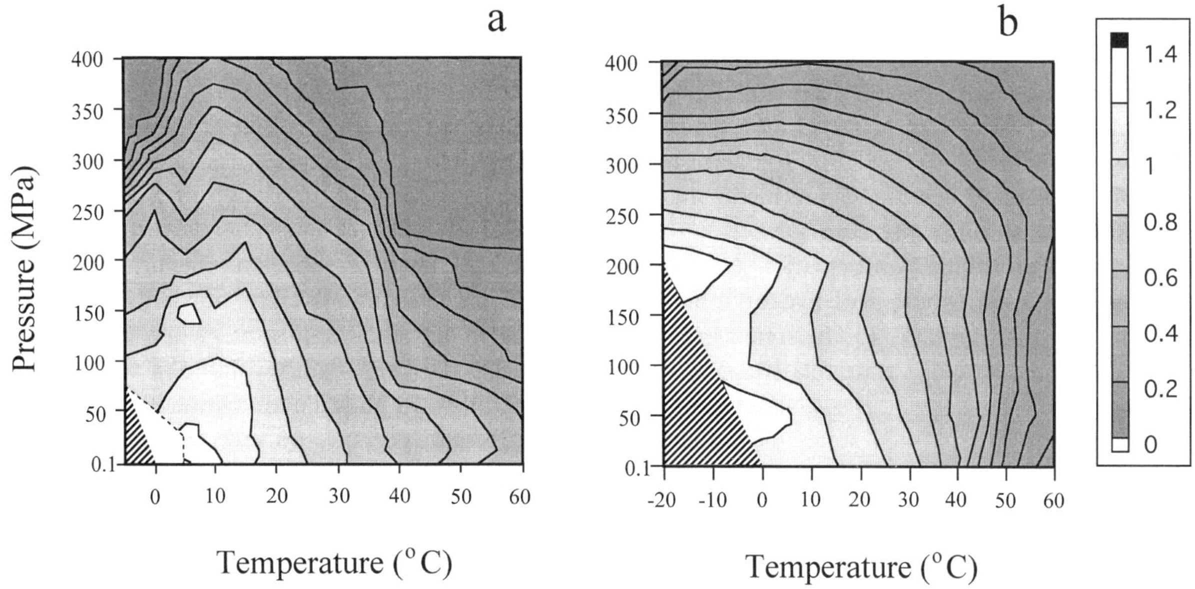

Fig. 11. Temperature and pressure dependence of the fluorescence peak intensity of thermolysin (a) and vimelysin (b). pH 6.0 (50 mM MES/NaOH) for a and pH 6.5 (50 mM HEPES/NaOH) for b. [E] = 0.4–8 μM. The peak intensity is presented relative to that at 25°C and 0.1 MPa. λ_{ex} = 290 nm. Taken from [48].

polymer and cold denaturation of protein is treated in the same framework as that of synthetic polymers.

One of the model systems is the so-called de novo proteins, structured synthetic polypeptides. Chapeaurouge et al. investigated the stability and equilibrium unfolding of a model three helix bundle protein, α_3-1, by GdnHCl, hydrostatic pressure, and temperature [50]. The combined use of these denaturing agents allowed detection of two partially folded states of α_3-1, as monitored by CD, intrinsic fluores-

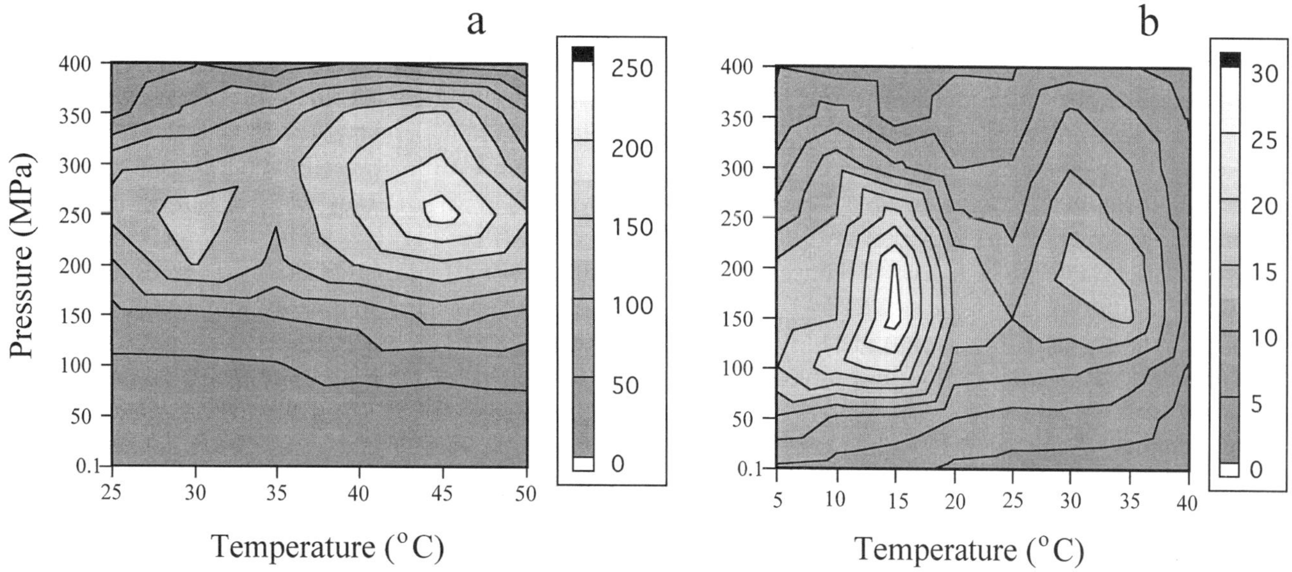

Fig. 12. Temperature and pressure dependence of the apparent second order rate constant ($k_{cat}/K_{m(app)}$ 10^4 M^{-1} s^{-1}) for the hydrolysis of Fua-Gly-LeuNH$_2$ by thermolysin (a) and vimelysin (b). pH 6.0 (50 mM MES/NaOH) for a and pH 6.5 (50 mM HEPES/NaOH) for b. [E] = 0.6–1.2 nM. [Substrate] = 0.2 mM. Taken from [48].

cence emission, and fluorescence of the hydrophobic probe bis-ANS. The overall free energy change for complete unfolding of α_3-1, determined from GdnHCl unfolding data, was +19.2 kJ/mol. The native state was stabilized by −5.9 kJ/mol relative to a partially folded pressure denatured intermediate (I_1). Cold denaturation at high pressure gave rise to a second partially (un)folded conformation (I_2), suggesting a significant contribution of hydrophobic interactions to the stability of α_3-1. The free energy of stabilization of the native-like state relative to I_2 was evaluated to be −10.5 kJ/mol. Bis-ANS binding to the pressure and cold denatured states indicated the existence of a significant residual hydrophobic structure in the partially (un)folded states of α_3-1. The demonstration of folding intermediates of α_3-1 lent experimental support to the protein folding simulation studies of other three helix bundle proteins, that predicted the existence of such intermediates [51,52].

There are several known synthetic polymers, including vinyl polymers, that show a more or less distinct change in their molecular level aggregational states (in solution) and volume phase transitions (in gel form) (thermoresponsive polymers) [53,54]. Poly(*N*-isopropylacrylamide) (pNIPAM) and other polyacrylamide derivatives with both hydrogen bonding properties and hydrophobic interactions in aqueous solutions are some of the best known examples. These polymers show a sharp reversible transition from their coiled (swollen) to collapsed (compact) states upon an increase in temperature: inverse temperature transition. The swelling and shrinking of gels prepared from such polymers have been extensively studied, both experimentally and theoretically. Besides by temperature, these thermoresponsive polymers can undergo transitions by changing other factors, such as salt concentration, surfactant addition, saccharide addition, solvent, etc. Furthermore, derivatives of pNIPAM undergo pressure induced transitions, both in solution and in gel forms.

These changes recall the behavior of proteins. Actually, pNIPAM can be considered as a simple but relevant protein model and its thermodynamic properties have been investigated from such a point of view [55,56]. The coil to collapse transition of these polymers upon increasing temperature corresponds to the transition from the unfolded protein to the folded structure, across the cold renaturation temperature, a model of the inverse process of cold denaturation.

Poly(*N*-vinylisobutyramide) (pNVIBA), an isomer of pNIPAM, shows more drastic thermoresponsive properties [57–59]. Basically, pNVIBA has amide bonds in its side chains in an inverted direction to pNIPAM. Both of these are chemically the isomers of polyleucine. Interestingly, polyleucine scarcely dissolves in aqueous solutions, while both pNIPAM and pNVIBA are readily soluble in water below the temperature of thermal transition. We have studied the effects of pressure on several types of thermoresponsive polymeric compounds [60–66].

The coil–collapse transition of pNIPAM or pNVIBA in aqueous solutions can be detected by, phenomenologically, observing the apparent transmittance or apparent light scattering of visible light, due to the formation of higher aggregates after the collapsed state being formed (cloud point measurement). More precisely, the static or dynamic light scattering from the solution can be used. The cloud point of these polymers showed very low concentration dependence and the dynamic scattering measurements indicated that they showed a coil–collapse change as a single molecular event at almost the same temperature as detected by the cloud point measurement, even in very dilute solutions, where actually no cloud point can be observed through ordinary methods. Thus, these changes are not driven by the aggregation of collapsed chains, but visible changes occur as a result of the formation of the collapsed chains.

The temperature–pressure diagrams of clouding of pNIPAM and pNVIBA were measured in aqueous medium (Fig. 13) [62]. Both T–P diagrams showed characteristic features. Near or at atmospheric pressure, the transition temperature increased with increasing pressure, but at higher pressures it decreased with increasing pressure. The diagrams seem to be explained by Eq. 11, based on the two state transition model. The experimentally obtained ΔC_P (in the direction from coil to collapse state) was negative, which is equivalent to the positive ΔC_P of cold denaturation (in the direction from native to denatured state). A study by numerical calculations revealed that ΔC_P cannot be constant anymore ($\partial\Delta C_P/\partial T < 0$), in order to observe a $T_m{}'$ higher than T_m

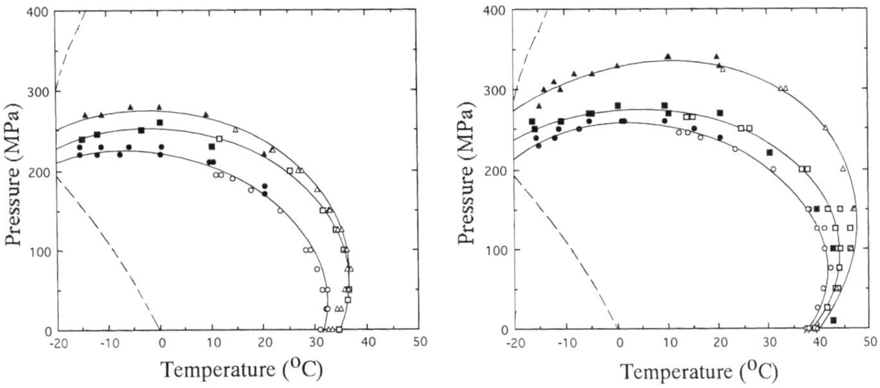

Fig. 13. Transition diagrams of 0.5% w/v pNIPAM left and 0.1% w/v pNVIBA right of various molecular masses in aqueous solution. – – –, freezing point of water under elevated pressures. (right) ○, 600 kDa; □, 40 kDa; △, 12 kDa. (left) ○, 460 kDa; □, 66 kDa; △, 11 kDa. Taken from [62].

and a negative ΔC_P at T_m' ($\Delta G(3)$ in Fig. 1) [60]. The polymers with higher molecular mass showed lower P_m' and T_m' due to larger values of ΔC_P at T_m'.

For these synthetic polymers, addition of third components can also alter their transition behavior. As is the case for the salt effects on proteins, so-called salting out ions can lower both T_m' and P_m', while salting in ions can shift them to higher values up to a certain salt concentration [62]. Furthermore, synthetic polymers can be designed so as to contain various kinds of functional groups in their molecules by adding various co-monomers during the synthetic process [63,64]. Fig. 14 shows the result of ionic monomers. Introduction of an anionic co-monomer such as acrylic acid (AA) into pNIPAM (pNIPAM-AA), or a cationic co-monomer vinylamine (VAm) to pNVIBA (pNVIBA-VAm) made the transition P–T curves highly pH dependent and P_m' was much more influenced than T_m'. In the DSC measurements of these copolymers in solution, the endothermic peak shifted to a higher temperature and the thermogram became broader with increasing (for anion) or decreasing (for cation) pH, while ΔH became smaller. These changes were attributed primarily to charge production in the side chains, causing electrostatic repulsive forces in the dehydration process and strengthening of the hydration process, which in turn hinders collapse of the polymer chains.

Apparently similar diagrams were obtained for the cloud point transition of the aqueous solutions of an elastin-like polypeptide with repeating pentapeptide Val-Pro-Gly-Ile-Gly sequence (MGLDGSMG(VPGIG)$_{40}$VPLE; (VPGIG)$_{40}$), prepared by bacterial expression of the corresponding artificial gene (Fig. 15) [65]. The (VPGIG)$_{40}$ solution exhibited a well-defined pressure induced cloud point (P_m') as well as a temperature induced one (T_m'). Near atmospheric pressure, up to 100 MPa, T_m' increased with an increase in pressure, but it decreased with increasing pressure above 200 MPa. A maximum T_m' appeared at 100–200 MPa. This transition has been shown to occur through the collapse from an extended chain to a β-spiral structure with three VPGVG units per turn, each pentamer adopting a type II β-turn conformation. (A recent molecular dynamics simulation study pointed out that the β-turn conformation is not complete for a longer polypeptide [66].) Different from the cases of proteins and thermoresponsive polymers, the changes were sub-

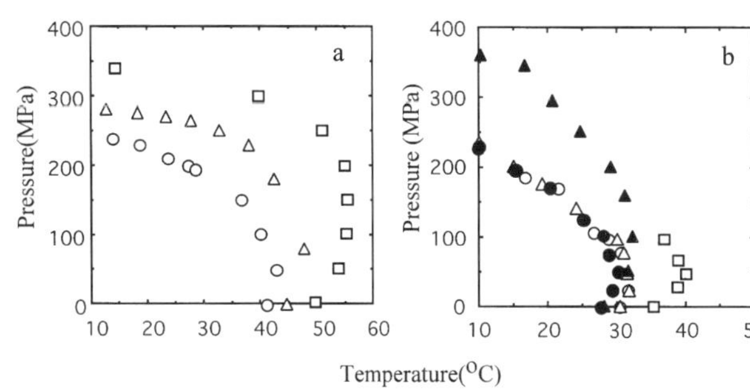

Fig. 14. Effects of ionic co-monomer insertion and pH on the T–P diagrams of pNVIBA (a) and pNIPAM (b) in aqueous solution (0.1% w/v). (a) NVIBA 92% and vinylamine 8%. ○, pH 13; △, pH 12; □, pH 11. (b) Open symbols, NIPAM 95% and acrylic acid 5%; closed symbols, NIPAM 62% and acrylic acid 38%. ○, pH 2; △, pH 3; □, pH 4. Taken from [64] (a) and [63] (b).

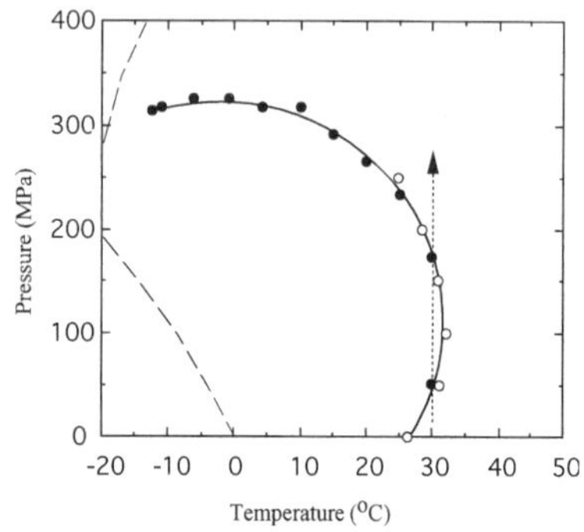

Fig. 15. *T–P* diagrams of $(VPGIG)_{40}$ in aqueous solution (0.05% w/v). Open symbols, temperature scanning; closed symbols, pressure scanning; arrow, reentry transition at 30°C with increasing pressure; ———, freezing point of water under elevated pressure. Taken from [65].

stantially concentration dependent, which infers that a different aggregation process is involved in $(VPGIG)_{40}$; an intermolecular interaction coupled with the chain conformational change is driving the transition of $(VPGIG)_{40}$.

Calorimetric measurements on this polypeptide indicated that ΔH and ΔS are around 12 kJ (mol pentamer)$^{-1}$ and 30 J (mol pentamer)$^{-1}$ K^{-1}, respectively (in the direction from coil to collapse state), and ΔC_P was small but negative (−0.5 kJ mol^{-1} K^{-1}). Such a convex *T–P* transition diagram is observed universally for other polymeric systems, such as (hydroxypropyl)cellulose microcrystallization [67] or micelle solutions, and is always accompanied by a large and positive enthalpy change and a negative ΔC_P (in the direction from coil (disordered) to collapse (ordered) state).

6. Conclusion

The cold denaturation phenomenon is a universal process, occurring not only in protein systems, and thus, proteins and synthetic polymers are to be treated in the same framework. High pressure is useful for studying this interesting behavior, in two senses: as destabilizing factor for the ordered (folded) structure ($T_m{}'$ shift), and lowering the freezing point of solvent water (antifreeze). Recently Zhou et al. found that yeast prion protein Ure2 showed cold denaturation below −5°C at 200 MPa and was more susceptible to cold denaturation at lower ionic strength [68].

For a deeper understanding of the cold denatured state of proteins, the in situ observation methodology is a critical issue. So far, various kinds of spectroscopic methods, such as fluorescence, UV absorbance, IR, and NMR, and scattering methods (including light scattering and SAXS), have been applied. Ex situ analysis coupled with modification processes, such as deuterium exchange and chemical modification, is also useful. However, these methods sometimes capture the behaviors of different parts and different levels. Thus, integration of the outputs from these various methodologies is inevitable, as well as the development of a new breakthrough in the in situ analytical methodology.

Furthermore, understanding of the generality of this phenomenon, on one side, by expanding the investigated subjects to a wide variety of polymeric substances, both experimentally and theoretically, is required, but, on the other side, the introduction of more specific interactions, which are characteristic to proteins, into these (model) polymeric systems is also very important for the comprehension of the cold denaturation phenomenon. In both these directions, perturbation by high pressure would be a very powerful tool.

References

[1] F. Franks, S.F. Mathias, R.H. Hatley, Water, temperature and life, Philos. Trans. R. Soc. London B Biol. Sci. 326 (1990) 517–531.

[2] R. Jaenicke, Protein structure and function at low temperatures, Philos. Trans. R. Soc. London B Biol. Sci. 326 (1990) 535–551.

[3] G. Weber, Protein Interactions, Chapman and Hall, New York, 1992.

[4] F. Franks, Protein destabilization at low temperatures, Adv. Protein Chem. 46 (1995) 105–139.

[5] L. King, G. Weber, Conformational drift of dissociated lactate dehydrogenases, Biochemistry 25 (1986) 3632–3637.

[6] L. King, G. Weber, Conformational drift and cryoinactivation of lactate dehydrogenase, Biochemistry 25 (1986) 3637–3640.

[7] J.L. Silva, Pressure induces folding intermediates that are crucial for protein-DNA recognition and virus assembly, this volume.

[8] P.L. Privalov, Cold denaturation of proteins, Crit. Rev. Biochem. Mol. Biol. 25 (1990) 281–305.

[9] P.L. Privalov, Physical basis of the stability of the folded conformations of proteins, in: T.E. Creighton (Ed.), Protein Folding, W.H. Freeman, New York, 1992, pp. 83–126.

[10] J. Jonas, Cold denaturation of proteins, ACS Symp. S 676 (1997) 310–323.
[11] G.P. Privalov, P.L. Privalov, Problems and prospects in microcalorimetry of biological macromolecules, Methods Enzymol. 323 (2000) 31–62.
[12] J.F. Brandts, The thermodynamics of protein denaturation. 1. The denaturation of chymotrypsinogen, J. Am. Chem. Soc. 86 (1964) 4291–4314.
[13] J.F. Brandts, J. Fu, J.H. Nordin, The low temperature denaturation of chymotrypsinogen in aqueous solution and in frozen aqueous solution, in: G.E.W. Wolstenholme, M. O'Connor (Eds.), The Frozen Cell, J. and A. Churchill, London, 1970, pp. 189–208.
[14] L. Smeller, Pressure-temperature phase diagrams of biomolecules, Biochim. Biophys. Acta 1595 (2002) 11–29.
[15] K.A. Dill, H.S. Chan, Multiple intermediates and transition states during protein unfolding, Nat. Struct. Biol. 4 (1997) 10–19.
[16] P.G. Wolynes, J.N. Onuchic, D. Thirumalai, Navigating the folding routes, Science 276 (1995) 1619–1620.
[17] A. Dinner, A. Sali, M. Karplus, E. Shakhnovich, Phase diagram of a model protein derived by exhaustive enumeration of the conformations, J. Chem. Phys. 101 (1994) 1444–1451.
[18] O. Collet, Warm and cold denaturation in the phase diagram of a protein lattice model, Europhys. Lett. 53 (2001) 93–99.
[19] P. De los Rios, G. Caldarelli, Putting proteins back into water, Phys. Rev. E 62 (2000) 8449–8452.
[20] A. Trovat, J. van Mourik, A. Marian, Swollen collapsed transition in random hetero-polymers, Eur. Phys. J. B 6 (1998) 63–73.
[21] R. Zwanzig, Simple model of protein folding kinetics, Proc. Natl. Acad. Sci. USA 92 (1995) 9801–9804.
[22] A. Hansen, M.H. Jensen, K. Sneppen, G. Zocchi, Hot and cold denaturation of proteins: critical aspects, Eur. Phys. J. B 10 (1999) 193–196.
[23] S.A. Hawley, Reversible pressure-temperature denaturation of chymotrypsinogen, Biochemistry 10 (1971) 2436–2442.
[24] A. Hawley, High pressure techniques, Methods Enzymol. 49 (1978) 14–24.
[25] J. Zipp, W. Kauzmann, Pressure denaturation of metmyoglobin, Biochemistry 12 (1973) 4217–4228.
[26] S. Kunugi, H. Yamamoto, M. Makino, T. Tada, Y.U. Kunugi, Pressure assisted cold denaturation of carboxypeptidase Y, Bull. Chem. Soc. Jpn. 72 (1999) 2803–2806.
[27] J. Jonas, A. Jonas, High pressure NMR spectroscopy of proteins and membranes, Annu. Rev. Biophys. Biomol. Struct. 23 (1994) 287–318.
[28] J. Zhang, X. Peng, A. Jonas, J. Jonas, NMR study of the cold, heat, and pressure unfolding of ribonuclease A, Biochemistry 34 (1995) 8631–8641.
[29] J. Jonas, L. Ballard, D. Nash, High resolution, high pressure NMR studies of proteins, Biophys. J. 75 (1998) 445–452.
[30] A.D. Blum, S.H. Smallcombe, R.L. Baldwin, Nuclear magnetic resonance evidence for a structural intermediate at an early stage in the refolding of ribonuclease A, J. Mol. Biol. 118 (1978) 305–316.
[31] D. Nash, B.S. Lee, J. Jonas, Hydrogen exchange kinetics in the cold denatured state of ribonuclease A, Biochim. Biophys. Acta 1297 (1996) 40–48.
[32] J.B. Udgaonkar, R.L. Baldwin, Early folding intermediate of ribonuclease A, Proc. Natl. Acad. Sci. USA 87 (1990) 8197–8201.
[33] W.A. Hourly, H.A. Scheraga, Structure of hydrophobically collapsed intermediate on the conformational folding pathway of ribonuclease A probed by hydrogen-deuterium exchange, Biochemistry 35 (1996) 11734–11746.
[34] G. Panick, G.J. Vidugiris, R. Malessa, G. Rapp, R. Winter, C.A. Royer, Exploring the temperature pressure phase diagram of staphylococcal nuclease, Biochemistry 38 (1999) 4157–4164.
[35] E. Mombelli, M. Afshar, P. Fusi, M. Mariani, P. Tortora, J.P. Connelly, R. Lange, The role of phenylalanine 31 in maintaining the conformational stability of ribonuclease P2 from *Sulfolobus solfataricus* under extreme conditions of temperature and pressure, Biochemistry 36 (1997) 8733–8742.
[36] D. Nash, J. Jonas, Structure of pressure assisted cold denatured lysozyme and comparison with lysozyme folding intermediates, Biochemistry 36 (1997) 14375–14383.
[37] S.T. Gladwin, P.A. Evans, Structure of very early protein folding intermediates: new insights through a variant of hydrogen exchange labelling, Fold. Des. 1 (1996) 407–417.
[38] K. Sasahara, K. Nitta, Pressure induced unfolding of lysozyme in aqueous guanidinium chloride solution, Protein Sci. 8 (1999) 1469–1474.
[39] D. Nash, J. Jonas, Structure of the pressure assisted cold denatured state of ubiquitin, Biochem. Biophys. Res. Commun. 238 (1997) 289–291.
[40] W. Dzwolak, M. Kato, A. Shimizu, Y. Taniguchi, FTIR study on heat induced and pressure assisted cold induced changes in structure of bovine alpha lactalbumin: stabilizing role of calcium ion, Biopolymers 62 (2001) 29–39.
[41] V.L. ValenteMesquita, M.M. Botelho, S.T. Ferreira, Pressure induced subunit dissociation and unfolding of dimeric lactoglobulin, Biophys. J. 75 (1998) 471–476.
[42] P. Kolakowski, E. Dumay, J.C. Cheftcl, Effects of high pressure and low temperature on β-lactoglobulin unfolding and aggregation, Food Hydrocolloids 15 (2001) 215–232.
[43] Yu. Griko, S.Yu. Venyaminov, P.L. Provalov, Heat and cold denaturation of phosphoglycerol kinase (interaction of domains), FEBS Lett. 244 (1989) 276–278.
[44] S. Kunugi, Y. Tanagi, M. Kitayaki, N. Tanaka, Y.U. Kunugi, Effects of high pressure on the activity and spectroscopic properties of carboxypeptidase Y, Bull. Chem. Soc. Jpn. 70 (1997) 1459–1463.
[45] M. Dumoulin, H. Ueno, R. Hayashi, C. Balny, Contribution of the carbohydrate moiety to conformational stability of the carboxypeptidase Y. High pressure study, Eur. J. Biochem. 262 (1999) 475–483.

[46] S. Kunugi, H. Hirohara, N. Ise, Pressure dependence of thermolysin catalysis, Eur. J. Biochem. 142 (1984) 565–570.

[47] S. Kunugi, M. Kitayaki, Y. Yanagi, N. Tanaka, R. Lange, C. Balny, The effect of high pressure on thermolysin, Eur. J. Biochem. 248 (1997) 567–574.

[48] H. Ikeuchi, S. Kunugi, H. Oyama, K. Oda, Activity and stability of a neutral protease from *Vibrio* sp. (vimelysin) in a pressure-temperature gradient, Eur. J. Biochem. 267 (2000) 979–983.

[49] F. Franks, R.H. Hatley, H.L. Friedman, The thermodynamics of protein stability. Cold destabilization as a general phenomenon, Biophys. Chem. 31 (1988) 307–315.

[50] A. Chapeaurouge, J.S. Johansson, S.T. Ferreira, Folding intermediates of a model three helix bundle protein. Pressure and cold denaturation studies, J. Biol. Chem. 276 (2001) 14861–14866.

[51] E.M. Boczko, C.L. Brooks III, First-principles calculation of the folding free energy of a three-helix bundle protein, Science 269 (1995) 393–396.

[52] Y. Zhou, M. Karplus, Interpreting the folding kinetics of helical proteins, Nature 401 (1999) 400–403.

[53] M. Shibayama, T. Tanaka, Volume phase transition and related phenomena of polymer gels, Adv. Polym. Sci. 109 (1993) 1–62.

[54] H.G. Schild, Poly(N-isopropylacrylamide) experiment, theory and application, Prog. Polym. Sci. 17 (1992) 163–249.

[55] E.I. Tiktopulo, V.E. Bychkova, J. Ricka, O.B. Ptitsyn, Cooperativity of the coil globule transition in a homopolymer; microcalorimetric study of poly(N-isopropylacrylamide), Macromolecules 27 (1994) 2879–2882.

[56] E.I. Tiktopulo, Y.N. Uversky, B.B. Lushchik, S.I. Klenin, V.E. Bychkova, O.B. Ptitsyn, Domain coil globule transition in homopolymers, Macromolecules 28 (1995) 7518–7524.

[57] M. Akashi, S. Nakano, A. Kishida, Synthesis of poly(N-vinylisobutyramide) from poly(N-vinylacetamide) and its thermosensitive property, J. Polym. Sci. A 34 (1996) 301–303.

[58] K. Suwa, Y. Wada, Y. Kikunaga, K. Morishita, A. Kishida, M. Akashi, Synthesis and functionalities of poly(N-vinylalkylamide). 4. synthesis and free radical polymerization of N-vinylisobutyramide and thermosensitive properties of the polymer, J. Polym. Sci. A 35 (1997) 1763–1768.

[59] K. Suwa, K. Morishita, A. Kishida, M. Akashi, Synthesis and functionalities of poly(N-vinylalkylamide). 5. Control of a lower critical solution temperature of poly(N-vinylalkylamide), J. Polym. Sci. A 35 (1997) 3087–3094.

[60] S. Kunugi, K. Takano, N. Tanaka, K. Suwa, M. Akashi, Effects of pressure on the behavior of the thermoresponsive polymer poly(N-vinylisobutyramide) (PNVIBA), Macromolecules 30 (1997) 4499–4501.

[61] K. Suwa, K. Yamamoto, M. Akashi, K. Takano, N. Tanaka, S. Kunugi, Effects of salt on the temperature and pressure responsive properties of poly(N-vinylisobutyramide) aqueous solutions, Colloid Polym. Sci. 276 (1998) 529–533.

[62] S. Kunugi, Y. Yamazaki, K. Takano, N. Tanaka, M. Akashi, Effects of ionic additives and ionic comonomers on the temperature and pressure responsive behavior of thermoresponsive polymers in aqueous solutions, Langmuir 15 (1999) 4056–4061.

[63] S. Kunugi, Y. Yamazaki, T. Tada, Effects of acrylic acid incorporation on pressure temperature behavior and calorimetric properties of PNIPAM in aqueous solutions, Colloid Polym. Sci. 278 (2000) 80–83.

[64] S. Kunugi, Y. Yamazaki, T. Tada, K. Yamamoto, M. Akashi, Thermodynamic studies on coil globule transitions of poly(N-vinylisobutyramide-co-vinylamine) (pNVIBA-co-VAm) in aqueous solutions, Langmuir 16 (2000) 2040–2042.

[65] T. Tamura, T. Yamaoka, S. Kunugi, A. Panitch, D.A. Tirrell, Effects of temperature and pressure on the aggregation properties of an engineered elastin model polypeptide in aqueous solution, Biomacromolecules 1 (2000) 552–555.

[66] B. Li, D.O. Alonso, V. Daggett, The molecular basis for the inverse temperature transition of elastin, J. Mol. Biol. 305 (2001) 581–592.

[67] S. Kunugi, D. Yoshida, H. Kiminami, Effects of pressure on the behavior of (hydroxypropyl)cellulose in aqueous solution, Colloid Polym. Sci. 279 (2001) 1139–1143.

[68] J.M. Zhou, L. Zhu, C. Balny, S. Perrett, Pressure denaturation of the yeast prion protein Ure2, Biochem. Biophys. Res. Commun. 287 (2001) 147–152.

Biochimica et Biophysica Acta 1595 (2002) 345–356

BIOCHIMICA ET BIOPHYSICA ACTA
BBA
www.bba-direct.com

Review

Protein crystallization under high pressure

Yoshihisa Suzuki [a,*], Gen Sazaki [b,c], Satoru Miyashita [d], Tsutomu Sawada [e], Katsuhiro Tamura [a], Hiroshi Komatsu [f]

[a] *Department of Chemical Science and Technology, Faculty of Engineering, University of Tokushima, 2-1 Minamijosanjima, Tokushima 770-8506, Japan*
[b] *Institute for Materials Research, Tohoku University, Sendai, Japan*
[c] *Center for Interdisciplinary Research, Tohoku University, Sendai, Japan*
[d] *Physics, Liberal Arts and Sciences, Toyama Medical and Pharmaceutical University, Toyama, Japan*
[e] *National Institute for Materials Science, Tsukuba, Japan*
[f] *Faculty of Policy Studies, Iwate Prefectural University, Morioka, Japan*

Received 7 September 2001; received in revised form 7 November 2001; accepted 8 November 2001

Abstract

Pressure is expected to be an important parameter to control protein crystallization, since hydrostatic pressure affects the whole system uniformly and can be changed very rapidly. So far, a lot of studies on protein crystallization have been done. Solubility of protein depends on pressure. For instance, the solubility of tetragonal lysozyme crystal increased with increasing pressure, while that of orthorhombic crystal decreased. The solubility of subtilisin increased with increasing pressure. Crystal growth rates of protein also depend on pressure. The growth rate of glucose isomerase was significantly enhanced with increasing pressure. The growth rate of tetragonal lysozyme crystal and subtilisin decreased with increasing pressure. To study the effects of pressure on the crystallization more precisely and systematically, hen egg white lysozyme is the most suitable protein at this stage, since a lot of data can be used. We focused on growth kinetics under high pressure, since extensive studies on growth kinetics have already been done at atmospheric pressure, and almost all of them have explained the growth mechanisms well. The growth rates of tetragonal lysozyme decreased with pressure under the same supersaturation. This means that the surface growth kinetics significantly depends on pressure. By analyzing the dependence of supersaturation on growth rate, it was found that the increase in average ledge surface energy of the two-dimensional nuclei with pressure explained the decrease in growth rate. At this stage, it is not clear whether the increase in surface energy with increasing pressure is the main reason or not. Fundamental studies on protein crystallization under high pressure will be useful for high pressure crystallography and high pressure protein science.

Keywords: High pressure; Crystallization; In situ observation; Growth kinetics; Protein

1. Introduction

For X-ray structure analysis of protein molecules, large and high quality single crystals are needed. However, the crystallization process is still a bottleneck. Pressure is a candidate to control protein crystallization well, since hydrostatic pressure affects the whole system uniformly and can be changed very rapidly. So far, a lot of crystallization studies under

* Corresponding author. Fax: +81-88-655-7025.
E-mail address: suzuki@chem.tokushima-u.ac.jp (Y. Suzuki).

0167-4838/02/$ – see front matter
PII: S0167-4838(01)00355-7

high pressure have been reported [1–12,14–17]. For the first time, Visuri et al. [9] reported that the growth rate of crystals of glucose isomerase was significantly enhanced with increasing pressure. Effects of pressure on protein crystallization have received much attention since then. The solubility of tetragonal lysozyme crystal increased with increasing pressure [1–6], while that of orthorhombic crystal decreased with increasing pressure [5]. That of subtilisin increased with increasing pressure [7,8]. The crystal growth rate of glucose isomerase increased with increasing pressure [8], while that of tetragonal lysozyme crystal [1–4,10–12] and subtilisin [8] decreased with increasing pressure. Effects of pressure on nucleation of protein crystal have also been studied, though the experiment and its analysis are difficult to get consistent results [2,14–17]. At the present stage, hen egg white lysozyme is the most suitable protein to study the effect of pressure on the crystallization mechanism more precisely and systematically, since a lot of data can be used. We focused on the growth kinetics of tetragonal lysozyme crystal under high pressure, since extensive studies on growth kinetics have been reported at atmospheric pressure (for a discussion, see for example [13]). The growth rate of tetragonal lysozyme crystal decreased with increasing pressure under the same supersaturation σ at each pressure [12]. The analysis of the growth rates by the birth and spread model of two-dimensional nucleation growth provided an increase in the average ledge surface energy of the two-dimensional nuclei with pressure.

In this chapter, we will focus on lysozyme, and special attention will be paid to the solubility and the crystal growth kinetics under high pressure.

2. Effect of pressure on solubility

In general, solubility is important and indispensable to study the equilibrium state of solution and crystal. Solubility is usually measured at the beginning of the crystallization research to determine the supersaturation σ ($\sigma = \ln(C/C_e)$, C: concentration of the solution, C_e: solubility), because the crystallization phenomena are often well described by using σ. Supersaturation is named the *driving force for crystallization*. For the studies at atmospheric pressure, σ has been useful for the discussion on protein crystallization [13]. Thus, for a high pressure study, the high pressure σ is also useful to discuss the mechanisms of protein crystallization. To determine the high pressure σ, the high pressure solubility is indispensable.

2.1. Method of solubility measurements

As mentioned above, a lot of researchers have measured the solubility under high pressure. However, the solubility varies according to the method, even though the composition of the solution is almost the same [6]. The variation prevents us from a quantitative discussion. Therefore, a more sophisticated method is expected to measure a more accurate solubility. Gross et al. and Lorber et al. reported the solubility of tetragonal lysozyme crystals under high pressure by ex situ measurement [1,3]. They incubated the supersaturated solution under high pressure for a certain period, then reduced the pressure and measured the concentration in the supernatant solution. They measured the solubility only from the supersaturated state, where it is always uncertain from 0 mg ml^{-1} to the asymptotic concentration. Takano et al. provided a much improved technique based on in situ observation [4]. They measured the equilibrium pressure in situ from supersaturation and undersaturation. They gradually increased the pressure of the sample over a period of a few days, and continuously recorded the images of the lysozyme crystal. The solubility was determined from changes in both the size of the crystal and the amount of the transmitted light through the crystal. However, the error of the equilibrium pressure was still large, and a long time period was still required to establish a solubility curve.

We have designed the following two in situ methods of solubility measurements. One (Section 2.1.1) is the precise method, and the other (Section 2.1.2) is the rapid method.

2.1.1. Precise method by continuous measurement of the transitional change in concentration using a Mach–Zehnder interferometer

We have designed an in situ precise method of solubility measurement using a Mach–Zehnder interferometer [18] and measured in situ the solubility of

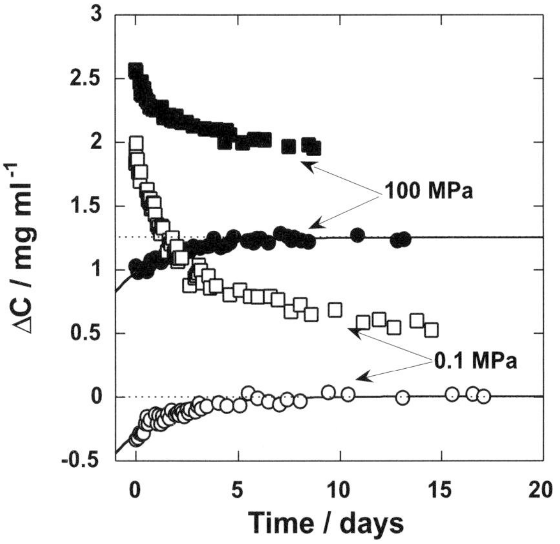

Fig. 1. Transitional change in concentration with time [6]. ΔC is the relative increase in concentration from the solubility at 0.1 MPa. Closed and open symbols indicate the data taken at 100 and 0.1 MPa, respectively. Circles and squares indicate the data for the dissolution and the growth relaxation, respectively. Conditions: 20°C; NaCl, 0.83 M; in 0.02 M sodium acetate buffer (pH 4.66).

lysozyme [6]. With this method, the relative change in concentration with time during equilibration was measured accurately and continuously, starting from a supersaturated state (growth relaxation) and an undersaturated state (dissolution relaxation). Further details are described elsewhere [6,18]. The transitional change in concentration with time is shown in Fig. 1. The asymptotic concentration for the dissolution relaxation was regarded as the solubility. We believe that this method is the most precise one for the high pressure solubility of lysozyme at the present stage, and the solubility will be used for the analysis of crystal growth rates under high pressure by a model function of σ in Section 3.2. However, a long time period is still required to establish a solubility curve by this method, and this remains one of the most serious barriers to further high pressure studies.

2.1.2. Rapid measurement of the equilibrium temperature using a Michelson interferometer

In order to decrease the time necessary for solubility measurement under high pressure, another interferometric technique has been developed, which can determine the solubility of lysozyme within at most 3 h [5,19]. This section introduces this method in detail.

In our interferometric method, an equilibrium temperature of a given concentration is determined by observing the concentration distribution around a crystal. The distribution can be visualized using a Michelson interferometer [20–22]. Fig. 2a shows the interferometer and interference cell placed in a high pressure cell. The interferometer was specially designed by Nikon. The working distance of the objective lens (25.5 mm) was long enough to observe the sample through the thick high pressure cell. A reference mirror could be translated over 10 mm to compensate for the large optical path through the interference cell. A spare sapphire window was also set in front of the reference mirror as a compensator. The interferometer was attached to an optical microscope of reflection type. The sample was illuminated by a white light source (halogen, 100 W) with a 547 nm band pass filter. The high pressure cell was made of stainless steel, and could be used to a maximum pressure of 120 MPa. Crystals in the interference cell could be observed in situ through a sapphire

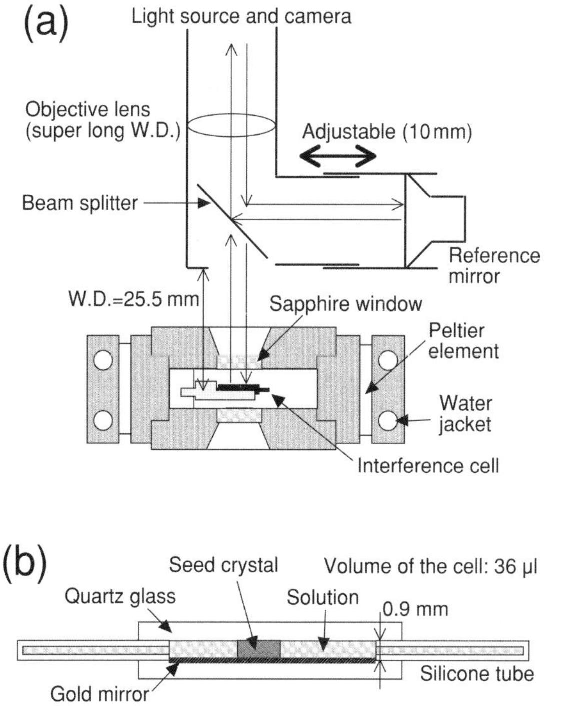

Fig. 2. Schematic drawing of the experimental setup [5]. (a) Michelson interferometer and an interference cell located in a high pressure cell. (b) Cross-sectional view of the interference cell.

window. The high pressure cell was filled with water and was connected to a hand pressure pump. The hydraulic fluid in the pump was Durasyn 166 (Amoco Chemicals). The pressure was monitored by a precise Bourdon-type pressure gauge (Heise, maximum 150 MPa, ±0.02 MPa) and was manually regulated to ±0.2 MPa. The temperature of the high pressure cell was controlled to an accuracy of ±0.1°C using Peltier elements. Fig. 2b shows a sectional view of the interference cell. The interference cell was made of quartz or glass plates. Gold was vacuum-deposited on the bottom of the cell to enhance the reflectivity. The thickness of the solution was 0.9 mm, and the volume of the cell was 36 μl. Pressure was applied to the crystals and solution in the interference cell via soft silicone tubes.

Under high pressure, the concentration distribution around the crystals was observed in situ with the Michelson interferometer. Fig. 3 shows interferograms of the solution around the tetragonal lysozyme crystal under 100 MPa. Here, the concentration of lysozyme was 40.0 mg ml^{-1}. When the crystal and the solution were almost equilibrated (21.2°C), the interference fringes were parallel straight lines, as shown in Fig. 3a. If the temperature of the sample was set lower than its equilibrium temperature, the crystal grew (10.2°C) and the fringes were bent in the vicinity of the crystal (Fig. 3b), because of the decrease in the concentration around the crystal. On the other hand, when the temperature was raised higher than its equilibrium temperature (27.0°C), the crystal dissolved and the fringes bent in the opposite direction (Fig. 3c). From observation of the fringes around the crystal, we determined the equilibrium temperature of the crystal and solution of a given concentration. Further details of this technique were reported in our previous study [19].

Six times recrystallized hen egg white lysozyme (Seikagaku Kogyo) was used without further purification. All other chemicals were of reagent grade. Lysozyme was dissolved in 50 mM sodium acetate buffer (pH 4.5) to a final lysozyme concentration of 120 mg ml^{-1}. Sodium chloride solution of 50 mg ml^{-1} was prepared in the same 50 mM acetate buffer. Supersaturated solution was prepared by mixing an equal volume of the lysozyme solution and the sodium chloride solution. The supersaturated solution was transferred to the interference cell and incubated for 3 days to prepare suitable seed crystals in the cell. The seed crystals of the orthorhombic form were grown in the sample cell at 40°C. After preparing the seed crystals, the solution in the cell was replaced with fresh lysozyme solution (15–60 mg ml^{-1}) for the interferometric observation.

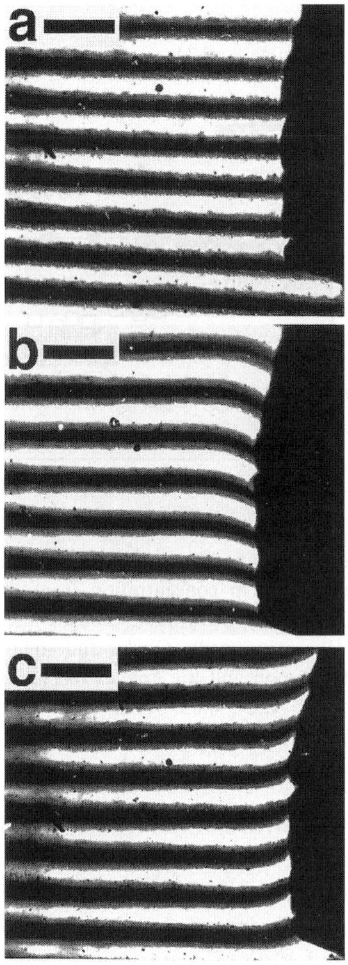

Fig. 3. Interferograms around the tetragonal lysozyme crystals under 100 MPa [5]. Conditions: lysozyme, 40.03 mg ml^{-1}; NaCl, 25 mg ml^{-1}, in 50 mM sodium acetate buffer (pH 4.5). (a) Equilibrium (21.2°C); (b) growth (10.2°C); (c) dissolution (27.0°C). The scale bar represents 200 μm.

Fig. 4 shows the solubility curves found for tetragonal and orthorhombic lysozyme crystals. The size of the error comes from the accuracy in measuring the straightness of the interference fringes. The solubility of both crystal forms increased with temperature at each pressure. In the case of the tetragonal form, the solubility increased with an increase of the hydrostatic pressure. In the case of the orthorhombic

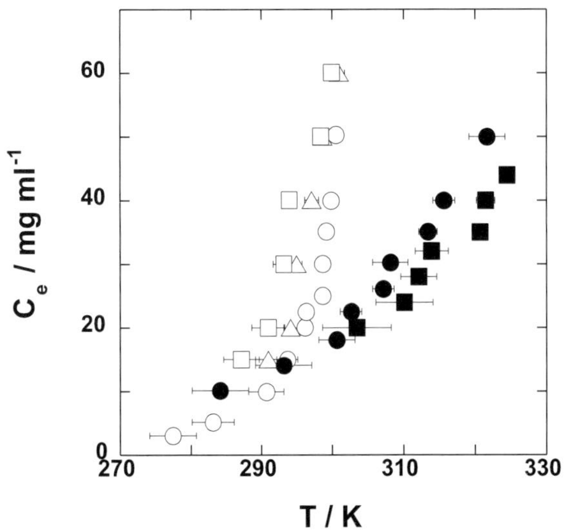

Fig. 4. Solubility curves of the tetragonal and orthorhombic lysozyme crystals measured by our technique [5]. Open symbols represent the solubility of tetragonal crystals. Pressure: 0.1 MPa (○), 50 MPa (△), 100 MPa (□). Closed symbols present the solubility of orthorhombic crystal. Pressure: 0.1 MPa (●), 100 MPa (■). Conditions: NaCl, 25 mg ml^{-1}; in 50 mM sodium acetate buffer (pH 4.5).

form, in contrast, the solubility was found to decrease with pressure.

2.2. Thermodynamic analysis of the solubility

From solubility data, thermodynamic parameters are often calculated using van't Hoff plots. If we assume that the effect of the activity coefficient is negligible, we can estimate the partial molar enthalpy of dissolution, ΔH, from Eq. 1 and the partial molar entropy of dissolution, ΔS, from Eq. 2.

$$\frac{\partial \ln C_e}{\partial (1/T)} = \frac{-\Delta H}{R} \tag{1}$$

$$\frac{\partial RT \ln C_e}{\partial T} = \Delta S \tag{2}$$

where C_e is the solubility (mg ml^{-1}) and R is the gas constant.

To estimate ΔH, ln C_e is plotted against $1/T$ in Fig. 5. If we assume that ΔH does not depend on temperature, ΔH is estimated from the slope by linear fitting of the plot. In this section, weighted fitting was done, because the temperature error was large at a lower concentration region. Fitted lines are also described in Fig. 5. ΔS is estimated from $T-RT \ln C_e$ plots.

From the dependence of pressure on solubility, if we assume that the effect of the activity coefficient is negligible, the volume change accompanying the dissolution, ΔV ($\Delta V \equiv \bar{V} - V_c$, $\bar{V}$: partial molar volume of the solute, V_c: molar volume of the crystal), is expressed as:

$$\Delta V = -RT \left[\frac{\partial \ln C_e}{\partial p} \right]_T \tag{3}$$

If ΔV does not depend on pressure up to p MPa, the molar volume change accompanying the dissolution at 0.1 MPa is expressed as:

$$\Delta V = -RT \frac{\ln C_{e,p} - \ln C_{e,0.1}}{p - 0.1} \tag{4}$$

where $C_{e,p}$ and $C_{e,0.1}$ indicate the solubility at p MPa and 0.1 MPa, respectively.

All thermodynamic parameters from our studies are listed in Table 1. ΔH and ΔS of tetragonal lysozyme crystal decreased with increasing pressure, while those of orthorhombic crystal did not change. ΔV of tetragonal lysozyme crystal shows a negative value, while that of orthorhombic crystal shows a positive value.

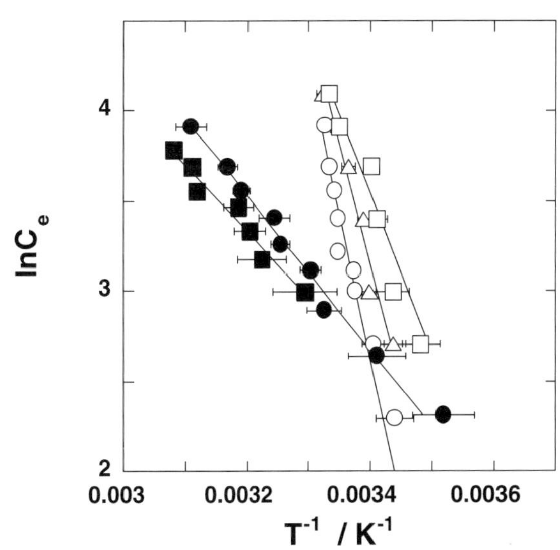

Fig. 5. van't Hoff plot of the solubility shown in Fig. 4 [5]. Open symbols represent the solubility of tetragonal crystals. Pressure: 0.1 MPa (○), 50 MPa (△), 100 MPa (□). Closed symbols represent the solubility of orthorhombic crystal. Pressure: 0.1 MPa (●), 100 MPa (■). Conditions: NaCl, 25 mg ml^{-1}; in 50 mM sodium acetate buffer (pH 4.5).

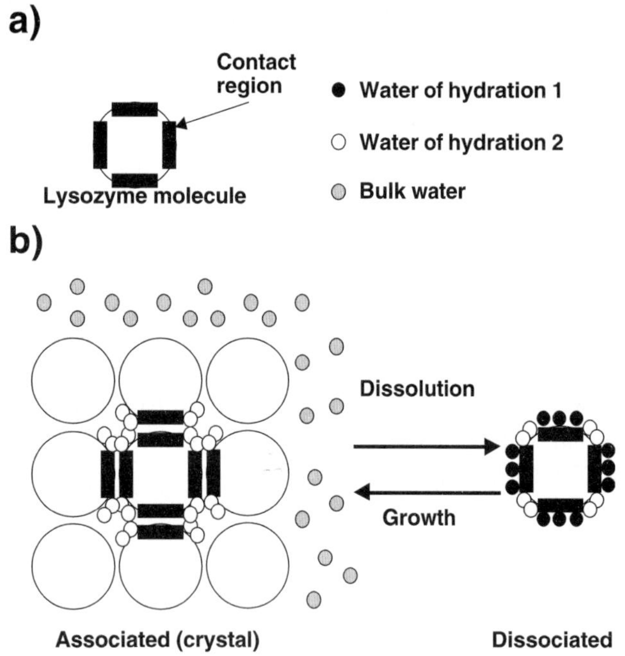

Fig. 6. Schematic diagrams of lysozyme molecules in the crystal and solution [5]. In the crystal, the molecules are in contact with each other at the regions indicated in black. Hydrated water 1 and 2 represent water molecules hydrating the contact and non-contact regions of the protein, respectively. The volume change on dissolution is given by the difference in volume between hydrated water 1 and bulk water.

The effect of pressure on the solubility of tetragonal crystal ($\Delta V < 0$) indicates that the partial molar volume of lysozyme, $\bar{V}$, is smaller than the molar volume of the tetragonal crystal [1–3,5], while $\bar{V}$ is larger than the molar volume of the orthorhombic crystal [5]. Fig. 6 represents a simplified model of the states of the protein in the crystal and in solution. Consider now the change from crystalline to the solution state. If we neglect any change in volume of the protein molecule, the bulk water, and the waters of hydration 2 (those around parts of the protein exposed in both crystal and solution), then ΔV is given by the volume of the waters of hydration 1 (around the contact surfaces of the protein) minus the volume occupied by these same water molecules as 'free' water when the protein is in the crystalline state. For this volume change to be negative, as found for the tetragonal form, the water molecules must be more tightly packed when hydrating the contact regions than when 'free' in the bulk water. This, in fact, is expected to be the case for contacts containing a large number of hydrophilic residues. Correspondingly, the positive volume change on dissolution of the orthorhombic form implies that the contact surfaces tend to structure the waters of hydration such that they occupy a larger volume than in the bulk. It predicts that the contacts in the orthorhombic form should be more hydrophobic than in the tetragonal form.

The decrease in ΔS of tetragonal lysozyme crystal with pressure can be explained by a decrease in ΔV with pressure. Since the solution is more compressible than the crystal, the magnitude of ΔV is smaller under high pressure than under atmospheric pressure. Smaller $|\Delta V|$ under high pressure can lead to a smaller change in the degree of freedom. However, in the case of orthorhombic crystal, ΔS is almost constant with increasing pressure. The reason is not clear. The change in ΔH is still not clear. Further crystallographic studies on the hydration of the intermolecular contact regions or the other independent measurements of ΔV, ΔH and ΔS may explain these phenomena.

Table 1
List of the thermodynamic parameters of lysozyme crystal (ΔV values are estimated at 300 K and 0.1 MPa)

Crystal form		Pressure		
		0.1 MPa	50 MPa	100 MPa
Tetragonal	ΔH (kJ mol^{-1})	130 ± 10	110 ± 20	70 ± 10
	ΔS (J mol^{-1} K^{-1})	460 ± 40	400 ± 60	280 ± 40
	ΔV (cm^3 mol^{-1})	−18 ± 46[a]		
Orthorhombic	ΔH (kJ mol^{-1})	35 ± 3	–	35 ± 5
	ΔS (J mol^{-1} K^{-1})	140 ± 10	–	140 ± 20
	ΔV (cm^3 mol^{-1})	5 ± 18[b]		

[a]Calculated from Eq. 4 at $p = 50$.
[b]Calculated from Eq. 4 at $p = 100$.

3. Effect of pressure on the growth kinetics of the crystal

Large and high quality single crystals are needed for X-ray structure analysis, as mentioned before. To obtain such crystals, a well-controlled crystallization process is expected. To achieve such a process, a fundamental study on the elementary process of protein crystallization is needed. The analysis of the growth rate by a growth mode function of σ is useful to clarify the elementary process of crystal growth [13,21,22]. Thus, the growth rate should be measured against σ at each pressure.

3.1. Measurements of the growth rate

Some nucleation and growth processes of tetragonal lysozyme crystal were analyzed using the transitional change in concentration of the supernatant with time [1,10,14,15]. However, pressure effects on the normal growth rate of the {110} face are different from that of the {101} face [2,3,12]. Thus, to study the growth mechanism in detail, a direct observation of the crystal is needed.

Lorber et al. observed the crystal with an optical microscope after releasing the pressure [3]. They found that the shape of tetragonal crystals becomes more elongated in a limited region of their phase diagram, but they could not observe the crystal continuously. To observe the crystal continuously and measure the growth rate precisely, an in situ observation is indispensable [2,4,12]. This section introduces the details of our method which is described in [12].

Our high pressure vessel is shown in Fig. 7. The vessel was filled with hydraulic oil. The hydrostatic pressure (up to 120 MPa) was applied to the vessel with a hand pressure pump through the oil. For in situ observation, the vessel has a pair of transparent sapphire windows. The pressure was monitored by a Bourdon-type pressure gauge (Heise) within an accuracy of ± 0.1 MPa. We could maintain the vessel pressure manually within ± 0.2 MPa. The temperature of the vessel was measured by a thermocouple. The vessel was placed in an air conditioned box and the temperature was kept constant within ± 0.1°C. The inner cell (Fig. 7b) for crystal growth was made of glass slides and set in the vessel. The crystals and the solution were put in the cell. The solution and the oil were separated by soft silicone tubes sealed with glue. The pressure was applied to the solution via these tubes. The crystallization process was observed by an optical microscope.

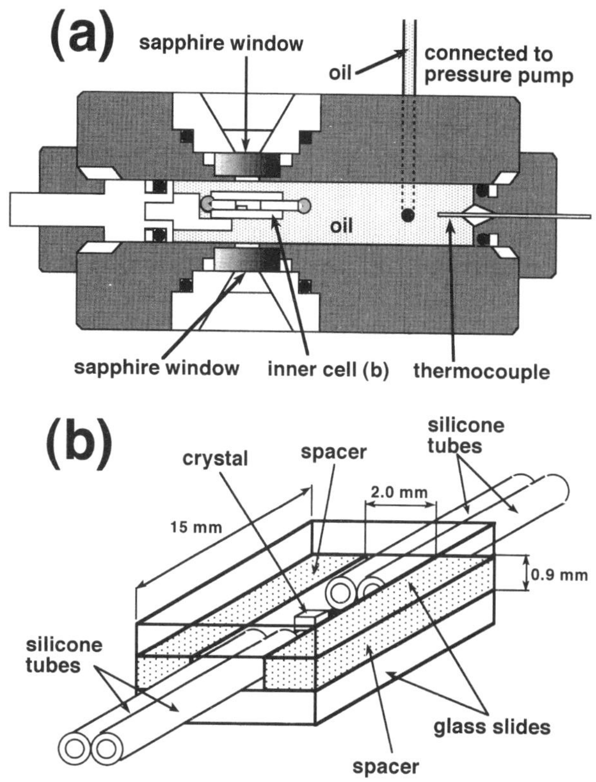

Fig. 7. Schematic illustrations of (a) the high pressure vessel and (b) the inner cell [12]. The vessel is directly connected to the high pressure pump, and pressure in the vessel is controlled precisely. The glass inner cell (b) is set between two sapphire windows, and the crystals in the cell are observed using an optical microscope through the windows. For the inner cell (b), the edges of the silicone tubes are sealed, and the crevices are filled up with glue.

The six times recrystallized lysozyme (Seikagaku Kogyo) was used without further purification. The buffer solution was 0.02 M sodium acetate (pH 4.66), and the precipitant was sodium chloride (0.83 M). The temperature was kept constant at $T = 20.0 \pm 0.1$°C. Under this crystallization condition, tetragonal form crystals bounded by {110} and {101} faces were obtained.

A supersaturated lysozyme solution of 50 mg ml^{-1} in concentration was incubated under atmospheric pressure (0.1 MPa) for about 10 h to get seed crystals 200–300 μm in size. The solution in the inner cell was replaced with fresh solution of a given lysozyme concentration through the silicone tubes with a peristaltic pump. Then the inner cell was inserted into the high pressure vessel, and pressure was applied. It took no more than 10 min to change the solution and to apply pressure. The normal growth rates of the {110} and {101} faces were measured by taking micrographs as Durbin et al. have done [23].

The solubility of lysozyme at $T = 20.0 \pm 0.1$°C under pressure p [MPa], $C_e(p)$ (mg ml^{-1}) was expressed as [12]:

$$C_e(p) = (4.7 \pm 0.2) + (1.8 \pm 0.2) \times 10^{-2} p - (5.7 \pm 1.4) \times 10^{-5} p^2 \quad (5)$$

The supersaturation σ was estimated by using Eq. 5 under each pressure.

3.2. Growth kinetics of tetragonal lysozyme crystals

Figs. 8 and 9 show the normal growth rates of the {110} and {101} faces with σ, respectively. Here, σ at each pressure was calculated by Eq. 5. The vertical error bars indicate the standard deviations of the growth rates measured by two or three crystals. The horizontal error bars are estimated by the error of the solubility measurements. Figs. 8 and 9 show

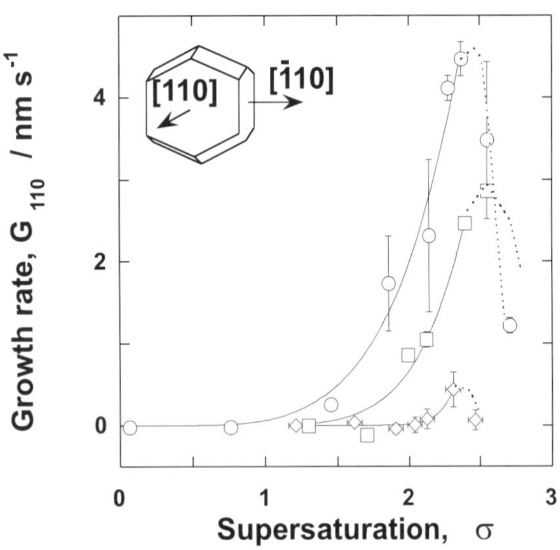

Fig. 8. Supersaturation dependence of the growth rates of the {110} face of a tetragonal lysozyme crystal [12]. Pressure: 0.1 MPa (○); 50 MPa (□); 100 MPa (◇). The solid lines are fittings by the birth and spread model. The broken lines are guides for the eye.

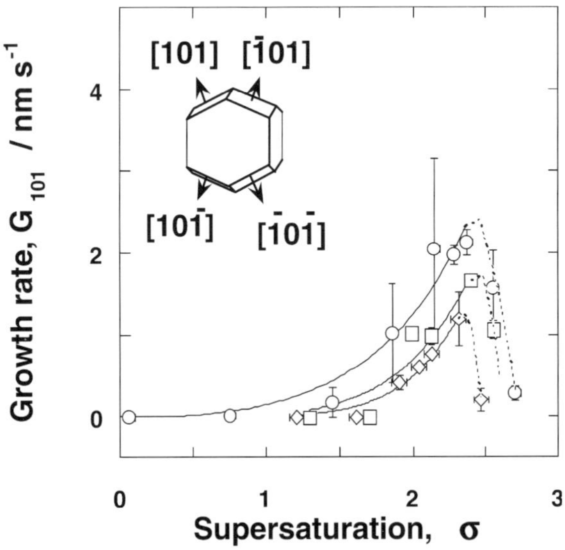

Fig. 9. Supersaturation dependence of the growth rates of the {101} face of a tetragonal lysozyme crystal [12]. Pressure: 0.1 MPa (○); 50 MPa (□); 100 MPa (◇). The solid lines are fittings by the birth and spread model. The broken lines are guides for the eye.

that: (1) as pressure increased, the normal growth rates of the {110} faces, G_{110}, and those of the {101} faces, G_{101}, decreased. (2) The growth rates reached almost zero for the small value of σ. No growth was observed till the supersaturation reached the critical value, σ_c, which suggests that the crystals grew after two-dimensional nuclei formed. (3) The growth rates had maxima at $\sigma = 2.3$–2.5. This phenomenon was observed under 0.1 MPa and explained by cluster formation [24] or impurity incorporation [13]. Here, we focused on (1) and (2), to elucidate the pressure effect on growth kinetics.

To find the growth mode, we fitted the obtained data by the empirical equation [25]

$$G \sim \sigma^n \quad (6)$$

where n is a constant. For n≦2, the growth is known to be governed by the spiral growth mode. For $n > 2$, the two-dimensional nucleation on the surface limits the growth rate. At each pressure, $n > 2$ holds and the two-dimensional nucleation was rate limiting in this study.

To confirm the growth mode, we observed the surface topography on the (110) face by an atomic force microscope (AFM) under 0.1 MPa [12]. The growth condition was $\sigma = 0.9$ and $T = 24$°C. The two-dimensional nucleation and the lateral growth of the two-dimensional islands were observed though the growth conditions were slightly different from those used for the high pressure experiments. Then we concluded that, in the range of σ from 0.9 to 2.4, crystals also grew by the birth and spread mechanism [26], which is expressed as [22]

$$G = k_1\left(\frac{C}{C_e}\right)^{\frac{2}{3}}\left(\frac{C-C_e}{C_e}\right)^{\frac{2}{3}}\sigma^{\frac{1}{6}}\exp\left(\frac{-k_2}{\sigma}\right) \tag{7}$$

where k_1 and k_2 are expressed as

$$k_1 = \left(\frac{\pi}{3}\right)^{\frac{1}{3}}a^2h^{\frac{7}{6}}\nu\lambda_0^{\frac{-2}{3}}C_e^{\frac{4}{3}}\exp\left(\frac{\varepsilon_{ad}-\varepsilon-2(E-\phi)}{3kT}\right) \tag{8}$$

$$k_2 = \frac{\pi\Omega\alpha^2h}{3k^2T^2} \tag{9}$$

In Eqs. 8 and 9, a is the distance between the molecules in the crystal ($=\Omega^{1/3}$, Ω: molecular volume in the crystal), h is the height of the step ($h = 5.7$ nm for the {110} face and $h = 3.4$ nm for the {101} face measured by AFM), ν is the thermal frequency of a solute, λ_0 is the average distance between the kinks on a step, ε_{ad} is the adhesive energy of a solute to the crystal surface, ε is the activation energy for a solute to be incorporated into a critical nucleus, $E-\phi$ is the activation energy for dehydration of a solute, α is the average surface energy on the step ledge, and k is the Boltzmann constant.

We fitted the obtained data by Eq. 7 with the least mean square method. The solid curves in Figs. 8 and 9 represent the fitted ones. They well reproduce the experimental data within the errors. The obtained values of k_1 and k_2 are listed in Table 2. On both faces, the values of k_1 and k_2 increased with pressure. An increase in k_1 means an increase in the growth rate while an increase in k_2 leads to a decrease in the growth rate. The results shown in Figs. 8 and 9 indicate that the change in k_2 dominated the pressure dependence of the growth rate. Eq. 9 shows that the increase in k_2 corresponds to the increase in α. However, the value of k_1 of the {110} face at 100 MPa for the best fit (Table 2) is extraordinarily large and this value has less physical meaning. This is due to the lack of data and the error of the growth rate. As in Eq. 8, there are too many factors to determine which dominate the increase in k_1 with pressure. To avoid such uncertainties, we fixed k_1 for each fitting. Since it is difficult to estimate a k_1 value under high pressure, we used the k_1 at 0.1 MPa as the fixed value. These fitting results are also listed in Table 2, and k_2 of both faces also increased with pressure. These results also support the fact that α increases with pressure.

If we regard the growth mechanism as the birth and spread type, the other possibility, which reduces the growth rate with pressure, is the decrease in k_1 (Eq. 8), but for the best fit, k_1 always increases with pressure. If we unreasonably fix k_2 for fitting, k_1 decreases with pressure, but the error of the fitting becomes significantly large. Thus, the most suitable reason why the growth rate decreases with pressure is the increase in α.

The evaluated values of the average surface energy on the step ledge, α, with Eq. 9 are plotted against pressure in Fig. 10. The values of α under 0.1 MPa are in the same order as those reported previously

Table 2
List of the fitted values of k_1 and k_2 according to Eq. 7

Coefficients	Pressure		
	0.1 MPa	50 MPa	100 MPa
{110} face			
k_1 (nm s^{-1})	0.84	3.5	7.1×10^6
k_2	3.6	8.7	45
k_2 (k_1 (nm s^{-1}) = 0.84 fix)	3.6	5.4	9.6
{101} face			
k_1 (nm s^{-1})	0.14	0.33	0.86
k_2	1.0	3.8	6.5
k_2 (k_1 (nm s^{-1}) = 0.14 fix)	1.0	2.0	2.6

[22,23]. The values of α of the {110} face are larger than those of the {101} face under each pressure. The values of α of both faces increased with pressure. The decrease in the lattice constants with pressure might cause an increase in the surface free energy. Kundrot et al., however, reported that the lattice constants, a and c, of tetragonal lysozyme crystals under 100 MPa were by 0.6% less and 0.1% more than those under 0.1 MPa, respectively [27]. As shown in Fig. 10, the values of α increased more than 2 times. These large changes in α could not be explained only by the changes in the lattice constants with pressure. Another evaluation of the change in α was also made. When pressure is applied, stress is applied to the crystal. The mechanical work by this stress is equal to the change in Helmholtz free energy of the crystal. We can estimate the mechanical work of compression per unit area of crystal surface, since Young's modulus of tetragonal lysozyme crystal has already been reported [28]. However, even if we use the dynamic Young's modulus (1.1 GPa; it is larger than the static Young's modulus (=0.2 GPa)), the previous works with increasing pressure up to 100 MPa (0.1 mJ/m^2) for the {110} face, 1.6×10^{-4} mJ/m^2 for the {101} face) still

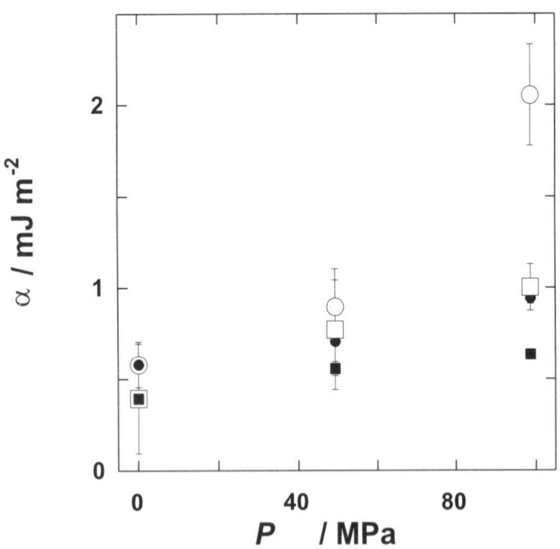

Fig. 10. Pressure dependence of the average surface energy of the step ledge, α, of two-dimensional islands on the {110} surface (circles) and the {101} surface (squares) [12]. Open symbols express the results of the best fits, and closed symbols express those fits when the value of k_1 of Eq. 7 is fixed. The vertical error bar was estimated by the standard deviation of the fitting.

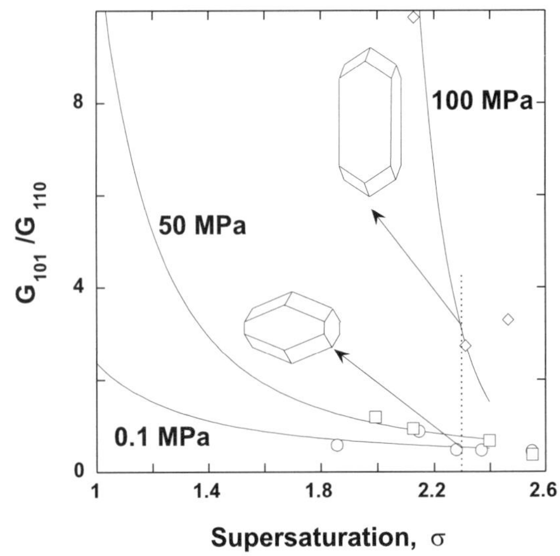

Fig. 11. Supersaturation dependence of the ratio of the normal growth rate of the {101} face, G_{101}, to that of the {110} face, G_{110}: ○, 0.1 MPa; □, 50 MPa; ◇, 100 MPa [12]. As pressure increased under the same supersaturation, the growth shape of the crystals was elongated along the crystallographic c-axis (see the inserts on the dashed line). Solid curves represent the fittings with Eq. 7.

show much smaller increases than those in α of this study (Fig. 10). At present, it is not clear why the values of α increased so much with pressure. Further studies, especially on hydration and dehydration, binding state between the molecules and so on, are needed.

Fig. 10 also shows that the change in α of the {110} face with pressure is larger than that of the {101} face. This larger increase in α of the {110} face leads to a larger decrease in the normal growth rate of the {110} face. The change in the growth rates of each face results in a morphological change of the crystal shape with pressure. In Fig. 11, the ratios of the growth rates at each pressure are plotted against the supersaturation, σ. The aspect ratio of the crystals along the c-axis was, indeed, drastically elongated as pressure increased. Thus, pressure, as well as supersaturation, is found to be one of the key parameters to control the crystal shape.

4. Concluding remarks

This chapter has introduced the present state of

protein crystallization under high pressure. So far, a lot of studies have been reported. Solubility, growth rate and nucleation rate of these studies depend on pressure, and the dependence of pressure on those differs from one protein to another. To study protein crystallization more precisely and systematically, we focused on the solubility and the growth kinetics of lysozyme crystal, since a lot of data under high pressure can be used and extensive studies on the growth kinetics at atmospheric pressure have already been done [13]. We have measured the dependence of temperature and pressure on solubility. The enthalpy and entropy of the dissolution (ΔH and ΔS) were calculated from van't Hoff plots. ΔH and ΔS of tetragonal crystal decreased with increasing pressure, while those of orthorhombic crystal were almost constant. The volume change accompanying the dissolution ΔV was negative for the tetragonal form, and positive for the orthorhombic form. The growth rate of tetragonal lysozyme crystal decreased with increasing pressure when it was plotted against the supersaturation σ at each pressure. The *birth and spread* model was applied to understand the pressure effects on growth kinetics. It was found that the increase in the average ledge surface energy of the two-dimensional nuclei with pressure explained the decrease in the growth rate.

This chapter has also introduced a precise in situ interferometric method of solubility measurement. High pressure measurements in combination with the in situ method will accelerate the study of protein crystallization under high pressure, because one plot of the high pressure solubility of lysozyme was determined within at most 3 h by our rapid method with a Michelson interferometer.

Needless to say, protein crystallization under high pressure is also useful for X-ray crystallography under high pressure [27,29,30]. The three-dimensional crystallographic structure of protein under high pressure should be useful to understand the enzymatic activity under high pressure [31], the pressure inactivation [32] or denaturation [33] of protein, and so on. The fundamental studies of protein crystallization under high pressure will be useful for such fields.

References

[1] M. Gross, R. Jaenicke, FEBS Lett. 284 (1991) 87–90.
[2] Y. Suzuki, S. Miyashita, H. Komatsu, K. Sato, T. Yagi, Jpn. J. Appl. Phys. 33 (1994) L1568–L1570.
[3] B. Lorber, G. Jenner, R. Giege, J. Cryst. Growth 158 (1996) 103–117.
[4] K.J. Takano, H. Harigae, Y. Kawamura, M. Ataka, J. Cryst. Growth 171 (1997) 554–558.
[5] G. Sazaki, Y. Nagatoshi, Y. Suzuki, S.D. Durbin, S. Miyashita, T. Nakada, H. Komatsu, J. Cryst. Growth 196 (1999) 204–209.
[6] Y. Suzuki, T. Sawada, S. Miyashita, H. Komatsu, G. Sazaki, T. Nakada, J. Cryst. Growth 209 (2000) 1018–1022.
[7] J.N. Webb, R.Y. Waghmare, J.F. Carpenter, C.E. Glatz, T.W. Randolph, J. Cryst. Growth 205 (1999) 563–574.
[8] R.Y. Waghmare, J.N. Webb, T.W. Randolph, M.A. Larson, C.E. Glatz, J. Cryst. Growth 208 (2000) 678–686.
[9] K. Visuri, E. Kaipainen, J. Kivimaki, H. Niemi, M. Leisola, S. Palosaari, Bio/Technology 8 (1990) 547–549.
[10] C.A. Schall, J.M. Wiencek, M. Yarmush, E. Arnold, J. Cryst. Growth 135 (1994) 548–554.
[11] M.V. Saikumar, C.E. Glatz, M.A. Larson, J. Cryst. Growth 151 (1995) 173–179.
[12] Y. Suzuki, S. Miyashita, G. Sazaki, T. Nakada, T. Sawada, H. Komatsu, J. Cryst. Growth 208 (2000) 638–644.
[13] F. Rosenberger, P.G. Vekilov, M. Muschol, B.R. Thomas, J. Cryst. Growth 168 (1996) 1–27.
[14] M. Gross, R. Jaenicke, Biophys. Chem. 45 (1993) 245–252.
[15] M.V. Saikumar, C.E. Glatz, M.A. Larson, J. Cryst. Growth 187 (1998) 277–288.
[16] E.H.P. Logtenberg, F. Meersman, P. Rubens, K. Heremans, J. Frank, High Press. Res. 19 (2000) 675–680.
[17] R.Y. Waghmare, X.J. Pan, C.E. Glatz, J. Cryst. Growth 210 (2000) 746–752.
[18] Y. Suzuki, T. Sawada, S. Miyashita, H. Komatsu, Rev. Sci. Instrum. 69 (1998) 2720–2724.
[19] G. Sazaki, K. Kurihara, T. Nakada, S. Miyashita, H. Komatsu, J. Cryst. Growth 169 (1996) 355–360.
[20] H. Komatsu, S. Miyashita, Y. Suzuki, Jpn. J. Appl. Phys. 32 (1993) L1855–L1857.
[21] S. Miyashita, H. Komatsu, Y. Suzuki, T. Nakada, J. Cryst. Growth 141 (1994) 419–424.
[22] K. Kurihara, S. Miyashita, G. Sazaki, T. Nakada, Y. Suzuki, H. Komatsu, J. Cryst. Growth 166 (1996) 904–908.
[23] S.D. Durbin, G. Feher, J. Cryst. Growth 76 (1986) 583–592.
[24] A. Nadarajah, E.L. Forsythe, M.L. Pusey, J. Cryst. Growth 151 (1995) 163–172.
[25] A.E. Nielsen, J.M. Toft, J. Cryst. Growth 67 (1984) 278–288.
[26] A.I. Malkin, A.A. Chernov, I.V. Alexeev, J. Cryst. Growth 97 (1989) 765–769.
[27] C.E. Kundrot, F.M. Richards, J. Mol. Biol. 193 (1987) 157–170.

[28] V.N. Morozov, T.Ya. Morozova, J. Biomol. Struct. Dyn. 11 (1993) 459–481.

[29] A. Katrusiak, Z. Dauter, Acta Crystallogr. Sect. D Biol. Crystallogr. 52 (1996) 607–608.

[30] R. Fourme, R. Kahn, M. Mezouar, E. Girard, C. Hoerentrup, T. Prange, I. Ascone, J. Synchrotron Radiat. 8 (2001) 1149–1156.

[31] S. Makimoto, H. Nishida, Y. Taniguchi, Biochim. Biophys. Acta 996 (1989) 233–239.

[32] Y. Taniguchi, K. Suzuki, J. Phys. Chem. 87 (1983) 5185–5193.

[33] K. Heremans, P.T.T. Wong, Chem. Phys. Lett. 118 (1985) 101–104.

ELSEVIER

Biochimica et Biophysica Acta 1595 (2002) 357–366

BIOCHIMICA ET BIOPHYSICA ACTA
BBA
www.bba-direct.com

Review

The use of high pressure for separation and production of bioactive molecules

Pierre Lemay *

INSA, Laboratoire de Biotechnologie et Bioprocédés, CNRS UMR5504, INRA UMR792, 135, Avenue de Rangueil, 31077 Toulouse Cedex 4, France

Received 31 August 2001; received in revised form 15 November 2001; accepted 15 November 2001

Abstract

Due to its action on the forces governing inter- and intramolecular interactions, the application of high pressure to biopurification or bio-elaboration of a product are of interest. The two closely thermodynamically related parameters, pressure and temperature, render processes based on their action clean, as no chemical reagents have to be added (and thus further removed) when they are applied. The use of high pressure in the development of desorption methods for the purification of bioactive molecules, particularly in the immunoaffinity field, is reviewed and discussed. Also mentioned is the application of the pressure parameter during the synthesis of a bioreagent. Finally, integrated processes relative to the synthesis and purification of these compounds are proposed.

Keywords: Immunoaffinity; Barodesorption; Chromatography; Bioseparation; Separation process; High pressure

1. Introduction

One of the most important challenges before using (or selling) a bioactive molecule is to purify it from the medium it is originating from. This step generally constitutes a bottleneck in the development of a process. Conventional purification procedures (e.g. molecular sieving, ion exchange chromatography etc.) are often cheap but have the great disadvantage not to be highly specific and must be repeated or combined to allow the attainment of highly pure products. By contrast, (immuno)affinity-based methods are more expensive, need reagents from biological origin but allow the rapid purification of products. Because these latter methods are highly specific, the interactions involved between the product and the support are often strong and imply the use of rather denaturing reagents (either for the product or the support) to attain an efficient desorption yield.

In this chapter, the application of high pressure to circumvent bioseparation problems, particularly in the affinity-based process field, will be discussed. Integrative procedures, including synthesis and purification of biologically active compounds, will also be addressed.

2. The fundamentals of (immuno)affinity-based separation

Biological recognition-based affinity chromatography was first described by Cuatrecasas and co-work-

* Fax: +33-5-61-55-94-00.
E-mail address: lemay@insa-tlse.fr (P. Lemay).

0167-4838/02/$ – see front matter
PII: S0167-4838(01)00356-9

ers in the late 60s [1,2]. Since this period, many reviews dealt with this subject: basic procedures for preparation and operation of affinity supports may be found in [3,4] for example.

The most widely used matrixes are agarose derivatives. These beads are rather sensitive to aggressive physical (particularly pressure and temperature) or chemical environments. In search for more resistant or inert matrixes, Vijayalakshmi [5] introduced the pseudobiospecific ligand affinity chromatography where inexpensive and robust substitutes (metal, dyes etc.) replace biospecific ligands. The design of polymerised imprinted molecules is reviewed in [6]; in this case, no ligand of biological origin is present on the support. Regarding the matrix, the grafting of ligands onto pressure-resistant silica supports was reported by the group of Mosbach [7,8]; also, a report specific to plane affinity filters was recently published by Kasper et al. [9].

2.1. Formation of affinity complexes

After the steric recognition between an antigen and its corresponding antibody, close intermolecular bonds stabilise the complex. These additional bonds are those currently occurring in inter- or intramolecular interactions ([10]; Boonyaratanakornkit et al., this issue), leading to binding energies ranging from 40 to 70 kJ mol^{-1}. The implementation of these bonds is accompanied by extrusion of ordered water (hydration water) from the complex [11]. The antigen–antibody complex formation is associated with a volume increase ranging from 150 to 800 ml mol^{-1}, in the case of polyclonal antibodies [12]. When haptens, such as the fluorescein molecule and various monoclonal antibodies form complexes, the volume change ranges from -12 to $+27$ ml mol^{-1} [13]. This is reminiscent of the variation of volume observed during dimer formation (50–150 ml mol^{-1}, reported by Silva and Weber [10]). These different behaviours might have some consequences on the development of baroprocesses (see below). The thermodynamic and the kinetic aspects of this reaction have also been reviewed (see [11,14], for example).

2.2. Disruption of affinity complexes

Many strategies have been proposed for antigen elution from immunoaffinity supports, such as extremes of pH, chaotropic salts, ionic strength and denaturants [15]. All these conditions lead to a rapid and irreversible loss of the properties of the immunoadsorbent. For example, a critical study of the long-term degradation mechanisms of six monoclonal antibodies grafted on a Sepharose 4B matrix by various chemical agents was achieved by Antonsen et al. [16]. In order to act on hydrophobic interactions some authors suggested the use of organic solvents (ethylene glycol, dioxane) combined with temperature control in a DNP-BSA anti-DNP model [17]. Nevertheless, in all cases the eluting reagent has to be further eliminated.

Physical methods have then been applied to overcome this drawback. The first application of electrophoresis to achieve antigen–antibody dissociation was reported by Grenot and Cuilleron [18] for the isolation of antibodies to steroid hormones from antigen-linked Sepharose. The major disadvantage was the slow release of antibodies and their high dilution. To remedy this point, isoelectrophoretic electrodesorption of monoclonal antibodies to transferrin linked to transferrin immobilised on polyethylene glycol tetraphthalate powder was performed by Kasicka and Prusik [19]. This method was developed on a nanoscale. Last, a review relative to principles, methodology, and applications of electrophoretic elution from biospecific adsorbents may be found in [20].

3. Baro-immunodesorption

A physical parameter susceptible to induce dissociation of affinity complexes is the application of high pressure to such entities. With regard to the nature of the bonds involved in the antigen–antibody reaction (see above) and the susceptibility of these bonds to pressure ([10]; Boonyaratanakornkit et al., this issue), it can be assumed that the application of high pressure combined with a defined temperature operating window would lead, for example, to the separation of an antigen from its corresponding antibody. Moreover, as mentioned earlier, the formation of affinity complexes is often accompanied by an increase in molecular volume. According to Le Châtelier's principle that governs the effects of pressure, application of pressure to these complexes favours the disruption of such entities. Briefly, if in a given physical environment a positive association volume is observed, then pressure-induced dissociation can occur and reciprocally.

The first report of baro-immunodesorption was based on the BSA–anti-BSA monoclonal filter-linked antibodies system [21]. A comparison between the stability of the immunoaffinity supports as a function of the elution procedure (i.e. high pressure or low pH) showed a greater stability of the affinity matrix after the application of pressures up to 200 MPa than after low pH treatments. Up to 90% of bound antigen were recovered after four cycles of pressure-induced desorption. Such immunodesorption experiments were performed with two other model systems: the prostate specific antigen and epiglycanin and their related antibodies [22]. In these cases, the antigen was adsorbed onto microtitration plates, and the amount of released antibodies was measured; here again, pressure acted favourably on the disruption of immune complexes.

The preceding methods dealt with monomeric antigens devoid of enzymatic activity. We thought to develop a baro-immunodesorption method whose recovery yield was based on the measure of an enzymatic activity. The tetrameric β-galactosidase of *Escherichia coli* was chosen as the model antigen. This enzyme is easily quantifiable, widely used in molecular cloning experiments, and often served as a tag in recombinant protein expression vectors. Moreover, its purification is easy either by conventional or affinity-based methods [23,24].

First the stability of the antigen under pressure was determined [25]. The binding ability of pressurised antibodies was assumed, either in a free state or bound to membranes [26]. These two latter determinations have to be done as others reported discrepancies, i.e. free or bound antibodies may behave differently with regard to pressure application [27]. The feasibility of the procedure was shown first with immunoaffinity membranes [28], and second with three-dimensional matrixes (i.e. agarose and silica beads) [29]. In addition, the recombinant β-galactosidase

Table 1
Recovery yields of various antigens following immuno-barodesorption experiments

Antigen	Antibody	Affinity matrix	Pressure (MPa)	Temperature (°C)	Recovery yield (%)	Ref.
^{125}I-BSA	Mouse monoclonal (homemade)	Cellulose membrane $Q_m = 0.435$ μg ml^{-1} $K = 2.4 \times 10^7$ M^{-1}	200	RT	90[a]	[21]
PSA	Mouse monoclonal (purchased)	Polystyrene microtiter plates	415	21	20	[22]
		Q_m = N.A. K = N.A.	550	21	30	
Chimerical *E. coli* β-galactosidase	Rabbit polyclonal (homemade)	Polyamide $Q_m = 0.44$ μg cm^{-2} $K = 0.4 \times 10^8$ M^{-1}	250	4	17.1 (13.5)[b]	[28,29]
		Cross-linked acrylamide/agarose $Q_m = 0.03$ μg mg^{-1} $K = 5.2 \times 10^8$ M^{-1}	250	4	13.4 (12.5)[b]	
		Silica beads $Q_m = 0.71$ μg mg^{-1} $K = 1.8 \times 10^8$ M^{-1}	250	4	25.5 (19.5)[b]	

BSA and PSA refer to bovine serum albumin and prostatic antigen respectively. RT stands for room temperature and N.A. for not available. Q_m corresponds to the maximal capacity of a matrix in terms of antigen binding ability. K represents the association constant.

[a]After 4 cycles of pressurisation.

[a]After 4 cycles of pressurisation and in terms of enzymatic activity; numbers in parentheses correspond to the value obtained with a high pH elution method; as a comparison, a 53% yield was obtained with a tag property-based elution method [30].

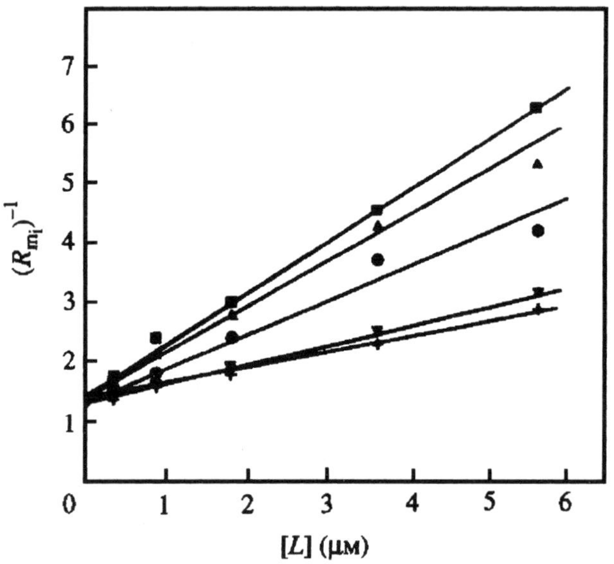

Fig. 1. Affinity electrophoresis. Affinity plot of relative mobility of protein (R_{mi} is the migration distance in the separating gel related to the tracking dye band) against the concentration of ligand ([L]) for the interactions between bovine serum albumin and an immobilised derivative of stearic acid at 35°C under high pressure. ■, atmospheric pressure; ▲, $P = 35$ MPa; ●, $P = 70$ MPa; ▼, $P = 120$ MPa; +, $P = 180$ MPa. The intercepts with the x-axis give the apparent dissociation constants of the complexes. (Reprinted from [31], with author's permission.)

used possesses a LytA tag, which allowed its fixation on a DEAE-cellulose resin; the high pressure process might then be compared to a tag-based desorption procedure [30]. Several examples of recovery yields from various immunodesorption experiments can be found in Table 1.

Protein–ligand interactions were also studied by affinity electrophoresis under pressure, and reviewed by Masson [31]. Many useful theoretical and experimental bases can be found in this paper. This method is funded on the retardation of migration of molecules in an electrophoretic field due to their interaction with ligands immobilised within the separating matrix. This can be carried out either on cylindrical or on slab gels. An illustration of the determination of the apparent dissociation constant of complexes formed between bovine serum albumin and an immobilised derivative of stearic acid is shown in Fig. 1.

All the above presented methods are fundamentally discontinuous processes and need many manual operations. Two groups have proposed apparatuses allowing high pressure bioseparations in a continuous way. The first one was tested on phage λDNA preparation, the separation of the RNA part of RNA/DNA hybrids, and the purification of an antibody following affinity capture by its corresponding antigen [32]. This apparatus is presented in Fig. 2. The second one is a versatile plant whose functioning is based on percolation of exchangeable cartridges. The design allows to perform bioseparations, but

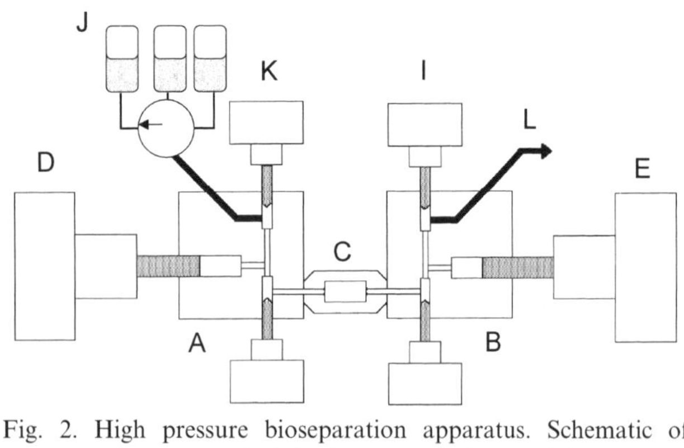

Fig. 2. High pressure bioseparation apparatus. Schematic of system and actuator. A, injection pump/valve module; B, drain pump/valve module; C, reaction chamber; D, injection pump pneumatic cylinder; E, drain pump pneumatic cylinder; F, injection inlet valve pneumatic cylinder; I, drain outlet valve pneumatic cylinder; J, reagent reservoirs; K, reagent select valve; L, effluent port. (Reprinted from [32], with permission.)

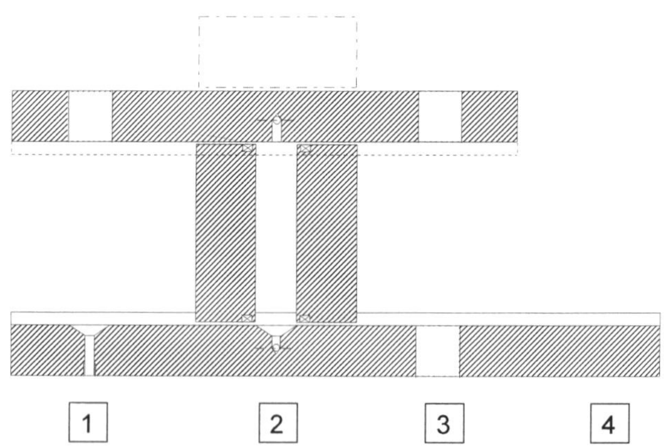

Fig. 3. Semi-continuous high pressure plant. This plant is computer driven and might be used either for (bio)synthesis and/or bioseparation processes. Its conception is based on a moving high pressure reactor containing removable cartridges. The reaction cylinder may occupy four positions permitting low (pos. 1), high percolation or reaction operations (pos. 2), change of cartridge (pos. 3), and maintenance (pos. 4). (Modified from [33].)

also reactions of (bio)synthesis (see below) [33]. This plant is shown in Fig. 3. The two devices are driven by LabView or LabWindows based computer software.

4. How to enhance the performance of baro-immunoaffinity separation?

In order to enhance the performance, in terms of specificity, safety, and yield, of a baro-immunodesorption process, one can act at different levels: the susceptibility of antigens to pressure, the generation of antibodies from non-animal origin, the development of new affinity matrixes, and the selection of (bioengineered) antibodies.

4.1. Susceptibility of antigens and antibodies to pressure

The first studies related to the influence of pressure on antigens and antibodies were performed on anaphylaxis, antitetanic and antidiphtherial models as early as 1933 [34,35]. An approach combining fluorescence (to detect protein conformational changes), FTIR (to determine secondary structure alterations), and ELISA (to check antibody recognition ability) was suggested to screen antigens and antibodies as to their usability in an immunodesorption process [36]. On the opposite, the application of high pressure was shown to speed the antigen–antibody reaction [37]; this would be of interest for the clinical diagnosis of the presence of HIV-1 p24, for example [37]. Also, the adjunction of a metatype antibody (which recognises antibodies in a liganded state) to immune complexes stabilises this latter against the action of pressure [38].

Besides the classical immunisation methods of animals with whole antigens, one of the 'oldest' ways to get specific antibodies looking like monoclonal ones is to raise antibodies against synthetic peptides after having determined potential antigenic sites on a given protein [39]. The great advantage is that this protein has not to be previously isolated; moreover, it may be a putative one revealed by a DNA coding sequence. On the opposite the generated antibody could either be only sequence or structure specific and then could not recognise the original whole native antigen or recognise many others.

4.2. Generation of antibodies from non-animal origin

Due to the possibility of contamination by pathogens or toxic agents some people are reluctant to use compounds of animal origin in the elaboration, production or purification of molecules, particularly in food or pharmaceutical science. To overcome this point, a recent approach consists in the selection of antibodies from phage display libraries. Strategies to design and optimise library selection have been reviewed and strategic flow charts have been elaborated [40]. Antibodies can also be produced in plants. Formerly developed to study plant metabolism or alter the expression of phytohormones (abscisic acid or gibberellin), the expression of antibodies (plantibodies) in plants presents many advantages: no use of high cost microbial fermentation techniques, no use of mammals, post-translational modifications achieved as in eukaryotic cells [41]. This latter property has to be emphasised, as it has been shown that the effects of glycosylation, for example, are of importance with regard to antibody function [42]. Finally, if 'clean' antibodies have to be obtained, immunoglobulins can be selected particularly with staphylococcal protein A-funded methods [43,44] or with peptides binding to the Fc portion and originating from a phage display peptide library [45]. Nevertheless, it should be kept in mind that these antibodies would obey to a pressure sensitivity criterion with regard to their barodesorption usability [21,26,46].

4.3. Development of new affinity matrixes

An improvement would also be found with the development of fully synthetic antibodies: the molecular imprinted polymers (MIP) [47]. These matrixes present physical as well as chemical stability, they resist storage and give recovery yields as high as 99% [48,49]. No living organisms are involved in their preparation, thus they may be considered as safe supports. Their properties render them of interest in view of the conception of new immuno-barodesorption processes.

4.4. Selection of (bioengineered) antibodies and matrixes for baroprocesses

The affinity constants, and also the kinetic parameters, have to be determined before using antibodies in a biotechnological process. Many biophysical methods for the determination of these constants, such as equilibrium dialysis, band shift, competition ELISA, fluorescence quench, isothermal titration calorimetry, or real-time interaction analysis, have been discussed [14]. A simple technique based on the Mancini immunodiffusion technique was proposed by Kennedy and Barnes [50] and applied to the selection of eluents for immunoadsorptive chromatography. The application of affinity electrophoresis in this field is of interest, as already mentioned [31]. As shown in Table 1, association constants in the range of 10^7–10^8 M^{-1} are suitable for barodesorption processes. All these determinations have to be made with respect to the reaction media; for example, another parameter involved in the sensibility of antigen–antibody complexes is ionic strength. A critical review on the combined effects of temperature, ionic strength and pressure on several antigen–antibody complexes is available [46]. A detailed analysis of the evolution of association volumes, salt bridges, aromatic stacking as a function of the physical parameters applied, as well as a thermodynamic analysis were reported [46]. All the above mentioned gaits would lead to the achievement of highly specific, pressure-sensitive antibodies (in terms of biodesorption).

5. New trends in bioproduction, biopurification and bioseparation

Until now mainly purification methods involving antigen–antibody recognition were evoked. The application of high pressure can also be considered in integrative schemes regarding the synthesis, production and purification of bioactive molecules.

5.1. High pressure-assisted design of bioactive molecules

Many reviews dealt with the stability of proteins as a function of pressure [10,51,52] and particularly with that of enzymes (see for example [53–58]). The affinity for substrate, reaction rates, and specificity of enzymes vary as a function of temperature, pressure, and solvent, thus it is advisable to use high pressure in the design or achievement of particular molecules. This can be achieved either with biocatalysts of mesophile [59] or extremophile [60] origin. The use of hyperthermophilic enzymes [61], as well as the characterisation of deep-sea microorganisms have been reviewed (see for example [62,63]). Studies have been done to explain the properties of thermozymes [64]. On the opposite, there also exist cold adapted enzymes of potential biotechnological interest [65]. Some other compounds such as natural catalytic monoclonal antibodies (abzymes) [66], or molecularly imprinted materials leading to synthesis or catalysis (plastizymes) [67,68] could have their activity(ies) modified or improved under high pressure. It should be kept in mind that the application of high pressure alters the specificity of organic reactions, leading generally to asymmetric syntheses [69]. This would be advantageous if this property was used in conjunction with the above mentioned biocatalysts to obtain a specific stereo-isomer of a given bioactive product.

5.2. Biopurification and bioseparation under high pressure

Biopurification can result from a direct precipitation of a protein from a crude extract, particularly with the use of antibodies. A review on immunoextraction applications was recently published [70]. It can be assumed that the use of high pressure to disrupt immune complexes thus obtained is of interest (see above).

Besides the immunochromatographic techniques previously described, there are more 'classical' methods of affinity chromatography separation, the performance of which can be improved by the application of high pressure as an elution protocol, for example the elution of proteins separated on the basis of their affinity for DEAE groups [30], dextran supports [71], metal affinity chromatography [72], or simple hydrophobic interactions [73].

Some new pressure-resistant chromatographic supports seem to be promising. These are non-porous copolymerised particles of styrene, methyl methacrylate and glycidyl methacrylate [74] or molecularly imprinted polymers (already successfully used in the separation of amino acids, peptides, proteins, receptors, and enantioselection of drugs) [75–78]. The precise understanding of the nature of recognition between molecular imprinted polymers and their 'substrates' [79], the rules governing the enantioselectivity of bioseparation [80], and the definition of the kinetic parameters involved in affinity interactions [81] would be very useful to enhance the performance in the bioseparation field by adding the potential positive effect of pressure to their intrinsic properties.

Finally, the ultimate stage required for determining the three-dimensional structure of a molecule is its obtention in a crystallised state. Here again, pressure was shown to act on the nucleation and growth of protein crystals (Suzuki et al., this issue; [82]).

5.3. Integrated processes

In the conception of a process leading to the production of a bioactive molecule under high pressure,

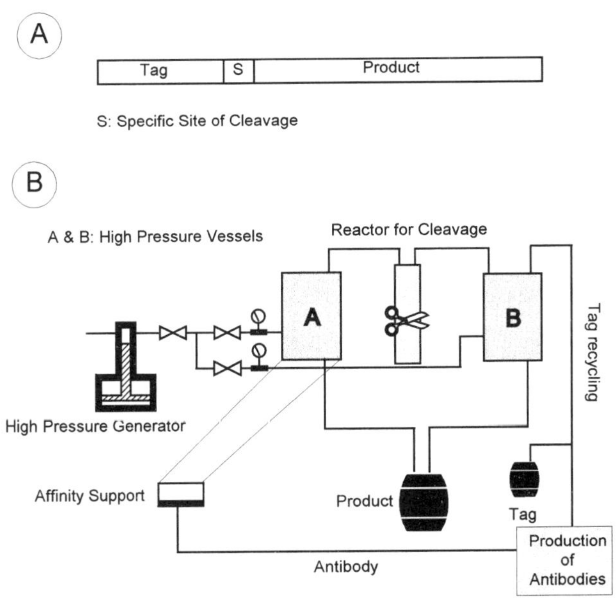

Fig. 4. Example of a biopurification process allowing the total recycling of all components. (A) Scheme of an expression vector comporting an immunoaffinity tag, a specific site of chemical or enzymatic cleavage and the product of interest. (B) Diagram showing a possible process allowing the recovery of the fused protein, the product, and the tag. All the steps might be done under high pressure. The antibodies are directed against the tag moiety of the fusion protein. In any case, it should be kept in mind that reusability of the various interveners is a fundamental criterion for their selection.

many things must be kept in mind (e.g. non-specific adsorption of proteins to matrixes [83]). In all cases, it is better to use immobilised reactants (i.e. enzymes are usually more active and stable when immobilised on a matrix [84–87]); they are also more convenient to proceed.

From the preceding, one can summarise as follows. Pressure may act at different steps in the development of integrated processes.

Elaboration of the product

- During in vitro synthesis by assuming enantioselectivity of organic synthesis, a modulation or modification of the specificities of natural (meso-, baro-, thermo-, psychro-) originating enzymes or abzymes, a modulation or modification of the properties of synthetic catalysts (plastizymes etc.).
- During in vivo production in bacteria, yeast, animals and plants, by operating on the metabolism of natural or engineered products, or allowing an overproduction of recombinant product (barosensitive promoters etc.).

Purification of the product

- Facilitation of crystallisation steps.
- Non-polluting way to assume desorption from ((immuno)affinity) matrixes.

A scheme presenting a process where all the compounds (including satellite products) may be recycled is presented in Fig. 4.

6. Concluding remarks

The contribution of the pressure parameter evidenced in the preceding example shows the great potentiality for the use of high pressure in the elaboration and purification of bioactive molecules. The effects of pressure on the growth of microorganisms and/or their inactivation, particularly those of viruses (i.e. [88,89]; Silva, this issue), were deliberately not mentioned in this chapter. These latter properties would speak in favour of baroprocesses for the production and purification of molecules of therapeutic or food interest either for man or animals.

Nevertheless, no predictive method with regard to the behaviour of biomolecules (in terms of function or interaction) under high pressure does exist, and this promises many happy years for the future in the field.

Acknowledgements

The helpful aid of Mrs Iris Marcos in the preparation of this manuscript is gratefully acknowledged. Drs C. Balny, J. Frank, K. Heremans, P. Masson, D. Combes and G. Goma brought their constant support during the evolution of the work done in my laboratory. Finally, I wish to thank Drs P. Degraeve, G. Demazeau, and A. Largeteau for their friendly and efficient collaboration. The work done in Toulouse was supported by grants from CNRS, INSA, INRA, ANVAR and the Région Midi-Pyrénées.

References

[1] P. Cuatrecasas, M. Wilchek, Single-step purification of avidin from egg white by affinity chromatography on biocytin-Sepharose columns, Biochem. Biophys. Res. Commun. 33 (1968) 235–239.

[2] P. Cuatrecasas, M. Wilchek, C.B. Anfinsen, Selective enzyme purification by affinity chromatography, Proc. Natl. Acad. Sci. USA 61 (1968) 636–643.

[3] M. Wilchek, T. Miron, J. Kohn, Affinity chromatography, Methods Enzymol. 104 (1984) 3–55.

[4] M.L. Yarmush, A.M. Weiss, K.P. Antonsen, D.J. Odde, D.M. Yarmush, Immunoaffinity purification: basic principles and operational considerations, Biotechnol. Adv. 10 (1992) 413–446.

[5] M.A. Vijayalakshmi, Pseudobiospecific ligand affinity chromatography, 7 (1989) 71–76.

[6] B. Ekberg, K. Mosbach, Molecular imprinting: a technique for producing separation materials, Trends Biotechnol. 7 (1989) 92–96.

[7] L. Ye, R. Weiss, K. Mosbach, Synthesis and characterization of molecularly imprinted microspheres, Macromolecules 33 (2000) 8239–8245.

[8] L. Ye, P.A.G. Cormack, K. Mosbach, Molecular imprinting on microgel spheres, Anal. Chim. Acta 435 (2001) 187–196.

[9] C. Kasper, O.W. Reif, R. Freitag, Evaluation of affinity filters for protein isolation, Bioseparation 6 (1996) 373–382.

[10] J. Silva, G. Weber, Pressure stability of proteins, Annu. Rev. Phys. Chem. 44 (1993) 89–113.

[11] D.R. Absolom, C.J. van Oss, The nature of the antigen-antibody bond and the factors affecting its association and dissociation, CRC Crit. Rev. Immunol. 6 (1986) 1–46.

[12] Y. Otha, T.J. Gill, C. Leung, Volume changes accompanying the antigen-antibody reaction, Biochemistry 9 (1970) 2708–2713.

[13] J.N. Herron, D.M. Kranz, D.M. Jameson, E.W. Voss Jr., Thermodynamic properties of ligand binding by monoclonal anti-fluoresceyl antibodies, Biochemistry 25 (1986) 4602–4609.

[14] D. Neri, S. Montagiani, P.M. Kirkham, Biophysical methods for the determination of antibody-antigen affinities, Trends Biotechnol. 14 (1996) 465–470.

[15] M.L. Yarmush, K.P. Antonsen, S. Sundaram, D.M. Yarmush, Immunoadsorption: strategies for antigen elution and production of reusable adsorbents, Biotechnol. Prog. 8 (1992) 168–178.

[16] K.P. Antonsen, C.K. Colton, M.L. Yarmush, Elution conditions and degradation mechanisms in long term immunoadsorbent use, Biotechnol. Prog. 7 (1991) 159–172.

[17] K. Andersson, Y. Benyamin, P. Douzou, C. Balny, Organic solvents and temperature effects on desorption from immunoadsorbents, DNP-BSA anti-DNP as a model, J. Immunol. Methods 23 (1978) 17–21.

[18] C. Grenot, C.Y. Cuilleron, Isolation of antibodies to steroid hormones by affinity chromatography on antigen-linked sepharose: an efficient electrophoretic elution procedure, Biochem. Biophys. Res. Commun. 79 (1977) 274–279.

[19] V. Kasicka, Z. Prusick, Isotachophoretic electrodesorption of proteins from an affinity adsorbent on a microscale, J. Chromatogr. 273 (1983) 117–128.

[20] M.L. Yarmush, W.C. Olson, Electrophoretic elution from biospecific adsorbents: principles, methodology, and applications, Electrophoresis 9 (1988) 111–120.

[21] W.C. Olson, S.K. Leung, M.L. Yarmush, Recovery of antigens from immunoadsorbents using high pressure, Biotechnology 7 (1989) 369–373.

[22] C.Y. Cheung, D.J. Green, G.J. Litt, J.A. Laugharn Jr., High-pressure-mediated dissociation of immune complexes demonstrated in model systems, Clin. Chem. 44 (1998) 299–303.

[23] J.H. Miller, Purification of β-galactosidase, in: J.H. Miller (Ed.), Experiments in Molecular Genetics, Cold Spring Harbor Laboratory Press, CSH, New York, 1972, pp. 398–404.

[24] E. Steers Jr., P. Cuatrecasas, H.B. Pollard, The purification of β-galactosidase from *Escherichia coli* by affinity chromatography, J. Biol. Chem. 246 (1971) 196–200.

[25] I. Heiber-Langer, N. Banzet, J.-L. Saldana, P. Lemay, C. Balny, Thermodynamic study of β-galactosidase for high pressure separation processes, Biotechnol. Tech. 7 (1993) 243–247.

[26] E. Gavalda, P. Degraeve, P. Lemay, High-pressure-induced modulation of the antigenic interactions between 2 β-galactosidases and anti-β-galactosidase antibodies, Enzyme Microb. Technol. 18 (1996) 10–17.

[27] S. Sundaram, C.M. Roth, M.L. Yarmush, Pressure-induced dissociation of antigen-antibody complexes, Biotechnol. Prog. 14 (1998) 773–781.

[28] L. Estevez, P. Degraeve, S. Espeillac, P. Lemay, High-pressure induced recovery of β-galactosidases from immunoad-

sorbents: stability of antigens and antibodies. Comparison with usual elution procedures, Biotechnol. Lett. 22 (2000) 1319–1329.

[29] L. Estevez-Burugorri, P. Lemay, Baro-immunodesorption: development of new affinity supports suitable for semi-continuous processes. Application to the purification of recombinant *E. coli* β-galactosidase, and comparison with a conventional purification method, High Press. Res. 19 (2000) 291–296.

[30] J.M. Sanchez-Puelles, J.M. Sanz, J.L. Garcia, E. Garcia, Immobilization and single-step purification of fusion proteins using DEAE-cellulose, Eur. J. Biochem. 203 (1992) 153–159.

[31] P. Masson, Electrophoresis of proteins under high hydrostatic pressure, in: W.H. Holzapfel, N.S. Isaacs (Eds.), High-pressure Techniques in Chemistry and Physics, Oxford University Press, Oxford, 1997, pp. 353–373.

[32] G.H. Dreier, F. Tao, R.A. Hess, C.Y. Cheung, L.E. Sciaba, D.J. Green, J.A. Laugharn Jr., A bioseparation apparatus with high-pressure fluid injection and fluid sampling, Anal. Biochem. 269 (1999) 223–229.

[33] G. Demazeau, A. Largeteau, P. Lemay, Conception et développement d'un bioréacteur hyperbar. Applications en bioséparations et en catalyse enzymatique, in: D. Tondeur (Ed.), Génie des Procédés Biotechnologiques, Centre de Perfectionnement des Industries Chimiques, Nancy, 2000, pp. 21–24.

[34] J. Basset, M. Macheboeuf, J.J. Perez, Etudes sur les effets biologiques des ultra-pressions. Modification de la spécificité antigénique des sérums sous l'influence des pressions très élevées, C. R. Acad. Sci. 200 (1935) 496–498.

[35] J. Basset, M. Macheboeuf, Etudes sur les effets biologiques des ultra-pressions. Etudes de l'immunité: influence des pressions très élevées sur certains antigènes et anticorps, C. R. Acad. Sci. 196 (1933) 67–69.

[36] J.R. Howlett, A.A. Ismail, D.W. Armstrong, P.T.T. Wong, Pressure-induced conformational changes in an antigen and an antibody and the implications on their use for hyperbaric immunoadsorption, Biochim. Biophys. Acta 1159 (1992) 227–236.

[37] D.J. Green, G.J. Litt, J.A. Laugharn Jr., Use of high pressure to accelerate antigen: antibody binding kinetic demonstrated in an HIV-1 p24: anti-HIV-1 p24 assay, Clin. Chem. 44 (1998) 341–342.

[38] T. Coelho-Sampaio, E.W. Woss Jr., Anti-metatype antibodies stabilize the fluorescein single-chain antibody 4-4-20 complex against dissociation by hydrostatic pressure, J. Biol. Chem. 269 (1994) 8146–8152.

[39] P.S. Stern, Predicting antigenic sites on proteins, Trends Biotechnol. 9 (1991) 163–169.

[40] H.R. Hoogenboom, Designing and optimizing library selection strategies for generating high-affinity antibodies, Trends Biotechnol. 15 (1997) 62–70.

[41] M.D. Smith, B.R. Glick, The production of antibodies in plants: an idea whose time has come?, Biotechnol. Adv. 18 (2000) 85–89.

[42] A. Wright, S.L. Morrison, Effect of glycosylation on antibody function: implications for genetic engineering, Trends Biotechnol. 15 (1997) 26–32.

[43] J.J. Langone, Applications of immobilized protein A in immunochemical techniques, J. Immunol. Methods 55 (1982) 277–296.

[44] J.J. Langone, Protein A of *Staphylococcus aureus* and related immunoglobulin receptors produced by streptococci and pneumococci, Adv. Immunol. 32 (1982) 157–252.

[45] M. Krook, K. Mosbach, K.O. Ramstrom, Novel peptides binding to the Fc-portion of immunoglobulins obtained from a combinatorial phage display peptide library, J. Immunol. Methods 221 (1998) 151–157.

[46] P. Roy, C.M. Roth, M.N. Margolies, M.L. Yarmush, Effect of pressure on antigen-antibody complexes: modulation by temperature and ionic strength, Mol. Immunol. 36 (1999) 1149–1158.

[47] O. Ramstrom, L. Ye, K. Mosbach, Artificial antibodies to corticosteroids prepared by molecular imprinting, Chem. Biol. 3 (1996) 471–477.

[48] R.J. Ansell, O. Ramstrom, K. Mosbach, Towards artificial antibodies prepared by molecular imprinting, Clin. Chem. 42 (1996) 1506–1512.

[49] K. Haupt, K. Mosbach, Plastic antibodies: developments and applications, Trends Biotechnol. 16 (1998) 468–475.

[50] J.F. Kennedy, J.A. Barnes, Affinity constants in immunoadsorptive chromatography: their simple determination and application to the selection of eluents in and the optimization of elution process, J. Chromatogr. 205 (1981) 31–37.

[51] K. Heremans, L. Smeller, Protein structure and dynamics at high pressure, Biochim. Biophys. Acta 1386 (1998) 353–370.

[52] K.E. Prehoda, E.S. Mooberry, J.L. Markley, High pressure effects on protein structure, in: Jardetzky et al. (Eds.), Protein Dynamics, Function, and Design, Plenum Press, New York, 1998, pp. 59–86.

[53] V.V. Mozhaev, R. Lange, E.V. Kudryashova, C. Balny, Application of high hydrostatic pressure for increasing activity and stability of enzymes, Biotechnol. Bioeng. 52 (1996) 320–331.

[54] V.V. Mozhaev, Mechanism-based strategies for protein thermostabilization, Trends Biotechnol. 11 (1993) 88–95.

[55] P. Degraeve, P. Delorme, P. Lemay, High hydrostatic pressure-induced deactivation of *E. coli* β-galactosidase: influence of pH and temperature, Biochim. Biophys. Acta 1292 (1996) 61–68.

[56] P. Degraeve, P. Lemay, High-pressure-induced modulation of the activity and stability of *Escherichia coli* (lacZ) β-galactosidase: potential applications, Enzyme Microb. Technol. 20 (1997) 550–557.

[57] V. Athes, R. Lange, D. Combes, Influence of polyols on the structural properties of *Kluyveromyces lactis* β-galactosidase under high hydrostatic pressure, Eur. J. Biochem. 255 (1998) 206–212.

[58] M. Noel Marie-Olive, V. Athes, D. Combes, Combined effects of pressure and temperature on enzyme stability, High Press. Res. 19 (2000) 317–322.

[59] V.V. Mozhaev, K. Heremans, J. Frank, P. Masson, C. Balny, Exploiting the effects of high hydrostatic pressure in biotechnological applications, Trends Biotechnol. 12 (1994) 493–501.
[60] R.A. Herbert, A perspective on the biotechnological potential of extremophiles, Trends Biotechnol. 10 (1992) 395–402.
[61] M.W. Adams, R.M. Kelly, Finding and using hyperthermophilic enzymes, Trends Biotechnol. 16 (1998) 329–332.
[62] D. Prieur, Microbiology of deep-sea hydrothermal vents, Trends Biotechnol. 15 (1997) 242–244.
[63] C. Kato, A. Inoue, K. Horikoshi, Isolating and characterizing deep-sea marine microorganisms, Trends Biotechnol. 14 (1996) 6–12.
[64] C. Vieille, J.G. Zeikus, Thermozymes: identifying molecular determinants of protein structural and functional stability, Trends Biotechnol. 14 (1996) 183–190.
[65] C.J. Marshall, Cold-adapted enzymes, Trends Biotechnol. 15 (1997) 359–364.
[66] B.S. Green, D.S. Tawfik, Catalytic monoclonal antibodies: taylor-made, enzyme-like catalysts for chemical reactions, Trends Biotechnol. 7 (1989) 304–310.
[67] K. Mosbach, Towards the next generation of molecular imprinting with emphasis on the formation, by direct molding, of compounds with biological activity (biomimetics), Anal. Chim. Acta 435 (2001) 3–8.
[68] O. Ramstrom, K. Mosbach, Synthesis and catalysis by molecularly imprinted materials, Curr. Opin. Chem. Biol. 3 (1999) 759–764.
[69] G. Jenner, High pressure and selectivity in organic reactions, Tetrahedron 53 (1997) 2669–2695.
[70] N. Delaunay-Bertoncini, V. Pichon, M.C. Hennion, Immunoextraction: a highly selective method for sample preparation, LCGC Eur. 14 (2001) 162–172.
[71] K. Kaseda, T. Kodama, K. Fukui, K. Hirose, A novel approach for purification of recombinant proteins using the dextran-binding domain, FEBS Lett. 500 (2001) 141–144.
[72] D.F. Westra, G.W. Welling, D.G.A.M. Koedijk, A.J. Scheffer, T.H. The, S. Welling-Wester, Immobilised metal-ion affinity chromatography purification of histidine-tagged recombinant proteins: a wash step with a low concentration of EDTA, J. Chromatogr. B Biomed. Sci. Appl. 760 (2001) 129–136.
[73] R. Ghosh, Separation of proteins using hydrophobic interaction membrane chromatography, J. Chromatogr. A 923 (2001) 59–64.
[74] C.-H. Chen, W.-C. Lee, Affinity chromatography of proteins on non-porous copolymerized particles of styrene, methyl methacrylate and glycidyl methacrylate, J. Chromatogr. A 921 (2001) 31–37.
[75] R.J. Ansell, D. Kriz, K. Mosbach, Molecularly imprinted polymers for bioanalysis: chromatography, binding assays and biomimetic sensors, Curr. Opin. Biotechnol. 7 (1996) 89–94.
[76] M. Kempe, K. Mosbach, Separation of aminoacids, peptides and proteins on molecularly imprinted stationary phases, J. Chromatogr. A 691 (1995) 317–323.
[77] O. Ramstrom, C. Yu, K. Mosbach, Chiral recognition in adrenergic receptor binding mimics prepared by molecular imprinting, J. Mol. Recognit. 6 (1996) 691–696.
[78] I.A. Nichols, L.I. Anderson, K. Mosbach, B. Ekberg, Recognition and enantioselection of drugs and biochemicals using molecularly imprinted polymer technology, Trends Biotechnol. 13 (1995) 47–51.
[79] H.S. Andersson, A.C. Koch-Schmidt, S. Ohlson, K. Mosbach, Study of the nature of recognition in molecularly imprinted polymers, J. Mol. Recognit. 9 (1996) 675–682.
[80] C. Yu, K. Mosbach, Influence of mobile phase composition and cross-linking density on the enantiomeric recognition properties of molecularly imprinted polymers, J. Chromatogr. A 888 (2000) 63–72.
[81] M.L. Yarmush, D.B. Pantakar, D.M. Yarmush, An analysis of transport resistances in the operation of BIAcore; implications for kinetics studies of biospecific interactions, Mol. Immunol. 33 (1996) 1203–1214.
[82] B. Lorber, G. Jenner, R. Giegé, Effect of high hydrostatic pressure on nucleation and growth of protein crystals, J. Cryst. Growth 158 (1996) 103–117.
[83] M. Wahlgren, T. Anerbrant, Protein adsorption to solid surfaces, Trends Biotechnol. 9 (1991) 201–208.
[84] E. Katchalski-Katzir, Immobilized enzymes-learning from past successes and failures, Trends Biotechnol. 11 (1993) 471–478.
[85] B. Solomon, E. Hadas, R. Koppel, F. Schwartz, G. Fleminger, Highly active enzyme preparations immobilized via matrix-conjugated anti-Fc antibodies, J. Chromatogr. 539 (1991) 335–341.
[86] L. Ye, O. Ramstrom, R.J. Ansell, M.O. Mansson, K. Mosbach, Use of molecularly imprinted polymers in a biotransformation process, Biotechnol. Bioeng. 64 (1999) 650–655.
[87] O. Bruggemann, K. Haupt, L. Ye, E. Yilmaz, K. Mosbach, New configurations and applications of molecularly imprinted polymers, J. Chromatogr. A 889 (2000) 15–24.
[88] S.-M. Tian, K.-C. Ruang, J.-F. Qian, G.-Q. Shao, C. Balny, Effects of hydrostatic pressure on the structure and biological activity of infectious bursal disease virus, Eur. J. Biochem. 267 (2000) 4486–4494.
[89] D.W. Bradley, R.A. Hess, F. Tao, L. Sciaba-Lentz, A.T. Remaley, A.J. Laugharn Jr., M. Manak, Pressure cycling technology: a novel approach to virus inactivation in plasma, Transfusion 40 (2000) 193–200.

ELSEVIER

Biochimica et Biophysica Acta 1595 (2002) 367–381

BIOCHIMICA ET BIOPHYSICA ACTA
BBA
www.bba-direct.com

Review

Pressure effects on in vivo microbial processes

D.H. Bartlett *

Center for Marine Biotechnology and Biomedicine, Marine Biology Research Division, Scripps Institution of Oceanography, University of California, San Diego, 8682 La Jolla Shores Drive, La Jolla, CA 92093-0202, USA

Received 25 September 2001; received in revised form 9 November 2001; accepted 12 November 2001

Abstract

Pressures between 10 and 100 MPa can exert powerful effects on the growth and viability of organisms. Here I describe the effects of elevated pressure in this range on mesophilic (atmospheric pressure adapted) and piezophilic (high-pressure adapted) microorganisms. Examination of pressure effects on mesophiles makes use of this unique physical parameter to aid in the characterization of fundamental cellular processes, while in the case of piezophiles it provides information on the essence of the adaptation of life to high-pressure environments, which comprise the bulk of our biosphere. Research is presented on the isolation of pressure-resistant mutants, high-pressure regulation of gene expression, the role of membrane lipids and proteins in determining growth ability at high pressure, pressure effects on DNA replication and topology as well as on cell division, and the role of extrinsic factors in modulating enzyme activity at high pressure.

Keywords: Deep sea; Extrinsic factor; High pressure adaptation; Piezophile; Pressure-altered mutant; Pressure regulated gene

1. Introduction

Pressure is a key physical parameter which has influenced the evolution and distribution of both microorganisms and macroorganisms [1,2]. The oceans, have an average depth of 3800 m and thus an average pressure of 38 megapascal (1 atmosphere ≈0.101 megapascal, MPa) as well as a maximal depth of approximately 11 000 m (~110 MPa). Additional high-pressure environments include deep lakes and the deep subsurface regions. The former includes Lake Baikal in Siberia and Lake Vostok located 3–4 km beneath the East Antarctic Ice Sheet (see for example, [3]). Subsurface communities of microorganisms have been detected as deep as 3500 m below the surface [4]. It is predicted that the largest number of prokaryotes in the biosphere are likely to reside in this realm below our feet [5]. Finally, while astrobiologists ponder the possibility of life on Jupiter's moon Europa [6], it is interesting to note that such life could be existing at pressures approximately twice that found in the deepest ocean trench on earth.

Pressures of different magnitudes exert different effects on organisms (Fig. 1). The xylem within trees can be at a negative pressure down to −2 MPa, which is necessary for water transport [7]. Planktonic crustaceans can display responses to pressure changes of less than 1 kPa [8]. Many fishes have evolved physiological responses to changes in hydrostatic pressure which influence the concentration of gases within a swim bladder and hence buoyancy [9].

* Fax: +1-858-534-7313.
E-mail address: dbartlett@ucsd.edu (D.H. Bartlett).

0167-4838/02/$ – see front matter
PII: S0167-4838(01)00357-0

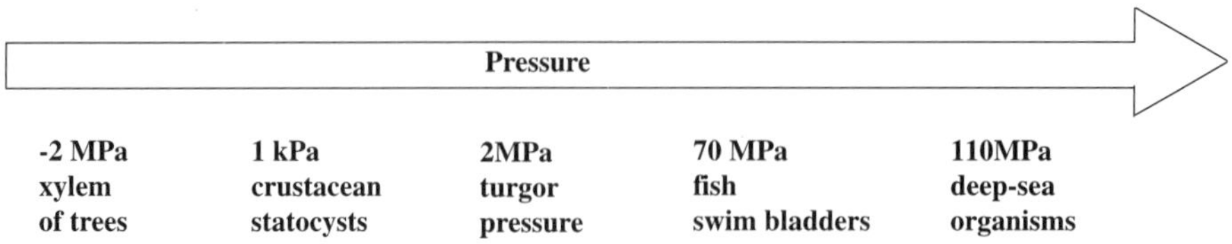

Fig. 1. Range of pressures present in various biological systems.

Turgor pressure must be sensed and regulated, and in microorganisms with large cell walls such as Gram-positive bacteria and certain species of algae it can be as high as 2 MPa [10]. Clearly the ability to adapt to pressure changes of one kind or another is a characteristic of all life. However, the capacity to respond and adapt to changes of tens of MPa is restricted to those organisms inhabiting high-pressure environments.

Evidence of high-pressure adaptation has been discovered in deep-sea organisms belonging to the lower Eukarya, [11], invertebrates and fishes [12], and even in deep diving marine mammals [13]. However, in this article the focus will be on the characteristics of high-pressure-adapted deep-sea prokaryotes. As can be seen in Table 1, prokaryotes belonging to both the *Archaea* (mostly from deep-sea hydrothermal vents) and *Bacteria* (from cold deep-sea habitats)

Table 1
Examples of piezophilic microorganisms

Microorganism	Pressure optimum (MPa)	Temperature optimum (°C)	Ref.
Bacteria			
Strain CNPT3	52	8	[128]
Strain PT64	90	9	[2]
Strain MT199	90	13	[2]
Colwellia MT41	103	8	[128,129]
Colwellia hadaliensis BNL1	85	10	[130]
Desulfovibrio profundus 500-1^T	15	25	[131]
Moritella PE36	41	15	[128]
Moritella japonica DSK1	50	15	[132]
Moritella yayanosi DB21MT-5	80	10	[133]
Photobacterium profundum strain DSJ4	10	10	[71]
Photobacterium profundum strain SS9	28	15	[128]
Shewanella SC2A	14	20	[123,134]
Shewanella violacea DSS12	30	10	[132]
Shewanella benthica WHB46	40	5	[135]
Shewanella benthica strain F1A	41	8	[128]
Shewanella benthica DB6101	50	10	[132]
Shewanella benthica strain PT48	62	8	[128]
Shewanella benthica strain PT99	62	8	[128]
Shewanella benthica DB5501	60	15	[132]
Shewanella benthica DB6705	60	15	[132]
Shewanella benthica DB6906	60	15	[132]
Shewanella benthica DB172R	60	10	[136]
Shewanella benthica DB172F	70	10	[136]
Shewanella benthica DB21MT-2	70	10	[133]
Archaea			
Methanococcus thermolithotrophicus	50	65	[137]
Methanococcus jannaschii	75	86	[138]
Pyrococcus abyssi	20	95	[139]
Thermococcus barophilus	40	85	[140]

domains of life exist which are adapted to pressures greater than 1 atmosphere. These microbes are referred to as piezophiles (literally 'pressure loving' [14], formerly described as barophiles). Piezophiles have optimal reproduction rates at pressures greater than 0.1 MPa, and hyperpiezophiles possess optimal growth rates at pressures >60 MPa. Although only a handful of high-pressure-adapted members of the *Archaea* have been reported, they span a phylogenetically broad collection of both the euryarchaeote and crenarchaeote kingdoms. In contrast, the members of the domain *Bacteria* so far obtained group primarily within a few closely related families of the gamma proteobacteria. Both the bacterial and archaeal piezophiles in culture are closely related to shallow-water microbes which are not piezophilic, indicating that high-pressure selection has not required the evolution of dramatically different lineages of life. Sequencing of small subunit ribosomal RNA genes from microorganisms collected from water samples in both the deep Pacific and Atlantic Oceans indicates that a subgroup within the crenarchaeotes are ubiquitous and highly abundant in these environments [15,16]. Although these microbes have not yet been cultured they are likely to include piezophilic members. This cluster of the *Archaea* could be among the most abundant organisms on Earth.

For those interested in additional aspects of environmentally relevant pressure effects on biological processes a number of other reviews should be considered: for deep-sea animals [1,17,18], for protein structure/function [19], for membrane-based adaptations [20], for physiology [21], for molecular biology, [20,22–25], for techniques and biophysics [14,26], and for reviews of older contributions to the field [14,26–31].

2. General effects of pressure on microbial cells

Despite the fact that the basis of all pressure effects arise from a single influence, namely the change in system volume that accompanies a physiological or biochemical process, high pressure exerts many effects on living organisms, making it difficult to pinpoint pressure-points in cell growth and viability. Making matters still more complex, pressure effects also depend on other physicochemical factors. For example, when the model mesophilic (atmospheric pressure-adapted) bacterium *Escherichia coli* is grown at a low temperature or a more acidic pH its growth is markedly more pressure-sensitive [32]. The types of nutrient used during culturing can also influence microbial pressure maxima by tens of MPa (reviewed in [33]). Some of the key pressure-sensitive processes described for *E. coli* are presented in Table 2.

An example of the unique nature of elevated pressure effects on microbial cells comes from investigation into the stress response of *E. coli* to pressure [34]. Exposure to 53 MPa resulted in the induction of 55 proteins, including 11 heat-shock proteins and four cold-shock proteins. Pressure is the only stressor known to simultaneously induce so many heat shock and cold-shock proteins, sets of proteins which are induced following exposure to opposing thermal regimes. The major pressure-inducible protein is an

Table 2
Pressure-sensitive processes in *Escherichia coli*

Process	Pressure abolishing process (MPa)	Ref.
Motility	10	[141]
Substrate transport[a]	26	[142]
Cell division	20–50	[31,90,107]
Growth	50	[91]
DNA replication	50	[91]
Translation	60	[91,143]
Transcription	77	[91]
Viability	100[b]	[144]

[a]Isopropylthiogalactopyranoside transport.
[b]~1 log decrease in viability of growing cells.

unknown small highly basic protein. High-pressure induction of heat-shock proteins has been confirmed in other studies with *E. coli* as well as with human cells [35,36]. Heat-shock proteins may also be induced in piezophiles upon decompression. A stress protein showing similarity to a heat-shock protein was found to be induced in the deep-sea piezophilic hyperthermophile *Thermococcus barophilus* subsequent to shift to atmospheric pressure [37].

Some of the known effects of pressure on biological systems may help explain the apparent paradox of simultaneous heat-shock and cold-shock stimulation (reviewed in [38]). High pressure and high temperature can destabilize protein quaternary and to a lesser extent tertiary structure. On the other hand, high pressure and low temperature share similar effects on protein synthesis and membrane structure. Thus, the induction of both heat-shock and cold-shock proteins may represent an attempt by *E. coli* to ameliorate the damaging effects of elevated pressure on membrane integrity, translation processes, and the stability of macromolecules.

3. Isolation of pressure-resistant mutants

Pressure could prove to be a useful tool in genetics in the isolation of conditional mutants, just as salt-sensitive and temperature-sensitive mutants have been used in the past. In this way, fundamental cellular processes in mesophilic microbes could be dissected and essential attributes of piezophilic organisms could be uncovered. A few hints of this potential have come from the isolation of mutants of mesophilic microbes which are resistant to elevated pressure. In subsequent sections I will describe some pressure-sensitive mutants that have been derived from piezophiles.

Mutants of the yeast *Saccharomyces cerevisiae*, have recently been isolated which are capable of growth at 25 MPa (F. Abe, unpublished results and [39]). High-pressure growth was found to result from increased expression of the TAT2 gene which encodes a high-affinity tryptophan permease. This alteration also improved growth at low temperature. The activation volume associated with tryptophan uptake was determined to be large (+46 ml mol^{-1}), further indicating that the uptake of this large aromatic amino acid across the membrane can be rate-limiting for growth at elevated pressure or low temperature. These results are consistent with transport studies with marine bacteria, which indicate that the rate of uptake of many substrates, including glutamate and acetate, are always greater at atmospheric pressure than that even at small increases in pressure up to 18 MPa [40]. Pressure inhibition of substrate transport can also limit the growth of the mesophile *Enterococcus faecalis* in the case of some substrates [29].

Marquis and colleagues have obtained mutants in *E. faecalis* following selection for growth at increased pressures (up to ~100 MPa, [41]). The mutants obtained were derepressed for catabolite repression. The result was the induction of additional ATP-generating catabolic enzymes as well as the accumulation of ammonia, due to the increased activity of arginine deiminase. Both of these factors could play a role in piezotolerance. Proton translocation by F_1F_0-ATPases from both *E. coli* and *E. faecalis* are highly sensitive to increases in pressure [42] [43] as is plasma membrane H^+-ATPase from *S. cerevisiae* [44]. Thus, enhanced growth at high pressure could have resulted from increased ATP availability for proton pumping. Alternatively, piezotolerance could have arisen from increased acid neutralization ability due to the increase in ammonia production.

Finally, it has also been possible to obtain mutants of *E. coli* which are capable of surviving lethal pressures [45,46]. This work stems from the interest in using high pressure for killing bacteria in food sterilization processes. Some of these mutants have been genetically characterized and found to possess mutations resulting in high RpoS activity [46]. This RNA polymerase sigma factor controls the transcription of numerous genes involved in surviving starvation and related stresses. These high-pressure-resistant mutants do not grow better than wild-type cells at sublethal elevated pressures [45], and therefore, the mechanisms responsible for facilitating high-pressure growth and high-pressure survival must be different.

4. High-pressure regulation of gene expression

Information on high-pressure signal transduction

has come from studies of pressure influences on gene expression in piezophiles. The moderate piezophile *Photobacterium profundum* species strain SS9 has become a workhorse for these and other studies of piezophile adaptation because of its genetic tractability and hyper-responsiveness to pressure changes. SS9 modulates the amount of outer membrane proteins such that when cells are shifted from atmospheric pressure to higher pressures the amount of the outer membrane protein OmpH increases and the amount of porin-like protein OmpL decreases [47–49]. The genes encoding these proteins are transcriptionally regulated by the ToxR and ToxS proteins [50]. These inner membrane proteins were first discovered in *Vibrio cholerae* where they function together as environmental sensors, regulating gene expression in response to changes in osmolarity, pH, temperature and to the levels of certain extracellular amino acids [51,52]. ToxR is an oligomeric protein whose cytoplasmic domain binds to genes directly under its control. ToxS modulates ToxR activity. Many if not all members of the *Vibrionaceae* possess *toxR* and *toxS* genes [50] and other bacteria can also possess the *toxR* gene alone (S. Kjelleberg, personal communication).

SS9 ToxR/S pressure sensing appears to depend on some aspect of the physical state of the inner membrane. Treatment of SS9 with local anesthetics at concentrations which are known to increase membrane fluidity results in a low pressure ToxR/S signaling phenotype (high OmpL, low OmpH abundance) even when the cells are grown at high pressure [50]. We hypothesize that at low pressure ToxR is capable of binding its cognate operator sequences, resulting in either activation or repression of gene transcription.

The SS9 ToxR/S system seems ideally suited for function as a piezometer. High pressure decreases the abundance as well as the activity of ToxR [50]. The former has been measured by Western blotting while the latter effect has been determined by artificially overproducing ToxR and ToxS in SS9 cells at high pressure. In contrast, at increased temperatures ToxR abundance drops but ToxR specific activity increases. Thus, temperature changes do not increase ToxR overall activity because of the cancelling effects of ToxR abundance and specific activity, whereas pressure has an additive effect on ToxR regulation due to the combined effects of pressure on both of these parameters.

In order to identify additional genes regulated by ToxR in SS9, we have used RNA arbitrarily primed PCR (RAP-PCR) with wild-type and *toxR* mutant strains of SS9 [53]. Seven ToxR-activated transcripts and one ToxR-repressed transcript were identified in this analysis. The cDNAs corresponding to these partial transcripts were cloned and sequenced. The products of these genes are all predicted to fall into one or both of two functional categories, those whose products alter membrane structure and/or those that are part of a starvation response. The connection between ToxR and starvation is further strengthened by the observation that the *ompH* gene is also inducible under conditions of carbon and energy limitation. Thus, the basic housekeeping function of the ToxR regulon in SS9 could be to maintain appropriate membrane structure as well as to help cells cope with various nutrient stresses. The transcript levels of all eight newly identified genes were also characterized as a function of hydrostatic pressure. Various patterns of pressure regulation were observed, indicating that ToxR activation or repression cannot be used to predict the influence of pressure on the expression of all ToxR-regulated genes in SS9, probably because many of the identified genes are controlled by multiple regulatory factors.

High-pressure activation of gene expression has also been analyzed among piezophiles belonging to the *Shewanella* piezophile taxonomic branch [54,55]. A promoter fragment was isolated which directed high-pressure-inducible transcription in *E. coli* as well as in the source *Shewanella*, *S. benthica* strain DB6705. This promoter directs the expression of two genes whose function is unknown, but which are highly conserved among the deep-sea shewanellas [56]. This operon was also cloned from the moderate piezophile *S. violaceae* strain DSS12 [57]. DNA-binding electrophoretic mobility shift assays performed on the identified DSS12 operator/promoter DNA fragment with cell lysates of DSS12 cells grown at low and high pressure indicate that a σ^{54}-like factor as well as another regulatory DNA-binding protein, possibly a pressure responsive DNA-binding protein, bind to the upstream region of this operator DNA [58]. RNA polymerase holoenzyme with σ^{54} is absolutely dependent on additional transcriptional factors for activity, and controls the expression of several ancillary processes in a variety of bacteria [59]. It would appear that this sigma factor in conjunction with a enhancer-binding pressure-responsive protein is required for high-pressure gene expression in piezophilic shewanellas. Downstream of this operon is a second pressure-regulated operon which includes a gene required for cytochrome *bd* assembly. This gene, *cydD*, appears to be important for high-pressure adaptation (see Section 6).

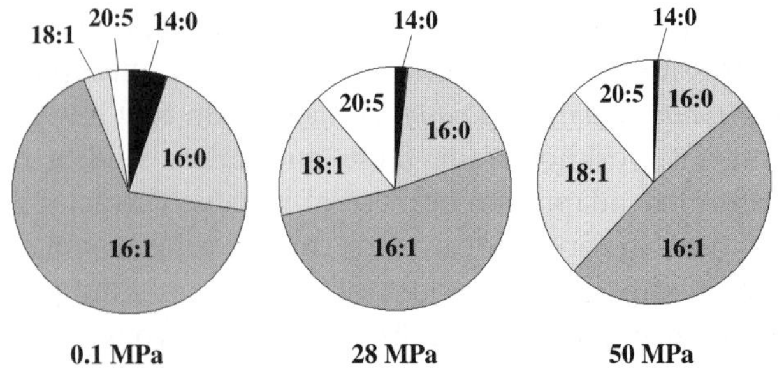

Fig. 2. Effect of culture pressure on the proportion of the major fatty acid species in the deep-sea bacterium *Photobacterium profundum* strain SS9.

5. Membranes

Lipids are particularly sensitive to pressure effects. They are on average an order of magnitude more compressible than proteins [60]. Increasing pressure enhances the order of hydrocarbon chains and raises the temperature of membrane phase transition from the gel state L_β to the liquid crystalline state L_α [61–63]. Neutron diffraction experiments indicate that pressure increases bilayer thickness by reducing the 'kinking' of acyl chains [64]. For many natural membranes the slope in a temperature pressure-phase diagram for the L_β–L_α main transition is about 20°C per 100 MPa at pressures less than 100 MPa. Based on this relationship the combined effect of temperature and pressure on the phase state of a membrane from a deep-sea organism at 100 MPa and 2°C (a typical temperature for the deep sea) is similar to a membrane at atmospheric pressure at a temperature of −18°C.

Many deep-sea organisms modulate their membrane fluidity and composition in response to pressure. This was first demonstrated by Cossins and Macdonald who measured the fluidity of fish brain myelin membranes from shallow-water and deep-sea fish [65]. Later they correlated increased proportions of membrane fluidizing unsaturated fatty acids in fish membranes with increased capture depth [66]. Since these studies many investigations have reported high levels of unsaturated fatty acids being present in deep-sea bacteria [55,67–73].

Studies with the genetically tractable deep-sea bacterium *Photobacterium profundum* strain SS9 have been used to demonstrate a role for unsaturated fatty acids in growth at high pressure. Fig. 2 shows that increased culture pressure increases the proportion of both the monounsaturated fatty acid 18:1, *cis*-vaccenic acid as well as that of the polyunsaturated fatty acid 20:5, eicosapentaenoic acid (EPA). Chemical mutants of SS9 which produce diminished amounts of monounsaturated fatty acids are pressure-sensitive [67] as is a strain constructed with a *fabF* mutation [74]. FabF encodes the enzyme β-ketoacyl-ACP synthase II, an enzyme catalyzing cis-vaccenic acid. We speculate that SS9 FabF activity is pressure-responsive just as the activity of the homologous enzyme in *E. coli* is low temperature-responsive, exhibiting increased substrate affinity and increased relative V_{max} at reduced temperatures [75].

Many deep-sea bacterial membranes also contain omega-3 polyunsaturated fatty acids (PUFAs) such as EPA or docosahexaenoic acid (DHA, 22:6). These fatty acids are of considerable biomedical interest, primarily because of their use in reducing the risk of human cardiovascular disease and certain cancers and their use as dietary supplements in mariculture and poultry farming [76]. The presence of PUFAs such as EPA and DHA is uncommon in most bacteria but present in a higher proportion of isolates from low temperature and deep-sea environments [73]. However, these fatty acids do not appear to contribute to piezoadaptation. A mutant in SS9 which does not produce EPA is not altered in its growth as a function of either pressure or temperature [67]. This result is surprising in view of the fact that the melting temperature of EPA is extremely low (−54°C versus 13.4°C for 18:1). One possibility is that PUFAs like EPA and DHA are produced by deep-sea bacteria for symbiotic interactions with higher deep-sea fauna, where they are needed as essential fatty acids. Alternatively, there may be key differences in the localization of these fatty acids within the membrane lipids. Recently bacterial production of omega-3 PUFAs has been found to proceed via a novel polyketide biosynthetic pathway [77].

What might be the biophysical explanation for the requirement for monounsaturated fatty acids (and certain other lipid modifications) in growth at high pressure? Three possibilities can be envisioned. Conformational changes in membrane physical structure as a function of increased pressure result primarily in tighter packing of fatty acyl chains within the membrane phospholipids. The close packing of acyl chains result in membranes assuming an ordered array where molecular motion is restricted. Thus, one explanation for increasing fatty acid unsaturation with pressure could be to maintain the membrane within a narrow range of viscosity. This type of process is termed homeoviscous adaptation [78]. Another possibility is that the selective pressure is rather to maintain the membrane within a liquid-crystalline phase, and thus the response of piezophiles is a homeophasic adaptive response. Most recently, it has been suggested that organisms primarily seek to optimize ion permeability across the membrane for bioenergetic purposes such as proton translocation and ATP synthesis [79]. Thus, permeability to ions such as protons or sodium ions could drive membrane restructuring in response to pressure. Whatever the reason, membranes are very susceptible to pressure changes. This is also true following decompression of membranes from piezophiles. Electron microscopic examination of the most piezophilic isolate yet obtained, *Colwellia* MT41, fixed during or shortly after decompression indicates that it responds to decompression by forming intracellular vesicles and releasing membrane fragments into the medium, followed later by cell lysis [80].

6. Membrane proteins

In addition to membrane lipids, membrane proteins have frequently been implicated as playing an important role in growth at high pressure. Work has already been alluded to implicating F_1F_0-ATPase and tryptophan permease in limiting the growth of some mesophiles at elevated pressure. Other studies have indicated a role for certain membrane proteins in the growth of piezophiles at high pressure. Two types of membrane proteins have been uncovered during the course of following high-pressure regulation of gene expression in piezophiles.

In SS9 transposon insertions within the *rpoE* operon have been found which decrease the expression of the pressure-regulated outer membrane protein encoding gene *ompH* [81]. RpoE is another alternative sigma factor. It regulates gene expression in response to stress within the membrane or periplasm [82,83]. The *E. coli rpoE* operon is comprised of *rpoE*, two downstream genes whose products modulate RpoE activity in response to environmental cues and the *rseC* gene whose product has a negligible effect on RpoE activity [84,85]. RseC and related proteins are believed to function as inner transmembrane proteins which anchor protein complexes to the membrane for the transfer of electrons to key enzymes such as those involved in thiamine biosynthesis [86]. Mutations in SS9 which decrease *rseC* expression result in both high-pressure-sensitive and cold-sensitive growth phenotypes, perhaps because of low temperature and high-pressure effects on membrane-based electron transport. Curiously, *rseC* mutants in mesophiles are not more pressure-sensitive or cold-sensitive, and SS9 mutants decreasing RpoE activity are not more temperature-sensitive as is the case with mesophiles. Lack of *rseA*, *rseB* or *rseC* expression decreases *ompH* transcript levels (Chi and Bartlett, unpublished results). The basis of this effect is unknown, however, it could depend on the activity of an additional regulatory gene which is under RpoE control.

During the course of searching for genes activated at high pressure Kato and colleagues uncovered a homologue of the *cydD* gene in the piezotolerant *Shewanella* species DSS12. In *E. coli*, CydD is required for assembly of the cytochrome *bd* complex. *E. coli cydD* mutants exhibit increased pressure sensitivity which can be complemented by introduction of the DSS12 *cydD* gene. The discovery that cytochrome formation could be a pressure-sensitive process then prompted an examination of cytochrome composition as a function of pressure in several deep-sea shewanellas. In strain DSS12 the cytochrome *bd* protein complex is only observed from cells grown at high pressure [87]. *Shewanella benthica* strain DB172F was found to produce two types of *c*-type cytochromes, one membrane bound and one cytoplasmic [88]. However, only the membrane bound *c*-type cytochrome was produced at high pressure (60 MPa). In addition, at high pressure a *ccb*-type quinol oxidase was increased in abundance [89]. These results suggest that when deep-sea shewanellas are shifted from a low to a high pressure there are major alterations in the respiratory chain and membrane bioenergetic properties.

7. DNA structure and function

In mesophiles such as yeast cells and *E. coli* the abundance of DNA per cell is substantially reduced at elevated pressure [39,90] and in one piezophile, *Colwellia* MT41, DNA content per cells decreases after culture decompression [80]. Measurements of *E. coli* DNA synthesis as a function of pressure during short term incubations suggest that it is less pressure-sensitive than protein synthesis. However, long-term incubations reveal that it is one of the most pressure-sensitive essential cell processes [91]. Pressures between 25 and 43 MPa promote synchronization of both cell division and also, apparently, DNA synthesis in *E. coli* [91]. The pressure-enhanced synchronization of DNA replication is consistent with an inhibition of either the initiation or termination of the replication process. One potential site of action of pressure on the initiation of DNA replication is the cell membrane. We have already alluded to the pressure-sensitive nature of membrane lipid structure. The initiation of DNA replication requires membrane fluidizing unsaturated fatty acids [92]. Also, the DnaA and SeqA proteins which are required for chromosome replication initiation also influence cell membrane structure [93].

Another critical feature of the high-pressure exposure in mesophiles is that there is no SOS response, despite the fact that sublethal pressures tend to induce filamentation (described in more detail in Section 8). The SOS response is a coordinated gene activation system which helps bacteria survive DNA damage or the inhibition of DNA replication and which (like high pressure) leads to dramatically filamentous cell morphologies [94]. The evidence indicating that there is no SOS response in *E. coli* following high-pressure exposure comes from two-dimensional gel studies of pressure-inducible proteins [34], the lack of a difference in pressure response between *recA*$^+$ or *recA* mutant cells [95] and the observation that SOS-inducible promoters are not induced (Bidle and Bartlett, unpublished results). The lack of SOS induction despite high-pressure inhibition of DNA replication is unusual but not without precedent. It has been observed that *E. coli* cells which overproduce the replication initiation protein DnaA. Such cells are filamentous with uncondensed chromosomes (Grigorian and Zyskind, personal communication). Although it is not known why DnaA overproducing cells do not mount a SOS response, this phenotype is consistent with the hypothesis that the basis of high-pressure filamentation in *E. coli* stems from an imbalance in DNA replication.

The DNA in *E. coli* cells incubated and fixed at moderate pressure is uncondensed [95]. The uncondensed state of mesophile bacterial DNA present during exposure to high pressure must result from direct or indirect effects on DNA supercoiling. These effects could include the inhibition of DNA gyrase or Muk protein action, or more generally, inhibition of replication which requires uncondensed parental DNA [96]. DNA gyrase catalyzed DNA supercoiling is not pressure-sensitive under physiologically relevant concentrations of enzyme and magnesium [97]. Indeed, in vitro experiments with topoisomerase I indicate that at high pressure a more supercoiled DNA form is favored [98], apparently because of its smaller spatial volume.

Genetic studies have implicated another DNA-binding protein in piezoadaptation [53]. A pressure-sensitive mutant of SS9 was found to possess a missense mutation in the *recD* gene. Subsequent *recD* insertion mutants were found to also be pressure-sensitive. RecD, along with RecB and RecC (exonuclease V), plays a major role in the homologous recombination pathway of bacteria such as *E. coli* [99]. RecBCD has been proposed to function as a destructive nuclease of linear DNA until it encounters the octameric sequence Chi [100]. At this sequence, RecD is hypothesized to dissociate from the complex, with the result that the RecBC enzyme is converted to a recombinogenic helicase, producing single-stranded DNA that serves as a substrate for RecA-catalyzed homologous recombination [101].

E. coli recD mutants are hyperrecombinogenic, decreased in plasmid stability, and increased in plasmid multimer formation [102]. SS9 *recD* mutants also display decreased plasmid stability as well as an enlarged, swollen cell morphology when incubated at elevated pressure (28 MPa).

The transfer of the SS9 *recD* gene into an *E. coli recD* mutant enables *E. coli* cells to divide normally at high pressure. Based on the well-documented role of RecD as a part of a major cellular nuclease, the basis of this effect could be in limiting DNA recombination at high pressure. This requirement would arise if high pressure either promotes increased extrachromosomal DNA recombination or decreased resolution of chromosome multimers. However, SS9 RecD contains an unusual domain of 64 amino acids not present in most other RecD proteins [53], and this could also be significant. This stretch of amino acids could confer an additional activity to the protein. At this juncture it seems fair to summarize the *recD* results as providing further evidence of a link between changes in DNA metabolism and cell division and piezophile adaptation to high pressure.

Another DNA-binding protein that has been studied from piezophiles is single-stranded DNA-binding protein (SSB). The gene for this conserved protein was isolated from four closely related bacteria one belonging to the genus *Shewanella*, but which differed in their extent of high-pressure adaptation. Then after overproduction and purification of the recombinant proteins the pressure-induced dissociation of the oligomeric SSB proteins was investigated using fluorescence anisotropy. SSB from the piezosensitive *Shewanella* strain *S. hanedai* was more sensitive to pressure than that of the SSB proteins from the piezotolerant or piezophilic bacteria. The volume change of association obtained from the pressure dependence of dissociation for a fixed protein concentration for SSB from *S. hanedai* was 400 ml mol^{-1}. The ΔV values for SSB from the deeper-living *Shewanellas* were smaller and ranged from 253 to 354 ml mol^{-1} (Chilukuri et al., in preparation). Differences between the primary structures of the SSB proteins that could correlate with differences in sensitivity to pressure-induced dissociation were examined and revealed that the deep-sea SSB has significantly less helix-destabilizing (glycine) and helix-breaking (proline) amino acid residues [103]. Therefore, increased SSB structural homogeneity could be favored by the high-pressure environments of the deep sea.

The SSB proteins DNA-binding characteristics as a function of pressure have not yet been examined. However, such studies have merit. Protein–DNA interactions can be extraordinarily sensitive to pressure with volume changes upon dissociation in the range of −1000 to −4000 ml mol^{-1} [104]. One DNA-binding protein whose interaction with DNA has been well studied is the lactose repressor. It is considerably destabilized at high pressure after binding to operator DNA [105].

8. Cell division

Many mesophilic microbes become filamentous when incubated at elevated pressures below those which prevent cell growth [31,90,106,107]. Thus, in these microbes cell division is more pressure-sensitive than cell growth. Piezophiles have been reported to be filamentous when grown at pressures below or above their pressure optima [108,109]. These results could result from pressure effects on DNA replication or condensation as described above.

In addition to pressure indirectly influencing cell division, more direct effects could also take place. The activity of several division proteins are possible targets of high pressure. Some of these proteins bear similarity to the cytoskeletal proteins of higher organisms which are well known to possess a pressure-sensitive assembly [110,111]. One of these proteins whose structure and localization has been studied as a function of pressure is FtsZ. FtsZ is a GTP-hydrolyzing tubulin-like protein which polymerizes into a ring at the middle of the cell during early stages of the septation process [112,113]. *E. coli* FtsZ rings are largely absent in cells incubated at high pressure but quickly form after decompression [95]. These results agree well with in vitro studies of pressure effects on tubulin filaments. The dissociation of these polymers has been reported to be accompanied by a volume change of −90 ml mol^{-1} and hence is very much promoted by elevated pressure [114].

Other cell division proteins of note include FtsA and FtsK, both of which are worthy of future attention in high-pressure studies, but which have not yet been studied in this context. FtsA is an ATP-hydrolyzing member of the actin/sugar kinase/hsp70 superfamily [115]. This is particularly significant because the stability of actin polymers has previously been studied in deep-sea animals (reviewed in [1]). It has been found that for skeletal muscle actins, the volume change that accompanies the assembly of the actin filament varies among species in apparently adaptive ways. For example, the volume change found for chicken actin is approximately 107 ml mol^{-1}, while that for the deep-sea fish *Coryphaenoides armatus* is only 2 ml mol^{-1}. The last division protein to mention here, FtsK, assists both chromosome partitioning and septation. It has a primary structure related to the plasmid transfer protein TrwB [116], whose three-dimensional structure has recently been reported to be similar to that of F_1-portion of the F_1F_0-ATPase [117]. Because proton translocation by this rotating pump is highly pressure-sensitive [42], FtsK activity could also be responsive to relatively small pressure changes.

9. Enzyme function and the role of extrinsic factors

Very little work has been done addressing the biochemical adaptations to high pressure in the proteins of piezophilic bacteria. However, D.S. Clark and colleagues have made considerable progress examining both enzyme stability and activity for a number of proteins obtained from piezophilic hyperthermophiles. Information on this topic can be found in the report by Clark in this issue as well as in a recent publication [118]. There has also been an impressive body of work on the enzyme adaptations of deep-sea animals. In this case adaptation is characterized by more pressure insensitive K_ms and reduced k_{cat}s [1,119].

An additional important consideration in protein function at high pressure is the presence of both organic and inorganic extrinsic factors. In the case of thermophily there is evidence that potassium salts of 2,3-diphosphoglycerate and di-*myo*-inositol-1,1′-phosphate function as thermostabilizers in some hyperthermophiles. [120,121]. Deep-sea fish, skates and crustaceans have been found to increase their levels of the organic osmolyte trimethylamine-*N*-oxide (TMAO) with depth [12,122]. TMAO has been shown to decrease pressure effects on enzyme substrate binding [122]. Stabilizing osmolytes such as TMAO are preferentially excluded from protein hydration layers, making it more entropically unfavorable for a protein to unfold and expose more hydrated surface area. Thus, osmolytes such as TMAO could favor a more compact protein structure.

Inorganic ions also can function as stabilizers of proteins at high pressure. The stability of malate dehydrogenase purified from the deep-sea bacteria SS9 and *Shewanella* SC2A was enhanced in the presence of various inorganic and organic salts [123]. The pressure sensitivity of both ribosome stability and DNA gyrase-mediated DNA supercoiling is sensitive to magnesium ion levels [97]. In vitro rates of protein synthesis using cell extracts obtained from *Pseudomonas bathycetes*, a deep-sea isolate obtained from the Mariana Trench, required supplementation of both magnesium and sodium ions in order for high-pressure rates of protein synthesis to approach the in vivo levels [124]. Magnesium and calcium can improve piezotolerance in at least one bacterial species [125]. Some of these effects could be dependent on osmotic pressure. Increased osmotic pressure often decreases protein hydration whereas increased hydrostatic pressure can promote the reverse [104]. One example of the role of hydration in protein stability is that the Arc repressor protein will not denature at high pressure in the absence of water [126].

10. Future prospects

Transitioning from a more descriptive to a more mechanistic understanding of pressure effects on biological systems is ongoing and will greatly benefit from some of the biophysical approaches described in other articles of this issue. This will include analyses of the structure and function of key proteins involved in high-pressure adaptation of piezophiles or pressure sensitivity in mesophiles, including enzymes catalyzing the production of unsaturated fatty acids and pressure-insensitive electron and proton transport, as well as proteins functioning in DNA binding and cytoskeletal structure. In many instances this progression will follow more global analyses of genes required for growth or survival at high pressure. And, where genome sequence information is available, genome level exploration of gene expression and protein abundance as a function of pressure will provide even more hints at the effects of elevated pressure on mesophile and piezophile cell physiology. Some genome level analyses of gene expression at high pressure have already begun [127]. All of this work should continue alongside the isolation and characterization of additional extremophiles from deep-sea and deep-subsurface habitats which display notable high-pressure physiological attributes and evolutionary origins.

Acknowledgements

I am grateful to the National Science Foundation for support.

References

[1] G.N. Somero, Life at low volume change: hydrostatic pressure as a selective factor in the aquatic environment, Am. Zool. 30 (1990) 123–135.
[2] A.A. Yayanos, Evolutional and ecological implications of the properties of deep-sea barophilic bacteria, Proc. Natl. Acad. Sci. USA 83 (1986) 9542–9546.
[3] D.M. Karl, D.F. Bird, K. Björkman, T. Houlihan, R. Shackelford, L. Tupas, Microorganisms in the accreted ice of Lake Vostok, Antarctica, Science 286 (1999) 2144–2147.
[4] U. Szewzyk, R. Szewzyk, T.A. Stenström, Thermophilic, anaerobic bacteria isolated from a deep borehole in granite in Sweden, Proc. Natl. Acad. Sci. USA 91 (1994) 1810–1813.
[5] W.B. Whitman, D.C. Coleman, W.J. Wiebe, Prokaryotes: the unseen majority, Proc. Natl. Acad. Sci. USA 95 (1998) 6578–6583.
[6] C.F. Chyba, C.B. Phillips, Possible ecosystems and the search for life on Europa, Proc. Natl. Acad. Sci. USA 98 (2001) 801–804.
[7] P.F. Scholander, H.T. Hammel, E.D. Bradstreet, E.A. Hemmingsen, Sap pressure in vascular plants, Science 148 (1965) 339–346.
[8] P.J. Fraser, A.G. Macdonald, Crab hydrostatic pressure sensors, Nature 371 (1994) 383–384.
[9] B. Pelster, P. Scheid, Countercurrent concentration and gas secretion in the fish swim bladder, Physiol. Zool. 65 (1992) 1–16.
[10] A.M. Whatmore, R.H. Reed, Determination of turgor pressure in *Bacillus subtilis* – a possible role for K^+ in turgor regulation, J. Gen. Microbiol. 136 (1990) 2521–2526.
[11] M.S. Atkins, A.P. Teske, C.D. Taylor, C.O. Wirsen, O.R. Anderson, Flagellate diversity and ecology at deep-sea hydrothermal vents, in: Int. Soc. for Microb. Ecol., 9th Meeting (2001), Amsterdam, The Netherlands, 2001, Abstract P.17.024, p. 291.
[12] R.H. Kelly, P.H. Yancey, High contents of trimethylamine oxide correlating with depth in deep-sea teleost fishes, skates, and decapod crustaceans, Biol. Bull. 196 (1999) 18–25.
[13] M.A. Castellini, J.M. Castellini, P.M. Rivera, Adaptations to pressure in the RBC metabolism of diving mammals, Comp. Biochem. Physiol. A 129 (2001) 751–757.
[14] A.A. Yayanos, Microbiology to 10,500 meters in the deep sea, Annu. Rev. Microbiol. 49 (1995) 777–805.
[15] E.F. DeLong, K.Y. Wu, B.B. Prezelin, R.V.M. Jovine, High abundance of archaea in Antarctic marine picoplankton, Nature 371 (1994) 695–696.
[16] J.A. Fuhrman, K. McCallum, A.A. Davis, Novel major archaebacterial group from marine plankton, Nature 356 (1992) 148–149.
[17] A.G. Macdonald, Effects of high pressure on cellular processes, in: N. Sperelakis (Ed.), Cell Physiology Sourcebook: A Molecular Approach, 3rd Edition, Academic Press, San Diego, CA, 2001, pp. 1003–1023.
[18] G.N. Somero, Biochemical ecology of deep-sea animals, Experientia 48 (1992) 537–543.
[19] M. Gross, R. Jaenicke, Proteins under pressure – the influence of high hydrostatic pressure on structure, function and assembly of protein complexes, Eur. J. Biochem. 221 (1994) 617–630.
[20] D.H. Bartlett, K.A. Bidle, Membrane-based adaptations of deep-sea piezophiles, in: J. Seckbach (Ed.) Enigmatic Microorganisms and Life in Extreme Environments Kluwer Academic Publishing, Dordrecht, The Netherlands, 1999, pp. 501–512.
[21] F. Abe, C. Kato, K. Horikoshi, Pressure-regulated metabolism in microorganisms, Trends Microbiol. 7 (1999) 447–453.
[22] D.H. Bartlett, Pressure sensing in deep-sea bacteria, Res. Microbiol. 142 (1991) 923–925.
[23] D.H. Bartlett, Microbial adaptations to the psychrosphere/piezosphere, J. Mol. Microbiol. Biotechnol. 1 (1999) 93–100.
[24] C. Kato, D.H. Bartlett, The molecular biology of barophilic bacteria, Extremophiles 1 (1997) 111–116.
[25] C. Kato, K. Nakasone, M.H. Qureshi, K. Horikoshi, How do deep-sea microoorganisms respond to changes in environmental pressure?, in: K.B. Storey, J. Storey (Eds.), Environmental Stressors and Gene Responses, Elsevier Science, Amsterdam, 2000, pp. 277–291.
[26] A.A. Yayanos, Empirical and theoretical aspects of life at high pressure in the deep sea, in: K. Horikoshi, W.D. Grant (Eds.), Extremophiles. Microbial Life in Extreme Environments, Wiley, New York, 1998, pp. 47–92.
[27] E.E. Allen, D.H. Bartlett, Piezophiles: microbial adaptation to the deep-sea environment, in: C. Gerday (Ed.), Extremophiles, Eolss Publishers, Oxford, 2001 (in press).
[28] D.H. Bartlett, Microbial life at high pressures, Sci. Prog. 76 (1992) 479–496.

[29] P. Matsumura, R.E. Marquis, Microbial life under pressure, in: D.J. Kushner (Ed.), Microbial Life in Extreme Environments, Academic Press, London, 1978, pp. 105–158.
[30] D. Prieur, Physiology and biotechnological potential of deep-sea bacteria, in: R.A. Herbert, R.J. Sharp (Eds.), Molecular Biology and Biotechnology of Extremophiles, Chapman and Hall, New York, 1992, pp. 163–202.
[31] C.E. ZoBell, Pressure effects on morphology and life processes of bacteria, in: A. Zimmerman (Ed.), High Pressure Effects on Cellular Processes, Academic Press, London, 1970, pp. 85–130.
[32] R.E. Marquis, High pressure microbiology, in: P.B. Bennett, R.E. Marquis (Eds.), Basic and Applied High Pressure Biology, University of Rochester Press, Rochester, NY, 1994, pp. 1–14.
[33] R.E. Marquis, Microbial barobiology, BioScience 32 (1982) 267–271.
[34] T.J. Welch, A. Farewell, F.C. Neidhardt, D.H. Bartlett, Stress response in *Escherichia coli* induced by elevated hydrostatic pressure, J. Bacteriol. 175 (1993) 7170–7177.
[35] M. Grob, I. Kosmowsky, R. Lorenz, H.P. Molitoris, R. Jaenicke, Response of bacteria and fungi to high pressure stress as investigated by two-dimensional polyacrylamide gel electrophoresis, Electrophoresis 15 (1994) 1559–1565.
[36] C.L. Haskin, K.A. Athanasiou, R. Klebe, I.L. Cameron, A heat-shock-like response with cytoskeletal disruption occurs following hydrostatic pressure in MG-63 osteosarcoma cells, Biochem. Cell Biol. 71 (1993) 361–371.
[37] V.T. Marteinsson, A.-L. Reysenbach, J.-L. Birrien, D. Prieur, A stress protein is induced in the deep-sea barophilic hyperthermophile *Thermococcus barophilus* when grown under atmospheric pressure, Extremophiles 3 (1999) 277–282.
[38] D.H. Bartlett, C. Kato, K. Horokoshi, High pressure influences on gene and protein expression, Res. Microbiol. 146 (1995) 697–706.
[39] F. Abe, K. Horikoshi, Tryptophan permease gene TAT2 confers high-pressure growth in *Saccharomyces cerevisiae*, Mol. Cell. Biol. 20 (2000) 8098–8102.
[40] H.W. Jannasch, C.D. Taylor, Deep-sea microbiology, Annu. Rev. Microbiol. 38 (1984) 487–514.
[41] R.E. Marquis, G.R. Bender, Isolation of a variant of *Streptococcus faecalis* with enhanced barotolerance, Can. J. Microbiol. 26 (1980) 371–376.
[42] R.E. Marqui, G.R. Bender, Barophysiology of Prokaryotes and Proton-translocating ATPases, Academic Press, London, 1987, pp. 65–73.
[43] P. Matsumura, R.E. Marquis, Energetics of Streptococcal growth inhibition by hydrostatic pressure, Appl. Environ. Microbiol. 33 (1977) 885–892.
[44] F. Abe, K. Horikoshi, Analysis of intracellular pH in the yeast Saccharomyces cerevisiae under elevated hydrostatic pressure: a study in baro- (piezo-) physiology, Extremophiles 2 (1998) 223–228.
[45] K.J.A. Hauben, D.H. Bartlett, C.C.F. Soontjens, K. Cornelis, E.Y. Wuytack, C.W. Michiels, *Escherichia coli* mutants resistant to inactivation by high hydrostatic pressure, Appl. Environ. Microbiol. 63 (1997) 945–950.
[46] M. Robey, A. Benito, R.H. Hutson, C. Pascual, S.F. Park, B.M. Mackey, Variation in resistance to high pressure and *rpoS* heterogeneity in natural isolates of *Escherichia coli* 0157:H7, Appl. Environ. Microbiol. 67 (2001) 4901–4907.
[47] D. Bartlett, M. Wright, A. Yayanos, M. Silverman, Isolation of a gene regulated by hydrostatic pressure, Nature 342 (1989) 572–574.
[48] E. Chi, D.H. Bartlett, Use of a reporter gene to follow high pressure signal transduction in the deep-sea bacterium *Photobacterium* SS9, J. Bacteriol. 175 (1993) 7533–7540.
[49] T.J. Welch, D.H. Bartlett, Isolation and characterization of the structural gene for OmpL, a pressure-regulated porin-like protein from the deep-sea bacterium *Photobacterium* species strain SS9, J. Bacteriol. 178 (1996) 5027–5031.
[50] T.J. Welch, D.H. Bartlett, Identification of a regulatory protein required for pressure-responsive gene expression in the deep-sea bacterium *Photobacterium* species strain SS9, Mol. Microbiol. 27 (1998) 977–985.
[51] J.J. Mekalanos, E.J. Rubin, M.K. Waldor, Cholera: molecular basis for emergence and pathogenesis, FEMS Immunol. Microbiol. 18 (1997) 241–248.
[52] K. Skorupski, R.K. Taylor, Control of the ToxR virulence regulon in *Vibrio cholerae* by environmental stimuli, Mol. Microbiol. 25 (1997) 1003–1009.
[53] K.A. Bidle, D.H. Bartlett, RecD function is required for high pressure growth in a deep-sea bacterium, J. Bacteriol. 181 (1999) 2330–2337.
[54] C. Kato, M. Smorawinska, T. Sato, K. Horikoshi, Cloning and expression in *Escherichia coli* of a pressure-regulated promoter region from a barophilic bacterium, strain DB6705, J. Mar. Biotechnol. 2 (1995) 125–129.
[55] Y. Nogi, N. Masui, C. Kato, Taxonomic studies of deep-sea barophilic *Shewanella* species, and *Shewanella violacea* sp. nov., a new moderately barophilic bacterial species, Arch. Microbiol. 170 (1998) 331–338.
[56] L. Li, C. Kato, Y. Nogi, K. Horikoshi, Distribution of the pressure-regulated operons in deep-sea bacteria, FEMS Microbiol. Lett. 159 (1998) 159–166.
[57] C. Kato, A. Ikegami, M. Smorawinska, R. Usami, K. Horikoshi, Structure of genes in a pressure-regulated operon and adjacent regions from a barotolerant bacterium strain DSS12, J. Mar. Biotechnol. 5 (1997) 210–218.
[58] K. Nakasone, A. Ikegami, C. Kato, R. Usami, K. Horikoshi, Mechanisms of gene expression controlled by pressure in deep-sea microorganisms, Extremophiles 2 (1998) 149–154.
[59] H. Barrios, B. Valderrama, E. Morett, Compilation and analysis of sigma(54)-dependent promoter sequences, Nucleic Acids Res. 27 (1999) 4305–4313.
[60] G. Weber, H.G. Drickamer, The effect of high pressure upon proteins and other biomolecules, Q. Rev. Biophys. 16 (1983) 89–112.
[61] D.A. Driscoll, J. Jonas, A. Jonas, High pressure ^{2}H nuclear

magnetic resonance study of the gel phases of dipalmitoylphosphatidylcholine, Chem. Phys. Lipids 58 (1991) 97–104.

[62] A.G. Macdonald, The role of membrane fluidity in complex processes under high pressure, in: H.W. Jannasch, R.E. Marquis, A.M. Zimmerman (Eds.), High Pressure Biology, Academic Press, London, 1987, pp. 207–223.

[63] P.T.T. Wong, D.J. Siminovitch, H.H. Mantsch, Structure and properties of model membranes: new knowledge from high-pressure vibrational spectroscopy, Biochim. Biophys. Acta 947 (1988) 139–171.

[64] L.F. Braganza, D.L. Worcester, Structural changes in lipid bilayers and biological membranes caused by hydrostatic pressure, Biochemistry 25 (1986) 7484–7488.

[65] A.R. Cossins, A.G. Macdonald, Homeoviscous theory under pressure. II. The molecular order of membranes from deep-sea fish, Biochim. Biophys. Acta 776 (1984) 144–150.

[66] A.R. Cossins, A.G. Macdonald, Homeoviscous adaptation under pressure. III. The fatty acid composition of liver mitochondrial phospholipids of deep-sea fish, Biochim. Biophys. Acta 860 (1986) 325–335.

[67] E.E. Allen, D. Facciotti, D.H. Bartlett, Monounsaturated but not polyunsaturated fatty acids are required for growth at high pressure and low temperature in the deep-sea bacterium Photobacterium profundum strain SS9, Appl. Environ. Microbiol. 65 (1999) 1710–1720.

[68] E.F. DeLong, A.A. Yayanos, Adaptation of the membrane lipids of a deep-sea bacterium to changes in hydrostatic pressure, Science 228 (1985) 1101–1103.

[69] E.F. DeLong, A.A. Yayanos, Novel bacterial lipids in deep-sea bacteria, Appl. Environ. Microbiol. 51 (1986) 730–737.

[70] K. Kamimura, H. Fuse, O. Takimura, Y. Yamaoka, Effects of growth pressure and temperature on fatty acid composition of a barotolerant deep-sea bacterium, Appl. Environ. Microbiol. 59 (1993) 924–926.

[71] Y. Nogi, N. Masui, C. Kato, *Photobacterium profundum* sp. nov., a new moderately barophilic bacterial species isolated from a deep-sea sediment, Extremophiles 2 (1998) 1–7.

[72] C.O. Wirsen, H.W. Jannasch, S.G. Wakeham, E.A. Canuel, Membrane lipids of a psychrophilic and barophilic deep-sea bacterium, Curr. Microbiol. 14 (1987) 319–322.

[73] Y. Yano, A. Nakayama, K. Yoshida, Distribution of polyunsaturated fatty acids in bacteria present in intestines of deep-sea fish and shallow-sea poikilothermic animals, Appl. Environ. Microbiol. 63 (1997) 2572–2577.

[74] E.E. Allen, D.H. Bartlett, FabF is required for Piezoregulation of cis-vaccenic acid levels and piezophilic growth of the deep-sea bacterium *Photobacterium profundum* strain SS9, J. Bacteriol. 182 (2000) 1264–1271.

[75] J.L. Garwin, A.L. Klages, J.E. Cronan, Structural, enzymatic, and genetic studies of β-ketoacyl-acyl carrier protein synthases I and II of *Escherichia coli*, J. Biol. Chem. 255 (1980) 11949–11956.

[76] A.P. Simopoulos, Omega-3 fatty acids in health and disease and in growth and development, Am. J. Clin. Nutr. 54 (1991) 438–463.

[77] J.G. Metz, P. Roessler, D. Facciotti, C. Levering, F. Dittrich, M. Lassner, R. Valentine, K. Lardizabal, F. Domergue, A. Yamada, K. Yazawa, V. Knauf, J. Browse, Production of polyunsaturated fatty acids by polyketide synthases in both prokaryotes and eukaryotes, Science 293 (2001) 290–293.

[78] A.G. Macdonald, A.R. Cossins, The theory of homeoviscous adaptation of membranes applied to deep-sea animals, Soc. Exp. Biol. Symp. 39 (1985) 301–322.

[79] J.L.C.M. Van de Vossenberg, T. Ubbink-Kok, M.G.L. Elferink, A.J.M. Driessen, W.N. Konings, Ion permeability of the cytoplasmic membrane limits the maximal growth temperature of bacteria and archaea, Mol. Microbiol. 18 (1995) 925–932.

[80] R.A. Chastain, A.A. Yayanos, Ultrastructural changes in an obligately barophilic marine bacterium after decompression, Appl. Environ. Microbiol. 57 (1991) 1489–1497.

[81] E. Chi, D.H. Bartlett, An *rpoE*-like locus controls outer membrane protein synthesis and growth at cold temperatures and high pressures in the deep-sea bacterium *Photobacterium* SS9, Mol. Microbiol. 17 (1995) 713–726.

[82] S. Raina, D. Missiakas, C. Georgopoulos, The *rpoE* gene encoding the σ^E (σ^{24}) heat shock sigma factor of *Escherichia coli*, EMBO J. 14 (1995) 1043–1055.

[83] P.E. Rouvière, A. De Las Peñas, J. Mecsas, C.Z. Lu, K.E. Rudd, C.A. Gross, *rpoE*, the gene encoding the second heat-shock sigma factor, σ^E, in *Escherichia coli*, EMBO J. 14 (1995) 1032–1042.

[84] A. De Las Peñas, L. Connolly, C.A. Gross, The σ^E-mediated response to extracytoplasmic stress in *Escherichia coli* is transduced by RseA and RseB, two negative regulators of σ^E, Mol. Microbiol. 24 (1997) 373–385.

[85] D. Missiakas, M.P. Mayer, M. Lemaire, C. Georgopoukos, S. Raina, Modulation of the *Escherichia coli* σ^E(RpoE) heat-shock transcription factor activity by the RseA, RseB and RseC proteins, Mol. Microbiol. 24 (1997) 355–371.

[86] B.J. Beck, L.E. Connolly, A. De Las Peñas, D.M. Downs, Evidence that *rseC*, a gene in the *rpoE* cluster, has a role in thiamine synthesis in *Salmonella typhimurium*, J. Bacteriol. 179 (1997) 6504–6508.

[87] H. Tamegai, C. Kato, K. Horikoshi, Pressure-regulated respiratory system in barotolerant bacterium *Shewanella* sp. strain DSS12, J. Biochem. Mol. Biol. Biophys. 1 (1998) 213–220.

[88] M.H. Qureshi, C. Kato, K. Horikoshi, Purification of two pressure-regulated c-type cytochromes from a deep-sea barophilic bacterium, *Shewanella* sp. strain DB-172F, FEMS Microbiol. Lett. 161 (1998) 301–309.

[89] M.H. Qureshi, C. Kato, K. Horikoshi, Purification of a *ccb*-type quinol oxidase specifically induced in a deep-sea barophilic bacterium, *Shewanella* sp. strain DB-172F, Extremophiles 2 (1998) 93–99.

[90] C.E. ZoBell, A.B. Cobet, Filament formation by *Escherichia coli* at increased hydrostatic pressures, J. Bacteriol. 87 (1963) 710–719.

[91] A.A. Yayanos, E.C. Pollard, A study of the effects of hydrostatic pressure on macromolecular synthesis in *Escherichia coli*, Biophys. J. 9 (1969) 1464–1482.

[92] J.A. Fralick, K.G. Lark, Evidence for the involvement of unsaturated fatty acids in initiating chromosome replication in *Escherichia coli*, J. Mol. Biol. 80 (1973) 459–475.

[93] A. Wegrzyn, B. Wrobel, G. Wegrzyn, Altered biological properties of cell membranes in *Escherichia coli dnaA* and *seqA* mutants, Mol. Gen. Genet. 261 (1999) 762–769.

[94] G.C. Walker, The SOS response of *Escherichia coli*, in: F.C. Neidhardt, R. Curtiss III, J.L. Ingraham, E.C.C. Lin, K.B. Low, B. Magasanik, W.S. Reznikoff, M. Schaechter, H.E. Umbarger (Eds.), *Escherichia coli* and *Salmonella* cellular and molecular biology, 2nd Edition, Vol. 1, ASM Press, Washington, DC, 1996, pp. 1400–1416.

[95] T. Sato, A. Ishii, C. Kato, M. Wachi, K. Nagai, K. Horikoshi, High hydrostatic pressure represses FtsZ-ring formation and chromosomal DNA condensation in *Escherichia coli*, in: Third Int. Congress on Extremophiles, Hamburg, Germany, 2000, Abstract P77, pp. 171–172.

[96] V.F. Holmes, N.R. Cozzerelli, Closing the ring: links between SMC proteins and chromosome partitioning, condensation, and supercoiling, Proc. Natl. Acad. Sci. USA 97 (2000) 1322–1324.

[97] L.N. Chilukuri, P.A.G. Fortes, D.H. Bartlett, High pressure modulation of DNA gyrase activity, Biochem. Biophys. Res. Commun. 239 (1995) 552–556.

[98] G.-Q. Tang, N. Tanaka, S. Kunugi, In vitro increases in plasmid DNA supercoiling by hydrostatic pressure, Biochim. Biophys. Acta 1443 (1998) 364–368.

[99] R.S. Myers, F.S. Stahl, Chi and the RecBCD enzyme of *Escherichia coli*, Annu. Rev. Genet. 28 (1994) 49–70.

[100] G.R. Smith, S.M. Kunes, D.W. Schultz, A. Taylor, K.L. Triman, Structure of chi hotspots of generalized recombination, Cell 24 (1981) 429–436.

[101] D.A. Dixon, J.J. Churchill, S.C. Kowalczykowski, Reversible inactivation of the *Escherichia coli* RecBCD enzyme by the recombination hotspot χ in vitro: evidence for functional inactivation or loss of the RecD subunit, Proc. Natl. Acad. Sci. USA 91 (1994) 2980–2984.

[102] D.P. Biek, S.N. Cohen, Identification and characterization of *recD*, a gene affecting plasmid maintenance and recombination in *Escherichia coli*, J. Bacteriol. 167 (1986) 594–603.

[103] L.N. Chilukuri, D.H. Bartlett, Isolation and characterization of the gene encoding single-stranded-DNA-binding protein (SSB) from four marine *Shewanella* strains that differ in their temperature and pressure optima for growth, Microbiology 143 (1997) 1163–1174.

[104] C.R. Robinson, S.G. Sligar, Hydrostatic and osmotic pressure as tools to study macromolecular recognition, Methods Enzymol. 259 (1995) 395–427.

[105] C.A. Royer, A.E. Chakerian, K.S. Mathews, Macromolecular binding equilibria in the *lac* repressor system: studies using high pressure fluorescence spectroscopy, Biochemistry 29 (1990) 4959–4966.

[106] C.E. ZoBell, A.B. Cobet, Some effects of hydrostatic pressure on the multiplication and morphology of marine bacteria, J. Bacteriol. 60 (1950) 771–781.

[107] C.E. ZoBell, A.B. Cobet, Growth, reproduction and death rates of *Escherichia coli* at increased hydrostatic pressures, J. Bacteriol. 84 (1962) 1228–1236.

[108] H.W. Jannasch, Effects of hydrostatic pressure on growth of marine bacteria, in: H.W. Jannasch, R.E. Marquis, A.M. Zimmerman (Eds.), Current Perspectives in High Pressure Biology, Academic Press, Toronto, 1987, pp. 1–15.

[109] A.A. Yayanos, E.F. DeLong, Deep-sea bacterial fitness to environmental temperatures and pressures, in: H.W. Jannasch, R.E. Marquis, A.M. Zimmerman (Eds.), Current Perspectives in High Pressure Biology, Academic Press, Toronto, 1987, pp. 17–32.

[110] H.C. Crenshaw, J.A. Allen, V. Skeen, A. Harris, E.D. Salmon, Hydrostatic pressure to 400 atm does not induce changes in the cytosolic concentration of $Ca2^{+}$ in mouse fibroblasts – measurements using fura-2 fluorescence, Exp. Cell Res. 227 (1996) 277–284.

[111] M. Sato, H. Kobori, S. Shimada, M. Osumi, Pressure-stress effects on the ultrastructure of cells of the dimorphic yeast candida tropicalis, FEMS Microbiol. Lett. 131 (1995) 11–15.

[112] H.P. Erickson, D. Stoffler, Protofilaments and rings, two conformations of the tubulin family conserved from bacterial FtsZ to alpha/beta and gamma tubulin, J. Cell Biol. 135 (1996) 5–8.

[113] A. Lockhart, J. Kendrick Jones, Interaction of the N-terminal domain of MukB with the bacterial tubulin homologue FtsZ, FEBS Lett. 430 (1998) 278–282.

[114] E. Morild, The theory of pressure effects on enzymes, Adv. Protein Chem. 34 (1981) 93–166.

[115] P. Bork, C. Sander, A. Valencia, An ATPase domain common to prokaryotic cell cycle proteins, sugar kinases, actin and hsp70 heat shock proteins, Proc. Natl. Acad. Sci. USA 89 (1992) 7290–7294.

[116] K.J. Beggs, S.J. Dewar, W.D. Donachie, A new *Escherichia coli* cell division gene, *ftsK*, J. Bacteriol. 177 (1995) 6211–6222.

[117] F.X. Gomis-Ruth, G. Moncalian, R. Perez-Luque, A. Gonzalez, E. Cabezón, F. de la Cruz, M. Coll, The bacterial conjugation protein TrwB resembles ring helicases and F-1-ATPase, Nature 409 (2001) 637–641.

[118] M.M. Sun, D.S. Clark, Pressure effects on activity and stability of hyperthermophilic enzymes, Methods Enzymol. 334 (2001) 316–327.

[119] G.N. Somero, Adaptations to high hydrostatic pressure, Annu. Rev. Physiol. 54 (1992) 557–577.

[120] R. Hensel, H. König, Thermoadaptation of methanogenic bacteria by intracellular ion concentration, FEMS Microbiol. Lett. 49 (1988) 75–79.

[121] S. Scholtz, J. Sonnenbichler, W. Schafer, R. Hensel, Di-myo-inositol-1,1′-phosphate: a new inositol phosphate isolated from *Pyrococcus woesei*, FEBS Lett. 306 (1992) 239–242.

[122] M.B. Gillett, J.R. Suko, F.O. Santoso, P.H. Yancey, Elevated levels of trimethylamine oxide in muscles of deep-sea gadiform teleosts: A high-pressure adaptation?, J. Exp. Zool. 279 (1997) 386–391.

[123] G.M. Tobal, Master's Thesis, Scripps Institute of Oceanography, University of California, San Diego, La Jolla, CA, 1993.

[124] J.V. Landau, D.H. Pope, Recent advances in the area of barotolerant protein synthesis in bacteria and implications concerning barotolerant and barophilic growth, Adv. Aquat. Microbiol. 2 (1980) 49–76.

[125] R.E. Marquis, C.E. ZoBell, Magnesium and calcium ions enhance barotolerance of streptococci, Arch. Microbiol. 79 (1971) 80–92.

[126] A.C. Oliveira, L.P. Gaspar, A.T. Da Poian, J.L. Silva, Arc repressor will not denature under pressure in the absence of water, J. Mol. Biol. 240 (1994) 181–187.

[127] M. Odani, H. Shimizu, Y. Komatsu, H. Iwahashi, Response of growing yeast cells in the hydrostatic pressure status monitoring by DNA microarray, in: R. Hayashi (Ed.), High Pressure Bioscience and Biotechnology (HPBB-2000), Kyoto, Japan, 2000, p. P-16.

[128] E.F. DeLong, D.G. Franks, A.A. Yayanos, Evolutionary relationships of cultivated psychrophilic and barophilic deep-sea bacteria, Appl. Environ. Microbiol. 63 (1997) 2105–2108.

[129] A.A. Yayanos, A.S. Dietz, R. Van Boxtel, Obligately barophilic bacterium from the Mariana Trench, Proc. Natl. Acad. Sci. USA 78 (1981) 5212–5215.

[130] J.W. Deming, L.K. Somers, W.L. Straube, D.G. Swartz, Macdonnel, M.T., Isolation of an obligately barophilic bacterium and description of a new genus, *Colwellia* gen. nov., Syst. Appl. Microbiol. 10 (1988) 152-160.

[131] S.J. Bale, K. Goodman, P.A. Rochelle, J.R. Marchesi, J.C. Fry, A.J. Weightman, R.J. Parkes, *Desulfovibrio profundus* sp nov, a novel barophilic sulfate-reducing bacterium from deep sediment layers in the Japan Sea, Int. J. Syst. Bacteriol. 47 (1997) 515–521.

[132] C. Kato, T. Sato, K. Horikoshi, Isolation and properties of barophilic and barotolerant bacteria from deep-sea mud samples, Biodivers. Conserv. 4 (1995) 1–9.

[133] C. Kato, L. Li, Y. Nogi, Y. Nakamura, J. Tamaoka, K. Horikoshi, Extremely barophilic bacteria isolated from the Mariana Trench, Challenger Deep, at a depth of 11,000 meters, Appl. Environ. Microbiol. 64 (1998) 1510–1513.

[134] A.A. Yayanos, A.S. Dietz, R. Van Boxtel, Dependence of reproduction rate on pressure as a hallmark of deep-sea bacteria, Appl. Environ. Microbiol. 44 (1982) 1356–1361.

[135] W. Liesack, H. Weyland, E. Stackebrandt, Potential risks of gene amplification by PCR as determined by 16S rDNA analysis of a mixed-culture of strict barophilic bacteria, Microb. Ecol. 21 (1991) 191–198.

[136] C. Kato, N. Masui, K. Horikoshi, Properties of obligately barophilic bacteria isolated from a sample of deep-sea sediment from the Izu-Bonin Trench, J. Mar. Biotech. 4 (1996) 96–99.

[137] G. Bernhardt, R. Jaenicke, H.-D. Ludemann, H. Koning, K.O. Stetter, High pressure enhances the growth rate of the thermophilic archaebacterium *Methanococcus thermolithotrophicus* without extending its temperature range, Appl. Environ. Microbiol. 54 (1988) 1258–1261.

[138] J.F. Miller, N.N. Shah, C.M. Nelson, J.M. Lulow, D.S. Clark, Pressure and temperature effects on growth and methane production of the extreme thermophile *Methanococcus jannaschii*, Appl. Environ. Microbiol. 54 (1988) 3039–3042.

[139] V.T. Marteinsson, P. Moulin, J.-L. Birrien, A. Gambacorta, M. Vernet, D. Prieur, Physiological responses to stree conditions and barophilic behavior of the hyperthermophilic vent archaeon *Pyrococcus abyssi*, Appl. Environ. Microbiol. 63 (1997) 1230–1236.

[140] V.T. Marteinsson, J.L. Birrien, A.L. Reysenbach, M. Vernet, D. Marie, A. Gambacorta, P. Messner, U. Sleytr, D. Prieur, *Thermococcus barophilus sp. nov.*, a new barophilic and hyperthermophilic archaeon isolated under high hydrostatic pressure from a deep-sea hydrothermal vent, Int. J. Syst. Bacteriol. 49 (1999) 351–359.

[141] R. Meganathan, R.E. Marquis, Loss of bacterial motility under pressure, Nature 246 (1973) 526–527.

[142] J.V. Landau, Induction, transcription, and translation in *Escherichia coli*: a hydrostatic pressure study, Biochim. Biophys. Acta 149 (1967) 506–512.

[143] M. Gross, K. Lehle, R. Jaenicke, K.H. Nierhaus, Pressure-induced dissociation of ribosomes and elongation cycle intermediates. Stabilizing conditions and identification of the most sensitive functional state, Eur. J. Biochem. 218 (1993) 463–468.

[144] R. Pagan, B. Mackey, Relationship between membrane damage and cell death in pressure-treated *Escherichia coli* cells: differences between exponential- and stationary-phase cells and variation among strains, Appl. Environ. Microbiol. 66 (2000) 2829–2834.

ELSEVIER

Biochimica et Biophysica Acta 1595 (2002) 382–386

BIOCHIMICA ET BIOPHYSICA ACTA
BBA
www.bba-direct.com

Review

Compressibility gives new insight into protein dynamics and enzyme function

Kunihiko Gekko *

Department of Mathematical and Life Sciences, Graduate School of Science, Hiroshima University, 1-3-1 Kagamiyama, Higashi-Hiroshima 739-8526, Japan

Received 16 November 2001; accepted 19 November 2001

Abstract

The adiabatic compressibility of enzyme is largely influenced by binding of coenzyme and substrate, due to the changes in atomic packing. Amino acid substitution also induces large changes in compressibility parallel to enzyme activity. These results demonstrate that a small alteration of local structure by ligand binding and mutation is dramatically magnified in the flexibility of protein molecule to affect the function. Compressibility gives new insight into protein dynamics and enzyme function from the aspect of atomic packing or cavity which cannot be obtained by other techniques.

Keywords: Adiabatic compressibility; Amino acid substitution; Cavity; Enzyme function; Ligand binding; Protein dynamics

1. Introduction

Compressibility is not only a basic physical quantity to analyze the pressure effects on protein structure, but also it gives new insight into protein dynamics which contributes to biological function of protein molecules. According to the X-ray crystal structure, globular proteins have a precisely defined compact structure and all atoms in the molecule seem to be closely packed as well as the crystalline amino acids. However, there are many packing defects or cavities in the interior of a protein molecule, which cause internal motions or flexibility response to thermal or mechanical forces. Such cavities are easily compressed by pressure, so compressibility is an important measure of protein flexibility or volume fluctuation. According to statistical thermodynamics, the volume fluctuation δV_{rms} of a protein with volume V_p is related to its isothermal compressibility β_T [1]:

$$\delta V_{\mathrm{rms}} = (kTV_{\mathrm{p}}\beta_T)^{1/2} \quad (1)$$

where k is the Boltzmann constant and T the absolute temperature.

Measurement of compressibility is possible, in principle, only for proteins in solution. Although volume fluctuation is directly linked to isothermal compressibility, most compressibility studies of proteins have been of adiabatic compressibility because of the accuracy and technical convenience [2–5]. The partial specific adiabatic compressibility $\beta_s°$ is defined as the change in the partial specific volume $v°$ with pressure. Since the constitutive atomic volume of a protein may be regarded as incompressible, the experimen-

* Fax: +81-824-247387.
E-mail address: gekko@sci.hiroshima-u.ac.jp (K. Gekko).

PII: S0167-4838(01)00358-2

tally determined $\beta_s°$ can be mainly attributed to its cavity v_{cav} and to changes in hydration Δv_{sol} [3],

$$\beta_s° = -(1/v°)(\partial v°/\partial P)$$
$$= -(1/v°)[\partial v_{cav}/\partial P + \partial \Delta v_{sol}/\partial P] \quad (2)$$

The cavity contributes positively and hydration negatively to $v°$ and $\beta_s°$, then $\beta_s°$ can be positive or negative depending on the magnitude of both terms. Most globular proteins have a positive $\beta_s°$ value of 0–14 $Mbar^{-1}$ at 25°C, indicating the large contribution of cavity overcoming the hydration effect [3]. The volume fluctuation is in the range from 30 to 200 cm^3 mol^{-1}, which correspond to only 0.3% of whole protein volume. But if concentrated in one area, it would produce sufficient cavities or channels in the protein molecule to allow penetration of solvent or probe molecules. Recent studies on the effects of ligand binding and amino acid substitution on $\beta_s°$ demonstrate that a small alteration of local structure is dramatically magnified in the overall protein dynamics to affect the enzyme function [6–9].

2. Effects of ligand binding

The enzyme–ligand complex formation generally proceeds in three states: formation of net bonds between enzyme and ligand, changes in hydration of the interacting species, and conformational changes of the enzyme. The solute-solute interaction in water should accompany dehydration of each solute molecule, resulting in the increase in $v°$ and $\beta_s°$. However, the observed changes in both parameters by ligand binding are not necessarily positive. A typical example is lysozyme–inhibitor complex formation. The $v°$ and $\beta_s°$ values of lysozyme decrease by addition of *N*-acetyl-D-glucosamine oligomers, in the order of monomer > dimer > trimer [6]. Assuming that the decrease in $\beta_s°$ is ascribed to the protein moiety only, the volume fluctuation is estimated to decrease by about 3%, 7%, and 14% on binding the monomer, dimer and trimer, respectively. The X-ray crystallographic study shows that the inhibitors are bound in the cleft of the active site of lysozyme. The effects of this 'induced-fit' type interaction would propagate throughout the entire protein molecule to enhance the internal atomic packing, resulting in the depressed fluctuation of the protein molecule. The hydrogen-exchange rate of lysozyme is also known to decrease in the same order of these oligomers [10]. It is noteworthy that the volume fluctuation changes comparably to the kinetically determined fluctuation.

More pronounced changes in $\beta_s°$ and $v°$ are found on ligand binding of *Escherichia coli* dihydrofolate reductase (DHFR). DHFR catalyzes the reduction of dihydrofolate (DHF) to tetrahydrofolate (THF) with the aid of coenzyme NADPH in the kinetic reactions cycling through five intermediates, DHFR·NADPH, DHFR·NADPH·DHF, DHFR·$NADP^+$·THF, DHFR·THF, and DHFR· NADPH·THF [11]. This enzymatic function is strongly repressed with an inhibitor, methotrexate (MTX). The X-ray structure of a DHFR·NADPH·MTX ternary complex is shown in Fig. 1 [12]. The $\beta_s°$ of these kinetic intermediates is shown as a function of the reaction coordinate in Fig. 2 [7], where the transient state DHFR· NADPH·DHF is substituted by a complex DHFR·NADPH·MTX since the X-ray structures of both complexes are quite similar. Evidently, the flexibility of DHFR changes alternatively by binding or releasing the coenzyme and substrate: the transient state DHFR·NADPH·DHF is expected to be most flexible and the state DHFR·$NADP^+$·THF most rigid in the intermediates. As illustrated in Fig. 3, the number, size and distribution of cavities also change on ligand binding, parallel to the changes in partial specific volume ($\Delta v°$) and compressibility ($\Delta K_s = \Delta v° \beta_s°$). On binding of these ligands, the total cavity volume is largely influenced (±40%) but the solvent accessible surface area only slightly decreases (at most 6%), indicating that the cavity plays dominant role in ligand-induced compressibility change overcoming hydration effects. Recent high pressure NMR study of folate-bound DHFR reveals that major conformation changes by pressure occur in the NADPH binding site and the flexible loops [13].

3. Effects of amino acid substitution

How does the single amino acid substitution affect the flexibility of overall protein structure? A key to answer this problem is found in the mutation effects on $\beta_s°$ of two *E. coli* enzymes, aspartate aminotrans-

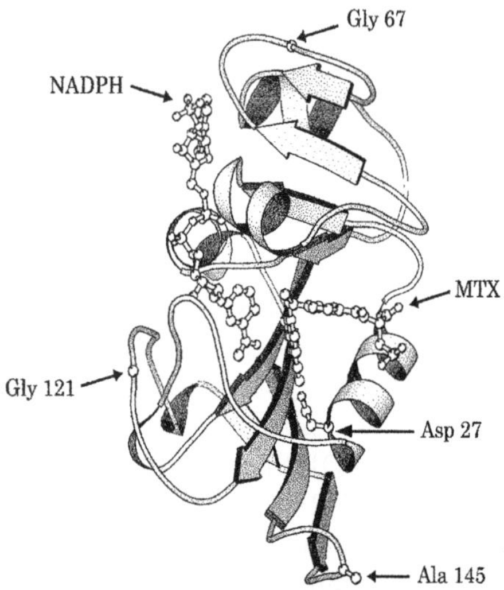

Fig. 1. The structure of a DHFR–MTX–NADPH ternary complex taken from Sawaya and Kraut [12]. Positions Gly67, Gly121, Ala145 (mutation sites) and Asp27 (catalytic site) are indicated by arrows.

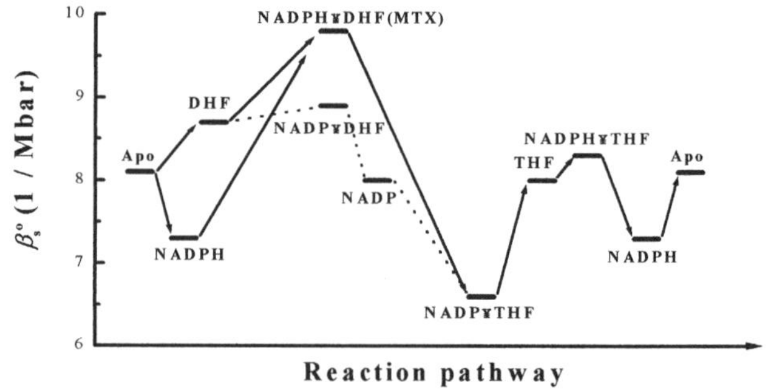

Fig. 2. The $\beta_s°$ level of the kinetic intermediates of DHFR as a function of the reaction pathway [7].

ferase (AspAT) and DHFR. AspAT has two identical 45 kDa subunits and the catalytic site is formed between the large and small domains in each subunit. Six mutant enzymes at Val39, which is located at the gate of the substrate binding site, have evidently different $\beta_s°$ values [8]. The observed positive linear correlation between $\beta_s°$ and van der Waals volume of the introduced amino acids suggests that the bulkiness of the side chain at position 39 makes the protein structure more flexible due to the overcrowding effect in atomic packing. The catalytic efficiency of AspAt is also modified by these mutations, where major changes occur on K_m but not on k_{cat}. Interestingly, there is a linear correlation between log (k_{cat}/K_m) and $\beta_s°$, showing that the amino acid substitution at Val39, which increases flexibility of overall structure, makes the enzyme more active.

The enzyme activity of DHFR is known to be influenced by mutations at Gly67, Gly121, and Ala145, which are located at the center of three different flexible loops (see Fig. 1), where K_m is only slightly influenced but k_{cat} remarkably changes [14–16]. This is very interesting because it is unlikely that the mutation sites directly participate in the enzyme reaction: α-carbons of Gly67, Gly121 and Ala145 are separated respectively by 29.3, 19.0 and 14.4 Å from the catalytic site Asp27, and at shortest by 8.5, 10.6 and 29.2 Å from the NADPH molecule. Furthermore, double mutants at Gly67 and Gly121, whose α-carbons are mutually separated by 27.7 Å, show non-additive effects on the function [17]. These findings suggest that these mutations would alter the

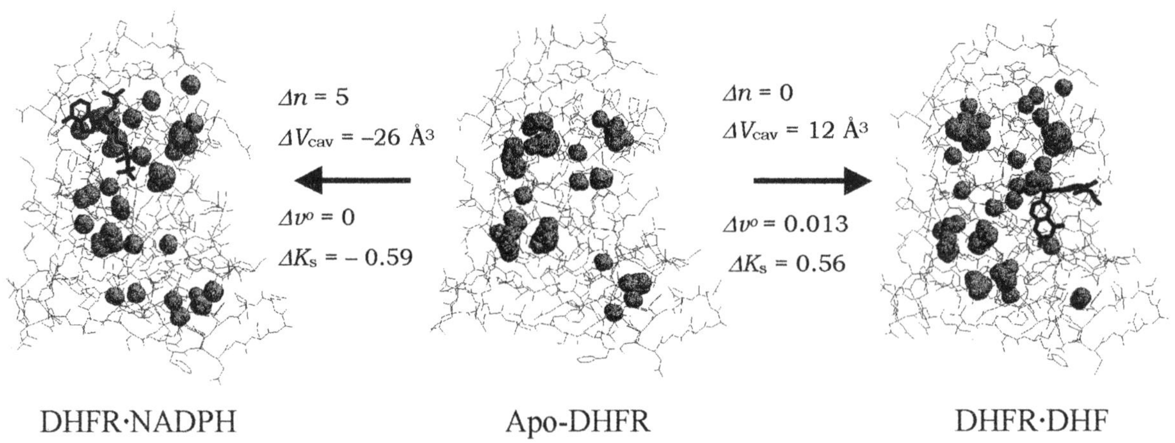

Fig. 3. The changes of number (n), position and total volume (ΔV_{cav}) of cavities in some kinetic intermediates of DHFR [7]. $\Delta K_s = \Delta v° \beta_s°$: $cm^3\ g^{-1}\ Mbar^{-1}$.

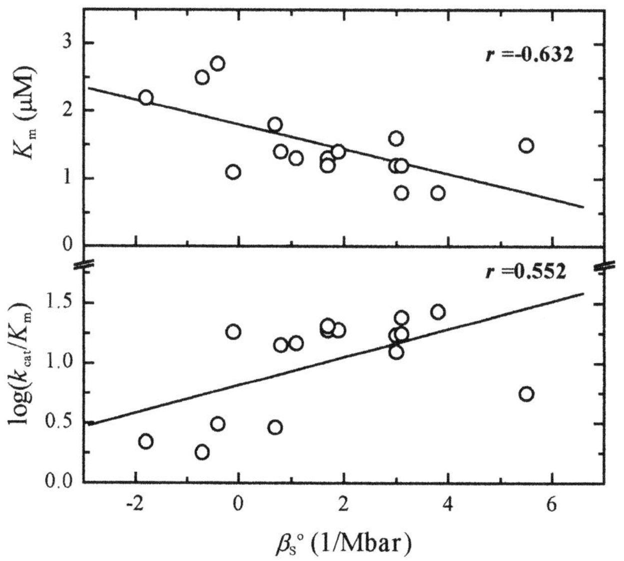

Fig. 4. Plots of K_m and log (k_{cat}/K_m) against $\beta_s°$ of DHFR mutants [9].

flexibility of the entire protein molecule. In fact, these mutations induce large changes in $v°$ (0.710–0.733 cm^3 g^{-1}) and $\beta_s°$ (−1.8–5.5 $Mbar^{-1}$) from the corresponding values of wild-type enzyme ($v°$ = 0.723 cm^3 g^{-1}, $\beta_s°$ = 1.7 $Mbar^{-1}$) [9]. As shown in Fig. 4, there is a definite correlation between $\beta_s°$ and K_m or log (k_{cat}/K_m), indicating that the structural flexibility positively contributes to the enzyme function, as is the case of AspAT, through an enhanced catalytic reaction rate and in part due to increased affinity for the substrate. It is important that the flexibility-mediated modification of enzyme activity can be derived by the mutation at the loop regions far from the catalytic site because this gives a new strategy to design more functional proteins. Our recent X-ray analyses reveal that the B-factors of main chain atoms and the cavities at positions far from the mutation sites are modified without any large change in solvent accessible surface area and the mutants with the large $\beta_s°$ values have large amount of cavity (K. Katayanagi et al., to be published). These results demonstrate that the effects of mutation extend to a long range and the mutation-induced $\beta_s°$ change is dominantly ascribed to modification of internal cavities.

As shown here, compressibility gives new insight into protein dynamics and enzyme function from the aspect of atomic packing or cavity which cannot be obtained by other techniques. A small alteration of local structure due to ligand binding and amino acid substitution is dramatically magnified in the overall protein dynamics, possibly via 'stepwise atomic reconstruction or long-range cooperative effect', leading to modification of catalytic activity of enzymes. Compressibility study in combination with X-ray crystallography and high-pressure NMR would be fruitful for elucidating the structure–flexibility–function relationship of enzymes.

References

[1] A. Cooper, Thermodynamic fluctuations in protein molecules, Proc. Natl. Acad. Sci. USA 73 (1976) 2740–2741.

[2] K. Gekko, H. Noguchi, Compressibility of globular proteins in water at 25°C, J. Phys. Chem. 83 (1979) 2706–2714.

[3] K. Gekko, Y. Hasegawa, Compressibility-structure relationship of globular proteins, Biochemistry 25 (1986) 6563–6571.

[4] T.V. Chalikian, V.S. Gindikin, K.J. Breslauer, Volumetric characterizations of the native, molten globule and unfolded states of cytochrome c at acidic pH, J. Mol. Biol. 250 (1995) 291–306.

[5] D.P. Kharakoz, Partial volumes and compressibilities of extended polypeptide chains in aqueous solution: additivity scheme and implication of protein unfolding at normal and high pressure, Biochemistry 36 (1997) 10276–10285.

[6] K. Gekko, K. Yamagami, Compressibility and volume changes of lysozyme due to inhibitor binding, Chem. Lett. (1998) 839-840.

[7] T. Kamiyama, K. Gekko, Effect of ligand binding on the flexibility of dihydrofolate reductase as revealed compressibility, Biochim. Biophys. Acta 1478 (2000) 257–266.

[8] K. Gekko, Y. Tamura, E. Ohmae, H. Hayashi, H. Kagamiyama, H. Ueno, A large compressibility change of protein induced by a single amino acid substitution, Protein Sci. 5 (1996) 542–545.

[9] K. Gekko, T. Kamiyama, E. Ohmae, K. Katayanagi, Single amino acid substitutions in flexible loops can induce large compressibility changes in dihydrofolate reductase, J. Biochem. 128 (2000) 21–27.

[10] M. Nakanishi, M. Tsuboi, A. Ikegami, Fluctuation of the lysozyme structure: (II) Effects of temperature and binding of inhibitors, J. Mol. Biol. 75 (1973) 673–682.

[11] C.A. Fierke, K.A. Johnson, S.J. Benkovic, Construction and evaluation of the kinetic scheme associated with dihydrofolate reductase from *Escherichia coli*, Biochemistry 26 (1987) 4085–4092.

[12] M.R. Sawaya, J. Kraut, Loop and subdomain movements in the mechanism of *Escherichia coli* dihydrofolate reductase: crystallographic evidence, Biochemistry 36 (1997) 586–603.

[13] R. Kitahara, S. Sareth, H. Yamada, E. Ohmae, K. Gekko, K. Akasaka, High pressure NMR reveals active-site hinge

motion of folate-bound *Escherichia coli* dihydrofolate reductase, Biochemistry 39 (2000) 12789–12795.

[14] K. Gekko, Y. Kunori, H. Takeuchi, S. Ichihara, M. Kodama, Point mutations at glycine-121 of *Escherichia coli* dihydrofolate reductase: important roles of a flexible loop in the stability and function, J. Biochem. 116 (1994) 34–41.

[15] E. Ohmae, K. Iruyama, K. Ichihara, K. Gekko, Effects of point mutations at the flexible loop of *Escherichia coli* dihydrofolate reductase on its stability and function, J. Biochem. 119 (1996) 703–710.

[16] E. Ohmae, K. Ishimura, M. Iwakura, K. Gekko, Effects of point mutations at the flexible loop alanine-145 of *Escherichia coli* dihydrofolate reductase on its stability and function, J. Biochem. 123 (1998) 839–846.

[17] E. Ohmae, K. Iriyama, S. Ichihara, K. Gekko, Nonadditive effects of double mutations at the flexible loops, glycine-67 and glycine-121, of *Escherichia coli* dihydrofolate reductase on its stability and function, J. Biochem. 123 (1998) 33–41.

ELSEVIER

Biochimica et Biophysica Acta 1595 (2002) 387–389

BIOCHIMICA ET BIOPHYSICA ACTA
BBA
www.bba-direct.com

Review

Experiments on ion channels at high pressure

Alister G. Macdonald *

Department of Biomedical Sciences, University of Aberdeen, Zoology Building, Tillydrone Avenue, Aberdeen AB24 2TZ, Scotland, UK

Received 31 October 2001; accepted 6 November 2001

Abstract

Ion channels are distinctive membrane proteins which provide a gated pathway for diffusing ions. High pressure (< 100 MPa) affects the kinetics of gating but not the conductance of the channel. Dynamic structural studies of channels at high pressure are, thus far, conspicuously absent but functional properties are studied at the single channel level with the patch clamp technique.

Keywords: Gating; Channel conductance; Inactivation; Porin; Patch clamp

This short review is intended for those who are unfamiliar with ion channels but knowledgeable about the biophysics of high pressure. Readers who, conversely, are knowledgeable about channels but unfamiliar with high pressure might wish to skip the first part and ponder the worth of the second.

Ion channels are polypeptides or proteins which traverse cell membranes, providing an aqueous pore through which ions diffuse down their electrochemical gradient. They select for cations or anions or more specifically for Na, Ca, etc. Diffusion is nevertheless rapid (10^7 monovalent ions s^{-1} per channel) and the overall transport of ions is regulated by the opening/closing (gating) of the channels' diffusion path [1–3]. Some channels conduct ions in one direction, ie rectify. Although channels can be regarded as enzymes, catalysing the diffusion of ions across membranes [4], most regard these macromolecules simply as distinctive membrane proteins, not least because of the specialized and powerful electrical recording techniques used in the study of their cellular functions. These underlie excitability, sensory transduction, much metabolic control and many other physiological processes.

Gating occurs on a μsec time scale and generally entails numerous state transitions en route to the functional open (ion conducting) or closed states, which usually exist on a millisecond scale. Inactivated states also occur on a longer time scale. Channels are gated by transmembrane voltage, or by ligand binding (the ligands being either extracellular, as with neurotransmitters, or intracellular, as with cGMP or Ca^{2+}) [1,2,5], or by mechanical energy, as in stretch-activated and -inactivated channels [6]. Temperature and high hydrostatic pressure also gate certain channels [7,8].

There are three main classes of channel. First, there are channels found in all prokaryote and eukaryote cells, which are based on a primary peptide sequence, in which α-helical transmembrane segments are connected by extra- and intracellular loops. The carboxyl and amino terminuses are both

* Fax: +44-1224-272396.
E-mail address: a.macdonald@abdn.ac.uk (A.G. Macdonald).

0167-4838/02/$ – see front matter
PII: S0167-4838(01)00359-4

intracellular. Channels selective for Na and Ca have four transmembrane segments [9], K-selective channels, the largest and most diverse group, vary [10,11]. Voltage-gated and calcium-activated K channels have six transmembrane segments, leakage K channels have four and inward rectifying channels have only two transmembrane segments. Specific functions are attributed to certain parts of the primary sequence. For example, transmembrane segment 4 (S4) is rich in positive charges and responds to voltage. Typically four primary units are aggregated to form voltage-gated channels, five in the case of ligand-gated channels and six form gap junctions. The water filled pore through which ions diffuse lies in the aggregate's axis of symmetry. Additional proteins accompany the multimeric channel and regulate some of its properties [9,10]. Another class of channels, the bacterial porins, are trimeric proteins which form water-filled pores traversing the outer membrane of Gram-negative bacteria. The three-dimensional structure of several has been resolved on the atomic scale. Their unit structure is a barrel of 16 or so anti-parallel β strands connected by helical loops, one of which projects into the pore and, probably by electrostatic means rather than occlusion, gates the pore [12–14].

Channel-forming peptides are the third category of channels. They are free agents, which partition into lipid membranes and aggregate to form a water-filled pore [15]. Examples include the dimer gramicidin and an antibiotic alamethicin, which forms decamer channels.

Channel sequences and structures are investigated by all the techniques used to study other proteins. Remarkable electron microscope images of the functioning acetylcholine-gated channel have been generated by Unwin [16]. Atomic force microscopy [13,17], electronparamagnetic [18,19] and fluorescence spectroscopy [20] are becoming used in structure/function studies. Channel function is studied by recording the electrical currents generated by the ion fluxes. In the last decade the patch-clamp recording technique has dominated the field, because it can resolve the transient currents passing through an individual channel (10^{-12} A; < 100 μs resolution), enabling the kinetics of channel gating to be observed free from the averaging of ensemble recording [17,21]. As Colquhoun and Hawkes point out ([17], Chapter 18) "The advent of single-channel recording has provided perhaps the strongest and most direct evidence for the existence of discrete states of large protein molecules". All three classes of channel can be studied at the single molecule level either using natural cell membranes or by reconstituting the channel in an artificial lipid bilayer [12,15,21]. Furthermore, the patch-clamp recording technique has been adapted for use at pressures up to 90 MPa [22,8], and there is potential for greatly extending the pressure range as well as the types of channel so far studied ([3], Chapter 59). As these electrical recordings give no direct information about channel structure there is also considerable scope for other pressure studies, especially using fluorescence spectroscopy [20].

Why experiment on ion channels at high pressure? The biological reason is that organisms live at pressures of up to 100 MPa so ion channels, like other cellular constituents such as enzymes are thought to be adapted, albeit in subtle ways, to such pressures [23]. Lesser pressures (< 10 MPa) elicit hyperexcitable symptoms in human divers and animals, and this too has stimulated interest in channels under pressure [24]. The main physical reason is the classical question of how high pressure might perturb the system in predictable, and to be candid, unpredictable, ways. Fast ionic conduction, ionic selectivity, protein–lipid and protein–protein interactions are all attractive targets. Gating has special interest, see below. The study of high-pressure effects on ion channels is in its infancy but there are sufficient published experiments to encourage the wary. Here are some examples.

Alamethicin, partitioning into a planar lipid bilayer, provided the first single-channel recordings at high pressure (100 MPa) [25]. The open state lifetime was prolonged by pressure. It also slowed the kinetics of the onset of conductance, probably affecting an intermediate non-conducting assembly stage. Pressure had no effect on the voltage sensitivity of monomer aggregation or on channel conductance. Similarly, the first high-pressure patch-clamp recordings, of voltage-gated Na and Ca channels, and acetylcholine-gated channels in cultured cells, showed the channel kinetics were slowed but their conductance was unaffected [22,26]. The latter is a general finding, suggesting that pore geometry is insensitive to the pressures used to date. Furthermore, the pressure-dependence of the Na channel gating matched that measured in earlier ensemble voltage-clamp measurements of squid giant axon Na currents [24].

Gating in general involves a sensing stage followed by an allosteric response. In ligand gating the free energy of binding provides the energy for the channels' structural response. In voltage-gated channels dipoles move in the electrical field, creating 'gating current'. Ensemble voltage-clamp recordings of this capacitative current generated by Na channels were achieved under high pressure by Conti and colleagues [27], using the squid giant axon. They found that gating current is less sensitive to pressure than the gating of the channel as a whole. This means that following the movement of dipoles there is a structural stage with a large positive activation volume. Contemporary thinking would equate the gating current with a rotation of the positive charge-carrying S4 helix [20]. This rotation is somehow transmitted to other transmembrane segments which open the channel, the process which involves a large positive activation volume.

The interpretation of pressure effects is always clarified by detailed knowledge of the molecular structures involved, and this maxim is exemplified by some recent channel experiments. A P-T study of the thermodynamics of N-type inactivation in mutant Shaker K channels by Meyer and Heinemann, for example, produced evidence that pressure dissociated a peptide from its channel binding site [28]. More speculatively, the high-pressure (90 MPa) activation of BK channels, manifest as a large increase in the open state probability, is also interpreted as pressure dissociating an endogenous inhibitor [29], first postulated by Salkoff and colleagues.

References

[1] B. Hille, Ionic Channels of Excitable Membranes, Sinauer Associates, Sunderland, MA, 1992.
[2] C. Miller, Curr. Opin. Chem. Biol. 4 (2000) 148–151.
[3] N. Sperelakis (Ed.), Cell Physiology Source Book, Academic Press, San Diego, CA, 2001.
[4] E. Marban, G.F. Tomaselli, Trends Neurosci. 20 (1997) 144–147.
[5] R.A. North (Ed.), Ligand and Voltage-gated Ion Channels, CRC Press, Boca Raton, FA, 1995.
[6] O.P. Hamill, B. Martinac, Physiol. Rev. 81 (2001) 685–740.
[7] P. Cesare, A. Moriondo, V. Vellani, P.A. McNaughton, Proc. Natl. Acad. Sci. USA 96 (1999) 7658–7663.
[8] A.G. Macdonald, Biophys. J. 73 (1997) 1866–1873.
[9] W.A. Catterall, Neuron 26 (2000) 13–25.
[10] W.A. Coetzee et al., Ann. N.Y. Acad. Sci. 868 (1999) 233–285.
[11] D.A. Doyle, J.M. Cabral, R.A. Pfuetzuer, A. Kno, J.M. Gulbis, S.L. Cohen, B.T. Chait, R. Mackinnon, Science 280 (1998) 69–76.
[12] A.H. Delcour, FEMS Microbiol. Lett. 151 (1997) 115–123.
[13] D.J. Müller, A. Engel, J. Mol. Biol. 285 (1999) 1347–1351.
[14] P. Van Gelder, F. Dumas, M. Winterhalter, Biophys. Chem. 85 (2000) 153–167.
[15] M.S.P. Sansom, Q. Rev. Biophys. 26 (1993) 365–421.
[16] N. Unwin, Phil. Trans. R. Soc. Lond. B 355 (2000) 1813–1829.
[17] B. Sakmann, E. Neher (Eds.), Single-channel Recording, Plenum, New York, 1995.
[18] E. Perozo, D.M. Cortes, L.G. Cuello, Nat. Struct. Biol. 5 (1998) 459–469.
[19] E. Perozo, A. Kloda, D.M. Cortes, B. Martinac, J. Gen. Physiol. 118 (2001) 193–205.
[20] L.M. Manuzzu, E.Y. Isacoff, J. Gen Physiol. 115 (2000) 257–268.
[21] D. Ogden (Ed.), Microelectrode Techniques, Company of Biologists, Cambridge, UK, 1994.
[22] S.H. Heinemann, W. Stühmer, F. Conti, Proc. Natl. Acad. Sci. USA 84 (1987) 3229–3233.
[23] A.G. Macdonald, Comp. Biochem. Physiol. (2001) in press.
[24] J.J. Kendig, Y. Grossman, S.H. Heinemann, in: A.G. Macdonald (Ed.), Effects of High Pressure on Biological Systems, Springer, Berlin, 1993, pp. 88–124.
[25] L.J. Brunner, J.E. Hall, Biophys. J. 44 (1983) 39–47.
[26] S.H. Heinemann, F. Conti, W. Stuhmer, E. Neher, J. Gen. Physiol. 90 (1987) 765–778.
[27] F. Conti, in: J.M. Richie, R.D. Keynes, L. Bolis (Eds.), Ion Channels in Neural Membranes, Liss, New York, 1986, pp. 25–41.
[28] R. Meyer, S.H. Heinemann, Eur. Biophys. J. 26 (1997) 433–445.
[29] A.G. Macdonald, Biophys. J. 80 (2001) 222a.

ELSEVIER

Biochimica et Biophysica Acta 1595 (2002) 390–391

BIOCHIMICA ET BIOPHYSICA ACTA
BBA

www.bba-direct.com

Review

Cell biology and high pressure: applications and risks

Horst Ludwig *

Institut für Pharmazeutische Technologie und Biopharmazie, Universität Heidelberg, Im Neuenheimer Feld 346, D-69120 Heidelberg, Germany

Received 8 November 2001; accepted 9 November 2001

High pressure is being used in more and more applications in the food industry. After one decade of enhanced basic research in mainly Japan and Europe, the breakthrough to the market seems to be occurring in the United States. In Europe, the Novel-Food regulations strongly impede the establishment of pressurized foods in the market. The only way out of this situation is to study the health risks associated with high-pressure treated foods. These risks include the formation of toxins or allergens during high-pressure treatment, and the development of pressure-resistant microorganism. The latter subject will be discussed in this brief article.

Microorganisms can live in quite different surroundings. They can be found at high salt concentrations, at low pH, at temperatures as high as 120°C, in the cold (deep freezer), and at high pressure (deep sea). These 'extremophiles' are the result of many years of adaptation. However, adaptation can also occur in a short time period. Hauben et al. succeeded in their efforts to obtain two pressure resistant *Escherichia coli* strains by repeatedly pressurizing bacteria and growing the survivors [1]. The authors explained their result as spontaneous mutations of the original strain of *E. coli*, and selection of the mutants by the often repeated pressurizing process. We applied the procedure of Hauben et al. to other bacterial strains and to moulds, but we could not find any acquisition of pressure resistance. This is not contradictory to Haubens results, since mutations occur by chance, and different species or strains differ strongly in their mutation frequency.

Recently, we experienced a very unusual modification of a microorganism under pressure. Massive quantities of spores of *Bacillus thuringiensis* subsp. *israelensis* were mutated into a new pressure-resistant strain by a single pressure treatment [2].

The modified bacteria have a different phenotype than the original strain. Their colonies on agar plates are perfectly circular and remain small in contrast to colonies of the mother strain (Fig. 1). Electron microscopy revealed that the ultrastructure of the new bacteria is different from that of the original strain as well [3]. The mutations in the new resistant bacteria

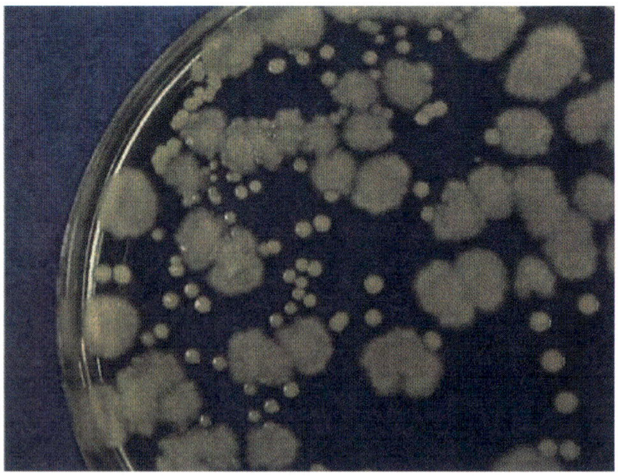

Fig. 1. Regular and HP-changed type of *Bacillus thuringiensis* ssp. *israelensis*.

* Fax: +49-6221-545960.
E-mail address: h.ludwig@urz.uni-heidelberg.de (H. Ludwig).

PII: S0167-4838(01)00360-0

are stable, so that all the changes are passed on to the progeny.

It should be emphasized that the mutation described above is only obtained by spores under pressure, not from the vegetative Bacilli. The reason for this may be that the very restricted metabolism in spores prevents repair mechanisms that are active in vegetative bacteria. If so, similar events could be expected with spores of other bacterial species. To our knowledge, this is the first example of such a cooperative mutation of microorganisms. Changes of the DNA and of proteins are currently being investigated.

References

[1] K.J.A. Hauben, D.H. Bartlett, C.C.F. Soontgens, K. Cornelis, E.Y. Wyuntack, C.W. Michiels, *Escherichia coli* mutants resistant to inactivation by high hydrostatic pressure, Appl. Environ. Microbiol. 63 (1997) 945–950.

[2] H. Ludwig, K.G. Werner, E. Schattmann, M. Schauer, High pressure induced alterations in morphology and cell characteristics of the bacterium *Bacillus thuringiensis*, in: R. Hayashi (Ed.), Trends in High Pressure Bioscience and Biotechnology, Elsevier Science, Amsterdam, 2002, pp. 303–310.

[3] K.G. Werner, H. Ludwig, A comparative electron microscopic study of cell growth and ultrastructure from a regular and a HP-changed type of *Bacillus thuringiensis ssp. israelensis*, in: R. Hayashi (Ed.), Trends in High Pressure Bioscience and Biotechnology, Elsevier Science, Amsterdam, 2002, pp. 337–342.

ELSEVIER

Biochimica et Biophysica Acta 1595 (2002) 392–396

BIOCHIMICA ET BIOPHYSICA ACTA
BBA

www.bba-direct.com

Review

Exploring hyperthermophilic proteins under pressure: theoretical aspects and experimental findings

Enrico Mombelli [a], Erlet Shehi [b], Paola Fusi [b], Paolo Tortora [b,*]

[a] *National Institute for Medical Research, The Ridgeway, Mill Hill, London NW7 1AA, UK*

[b] *Dipartimento di Biotecnologie e Bioscienze, Università di Milano-Bicocca, Piazza della Scienza 2, I-20126 Milan, Italy*

Received 23 November 2001; accepted 26 November 2001

Abstract

Proteins from hyperthermophilic microorganisms are generally capable of withstanding temperatures close to, or even higher than the boiling point. As a rule, these proteins are strongly piezostable as well, although exceptions have been also reported. This observation has a theoretical relevance, as the understanding of the effects of pressure and temperature on protein stability is equally important to develop a comprehensive model for their thermodynamic stability. Nevertheless, the structural features justifying the correlation between heat resistance and pressure resistance are poorly understood. Actually, most reports do not exceed the phenomenological level. Only in the case of the small protein Sso7d from *Sulfolobus solfataricus*, characterisation of wild-type and some mutants showed that both properties are largely accounted for by a network of aromatic residues found in the hydrophobic core of the molecule. Current knowledge, however, does not allow to establish to what extent this finding may be generalised. In a biotechnological perspective, hyperthermophilic enzymes seem to be more suitable for bioprocesses at high pressure with respect to their mesophilic counterparts. Indeed, thanks to their higher resistance towards pressure and temperature, they may be exploited in a much broader range of working conditions for tuning activity and specificity. Furthermore, they are often activated by increasing pressure, although it cannot be established, to date, to what extent this is a common feature.

Keywords: Hyperthermophile; Archaea; Protein thermostability; Protein piezostability; Enzyme catalysis; Hydrostatic pressure

Microorganisms thriving in hot environments (mostly but not exclusively archaeal) have developed a set of adaptation mechanisms to withstand such extreme conditions, including an inventory of heat-resistant macromolecules. In particular, bacterial forms referred to as hyperthermophiles grow optimally in the range 80–110°C. As a result, not only proteins from these organisms are capable of withstanding temperatures close to, and even higher than the boiling point of water [1–3], but also require them to exert their biological activity.

There are not unique structural requirements for making a protein thermostable. Indeed, the exceptional stability of hyperthermophilic proteins results from a subtle balance of stabilising and destabilising factors, the latter being mainly accounted for by conformational entropy [4]. Thermostabilising interactions include hydrophobic effect, van der Waals contacts, intramolecular hydrogen bonding, salt and disulfide bridges. Generally, at the physiological tem-

* Corresponding author. Fax: +39-02-6448-3565.
E-mail address: paolo.tortora@unimib.it (P. Tortora).

0167-4838/02/$ – see front matter
PII: S0167-4838(01)00361-2

perature stabilising interactions only slightly exceed destabilising ones, thus proteins are only marginally stable [4].

In recent times, the growing attention for the effect of hydrostatic pressure on protein stability [5–7] raised the issue of whether protein piezo- and thermostability are correlated. This issue has both theoretical and biotechnological relevance. In a theoretical perspective, the understanding of the effects of pressure and temperature are equally important to develop a comprehensive model for protein thermodynamic stability [8]. Indeed, a complete phase diagram is not accomplished if the effect of pressure is not taken into account. In a biotechnological perspective, the availability of enzymes both stable and active at high pressure may help develop enzyme-catalysed processes in which either selectivity is modulated, or reaction rate is increased by pressure [7].

Nevertheless, the relationship between protein piezo- and thermostability is not yet fully understood. Actually, pressure and temperature do not affect in the same way all non-covalent interactions. Although intramolecular ionic interactions are destabilised by increasing both pressure and temperature, in contrast, intramolecular hydrogen bonds are impaired on heating [4,8], whereas they are virtually pressure-insensitive [5–7]. Furthermore, the exposure of hydrophobic residues to the solvent is believed to be exothermic in itself and thus favoured by decreasing temperature, irrespective of whether contacts are between aromatics or aliphatics [9]. On the other hand, interactions between aromatics in the protein core are even reinforced by pressure [6,7]. As regards the effects of pressure on hydrophobic contacts between aliphatics, available data are not straightforward. Kauzmann reported that the volume change (ΔV) upon protein unfolding (largely accounted for in terms of exposure of apolar residues to the aqueous medium) is positive at low pressures but negative at pressures higher than 1–2 kbar [8], and that the transfer of hydrocarbons into water is governed by an opposite behaviour, i.e., it shows a negative ΔV at low pressures and a positive one at high pressures. This issue has been recently addressed by Hummer et al. [10], who pointed out that pressure and heat denaturation represent processes underlying distinct mechanisms, the first corresponding to the incorporation of water into protein, the second to the transfer of non-polar groups into water. Thus, pressure-denatured proteins have a reduced compactness, yet they retain a considerably more ordered structure with respect to heat-denatured proteins [10,11].

The above considerations make it apparent that one can hardly infer, on a theoretical basis, a protein's pressure resistance from data regarding its heat resistance. Nevertheless, the majority of experimental findings shows that a thermostable protein is piezostable as well. It has been found, in fact, that in many cases thermophilic proteins are also stabilised by pressure values which lead to inactivation of the mesophilic counterparts. This holds true, for instance, for hydrogenases from *Methanococcus* strains [12], *Sulfolobus solfataricus* carboxypeptidase [13] cytochrome P450 from the same microorganism [14], DNA polymerases from *Thermus aquaticus* and from different *Pyrococcus* strains [15]. This is not usually observed in mesophilic proteins. Therefore, thermostable enzymes represent an optimal pool to choose from for possible biotechnological applications under pressure. The increase in protein stability on increasing pressure may be accounted for on the basis of the well-known pressure–temperature phase diagram, first depicted by Bridgman [7,16]. This predicts that, over a broad temperature range, proteins are first stabilised and then destabilised by increasing pressure. Thus, the described stabilising effect should correspond, for each tested protein, to the lower pressure range of the diagram. This implies that for thermophilic proteins both stabilisation and destabilisation occurs in a higher range of pressure as compared to mesophilic counterparts. These observations, however, leave many questions unanswered. First, this correlation is not always observed as shown, for instance, in the case of adenylate kinases from *Methanococcus* strains [17]. Second, in the different cases reported, working pressures spanned different ranges, in some cases up to 4 kbar [13], in others barely exceeding 0.5 kbar [12]. Third, and more importantly, rarely did these reports exceed the phenomenological level. In the case of *Sulfolobus* carboxypeptidase, intramolecular electrostatic interactions were formerly shown to be a major determinant of thermostability [18]. This fits well with previous papers which, comparing the structure of hyperthermophilic proteins with mesophilic counterparts, concluded that ion pairs often play a major role in hyperthermostability [19], and also point out that these occur in form of large networks rather than single salt bridges [19]. However, no clear-cut explanation for carboxypeptidase piezostability could be proposed [13].

A significant clue to the understanding of major determinants of piezostability came instead from the characterisation of the 7-kDa DNA-binding protein Sso7d from *S. solfataricus* (formerly referred to as P2) [20]. Sso7d is rich in β-sheet and has an aromatic cluster consisting of several residues, Phe31 being located at the core of the cluster itself. Protein piezostability of wild-type and F31A mutant was assayed by FTIR and fourth derivative spectroscopy [20,21]. This showed a midpoint transition beyond 14 kbar (a pressure at which water undergoes a phase transition) for the former and of 4 kbar for the latter. Likewise, wild-type Sso7d displayed a midpoint transition temperature of 361 K and the mutant of 334 K, as determined by both NMR and CD [20]. The impressive example of Sso7d is unprecedented as regards both piezostability and the dramatic loss in piezo- and thermostability resulting from a single-point mutation. Furthermore, the correlation between the two physical parameters is quite apparent as well. Work performed by Burley and Petsko [22] provides a rationale to account for protein thermostability. On examination of a set of 34 proteins they found that several aromatic side chains are involved in aromatic pairs which often also form networks of three or more interacting side-chains, where phenyl ring centroids are separated by a preferential distance ranging from 4.5 to 7 Å, and dihedral angles approaching 90° are most common. The aromatic cluster found in Sso7d fits well this pattern: thus, removal of Phe31 abolished the interactions of this residue with three other aromatics (Phe5, Tyr7 and Tyr33), which resulted in a substantial destabilisation [20,21]. In this case, both thermo- and piezostability could be largely accounted for by the same, well-defined structural features. Furthermore, the fact that stacking interactions between aromatics are substantially pressure insensitive, as above mentioned [7], fits well with the observed behaviour of wild-type and the resulting destabilisation of the mutant. Probably, the small size of this protein also justifies the magnitude of the mutational effect, as well as the two-state, highly cooperative transition and, to some extent, its unusual thermostability [23]. This is not always observed in pressure-induced protein denaturation. For instance, in the case of hen egg-white lysozyme, a biphasic transition was demonstrated, with the α-domain unfolding first, followed by the β-domain [5].

In recent years, the possibility to exploit hydrostatic pressure in enzyme-catalysed processes of biotechnological relevance has been investigated [7,24,25]. Pressure influences enzyme behaviour mainly (but not exclusively) by affecting the individual rate constants involved in the kinetic mechanism and that concur to the definition of k_{cat} and K_m. Although the prediction of these effects on K_m is not straightforward, it is expected that an in increase in pressure would increase the k_{cat} in reactions for which the $\Delta V^{\ddagger}$ (i.e., the difference in volume between the transition state and the ground state) of the limiting step has a negative value [5–7]. Thus, in some cases pressure allows to enhance enzyme activity and/or to modulate enzyme specificity, whatever may be the kinetic mechanism. This has been observed, for instance, when α-chymotrypsin activity in reversed micelles was assayed [24]: actually, enzyme-catalysed hydrolysis of an anilide underwent a 7-fold increase on increasing pressure from 1 bar to 2 kbar, whereas hydrolysis of an ester was decreased 10-fold under the same conditions. Likewise, *Sulfolobus* carboxypeptidase, an enzyme hydrolysing almost any N-blocked amino acid [26] displayed impressive differences in the pressure dependence of k_{cat}, with $\Delta V^{\ddagger}$ values ranging from -45 ml mol^{-1} (using benzoyl-methionine as a substrate) up to $+10$ ml mol^{-1} (using benzoyl-histidine) (R. Lange, N. Bec, unpublished results).

The relevance of hyperthermophilic enzymes to the field of biocatalysis under pressure stems from their generally different behaviour with respect to the mesophilic counterparts, in terms of both stability and catalytic activity. This justifies both a theoretical and a biotechnological interest. For instance, unlike the aforementioned hyperthermophilic carboxypeptidase, the activity of the mesophilic carboxypeptidase A was virtually pressure insensitive [27]. Likewise, a comparison between yeast and *Thermoanaerobium brockii* alcohol dehydrogenases showed that the former was continuously inhibited by increasing pressure, whereas the latter underwent an increase in

k_{cat} and a concomitant decrease in K_m up to 1 kbar, with no dissociation of the oligomer [28]. Furthermore, activities of almond and *S. solfataricus* β-glucosidase were assayed as a function of pressure, which showed a steady decrease in activity of the mesophilic enzyme up to 2 kbar, whereas the activity of *Sulfolobus* enzyme underwent an increase up to 1.25 kbar, exclusively accounted for by a rise in the k_{cat} [29]. However, the catalytic efficiency of both thermophilic alcohol dehydrogenase and β-glucosidase declined after reaching the respective maxima. One of the most impressive examples of pressure-promoted enzyme activation is that of thermolysin, a moderately thermophilic, zinc-dependent protease, which displayed an about 10-fold increase in the catalytic efficiency (expressed as the second-order rate constant k_{cat}/K_m) when pressure was increased from 1 bar to 2.5 kbar at 40°C [25]. Remarkably, the activating effect of pressure was magnified by increasing temperature up to at least 40°C [25]. However, at higher pressure the activity declined, mainly due to zinc dissociation.

At present, no common paradigm can be devised to justify the differences frequently found in the catalytic behaviour under pressure between thermophilic and mesophilic enzymes. Nevertheless, current knowledge highlights that the former might offer some significant advantage with respect to their mesophilic counterparts, as far as their exploitation in bioprocesses at high pressure is concerned. First, their higher stability towards pressure, temperature and generally towards organic solvents as well [30], involves that they are less prone to inactivation during bioprocesses developed by combining these chemical–physical parameters. Second, higher stability also involves that a much broader range of working conditions is made accessible for tuning activity and specificity. Third, it is frequently observed, as above outlined, that thermophilic enzymes are often activated by increasing pressure, although it cannot be established, to date, to what extent this is a common feature.

In conclusion, this survey makes it apparent that thermophilic enzymes have an intrinsically better potential in bioprocesses under extreme conditions, which has been exploited in specific applications. Quite likely, future investigations will substantially broaden the scope of applications in high-pressure technology of these unusually resistant molecules.

Acknowledgements

This work was supported by grants from the Italian Ministry for University and Scientific and Technological Research (PRIN 1999, Structural, Functional and Applicative Prospects of Proteins from Thermophiles). E.M. gratefully acknowledges the French Institute for Medical Research (INSERM), which has supported his research through a 'Poste verte' fellowship.

References

[1] M.J. Danson, D.W. Hough, Trends Microbiol. 6 (1998) 307–314.
[2] C. Vieille, G.J. Zeikus, Microbiol. Mol. Biol. Rev. 65 (2001) 1–43.
[3] R. Sterner, W. Liebl, Crit. Rev. Biochem. Mol. Biol. 36 (2001) 39–106.
[4] A. Fersht, Structure and Mechanism in Protein Science, Freeeman, New York, 1977.
[5] M. Gross, R. Jaenicke, Eur. J. Biochem. 221 (1994) 617–630.
[6] V.V. Mozhaev, K. Heremans, J. Frank, P. Masson, C. Balny, Proteins Struct. Funct. Genet. 24 (1996) 81–91.
[7] V.V. Mozhaev, K. Heremans, J. Frank, P. Masson, C. Balny, Trends Biotechnol. 12 (1994) 493–501.
[8] W. Kauzmann, Nature 325 (1987) 763–764.
[9] T.E. Creighton, Proteins. Structure and Molecular Properties, Freeeman, New York, 1993.
[10] G. Hummer, S. Garde, A.E. Garcia, M.E. Paulaitis, Proc. Natl. Acad. Sci. USA 95 (1998) 1552–1555.
[11] J. Zhang, X. Peng, J. Jonas, Biochemistry 34 (1995) 8631–8641.
[12] D.J. Hei, D.S. Clark, Appl. Environ. Microbiol. 60 (1994) 932–939.
[13] N. Bec, A. Villa, P. Tortora, V.V. Mozhaev, C. Balny, R. Lange, Biotechnol. Lett. 18 (1996) 483–488.
[14] M.A. Mc Lean, S.A. Maves, K.E. Weiss, S. Krepich, S.G. Sligar, Biochem. Biophys. Res. Commun. 252 (1998) 166–172.
[15] M. Summit, B. Scott, K. Nielson, E. Mathur, J. Baross, Extremophiles 2 (1998) 339–345.
[16] P.W. Bridgman, J. Biol. Chem. 19 (1914) 511–512.
[17] J.K. Konisky, P.C. Michels, D.S. Clark, Appl. Environ. Microbiol. 61 (1995) 2762–2764.
[18] A. Villa, L. Zecca, P. Fusi, S. Colombo, G. Tedeschi, P. Tortora, Biochem. J. 295 (1993) 827–831.

[19] A. Goldman, Structure 3 (1995) 1277–1279.
[20] P. Fusi, K. Goossens, P. Consonni, M. Grisa, P. Puricelli, G. Vecchio, M. Vanoni, L. Zetta, K. Heremans, P. Tortora, Proteins Struct. Funct. Genet. 29 (1997) 381–390.
[21] E. Mombelli, M. Afshar, P. Fusi, M. Mariani, P. Tortora, J. Connelly, R. Lange, Biochemistry 36 (1997) 8733–8742.
[22] S.K. Burley, G.A. Petsko, Science 229 (1985) 23–28.
[23] P. Alexander, S. Fahnestock, T. Lee, J. Orban, P. Bryan, Biochemistry 31 (1992) 3597–3603.
[24] V.V. Mozhaev, N. Bec, C. Balny, Biochem. Mol. Biol. Int. 34 (1994) 191–199.
[25] E.V. Kudryashova, V.V. Mozhaev, C. Balny, Biochim. Biophys. Acta 1386 (1998) 199–210.
[26] S. Colombo, S. D'Auria, P. Fusi, L. Zecca, C.A. Raia, P. Tortora, Eur. J. Biochem. 206 (1992) 349–357.
[27] E. Mombelli, N. Bec, P. Tortora, C. Balny, R. Lange, Food Biotechnol. 10 (1996) 131–142.
[28] S. Dallet, M.-D. Legoy, Biochim. Biophys. Acta 1294 (1996) 15–24.
[29] V. Hamon, S. Dallet, M.-D. Legoy, Biochim. Biophys. Acta 1294 (1996) 195–203.
[30] E. Keinan, E.K. Hafeli, K.K. Seth, R. Lamed, J. Am. Chem. Soc. 108 (1986) 162–169.

ELSEVIER

Biochimica et Biophysica Acta 1595 (2002) 397–399

BIOCHIMICA ET BIOPHYSICA ACTA
BBA
www.bba-direct.com

Review

High pressure in bioscience and biotechnology: pure science encompassed in pursuit of value

Rikimaru Hayashi *

Division of Applied Life Sciences, Graduate School of Agriculture, Kyoto University, Sakyo-ku, Kyoto 606-8502, Japan

Received 7 November 2001; accepted 8 November 2001

Abstract

A fundamental factors, pressure (P), is indispensable to develop and support applications in the field of bioscience and biotechnology. This short sentence describes an example how high pressure bioscience and biotechnology, which started from applied science, stimulates challenges of basic science and pure science in the biology-related fields including not only food science, medicine, and pharmacology but also biochemistry, molecular biology, cell biology, physical chemistry, and angineering.

1. Introduction

Since Royer first reported, in 1895, that hydrostatic high pressure treatment kills bacteria, Hite and coworkers reported on the effect of this on milk preservation in 1899 and Bridgman observed the coagulation of egg white as the result of high pressure treatment in 1914. With the passing of time, considerable advances in the utilization of high pressure in food science and technology have been made, especially in the last decade of the 20th century; a major breakthrough occurred in Japan where high pressure-processed jams, prepared by non-heating food processing appeared on the market in 1992. Since then, pressure processing of food, in addition to heat processing or a combination thereof, has attracted the interest of food industries on a world-wide scale. With the finding that high-pressure techniques are useful in food science and technology, it has naturally been extended to medical and pharmaceutical fields as a general method in bioscience and biotechnology, since high-temperature processing has been used in various aspects of human welfare.

In order to develop and support these applications, basic research, not only in the areas of food science, medicine, and pharmacology, but other disciplines including biochemistry, molecular biology, cell biology, physical chemistry, and engineering, are indispensable, based on the expectation that the two fundamental factors, pressure (P) and temperature (T), could be widely applied to our daily lives. Recent developments in these areas of basic research are summarized in a recent book [1].

This short sentence describes an example of how high-pressure bioscience and biotechnology, the origins of which are in applied science, has been applied to research in basic and pure science in biology-related fields.

* Fax: +81-75-753-6128.
E-mail address: hayashi@kais.kyoto-u.ac.jp (R. Hayashi).

0167-4838/02/$ – see front matter
PII: S0167-4838(01)00362-4

2. Applied science for food processing

When food materials, including microbial cells and plant and animal tissues, are pressurized at 100–400 MPa as a pre-treatment for food processing, cellular components such as metals, amino acids, and proteins become solubilized and extracellular compounds such as salts are permeated into cells and tissues. After pressure treatment, an electron microscopic observation of yeast cells and a biochemical analysis of animal tissues revealed that nuclear membranes, lysosomes, and vacuoles undergo significant disruption, in addition to a small amount of damage to cell membranes and cell walls.

Observations obtained from related applications raise the question of why small molecules such as salts and amino acids become solubilized from or penetrate into living cells under the conditions of restricted diffusion which are associated with high pressure.

3. Basic science for applications

In order to understand these morphological and biochemical phenomena, pressure-induced functional and structural changes in membrane-bound Na^+,K^+-ATPase from pig kidney has been studied as one model system of the cell membrane which consists of lipid bilayers and transmembrane proteins [2]. The enzyme consists of two subunits, an α-subunit (molecular mass 94 000–120 000) and a β-subunit (molecular mass 40 000–57 000) and approximately two thirds of the total enzyme protein protrudes from the lipid bilayer as a water-soluble domain and is in contact with the aqueous environment, while the remainder, in the form of a transmembrane segment, is surrounded by the lipid bilayer.

The activity of the enzyme showed at least a three-step change, depending on the pressure applied at 0.1–100 MPa, 100–220 MPa, and 220 MPa or higher. Pressures of 100 MPa or lower induce a decrease in the fluidity of the lipid bilayer and reversible conformational changes in the transmembrane protein, resulting in the functional disorder of the enzymes. Pressures of 100–220 MPa cause a reversible phase transition of the lipid bilayer and dissociation of the protein subunits. These changes bring about the separation of the protein and the lipid bilayer, producing transmembrane tunnels. Pressures of 220 MPa or higher lead to irreversible protein unfolding and fragmentation of the lipid bilayer, thus destroying the gross membrane structure.

The perspective of the pressure-induced change at the molecular level provides fundamental information that can be applied to practical applications.

4. Pure science

A high hydrostatic pressure causes the dissociation of the protein subunit in aqueous solution at pressures of 200 MPa or less. However, the issue of how the lipid bilayer affects the pressure-induced dissociation of protein subunits is not currently understood.

The phase transition of a lipid bilayer from the liquid-crystalline phase to a gel phase is accompanied by an increase in the thickness of the lipid bilayer and a decrease in its cross-sectional area, due to the ordering of hydrocarbon chains (see [3] for an example). These changes would be expected to alter the environment of transmembrane proteins at the contact surfaces of proteins and water, and/or protein and the lipid bilayer: an increase in the thickness of the lipid bilayer partially covers the transmembrane proteins and, thus, decreases the water-soluble region, or its lateral shrinkage separates protein from lipid. Useful information can also be obtained concerning membrane dynamics from studies of the effects of pressure on membrane-associated proteins.

The introduction of P in addition to T is generally accepted as an important issue in biophysical sciences. P itself is also important as an environmental factor in biochemistry and microbiology, including molecular and cell biology, because microorganisms, which live under extreme conditions including high pressure, are able to thrive. Thus, the example of pressure-induced biochemical phenomena, as described above, involves issues of pure science.

In conclusion, as far as interest in high pressure is focused on its use in biology-related fields including food science, pharmacy, and medicine, high-pressure bioscience and biotechnology represents a cultural science or *Kulturwissenschaft* in which 'value' is ultimately pursued as described by Heinrich Rickert [4]. Basic science precedes applications, and applied science requires basic science in order to achieve further developments. Thus, basic science and applied science mutually affect one another and are interdependent. Concerning pure science, which pursues no value, it is discriminated from basic science, the intent of which is eventually provide support and methodology for applications in the current world of advanced science and technology. High-pressure bioscience and biotechnology encompasses not only applied and basic sciences but also pure science, which serves to stimulate the interest of young scientists.

References

[1] R. Hayashi (Ed.), Trends in High Pressure Bioscience and Biotechnology, Elsevier, 2002.

[2] M. Kato, R. Hayashi, T. Tsuda, K. Taniguchi, High pressure-induced changes of biological membrane: Study on the membrane-bound Na^+, K^+-ATPase as a model system, Eur. J. Biochem. 269 (2002) 110–118.

[3] T. Hata, H. Matsuki, S. Kaneshina, Effect of local anesthetics on the bilayer membrane of dipalmitoylphosphatidylcholine: interdigitation of lipid bilayer and vesicle–micelle transition, Biophys. Chem. 87 (2000) 25–36.

[4] H. Rickert, Kulturwissenschaft und Naturwissenschaft (Japanese translation), Iwanami Publishing, 1939.

ELSEVIER

Biochimica et Biophysica Acta 1595 (2002) 400–401

BIOCHIMICA ET BIOPHYSICA ACTA
BBA

www.bba-direct.com

Protein Structure and Molecular Enzymology

Author Index

PII: S0167-4838(02)00245-5

Biochimica et Biophysica Acta 1595 (2002) 402–403
www.bba-direct.com

Protein Structure and Molecular Enzymology
Subject Index

PII: S0167-4838(02)00245-5

SpringerWienNewYork

Acta Neurochirurgica
Supplements

Editor: H.-J. Steiger

Joint Residency Advisory and Accreditation Commitee (JRAAC)
A Joint Commitee of the the U.E.M.S. and E.A.N.S.

Training in Neurosurgery in the Countries of the EU

A Guide to Organize a Training Programme

Edited by
H.-J. Reulen

in collaboration with
J. Brennum, Fl. Gjerris, J. Haase, W. Kanpolat, D. Lang, K. W. Lindsay, J. Lobo Antunes, D. W. Long, G. Neil Dwyer, G. Schackert, J. Steers, H.-J. Steiger, J. C. Tonn, T. Trojanowski, and P. A. Winkler

Acta Neurochirurgica
Supplement 90

SpringerWienNewYork

Univ.-Prof. Dr. H.-J. Reulen
Neurochirurgische Uni.-Klinik, Klinikum Großhadern, Munich, Germany

Printed in Austria
SpringerWienNewYork is a part of Springer Science+Business Media
springeronline.com

Typesetting: Asco Typesetters, Hong Kong
Printing and Binding: Druckerei Theiss GmbH, 9431 St. Stefan, Austria

Printed on acid-free and chlorine-free bleached paper

SPIN: 10992920

Library of Congress Control Number: 2004109046

With 4 Figures

ISSN 0065-1419
ISBN 3-211-21322-8 SpringerWienNewYork

Preface

The idea for this book was born during the discussions of JRAAC following certification visits to a number of different neurosurgical departments all over Europe. The members of JRAAC realized recurrent problems in the structure of training programmes on the one side an many positive solutions on the other side. The idea evolved to compile a manual or practical guide, respectively, how to solve such typical problems. Hence this book is a common effort of the members of JRAAC to provide advice and support to those involved in the organisation of neurosurgical education.

The new Neurosurgical Training Charter of the UEMS, as presented in Chapter II, defines now a "European Model" to *structure* a training programme with a clear set of standards and requirements. This is inevitably necessary if we want to move toward a steady improvement of neurosurgical education in Europe. It should be emphasized that this concerns the *external* structure, the skeleton. The *internal* structure of a programme, the curriculum, the content of the various training years, the philosophy of a department, the formation of an educational environment, etc. remain still in the responsibility of the department. Both together determine the charm and attractiveness of a department.

The chapters in this volume cover most of the mentioned "standards and requirements" asked for in the new UEMS Training Charter. The solutions offered are based on many years of personal experience of the authors as well as good examples encountered over the years in the different departments. The book may be used as a guide by those departments that are in the process of developing or improving their training programme. But also departments with an already advanced programme will find a variety of important information. The solutions offered may be used as they are or be modified and adjusted to the local situation.

We wish good luck and success in your endeavours.

For the members of JRAAC

H.-J. Reulen

Contents

Listed in Current Contents

Acta Neurochir (2004) [Suppl] 90: 1–2

Printed in Austria

Introduction

J. Lobo Antunes

Serviço de Neurocirurgia, Hospital de Santa Maria, Lisboa, Portugal

One of the main concerns of the founding fathers of the European Association of Neurosurgical Societies (EANS) was certainly to contribute to improving toward the quality of neurosurgical education in Europe. We were then living in a politically divided continent with striking asymmetries in the quality of care delivered to the various populations.

Today we are experiencing the dawn of a new period, not only because the European Union will soon encompass twenty five countries, but because we have witnessed the progressive disappearance of the political, economical, and educational barriers and a totally new mobility of students and professionals, although it is true that from its inception, EANS knew no frontiers. But at the same time we recognize the need to face the new challenges that the increasing sophistication of training in our speciality definitely pose.

Some of us, who for many years have been members of the juries of the European Examinations, have recognized that there still remains remarkable disparity in the quality of training in the various countries, which is certainly, at least in part, due to different educational methodologies and a rather incipient concern about the need to assess what is being accomplished. We are fully aware that it is crucial for each country to preserve its own culture and tradition, and that neurosurgeons carry on their work in a specific social, economic, and professional context. But in a Europe without frontiers, it is important to define a set of requirements of training and professional experience that may allow us to define a "European model", which inevitably will have to be drawn by the wise contribution of all countries involved.

Hans-Jürgen Reulen who was, until recently, the Chairman of the Joint Residency and Advisory and Accrediation Committee had the foresight to recognise these new paths in neurosurgical education and the need to join efforts with the Union Européenne des Médecins Specialistes and set up the criteria that would allow the recognition of excellence of the various European services, by a demanding and rigorous process of certification.

The present guide is the logical development of that effort and will be most helpful to any person involved in the education and training of young neurosurgeons. It is devoid of the usual verbiage of educational handbooks: it is a practical guidebook addressing the needs of anybody in charge of this wonderful, almost magic, task of making a neurosurgeon with enough cognitive information and technical skills but also with all the other ingredients that define us as a unique brand of doctors including what is part of the so-called "hidden curriculum".

This publication is particularly interesting also because it collects the experiences of people who have been working in these areas for many years and have different backgrounds and cultural heritages. But this is the fascination of our old continent!

As I mentioned, the basic goal of this publication was to make it a practical guide, so naturally some topics which are complementary to the education of a young neurosurgeon, such as uncertainty and error, risk management or dealing with conflicts of interest, were deliberately left out. But it will give the reader the basic tools to structure a neurosurgical service, following simple principles and a common language which will help to achieve one of the aspiring goals of the EANS: excellence in what we do.

I further believe we are anticipating what will soon become a political mandate – to harmonize specialty training in Europe, to allow the development of uniform criteria for a European accreditation, and guarantee the free circulation of skilled professionals. And this was what our founding fathers envisioned many years ago.

J. Lobo Antunes
Past President of the EANS

Acta Neurochir (2004) [Suppl] 90: 3–11

Printed in Austria

I UEMS charter on training of medical specialists in the EU – the new neurosurgical training charter

J. Steers[1], **H.-J. Reulen**[2], and **K. W. Lindsay**[3]

[1] Department of Clinical Neurosciences, Western General Hospital, Edingburgh, UK
[2] Department of Neurosurgery, Klinikum Großhadern, Ludwig-Maximilians-University, Munich, Germany
[3] Department of Neurosurgery, Southern General Hospital, Glasgow, UK

In the Directive 93/16 EEC of 5.04.1993, the European Commission laid down the general guidelines for the training of medical specialists in the member states of the EU, EFTA and associated member states. On the European level, the UEMS (European Union of Medical Specialists) is responsible for harmonisation and improvement of the quality of medical specialist practice in the EU. Its statutory purpose is the formulation of a common policy in the field of training.

In 1995 the "Charter on Training of Medical Specialists in the European Community" was published, outlining the general requirements and guidelines of postgraduate training as well as the special requirements for the various medical specialities including neurosurgery (www.uems.be → Training/Formation). Responsible national organisations have strongly been recommended to implement these requirements and guidelines in their national training programmes.

On request of the UEMS, a novellation was formulated in 2002–2004 for all medical specialities which is now available also for the field of neurosurgery. It describes more precisely then so far the structures of neurosurgical training including the external audits required now. The present version at the time of publication of this book has not definitely been ratified yet so that minor alterations are still possible. Also, the operative figures required at the end of training have not yet been finalised. Therefore the chapter does not contain the appendix 1+2, mentioned on page 11/12. However, we thought it necessary to include the new *Neurosurgical Training Charter* since it will be the working basis for the JRAAC and above all for the national societies and national authorities.

This document will be the professional position paper needed by the European Commission when further EU directives in the field of medical specialist practice are being contemplated.

Correspondence: J. Steers, Department of Clinical Neurosciences, Western General Hospital, Crewe Road, Edingburgh, EH4 2SU, UK. e-mail: js@skull.dcn.ed.ac.uk

Charter on training of medical specialists in the EU – Requirements for the speciality of neurosurgery (as of February 28th, 2004). (UEMS Specialist Section of Neurosurgery)

Foreword

This document sets out standards and guidelines for neurosurgical residency training and for approval of training programmes in the countries of the EU/ EFTA and associated member states. It is recognised that there are a number of structural and operational differences in the health care systems, appointment procedures and training systems in these different countries. This document provides the basis for the development of a harmonised, comprehensive, structured and balanced training programme in Neurosurgery.

The future of European neurosurgery will depend on the quality of training offered to our trainees. Apprentice style training, which has been at the heart of traditional training, is increasingly being threatened by regulation and legislation. Hours of work for both trainers and trainees is coming under increasing pressure from many competing demands, many of which further fragment the training opportunities available.

It is recognized that in future there may be a need to prolong the training time in clinical neurosurgery from 4 to 5 years.

Goal of training programme

The primary goal of a training programme is to provide the trainee with a broad knowledge base, the necessary operative and procedural skills and experience as well as professional judgement for independent neurosurgical practice; a further goal is to teach him/her self-criticism, critical assessment of his/her results, the ability to self-directed learning which will eventually lead to continued growth, expert practice and professionalism.

Definition of speciality

Neurosurgery is a discipline that provides the diagnosis, the operative, and non-operative management (i.e. prevention, diagnosis, evaluation, treatment, intensive care, and rehabilitation) of patients with pathological processes that affect the central, peripheral (and autonomic) nervous system, including their supporting structures and vascular supply as well as the operative and non-operative management of pain. This encompasses the modern treatment of disorders of the brain, meninges, skull and their blood supply including the extracranial carotid and vertebral arteries; disorders of the pituitary gland, disorders of the cranial, spinal nerves, peripheral nerves and disorders of the autonomic nervous system, disorders of the spinal cord, meninges and spine including those which may require treatment by spinal fusion or instrumentation.

Article 1: General rules on monitoring and accreditation

1.1 Manpower planning

Manpower planning should be developed, based on the demands and provision of safe neurosurgical care across the countries of the EU and associated member states. Planners will have to take into consideration demographic changes in any given population such as its growth and ageing, changing treatment modalities and actual workload, the possible effects of legislation on working hours and, in some centers, the involvement in education of medical professionals. Whilst many countries intend to increase the number of trained neurosurgeons, there is a recognised risk that overproduction leads to the dilution of experience and consequent difficulties in maintaining competencies.

1.2 Monitoring authority at European level

Harmonisation of neurosurgical training throughout Europe will require standards of training and monitoring, and centralised registration of approval of neurosurgical training programmes in the EU and associated countries. The central monitoring body is the **J**oint **R**esidency **A**dvisory and **A**ccreditation **C**ommittee (JRAAC); a joint committee of the Section of Neurosurgery of the UEMS and the European Association of Neurosurgical Societies (EANS).

National professional bodies (responsible for the recognition of medical specialists in individual countries) can monitor and recognise neurosurgical training programmes using UEMS standards based on the training charter. In the interest of developing common standards, cooperation with JRAAC is recommended.

1.3 Accreditation of training institutions

The visiting and evaluation of training institutions is an important feed back mechanism for quality improvement.

1.3.1 A training institution or training programme must have European or national recognition/accreditation, respectively, according to UEMS/national standards.

1.3.2 To be accredited, an educational programme must substantially comply with the special requirements for residency training in neurosurgery as set down by the UEMS Training Charter.

Programmes must demonstrate their compliance with these requirements at the time of the site visit. Site visits will be carried out in accordance with the Charter on Site Visits.

1.3.4 Nationally accredited training programmes in neurosurgery may also be approved by the Joint Residency Advisory and Accreditation Committee.

1.3.5 A training programme must be reviewed every 5 years, or within 12 months following the appointment of a new director of training programme.

1.4 Accreditation of trainers

Trainers must be certified neurosurgeons and the Programme Director must be registered in accordance with the medical licensing authority of the country of the training programme and possess the necessary administrative, teaching and clinical skills required to conduct the programme.

Article 2: General aspects of training in the speciality

2.1 Selection for and access to the speciality

2.1.1 Applicants should have a valid license to practice medicine within a training programme in EU and associated countries; this license has to be recognised by the country where he/she will be trained.

Training institutions or, if present, responsible bodies should select or appoint trainees suitable for the speciality in accordance with an established selection procedure. This selection procedure should be transparent and fair, and application should be open to all eligible persons. The candidates should be aware of these requirements.

2.1.2 After appointment of a trainee, a training agreement should be entered into by the Director of the Programme and trainee and duly signed by the trainee and the Director. The agreement should define – in terms of education and training – the relationships, duties and obligations on each side.

2.2 Duration of training

Training must cover the full range of the speciality and lead to the ability for independent practice on completion of training.

2.2.1 Neurosurgical training is of 6 years minimum duration and consists of a minimum of 4 years' training in clinical neurosurgery in an accredited programme. Of these 4 years dedicated to neurosurgery, at least 3 years should be spent in a UEMS member state and not less than 3 years in the same recognised programme. Training must include adequate exposure to intensive care and to pediatric neurosurgery (see 3.2.1).

2.2.2 Up to a total of two years may be spent in related disciplines (in a surgical discipline, in neurology, in neuropediatrics, in neuroradiology, in neuropathology, in neurophysiology) and/or activities including research in neurosciences.

2.3 Curriculum of general and specific training periods

2.3.1 A written *Training Curriculum* must be designed to provide a diversified and balanced quality (theoretical and practical) of neurosurgical education describing the contents and aims in each year of training. This must be available to trainees and the faculty. Emphasis should be placed on adequate time allocated for study and tuition independent of clinical duties. It may be necessary for some units to formally organize specific training periods in associated neurosurgical units, if adequate experience cannot be provided internally.

2.3.2 There should be established *Rotation Periods* covering all main areas of neurosurgery (including pediatric neurosurgery). These rotations should be organized in such a way as to give trainees increasing responsibility as they progress through their training with regard to patient care and surgical experience. Rotations may include neurology, surgical disciplines, neuroradiology, neuropathology, radiosurgery, etc., neurosciences or research, depending on requirements, local availability and the department's emphasis.

2.3.3 Some institutions may wish to use a structured *Surgical Training Plan*. The main idea of this is a continuous and systemic escalation of surgical responsibilities and competence through training years 1–6.

2.3.4 Education programme

There should be a documented, continuous *Education Programme* throughout the training, which should include daily, weekly and monthly conferences, etc. There must be protected time for study and tuition.

This Education Programme should consist of

- A programme of lectures including visiting speakers
- Clinical presentations from all neuroscience disciplines
- Neuropathological and neuroradiological conferences
- Journal club
- Mortality and morbidity conference (with audited attendance)
- Research meetings
- Regular teaching conferences including subspecialities (residents should take increasing responsibility in the conferences and in the teaching of junior trainees, nurses, students)
- Teaching in ethics, administration, management and economics

2.3.5 Exposure to research

Trainees should be encouraged to undertake research and would be expected to develop an understanding of research methodology. An appropriately qualified person should supervise specific research projects. There should be a protected period of time within a 6-year-programme where a trainee can participate in a specific research project. All trainees will be expected to be able to assess published work. In academic programmes, clinical and/or basic research opportunities must be available to the trainee with appropriate faculty supervision.

2.3.6 Participation in meetings/courses

It is recommended that trainees attend the meetings of the national neurosurgical society once a year (or an equivalent meeting). If possible trainees should participate in the European Association of Neurosurgical Societies' training courses or equivalent national training courses. During their training, they should also attend a subspecialty course/meeting (spinal, stereotaxy, paediatric, peripheral nerves, neurooncology, neurovascular, etc.) and if possible a hands-on-course in anatomy or surgical techniques.

2.3.7 Trainees should keep a *Trainee Portfolio* containing details of previous training posts, examinations passed, lists of publications and presentations at meetings, courses attended, cumulative operative totals, copies of assessment forms of these different training periods.

2.4 Training log-book and periodic progress assessment of trainees

2.4.1 Log-Book

Each trainee must keep an authorised *Log-Book* that meets the standards of the EANS/UEMS log-book for documentation of operative experience. The trainee will have to demonstrate that he/she has assisted a wide range of cases which should include a balance of trainer assisted and personal cases under supervision. Log-book entries must be monitored by regular inspection and signed off by the appropriate trainer. The log-book must be available at Board examination.

2.4.2 Periodic progress assessment

The purpose of assessment is to ensure continuing progress in the trainee's knowledge and skills as well as professional conduct and ethics.

Trainees have to meet the agreed standards and requirements of the planned programme. Assessment must be performed on a six monthly basis or at the end of each rotation period by the appropriate trainer in writing using an evaluation sheet. The log-book is used as supporting documentation. The result of the evaluation must be discussed with each resident. Failure to meet agreed targets must be brought to the attention of the training programme director.

It is the responsibility of the training programme director to identify any failure in a trainee's progress, to conduct and to provide appropriate advice, and to take remedial action.

In the event of trainees not progressing as required, there are three stages of remedial action. Targeted training: closer monitoring and supervision to address particular needs; intensified supervision and, if necessary, repetition of the appropriate part of the programme; and finally the potential to withdraw a trainee from the programme.

It is of the greatest importance that accurate records of the trainees' progress are kept (trainee's portfolio).

In future a parallel assessment for trainees to assess their training may be introduced to monitor the effectiveness of the training programme.

Article 3: Requirements for training institutions

3.1 Process for recognition as a training institution

In order to be recognised, the training institution must substantially comply with the special requirements for Residency Training in Neurosurgery and the General Requirements in Graduate Medical Education of the UEMS Training Charter. The training institution (programme) must be able to demonstrate its compliance with these requirements at the time of a site visit conducted by the JRAAC or equivalent national body.

3.1.1 The application

The Programme Director must submit a Programme Application Form to JRAAC describing the personnel, space, technical facilities, and in particular the Residency Training Programme. The Application Form can be obtained from the secretary of JRAAC or the web site of EANS (www.eans.org → What's new → Accreditation of Training Centers).

3.1.2 The site visit

The next step will be a site visit of the applicant institution, conducted by two independent visitors nominated by JRAAC and a third national observer/visitor appointed by the applicant institution. The date of the site visit will be arranged between the Programme Director and the site visitors. The Programme Director will receive the necessary information to prepare the site visit in due time. The site visit will be performed in accordance with the guidelines of the UEMS Charter on Visitation of Training Centers.

The site visit serves to explore in detail the training programme, the educational and scientific environment, by holding discussions with the Chairman, the teachers, the trainees, and administration of the unit. A report will be prepared by the site visitors and will be part of the final decision of the unit on the accreditation status of the programme. All information obtained during the interviews with trainers and trainees remains absolutely confidential.

The accreditation status as decided by the JRAAC will be reported to the Programme Director by formal Letter of Notification. Together with the site visit report, additional advice and recommendations – if necessary – will be given to further improve the Training Programme.

3.1.3 The accreditation

The following decisions may be taken by the JRAAC with regard to the accreditation status of a Training Institution and Programme:

Full accreditation may be granted if the programme has demonstrated its full compliance with the European Training Charter. The Department will receive a certificate indicating that the Department and the Training Programme fulfill the criteria meeting European Standards of Excellence for Education in Neurosurgery. The accreditation shall be re-assessed after 5 years or within one year after change of Programme Director.

Provisional accreditation indicates that the programme is basically in line (but not in compliance) with the requirements and standards. It is considered to be at the development stage of its training programme. The Programme Director will be requested to submit a so-called Progress Report within one or two years of notification. The Committee shall specify precisely the information to be provided. When a Progress Report is requested, a specific due date should be included in the request.

Accreditation may be withheld if the programme does not substantially comply with the requirements and standards. The Committee will cite those areas in which the reviewed programme does not comply with the standards. A new application can be submitted when the areas indicated are brought into compliance with the requirements and standards.

Accreditation may be discontinued if a programme for some reason is no longer in compliance with the requirements and standards. A new application can be submitted if the requirements are again fulfilled.

Reassessment of a Programme is usually done after 5 years or within one year after a change of Programme Director.

3.2 Requirements on equipment and educational facilities

3.2.1 The training programme

- There must be a referral base sufficient to provide an adequate case volume and mixture to support the training programme.
- There must be a minimum of 4 trainers (including chairman/programme director).
- There must be at least 30 neurosurgical beds, and in addition critical care beds (7–10 per million).
- There must be at least two designated fully staffed (neurosurgically trained staff) and appropriately equipped operating theatres with availability of a 24 hour operating theatre.
- There must be an operating microscope with CCTV for each theatre. The following are deemed to be essential equipment: ultrasonic aspirator, image guidance, a stereotactic system, radiological imaging, endoscopy equipment, ultrasound equipment as standard. Laser equipment may be necessary for specialised procedures.
- Neurosurgical theatres should be covered by anaesthetists with a special interest in neuroanaesthesia. Anaesthesia cover should be available at all times for neurosurgery.
- There must be designated and fully staffed neurosurgical intensive care beds. Neurosurgical intensive care may be managed by neurosurgery or there may be joint responsibility between neurosurgery and anaesthesia.
- There must be an emergency unit with 24 hrs admission.
- There must be outpatient clinics where non-emergency patients are seen before and after surgical procedures.
- There must be exposure to paediatric neurosurgery as a mandatory component of a training programme. Where this does not form part of routine work of the neurosurgical department, a six-month-secondment to an appropriate paediatric programme should be arranged.
  It must be recognised that in some European states paediatrics requires special training and a protected environment.
- There should be opportunity to obtain experience of functional neurosurgery either within the department or in another neurosurgical department specialised in this field.

Highly specialised centres not covering the whole neurosurgical field can be included in rotational systems but cannot be training centres in their own right.

3.2.2 Associations and access to other relevant specialities

Allied specialities must be present to a sufficient extent to provide the trainee with the opportunity of developing his/her skills in a team approach to patient care. The training programme should be closely associated with the following departments or units officially certified for training:

a department of neurology
a department of surgery and traumatology to support neurosurgical involvement in cranial and spinal trauma
a department of anaesthesiology with special responsibility for neuroanaesthesiology
a department of radiology
a department or unit of neuroradiology which has: imaging techniques with dedicated CT-scanning, access to MR-scanning on site and appropriate angiography equipment for diagnostic procedures including availability of interventional radiology
a department of pathology
a department or unit of neuropathology
a department of radiotherapy
a department of internal medicine and/or oncology
a department of paediatrics

3.2.3 Educational facilities

Easily accessible library with an adequate selection of books and journals on neurosurgery (mandatory)
Office space for both faculty and trainees
Facilities for computer literature searches
Space and opportunity for practical and theoretical study (mandatory)
Space and equipment for practical training of techniques in a laboratory setting
Space, equipment and supporting personnel for clinical and/or basic research in academic programmes.

3.3 Institutional quality management provisions

A training institution must have an internal system of medical audit or quality assurance.

3.3.1 Internal regulations. There should be written general guidelines of the training institution concerning patient care and patient information (patient's consent), referrals, medical records, documentation, on-call and back-up schedules, days off, residents' work schedules, attendance at conferences and educational activities. These should be available to staff and trainees.
3.3.2 Internal medical quality assurance. There must be an internal system of medical audit, such as mortality and morbidity conferences, together with a structured procedure for reporting of accidents.
3.3.3 The hospital should have measures in place (perhaps in the form of a committee) in relation to quality control such as infection control and drugs and therapeutics committee should exist.
3.3.4 A programme and training in risk management should be in place.
3.3.5 The hospital or the training institution should have an annual activities report.

Article 4: Requirements for training programme director and trainers

4.1 Criteria for training programme director

4.1.1 A Training Programme requires the appointment of a Training Programme Director to coordinate the training activities of the unit.
4.1.2 The Training Programme Director does not need to be the head of the training institution.
4.1.3 He/she must be a certified specialist of a minimum of 5 years. His/her substantial working contract must be with the training institution.
4.1.4 The CV of the Programme Director should provide evidence of continuing professional development (CPD).
4.1.5 The Programme Director must have full secretarial and administrative support and there must be sufficient protected time to carry out his/her responsibilities.

4.2 Responsibilities of training programme director

He has to
4.2.1 establish a transparent and fair selection and appointment process for trainees.
4.2.2 arrange a balanced training programme with established rotations ensuring that the trainee will have complete exposure to all aspects of neurosurgery. The programme must be written and available to trainers and trainees.
4.2.3 ensure that there is dedicated time allocated to the trainers for training and that the trainers are fulfilling their training responsibilities.
4.2.4 oversee the process of periodic assessment and review of the trainees.
4.2.5 ensure that the individual trainees' documentation (training portfolios) are up to date.
4.2.6 ensure that trainees attend appropriate and approved courses.
4.2.7 provide valid documentation as to the satisfactory completion of training.
4.2.8 ensure the annual collection and compilation of the number and types of neurosurgical operative procedures performed in the department and also in participating units connected with the training programme.
4.2.9 provide opportunity for research, audit and other educationally valid activities such as opportunities to attend training courses and scientific meetings.

4.3 Criteria for neurosurgical trainers

4.3.1 Trainers should be certified neurosurgeons who can demonstrate that they are in compliance with the requirements of continuing professional development.
4.3.2 Trainers should possess the necessary administrative, teaching and clinical skills, and commitment to conduct the programme.

4.3.3 Trainers should have undertaken instruction in training (learning needs and teaching objectives) and assessment of trainees.
4.3.4 Trainers should provide evidence of scholarly activities (clinical and/or basic research, publications in recognised journals and participation in neurosurgical scientific meetings).
4.3.5 Trainers will require secretarial and administrative support.

4.4 Responsibilities of neurosurgical trainers

They have to
4.4.1 set realistic aims and objectives for a rotation or training period
4.4.2 supervise the day to day work of the trainee on the ward, in clinic and in the operating theatre
4.4.3 support and assess trainees' surgical progress and supervise trainees' log-books
4.4.4 assess and report on trainees' progress at the end of each rotation or training period
4.4.5 ensure that the assessments and reports are documented and signed both by the trainer and the trainee
4.4.6 inform the programme director of problems at an early stage.

Article 5: Requirements for trainees

5.1 Minimal/optimal numbers

Trainees during their training must be exposed to at least 4 different trainers and the full spectrum of neurosurgical procedures.

Appendix 1 lists the minimal and optimal numbers of procedures that trainees should have performed at completion of training.

Trainees should have been directly involved in the pre- and post-operative management of these patients and should have a detailed understanding of the pre-operative investigation.

Many of the above procedures demand the use of the operating microscope with which the trainee must be fully familiar.

In addition to the above list of operative procedures, the trainees should have assisted in or partly performed operations for:

- pituitary microadenomas
- complex basal meningiomas
- arterio-venous malformations
- paediatric procedures – supra and infra-tentorial tumours
- spinal conditions

5.2 Communication abilities

- the trainee must demonstrate ability to record and convey patient details of history, examination and investigative findings to senior staff
- consent patients for operative procedures listed in 5.1 clearly detailing the reasons for performing the procedure and the risks involved
- communicate with patients and relatives and of passing on distressing information (e.g. malignancy or bereavement) in a sensitive and caring manner.

5.3 Log-book and assessment

5.3.1 The trainee must maintain an operative log book detailing his/her involvement in all cases.
5.3.2 The trainee should keep a training portfolio, which should include an up-to-date curriculum vitae incorporating
- details of previous training posts, dates, duration and trainers
- details of examinations passed
- list of publications with copies of published first page (abstract)
- list of research presentations at local, national and international meetings
- list of courses attended
- cumulative operative totals
- copies of assessment forms for each training period, completed and signed by trainers for that period

5.4 Specifications of training

The formal basis is the Training Curriculum of the department with training periods covering all main areas of neurosurgery. During his/her training, a trainee may wish to emphasise academic or research exposure or a particular area of subspecialisation. This can be organised with the Programme Director if the trainee's progress and performance allows for this, and the rotation may be adapted correspondingly. In future more trainees may wish to acquire higher competence in a subspeciality area after finishing their formal 6-year-training. This may be organised through so-called fellowship programmes.

5.5 Certifications for individual techniques/ procedures

The training record (Appendix 2) lists procedures required for neurosurgical training and indicates the minimal competency required at the end of six years' training. On completion of training, the trainee must tabulate his/her cumulative operative totals and indicate his/her level of competence. The training programme may require completion of this form at the end of each year of training.

At the end of neurosurgical training, the Training Director will certify as to the attainment of:

- satisfactory operative totals (in conjunction with Appendix 1)
- adequate competency level for each procedure (Appendix 2)
- satisfactory assessment forms for each year of training

Appendix 1 and 2 are not yet finalised.

Article 6: Certification of completion of training

The National Authority is the responsible body for recognition/certification of medical specialities in each member state of the EC/EFTA. Several of these countries now have a compulsory Board Examination. This is usually an oral examination, which is one method of assessing knowledge, clinical judgement and the candidates' thought processes. This is however not a standardized test. An examination at the end of training will become compulsory in the near future due to EU legislation. National bodies should therefore note the existence of the EANS 2-part examination (written and oral) with sessions twice a year, which leads to European certification. This may be a useful tool which could be assimilated by countries which do not have board certification examination arrangements in place, although at the time of writing, European certification is not recognised as being equivalent to national certification.

Subspecialisation

Training is a continuing process. Competence in complex procedures exceeding the operative totals and competence levels of appendices 1 and 2 can be developed either during the subspecialisation year or more likely after completion of training within the frame of a $1 - 1\frac{1}{2}$ year subspecialisation fellowship.

Correspondence: Prof. Dr. H.-J. Reulen, Schwojer Str. 19, 81249 München, Germany. e-mail: Ilona.Anders@med.uni-muenchen.de

Acta Neurochir (2004) [Suppl] 90: 13–16

Printed in Austria

II Neurosurgical training programme director position and responsibilities

G. Neil-Dwyer[1], **D. A Lang**[1], and **T. Trojanowski**[2]

[1] Wessex Neurological Centre, Southampton, UK
[2] Department of Neurosurgery, University Medical School, Lublin, Poland

In considering this key post in a training programme a few introductory remarks need to be made. The training programme needs to have an aim and a direction so that it can produce a well qualified, well trained and accredited Neurosurgeon capable of working in a multi professional environment. The training programme needs to be structured with progress monitored against clear objectives with trainees aware of their career pathways and goals. The training programme should be fair, flexible, transparent and there should be quality assurance of the programme. To achieve such a training programme requires commitment, vision and above all patience.

The appointment of a programme director for neurosurgical training

A Training Programme Director needs to be appointed to each neurosurgical training scheme. In many European countries the appointment as Programme Director is made by National Medical Authorities responsible for postgraduate education while in some it is made by the Dean of the University or the Postgraduate Dean who is responsible for postgraduate education. In many countries the Chairman automatically is considered to be the Programme Director. However, this linkage is not necessary and the appointment may take place with a nomination from the certified trainers who take part in the training programme. The appointment will normally be for a period of five years with possible re-election.

The criteria for a neurosurgical programme director

It is important to recognise that the Director of Training need not be the Head of the Department. In fact in many ways it is advisable to separate the two functions because their commitments are different and having one individual to perform both tasks will inevitably produce areas of conflict. The proposed Director should be a certified specialist for a minimum of five years, should have worked full-time within the department (s) central to the training programme. The CV of the proposed Director should provide evidence of current and continuing scientific work with some indication as to his/her commitment to training.

Importantly an Assistant Training Director must be appointed and should be a certified specialist for a minimum of five years.

The Director must have full secretarial and administrative support for this task and it should be quite separate from his/her clinical commitment. It is vital that designated time be allocated to ensure that Programme Directors have sufficient protected time to carry out their responsibilities.

The responsibilities of a training programme director

One of the principal responsibilities of the Programme Director will be to establish a transparent and fair selection and appointment process for trainees. In setting up this process the Director needs to have a clear idea of the profile of the type of candidate suitable for the training programme and in coming to this conclusion he/she will need to have discussed this with the other trainers. The Director will have to decide on basic qualifications, the mandatory educational, personal and character requirements as well as important and contributory requirements which may be helpful in a trainee if the aims of the training programme are to be achieved. It is also important that those applying for the training post are aware of the requirements.

It is the inherent responsibility of the Director to arrange and have written a carefully considered and balanced training programme which addresses the changing and progressive educational and training needs of each trainee throughout their training period.

The Director will need to organise an academic programme (lectures, research meetings, demonstrations, audit reviews, mortality and morbidity meetings) which is a central part of the trainee's educational programme. Trainees should be encouraged to be involved in the organisation of these academic sessions.

A period of time in research may benefit the trainee. In deciding on whether this should be part of an individual trainee's programme a Director will need to be certain of a number of factors. These are the trainee's ability not only academically but also in terms of being able to complete a project in time. The Director needs to know that the project is well designed, will be supervised effectively, has clearly defined and timed end-points and is properly funded. It goes without saying that the trainee will need to be enthusiastic.

After the appointment of a trainee a training agreement must be completed by the Director and signed by the trainee and the Director. This is an educational contract and should provide the trainee with an overview of his/her training.

The responsibilities of a programme directors to trainees

Their main responsibility must be to organise and co-ordinate the training programme. The trainees must be aware of all the processes and requirements of their training. The organisation and progression throughout training will need to recognise the needs of the trainee and a degree of flexibility will be required. Directors will need to provide leadership and assist in the assessment of trainees on a six monthly basis. In achieving this they will need to organise the trainees' appraisal processes, assess the competence of trainees and ensure that all trainees training portfolios are up to date. This will include signed reports, evaluation records and completed and signed logbooks. There is a European Evaluation Sheet and a logbook developed by the Joint Residency Advisory and Accreditation Committee (JRAAC) – a Joint Committee of the UEMS and EANS. The trainee will need to provide valid documentary evidence of the satisfactory completion of training. Throughout all this process the Programme Director needs to monitor, counsel and inform both the trainees and trainers.

It would be advisable, as a means of enhancing the evaluation of the effectiveness of the training programme, to have a parallel assessment form for trainees to assess their training. This assessment should be monitored by both the Director and the Authority responsible for the recognition of the programme.

The promotion of an ethos of a high level of professional conduct and ethics is mandatory within the training programme and although this lies within the remit of the Director, it must be a shared responsibility between the Director, the Trainers and the Trainees.

The responsibilities of a programme director to trainers

The Programme Director will need to ensure that all trainers have the appropriate training and support. If necessary they should arrange for trainers to attend "Training the Trainers" courses and other educational courses considered to be relevant. The Director has to ensure that trainers are fulfilling their training obligations, and, importantly, make certain that relevant documentation (assessments, logbooks) is completed in a timely (contemporaneous) fashion. The Director will also need to evaluate the effectiveness of education provision and ensure that the necessary resource elements are in place.

The responsibilities of a programme director to the training board

The Training Programme Director will need to make certain that the training programme's functions are agreed and to inform the Authority responsible for the recognition of the programme (JRAAC, National Board).

In this respect the Director will need to ensure that each training post delivers the education and training expected and agreed for that period.

The Director needs to advise the local administration or authorities on the facilities required for training and to point out any deficiency that will effect their ability to deliver the training programme.

There is also the major responsibility of organising and preparing the Unit for site inspections by authorised visitors. The Director will have the onerous task of identifying and reporting failing trainees and trainers to the responsible Training Board. The Director will be aware that for trainees who are not progressing there are recognised remedial options. These are targeted training, closer monitoring and supervision to address particular needs, intensified supervision and, if necessary, a repeat of the appropriate part of the programme and finally withdrawal from the programme. It will need to be established that any failing trainee or trainer has been made aware of the reasons for their failure, there should be evidence to demonstrate that attempts have been made to correct the problems, bearing in mind the possible remedial options, and in spite of these efforts the situation has not changed. Under these circumstances the Director has no alternative but to take the matter to the Dean and/or the responsible Training Authority.

The criteria for neurosurgical trainers

The Programme Director will need to develop with the help of colleagues and the Dean the criteria for neurosurgical trainers. A trainer should be a certified specialist, be committed to and familiar with the requirement of the training programme. The Director needs to ensure that trainers recognise their responsibilities and can reconcile them with service pressures. Directors should also recognise that trainers will require secretarial and administrative support to carry out the tasks and they should ensure that the necessary resources are made available.

The responsibilities of neurosurgical trainers

The responsibilities of the trainer should be very clearly laid out and the training programme should be available to the trainee.

It is essential that a trainer meets at an early stage their trainee and establishes their experience and identifies their learning needs. The trainer and the trainee should set realistic aims and objectives for the next six months or rotation. These should be documented and signed by both the trainer and trainee. The trainer needs to have the time to supervise the day to day work of the trainee on the ward, in the clinic and in the operating theatre. It needs to be recognised that training trainees takes up time and this must be taken into consideration in balancing the service requirements. The trainer will assess the trainee's surgical competency and knowledge on a continual basis and at the end of the six month training period or rotation, either by writing a report and/or by completing an evaluation sheet on the trainee's progress. This must be documented and discussed in detail with the trainee and signed by both the trainer and the trainee. Importantly this does not necessarily mean that the trainee agrees with the assessment but that he/she has had the opportunity to see and discuss it. Clearly if there is a problem with the assessment then this is an area for discussion with the Programme Director. The trainer will need to ensure that all assessments and reports are documented and signed by both the trainer and trainee.

An area in which most trainers will require support and guidance is in carrying out appraisal and assessment and giving constructive feedback. A Director will need to review regularly these processes to ensure not only that reports are being done but that they are being done appropriately.

Finally trainers will need to inform the Programme Directors of problems with the trainees at an early stage so that a failing trainee can be identified at an early stage.

The responsibility of the neurosurgical trainee

Trainees will need to be aware of an important fact-their training is their responsibility. While a programme may be well designed with trainers who are committed, effective and efficient and a Director committed to training, it is entirely up to the individual trainee to get the best out of their training and, if they do, this will inevitably improve the programme.

Appendix I

Criteria for a neurosurgical training programme director

1. The Director of Training need not be the Head of Department.
2. He/She should be a certified specialist for a minimum of 5 years.
3. He/She should work full-time within the Department(s) central to the Training Programme.
4. The CV of the Director should provide evidence of current and continuing scientific work.
5. An Assistant Director must be appointed and should be a certified specialist for a minimum of 5 years.

6. The Director must have full secretarial and administrative support.
7. Designated time must be allocated to ensure Programme Directors have sufficient protected time to carry out their responsibilities.

Appendix II

Responsibilities of a training programme director

The principal responsibilities include establishing a transparent and fair selection and appointment process for trainees.

After appointment of a trainee a Training Agreement must be completed by the Director and signed by the trainee and the Director. This is an educational contract and should provide the trainee with an overview of his/her training.

The promotion of an ethos of a high level of professional conduct and ethics within the training programme is essential.

To trainees

a. Organise and co-ordinate a fully balanced Training Programme.
b. Monitor, counsel and inform.
c. Ensure trainees attend appropriate and approved courses.
d. Assist and provide leadership for the assessment of trainees on a six monthly basis.
e. Organise trainees appraisal processes.
f. Assess the competence and knowledge of trainees.
g. Ensure Training Portfolios are up to date.
h. Provide valid documentary evidence of the satisfactory completion of training.

To trainers

a. To evaluate the effectiveness of education and academic provision.
b. To ensure that trainers have the appropriate training and support.
c. To ensure that relevant documentation is completed in a timely (contemporaneous) fashion (assessment reports, evaluation sheet, logbooks).
d. To ensure trainers are fulfilling their training obligations.

To the authority responsible for recognition of the training programme

a. To ensure that the Training Programme functions as agreed.
b. Inform the Authority of major changes in facilities, trainers, trainees.
c. Report failing trainees and trainers to the Authority.
d. To organise and prepare their Unit for site inspections.
e. To advise on the facilities needed for training.

Appendix III

Criteria for neurosurgical trainers

1. A Trainer should be a certified specialist.
2. Trainers should be committed to and familiar with the requirements of the Training Programme.
3. Trainers need to recognise their responsibilities and reconcile them with service pressures.
4. Trainers will require secretarial and administrative support.

Responsibilities of neurosurgical trainers

1. Meet at an early stage with the trainee and establish their experience and identify training needs.
2. Set realistic aims and objective for 6 months.
3. Possess the skills to carry out appraisal and assessment and give constructive feedback.
4. Supervise the day to day work of the trainee on the ward, in the clinic and in the operating theatre.
5. Assess trainee's surgical competency on a continual basis.
6. Supervise, check and sign the trainee's log book at regular intervals.
7. At the end of the 6-month-training period assess and report on the trainee's progress(evaluation form).
8. Ensure that all assessments and reports are documented and signed by both the trainer and the trainee.
9. To inform the Programme Director of problems at an early stage.

Correspondence: G. Neil-Dwyer, Wessex Neurological Centre, Southampton University Hospitals, Tremona Road, Southampton SO16 6YD, Hampshire, UK. e-mail: glenn@neil-dwyer.com

Acta Neurochir (2004) [Suppl] 90: 17–19

Printed in Austria

III Teaching staff

J. Lobo Antunes

Serviço de Neurocirurgia, Hospital de Santa Maria, Lisboa, Portugal

Education of a surgeon is a lifelong process where two components can be distinguished: one is what may be called the *formal curriculum* and is analyzed according to the current principles of medical education using a somewhat reductionist and predominantly quantitative approach. It often begins with definition of the educational goals which, in our discipline, are usually aimed at training a "general neurosurgeon". It should be stated, however, that the extent to which it is permissible to demand from those who have just finished their training complete technical autonomy in dealing with the most complex pathologies that constitute general practice in our specialty, is not at all clear. In the United States, residents who have finished a training program usually have to move elsewhere, joining other surgeons in group practices or go to specific areas for additional training. In Europe, in many instances, the "training period" does not end with conclusion of a six-year-program since trainees remain in the same clinical environment for a few additional years. But there is necessarily a fair amount of variation among the different countries in Europe concerning the very definition of the training goals and methodology that is required for its fulfilment.

There is general agreement, however, that there should be a clear definition of a timetable of rotations, of a quota of surgical procedures that may guarantee an acceptable degree of technical competence, a structured learning program which should preferably include a period in a research laboratory, and explicit criteria of evaluation, since any formal curriculum is closely tied to an evaluation process.

When I started to reflect on my own education as a neurosurgeon and searched for the intangible components of this mixture of theoretical knowledge learned from books and from the experiences in my own life, of good and bad examples, courage and cowardice, altruism and egoism, triumph and defeat, I began to appreciate the concept of the "*hidden curriculum*" [1]. This is engraved in the bricks, the mortar, and the memory of a school or a department and is preserved and transmitted from generation to generation. This curriculum is much more difficult to evaluate, as its appraisal slowly emerges within ourselves and gains its voice throughout life. This curriculum takes form in the wards, in the locker room, in the cafeteria, in the operating room, in the emergency room, everywhere we are called to do our job. It has no timetable, gives no credits or diplomas, is often silent or rooted in an oral tradition, and is made up of stories, anecdotes, personal experiences or aphorisms.

The *formal curriculum* is particularly concerned about the training of trainers. The *hidden curriculum* is fed by models [6] which, in their essence, are the purest expression of professional solidarity. Wilder Penfield once wrote that "no man goes alone on his eventful journey through medical school" [3]. Models are defined by qualities that include compassion, sense of humour, integrity, ability to teach, capacity to explain clearly in a non intimidating fashion complex matters, technical competence and tact in our relationship with patients and their families. All this is crucial for a well balanced teaching staff.

I would also like to emphasize that there is much to be learned from non-medical professionals, and during my residency in New York, I learned a lot from the experienced nurses working there. Throughout my life as a neurosurgeon I have learned even more from my patients.

So it is clear to me that the "skeleton" of this hidden curriculum is made up of values that are easy to recognize but hard or even impossible to quantify: sense of duty, curiosity, diligence, integrity and altruism. But it also requires the understanding of ambiguity, uncertainty and error. My point is simply this: education of any physician needs to find a proper balance between two indispensable forms: the *formal* and the *hidden*.

This somewhat long introduction was necessary in my view to put forth what I believe are the crucial issues in the matter I am supposed to discuss – the training staff of a department. I dealt with them already some years ago in the European Lecture of the EANS [2].

The ideal setting for teaching may take different forms. Many of us are convinced that we run nearly perfect services which we strongly dominate by the lustre of our personalities. Alone we can do very little, however, as we need people with various intellectual abilities, technical skills, and scientific interests, and it is the tapestry of interwoven talents and temperaments which constitutes a true *school of surgery*.

A school of surgery needs different actors to play different roles. It requires a *leader* who preferably should be the Chief of Service, who represents the repository of experience, and who should be an acknowledged expert within a specific clinical area. In my view it is important that the "Chief" is someone with an outstanding technical or scientific reputation, a truism which in fact is not always understood.

It is certainly crucial to have among the staff group a "*scientist*", someone who can help to set a research project, to teach the scientific method, to criticize the results, to introduce objectivity and precision into clinical reasoning.

It is equally helpful to have a "*scholar*", and true scholarship is sadly dying out in our profession. He or she will be able to give the correct reference, to teach how to write an article, to make trainees acquainted with the history of our discipline, to develop a sense of healthy scepticism about what they read. This is probably the person who should also be in charge of the "Journal Club" and of the library of the department.

Of course it is also indispensable to have a number of "*teachers of surgery*". Although we all would like to shape artists, the truth is that we have mostly to form artisans. Many of us when working will just move along, not sensing the need to explain, happy to be watched like any great performer. It is true that surgical teaching places emphasis on decision making accounting for about 75% of any operation. This is perhaps one reason why teaching dexterity has been so neglected. Other reasons are that teaching technique is very time consuming and many people don't know how to do it, anyway. So it is absolutely crucial to have people with ability and taste to teach the craft in all its minute details, from the correct position of the patient to the strategic placement of the bone flap. They will show the trainee how to handle the tissues gently and care for the tools with affection. Those, like me, who were fortunate to have had that kind of master will always feel their presence in the operating room, standing behind them observing and guiding.

I believe that it is quite useful to have a "*confessor*", someone the trainees can appeal to to find solace, and upon whose shoulder they can shed their tears. He will absolve the mistakes, confiding that he made them himself, but at the same time assuring that all will end well, even when the situation seems hopeless. They play an irreplaceable stabilizing role, and this is particularly important as it has recently been emphasized by Volpp et al. [5]. It is essential to have a frank discussion of the errors committed, otherwise silence prevails and errors tend to be concealed. In a study by Wu et al. [7] only 54 percent of the house officers told their attending physician about the most serious errors they had committed in the previous year. As noticed, trainees who are willing to accept responsibility for their errors and to discuss them are more likely to report constructive changes in practice than trainees who do not openly acknowledge their errors.

It also helps to have a "*bureaucrat*" in a positive sense of the word, somebody who can help the Chief of the Department often too busy to run the daily teaching affairs. He coordinates the various educational activities, checks the surgical logbook, relates to the hospital officers, organizes and programs the rotations through the various departments, etc. It may be relevant that each trainee has an assigned *tutor*, with whom he may develop a more intimate professional and personal relationship.

I think it is important to emphasize that in a teaching hospital everybody should be involved in the process of teaching and that includes, of course, the trainees themselves. In other words, it is essential that a teaching philosophy permeates the daily activities of a service, and the more advanced resident should assist and help the juniors. In my experience in the United States, this contributed to creating a sense of hierarchy and discipline that was quite useful for the daily running of the service and a growing sense of leadership.

Conclusion

In conclusion, I would like to insist that surgical education should always be a molding process tailored to the needs of each trainee. This requires a teaching staff composed of different personalities and talent. And returning to the concluding remarks from my European Lecture, with all this in mind, we will be prepared to teach and study a profession that still belongs, in the words of Trotter [4], to the "very small class of professions [that] in this tame world can be called jobs for men (...). By it I mean professions in which it is possible for people – men or women – to pursue the dying ideal that an occupation for adults should allow for intellectual freedom, should give character as much chance as cleverness, and should be subject to the tonic of difficulty and the spice of danger."

References

1. Hafferty FW (1998) Beyond curriculum reform: confronting medicine's hidden curriculum. Acad Med 73: 403–407
2. Lobo Antunes J (1987) Teaching and learning neurosurgery. Acta Neurochir (Wien) 86: 69–74
3. Penfield W (1977) No man alone: a neurosurgeon's life. Little Brown, Boston, pp 31–32
4. Trotter W (1946) Art and science in medicine. The collected papers. Oxford University Press, p 85
5. Volpp KGM, Grande D. (2003) Residents' suggestions for reducing errors in teaching Hospitals. N Engl J Med 348: 851–855
6. Wright SM, Kern DE, Kolodener K, Howard DM et al. (1998) Attributes of excellent attending – physician role models. N Engl J Med 339: 1986–1993
7. Wu AW, Folkman S, Mc Phee SI, Lo B. (1991) Do house officers learn from their mistakes? JAMA 265: 2089–2094

Correspondence: J. Lobo Antunes, M.D. Ph.D., Department of Neurosurgery, Hospital Santa Maria, Av. Prof. Egas Moniz, 1649-035 Lisbon, Portugal. e-mail: jlobo.antunes@mail.telepac.pt

Acta Neurochir (2004) [Suppl] 90: 21–31

Printed in Austria

IV The ideal neurosurgical training curriculum

A. General aspects

D. M. Long

Department of Neurosurgery, Johns Hopkins Medicine, Baltimore, USA

Introduction

There are two aspects to the development of an ideal curriculum for neurosurgical training which are fundamental to curriculum structure. Of course, local considerations may change specific sequences so the ideal curriculum is always more theoretical than practical. What is important is not to slavishly follow any guidelines for sequencing material but to understand the concepts of competency-based training as a way to assure the competency of training program graduates and to prepare them for the next phase of post-graduate medical education which leads to mastery of our field [9].

There are some theoretical teaching considerations which underlie the development of any curriculum [7]. Much is known about perceptual learning [1]. The general agreement is that the students learn best through memorization and understanding rules as a first step, followed by externally directed learning, usually in the form of lectures, conferences or core curricula. Then students progress to self-directed learning in which the trainees take primary responsibility for their education with guidance from faculty and mentors. How we learn and how we apply that knowledge are important issues in any educational system. One schema which is particular germane to neurosurgery proposes that we first begin by acquiring necessary skills such as history taking, the neurological examination, interpretation of imaging studies, and how to do procedures. We then learn to apply these skills through following sets of rules first laid down by experts and then developed for ourselves. From rules following the master surgeon should progress to knowledge-based practice in which the product of all knowledge and experience is correctly applied often without a specific sequential thought process [2, 10, 11]. The synthesis of much material has occurred within the brain automatically and is then applied effectively, often without specific steps [4].

There is much information concerning various stages of learning and what it takes to progress from each stage [3]. In surgical training we take beginners with very little previous practical experience in neurosurgery and very little specific neurosurgical information to enter training. From the beginning the goal of neurosurgery training is the progression to competency to practice the field. It is of interest that there is very little information about how long it takes or what kind of application is required in terms of hours of training and practice to move from the beginning stage to competency [5]. Much more is known about the next phase in which the competent practitioner progresses to mastery of the field. Comparison of a number of diverse occupations indicates that it takes about ten thousand hours of practice and approximately ten years to move to clear mastery of a field. Therefore we have to always remember that the training program has a twin goal of producing a competent practitioner and preparing that practitioner for self-directed education which leads to the desired goal of master surgeon. Core curricula are fundamental to a competency-based training program, but in themselves do not represent competency. In order to become competent it is necessary to know the body of skills and knowledge which define competency. The American College of Surgeons, The American Association of Neurological Surgeons/Congress of Neurological Surgeons, and the Society of Neurological Surgeons have all developed core curricula for different phases of training. Whenever a curriculum is developed there is then a great tendency to treat that information as the maximum which is to be learned. In fact, core curricula always should represent the minimum and only lead the perceptual learner through externally directed learning. The next phase of self-directed learning is key to development of competency and for progressing to mastery after competency [10].

There is one other concept which is important to think about. How material to be learned is presented is very important. In the first phases of perceptual learning when memorization and rules following are important, it is essential to have didactic presentations which present both the minimum material to be learned and the rules to be followed. This can be achieved for example by rules presenting how the information should be applied to obtain a desirable outcome most of the time. However, these formal presentations have little value once learning becomes self-directed. There is good information that they do not impart new information effectively. The traditional lectures, grand rounds presentations, teaching conferences conducted only by senior neurosurgeons, and all other non-participatory conferences have limited impact on education, and cannot prepare the trainee for lifelong learning [10].

The ideal training program year by year

Ideally I believe that the concept of numbered postgraduate years should be eliminated and all students should progress within the neurosurgery training program as rapidly as their individual skills and acquisition of knowledge allow. However, we have to start somewhere in making a transition so it is reasonable to use the current post graduate year (PGY) concept as a format (Table 1). We should realize that in an ideal training program these designations should have no meaning. Rather we should think about what needs to be learned, what applications are required, and what independent judgements the trainee should be able to exercise. These are likely to be sequenced according to local needs and patterns of practice rather than conform to any idealized program. What is important is that the core curricula be mastered at a minimum and that residents proceed in training with escalating responsibilities as rapidly as their individual skills allow.

Table 1. *The current Johns Hopkins residency program*

PGY 1	12 months	General Surgery, Neurology (3 mos), Surgical Specialties
PGY 2	12 months	General, Tumor, Vascular, Spine Neurosurgery
PGY 3	12 months	Pediatrics, Functional, General Neurosurgery
PGY 4	12 months	Chief Residency, 4 months of each-Tumor, Vascular, Spine
PGY 5, 6	24 months	Required Research Training
PGY 7	12 months	Supervised Practice- 4 months each of Tumor, Vascular, Spine, Subspecialization

The current training program

PGY 1-the internship

What do trainees need to learn in the internship as it currently is structured in the United States? Trainees need to learn conduct in the operating room and fundamental operative skills. They need to understand fluid and electrolyte balance and nutrition. They should become familiar with the management of patients suffering trauma. They should learn to recognize and treat surgical complications and they should have experience in those surgical specialities which are most inter-related with neurosurgical practice later. In the United States it has been decided that the neurology experience should also occur in the first year. I personally do not believe that three months of neurology in the first year of training is sufficient to provide the ideal of the surgical neurologist which was a part of the founding concept of neurosurgery. However, it is quite reasonable that the management of stroke, central nervous system infections, and neurological intensive care all be learned in the first year. Extensive experience in general surgery beyond what is needed to master fundamental surgical skills is not required and there is little educational purpose to extended periods of patient care dealing with diseases which are not pertinent to neurosurgery. PGY 1 has been modified to provide six months of general surgery and 6 months of neurology/surgical specialties.

PGY 2, 3, 4

Current requirements in the United States are for 36 additional months of clinical training in neurosurgery.[1] These 36 months must culminate in a chief residency in which quasi-independent practice is possible. Of course, legal requirements mean that this experience can never truly be independent. In our training program we have broken down the 36 months into junior residency (12months), senior residency (12months), and chief residency (12months). In most training programs of the past there was very little sub-specialization and assignments were generally made on the basis of assisting individual surgeons or general experience in a specific hospital. Sub-specialty practice is now common in academic centers and in larger group practices throughout the world. Training programs should reflect this trend to sub-specialization. Since the concept of junior and senior resident disappears in a competency-based program it is not necessary to stratify training. Thus for the 24 months required before chief residency we now assign residents according to broad sub-specialty categories. These include: 1. Tumor neurosurgery 2. Vascular neurosurgery 3. Surgery of the spine 4. Pediatric neurosurgery (including pediatric trauma) 5. Peripheral nerve surgery 6. Pain 7. Functional neurosurgery including stereotaxis for all purposes 8. Epilepsy surgery. At Johns Hopkins trauma is not a separate service and trauma experience is primarily in pediatrics and spread throughout the residency.

In addition we have one affiliated hospital which is a general hospital. A general practice of neurosurgery which is not specialized is the core of the experience in this affiliate.

In addition residents return for at least one month of neurological intensive care to supplement the two-three month experience of the first year.

Tumor, vascular, and spinal neurosurgery are experienced twice during the 24 month period. All of the other rotations are a single experience. All these rotations are currently of a specific length and content.

It is the responsibility of the director of each service to define for the trainees what should be learned in terms of knowledge and skills. Knowledge is assessed by an examination at the end of each rotation. At present the assessment of skills and the application of both skills and knowledge is left to the judgement of the faculty. We currently have no better way and there is good evidence that the judgement of experts is reliable [12]. In the ideal program no resident should complete one of these rotations without having satisfactorily demonstrated competence in all of the minimum skills defined by the faculty. Furthermore they should progress beyond these minimums as rapidly and as far as their personal abilities allow. By defining the sequence of skills acquisition in the operating room and allowing the residents to move ahead independently in applying these skills we demonstrated acceleration of competency to do specific neurosurgical operations by 6 to 18 months. This strongly suggests that our current training programs retard the speed with which many residents can progress in training. In competency-based training residents should move ahead as rapidly as they demonstrate competency and this simply means moving from one set of skills to another so that much more is learned for any given period of assignment. No resident should complete one of these rotations until all the faculty involved agree that competency has been gained in at least the minimum skills defined as the requirement of the rotation [8, 9, 10].

The core curricula which underlie these rotations are being developed from a variety of sources. The specific sequence in which material is learned is going to vary from institution to institution. What is important is that the minimums should be uniform throughout the world and that residents should have the opportunity to pass beyond the minimums as rapidly as their individual talents allow.

There is another fundamental premise in the idea of competency-based training. Residents should be able to make independent patient management judgements when they are deemed competent to do so and the independence of these judgements should escalate according to experience and demonstrated competence. In today's legal environment in the United States it is not possible for residents to truly function independently but they should be given the ability to demonstrate that they could be independent at each level before competency is verified [10].

PGY 4 – the chief residency

Chief residency experience is now 12 months. During that time residents are expected to take on major responsibilities for evaluation of patients, implementation of treatment plans, and the surgical experience. This traditional training concept of quasi-indepen-

[1] According to the European Training Charter a minimum of 4 years must be spent in clinical neurosurgery. The curriculum can be organized appropriately. (H.J.R.)

dence which has been fundamental in the United States for the past hundred years is under attack from the legal profession, but persists as a requirement for certification. Independent practice is not possible so supervision must occur in ways that assure patient safety without compromising the educational experience for chief residents. During the chief residency trainees should have the opportunity to examine patients independently without being influenced by the prior information from senior neurosurgeons. The training program should require that they develop treatment plans including prognosis and treatment alternatives, interact with patients, participate in surgical procedures to the extent of their abilities, and do all of this in a progressively escalating fashion in which they achieve greater and greater independence throughout the year. This should culminate in complete independence. Specific numbers of operative procedures are not required in American training programs. There are minimums for number of craniotomies, and examining board members and residency reviews make arbitrary judgements about the adequacy of numbers of procedures done, but there are no specified numbers required, and therefore the experience of chief residents may vary enormously from one training program to another. Our trainees at Johns Hopkins typically have 350–500 operations in the chief year. Training to competency is the goal of the chief year. Now we clearly define the end of training as that degree of competency which will allow independent practice the following year without endangering patients. This means that every chief resident must be competent to do most procedures of neurosurgery. Remember competency does not imply mastery [10]. Competency means that the individual will carry out all aspects of patient care at the level of his or her peers and in a fashion that does not endanger patients. It does not imply that every trainee will be a master of every field or that every trainee can do everything in neurosurgery irrespective of how difficult it may be. Competency means that individuals can practice independently while knowing their own limitations, and that they are prepared for the ongoing education which will lead to mastery of those parts of neurosurgery he/she chooses to emphasize. I do not believe that competency can be assured by exposure to specific numbers of patients, by independent practice without supervision, by observation without experience, or through acquisition of a knowledge base alone.

A year of supervised practice

For many years at Johns Hopkins we have supplemented traditional residency training with a year of supervised, though independent, practice as originally envisioned by Halstead [7]. After completion of training for the required 48 months of neurosurgery all Johns Hopkins trainees stay one additional year serving as *instructors* in the medical school and independent members of the faculty. This year supplements the traditional chief residency with a year of busy practice in which trainees maintain independent clinics, evaluate patients, plan treatment, provide prognosis, carry out surgery, and administer all peri-operative care. They are required to verify all of this information with senior faculty in the pertinent sub-specialties and senior faculty members serve as assistants during virtually all surgical procedures. A typical year of independent practice sees the trainee adding about 350–400 major neurosurgical operations to their experience. Sub-specialization is possible, if desired, and may substitute for a fellowship year.

Fellowship years

It is now a common practice for trainees to progress from residency to fellowships in specialized areas of neurosurgery. I personally believe that the introduction of true competency-based training concepts will allow the enfolding of much specialty training into our current programs. As residents accelerate learning and skills application, they will progress to areas of sub-specialization more rapidly. I think this will reduce the amount of time required for post-graduate sub-specialty training. It is my view that competency to practice a sub-specialty should be demonstrated in the same way that competencies are assessed throughout residency program and should not be dependent upon time spent in fellowship training. Rather it is the acquisition and application of knowledge and skills which should define sub-specialty competency [12].

Laboratory training

Board requirements in the United States now include one year of research experience. There is very strong evidence that very few neurosurgeons continue research, and it is my view that research training should become an elective as any other sub-specialty. At Johns Hopkins we have had a 2 year requirement for research for many years. I believe that all neurosurgeons would benefit from an experience in clinical research which could be possibly as short as one year. It is unlikely that even 2 years of laboratory research are enough to prepare the trainee for competitive research, and those who wish to do bench research as a part of their career should have the opportunity to be adequately prepared. Such preparation might involve 3 or more years of training. The possibility of a Ph.D. experience should be a part of neurosurgery training. However, these options should be available only for those who wish to take advantage of them and there is little value in all trainees continuing to receive basic research training.

Close reading, statistical methods, study design, and the related issues can be taught to competency by specific programs rather than learned haphazardly in the course of less structured research training. Clinical research training would be of value for all neurosurgeons. I believe basic research training should be limited to those with a serious interest in pursuing research as a part of an academic career and that training should be expanded beyond what is currently available. An obligatory year of basic research serves little educational purpose [6].

Professionalism

The competent practice of medicine requires certain behavioral patterns which are loosely termed professionalism. These include relations with and dedication to patients, relations with peers, meeting documentation standards, personal behavior that never compromises patient care, and a dedication to life-long learning as examples. There is excellent evidence that students and residents who deviate from standards of professional behavior often have difficulty in practice and are disciplined frequently. No one knows how to impart these ideals of professional behavior but excellent mentors are thought to be extremely important. In today's complex medical practice it is also important that didactic educational programs which detail the many administrative responsibilities of physicians are presented to trainees.

Case management as an educational technique

The traditional didactic forums in residency training are Grand Rounds which usually are in lecture form, patient conferences in which ad hoc presentations are the rule, pathology and radiology conferences where ad hoc review of recent cases is carried out, walking rounds which depend upon the nature of hospitalized patients, and journal clubs which review important current papers. All of these forms have significant short-comings though all are valuable for different reasons.

The *formal lecture* is most valuable to the person giving it for all need to learn how to synthesize material, to present it effectively, and to incorporate visual aides for teaching purposes. A summary lecture given by an expert can present material not easily available in any other way. However, when conferences are not participatory they are unlikely to be remembered, and an emphasis upon lectures as an educational form probably retards the acquisition of knowledge. *Patient conferences* and *rounds* which are ad hoc suffer from the fact they are not comprehensive or rank ordered by importance, and typically do not expose the trainee to the spectrum of diseases. *Didactic conferences* in neuro-pathology and neuro-radiology have the same problems if they depend upon current patients for material and can be better organized if they are comprehensive and sequential. *Journal clubs* are of value if they teach close reading, study design, assessment of statistical treatments, and look seriously at the value of the reports being reviewed. If they simply substitute for reading current literature they have little merit.

Case management conferences offer a significant improvement over the conventional forms and are particularly amenable to computerization and standardization. In an appropriately designed case management format patients are sequentially presented beginning with the most typical and progressing to the most unusual presentations of any specific disease. Diagnostic possibilities are outlined; prognosis is determined; alternative therapies described and the outcomes of these therapies presented; moral, ethical and research issues where pertinent can be included. I believe that development of fundamental case management system covering all of neurosurgical disease will be an extremely important next step in educational evolution. The organization of these conferences is also important. They should be participatory and symposium in style. The patient should be presented and the trainees challenged to make the appropriate diagnosis based upon history and physical examination, to interpret the imaging and other diagnostic studies, and to discuss therapeutic options and their outcomes. (See also chapter P. Winkler/J. C. Tonn)

There is also excellent evidence that individuals remember and synthesize material best if it is repeated multiple times and if it is encountered in multiple venues. Neurosurgical trainees should encounter a broad spectrum of diseases in the clinic. A lack of clinic experience is a major deficit in neurosurgical training throughout the world currently. Trainees will then encounter the same problems in the hospital, in the intensive care unit, and in the operating rooms, as well as through related case management conferences. This combination of approaches improves retention and recall [10].

Clinic experience

The issue of clinic experience is an extremely important one. Increasingly in the United States, all surgical specialties find their training focused upon inpatients who come to surgery or suffer significant complications. In most diseases there will be large numbers of patients who do not come to surgery. Some patients will not even have the disease as suspected. The outcome of procedures can only be known with long-term follow-up and trainees are usually excluded from this follow-up. It is extremely important that trainees have a defined outpatient experience which parallels their operative experience and escalates in responsibility in the same way that procedural skills do. This outpatient experience can be correlated with formal case management training to provide a much broader analysis of disease than is possible when one focuses only upon inpatients who are having procedures.

Summary

The current training program at Johns Hopkins is 7 years in length. The first year is equally divided by general surgery and neurology/neurosurgery-related sub-specialties. This year needs to be restructured so that it meets all the specific education required by the beginning neurosurgeon.

In our program the next three years are spent in residency escalating from junior to senior to chief resident to complete the additional required 36 months of training for board eligibility. It is this phase of training that is most amenable to change through competency-based teaching. No one knows how long this residency training should be, and there is very little information about how rapidly neurosurgeons can acquire the requisite knowledge and skills to practice independently. It is important that these years escalate in responsibility and culminate in supervised quasi-independent practice. It is my belief that with escalation based upon demonstrated competencies most residents will progress through training much faster then they do currently. This will allow enfolding of much sub-speciality training into existing programs, especially in the chief year [10].

Years 5 and 6 at Johns Hopkins are spent in a required laboratory rotation though clinical research and sub-specialization which leads to a certification have been reasonable alternatives. It is my belief that obligatory laboratory time serves little useful purpose for most residents. I think basic laboratory research should be reserved for the small number of neurosurgeons who are actually committed to a research career. Most neurosurgeons would benefit from participation in structured clinical research. Research should be treated like any other sub-specialty rather than viewed as a requirement in my opinion.

The final required year at Johns Hopkins has been spent as a faculty member with an independent but supervised practice. This is the key year for competency training in my opinion. Trainees should be allowed to sub-specialize as they continue to expand experience in all fields. This year may be enough to meet any sub-specialty requirements if those are defined by competencies rather than by time. Our residents consider this final year the most important to their successful training and I believe the concept of a final year of supervised practice should be expanded to become a requirement in all training programs. It is in this final year that the acquisition of all competencies should be proven for certification.

An ideal training program utilizes competency as the measure of progression through the program and ignores years in training. The acquisition of knowledge is emphasized progressively to self-directed learning. The application of skills should be carefully monitored from rules following to knowledge-based practice and educational techniques should reflect individual learning styles and the new information available to direct perceptual learning. In the ideal training program the core of knowledge and skills required for each aspect of neurosurgery is defined in advance. Beginners are guided by external directed learning techniques through the minimums to self-directed learning which should occur at whatever rate each trainee can accomplish. Skills are acquired in the same way and no trainee should advance through the training program without demonstrated mastery of the minimums of both knowledge and skills at each level. These concepts mean major changes in our training programs. Education becomes the focus of training and service is relegated to those issues which support the educational mission. That does not mean that service ceases to exist. A part of the educational mission is giving senior faculty members time to teach. However, it does mean that service alone is not a justification for any trainee activity. Trainees will have to demonstrate that acquired skills can be applied effectively at levels deemed acceptable by expert practitioners of neurosurgery. No one will complete any part of a training program without having demonstrated both the knowledge and the skills sufficient to assure competency required for that level [10].

National and international verification of competency is important. Knowledge can be assessed by written examination and is done so quite well now in the United States. Skills are probably best judged by other expert practitioners in the field. The application of knowledge can be assessed by case management techniques. If we are really going to be serious about examining the application of knowledge we need to develop much better methods than the current brief oral ad hoc assessment by our board examinations. This is a field where computerized standard examinations have enormous promise for assessment and educational purposes. Still careful evaluation of application of knowledge and skills by personal contact with master neurosurgeons must remain a part of the assessment process.

Competency-based training offers an enormous opportunity for research in neurosurgical education and in how to assess competency. Self-assessment should be a part of the requirements for every training program, and for every neurosurgeon. Outcomes will eventually be used to define training programs and the practice of their trainees. Competency-based training focuses upon product, not process, so competency, not years of training, becomes the goal.

The original goal of the liberal education was to prepare students to be responsible citizens. Our own training programs are in danger of becoming purely technical. As we change to competency as the primary goal of residency training, we must assure that the liberal aspects of that training are also improved. Deviation to patient welfare, lifelong learning, and effective neurosurgery citizenship all characterize the competent neurosurgeon who will proceed to mastery of the field. Assuring all those desirable characteristics of professionalism is another challenge of competency-based training.

B. Curriculum development in neurosurgery

Introduction

The goal of neurosurgery training is the assurance of competency. Therefore, the first step has to be to understand what is meant by competency in order to define what it will take to make the aspiring neurosurgeon competent. Competency has a specific definition within a theoretical framework. The categories generally accepted are Novice, Beginner, Advanced Beginner, and Competent, beyond competency stand Expert and Master. In general terms we can think of these categories as they relate to the current concept of post-graduate years in training (PGY). Year 1, or the internship, is the Novice year, year 2 is the Beginner year, years 3 and 4 are the Advanced Beginner years and the progression of understanding responsibility and skills should bring the trainee from Advanced Beginner in year 2 to Competent by the end of year 4. These goals are summarized in Table A. A competent physician should have the knowledge, surgical techniques, and patient management skills adequate to practice independent neurosurgery, be prepared for lifetime learning, and be prepared to mature to Expert and Master of the field over the next 10 to 20 years. The program must define the knowledge required, examine the application of that knowledge, define the surgical skills required and assess proficiency, and

Table A.

What must be done to assure Competency

1. Define competency?
   - A. Knowledge
   - B. Application of knowledge
   - C. Skills (Surgical Proficiency)
   - D. Application of Skills

The competent physician should manage patients and their problems at a level expected of others of equivalent experience.

2. Define the knowledge and skills required and the sequence of learning
3. Define measurement of competency
4. Develop the curriculum
5. Define outcome measures

Table B.

1. Medical Knowledge
2. Patient Care- Application of Knowledge
3. Medical Skills
4. Procedural Skills
5. Application of Skills to Patient Care
6. Practice Improvement- Use of Texts and Literature- Self-Directed Learning
7. Interpersonal and Communication Skills
8. Professionalism

Table D.

1. The Doctor- Patient Relationship
2. Conduct of Medical Research
3. Valid Study Design and Research
4. Responsible Medical Writing/Speaking
5. Communicating with Patients
6. Close Reading, Critical Thinking
7. Conflict of Interest
8. Relations with Peers
9. Quality Assurance and Outcomes Assessment
10. Physician Impairment and Discipline

determine the application of those skills to patient care. In addition, this sequence of acquisition must be determined independently for each training program. Organized neurosurgery should define measures of competence, devise a minimum core curriculum, and examine outcome measures for the educational process.

General competencies

There are a group of general competencies that every physician must have and every neurosurgical trainee certainly must acquire these skills. These are listed in Table B. The training program has to assess technical competency to perform procedures and the application of knowledge and surgical skills to patient care on a continuous basis. Progress and acquisition of knowledge can be assessed by written or oral examination such as those used by the American Board of Neurological Surgery. It is important to assess professional behavior and require moral, ethical, and personal competence in all aspects of professionalism throughout training.

Surgical skills curriculum

Acquisition of surgical skills should be sequenced from least complex to most complex and sequenced throughout the training program. Table C demonstrates the general progression. The sequence actual procedures can be determined independently in each program in accordance with the overall goals of the curriculum.

Table C.

1. Assisting a qualified surgeon until knowledge of anatomy and instrument use is verified.
2. Positioning and Preparing
3. Closure
4. Opening and Exposure
5. Progressively performing the procedure with supervision
6. Independent Surgery

Professionalism in the ideal curriculum

Teaching neurosurgery trainees to be good doctors, good colleagues, and exemplary human beings is important. This is probably best done by example and mentors who behave correctly are the most important influences for young physicians. Some aspects of professionalism can be taught and discussed in a formal way. Table D lists some of these more important aspects of professional behavior.

The ideal curriculum

Specific curriculum development is being addressed by a number of professional organizations in the United States. The American College of Surgeons and the Society of Neurological Surgeons both have major efforts and core curricula are available or soon will be to supplement those already proposed by our national neurosurgery organizations.

The educational requirements of the Novice (PGY-1) year are found in Table E. During this year the neurosurgical trainees should learn the fundamentals of operative and peri-operative care. They should learn how to deal with the major complications which beset neurosurgery and they should be able to deal with serious multi-system trauma. This is a good time to learn the fundamentals of neurology, particularly acute care neurology. The goal of the year is to prepare the young neurosurgeon with all of the fundamentals which will be needed as the trainee becomes involved exclusively with neurosurgical patients.

Table E. *The ideal curriculum – year one**

1. Peri-operative Evaluation and Management
2. Management of Pain
3. Management of Cardiac Disease
4. Management of Pulmonary Disease
5. Principles of Fluid Balance and Nutrition
6. Cardiopulmonary Resuscitation
7. Fundamental Surgical Skills
8. Fundamentals of Neurology
9. Principles of Acute Care Management
   - Acute Neurological Deterioration
   - Pulmonary Distress
   - Anesthetic Emergencies
   - Multi-system Trauma
   - Coma and Stroke

(See the core curriculum details available from the American College of Surgeons for all recommendations).

Table F. *The ideal curriculum – years two and three general beginning competencies*

1. Fundamentals of Neuro Imaging
2. The Neurological Examination
3. Acute Neurological Deterioration
4. Spinal Emergencies
5. Peripheral Nerve Injury and Repair
6. Principles of Neuroanesthesia

During the Beginner and Advanced Beginner phases of training (PGY-2,3) the trainee becomes exclusively involved in neurological issues. The fundamentals which should be mastered early in the beginning phase are found in Table F. Obviously all of these will be expanded during the entire neurosurgery training. But these topics should all be covered in the first few months of the first year of specific neurosurgery training.

Core curriculum

A draft core curriculum has been presented to the Society of Neurological Surgeons. I present one of these proposals as an example of how a core curriculum should be structured initially. There are two important issues to remember. The first is that the core curriculum is the minimum not the maximum required and the second is that the core curriculum is simply for the introduction of these topics. It is expected that each topic will be developed as a continuum of learning throughout the next three years of training.

There is another important point for all core curricula. They can be as specific or as broad as thought appropriate by each individual program director. The sequence in which these topics are introduced will also be program specific and will vary from program to core program according to the types of patients seen. The topics can be equally broad or specifically defined. The program director may choose to list everything an individual is supposed to know while another may define only broad categories and let the trainees develop the specifics themselves. The importance of the core curriculum lies in the fact that every general topic to be mastered is listed. The general sequence in which these topics should be studied and mastered is defined. The core curriculum should have some relationship to the patient management and surgical experience of the trainee. However, individual programs should take much responsibility for developing their own curricula and integrating the curricula available from organized sources into a curriculum specifically developed for their own purposes. General examples of core curricula are found in Table G sub-topics I–VIII.

General principles of education in patient care

Education in patient care is extremely important. The general principles to be followed are found in Table H. The most important issue is that large numbers of patients must be seen in multiple venues and that trainees must be given the responsibility of formulating independent patient management schema.

Education to competency

The principle of education to competency will eventually eliminate the concept of specific years of postgraduate training. However, in order to make the transition it is necessary to use the PGY concept to have some structure in which the acquisition and application of knowledge can be judged. It is to be expected that with time training programs will gradually change to allow trainees to proceed at their individual rate in both acquisition of knowledge, mastery of surgical skills, and the application of both to the management of patients with neurosurgical diseases.

Table G. *Curriculum topics-years two and three**

Trauma
1. Head trauma
   - Diagnostic and surgical management of depressed skull fractures
   - Diagnostic and surgical management of acute and chronic subdural hematomas
   - Diagnostic and surgical management of epidural hematomas
   - Diagnostic and surgical management of intracerebral hematomas
   - Management of increased intracranial pressure including ICP measurement
   - Management of DAI and cerebral contusion
   - General trauma management
   - Management of CSF otorrhea and rhinorrhea
   - Recognition of brain death including legal implications
   - Familiarity with AANS/CNS guidelines for management of severe head injury
2. Spine trauma
   - Diagnostic and surgical management of cervical spine injuries
   - Diagnostic and surgical management of thoracic spine injuries
   - Diagnostic and surgical management of lumbar and sacral injuries
   - Fundamental instrumentation techniques
3. Peripheral nerve trauma
   - Recognition and initial management of peripheral nerve trauma
   - Major nerve exploration and repair

Vascular disease
1. Extracranial vascular disease
   - Extra and intracranial occlusive disease
   - Carotid endarterectomy
   - Management of acute infarction and TIA
2. Intracranial vascular disease
   - Intracerebral hematomas
   - Anterior circulation aneurysms
   - Straight forward AVMs
   - Dissections
   - Cavernous malformations
   - Vasospasm management

Tumors
- All glial tumors
- Convexity meningiomas
- Parasellar Tumors
- Neuronavigation
- Cortical localization
- All metastatic tumors
- Pituitary Tumors – functional and nonfunctional

Spine
- Diagnosis and management of degenerative disease
- Management of epidural tumors and infection
- Management of bone infection
- Management of intradural extra axial tumors
- Management of intra axial tumors
- Management of dural AVM
- Basic stabilizing instrumentation

Table G (continued)

Peripheral nerve and functional
- Diagnosis and surgical management of upper and lower extremity compression syndromes
- Diagnosis and management of peripheral nerve tumors
- Diagnosis of Brachial plexus tumors, compressive syndromes and inflammatory disease
- Balloon compression or radiofrequency Rx of trigeminal neuralgia
- Dorsal column stimulator
- MVD for trigeminal neuralgia or other cranial neuralgias
- Cordotomy
- Drug delivery pump (intrathecal)
- Vagal nerve stimulator for seizures
- Temporal lobectomy for seizures
- Sterotactics
- Epilepsy Surgery

Pediatrics and shunting
- Medical and surgical management of common congenital abnormalities
- Medical and surgical management of common brain tumors
- All cranial trauma
- All shunts

Other skills
- Ventriculostomy
- Tongs application
- Halo application
- Muscle and nerve biopsy
- Lumbar CSF drainage and LP
- Peripheral nerve blocks

Skull base surgery
- Acoustic tumors
- Basal meningiomas
- Other skull base training
- Complex approaches to the skull base

*Tables G-I–VIII are a synopsis of draft recommendations presented to the Society of Neurological Surgeons, 2003.

Table H.

1. Large numbers of diverse patients
2. Multiple venues: clinic, ward, emergency, operating room
3. Independent assessment and formulation of management plan
4. Interaction with senior faculty to critique management
5. Large diverse surgical volume
6. Repetition

References

1. Anderson JR (ed) (1981) Cognitive skills and their acquisition. Lawrence Erlbaum, Hillsdale NJ
2. Blum BI, Sigilliot VG (1986) An expert system for designating information systems. Johns Hopkins APL Tech Dig; 7: 23–31
3. Edelman S, Intratar N (2002) Models of perceptual learning. chap. 18 n Perceptual Learning (eds) Poggio and Fahle MIT Press, Campbridge, London
4. Hamill BW, Stewart RI (1986) Modeling the acquisition and representation of knowledge for distributing tactical decision making. Johns Hopkins APL Tech Dig 17:31–38

5. Harden RM, Hart IR, Mulholland H (eds) (1992) Current developments in assessing clinical competence. Page Brothers, Norwich, England
6. Herbert RS, Levine RB, Smith CG, Wright SM (2003) A systematic review of resident research curricula. Acad Med 78 (1): 61–67
7. Long DM (1996) Educating neurosurgeons for the 21st century. Neurosurg Quart 62 (2): 78–88
8. Long DM (1998) Assurance of competency in residency training: neurosurgical education in the 21st Century. Neurosurgery in transition. The socioeconomic transformation of neurological surgery, concepts in neurosurgery, vol 9. In: Bean JR (ed) Williams & Wilkins, Chap 10, 99: 147–156
9. Long DM (2001) Competency based residency training: the next advance in graduate medical education. Risk control and quality management in neurosurgery. In: Steiger HJ, Uhl E (eds) Acta Neurochir (Wien) [Suppl] 78: 153–158
10. Long DM (2004) Competency based training in neurosurgery: the next revolution in medical education. Surg Neurol 61 (1): 5–15
11. Rasmussen J, Lind M (1982) A model of human decision making in complex systems and its use for design of system control strategies. Riso National Laboratory Report. Riso-M-2349. Roskilde, Denmark
12. Williams RG et al (1987) Direct, standardized assessment of clinical competence. Med Educ 21: 482–489

Correspondence: Donlin M. Long, Department of Neurosurgery, Johns Hopkins Medicine, 600 North Wolfe Str./Carnegie 466, Baltimore, MD 21287-7709, USA. e-mail: dmlong@jhmi.edu

Acta Neurochir (2004) [Suppl] 90: 33–38

Printed in Austria

V Internal regulations and general guidelines of a neurosurgical department and training programme

H.-J. Reulen[1] and **H.-J. Steiger**[2]

[1] Department of Neurosurgery, Ludwig-Maximilians-University, Munich, Germany
[2] Department of Neurosurgery, Heinrich-Heine University, Düsseldorf, Germany

Summary

It is strongly recommended that general guidelines and internal regulations be laid down in written form by the department. The guidelines regulate the general workflow and related trainee duties and responsibilities, education and research, special procedures for particular diseases and conditions (standards of care). Regarding general workflow of the department, the following items should be laid down once in written form: time table of department, working hours, admission procedures, medical records directives, responsibilities on wards and in the emergency unit, patient information, discharge procedures, outpatient consultations, call schedule and on-call duty plan, as well as week-end and holiday regulations for rounds on wards and ICU. Regarding education and research, the following items should be written down: conferences and teaching rounds, policies on presentations and publications, policies regarding meetings and courses, research leave and vacation plan. The written definition of standards of care is still considered optional in neurosurgery at the present time.

Keywords: Neurosurgical training programme; regulations and procedures; residency training; standards of care; quality management.

Introduction

It is strongly recommended that internal regulations and general guidelines of a department be laid down in written form. Defined procedures and punctuality are necessary for an efficient workflow and to ensure principal standards in patient care, last but not least also from a medico-legal point of view. Written procedures are the heart of any quality management system. Such "internal regulations" also facilitate daily life and collaboration among doctors, nurses, and other personnel. It is important that a new trainee receives a written hand-out when entering the programme.

This chapter has been designed as a possible guide for all those departments that have not yet developed such internal regulations. It is based on experiences of many years and simultaneously is a collection of good examples seen over the years in different departments. The model presented here in the following sections can either be used as they are or be modified and adjusted to the local situation. Some references are given for further reading.

Internal regulations and guidelines of the department objectives of the training programme

The goal of a training programme is to bring the rising generation efficiently to the necessary level of competent patient care and conduction of scientific research to produce a small group of fine clinicians and scientists. The programme needs trainees who are interested in becoming outstanding clinicians and surgeons devoted to patient care, who are also interested in acquiring basic laboratory and/or clinical research skills that will allow them to contribute to new knowledge and future developments. The training programme emphasises a team approach in the major areas of our specialty. The programme expects outstanding performance from every trainee and in return we are committed to provide an outstanding educational environment for our trainees. The training programme constitutes essentially a contract between two generations of neurosurgeons. The younger generation helps to run the service and receives in exchange devoted formation by the older generation.

List of guidelines

The following items should be laid down once in written form and be revised from time to time. They should be handed to the resident when entering the programme.

A) Workflow and related trainee duties and responsibilities
   1) time table of department
   2) working hours
   3) admission procedures
   4) medical records directives
   5) responsibilities on wards and in the emergency unit
   6) patient information
   7) discharge procedures
   8) outpatient consultations
   9) call schedule and on-call duty plan
   10) weekend and holiday regulation for rounds on wards and ICU

B) Education and research
   1) conferences, teaching rounds and education
   2) presentations and publications
   3) policies regarding meetings and courses
   4) research leave
   5) vacation policies and plan

C) Special management guidelines (standards of care, optional)

In the following examples are given for some of the items from the given table of contents. They can be modified and adjusted to the local situation. Local circumstances determine whether certain concepts can be implemented or not. For example, in our opinion the teaching effect of the educational conferences is optimal if an entire morning of the week is devoted completely to teaching, possibly in combination with ward rounds with the chairman. No elective procedures are planned for this weekday. Implementation of this concept is only possible if the operating theatres are shared with other disciplines to allow for an even utilisation of these resources. Not all items mentioned in the survey are covered.

Workflow and related trainee duties and responsibilities

1. Time Table of Department
Punctuality is critical for efficient functioning of the service

a) *Morning rounds [7, 9]*: Monday to Friday, beginning at (07^{00}) hrs with the respective staff members/consultants.
b) *Morning conference [4]*: Monday to Friday, beginning at (07^{45}) hrs in the conference room.[a] A short report of emergencies and admissions during the night is presented by the trainee on call, a status report on patients of the ICU and intermediate ward by the respective residents, particular occurrences and problems on the wards are reported by the ward residents.
c) *Operating theatres*: OR begin at (08^{15}) hrs according to the OP programme of the day. The surgeon or the assigned trainee must be present in the OR for positioning. Any changes in the OP programme must be made known to the anaesthesiologist and the responsible scrub nurse in due time.
d) *Neuroradiological conference and/or OP conference* Monday to Friday at (14^{00}) hrs in conference room. All newly admitted patients are presented with previous radiological examinations etc., so that the patient management can be planned conclusively, if possible. All patients to be operated the following day are discussed regarding the surgical approach and details of the procedure.
e) *Afternoon ward rounds* (if any)
f) *Intensive care round*: Surgeons must see their patients on the ICU at least once per day and discuss the treatment protocol with the responsible colleagues.[b] Notes are to be written on all neurosurgical ICU patients either by the surgeon or by the trainee.
g) *Briefing of trainee on call*: Important information shall be transmitted to the trainee on call within the frame of a formal transfer. A list of all patients has to be generated to update the trainee on call on the status of all patient problems on the service.

Working hours[c]

Admission procedures

a) Patients are admitted either to the department in general or to one of the consultants/staff specifi-

[a] Most European departments have morning conferences where all staff members and trainees participate.

[b] The personal care by the designated attending surgeon is a traditional concept in the USA. In Europe the personal care by the responsible surgeon is also the concept which meets best the expectations of the patients and relatives. Furthermore, specific knowledge of the details of the pathology and the surgical procedure is necessary for an appropriate assessment and management of surgical ICU patients.

[c] There is no general standard for the working hours and changes are to be expected in the light of new European and national regulations. Still, it is advisable to define the working hours and also the dependence of working hours on specific needs of the patients and the department [5].

cally. Patients in the outpatient clinic are seen by the assigned trainee together with the respective consultant/staff.

b) Patients arriving directly on the ward are seen by the assigned trainee as soon as possible. History taking, physical examination, and the organisation of further laboratory, radiological or – if necessary – additional examinations are indispensably required on day of admission. All patients should have a full physical and neurological examination and a documented plan for necessary lab and x-ray work-up. New admissions are to be discussed with the respective consultant/staff and the further procedure is determined.

c) Emergencies admitted through the emergency unit are seen by the trainee on call, who discusses the case and further procedures after primary assessment with the consultant/staff on call.

Medical record chart (patient file)

a) Trainees are expected to keep all charts current. All admission data (patient histories, physical and neurological exams), operative notes of trainee's cases[d], discharge summaries as well as consultation notes must be dictated or written down without delay. A daily progress note must be entered in the chart. Additional entries are required in case of particular events. Document the findings of all urgent diagnostic studies, especially MRT, CT, angiography immediately on the chart!

b) Postoperative orders and notes: At the end of surgery the trainee must write all orders for the ward/ ICU on the respective order sheet, in particular arterial pressure and other limits, directives with regard to drains, positioning and mobilization, infusions and medications, and control examinations such as x-rays, MRI etc. It is expected that the trainee writes a short postoperative note in the chart regarding the result of surgery.

c) On Fridays, please document a neurological exam and outline a week-end plan! This will help the weekend trainees on call. Document all entries and orders on the chart with date, time and your name! Remember to order the drugs patients come with as well as drug levels (phenytoin, valproate, etc.). Address and inform the nurses also directly! This will help the nurses.

Responsibilities on wards and emergency unit

a) OP preparation
The trainee is responsible that for surgical procedures of the next day, the complete patient chart including all pertinent neuroradiological material for the operation is available on the ward on the evening before the operation. He will check and complete preoperative orders.

b) Emergencies and on call duties
The trainee on call should not be assigned to the routine OP programme of this day, so that he is available any time for emergencies[e]. Night duty starts at (16^{00}) hrs and lasts until (08^{00}) hrs the following morning.

Accurate lists concerning patients with open problems must be passed on to the next trainee on call. The patient list must be updated throughout the night or weekend on call.

On weekends there may be sometimes a more junior or a senior trainee and tasks have to be specified accordingly. Emergencies are primarily seen and examined by the trainee on call. He informs the respective consultant/staff on call and all further decisions are made together.

Problems arising with patients on the ICU or on the ward will be evaluated first by the trainee on call who will then rectify the problem or contact the consultant/staff on call. In case of major patient problems the trainee has to contact the consultant/ staff immediately.

Information of patients and their relatives

The personal guidance and care of our patients is most important. Examinations and procedures should be explained in detail to the patient. The surgery has to be fully explained to the patient minimally 24 hours before a planned operation (Germany!), if necessary by means of a form or a schematic drawing. Risks have to be discussed and consent to the operation has to be obtained in written form [6]. The patient and the doctor and, in case of a major operation, preferably

[d] Consultant and trainee must discuss who will take care of the operative note. In general trainees will dictate the note if they have performed the majority of the procedure and consultants/staff will take care of the operative note for complex cases.

[e] This rule depends on local situation. If a junior and a senior house officer are assigned to on call duty, one of them can be involved in elective surgery to a limited degree.

another witness have to sign the form. When complex surgery is imminent, it is recommended to inform the next of kin. For children both parents have to agree with the operation and have to give their written consent. Information is usually given by the operating surgeon (or an assigned trainee).

Since nowadays patients are often admitted the evening prior to the operation only, it is advisable to inform the patient and obtain his consent already before, for example in the outpatient clinic.

Patients should be informed in a sensitive and caring manner about their disease, especially in case of an unfavourable prognosis. It is advisable to first interview the relatives and discuss with them in how much detail the patient can and wants to be informed. Information on the phone should be given only in exceptional cases or to personally well known relatives.

Discharge Procedures

The discharge summary must be ready at the time of discharge and a copy should be given to the patient. The discharge summary should be addressed to the referring physician. Family doctor and other involved colleagues also receive a copy. The discharge summary must contain the relevant diagnoses, procedures performed and a summary of the hospital course and outcome. Medications at the time of discharge must be listed and weaning instructions, if applicable, must be detailed. A follow-up date and time in the outpatient clinic should be scheduled prior to discharge and be mentioned in the discharge summary.

Prescriptions for necessary medications are handed to the patient upon discharge. In case of evening or weekend discharge, a supply of necessary medication for 24 hours is given to the patient.

Available medical reports and data sheets are put into the medical record chart together with the discharge summary. The completed chart is given to the ward secretary for final coding and archiving.

Outpatient consultations

On-call duty schedule and plan

Weekend and holiday regulations for ward and ICU rounds

Weekend and holiday ward rounds by the house staff on call are at (08^{30}) hrs. A briefing between the trainee leaving duty and the trainee beginning work is necessary (see 1.g.). Rounds on ICU start at (09^{00}). Problems on the ward are discussed with the chief trainee or the staff member on back-up call.[f]

Education, research and absences

Conferences, teaching rounds and education

a) *Journal club* (conference room, Tuesday 16^{00} till 17^{00}) [1, 7, 9, 10]
A journal club will be held twice monthly. The sessions consist of either current articles or a group of articles dealing with a common theme. The journal club has the purpose to review and discuss selected articles from the recent literature. Each trainee should review and present at least two articles per year. Articles can be selected with either a staff member or with the chief resident. The trainee learns to present the essence or "take home message", respectively, of the article but he should also learn to critically analyse whether the topic has been approached appropriately, whether methods applied are adequate, whether conclusions drawn from the results are balanced and not over- or under-estimated, and also whether they fit into the pertinent literature. Such a systematic approach is important for a trainee so that he acquires the ability to weigh the literature critically for its relation to his practice in neurosurgery.

b) *Afternoon teaching conference* (conference room, Tuesday 16^{00} till 17^{00} alternating with journal club)
Afternoon teaching rounds are mandatory and shall be attended by all trainees, students, and fellows unless they are in the OR.

c) *Weekly clinical neurosurgery conference (CNC)* (conference room, Friday 08^{15} till 10^{00})[g]
The CNC is dedicated to neurosurgical topics pertinent to the clinical service. All trainees, students, and fellows are expected to attend and to participate actively in this conference.

[f] Weekend procedures depend very much on local circumstances. A formal in-house briefing with the consultant/staff appears advisable if junior trainees are on call without back-up of a senior trainee.

[g] The classical grand rounds conference can be modified in many directions according to local circumstances. The here cited Friday morning model may not be appropriate in hospitals where Friday afternoon is already considered as part of the weekend. Upcoming working hour's regulations will also have a major impact on this traditional common teaching event.

d) *Monthly morbidity & mortality conference (MMC) [2, 3]* (conference room, Friday 08^{15} till 10^{00} instead of CNC)
The MMC reviews all mortalities, morbidities and complications of the service. The cases are selected by the consultant/staff and the trainees present the case. The consultant or trainee review the pertinent recent literature. The purpose of the MMC is future avoidance of occurred complications.

Presentations and Publications

Trainees should be well supervised during clinical and/or experimental research. They have to acquire skills to study and judge clinical case material. The following is a recommendation for presentations and publications

Y1: at least one poster presentation at a national congress/meeting
Y2+3: at least one poster and one oral presentation at a national congress or meeting
Y4–6: at least one poster and one oral presentation as well as one paper sent to a national or international journal for publication.

If a trainee presents a paper, poster or abstract at a meeting, expenses are/can be paid by the department according to prior individual agreement with the programme director.

Travel to meetings and courses

Trainees are encouraged to participate in meetings of the national society, meetings with special topics, or a neurosurgical course (European or national) and during year 4–6 also in some international meetings. Travel to meetings needs approval of the programme director and hospital administration. A travel request must be made at least 2 weeks prior to departure. Travel expenses are generally reimbursed if the resident contributes actively to the meeting. Financial support for courses is given according to prior individual agreement with the chairman.

Research leave

Concentrated periods of research in the lab or at another institution are encouraged and supported for trainees during the second half of training. The periods should range from 3 to 4 weeks and are generally allowed once or twice per year.[h] A fully detailed research project is a necessary condition. Co-ordination of the absence follows the procedures for vacations.

[h] Other models are a research rotation of 6–12 month's.

Vacation plan

Planned vacations must be permitted by the designated coordinator. Requests must be filed generally as early as possible. Requests will be generally accepted if at least 1 senior and 1 junior trainee remain on the respective ward during the absence. Replacement for other special duties, such as outpatient clinics and scheduled participation in conferences must be arranged by the trainee seeking leave of absence. No vacations are generally permitted during the national neurosurgical meeting.

Special management guidelines (standards of care, optional)

Guidelines for the management of all diseases are increasingly published by national and international organisations of medical specialists [12]. These guidelines focus usually on the available evidence and are rarely detailed enough to guide a resident. If standards of care are written for the department, the pertinent guidelines need to be concretised according to the tradition of the hospital. The authors have published such a manual in the German language [13].

Final remarks

These internal regulations have been designed with the intention to clearly define the duties and competencies of the trainees in order to avoid misunderstandings and mistakes as much as possible and to ascertain optimal patient management. They are not meant to be a rigid scheme for each single case. Initiative of one's own should be developed and be guided by personal conscience and knowledge.
Signature: Programme Director

References

1. Alguire PC (1998) A review of journal clubs in postgraduate medical education. J Gen Intern Med 13: 347–353
2. Bonsanto MM, Hamer J, Tronnier V, Kunze S (2001) A complication conference for internal quality control at the Neurosurgical Department of the University of Heidelberg. Acta Neurochir (Wien) [Suppl] 78: 139–145

3. Brennum J, Gjerris F Morbidity and mortality conferences. How we do it? This volume
4. Durning SJ, Sweet JM, Cation LJ (2003) Morning report: an analysis of curricular content and comparison to national guidelines. Teach Learn Med 15: 40–44
5. Li J, Tabor R, Martinez M (2000) Survey of moonlighting practices and work requirements of emergency medicine residents. Am J Emerg Med 18: 147–151
6. McClean KL, Card SE (2004) Informed consent skills in internal medicine residency: how are residents taught, and what do they learn? Acad Med 79: 128–133
7. McLeod PJ (1986) A successful formula for ward rounds. Cmaj 134: 902–904
8. Moberg-Wolff EA, Kosasih JB (1995) Journal clubs. Prevalence, format, and efficacy in PM&R. Am J Phys Med Rehab 74: 224–229
9. Pathak A, Pathak N, VKak K (2000) Ideal ward round making in neurosurgical practice. Neurol India 48: 216–222
10. Seelig CB (1991) Affecting residents' literature reading attitudes, behaviors, and knowledge through a journal club intervention. J Gen Intern Med 6: 330–334
11. Seelig CB (1993) Quantitating qualitative issues in residency training: development and testing of a scaled program evaluation questionnaire. J Gen Intern Med 8: 610–613
12. Steiger HJ (2001) Standards of neurosurgical procedures. Acta Neurochir (Wien) [Suppl] 78: 89–92
13. Steiger HJ (1999) (ed) Manual Neurochirurgie. Ecomed-Verlag, Landsberg

Correspondence: Prof. Dr. H.-J. Reulen, Schwojerstr. 19, 81249 Munich, Germany. e-mail: Ilona. Anders@med.uni-muenchen.de

Acta Neurochir (2004) [Suppl] 90: 39–50

Printed in Austria

VI Principles of teaching in a structured training programme, the rotations, and the surgical training plan

H.-J. Steiger

Department of Neurosurgery, Heinrich-Heine-University, Düsseldorf, Germany

Summary

The present paper is divided into two parts. In the first part it summarises *the essentials* of transfer of knowledge and personality from trainer to trainee. The training capacity of a programme should be identified first since the number of residents in training has implications on the structure of the programme. The capacity is limited by the number of surgical "resident cases", which should at least amount to 70 per year and resident. For the future, more emphasis should be laid on the acquirement of methods of self-teaching and continuous self-education, in order to provide the trainees with the ability to cope with changes during their later career more easily. In part two the organisation of rotations as well as the structured surgical training plan are discussed. Teaching can be organised either as a gradual exposure to more and more complex procedures or as a sequence of speciality rotations. Structured teaching of theoretical contents should accompany practical teaching and in academic programmes a scientific rotation should be integrated. The ongoing subspecialisation is currently exceeding the limits of complete coverage during a six-year-programme.

Keywords: Competence; neurosurgical residency programme; paedagogic principles; subspeciality training, structured curriculum, rotations, surgical training plan, self-directed learning.

Introduction

In a surgical speciality the transfer of practical skills in conjunction with the theoretical background allowing differential diagnosis and correct indication of procedures is of primordial importance. During the last decades, the number of procedures that need to be learned during residency training has steadily increased, and an unstructured programme cannot cover all necessary aspects within the available time. A sequence of well-defined clinical rotations with dedicated tutors is the simplest and most frequently used way and can presently be considered the standard way of achieving an efficient training. In these chapters the key aspects of organising such a training plan are dealt with. However, even with the advantages of a well-defined frame of rotations, it will not be possible anymore to accommodate up-coming new specialities within the available time. Therefore more flexible training plans shall be discussed, which can be implemented as alternatives to the standard system.

A. The transfer of knowledge and skills from trainer to trainee

Embedding of training in the hospital system

Training has always been an integral part of hospital life, but now the boundary conditions become more clearly defined in the context of hospital quality management programmes. As an example, the German KTQ® system (Kooperation für Transparenz und Qualität im Krankenhaus) shall be mentioned. In this system six *categories* are considered:

1. patient information
2. personnel management
3. hospital safety
4. data management
5. hospital planning
6. quality management

Category 2, personnel management, contains:

2.1. planning of personnel recruitment
2.2. personnel development
2.2.1. systematic concepts
2.2.2. definition of qualifications
2.2.3. training and education
2.2.4. financing of training and teaching
2.2.5. availability of training instruments
2.2.6. quality control of training

It is recommended that responsibilities of specific tasks are best delegated to the individual departments. The hospital or the university should provide the resources for the organisation of such tasks and the financing, respectively.

How many residents can be trained?

In general, comparing the number of resident cases can assess the training capacity of a unit and the required number of procedures required for board certification. If, for example, the national guidelines prescribe a number of 400 procedures done by each resident during a six-year programme, grossly 80 cases per year must be calculated for each resident in training. Depending on to which degree the staff members are committed to resident training, between 30 and 50% of all procedures can be resident cases (with a trainer supervising the procedure). Thus, there is a general recommendation that a total of about 250 cases will be needed per year to train one resident. If, for instance, a department has 1500 operations per year, between 5 and 6 residents can be trained at different levels. If the available number of resident cases is insufficient, it is better to limit the number of neurosurgical residents and fill the remaining assistant positions with rotating residents from other specialities.

The spectrum of cases in the training centre must also be considered. Although the total number of cases may be sufficient, deficiencies in specifically required areas of our speciality may limit the number of residents passing through the programme. If, for example, the number of cases in spinal surgery is very small at a university hospital specialised in complex intracranial surgery, the spinal sector limits the number of residents unless a training alliance is formed with a spinal centre, allowing the residents to spend a rotation in this unit.

The board certification requirements in most countries specify besides surgical experience, additional general and diagnostic experience, such as treatment plans, electrophysiological, ultrasound and radiological examinations, intensive care procedures and writing reports and expertises. Under most local circumstances, these aspects do not limit the training capacity, but may do so in special situations.

Numerical relations of trainer to trainees and who is teaching whom?

In the surgical disciplines learning by supervised doing is still prevailing due to the lack of surgical simulators or alternative methods for the transfer of surgical skills. Competence is measured in terms of experience. The ancient master-apprentice relation is still the key of surgical training.

Certain consideration must be given to the numerical relations between trainers and trainees. During a surgical procedure, practical knowledge can be transferred only to one resident while more theoretical knowledge can be transferred to small groups, for example during the planning conference for next day's procedures. In programmes with a well-defined rotation plan, a one-to-one relation (the tandem model) is ideal from the training point of view and ensures intensive transfer of practical and theoretical skills. If the trainee spends the day with one trainer, he participates intensively in patient care, decision making, planning of procedures, reflections concerning risks, the operation, the postoperative care and the follow-up.

In the one-to-one-system, the consultant primarily teaches the basic procedures and the openings. In a one-to-one system usually only one house staff is available for night on-call duty and the consultant needs to come in for surgical procedures. This close relationship ensures that procedures are learned in a specific manner according to the general agreements of the attending staff-group. Since learning surgical skills during a residency programme fills only part of the daily activities, it also appears conceivable that one trainer serves two residents (double-tandem model), a senior and a junior resident. The senior resident is directly engaged in the more complex cases and also can supervise the junior resident in the so-called component parts of the operation, for example the opening and closure. From the training point of view this may be less ideal, but two residents per trainer accommodates well the realities and necessities of hospital life.

As indicated, the one staff per two residents principle works well for neurosurgery. This principle allows also that the senior resident is involved in training the junior resident. During elective cases the senior resident will instruct the junior resident how to perform the opening or parts of the procedure, and if the on-call service is provided by a team of two, also the senior resident will teach the junior in the area of emergency procedures during the night, provided the local legal regulations allow such a system. If this principle is accepted for the training programme, one must be aware that the style of the basic procedures evades to a large degree the influence of the attending staff. The style of craniotomies is transferred from one senior resident generation to the next. Corrective influence from the staff becomes apparent only in the course of working up complications. The positive aspect of this system is that all three participants benefit. For the junior resident the effect is the same, whether instructed by the senior resident or by attending staff, the senior resident can develop teaching skills and the attending staff gets some time off for other duties.

The total number of trainers/consultants and residents certainly is influenced by a number of factors, i.e. size of department, number of procedures, presence of subspecialties, academic involvement, etc. In programmes with a large case load usually a relation of one to one is given but in some smaller programmes residents may represent under the present conditions a minority.

Up-coming legal regulations with the new European working hours legislation, especially "off" time after on-call duty, in the future will have an increasing and severe impact on the organisation of training. The requirement to go home after a night on call or to have some days off following a weekend on call, prolongs the necessary time span to obtain the required numbers of procedures performed and disturbs the rotations. Eventually it may lead to prolongation of the presently obligatory 6 years of training. A major additional effect is the uncoupling of the teacher/trainee relation if the trainee cannot spend anymore his whole time with his trainer. There is no question that training is less intensive or effective if the resident works with different trainers at the same time. Although there is presently not enough experience of how the training can be organised under such aggravated circumstances, the double tandem model may offer some advantages. On-call duties or time off respectively may be organised in a way that always two of the trio are working together, thus maintaining the necessary continuity for the physicians as well as the patients.

Written contents of training rotations and learning objectives

The aims and the contents of the individual rotations including theoretical knowledge and the type and number of diagnostic and operative procedures should be available in written form. This is important to frame the expectation of the trainee and to guide the teacher as well. Furthermore only written curricula allow internal or external quality control of the programme and only written learning objectives allow assessment of the trainee. An example of the curriculum for the rotation "Stereotactic and functional neurosurgery" could look as described in appendix 1.

When such curricula are developed, there is a trend for inclusion of everything in the speciality. It must be accepted that at the more junior or middle level the procedures to be performed cannot exceed the simple routine procedures. The subspeciality rotation shall create a platform for the trainee allowing for self-development thereafter.

Balance of theoretical and practical teaching

Although practical skills are of overwhelming importance in a surgical discipline, the theoretical background and knowledge should not be neglected and also be offered in a structured fashion. This is particularly important since the resident cannot autonomously recognise the importance of theoretical knowledge, and the programme director and the teaching staff must foster this notion systematically. In today's evidence driven medical world only knowledge of the pertinent literature allows for critical and sound settling of surgical indications. Theoretical knowledge needs to be subdivided into textbook material and recent literature. Besides recommending specific textbooks, the textbook material can be periodically covered in a dedicated conference.

However, this usually exceeds the local potential. A more effective way is to encourage trainees to participate in either the European training courses or comparable national courses, which cover most of the relevant areas in 4 or 5 dedicated courses. The residents must also be assisted with the acquisition of recent material. A "Journal Club" is a time proved forum to point at important new papers and to sharpen critical reading of scientific papers. The residents should also be encouraged to read the important neurosurgical journals. The increasing number of journals renders a clear recommendation unavoidable with regard to the expected reading. It is realistic for residents to read 2 or 3 core journals. A debatable question is whether the residents should be incited to read the journals already at the beginning of the residency or only later. Only if they begin reading early, they will survey the relevant literature at the end of their training. They will be able to discuss with experts at that point and they will be accepted. It is one of the important (but also difficult) duties of the Programme Director and the attending staff to create an educational environment with discussions on approaches, etc: readiness to learn new facts, methods and techniques becomes an integral part of clinical life.

It is an illusion to expect that anybody is able to survey the vast amount of new and potentially important publications in all the subspecialities of our field. The horizon can be somewhat enlarged beyond the core journals if a culture of communication with the subspecialists is encouraged. If, for example, the neurooncologist tracks a potentially important paper in a subspeciality journal, he should discuss this with the residents either on rounds or in the journal club. As soon as such a culture is established, it will be beneficial for all sides, and also the consultants will begin to learn from information coming from the residents.

External courses

Some important practical concepts cannot be conveyed locally without inadequate efforts. Trainees should be supported to attend specific hands-on training courses of their respective level of training. Examples are the following:

- microvascular anastomosis course
- anatomical dissection courses
- hands-on courses on specific approaches and techniques
- rhetoric courses, etc.

B. The rotations and the surgical training plan

According to the European Training Charter a department has to develop a Rotation Plan describing the training periods during the 6 years of training. Rotations in general are of 6 months' duration although some may be shorter. The idea is to structure these 6 years and not leave them at random. It is recognised today that covering all main areas of our speciality and teaching the many procedures is only possible within the frame of a strict organisation. For some departments it will be necessary to organise specific training periods in another neurosurgical unit, if adequate exposure in one area cannot be provided. It is also important to keep in mind that the allocation of rotations should preferentially be governed by the requirements of the training programme and not mainly by the needs of work on the wards! Experience from many institutions teaches that often conflicts may arise between the interests of clinical work and maintenance of the rotation schedule.

When planning a rotation schedule it should primarily be defined how much time shall be spent with non-neurosurgical activities (neurology, surgical disciplines, neuropathology, neuroradiology, research, etc.). The mandatory 48 months to be spent with clinical neurosurgery are an absolute minimum to teach the full variety of modern neurosurgery. In order to ascertain that all trainees have the same exposure, a rigid sequence of rotations is preferable. This allows an objective assessment and comparison of trainees' progress after each of the rotation periods.

A rotation plan can be developed in increasing order of comprehensiveness according to local possibilities and facts. A few examples shall be given.

The traditional model

In the traditional model that is dominated by generalist neurosurgeons, residents are involved in all cases on a ward and thus see always the broad spectrum of the speciality. They gradually grow into more responsibility according to the curriculum of the department.

It is recommended to assign trainees to a certain trainer during each rotation. This is more effective than working together with several staff members of whom no one has any defined responsibility. It also avoids the widely known problem where trainers give preference to active trainees and try to work together

Table 1. *Example A: the traditional model*

	Period 1	Period 2
Year 1	neurology or surgical discipline	neuropathol./ neuroradiology
Year 2	NS (Ward A)	NS (Ward B)
Year 3	intensive care	NS (Ward C)
Year 4	paediatrics (outside)	research
Year 5	NS (ward A)	NS (Ward B)
Year 6	NS (ward C)	subspeciality interest
Year 7	instructor, subspeciality interest or change to other department	

Example A is often used in smaller departments with 2–4 neurosurgical trainees and generalist neurosurgeons as trainers. In this example there are 2 or 3 wards. Trainee spends 6 months in intensive care and spends a 6 months' rotation in paediatrics (or any other missing area) in another unit outside. The last rotation is spent in one of the subspeciality areas of personal interest. In this model, residents of different stages manage patient care on the ward together. Allocation to surgical procedures is usually done by a staff member.

with them as long as possible, which may leave the trainee with deficiencies in other areas. During an assignment period (rotation) the trainer should have a clear responsibility to instruct and supervise a trainee to perform the procedures of the respective level (of the surgical training plan).

In this example it is further recommended to assign certain operative procedures, activities, and teaching to each one of the rotations. A structured surgical training plan allows for a continuous and systematic escalation of surgical responsibilities and competence through training years 1–6. A typical and widely used plan is presented in Appendix 2. Required minimum and/or average operative case numbers for each category can be assigned for each training year. A trainee is expected to fulfil the minimum numbers listed for the respective training period. This allows to detect deficiencies very early and correct them.

The subspecialist oriented rotation plan

In a department where several subspecialists are present, residents can work together with the consultants in specific subspeciality rotations. Training is provided through successive subspecialities according to a rotation schedule. The subspecialist groups can be organised with one staff and one or two residents or a fellow, respectively. If a department has a particular focus, e.g. functional or vascular neurosurgery, more than one staff member may share the workload in these subspecialities (Table 2).

Table 2. *Example B: subspeciality oriented rotation plan*

	Period 1	Period 2
Year 1	neuroradiology/ICU and neurotraumatology	neurology or surgical speciality
Year 2	functional and stereotactic NS, pain	neuro-oncology and neuropathology
Year 3	spinal NS	paediatric NS, epilepsy surgery
Year 4	research	research
Year 5	skull base and pituitary or spine and peripheral nerves	vascular NS
Year 6	neuro-oncology	free for subspeciality interest
Year 7	instructor or subspeciality interest, or change to other department	

In this example the trainee has rotations in all areas of neurosurgery. During neurotraumatology he is assigned to ICU, during neuro-oncology he has exposure to neuropathology, etc. Allocations to surgical procedures are done by his trainers. This is an example and can be changed in any way.

In this system it is necessary that all staff members and residents provide basic care and general neurosurgery. In practice a call schedule consisting of a staff person and one or several residents covers traumatology and other emergencies. In addition all staff members and residents care for elective cases not assigned to a specific subspeciality. This general root is a slim variant of the generalist-dominated programme. Knowledge and skills in the general aspects are also acquired over the years of training in a stepwise fashion. An example of a surgical training plan is presented in Appendix 3.

The subspeciality-dominated rotations have some advantages. The concentration on one subspeciality allows to see and perform more of the same procedures in a row (repetition effect) and focuses reading and acquisition of theoretical knowledge. Working with a competent subspecialist provides a deep and comprehensive insight and also renders account of the increasingly complex field of neurosurgery.

The limited time in the rotations certainly is not sufficient for a resident to acquire a level of expertise in the respective area. However, the resident is able to select one of these subspecialties for an additional fellowship. If neurosurgery shall survive in the border areas to other disciplines, such as carotid surgery or spinal instrumentation, specific subspeciality training is unavoidable.

Organisation of the rotation schedule in this example on the other side is not without problems. One question that needs to be solved with the staff members is how the sequence of rotations shall be established. It certainly would be problematic if a resident were exposed to clip an aneurysm during the first year of training. In general, all subspecialists argue that their discipline be reserved to the last two years.

Another problem is how many subspecialities can be accommodated in the six years of training. If all possible areas are included as six-month-rotations, and if a scientific year is integrated, the available time will be exceeded. We have made the experience that some rotations which are not too busy can be combined, which means that one resident works together with two staff members. When planning the rotation schedule, it is conceivable to divide the subspecialities into mandatory and optional rotations. The optional rotations should be done at the end of training in order to preclude premature commitment. A model could be to organise the four mandatory clinical years in a standard manner, have one year free for research or

Table 3. *Example C: variable subspeciality oriented program*

Subject	Total months (weeks)	Year of training					
		I	II	III	IV	V	VI
General surgery or traumatology							
Neurology							
Neurophysiology							
Neurotraumatology							
Neuroradiology							
Microsurgery (course)							
Neuropathology							
Spinal neurosurgery I							
Spinal neurosurgery II							
Peripheral nerves							
Rehabilitation							
Stereotactic and functional, pain							
Epilepsy surgery							
Intensive care							
Neuro-oncology							
Paediatric neurosurgery							
Vascular neurosurgery							
Research							
Skull base and pituitary							
Hands-on courses							
European or national training courses							
Medical laws							
Hospital economics							
No of months							
No of examinations							

In this example the various possible rotations and courses are listed (and can be completed), the total time in months or weeks is determined and finally the rotations and courses are assigned to the year of training. This example allows organising a core training programme of the 4 mandatory clinical years and then an individual shaping of the residual time. Instead of performing neurology or research, a second rotation in spinal neurosurgery can be chosen.

other commitments, and have 6–12 months as flexible rotations.

A variant of a subspeciality oriented rotation schedule is presented in Example C.

Rigid or flexible rotations?

The goal of a training programme is to assure that residents completing training have achieved the highest possible level of competency. The easiest way to ascertain that all residents have the same chance to reach this ideal is to define a rigid sequence of rotations such as in the previous example. All residents pass through all rotations.

In addition, the rigid schedule is the most simple and practical model to organise. The introduction at our institution proved to be a big step ahead with regard to resident competency at each level, particularly at the end of training. There was a constantly growing acceptance of responsibility and of independence. Also some aspects of competency-based training [4] can be included, thus allowing a certain amount of flexibility. Talented residents in each rotation could do numerically more operations as well as advance faster to complex procedures without disturbing the whole programme. This kind of flexibility exists in fact in most institutions.

A more *flexible curriculum* is certainly conceivable, particularly toward the end when, for instance, the last year can be tailored to the individual trainee's interest. The last year may be used either as an intensified academic training with research, as a rather practical training for residents aiming at private practice, or as a preparation for a subspeciality fellowship. A separation between academic neurosurgery and non-academic neurosurgery at an early stage is not desirable. On the other side, even academic programmes should not consider their research rotations mandatory but offer a more clinically oriented alternative for residents not interested in basic research [3]. Moreover, crossover should be accepted and structurally included. If a resident does not succeed in basic or clinical research or realises that he does not enjoy lab work, a crossover to a purely clinical curriculum should be welcome.

D. Long proposed a flexible way through the stages of residency according to competency [4]. In his competency based training scheme a talented resident could advance much faster to complex procedures than less gifted colleagues. From the theoretical point of view the competency-based scheme is convincing. There is little doubt, however, that organising a competency-based programme represents a considerable challenge for the programme director. It can happen that two or more residents end up at the same level simultaneously, which can exceed the case resources of a training programme. Collisions are certainly less likely if the number of residents is small in comparison with the number of staff.

Assessment of rotations

At the end of each training period or rotation an assessment of the trainees' progress is obligatory and performed by the supervising consultant [6]. The most practical way to assess surgical progress is the Logbook, the general progress of knowledge and skills by using an Evaluation Form. Details of the proceedings are described in the respective chapters. The assessment process allows not only an assessment of the trainee but also of the responsible tutor, and therefore is an instrument for programme optimisation.

In a teaching programme with a distinct junior (first 2 or 3 years) and senior level with different responsibilities, it may be appropriate or desirable to make an additional assessment prior to entering the senior resident level. Such midterm evaluation should include the following key points:

Midterm evaluation

Knowledge
- familiarity with the management standards of the service [9]
- text book material

Practical skills
- competence with lumbar discectomy and laminectomies
- competence with external cervical stabilisation (Halo etc.)
- competence with ventriculostomies and ventricular shunts
- competence with craniotomies
- competence with evacuation of intracranial haematomas

Scientific assessment (optional)
- publications
- actual projects

In order to prevent failure at the board exams and also to be able to guarantee for the performance of the residents released into independent practice, some departments organise an internal practical exam where residents have to demonstrate their competence in some procedures including the preparation and postoperative care of these patients. Assessment of theoretical knowledge in an objective and reproducible way is not possible for the individual training programmes and therefore programmes increasingly recommend their trainees to participate in the EANS written examination (see specific chapter).

Ways to go

The steadily growing amount of knowledge that needs to be acquired during residency training requires in each training centre the development of a well-structured written training curriculum. However, the old fashioned apprentice and master relationship remains the most important part of residency training, particularly in the subspeciality rotations, where the trainee collaborates with a specific tutor very intensely for a period of several months. During these rotations a major part of the knowledge and practical skills, and maybe also a portion of the personality of the specific tutor, is adopted. The responsibility of the trainer during these rotations cannot be overemphasised. The trainer cannot teach something different from what he is doing himself. To talk with Rudolf Steiner: the teacher must be a personification of the good, the true, and the beautiful. The child will accept what he really is and what he says [10].

In neurosurgery learning by doing still dominates despite efforts to develop simulators as used in aviation. The number of procedures performed as documented in the logbook is still the best way to estimate competence. In contrast, a pilot can be fully trained in a simulator and fly the real plane competently already at the first time. The availability of comparable neurosurgical simulators would allow a major step ahead in residency training. The crude "virtual manipulators" available today, however are rather toys than effective instruments. The reason for the insufficiency of "surgical trainers" appears not to be a principal one but a question of financial resources. In some surgical fields more progress has been made as compared to neurosurgery [8], and some specialities like anaesthesiology are easier objects for simulation [7]. Design and construction of a common platform of a surgical simulator allowing specification for the surgical specialities could be an affordable strategy for a company in the near future.

There are some principal questions with regard to the aims and methods of residency training. Major emphasis is currently put on conveying knowledge and skills. The fact that most knowledge and skills in medicine is relatively short-lived is not adequately accounted for. Many staff physicians stick for their entire career to what they have learned during residency. If certain methods become outdated, they just give them up without acquiring new ones. Therefore more weight should be put during residency on techniques of self-teaching and self-development. The paedagogic problems of self-directed learning are not trivial and are widely ignored in current continuous medical education (CME) programmes. One aspect is an intrinsic resistance to adult self-education. Hiemstra gives some practical hints with regard to overcoming this problem [2]. For learners, there are at least two factors that can be linked with resistance: self-concept and self-awareness. Many adults enter a teaching/learning transaction with low confidence and poor self-concept, making it initially difficult to take a high degree of responsibility for one's learning. Other learners, perhaps because of previous experiences with education, are simply not aware of the power they possess as learners and thus make the assumption that a highly teacher-directed approach is the way education "should" happen. Some of the strategies to address this concern include: self-reflection; peer reflection/judgements; interviewing techniques that allow individuals to learn from one another; generating lists of possible learning resources; portfolio review and assessment; journal writing; proactive reading; discussions (sharing with colleagues/peers); using learning contracts; and obtaining feedback from many different sources. The teachers can encourage self-education by several methods:

- Teach learners how to be self-reflective.
- Develop recognition of rewards for self-directed learning.
- Provide guidelines for organising and conducting self-directed learning projects.
- Help learners develop skill in using technology.
- Use technology for advisement and learner feedback.
- Help learners learn how to investigate options, opportunities, and resources.
- Help learners learn how to match individual strengths with interests.
- Help learners develop education/learning plans.
- Help learners develop good technical learning skills.
- Help learners feel comfortable with new content.
- Help learners enhance their sense of personal learning competence.
- Help learners develop confidence and skill in taking control of elements of the teaching/learning transaction (for example, needs assessment, goal

setting, selection of content and process, and self-evaluation).
- Help learners create and control effective learning environments.

Adult self-education does not principally differ from learning during childhood. Already at the end of the 19th century, John Dewey expressed a number of paedagogic concepts that are still valid [1]. The arising interests should be watched and fostered, not be repressed. To repress interest is to substitute the adult for the child, and so to weaken intellectual curiosity and alertness, to suppress initiative, and to deaden interest. The interest is always the sign of some power below; the important thing is to discover this power. This concept is easily applicable to residency training.

Appendix 1

Example of a curriculum of a subspeciality training rotation. Contents of training rotation "stereotactic and functional neurosurgery"

a) *Procedures and number to be performed under supervision*
   - stereotactic biopsy 10
   - stereotactic drainage of intracranial cysts/abscesses/haemorrhage 3
   - radiosurgical planning 5
   - functional electrostimulation of the spinal cord 5
   - functional deep brain stimulation 2
   - procedures for intrathecal morphine (baclofen) 5
   - Percutaneous selective trigeminal rhizotomy, or glycerol injection, respectively 5
   - microvascular decompression (Jannetta) 3

b) *Learning objectives [5]*
   - Choice of therapy for trigeminal neuralgia.
   - Indications and patient selection for spinal cord stimulation.
   - Benefits and disadvantages of infusion pumps.
   - Definition of factors guiding the choice of neuroimaging (CT, MRI, angiography) for stereotactic procedures.
   - The rationale for various MRI sequences used for tumour localisation and functional procedures.
   - The benefits and limitations of frame-based stereotactic procedures.
   - Patient selection for VL thalamotomy and pallidotomy and stimulation.
   - Advantages and disadvantages of ablative procedures.
   - Technical considerations to minimise the potential for an intracranial haemorrhage after a stereotactic biopsy.
   - Technical considerations to maximise the potential of diagnostic stereotactic biopsy.
   - Trajectories to biopsy a lesion in the pineal region, midbrain, pons, and medulla.
   - Advantages and disadvantages of radiosurgery and surgical resection for tumours and vascular malformations.

Appendix 2

Example of a surgical training plan

Year	Mandatory procedures*	Conditional procedures*
1	– spinal and ventricular taps, wound care – assistance at craniotomies – burr holes – opening and closing of procedures – nerve and muscle biopsies	– nerve decompression – lumbar discectomy – simple craniotomies – HALO
2	– elective convexity craniotomy – evacuation of epi- and subdural haematomas – cranioplasty – lumbar discectomy – impression fractures	– carpal and ulnar tunnel decompression – ventricular shunts
3	– posterior fossa and skull base craniotomies – lumbar and thoracic laminotomies – intracerebral haematomas – superficial meningiomas and gliomas – ventricular shunts – more complex traumatology (simple basal CSF leak) – spinal metastasis	– cerebellar haematoma – ventral approach to cervical spine – cervical laminotomies
4	– cerebellar haematomas – ventral and dorsal approach to cervical spine – spinal stenosis – cerebral metastases	– cervical discectomy – extraforaminal lumbar discectomy
5	– extraforaminal lumbar discectomy – cervical discectomy – intradural extramedullarytumours	– cervical fusion with plates – peripheral nerve grafts
6	– posterior fossa tumours – skull base meningioma – peripheral nerve grafts – spinal instrumentation	– aneurysm clipping – transsphenoidal procedures – stereotactic operations

* It is important to emphasise that in this scheme the exposure to the mentioned problems and procedures is not limited to the year of primary exposure but that the experience in these areas is gradually expanded during the subsequent years. Required minimum or average operative case numbers for each category can be assigned for each training year. Any change in the sequence of procedures is possible. This is an example.

Appendix 3

Example of a subspeciality oriented surgical training plan

Year	General procedures[1]	Speciality procedures[2]
1	– spinal and ventricular taps, wound care – assistance at craniotomies – burr holes – opening and closing of superficial procedures – nerve and muscle biopsies	a) *ICU rotation* – invasive monitoring – intubation, ventilation
2	– elective convexity craniotomy – evacuation of epi- and subdural haematomas – trigeminal thermorhizotomies – cranioplasty – lumbar discectomy – impression fractures	b) *stereotactic & functional* – stereotactic biopsies – radiosurgical planning – microvascular decompression (Jannetta) c) *neuro-oncology + neuropathology* – removal of gliomas – craniotomies, posterior fossa craniotomies – removal of convexity meningiomas
3	– posterior fossa craniotomy – intracerebral haematomas – ventricular shunts – superficial brain tumours – laminotomy for spinal metastasis	d) *spinal rotation* – lumbar and thoracic laminotomies – ventral and dorsal approach to cervical spine – cervical discectomies – extraforaminal lumbar discectomy – cervical fusion with plates – spinal instrumentation e) *paediatric and epilepsy* – invasive monitoring – temporal resections – paediatric shunts – simple paediatric tumours
4	– cerebellar haematomas – spinal stenosis decompression – skull base craniotomies – cerebral metastases	
5	– extraforaminal lumbar discectomies – intradural extramedullary tumours	f) *pituitary & skull base rotation* – skull base craniotomies – more complex traumatology (simple basal CSF leak) – skull base tumours transsphenoidal procedures g) *neurovascular rotation* – aneurysm clipping – carotid endarterectomy – resection of cavernomas – EC-IC bypass
6	– posterior fossa tumours – peripheral nerve grafts	h) *neuro-oncology rotation (elective subspeciality interest)* – removal of gliomas, metastases, meningiomas, etc – craniotomy under local anaesthesia free for subspeciality interest

[1] The exposure to the mentioned procedures is not limited to the year of first exposure, but experience in these areas is gradually expanded during the subsequent years.
[2] In contrast in the subspeciality areas, exposure to the procedures is generally limited to the rotation. Any change in the sequence of procedures is possible. This is an example.

References

1. Dewey J (1897) My pedagogic creed. School 54: 77–80 (*http://www.pragmatism.org/genealogy/dewey/My_pedagogic_creed.htm*)
2. Hiemstra R, Brockett RG (eds) (1994) Overcoming resistance to self-direction in adult learning (No. 64) Jossey-Bass, San Francisco (*http://www-distance.syr.edu/learning.html*)
3. Kempski O (2001) Research and clinical routine in the next century: segregation or co-operative networks? Acta Neurochir (Wien) [Suppl] 78 : 213–216

4. Long DM (2001) Competency based residency training. The next advance in graduate medical education. Acta Neurochir (Wien) [Suppl] 78: 153–158
5. Neurosurgery://on-call (AANS-CNS website), professional pages, residency curriculum (*http://www.neurosurgery.org/resident/curriculum/index.html*)
6. Reulen HJ (2001) The semi-annual residency rotation summary: a means to assess the quality of neurosurgical training. Acta Neurochir (Wien) [Suppl] 78: 159–164
7. Schwid HA, Rooke GA, Carline J, Steadman RH, Murray WB, Olympio M, Tarver S, Steckner K, Wetstone S (2002) Evaluation of anesthesia residents using mannequin-based simulation: a multiinstitutional study. Anesthesiology 97: 1434–1444
8. Seymour NE, Gallagher AG, Roman SA, O'Brien MK, Bansal VK, Andersen DK, Satava RM (2002) Virtual reality training improves operating room performance: results of a randomized, double-blinded study. Ann Surg 236: 458–463
9. Steiger HJ, Reulen HJ (1999) Manual Neurochirurgie. Landsberg, Ecomed-Verlag
10. Steiner R (1923) Die Erziehung des Kindes vom Gesichtspunkte der Geisteswissenschaft. Ein Vortrag über Pädagogik. Rudolf-Steiner-Verlag (edn 1992)

Correspondences: H.-J. Steiger, M.D., Neurochirurgische Klinik, Universitätsklinikum, Moorenstraße 5, 40225-Düsseldorf, Germany. e-mail: Steiger@uni-duesseldorf.de

Acta Neurochir (2004) [Suppl] 90: 51–57

Printed in Austria

VII A structured neurosurgical training plan and the neurosurgical logbook in the UK

K. W. Lindsay

Department of Neurosurgery, Southern General Hospital, Glasgow, UK

Introduction

Although there are many similarities in neurosurgical training methods throughout Europe, some differences exist in training programmes and the extent of practical experience trainees gain in different countries.

In the UK a structured neurosurgical training plan and a log-book have a long tradition for over 20 years and are mandatory for any training programme. The UK training programme aims to produce a neurosurgeon capable of independent practice, wholly responsible for patients admitted under his/her care. This requirement perhaps differs from some other European countries where a hierarchical system exists and the newly trained neurosurgeon has limited autonomy. Whatever system is present, learning does not stop on gaining accreditation, but should persist throughout the neurosurgeon's entire working career.

A structured training plan should ensure that the trainee has gained adequate exposure to a wide variety of neurosurgical conditions in a training unit with appropriate facilities and has attended a weekly educational programme covering the necessary range of academic disciplines. It is essential that such a training programme be combined with a robust method of assessing progress. Ensuring that the trainee completes a specific number of years within a training programme is straightforward, but determining whether or not he/she has achieved the required level of knowledge and the required level of clinical and operative competence at each stage of training presents challenging difficulties.

To become competent in performing an operative technique begins with observation, followed by supervised guidance by a patient teacher. Finally experience by repetition improves ability, increases confidence and provides increasing exposure to anatomical variations and to potential complications. The maintenance of an operation logbook records such experience and is an important aspect of a structured training programme.

This chapter sets out guidelines to establish a comprehensive, structured and balanced neurosurgical training programme, based primarily on the system currently used in the United Kingdom and Ireland. It also describes the neurosurgical logbook and reviews the cumulative operative totals achieved by UK/Irish trainees over the last decade. The use of operative totals in assessing progress and attainment is reviewed along with other methods of determining competence in chapter XI on page 77. These continue to evolve in the need to maintain standards and to ensure that trainees achieve satisfactory training.

It is hoped that those countries in which training is less well structured may take up these suggestions and provide an opportunity to establish a more uniform approach to training throughout Europe according to the UEMS Training Charter, the respective recommendations of the UEMS Section of Neurosurgery and the Joint Residency Advisory and Accreditation Committee.

The structured training plan

Objectives

Neurosurgery is a specialty dealing with the diagnosis, evaluation and treatment of patients with disorders of the cranium, spine and nervous system. On completing training, the trainee must be capable of practising general neurosurgery competently and independently and should develop a special expertise in at least one of the subspecialties and/or research.

It is not expected that trainees on completion of training, should attain competence in every operative procedures (see Table chapter XI). When not competent to deal with all aspects of a particular condition, the trainee should have sufficient knowledge and expertise to care for such a patient until referral to an appropriate neurosurgeon is arranged.

Requirements for entry into neurosurgical training

Although such requirements will vary from country to country these should be clearly specified. It is suggested that trainees should not enter a neurosurgical training programme immediately after gaining a medical degree, but should obtain experience at a junior level in other surgical specialties as well as in neurosurgery. Such exposure may help convince a trainee that he/she has made the correct career choice before embarking on a prolonged period of training. During this time the trainee should acquire knowledge of the management of surgical patients in general e.g. – wound healing, fluid balance, coagulation and thrombotic disorders, pain control, triage, assessment and resuscitation after trauma, airway control, respiratory problems and ventilation, chest, abdominal and limb injuries, blood pressure control, sepsis and shock, multi-system failure, principles of oncology and the principles of outcome assessment after surgery.

In the UK such knowledge is tested by an MCQ, an oral and a clinical examination on basic surgical training. In addition trainees must attend and satisfactorily complete a five day course on basic surgical skills.

Recommendations

- Six months general surgery
- Six months orthopaedic surgery (including emergencies)
- Six months accident and emergency
- Six months neurosurgery

Duration of neurosurgical training in the UK

In most instances it is expected that neurosurgical training should last a *minimum of six years* consisting of

- *Five years in clinical neurosurgery* which must incorporate six months of *paediatric neurosurgery*. During this training period, the trainee must acquire a good working knowledge of neuroradiology, neuropathology and neurophysiology, although he/she need not have worked in these specialties.
- *One further year of training* should be completed in any related branch of the specialty, e.g.,
  - laboratory or clinical research,
  - subspecialisation,
  - training overseas,
  - up to 6 months in neurology, neuroradiology, neuropathology or neurophysiology (more than one could be taken).

Alternatively, the trainee could complete a further year of clinical neurosurgery within his training unit. The trainee need not take this year at the end of training, particularly if undertaking research. Some may wish to extend a period in research to two or three years to attain a higher degree.

Rotational programmes

Since it may not be possible to incorporate all training requirements within a single neurosurgical unit, it may be necessary to set up rotational programmes incorporating two, three or even four units through which the trainees would rotate for specific periods. Such rotational arrangements would be essential

- in units with no (or limited) paediatric neurosurgery.
- in small units with limited bed numbers where less than four specialists/consultant neurosurgeons practice
- in units without access to research facilities.

Content of training programme

Because hours available for training vary, the content is *not prescriptive* with respect to each individual year of training. General principles apply and include –

In the first 2 years, the trainee must become competent at history taking, neurological examination, assessment of clinical problems, non operative management, neurosurgical intensive care, interpretation of neuroradiological, neuropathological and other neuroscience investigations and the basics of operating, e.g., should be able to perform craniotomies, for cranial trauma and malignant tumours, shunting procedures, procedures for lumbar disc disease and cervical degenerative disease. Reading should include standard neurosurgical texts and the major neurosurgical journals.

In years 3 and 4 the range of conditions operated upon should widen to include benign tumours and posterior fossa procedures.

In years 5 and 6 the trainee should increase his/her experience of more major cases, e.g., benign tumours and aneurysms and develop some experience in more specialised fields, e.g., skull base surgery, transphenoidal pituitary surgery and pain.

Trainees especially during 5th and 6th years would be expected to be involved in the training of more junior neurosurgical trainees.

Neuroradiological, neuropathological and neurophysiological experience should be gained in relation to each clinical case, in addition to any coverage in the academic programme. Throughout the period of training, reading of journals should continue.

Research

All trainees are encouraged to undertake research and are expected to develop an understanding of research methodology during neurosurgical training. Up to one year of full-time research can count towards the six years of required neurosurgical training. Some trainees may wish to extend their period in research for a further year or two to obtain a higher degree. Research during the training programme should be supervised by a specifically named trainer with appropriate qualifications.

All trainees including those who do not undertake a specific piece of research will be expected to gain sufficient understanding of research and statistical techniques to permit assessment of published work.

Subspecialty training in neurosurgery

Trainees should be encouraged to develop a subspecialty interest provided this does not adversely affect training in general. The trainee may wish to spend up to one of the six years focusing on one or more subspecialty interests, e.g., skull base procedures, complex spinal procedures, functional neurosurgery, paediatric neurosurgery or complex vascular procedures. It is expected that most trainees would continue to develop a subspecialty interest beyond completion of their CCST.

Study leave

Trainees should have the opportunity to attend both national and international meetings in neurosurgery or in a neurosurgical subspecialty throughout their training. Trainees would be expected to attend at least one national neurosurgical society meeting per year. Where possible, expenses should be provided either wholly or partly to enable the trainee to attend. Trainees should also be encouraged to apply for EANS courses during the 2nd or 3rd year of training.

Training agreements

Some countries may require that a formal training agreement is drawn up between the body responsible for training and the trainee. This should define in terms of education and training, the relationship, duties and obligations on each side.

The training unit

It is expected that the facilities of any training unit would include

- at least 25 neurosurgical beds,
- either dedicated neurosurgical intensive care beds or access to intensive care beds within a general hospital,
- a 24 hour operating theatre with operating microscope,
- access to CT scanning and magnetic resonance imaging,
- access to other specialties including neurology, neurophysiology, neuroradiology, neuropathology, neuroanaesthesiology, neuro-otology, neuro-ophthalmology, maxillofacial surgery, general surgery, orthopaedic surgery, internal medicine,
- an accessible library with books and journals on neurosurgery,
- computer search facilities,
- an academic programme as specified below.

Academic programme (for UEMS-recommendations see p. 8, chapter I)

Each training unit should maintain an academic programme of neurosurgical/neuroscience education. This should be continuously available throughout the training period and should include

- Lectures, including visiting speakers and incorporating basic neurosciences
- Clinical presentations with all neuroscience disciplines
- Neuropathology including clinicopathological conferences
- Neuroradiological conferences
- Journal club
- Audit
- Research meetings.

Trainee assessment

No structured training programme is complete without a method of assessing trainee progress and attainment at regular intervals, followed by a final assessment leading to the award of the Certificate of Completion of Specialist Training (CCST). The methods currently used in the UK and Ireland are described in Chapter XI along with newer concepts for competence assessment.

Training programme assessment

In the UK, each training programme must be approved by the Specialist Advisory Committee (SAC) in Neurosurgery. This committee works on behalf of the Joint Committee for Higher Surgical Training (JCHST), supervising all surgical specialties for the four Surgical Royal Colleges, the bodies responsible for the standard of training. SAC members visit every training programme on a four-year basis to monitor standards and the quality of training in individual neurosurgical units. Both trainers and trainees are interviewed and facilities are inspected. The findings of the visit are documented in detail and reported to the whole SAC committee. The SAC decides on whether the volume of work is sufficient to provide adequate training material for the number of trainees in the unit and compares this with the actual experience gained. Such visits led to the withdrawal of recognition of training from six units over the last decade. All eventually regained recognition after introducing the recommended improvements.

The Joint Residency Advisory and Accreditation Committee (JRAAC), under the auspices of the UEMS and the EANS, now offer an equivalent accreditation facility for any European centre wishing to participate. Visits began in 2000. A number of centres from different countries have already taken up this offer.

The neurosurgical logbook

For over 20 years, trainees in the UK and Ireland have recorded operative procedures in which they have participated, either as assistant, or as the surgeon (with or without assistance) in a standardised logbook. From this they derive cumulative totals categorised under the headings listed in Table 1. These are reviewed by the Training Programme Director as part of the trainee's annual assessment. At the end of six years of training, each trainee submits his or her cumulative totals to the SAC as part of the accreditation process (see chapter XI).

The logbook categorises each procedure into the following groups:

P – The trainee has performed the operation. The consultant may have scrubbed to either see the pathology or because the trainee required an assistant, but the consultant must not have made a contribution/intervention significantly affecting the execution of the operation.

T – Trainee training a more junior trainee and assisting with the operation.

C – The trainee has done the operation with a consultant or a more senior trainee assisting or standing-by in the operating theatre and the supervisor making a significant decision or practical manoeuvre during the operation.

A – The trainee is the principal assistant at the operation.

Between 1990 and 2000, as both a member then chairman of the SAC in neurosurgery, I have compiled a database of operative totals for each trainee, permitting analysis of the range, mean and median totals for each procedure. For simplicity, any "T" procedure was reclassified as a "P".

Over the eleven year period, 109 trainees achieved accreditation or the CCST (introduced in 1996). Paediatric procedures were classified separately in 1997, increasing the operative categories from 50 to 74 (Table 1). Thirty-two trainees categorised paediatric procedures separately.

Operative totals

Table 1 shows the range and median values of operative totals over the decade for procedures carried out "personally" i.e. where the consultant may or may not be present, but if present, not making a contribu-

Table 1. *The median, mean and range of trainee operative totals performed 'personally' on completion of training**

Operative experience performed personally

Adults		Min	Max	Mean	Median
Head injuries	burr holes – ICP monitoring	0	131	35	32
	burr holes – ch. subdural	14	140	53	49
	craniotomy – extradural	8	60	25	23
	craniotomy – acute subdural	7	64	28	26
	craniotomy – intracerebral	2	37	12	9
	craniotomy – other	1	21	5	4
	depressed fractures	2	60	18	16
	dural repair	1	25	7	6
	combined craniofacial repair	1	28	4	3
	cranioplasty	0	46	11	9
Supratentorial tumours	biopsy	1	345	37	33
	stereotactic/Ids biopsy	0	142	28	23
	craniotomy – intrinsic tumour	11	162	52	49
	meningioma	8	38	18	17
	other benign	1	14	4	4
Midline tumours	transphenoidal	0	26	7	6
	other	0	9	2	1
Posterior fossa tumours	malignant	2	28	8	8
	acoustic	0	22	2	1
	other benign	1	33	7	6
Hydrocephalus	external ventricular drain	13	128	50	43
	shunts insertion	27	230	95	89
	shunt revision	5	171	43	35
	Arnold Chiari	1	11	4	4
Vascular	craniotomy – aneurysm	17	108	39	36
	craniotomy – AVM	0	11	3	3
	craniotomy – intracerebral haematoma	1	43	16	15
	posterior fossa haematoma	1	24	7	6
	craniotomy/craniectomy – other	0	3	1	1
	endovascular – aneurysm	0	0	0	0
	endovascular – AVM	0	0	0	0
	endovascular – tumour circulation	0	0	0	0
	carotid endarterectomy	0	26	2	0
Infection	abscess	2	66	11	10
	subdural empyema	0	14	5	4
	spinal abscess	0	8	3	3
Spine	cervical disc disease/spondylosis	21	149	59	54
	thoracic disc disease	0	11	1	0
	lumbar disc disease/spondylosis	5	343	101	88
	transoral	0	6	1	0
	syrinx	0	8	2	2
	dysraphism	0	8	2	1
	tumours – extradural	2	59	17	14
	tumours – intradural extramedullary	0	16	6	6
	tumours – intradural intramedullary	0	12	2	2
	spinal fixation	0	60	11	7
Pain:	trigeminal neuralgia – lesion	0	59	9	7
	trigeminal neuralgia – MVD	0	23	4	3
Functional	stereotaxy	0	46	3	0
	surgery for epilepsy	0	31	2	1

tion or intervention significantly affecting the execution of the operation. Inevitably numbers vary considerably depending on both the complexity of the procedure and its frequency of occurrence. For example, the highest median values exist for shunt insertion (89) and operations for lumbar disc disease (88). For relatively common but more complex procedures such as craniotomy for aneurysm repair and meningioma removal, trainees achieved median operative totals of 36 and 16 respectively. Few trainees performed per-

Table 1 (continued)

Operative experience performed personally					
Paediatric		Min	Max	Mean	Median
Head injuries	extradural	0	25	5	3
	other acute haematoma	0	12	3	3
	chronic subdural/hygroma	0	14	4	3
	depressed fracture	0	12	5	4
	ICP monitoring	0	80	16	12
Tumours	supratentorial	0	6	2	2
	infratentorial	0	9	3	2
	stereotactic/ids biopsy	0	16	3	2
Hydrocephalus	external ventricular drain	0	32	11	9
	shunt insertion	0	129	34	27
	shunt revision	0	102	29	26
	Arnold Chiari	0	2	1	0
	encephalocele	0	4	1	1
Spine	mening/meningomyelocele	0	70	3	1
	tethering syndromes	0	12	2	1
	tumours	0	5	1	1
Paediatric functional	spasticity	0	3	0	0
	surgery for epilepsy	0	5	1	0
Neuroendoscopy	adult	0	25	1	0
	paediatric	0	6	2	1
Craniostosis	sutural craniectomy	0	18	3	1
	combined craniofacial	0	4	1	1
Other	vascular	0	2	1	1
	infection	0	20	4	3

* From 1990 to 2000, 109 trainees were assessed. *ICP* Intracranial pressure; *AVM* Arteriovenous malformation; *MVD* Microvascular decompression.

sonally complex procedures such as for intradural intramedullary tumours or arteriovenous malformation, (median values of 2 and 3 respectively). No trainee performed unassisted a carotid endarterectomy, a thoracic disc removal or a transoral approach. A similar table was produced giving the range, mean and median totals for procedures carried out by the trainee, but assisted by the consultant, thus providing a guide to the extent of direct supervision.

It is important to note that the figures presented in Table 1 are the median totals actually achieved by trainees on completing their neurosurgical training. These reflect the service aspect of the training period as well as the training itself. They do *not* reflect the training committee's views of the minimum number required to achieve competence to perform a specific procedure. No one would claim that it is necessary to perform 89 adult shunt insertions to achieve competence for that specific procedure. No attempt has been made to define such numbers. Tables of operative totals were fed back to training programme directors from 1992 onwards. It was explained that the SAC expected trainee operative totals to 'approach the median value for the majority of procedures'.

Over the decade, no significant change occurred in the median number of operative procedures performed for each condition from one year to the next. Figure 1 indicates the median totals achieved each year for five

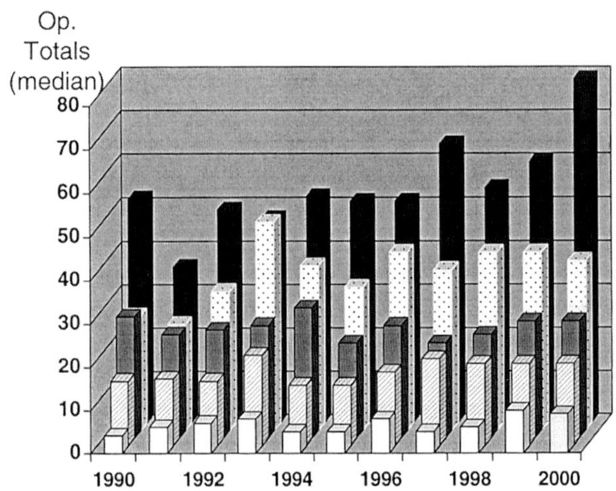

Fig. 1. The median operative totals achieved per year for five procedures from 1990–2000. The number of trainees achieving accreditation /CCST per year was respectively –6, 12, 7, 5, 13, 14, 8, 9, 11, 13 and 11. ■ Cervical disc/spondylosis; ▨ Aneurysm; ▤ Subdural; ▧ Meningioma; □ Transsphenoidal

key procedures. In particular, it is noted that the number of craniotomies for aneurysm and traumatic subdural haematoma remain constant despite the recent trend for coil embolisation and the reduction in trauma from road traffic accidents. With release of results from the ISAT trial (4), the number of patients undergoing direct aneurysm repair and subsequently the number of procedures performed by trainees will inevitably fall. In the future, trainees may only perform this procedure if they have expressed a subspecialty interest in vascular surgery.

Recording cumulative operative totals from the neurosurgical logbook provides an objective method of monitoring operative experience. To use this as a method of evaluating progress requires some caution. Although trainers check and countersign pages within the logbook, the categorisation of the procedure i.e., "P", "C", or "A" depends to some extent on the integrity of the trainee. Even with the best intentions, some variability must exist in deciding the extent of the trainers' participation. The "P" category has recently been sub-divided into "P = consultant available but not in the operating room" and "SU = consultant on standby in the operating theatre but unscrubbed". The "C" category has been redefined to mean, "surgeon scrubbed". These measures may help to reduce variability in recording the data and help identify the degree of supervision within theatre. Other pitfalls will be described in chapter XI.

References

1. Lindsay KW (2002) Neurosurgical training in the UK and Ireland: assessing progress and attainment. Neurosurgery 50: 1103–1113
2. Joint Committee on Higher Surgical Training (1996) A manual of higher surgical training in the United Kingdom and Ireland. Royal College of Surgeons of England London, (http://www.jchst.org/)
3. Joint Committee on Higher Surgical Training. (Supplement) (1996) A curriculum for higher training in neurosurgery. Royal College of Surgeons of England London (http://www.jchst.org/)
4. International Subarachnoid Aneurysm Trial (IASA) Collaborators Group (2002) Lancet 360: 1267–1274

Correspondence: K.W. Lindsay, Department of Neurosurgery, Southern General Hospital, Govan Road, Glasgow G51 4TF, UK. e-mail: kwlindsay@sgh.scot.nhs.uk

Acta Neurochir (2004) [Suppl] 90: 59–66

Printed in Austria

VIII The European Neurosurgical Log-Book (UEMS/EANS)

H.-J. Reulen

Department of Neurosurgery, Klinikum Großhadern, Ludwig-Maximilians-University, Munich, Germany

In the European Training Charter (1), Chapter 6, 1.B.1, the use of a log-book for neurosurgery is specified as mandatory: "The candidate neurosurgeon should keep an authorized log-book of operative experience". Chapter 5.3 formulates the requirements for trainees: "The trainee should keep his/her personal log-book or equivalent up-to-date according to national rules and EC directives as well as considering UEMS/European Board recommendations."

Written documentation of the operative procedures done by a resident and attested by the signature of the supervising trainer/programme chairman is presently required in more or less all European countries. A compilation in tabular form of all surgical procedures is also called for in the national board examination and the formal recognition as a specialist, respectively. However, the many European Neurosurgical Societies, as a consequence of their different historical background and development, employ a variety of systems and categories of procedures.

With growing consciousness for a uniform European Resident Training Programme of high standard, it became necessary to develop a modern edition of a Neurosurgical Log-book to allow for documentation of the broad variety of surgical procedures of today's neurosurgery with its various subspecialties.

The present Neurosurgical Log-Book was developed in 1997/98 by the Residency Advisory Committee of the EANS (later JRAAC), based on the long-term experience of our colleagues in the UK, the USA, and other national societies where such a system for documentation of surgical procedures is used regularly. Experienced individuals from many national societies have contributed to the new European log-book with their proposals and advice. The Log-Book has been approved by the Administration Council of the EANS (June 1998) and by the Section of Neurosurgery of the UEMS as the official European Log-Book in Neurosurgery.

In this European Log-book procedures have been included that are of relevance in a modern neurosurgical training programme. The log-book really represents the common European suggestion for a core education in our specialty. Thus trainees during their 6 years of training should be adequately exposed to the respective diseases and operative procedures, and they should demonstrate that they have acquired the necessary competency called for.

If a training site cannot offer exposure in one or more fields of our specialty, they must seek collaboration with another training site so that residents will also receive that missing part. In future two or even three training sites will have to organise one common training programme. This will require a certain degree of flexibility, but some countries have already demonstrated that it is possible.

Explanations how to use the Log-Book

The Neurosurgical Log-Book written in English is intended to be a record of all operations the trainee performs or participates in during his training. In consecutive order the date of operation, patient's name, and nature of the operation should be entered on the pages as indicated (Table 1). The last column of the page serves to designate the exact role of the trainee during the operation. Operations are to be classified into *one* of the four following groups. For the national or European Board examination, the two groups T and TS are to be considered when preparing the consolidated list of operation.

Table 1.

Date of operation	Patients number	Patients name	Nature of operation	S, TS, C, A

T The trainee has done the operation. The supervising consultant must not have made a decision/ practical manœuvre during the operation.
TS The trainee has done the operation, but the supervising consultant has made a significant decision/ practical manœuvre during the operation.
C The trainee has performed component parts during the operation under supervision of a senior surgeon: positioning, operative approach (i.e. craniotomy, opening, closure, drainage, draping, written instructions for postoperative care).
A The trainee is the principal assistant during the operation.

At the end of the log-book there is a consolidated list of all procedures which are of relevance in a modern training programme, divided into "adults" and "paediatrics" (Table 2).

Categorisation of groups T and TS depends to some extent on the integrity of the trainee. Even with the best intentions some variability will exist between trainees and centres. Also in some countries, legislation strictly requires trainees to be closely supervised during a procedure while in others it is sufficient for the trainer to be available but not necessarily present in the operation theatre. It is wise, therefore, to add T and TS groups when recording cumulative operative totals.

The log-book must be supervised and signed by the Programme Director or the responsible tutor at regular intervals.

The log-book can be conducted in form of a booklet and can be obtained free of charge through Aesculap AG + Co KG, P.O. Box 40, D – 78501 Tuttlingen (attn. Mrs. Eva Streit), who have generously financed the print. Recently an electronic version has been developed by the Joint Residency Advisory and Accreditation Committee with the financial support of the EANS. This electronic version can be downloaded from the EANS web site (*www.EANS.org*) and managed by oneself. The version is self-explanatory and leads the user in a simple and logic sequence through the programme. The booklet, and more easily the electronic version, allows to establish a consolidated list of operations performed during any desired period, for instance during a specific rotation, on an annual or semi-annual basis, or for the entire 6 years of training. It permits the trainer to assess a trainee's progress at regular intervals for evaluating a rotation and his growing autonomy. It also allows to detect any possible deficiencies in the operative training very early so that they can be corrected. Furthermore it permits to register components of an intervention and the degree to which the trainee has been involved. This offers the possibility to observe the progress of a trainee as well as his growing autonomy.

It allows also a comparison between different trainees as well as between different training centres. Finally it can be used, optionally or already obligatory as in some European countries, as the required documentation for national or European Board examinations.

In future this official European Neurosurgical Log-book will be requested for departments who wish to participate in the European Accreditation and Certification Process.

Conclusions

The introduction of a common European Log-book was certainly a big step forward in the endeavour to organise a well-structured European residency training programme. Its advantage is to have a clear and comparable system for documentation of all surgical interventions in which a trainee has been involved. It precisely fulfils the requirements of the European Training Charter. Following its introduction, several national societies and many departments have already made its use mandatory for their trainees, and in a not too distant future we hope that all member societies of the EC formally will have implemented this Log-book.

Some national societies may wish to use the Log-book in their own language. If the electronic version in a national language is desired, the JRAAC can help to supply this at a reasonable price.

In the previous chapter the experiences with a fairly similar log-book used in the United Kingdom and Ireland are presented, where a structured training programme has already been in existence for many years. In this article many interesting and essential data are reported which are of importance for everyone who is to organize or improve a training programme.

An important question is whether each trainee needs to be competent in all of the very complex procedures of the entire speciality. Must each one of the trainees have done a large amount of aneurysms, acoustic neurinomas or insertion of a fixateur in a traumatic lumbar spine fracture, to name some of the typical, complex interventions? It is recognised that our specialty has grown rapidly to a point where one individual probably no longer is in a position to cover all

Table 2. *Consolidated list of operations performed*

ADULT

	NATURE OF OPERATION	T	TS	C	A
1.	**Head Injuries**				
	Burr holes: ext. ventricular drainage/ICP-monitoring/reservoir				
	exploration/chronic subdural				
	Extradural haematoma				
	Acute subdural haematoma/contusion				
	Intracerebral haematoma				
	Depressed skull fracture				
	Gunshot / Penetrating wound				
	Dural repair (CSF fistula)				
	Combined craniofacial repair				
	Cranioplasty				
	Others				
2.	**Supratentorial Tumours and Lesions (excl. stereotactic procedure)**				
	Primary/intrinsic tumours				
	Metastatic lesions				
	Meningioma				
	Other benign lesions (Craniopharyngioma)				
	Pituitary - transphenoidal approach				
	Pituitary - transcranial approach				
	"Craniotomy Flap"				
	Others				
3.	**Posterior Fossa Lesions**				
	Primary and metastatic tumours				
	Acoustic neurinoma				
	Meningioma				
	Other benign lesions (abscess, epidermoid, etc.)				
	Arnold Chiari malformation				
	Others				
4.	**Infection**				
	Abscess				
	Subdural empyema				
	Others				

Table 2 (continued)

ADULT

	NATURE OF OPERATION			T	TS	C	A
5.	**Vascular**						
	Craniotomy:	Aneurysm					
		AVM					
		Intracerebral haematoma					
		Cavernous angioma					
	Endovascular:	Aneurysm					
		AVM					
		Tumour embolisation					
	Occlusive:	Bypass					
		Endarterectomy					
	Haematoma (spontaneous intracerebral)						
	Others						
6.	**Hydrocephalus** (≥16 years)						
	Shunting procedure, initial						
	Shunt-revision						
	Endoscopic fenestrations						
	External ventricular drainage						
7.	**Spine**						
	Cervical Disc Disease or Spondylosis						
		Anterior approach with (bone) graft					
		with (out) graft					
		with instrumentation					
		Posterior approach (foraminotomy)					
		with instrumentation					
		Laminotomy/Laminoplasty					
	Thoracic Disc Disease						
	Lumbar Disc Disease or Spondylosis						
		Lumbar disc					
		with instrumentation					
		Laminotomy/Laminectomy					
	Spinal Tumours						
		Extradural					
		Intradural:	extramedullary				
			intramedullary				
		Instrumentation in vertebral tumours					
	Spinal Trauma						
		Decompression					
		Instrumentation					
	Others						

Table 2 (continued)

ADULT

	NATURE OF OPERATION	T	TS	C	A
8.	**Trigeminal and other Neuralgias**				
	Injection techniques/RF-lesion				
	Microvascular decompression				
	Others				
9.	**Stereotactic and Functional Neurosurgery**				
	Tumour biopsy				
	Seed implantation				
	Thalamotomy, Pallidotomy/Stimulation technique				
	Neuroablative procedures (chordotomy, DREZ)				
	Therapeutic electrostimulation (peripheral nerve, spinal)				
	Implantation of ports/pumps for intrathecal drug delivery				
	Others				
10.	**Surgery for Epilepsy**				
	Invasive diagnostic procedures (deep electrodes,grids,plates)				
	Temporal lobe resection				
	Extratemporal resection				
	Others				
11.	**Peripheral Nerve**				
	Entrapment decompression/transposition				
	Sutures (with graft)				
	Neurorhaphy				
	Sympathectomy				
	Brachial plexus repair				
	Tumour				
	Others				
12.	**Computer-aided Neurosurgical Planning (not the intervention)**				
13.	**Other Procedures (Please specify)**				

Table 2 (continued)

PAEDIATRIC (through 15 years)

	NATURE OF OPERATION	T	TS	C	A
1.	**Hydrocephalus and Congenital Malformation**				
	External ventricular drainage				
	Shunting procedure: initial				
	revision				
	Endoscopic fenestration				
	Arnold-Chiari/Dandy Walker				
	Encephalocele				
	Single sutural craniosynostosis				
	Complex craniosynostosis/ Craniofacial reconstruction				
2.	**Head and Spine Injuries**				
	Burr holes, ICP-monitoring/drainage/reservoir				
	Chronic subdural haematoma/hygroma				
	Extradural haematoma				
	Acute subdural haematoma				
	Acute intracerebral haematoma				
	Depressed fracture				
	Spinal trauma				
	Others				
3.	**Brain Tumours**				
	Supratentorial hemispheric tumours				
	Supratentorial midline tumours				
	Infratentorial tumours				
	Others				
4.	**Spine**				
	Meningo/meningomyelocele				
	Tethering syndromes				
	Tumour				
	Spinal dysraphism				
5.	**Functional**				
	Spasticity				
	Surgery for epilepsy				
6.	**Other Procedures (Please specify)**				

subspecialty areas of our specialty, particularly if one has to overlook the literature and work to scientific standards. However, a trainee during his training should see enough patients in all categories of the log-book to be able in recognition, in the diagnostic work-up, the differential diagnosis and the various forms of treatments of these diseases. He should be capable to inform the patient about the risks, the advantages, etc. of a certain treatment and he should be able and willing to refer the patient – if he does not perform the respective procedure with "adequate routine" – to a more experienced colleague or department.

One possible and recommended solution for the future is with all probability the development of subspecialty areas (as described in Chapter 16, G. Schackert) which may start during the last 1 or $1\frac{1}{2}$ years of the training programme.

During his training the resident should become familiar with all the techniques and procedures and should be able to handle those complex cases which he needs to know when choosing a subspecialty. A further deepening of the experiences and knowledge can then be acquired within the scope of an additional fellowship programme.

A second important question is how many procedures of each category must be performed during the training to reach "adequate routine" and competence.

The figures published in the UEMS neurosurgical part of the Training Charter are minimum figures and should be attained by every resident. Since residents have different abilities, ambition und capabilities, a competence range should be introduced. Thus a resident must at least receive the minimum competency in the respective area but may result in a much higher competence range. A very good description of the discussion on the operative total figures and the competence range is presented in one of the following chapters (K. W. Lindsay, Periodic Progress Evaluation).

Learning certainly does not stop at the end of a training programme but rather has to continue for the rest of life.

Correspondence: Prof. Dr. H.-J. Reulen, Schwojer Str. 19, 81249 München, Germany. e-mail: Ilona.Anders@med.uni-muenchen.de

Acta Neurochir (2004) [Suppl] 90: 67–71

Printed in Austria

IX Morbidity & mortality conferences – How can we do it?

J. Brennum[1] and **F. Gjerris**[2]

[1] The University Clinics of Neurosurgery, Copenhagen County Hospital, Glostrup, Denmark
[2] Rigshospitalet, Neuroscience Centre, Copenhagen, Denmark

Introduction

Quality assurance is a very important aspect of daily life in a neurosurgical department as well as an important parameter for the neurosurgical patient. Quality assurance programmes of some sort are common but still not implemented in all units. Some quality assurance programmes have been formulated as accreditation programmes to set a measure of good quality surgery [2], and have ended up in web site programmes [1]. In the near future such programmes [1] will play an increasing role, as the accreditation programmes sweep across Europe. However, quality assurance programmes, in which the performance of both the department in general and its individual members are analysed, are far more seldom. Imagine if airline pilots did not conduct debriefings of so-called near misses, situations that might have developed into a disaster. Or for that matter, if plane crashes did not automatically result in a full investigation. Would it make you feel safe, as a frequent flyer, if the pilot organisations and the airlines declared that these measures were no longer necessary due to the high ethical and moral standard of pilots in general, which in itself guarantees the best quality possible.

Some departments have implemented quality assurance or auditing systems [3, 4, 6, 7, 14, 16, 18] similar to those known in the airlines – *morbidity & mortality conferences*. Morbidity & mortality conferences seem to be more common in North America than in Europe. Morbidity & mortality conferences are well suited for any kind of department, be that University or general neurosurgical units [9, 12, 16, 18].

We have used morbidity & mortality conferences as an intricate part of our training programmes for a number of years, and found them to be an extremely helpful teaching tool.

Definition

By a morbidity & mortality conference we understand a conference within the individual department, or including a number of departments, as may for example be relevant for a trauma-receiving centre [7]. The conferences are based on real cases, i.e. near misses or plane crashes, from which it is possible for the individual staff member to learn and for the department to reflect upon its procedure relevant to the pertinent case. This means that the discussion covers "*what went wrong*" in the single case, "*what can we learn*" from the present case, where the problem may stem from part of the handling of the patient, i.e. history, diagnosis or operative activity. Now, having identified the source of the mishap, the next part of the learning process is to reflect upon "*what can we do better*" and "*how can we avoid the same mistake*" in the future?

Objective

Morbidity & mortality conferences may have many forms, but the common objective is to continuously improve the performance of the department, by improving the performance of its individual staff members and their collaboration. A way point towards this goal is to develop a foundation of trust and mutual confidence [6, 9, 10, 15, 17]. Hence, everybody should be encouraged to participate actively and open minded in these conferences. Even the slightest hint of a witch-hunt is extremely counterproductive and should be avoided [10].

Form

It is very important that the form of the morbidity & mortality conferences reflects the objective [11, 13, 14, 16]. The aim is not to denounce any members of the team or the staff but to allow everyone to take part in the discussion and learn from individual members mistakes or misjudgements. The trainees in neurosurgery, as in any other speciality, are for the vast majority young colleagues who are very eager to learn, and who aim to treat their patients as expertly as possible. There are lots of anxiety in being a young trainee and they need to know that their department will provide adequate supervision and training. They should not be harassed for not being as theoretically well founded as their trainers believe they themselves are. The trainees need a) to feel safe, b) to know that it is always acceptable to ask questions, c) that teaching is an elaborate part of the daily life in the department, and d) to know that in any patient's outcome, even a less favourable than expected, there may well be a lesson to learn.

Morbidity & mortality conferences may be viewed as a multipurpose tool that can be used as a quality assurance measurement, a teaching device and an opportunity for reflection, where protocols and procedures are reviewed, developed and improved. A good teaching neurosurgical department is not only a department full of theoretical and technically talented neurosurgeons but also a place with a scheduled and balanced training programme and regular teaching sessions and conferences. These are all valuable ingredients but in our way of thinking the most important foundation is the teaching culture and spirit within the department. The talented neurosurgical consultant may sometimes be the one in need of training, when it comes to the trade of teaching.

Structure

Morbidity & mortality conferences should be held on a regular basis. In our departments morbidity & mortality conferences are conducted once or twice a month. Morbidity & mortality conferences may have a variety of configurations:

- Review of all recent mortality cases since the last conference
- Review of all severe morbidity cases since the last conference
- Review of single cases
- Review of a couple of similar cases

We prefer one of the two latter forms because they allow us to choose different clinical subjects for the individual conferences. As stated previously, it is of utmost importance that the atmosphere of the morbidity & mortality conferences is that of a forum for mutual learning and not a place of finger pointing and punishment [7, 10]. This is not different from how the atmosphere and the teaching environment should be in the neurosurgical department in general.

One member of the staff, which may well be one of the trainees, presents a relevant summary of the pertinent case or cases, including demonstration of relevant images. Especially when morbidity & mortality conferences are introduced into the teaching programme of a department, it is prudent, at least in the beginning, to use anonymous cases [10]. This is simply done by using cases that are not recent, and by using transcripts from the patient charts, where any patient identifiers, dates and names or initials of doctors or nurses involved in the treatment are omitted. All this is done to get an objective, unbiased review of the case that is not contaminated with the urge to blame an individual or the urge to defend one's own action in the case.

The two first forms are also important, but they might be extremely time-consuming in very busy departments, for instance as in our departments with altogether up to 7,500 admissions and 4,500 operations per year.

It is our experience that the usage of anonymous cases makes everybody more open minded and receptive to learning. Once the morbidity and mortality conferences have become an accepted part of the teaching programme of the department and everybody recognises the merits of the conferences as mutual teaching experiences, there is no longer any need to maintain the use of anonymous cases. Furthermore, when the department has developed to a stage where it is generally accepted and expected that everybody comments on the performance of others with the aim to improve the general performance of the department, the resulting effect will be even greater. However, if you do not choose to use cases that are anonymous, it is important that the cases chosen for morbidity and mortality conferences are not selected only to reflect the practice of the more junior staff. The selection of cases should demonstrate that it is human and acceptable not to be perfect. What is not acceptable is not to learn from experience. If the selection of cases is biased towards demonstrating the mishaps of the junior staff, soon the feeling of being exposed and blamed is going to devaluate the teaching experience and replace it with fear [10].

In our department all head staff members participate in the identification of cases that are suitable for mortality & morbidity conferences. The director, the professor, or the head staff member responsible for the particular mortality & morbidity meeting, selects the cases to de demonstrated. Confidentiality is obviously very important. If the discussion of a case during a morbidity & mortality conference could have legal or other repercussions, the teaching value would be degraded, and it is likely that the morbidity & mortality conferences would altogether cease to exist.

To improve or elevate the learning experience, the staff member presenting the case should submit a short literature review or a specific scientific article relevant to the presented case or cases. The presentation itself should normally be brief, typically between 5 to 10 minutes, and the following 20 minutes should be used for open and free minded discussion of whether the diagnostic work up, treatment and care of the patient lived up to the expected standard of the department or whether the performance could have been better. If the consensus is that the performance of the department in the particular case did leave room for improvement, this then leads the discussion into the next phase. How can we improve the performance in future and avoid similar suboptimal performance? Should the case be reported to the Danish Patient Compensation Assurance System [8]? Sometimes the result of a morbidity & mortality conference is that a task force is formed to update the department's routine procedure book, based on the lessons learned. All doctors are informed of the final decisions. It is by no means necessary that the cases presented at a morbidity & mortality conference reflect mortality or severe morbidity, it may also be a brilliant way of demonstrating or exposing inefficient or obsolete procedures.

When morbidity & mortality conferences work at their best, they function both as a quality assurance process, a learning process, and as a cultural process in which the spirit of aiming towards mutual learning and improvement by openly sharing not only the success stories but also the problems and mistakes. Far more can be learned by analysing cases with morbidity and mortality than by analysing successful stories. The morbidity and mortality conferences are popular because they offer learning and opportunity to participate actively, independent of one's training level. Furthermore, they are popular because they are designed to be interactive and relevant, in contrast to the passive learning experience of a lecture.

Examples

In the following a couple of examples of recent morbidity & mortality conference cases are presented in abbreviated form, and the outcomes of those conferences are commented.

Example A

A 67-year-old man was admitted with a history typical of lumbar spinal canal stenosis, decreasing gait distance, and no paresis or sphincter disturbances. A stenosis at the L4/5 level was demonstrated by myelography, and he was treated with a laminectomy. Postoperatively, on the same day as the procedure, normal neurological function of the lower extremities was noted, but intermittent catheterisation was necessary. On the second day, the patient complained of a strange feeling in his feet, and intermittent catheterisation was still needed. The patient notes did not reveal whether any objective evaluation of patient was performed at this point in time. On the 8th day after surgery the patient was discharged with a catheter and a scheduled follow-up appointment at the department of urology. On the 12th post-surgical day the patient was readmitted because he still had a strange sensation in his legs and then had suffered several falls at home. A clinical investigation demonstrated that he had increased tone in his legs, bilateral extensive plantar responses, normal motor power, and slightly reduced touch sensation diffusely in his legs. An acute MRI demonstrated an extensive subdural haematoma reaching up to the mid-thoracic level.

Comments

Nobody at the morbidity & mortality conference had any trouble recognising the mistakes made in this case: On the second day a suspicion of a haematoma should have been raised, a clinical examination should have been performed and eventually relevant investigations made, i.e. an MRI. However, in our mind such a case serves as a far more powerful reminder of why all staff members have to be conscious in their evaluation of patients postoperatively than a scalding by the department chairman would have been.

Example B

A 72-year-old man was referred from the department of oncology with slight gait disturbances and back pain. He had a known history of colorectal cancer, operated 5 years previously. He had bilateral lung metastases and possible retroperitoneal metastases. A single level columnal pathological fracture at Th8 was demonstrated on MRI, with a wedge-shaped vertebra, with involvement of the most anterior part of the pedicles on both sides. The subarachnoidal space was obliterated at this leve, but the spinal cord was not deformed. The patient was found to have increased tone in his legs, normal sensation for touch, moderately reduced position sense and normal motor power and sphincter function. The patient was referred back to the department of oncology for acute radiotherapy. Over the following weeks the patient's lower extremity function deteriorated, and at this time a decompression and stabilization procedure was performed. This did not improve the function of the lower extremities and a few weeks later the patient was found to have multiple brain metastases.

Comments

This case led to a very interesting discussion of ethics and the role of the neurosurgeon in palliative treatment of the inevitably dying patient. Furthermore, this case led us to form a task force with the assignment to review and update our procedures for treatment of metastatic disease of spinal column, which was presented and discussed at a conference a month later.

Example C

A 37-year-old woman was admitted with a history typical of a one-sided acoustic schwannoma, confirmed by MRI. She was operated by a translabyrinthine approach without any operative complications. During the first day after operation she had no complaints at all but on the second day she slowly developed headache and vomiting, treated by medicine. The third day she demonstrated beginning unconsciousness, which was interpreted as caused by the given medicine. Early in the morning on the fourth day she became unconscious and an acute MRI disclosed a haematoma in the operative field and severe hydrocephalus. She was re-operated, but a fatal herniation had already occurred.

Comments

The case led to a many-sided discussion of the problems of post-operative observation by "too many different" doctors. Furthermore, the case led to an updated education of house officers in postoperative observation and repetitive courses in the most important symptoms and signs to be observed in patients following intracranial procedures.

Example D

A 42-year-old man was admitted with a typical history of long lasting lumbar pain and a shorter one of one-sided radicular symptoms and signs of a lumbar disc herniation, confirmed by MRI at the level of the 4th lumbar disc. He was operated with a classical one-sided lumbar approach. After the operation the symptoms were only slightly improved but in spite of this he was dismissed. At the postoperative control in the outpatient clinic three months later, the patient said that the preoperative symptoms had returned within two weeks of the operation, and had been unchanged since. A control X-ray disclosed, that the patient had been operated at a wrong level. The patient was re-operated, but some of the symptoms remained.

Comments

During the mortality and morbidity conference the general rules for spinal operations were discussed. The internal investigations showed that the surgeon did not follow the instructions of the clinics, which clearly stated that a peri-operative X-ray confirmation of the height is the gold standard. A discussion followed concerning the problem of casual post-operative observation by "too many different doctors" versus the problem of potentially biased postoperative observations, when these only are performed by the surgeon. It was decided to make a notification to the Danish patient compensation insurance company [8]. Finally, it was debated how procedures at a wrong level could be avoided and later again this was discussed in a more scientific forum [5].

Conclusion

We regularly use such cases to adjust behaviour. The insight gained by the individual member of the staff of their role or potential role in the outcome of cases like the ones above is an eye-opener. Of course the first one was not the result of everybody "not bothering" but reflected an extremely busy week in the department during a holiday, where corners were cut to make ends meet. If cases like these are addressed by lecturing the staff about "their irresponsible and poor performance", this will put everybody in a defensive mode, justifying their actions with the workload etc. When cases like number one and three were presented at the morbidity & mortality conference, the result was that everybody felt sorry for the patients and acknowledged the need to improve and maintain a high standard of postoperative evaluation of patients. To maintain a high standard of postoperative evaluation, another morbidity & mortality conference may be used as a booster, if the quality of postoperative evaluations declines again.

We have found the morbidity & mortality conference to be of great value in our departments, both as a teaching tool and as a tool to modify clinical behaviour and the culture of the department. The morbidity & mortality conferences have made it more acceptable to comment on and discuss each other's performance for the benefit of the patients without antagonising each other. In our opinion morbidity and mortality conferences should be part of the daily life in a neurosurgical department.

References

1. Agency for Healthcare Research and Quality. AHRQ WebM&M. *www.webmm.ahrq.gov*
2. Ashbaugh DG, Lung JA (1977) Benefits of an intensive community surgical audit program. Arch Surg 112: 389–390
3. Berthold LD, Hoppe M, Konig H, Saar B, Wagner HJ, Lorenz W, Klose KJ (1998) Mortalitats- und Morbiditatskonferenz in der Radiologie. [Mortality and morbidity conference in radiology]. Rofo Fortschr Geb Rontgenstr Neuen Bildgeb Verfahr 169: 585–589
4. Biddle C (1991) The morbidity and mortality conference: a prolegomenon on its role as an anesthesia risk-management tool. Nurse Anesth 2: 165–171
5. Degn J, Gjerris F (2004) Operation against lumbar disc prolapse. Is it possible to avoid a wrong level? (Operation for lumbal diskusprolaps. Kan indgreb på forkert niveau forebygges? (In Danish)) Ugeskr Læger 166: 1757–1759
6. Eckel HE, Streppel M, Schmalenbach K, Volling P, Schrappe M, Dietz A, Bootz F (2000) Qualitätsicherung in der HNO-Tumorchirurgie. HNO 48: 902–910
7. Esselman PC, Dillman Long J (2002) Morbidity and management conference: an approach to quality improvement in brain injury rehabilitation. J Head Trauma Rehab 17: 257–262
8. Gjerris F, Grünfeld A (1996) Patients insurance and compensation policy in Denmark and other Scandinavian countries. Invited Lecture at the "European Conference on Neurosurgery and Public Health". WHO, Madrid, Spain
9. Harbison SP, Regehr G (1999) Faculty and resident opinions regarding the role of morbidity and mortality conference. Am J Surg 177: 136–139
10. Hobgood CD, Ma OJ, Swart GL (2000) Emergency medicine resident errors: identification and educational utilization. Acad Emerg Med 7: 1317–1320
11. Hoyt DB (1997) Clinical practice guidelines. Am J Surg 173: 32–34
12. Ku Y, Tominaga M, Iwasaki T, Fukumoto T, Suzuki Y, Fujino Y, Hori Y, Kuroda Y (2000) Chairpersons' opinions regarding quality control of surgical faculty performance in Japanese academic surgery departments. Am J Surg 180: 46–50
13. Murayama KM, Derossis AM, DaRosa DA, Sherman HB, Fryer JP (2002) A critical evaluation of the morbidity and mortality conference. Am J Surg 183: 246–250
14. O'Connell PR, Kirwan WO, Brady MP, O'Donnell JA (1988) Surgical audit: the value of a morbidity and mortality conference. Ir J Med Sci 157: 100–103
15. Orlander JD, Barber TW, Fincke BG (2002) The morbidity and mortality conference: the delicate nature of learning from error. Acad Med 77: 1001–1006
16. Seiler RW (2001) Principles of the morbidity and mortality conference. Acta Neurochir (Wien) [Suppl] 78: 125–126
17. Thompson JS, Prior MA (1992) Quality assurance and morbidity and mortality conference. J Surg Res 52: 97–100
18. Zubarik R, Eisen G, Mastropietro C, Lopez J, Carroll J, Benjamin S, Fleischer DE (1999) Prospective analysis of complications 30 days after outpatient upper endoscopy. Am J Gastroenterol 94: 1539–1545

Correspondence: J. Brennum, The University Clinic of Neurosurgery, Copenhagen County Hospital, DK-2600, Glostrup, Denmark. e-mail: brennum@dadlnet.dk

Acta Neurochir (2004) [Suppl] 90: 73–76

Printed in Austria

X Problem-based learning in residency training and the tutorial process for training and education in neurosurgery

P. A. Winkler and **J.-C. Tonn**

Department of Neurosurgery, Klinikum Großhadern, Ludwig-Maximilians-University of Munich, Munich, Germany

Introduction

The educational climate in more traditional, didactic educational settings in neurosurgery, at least in Europe, works against "facilitatory tutorial teaching" of the residents in neurosurgery. In these settings, the resident and fellow is typically expected to know the right answer. S/he learned a long time ago that it is best to stay silent if s/he does not know the answer, or is unsure of it, for s/he fears that s/he may blame himself when being wrong with an answer. An admission of not knowing something would be used as evidence of inadequate study or lack of knowledge. Unless a resident has been in a more liberal or alternative high school or university, this climate has usually been in existence throughout his educational career, from kindergarten on [3].

The actual personnel and financial requirements necessitate a careful and economical application of the available resources, in this particular situation the teaching resources. "Facilitatory tutorial teaching" describes a culture of teaching in which a constructive atmosphere under the supervision of an experienced tutor is created.

At the anglo-saxon universities the title "tutor" is usually associated with someone who takes up the position of a teacher or master. Although the term "tutor" may therefore not be ideal for a teacher employing a facilitatory or Socratic teaching, respectively, style, it has become commonly used in this way. As "tutorial process" the interaction between the tutor and a "small group" of trainees with a maximum of eight group members is defined [1, 2]. The ability and, with increasing experience the skills of the tutor to use facilitatory teaching during the group learning process is the major determinant of the quality and the success of any educational method. Tutoring is a teaching skill central to problem-based, self-directed learning. Several universities and national societies offer "tutoring" and "facilitatory teaching" in form of a course for trainers to prepare them in modern teaching. It would be wise, if the EANS in future would also develop courses for training in modern teaching techniques.

How can we apply such teaching techniques in the clinical environment? In the daily practice of neurosurgery it is logical to combine the admission examination and the discussion about the best management of the patient with a problem-based, case-oriented learning for the residents and fellows. Basically, this technique can be applied during all forms of teaching, in conferences, seminars etc.

The intent of this book chapter is to present a conceptual basis for facilitatory teaching, making this new method of learning easier to understand, as well as to present practical suggestions how to develop and practise these skills with time. Over the last ten years, the first author of this book chapter had the opportunity to practise this style of teaching in seminars of neuroanatomy for second-year-students (here combined with conventional didactic methods), in courses of clinical neurosurgery for fifth-year-students and in a residency-training program.

The case presentation

As a vehicle of learning, the daily case should be used to initiate and drive the teaching process. Depending on the level of knowledge a straight forward case or a more complex case can be selected. This decision preferably is made by the tutor. Decision points elicit intellectual commitment, and help to raise the level of inquiry [4]. Given the busy daily time schedule a specific time period should be preserved each day for case presentation and discussion. It seems that the possibly best time is after the "core activities" in the operating theatre and the work on the ward. "The case" is best presented by a member of the small group, a trainee or a student, who during the day had the opportunity to interview and examine the patient meticulously under supervision. Beginning with the vital characteristics of the patient, (i.e. vital status, chief complaint, history of the present illness, pertinent past medical history, social and family history, physical exam, labs and diagnostic studies, treatments, etc.), actual problems of the patient and neuroradiological imaging-presentation, a whole picture of the suspected disease has to be described. For this procedure 8 out of 40 minutes of the whole teaching session are reserved.

Only if the tutor and the group have received all relevant data, the process can continue. If not, the tutor has to take care that this information is supplied. Next, the trainee continues with the discussion of the differential diagnosis, the disease etiology, its natural history, appropriate therapeutic options, neurosurgical procedures and approaches, its outcome and complications. At this point, the skilled tutor animates the other group members to take part in the discussion:

- Does everybody agree with the statements made?
- Are the conclusions adequate?
- Are there alternative therapeutic options, etc.?

The tasks of the tutor

The tutor has to find the right balance as a facilitator of the learning process. He asks questions, activates the group members to become involved, but he should not provide the desired information himself. The tutor's function aims at developing the residents' thinking or reasoning skills (problem solving, metacognition, critical thinking) and at helping to become independent, self-directed learners.

The term "*metacognition*" describes aspects like thought, reflection and deliberation. Metacognition is the executive function in thinking: pondering, deliberating, or reflecting on a problem or situation; reviewing what is known about the problem at hand; creating hypotheses; making decisions about what observations to make and what questions to ask; questioning the meaning of new information obtained from inquiry; evaluating other sources of information; reflecting on and reviewing what has been learned, what it all may mean and what needs to be done next.

Anyone who does not use metacognitive skills when faced with new and difficult problems is an impulsive thinker, someone who operates on reflex behaviours, guesses or hunches (i.e. without thinking). This is the trainee we all dread and the professional from whom we would prefer not to receive a service.

The tutor must carefully guide the respective trainee but also the whole group through all the steps of the particular learning process. The tutor should carefully guide trainee and group through analysis and synthesis. If action steps or solutions for a problem are to be undertaken, the group must entertain alternatives and consider the pros and cons, for instance the pros and cons for a stabilisation/instrumentation in a specific case. The tutor should be certain that all the appropriate steps are thoroughly covered. Particularly if a complex problem is to be analysed, the tutor must be sure the trainee structures the problem and goes through all stages of the hypothetico-deductive process.

A further task of the tutor is to keep the discussion process moving, to make sure that no essential part of the case evaluation is passed over or neglected and that the phases are addressed in the right sequence. For example, the tutor should be certain that all possible hypotheses or explanations for the cause of the problem are explored before he allows the group to go on with the next step.

The tutor should never let ideas, terms, explanations or comments go unchallenged or undefined. Anytime a scientific term, label or eponym is used, the trainee should be asked to define the term. The tutor cannot assume that a trainee correctly understands a concept or entity because he can use the label correctly. With the discussion going on the tutor gains a pretty good impression of the trainee's knowledge and understanding of the particular case. If necessary, he may ask some questions until the trainee has come to the depth of knowledge expected of him or her.

An important task is to involve all trainees in the group work. If the final solutions and decisions are made more or less as a consensus in the group, this is not only a reward but also an important experience for the participants. To facilitate the interchange between trainees, the tutor has to avoid to become the focal point of the discussions. If s/he is not careful, s/he will find that each trainee is answering only his/her questions and that each trainee is addressing the tutor in the discussions and not each other. Instead of a working group, the tutor will have five or six individuals, each interacting only with the tutor.

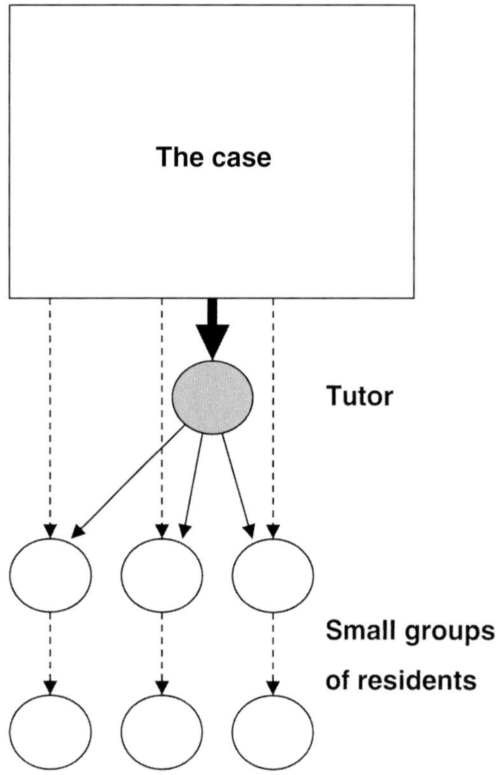

Fig. 1. Conventional teaching style

Interpersonal dynamics as a challenge for the tutor

Occasionally "*interpersonal dynamics*" may arise in the group. Silence, late arrival, sarcasm, lack of individual productivity, lack of initiative, personal instead of objective arguments, trainees taking sides on an issue, expressions of dissatisfaction with learning or attempts by a trainee to take over in the group should be recognized as important symptoms. When such problems become apparent to the tutor, it is best to approach them at a metacognitive level, i.e. at the level of objective, unprejudiced and unbiased decision makings. The two basic challenges for a tutor in small group learning are attending the group's educational progress at the metacognitive level and the group's interpersonal dynamics. If the tutor appreciates that s/he is the trainee's metacognitive conscience, his or her specific teaching role becomes more understandable.

Architecture of the small group learning process

It is important to form small groups consisting of the responsible tutor (consultants, subspecialists) and the trainees (residents, fellows, students, guests). The number of participants, however, should not exceed eight (8!) persons (Figs. 1 and 2).

In the clinical setting, either the assigned trainer works with his team on a regular schedule or a subspecialist works with all residents once a week and teaches a particular pathology. Even a combination of both is possible. If this is done on a regular basis, the trainees certainly will acquire a high competence in problem solving.

Summary of the general principles for tutorial teaching

Every responsible consultant or staff member who functions as a tutor should follow the subsequent "ten laws":

1. It is essential that the tutor avoids expressing a deprecatory opinion concerning the correctness or quality of any trainee's comments or contributions.
2. The tutor should push the trainee to a deeper level of understanding.
3. Discussions between trainees, comments and critical analysis of each other's ideas or knowledge must always be encouraged.
4. All decisions should be by group consensus. The tutor must be certain that all trainees contribute to the final decision.
5. The tutor should do whatever is necessary to get the trainees to talk and discuss among themselves.
6. Questions, particularly like "Are you sure you're right?" or "Are you comfortable with that decision?" should be asked to trainees when they are correct as often as when, in the mind of the tutor, they are incorrect in their opinions and statements.
7. The tutor should determine the goal of the specific session.
8. The tutor must monitor whether this goal has been reached and whether a trainee has made an educational progress.

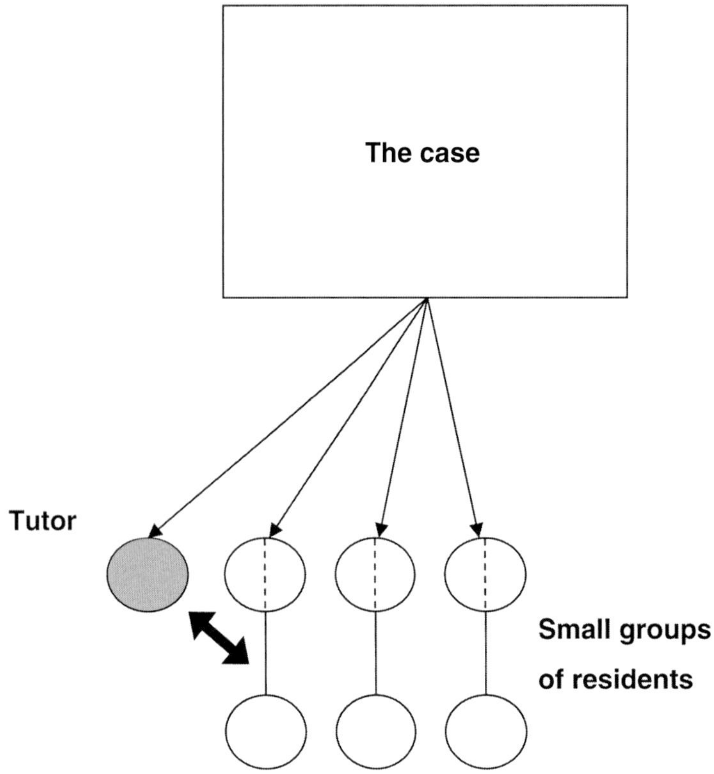

Fig. 2. Problem based learning and the tutorial process

9. The tutor needs to be aware of potential interpersonal problems in the group and make the interventions necessary to maintain an effective group process.
10. None of these tutorial activities should become the sole task of the tutor. The tutor must constantly work toward getting the group of trainees to take on the responsibility for the group learning process.

Conclusions

The tutor is responsible for the process of neurosurgical trainees' learning. S/he facilitates the trainees' learning through guidance at the metacognitive level. The small group-tutorial process should provide trainees with the metacognitive skills they will need to do this – as they acquire the skills to learn in contexts and practices relevant to their future.

Acknowledgement

The authors deeply acknowledge and thank Prof. Tom Aretz from Harvard-Medical-School in Boston for his valuable help in correcting and revising the manuscript.

References

1. Albanese MA, Mitchell S (1993) Problem based learning: (a review of literature on its outcomes and implementation issues. Acad Med 68: 1–48
2. Armstrong EG (1997) A hybrid model of problem based learning. In: Boud D, Feletti G (eds) The challenge of problem based learning, 2nd edn. London, Kogan Page 14: 137–150
3. Barrows HS (1992) The tutorial process, revised edn. Southern Illinois University School of Medicine, Springfield, Illinois
4. Glick TH, Armstrong EG (1996) Crafting cases for problem-based learning: experience in a neuroscience course. Med Educ 30: 24–30

Correspondence: Peter A. Winkler, Department of Neurosurgery, Klinikum Großhadern, Ludwig-Maximilians-University of Munich, Marchioninistr. 15, Munich, Germany. e-mail: pwinkler@nc.med.uni-muenchen.de

Acta Neurochir (2004) [Suppl] 90: 77–95

Printed in Austria

XI Periodic progress evaluation

K. W. Lindsay

Department of Neurosurgery, Southern General Hospital, Glasgow, UK

Introduction

Successful completion of neurosurgical training requires that trainees attain a variety of skills (Table 1) of which technical competence and judgement are perhaps the most critically important. But how do we evaluate these skills? How do we ensure that trainees satisfactorily progress throughout their training? How do we ensure that after a specified number of years trainees have achieved sufficient competence to permit safe neurosurgical practice? Too often such decisions depend on the subjective impressions of the clinical teachers. These may be reliable, particularly if based upon the consensus view of a large number of observers – but does this always apply? Does adequate documentation of the trainees' progress or failure always exist? It is no longer acceptable for trainees to complete training and be deemed 'competent' without undergoing some form of formal assessment. Yet to obtain a valid, objective and reliable method of assessing competence in these numerous skills, which remains feasible in terms of human and physical resources, presents a formidable challenge.

Table 1. *Competencies required in neurosurgical training*

Medical skills
Knowledge of basic neurosciences
Knowledge of clinical skills – examination
interpretation of symptoms/signs
making a diagnosis
treatment options
Decision making
Surgical skills/manual dexterity
Application of knowledge to clinical practice
Communication skills
Establish patient/clinician relationship
History taking
Relay information to the health care team
Obtain consent
Communicate bad news/bereavement
Teaching/learning skills
Critically appraise medical information incl. research
Develop continual self education
Pass on knowledge and skills to others
Personal effectiveness
Consult effectively with other clinicians
Contribute to teamwork
Use time effectively
Management
Utilise and prioritise available resources within healthcare organisation
Utilise information technology

In 1997, a task force charged with reporting on "Clinical Academic Careers" recommended that the Royal Colleges "give serious consideration to establishing innovated procedures, other than the written examination, to assess clinical competence of candidates for the Certificate of Completion of Specialist Training (CCST), rather than just time spent" [8]. In response to this the Surgical Royal Colleges set up working parties and committees to look into the issues involved, but progress has been slow.

In this chapter I will describe the methods used in the UK in the last decade to assess progress through neurosurgical training and will discuss their limitations. I will also review other methods, presently under consideration, which attempt to determine more objectively whether or not the trainee has achieved a satisfactory level of competence.

Progress evaluation – process and timing

Evaluation of progress of a trainee takes place at two levels, firstly at a local level by the trainers, by the Training Programme Director and by the Postgraduate Dean (the statutory source of trainee funding) and secondly, at a National level (for the UK and Ireland) by the Specialist Advisory Committee (SAC) in Neurosurgery acting on behalf of the Joint Committee for Higher Surgical Training (JCHST) of the Surgical Royal Colleges.

Local evaluation

Assessments occur at *6 months, one year and annually* thereafter. The *trainee assessment form* (Appendix I and II) forms the basis of these progress assessments. All trainers supervising training complete a form at the above periods or on completion of a training placement, if this is of less than one year's duration, taking into account the years of training already completed. The forms attempt to cover the range of required skills and include the trainees' capability of recording and conveying patient details of history, examination and investigative findings to senior staff, consenting patients for operative procedures, communicating with patients and relatives and passing on distressing information (e.g., malignancy or death) in a sensitive, caring manner. For each criterion, they categorise the trainee as – satisfactory, *or* needing targeted training in an area *or* unsatisfactory, needing to repeat training in this area.

The trainer produces a written summary and indicates whether

- training is satisfactory in all aspects permitting the trainee to proceed to the next stage of the programme,
- any areas of weakness have been identified requiring correction at the next placement,
- training must be repeated *or*
- the trainee should be counselled by the Postgraduate Dean/Programme Director about choosing an alternative career pathway.

The trainers discuss their conclusions with the trainee, who is asked to add comments and to sign the form (to confirm that he has seen, but not necessarily agreed with the conclusions).

A very similar form, modified by H. J. Reulen is available from the Joint Residency Advisory and Accreditation Committee of the EANS/UEMS (see web site EANS).

The *assessment panel*, composed of at least four members including the Training Programme Director, the Postgraduate Dean's representative and at least one other trainer, meets annually and interviews all trainees on the training programme to ascertain achievements and progress. The panel reviews the

- Training portfolio which includes an up to date *curriculum vitae* incorporating –
  - (i) details of previous training posts, dates, duration and trainers.
  - (ii) details of examinations passed e.g. MRCS
  - (iii) list of publications with copies of published first page/abstract.
  - (iv) list of research presentations at local, national and international meetings.
  - (v) list of courses attended.
  - (vi) updated cumulative operative totals (see Chapter VI)
  - (vii) copies of assessment forms for the end of each training year, completed and signed by both trainers and trainee.
- Trainee assessment form
- Neurosurgical logbook

and makes a final decision regarding progression to the next training placement.

Prior to the interview, the trainee also completes and submits a *training post assessment form* (Appendix III) commenting on any perceived deficiencies in the training programme related to clinical training, the workload, teaching and supervision in theatre, research support, feedback of progress and career advice. The panel can discuss any items of concern with the trainee. Where possible, the Training Programme Director would be responsible for correcting any such deficiencies. Completed assessment forms are sent to the central Specialist Advisory Committee in Neurosurgery. Forms indicating a significant problem in training may trigger an interview with this committee (see below).

Right of appeal

If the trainee disagrees with any decision, he or she has the right of appeal to the assessment panel. If asked to repeat training or to withdraw from the programme, the trainee has the right of appeal to an appeals panel, chaired by the Postgraduate Dean, that excludes any member of the original assessment panel.

National level evaluation

Two members of the Specialist Advisory Committee (SAC) interview trainees in their third year of training, or more frequently if any significant problems arise. Before each interview, trainees complete a short structured information sheet providing details on weekly timetable, operative experience and supervision, the on-call rota, the academic programme, research, and administrative duties. They also present their cumulative operative totals for review. The interviewers report their findings back to the whole committee. If the SAC believes any deficiencies exist the Training Programme Director is notified.

Two SAC members visit every training programme once every four years to monitor standards and quality of training. As well as inspecting facilities and reviewing the trainees' workload, the visitors interview all trainees on the programme, irrespective of their year of training.

End of training assessment

At the end of a *minimum of 6 years* of training the trainee applies to the SAC for recommendation for the Certificate of Completion of Specialist Training (CCST). To succeed, the trainee should have

- completed 5 years of clinical neurosurgical training and one in research or any other branch of the specialty,
- obtained satisfactory assessments at the end of each year,
- maintained a logbook (Chapter VI), validated by his trainers, showing that he or she has assisted in a wide variety of cases, including those performed personally, those assisted by a trainer or more senior trainee, and those where the trainee acted as assistant,
- a cumulative list of operative totals for each procedure, indicating that these approach the median figure for the majority of procedures (see chapter VII, Table 1),
- passed the Intercollegiate Board Examination in Neurosurgery,
- received a reference from all trainers indicating a satisfactory performance.

The SAC scrutinises the above information and tries to ensure that no deficiencies exist. In a proportion, recommendation for the CCST award is deferred.

Progress evaluation – methods

In-training evaluation

Trainers (i.e. consultants or specialists) working directly with the trainee for months or years should be capable of providing a meaningful judgement of his or her abilities. But such observations, although perhaps accurate, are inevitably based on subjective assessment rather than objective facts. The *trainee assessment form* provides a structured format, compelling trainers to consider trainees' abilities for individual skills, but the available categories may fail to discriminate between trainees of varying ability. Completion of assessment forms by multiple trainers for any one trainee must help to improve validity but other methods should be sought.

Operative totals

The trainee logbooks and summative operative totals go some way toward providing a more objective method of assessment. These detail operative experience, and this in turn should indirectly reflect operative skills. Trainers will more readily permit trainees to perform procedures if they are confident of the trainees' abilities. The number of procedures performed for any one condition will reflect the likelihood that the trainee will have experienced complications and difficulties that might arise for that condition. Operative totals are simple to monitor by the trainer and by the local and national training committees. The national training committee (the SAC) can set minimal targets for certain key procedures and encourage trainees to achieve median values for the majority of procedures.

Using operative totals to monitor competence requires some caution –

- The numbers of procedures performed personally by the trainees and the numbers when the consultant has actively assisted may depend to some extent on the personality of the trainer and the extent of supervision he or she imposes.
- A degree of variability may exist in the way the trainee categorizes his or her participation in the

procedure. This depends to some extent on the integrity of the trainee, unless the trainer checks and countersigns every procedure (for categories and redefinitions see Chapter VII).

- Performing many operative procedures does not necessarily mean that the trainee has received good training; nor does it ensure that the procedures were performed well. Operative training should progress through various stages – the trainee assisting the trainer, the trainer assisting, the trainer observing scrubbed, the trainer observing unscrubbed, and finally with the trainer available but not in the operating theatre. Some countries (e.g., Germany) require that the trainer be present in the operating theatre for the whole procedure, but this deprives the experienced trainee of the opportunity to gain the self-confidence required for independent practice. Scrutiny of the logbook for numbers of procedures "assisted by the trainer" ensures that the appropriate training steps have been followed.
- If a training committee set a target number of procedures, the danger would be that trainees spend their time chasing operative numbers rather than ensuring that they obtain a balanced neurosurgical training. This criticism may apply, but most would accept that, along with judgment, operative ability is the most crucial aspect of neurosurgical training.

Operative competencies

Recording operative totals provides a limited measure of competence. To seek alternative methods of evaluating progress, the SAC in Neurosurgery proposed taking operative competencies a step further. Trainees in conjunction with their trainers would categorize their operative abilities for each procedure into one of four grades, as follows: *(A) not able to perform the procedure; (B) competent to perform the procedure with direct supervision; (C) competent to perform the procedure without direct supervision; and (D) competent to deal with complications and difficulties that may arise.* These competencies would be recorded at three stages in training: at the end of the second year, at the end of the fourth year, and at the end of the sixth year [4]. The SAC has drawn up three tables indicating the minimal competency expected at these three stages (Table 2 a,b,c). At each stage, the trainee would tabulate his/her cumulative operative totals and perceived level of competence and the trainers and the Training Programme Director would validate the completed forms.

Although training in the UK aims to produce a neurosurgeon 'capable of independent practice', it is not expected that trainees attain competence in every operative procedure (although trainees should have sufficient expertise to care for the patient until referral to an appropriate neurosurgeon is arranged). Table 2c illustrates those procedures where the SAC would expect operative competence to be developed *after* completion of training (even though the trainee undertook a subspecialisation year in the field). Such procedures would include

- complex basal meningiomas,
- transphenoidal micro-adenomas,
- acoustic neuromas
- complex arteriovenous malformations,
- paediatric neurosurgical procedures other than shunt insertion/revision.
- intradural intramedullary spinal tumours
- functional neurosurgery

Another method of assessing technical skills under consideration, is that described for general surgical procedures by Winckel et al. [9]. This involves defining a few core procedures and creating a Structured Technical Skills Assessment Form (STSAF) which consists of two parts. In the first part a checklist divides the procedure into numerous components e.g. positioning on the table, incision, use of cautery, exposure, use of the microscope, dural opening, achievement of the procedure, closure, postoperative care etc. A trained evaluator assesses the trainee's capability of performing each step. The second part of the form assesses global skills e.g. respect for tissues, instrument handling, flow of procedure, understanding of surgical anatomy, use of assistants etc. Winckel et al. found high inter-rater reliability with this technique. The checklists also competently distinguished between junior and senior trainees.

Reznick's team also developed an objective method to test surgical skills using 'bench models' to simulate different stages within a surgical procedures (e.g. bowel anastomosis) [5, 7]. Again the authors created a structured form – the Objective Structured Assessment of Surgical Skills (OSATS) has been created with checklists evaluating ability at performing individual steps and a 'global rating scale' including the skills listed above, looking at surgical 'behaviour'. These

Table 2 (a). *Expected competencies – end of 2nd year*

NEUROSURGICAL TRAINING Name: Date:

OPERATIVE COMPETENCIES - END OF 2nd YEAR

- A = Not able to perform procedure
- B = Competent to perform procedure under direct supervision
- C = Competent to perform procedure without direct supervision
- D = Competent to deal with complications and difficulties which may arise

Cumulative operative totals

- P = No direct supervision
- SU = Trainer in theatre
- SS = Trainer scrubbed
- A = Assisting trainer

		A	B	C	D	P	SU	SS	A
Basic techniques	Burrhole insertion								
	Craniotomy								
Head Injuries:	craniotomy - extradural								
	craniotomy - acute subdural								
	craniotomy - intracerebral								
	depressed fractures								
	burr hole - chronic subdural								
Supratentorial tumours:	stereotactic/lds biopsy								
	craniotomy - intrinsic tumour								
	meningioma - vault								
	meningioma - parasagittal								
	meningioma - complex basal								
Midline tumours	transphenoidal macroadenoma								
	transphenoidal microadenoma								
Post. fossa tumours	malignant								
	acoustic								
Hydrocephalus	external ventricular drain								
	shunts insertion								
	shunt revision								
Neuroendoscopy									
Vascular	craniotomy - aneurysm								
	craniotomy - AVM								
	craniotomy - intracerebral haematoma								
	posterior fossa haematoma								
Spine	cervical disc disease / spondylosis								
	thoracic disc disease								
	lumbar disc disease/spondylosis								
	tumours - extradural								
	tumours - intradural extramedullary								
	tumours - intradural intramedullary								
	spinal fixation								
Pain:	trigeminal neuralgia - lesion								
	trigeminal neuralgia - MVD								
PAEDIATRIC									
Head injuries	haematoma								
	chronic subdural/hygroma								
	depressed fracture								
	ICP monitoring								
Tumours	supratentorial								
	infratentorial								
	stereotactic / ids biopsy								
Hydrocephalus	external ventricular drain								
	shunt insertion/revision								
	Arnold Chiari								
	Neuroendoscopy								
Spine	mening/meningomyelocele								
	tethering syndromes								
Paediatric functional	spasticity								
	surgery for epilepsy								
Craniofacial	sutural craniectomy								
	combined craniofacial								
	encephalocele								

[shaded box] - minimal competency

Table 2 (b). *Expected competencies – end of 4th year*

NEUROSURGICAL TRAINING **Name:** **Date:**

OPERATIVE COMPETENCIES - END OF 4th YEAR

A = Not able to perform procedure
B = Competent to perform procedure under direct supervision
C = Competent to perform procedure without direct supervision
D = Competent to deal with complications and difficulties which may arise

Cumulative operative totals

P = No direct supervision
SU = Trainer in theatre
SS = Trainer scrubbed
A = Assisting trainer

		A	B	C	D	P	SU	SS	A
Basic techniques	Burrhole insertion								
	Craniotomy								
Head injuries:	craniotomy - extradural								
	craniotomy - acute subdural								
	craniotomy - intracerebral								
	depressed fractures								
	burr hole - chronic subdural								
Supratentorial tumours:	stereotactic/ids biopsy								
	craniotomy - intrinsic tumour								
	meningioma - vault								
	meningioma - parasagittal								
	meningioma - complex basal								
Midline tumours	transphenoidal macroadenoma								
	transphenoidal microadenoma								
Post. fossa tumours	malignant								
	acoustic								
Hydrocephalus	external ventricular drain								
	shunts insertion								
	shunt revision								
Neuroendoscopy									
Vascular	craniotomy - aneurysm								
	craniotomy - AVM								
	craniotomy - intracerebral haematoma								
	posterior fossa haematoma								
Spine	cervical disc disease / spondylosis								
	thoracic disc disease								
	lumbar disc disease/spondylosis								
	tumours - extradural								
	tumours - intradural extramedullary								
	tumours - intradural intramedullary								
	spinal fixation								
Pain:	trigeminal neuralgia - lesion								
	trigeminal neuralgia - MVD								
PAEDIATRIC									
Head injuries	haematoma								
	chronic subdural/hygroma								
	depressed fracture								
	ICP monitoring								
Tumours	supratentorial								
	infratentorial								
	stereotactic / ids biopsy								
Hydrocephalus	external ventricular drain								
	shunt insertion/revision								
	Arnold Chiari								
	Neuroendoscopy								
Spine	mening/meningomyelocele								
	tethering syndromes								
Paediatric functional	spasticity								
	surgery for epilepsy								
Craniofacial	sutural craniectomy								
	combined craniofacial								
	encephalocele								

(shaded box) - minimal competency

Table 2 (c). *Expected competencies – end of 6th year*

NEUROSURGICAL TRAINING **Name:** **Date:**

OPERATIVE COMPETENCIES - END OF 6th YEAR

A = Not able to perform procedure
B = Competent to perform procedure under direct supervision
C = Competent to perform procedure without direct supervision
D = Competent to deal with complications and difficulties which may arise

Cumulative operative totals

P = No direct supervision
SU = Trainer in theatre
SS = Trainer scrubbed
A = Assisting trainer

		A	B	C	D	P	SU	SS	A
Basic techniques	Burrhole insertion								
	Craniotomy								
Head Injuries:	craniotomy - extradural								
	craniotomy - acute subdural								
	craniotomy - intracerebral								
	depressed fractures								
	burr hole - chronic subdural								
Supratentorial tumours:	stereotactic/ids biopsy								
	craniotomy - intrinsic tumour								
	meningioma - vault								
	meningioma - parasagittal								
	meningioma - complex basal								
Midline tumours	transphenoidal macroadenoma								
	transphenoidal microadenoma								
Post. fossa tumours	malignant								
	acoustic								
Hydrocephalus	external ventricular drain								
	shunts insertion								
	shunt revision								
Neuroendoscopy									
Vascular	craniotomy - aneurysm								
	craniotomy - AVM								
	craniotomy - intracerebral haematoma								
	posterior fossa haematoma								
Spine	cervical disc disease / spondylosis								
	thoracic disc disease								
	lumbar disc disease/spondylosis								
	tumours - extradural								
	tumours - intradural extramedullary								
	tumours - intradural intramedullary								
	spinal fixation								
Pain:	trigeminal neuralgia - lesion								
	trigeminal neuralgia - MVD								
PAEDIATRIC									
Head injuries	haematoma								
	chronic subdural/hygroma								
	depressed fracture								
	ICP monitoring								
Tumours	supratentorial								
	infratentorial								
	stereotactic / ids biopsy								
Hydrocephalus	external ventricular drain								
	shunt insertion/revision								
	Arnold Chiari								
	Neuroendoscopy								
Spine	mening/meningomyelocele								
	tethering syndromes								
Paediatric functional	spasticity								
	surgery for epilepsy								
Craniofacial	sutural craniectomy								
	combined craniofacial								
	encephalocele								

(shaded) - minimal competency

appear to correlate with the trainers' opinions about trainees' progress in the operating theatre. However to create such facilities on a wide scale basis would be labour-intensive and expensive and Reznick's techniques do not apply to neurosurgical practice. Similar assessment devices provide objective measurements of economy of motion and number of movements of laparoscopic surgical instruments using both real and virtual reality [2], but again such techniques are not applicable to neurosurgical practice. Much work is underway to develop *virtual reality* models of neurosurgical procedures with the aim of enabling trainees to acquire operative experience and to learn operative techniques without the need to 'practice' on patients [3]. The extensive range of technical manoeuvers and the variety of conditions met during a single operative procedure make this a daunting task. In theory, such computer based programmes could provide a method of assessment as well as training, but I suspect that we are many years away from using this as a feasible alternative.

Examination

After at least 4 years of clinical neurosurgical training in the UK, trainees sit an Intercollegiate Board Examination in Neurosurgery. This is comprised of two parts, the first, a MCQ on the applied neurosciences can be taken at any stage after appointment to a training programme. The second part is composed of a clinical exam of one hour duration, a 'spot' examination of one hour duration where slides of CT/MR images, histopathology, clinical features, operative anatomy etc. are shown to all candidates along with questions requiring short answers, and three $\times \frac{1}{2}$ hour vivas on 'operative surgery and surgical anatomy', 'the non-operative practice of neurosurgery', and' investigation of the neurosurgical patient'.

Such an examination tests clinical and basic scientific knowledge over a relatively wide range. The clinical and oral section tests history taking, patient examination, investigations, judgement and to a limited extent, communication, but time limitation restricts the range of topics. The examination does not evaluate decision-making, surgical skills and manual dexterity or experience; nor does it assess teaching, learning and management skills, personal effectiveness, or the ability to work in a team.

Other forms of examination such as the Objective Structured Clinical Examination (OSCE), consisting of short observations of clinical scenarios, utilising checklists to assess many clinical skills, can provide valid, reliable methods of testing many clinical skills, but these take considerable time and effort to develop and run [6].

Few other European countries conduct examinations in neurosurgery, but an examination of a similar standard is run by the European Association of Neurosurgical Societies (EANS). This EANS Board Examination is available in two parts. The first, an MCQ, is taken at any stage of training and the final part composed of three x 1 hour oral examinations, after completing one year in practice beyond gaining a CCST. Trainees from countries where no local examination exists should be encouraged to sit the EANS MCQ examination. Two countries have already made this a compulsory examination for their trainees.

Conclusions

Over the last 10 years in the United Kingdom and Ireland, we have established systematic procedures both locally and on a National level to evaluate trainees' progress throughout neurosurgical training. Present methods use a combination of in-training evaluation by the local trainers and the Training Programme Director, monitoring cumulative operative totals and an Intercollegiate examination. But operative totals reflect competence indirectly and could prove misleading. In-training evaluation by the trainers may give reliable information, particularly when multiple observers are involved [1], but at present this still lacks objectivity and validity. The suggested method of using competence tables or the structured technical skills assessment form may provide a useful addition to evaluating progress, but these still require validation in neurosurgical practice.

Despite the competitive selection process encountered before entering neurosurgical training, trainees progress at a variable rate. If it were feasible to measure clinical competence, then training could become "competence-based" rather than "time-based" – that is, the duration of training would depend on the time taken to achieve a certain level of competence rather than lasting a prescriptive minimum of 6 years. Whether or not these ideals are ever achieved, efforts to provide a more objective means of assessing competence can only improve current methods and raise the standards of training in general.

Appendix I. Joint committee on higher surgical training: trainee assessment form

This is an official document. A separate form is to be completed at the end of a placement by each trainer (forms are to be completed every 6 months and must be completed within a month of finishing the placement). The original is the property of the JCHST. Signed and completed forms are to be returned to the JCHST offices with a copy going to the Programme Director and Postgraduate Dean.

Guidance notes on the completion of this form are available on the JCHST web site, www.jchst.org from the Postgraduate Dean or Programme Director.

General Information

(to be completed in block capitals by trainee before handing to trainer for completion)

Form completed by (Name of trainer) ________________________________

Name of trainee ________________________________

Programme Director ________________________________

Specialty ________________________________

Training Number ________________________________
(NTN/VTN/FTN or LAT)

Expected CCST date (if applicable) ________________________________

GMC Number ________________________________

Post and Trust/Hospital ________________________________
(Number if known)

Assessment period From: ________________ To: ________________________

Sick Leave
Duration of absence due to sick leave during this period ________

Duration of sick leave since entering SpR training ________

Trainer Signature ____________________________ Date____________________

Trainee Signature ____________________________ Date____________________

September, 2001

To be completed by trainer

Year of SpR training (please circle) 6 mths 1 2 3 4 5 6

CRITERIA	unsatisfactory needs to repeat training in this area	needs targeted training in this area	satisfactory	COMMENTS
A. Clinical Skills				
History Taking				
Physical Exam				
Investigations				
Diagnosis				
Judgement				
Operative skill				
After care				

B Knowledge				
Basic Science				
Clinical				

C. Postgraduate Activities				
Teaching				
Lecturing				
Case presentation				
Publications				
Learning skills				
Research				
Audit				

CRITERIA	unsatisfactory needs to repeat training in this area	needs targeted training in this area	Satisfactory	COMMENTS
D. Attitudes				
Reliability				
Self Motivation				
Leadership				
Team working				
Administration				
Relationships & Communication With:				
a) Colleagues				
b) Patients				
c) Other staff				
Communication skills:-				
a) Informed consent				
b) Bereavement				
c) Breaking bad news				

Comments from trainer (please extend to an attached sheet if necessary – each attached sheet must be signed and dated by trainer and trainee)

____

____

____

____

____

____

____

September, 2001

Summary conclusion

❒ Satisfactory in all respects to proceed

❒ Satisfactory to proceed, but the following areas for improvement have been identified and must be addressed in the next placement (detail areas for improvement - please extend to an attached sheet if necessary –attached sheets must be signed and dated by trainer and trainee)

____

____

____

____

____

____

____

____

❒ Unsatisfactory to proceed and the training placement needs to be repeated

❒ Unsatisfactory and should be referred for advice to Postgraduate Dean / Programme Director about choosing an alternative career pathway

Comments from Trainee (please extend to an attached sheet if necessary – each attached sheet must be signed and dated by trainer and trainee)

____

____

____

____

____

____

____

____

Appendix II. Guidance notes on the completion of the JCHST Trainee Assessment form

1. The assessment form is CONFIDENTIAL once completed, and must be handled accordingly
2. Trainees are to complete the first section of the form **BEFORE** handing to their trainers.
3. It is a trainees responsibility to ensure that a separate form is completed by each of their consultant trainers
4. It is a trainees responsibility to ensure that a form (or forms) is completed at least every 6 months.
5. THE FORM WILL BE RETURNED IF INCOMPLETE

Assessors must take into account the year / level of training that the trainee has reached and assess accordingly.

Assessors are asked to put a X orin the appropriate column.

Assessors are to indicate any areas where improvements are necessary (requiring improvements does not necessarily mean a trainee will have to repeat training).

The columns below are not prescriptive but for guidance to how an assessment may differentiate between 'satisfactory' or 'in need of targeted training'. It is of overriding importance when completing this form and considering the performance of a trainee, that the trainee is judged in conjunction with the requirements of their specialty curriculum for their year of training and against the standards you would expect of their contemporaries. It is important that if deficiencies are identified these are evidence based and clearly noted.

Trainers are recommended to complete the form in draft PRIOR to meeting with the trainee.

CRITERIA	Examples of performance which would be considered unsatisfactory	Examples of performance which would be considered satisfactory
A. Clinical Skills		
History Taking	Incomplete, inaccurate. Poorly recorded	Complete, orderly, perceptive
Physical Exam	Misses important physical signs, relies unnecessarily on investigations	Thorough, accurate, can elicit correct signs, recognises most significant findings
Investigations	Inappropraite, random, inability to interpret tests	Usually appropriate, good knowledge on interpreting tests
Diagnosis	Fails to interpret and synthesise symptoms, signs and investigations	Good knowledge with an orderly logical approach to differential diagnosis, good clinical memory
Judgement	Difficulty in managing emergencies. Fails to take appropriate action	Reliable, competent under pressure, asks for advice appropriately
Operative skill	Not at competency level for year of training, treats tissues roughly, requires significant guidance	Competent to carry out procedures as required by the level of training
After care	Fails to notice complications and acts inappropriately	Good awareness of complications, takes appropriate action
B. Knowledge		
Basic Science	Not aware of the basic sciences applicable to the clinical practice	Adequate knowledge of the basic science related to the relevant surgical specialty
Clinical	Does not have necessary clinical knowledge	Adequate clinical knowledge to make decisions appropriate to year of training

September 2001

CRITERIA	Examples of performance which would be considered unsatisfactory	Examples of performance which would be considered satisfactory
C. Postgraduate Activities		
Teaching	Appears uninterested in teaching. Contributes little to the education of juniors. Poor style.	Competent and conscientious in teaching others, good style
Lecturing	Inadequate preparation, poor style	A good speaker, well prepared and has prepared for questions
Case presentation	Needs to be better organised and/or better prepared. Poor on history, signs, diagnosis and discussion	History and signs correct, good deductions
Publications	Needs to complete writing up of a paper or thesis and to augment c.v. - no publications/presentations during period	Paper published or accepted for publication in a peer review journal or presentations at learned society during assessment period
Research	Not involved in research, poor graps of statistic and research methods	Able to critically analyse, involved in research and publishing results
Audit	Avoids audit, inadequare presentation	Involved in audit, good audit presentation
D. Attitudes		
Reliability	Poor time keeping, attention to detail and continuing patient care	Conscientious time keeper and in respect of patient care
Self Motivation	Disorganised in work practices, needing constant organisation and does not seek opportunities to learn	Able to organise working routine without supervision, looks for opportunities to learn
Leadership	Gives mixed and/or unclear instructions. Others not willing to be guided by the trainee	Gives clear instructions, able to motivate others. Setting a good example. Guide actions or opinions of others
Team working	Does not work well in a team: works alone and/or for own goals rather than team ones	Good relationships with team members, strive towards a common goal by co-operation and combined effort
Administration	Did not carry out the range of administrative duties required of post. Production of documentation not timely.	Carried out routine administration promptly and well. Has a good grasp of hospital management
Record Keeping	Inaccurate, insufficient information recorded, not timely, illegible	Accurate, inclusive and timely

Relationships & Communication with:

a) Colleagues	Poor rapport with both senior and junior colleagues: difficult to work with	Willing to help even if personally inconvenienced, able to defuse problems in the surgical team. Easy to work with
b) Patients	Does not establish a rapport, increases patients anxieties	Caring attitude. Listens well, explains well and can allay patients fears
c) Other Staff	Does not treat staff with respect. Is not approachable	Sound and professional yet approachable. Treats others with respect and is respected in return

Communication skills:

a) Informed consent	Gives inadequate details of procedures, risks and or alternatives. Is inappropriately persuasive	Able to obtain full informed consent, risks and alternatives explained clearly and well
b) Bereavement	Appears Casual: unfeeling	Sympathetic: empathetic
c) Breaking bad news	Inadequate explanations given: appears unsympathetic	Does in a sympathetic manner

The form is to be signed by trainer and trainee on completion

Appendix III. Confidential – joint committee on higher surgical training training post assessment form (for completion by higher surgical trainees)

This is an official document. The original is the property of the JCHST. After completion it should be passed to the Training Programme Director/Chair of the Regional Training Committee who will collate and scrutinise all reports relating to the programme, before making them available to the Regional Postgraduate Dean. The Training Programme Director/Chair of the Regional Training Committee will retain copies, submitting originals to the JCHST Office at The Royal College of Surgeons of England, 35/43 Lincoln's Inn Fields, London WC2A 3PE for scrutiny by the SACs.

TRAINEE NAME: .... **DATE:** ....

HOSPITAL BEING ASSESSED: .... **DATE STARTED:** ....

REGION: .... **NTN/VTN/FTN or LAT** ....

CONSULTANTS: .... **FROM:** .... **TO:** ....

SPECIALTY: .... **SPECIAL INTEREST(S):** ....

CLINICAL TRAINING	**Deficient**	**Satisfactory**	**Good**	**Comments**
Out Patients				
Special Clinics				
Ward Rounds				
Surgical Meeting				
Audit				
Journal Review				
OPERATIVE TEACHING				
Adequate Opportunity to Operate				
Demonstration of Techniques				
Supervision in Theatre				
Communication / Rapport with Consultant				
RESEARCH				
Opportunity				
(Detail Sessions............)				
Encouragement				
CAREER ADVICE				
CLINICAL MANAGEMENT				
1) Did the consultants allow adequate responsibility for patient management?				
2) Did you have adequate support with Emergency cases?				
a) in theatre				
b) advice				
FEEDBACK				
Did the Consultant provide you with appropriate feedback of your performance?				

GENERAL				

1) Strengths of firm:

2) Weaknesses of firm:

3) Suggestions for improvement:

4) Did this placement fulfil your expectations?

Deficient	Satisfactory	Good

OVERALL RATING

TRAINEE'S TIMETABLE

	Monday	Tuesday	Wednesday	Thursday	Friday	Saturday
Timetable (current)						

On Call – ROTA Tiers e.g. HST
ADH's SHO
PRHO

COURSES & MEETINGS ATTENDED IN LAST 12 MONTHS

Title	Date	Location

Number of days Study Leave granted ....

Course/meeting fee paid? **YES/NO**

Incidental expenses paid? **YES/NO**

These forms are strictly confidential. Completion and return will greatly facilitate the Department's activities.

NOTES TO ACCOMPANY JCHST TRAINING POST ASSESSMENT FORM

1. Assessment Form is **CONFIDENTIAL** once completed, and must be handled accordingly.
2. The following guidelines are for trainees completing the form.
   a. Complete as fully as possible the post details at the top of the form.
   b. Complete assessment by placing an '**X**' in one box only against each criterion, with comments if desired. The following guidelines are offered for use in grading criteria.

	DEFICIENT	SATISFACTORY	GOOD
CLINICAL TRAINING			
Out patients	Do not see new patients. No time for / interest in discussion with consultant. Large number of patients. Poor organisation.	◄──────►	See new & old patients. Time for discussion with consultant. Reasonable time with patient. Well organised.
Special Clinics	As above. Do not learn / use any special investigations / techniques. Often work alone.	◄──────►	As above. Opportunity to learn special investigations / techniques. Often work with consultant. Multi disciplinary.
Ward Rounds	Rarely consultant led. Rapid decisions, little discussion. Junior views not listened to.	◄──────►	Usually consultant led. In-depth presentation / discussion of patients. Adequate time allowed.
Surgical Meetings	Poor consultant support. Badly attended. Rigid non-innovative programme. Not multi-disciplinary. Held outside normal working hours. Little input from consultants.	◄──────►	Consultant led. Well attended by all grades. Varied programme. Often multi-disciplinary. Regularly held in normal session time. Juniors encouraged to present / take part.
Audit	Morbidity / mortality only. No in-depth review of clinical practice / problems. Does not lead to change in clinical practice. Retrospective data. Juniors expected to collect all data. Non constructive / threatening atmosphere.	◄──────►	Proper audit cycle utilised. Leads to change in clinical practice. Prospective data collection. Juniors assisted with data collection. Friendly, non-confrontational atmosphere.
Journal Review	Juniors expected to do all reviewing. Poor consultant attendance. Didactic discussion?	◄──────►	Equal consultant / junior participation. Articles precied and discussed.
OPERATIVE TEACHING			
Opportunity	Usually left to do minor surgery. More than 5 elective sessions / week. Only assists and rarely performs more major cases.	◄──────►	Mix of Major & minor elective surgery. At least 3 elective sessions / week. Exposure to day surgery, and minimal invasive surgery.

Teaching	Works on own. Poor senior support. Not shown / taught new or more advanced techniques.	←→	Taken through procedures. Shares cases with consultant. Video teaching films. Anastomotic and new technique workshops / courses encouraged.
Supervision	Consultant rarely present in same or adjoining theatre. Own lists. Cannot readily summon senior assistance if in difficulty. No clear guidelines.	←→	Consultant usually present in same or adjoining theatre. Assistance at senior level readily available. Given clear guidelines as to when to call / inform / discuss with consultant.
RESEARCH			
Opportunity	No fixed time allowed. Any identified time often not taken due to other pressures. Clinical work precludes time for research.	←→	Fixed session / protected time allocated. Arrangements made to free trainee of some clinical work to allow research activity.
Encouragement	No interest shown by consultants. No ideas or stimulation.	←→	Able to discuss / plan ideas with consultants. Directed to appropriate sources for information / opportunities / funding.
CAREER ADVICE			
	Consultant not interested in trainee or his career.	←→	Consultant offers advice / help. Directs trainee to source of advice / help.
CLINICAL MANAGEMENT			
Patient Management	No guidelines. No trust. Consultant questions all decisions. Consultant does not back trainee.	←→	Consultant readily offers help / advice. Trainee given guidelines. Trusted to use own initiative / judgement. Consultant backs trainee.
Emergency Operating	Advice / help not easy to obtain. Consultant difficult to find / contact. Also not keen to come in / assist.	←→	Advice / help readily available. Consultant always happy to be phoned / consulted / give advice.
Feedback	Poor or absent appraisal. No specified protected time for discussion of trainee's performance. Consultant not frank about performance. Mainly critical. Rarely praises.	←→	Regular appraisals sessions in clearly specified time. Consultants open about strengths / weaknesses / areas for improvement.
GENERAL			
	No objectives. All clinical work. Poor education / learning.	←→	Clear objectives for trainee. Good balance / clinical / teaching / learning / research.

References

1. CanMeds 2000 Project. Skills for the new millennium: report of the social needs working group. Sept. 1996. Royal College of Physicians and Surgeons of Canada
2. Darzi A, Smith S, Taffinder N (1999) Assessing operative skill. *Brit Med J* 318: 888–889
3. Haase J, Musaeus P, Boisen E (2001) Virtual reality and habitats for learning microsurgical skills. (in press)
4. Lindsay K.W (2002) Neurosurgical training in the UK and Ireland: assessing progress and attainment. *Neurosurgery*. 50: 1103–1113
5. Martin JA, Regehr G, Reznick RK *et al* (1997) Objective structural assessment of technical skill (OSAT) for surgical residents. Br J Surg 84: 273–278
6. Merrick HW, Nowacek GA, Padgett B, Francis B, Gohara SF, Staren ED (2002) Ability of the objective structured clinical examination to differentiate surgical residents, medical students, and physician assistant students. *J Surg Res* 106: 319–322
7. Reznick RK, Regehr G, Martin J *et al* (1997) Testing technical skills via an innovative 'Bench Station' examination. *Am J Surg* 173: 226–230
8. Richards R (1999) Why do we need to assess competence? In *Surgical Competence*. Royal College of Surgeons of England, London, pp 6–15
9. Winckel CP, Reznick RK, Cohen R, Taylor B (1994) Reliability and construct validity of a structured technical skills assessment form. *Am J Surg* 167: 423–427.

Correspondence: K. W. Lindsay, Department of Neurosurgery, Southern General Hospital, Govan Road, Glasgow G51 4TF, UK. e-mail: kwlindsay@sgh.scot.nhs.uk

Acta Neurochir (2004) [Suppl] 90: 97–101

Printed in Austria

XII Research rotation in a trainee's curriculum

H.-J. Reulen[1] and **Y. Kanpolat**[2]

[1] Department of Neurosurgery, Ludwig-Maximilians-University, Klinikum Großhadern, Munich, Germany
[2] Yücel Kanpolat, Neurosurgical Department, Ankara University, Turkey

Introduction

All over the world, residency programmes for neurosurgery emphasize heavily technical issues. This training, more or less successful, teaches surgical techniques and procedures to neurosurgery trainees. Other issues, for instance training of administrative capabilities, despite their obvious importance in the life of a neurosurgeon, do not receive the merit they deserve. Research constitutes another important element in the training programme. For some, research is a matter of science to be conducted only by academicians in centers of quality. However, research, as a key component of occupational practice and indeed of life in general, should be stressed. This article aims at discussing the role and methods of application of research within neurosurgical practice.

The goals of a research rotation

What do we expect and what are the goals of a research rotation? Most of the academic programme directors or chairmen consider a research period an essential component of education, although most agree that not every resident needs such a rotation [4, 5, 7, 10, 12, 13, 14, 15, 16]. If we look at the background of our successful neurosurgical teachers, most of them had such an elective exposure to research. The goal of a well-organized research rotation is to gain an understanding of the intellectual processes involved in collecting, analyzing and interpreting clinical observations in a systematic way [1, 2, 8, 9]. Such experience will significantly improve the individual's rational approach to the management of patients, and the critical judgment of his own work and, for instance, the literature [10], when later he works outside a clinical environment.

George Ojeman in 1985 [10, 11] wrote about the role of research training in a neurosurgical residency: "The essential features of research are making systematic observations and organizing these into hypothesis and written documents. Often, but not always, this is done in a planned experiment. With this general definition, research training has a place in every neurosurgical residency programme regardless of the trainee's ultimate career goals and even if they do not continue with research, for opportunities to make new observations – of unique cases, the effects of therapy, pathophysiology of neurosurgical diseases, or the function of the nervous system- will occur in every neurosurgical career."

Additional goals could be added which may represent a second level. The resident should be introduced to the art of science in a way that he learns and experiences the real excitement of uncovering and describing new knowledge. Also he should come to a point where he is able to start his own research project, raise funds, and teach students and younger colleagues.

It is a matter of fact that in our training institutions there are residents who are planning an academic career, and others who plan to practise, although it must be stressed that many trainees at the early stage of their careers are unsure of their interests. With the above definitions a wide range of options can be offered to both groups, according to their specific interests.

Structures and categories of research

Different structures are used to organize research rotations. Some departments focus exclusively on re-

Table 1. *Categories of Research in Neurosurgery*

A.	Fundamental basic neurosciences (blue sky research)
B.	Laboratory studies simulating clinical diseases (experimental cerebral vasospasm, brain edema, experimental brain tumors, etc.)
C.	Applied clinical research (application of basic research techniques to a clinical problem)
D.	Research related to technical innovations and to improving surgical techniques (neuronavigation, functional mapping, laser-teachnique, instrumentation and biomechanics)
E.	Clinical observational studies (restrospective studies, case reports, prospective non-randomized studies, studies on natural course or outcome)
F.	Randomized prospective controlled trials

search training in a laboratory within the department under the guidance of neurosurgical faculty members experienced in research. Others prefer to send their trainees to a different institution where applied clinical research is performed, particularly if the department in question does not have appropriate conditions, and finally a third group of departments prefer rotation in an area of basic research in one of the basic neuroscience institutions. Thus, as shown in Table 1, a wide range of possibilities exists today, extending from fundamental basic research in neurosciences to patient-oriented studies and finally to randomized controlled trials (Table 1).

The important question is whether all these options do have the same rating in the context of neurosurgical training. Can we achieve the above-defined goals of research both with a rotation in basic research for instance tissue slices- or by participating in a prospective clinical trial?

Selection of a research area

There are certainly arguments that strongly support clinically oriented research. This is advocated by most of the experienced programme directors. A successful academic career depends on two columns: a) performance of competent clinical patient care and b) generation of high-level research [15]. Thus clinical practice and research projects regularly compete for the academic neurosurgeon's time, particularly if other tasks like administration, teaching, etc. have also to be performed. In contrast to colleagues in neurology or internal medicine, surgeons spend a large part of their day in the operating theater. The time remaining for research projects is always limited. Therefore when choosing a research area, there are definite advantages of combining research with the individual's field of clinical interest, for instance working on vasospasm if the main clinical interest is vascular neurosurgery. Thus the neurosurgeon may use questions posed on a ward round as subject for an adequate study, be it experimental or clinical. Such studies are likely to influence the neurosurgeon's future practice.

During the past 25 years a deep shift in the understanding and execution of neurosurgery has occurred. Between 1960 and 1970, it was still possible for one person to have an overview of more or less all technical procedures as well as the theoretical knowledge. Most of the scientific neurosurgical publications at that time were understandable and of interest to the majority of neurosurgeons.

In the last 10 to 15 years neurosurgeons have tried to increasingly concentrate on specific areas as for instance neurooncology, functional and stereotactic surgery, spinal neurosurgery, skull base surgery, posterior fossa surgery, pediatric neurosurgery, etc. and consequently in all those fields a more profound knowledge has built up involving new technical operative skills and procedures, and improvement of the technical tools, etc. This finally led to the development of subspecialty areas as seen today in many departments, a development which offers to patients a higher competence in a specific section and which cannot be reversed. In any of those subspecialties the literature has increased considerably so that today it has become impossible to comprehend or even obtain all the new information. Furthermore a growing inclination can be observed to plan and organize research projects in a way that the criteria for class I evidence are met [8, 15].

This change in paradigm coincided with a change in fields of research. There are now many more research areas than in the 60s or 70s. Each one of the subspecialty areas comprises many responsibilities, again ranging from basic research to randomized controlled clinical trials, as shown in Table 1.

As a consequence and taking into account the obvious tendency of our young colleagues to focus their clinical work on one of the various subspecialty areas of neurosurgery, the choice of the research topic will become even more important. Two examples are given below:

a) A resident in his/her third or fourth year of training received a fairly good overview of most clinical areas. The resident decides that his/her major interest is spinal neurosurgery, and, after finishing his/her regular training programme, wishes to acquire special knowledge in this field by entering a respective fellowship in spinal neurosurgery. It would be logical to choose as research rotation a project closely related to the resident's future work, in this case for instance spinal neuronavigation, spinal surgical anatomy, spinal biomechanics, or instrumentation, etc.

It would not be logical or even a waste of time, however, if this resident got involved for example in studies on hippocampus slices, in experimental studies on vasospasm, or in any other project remote from his/her clinical interest. Even with the best introduction into the art of science, the final result of such studies would not likely substantially influence his/her practice in spinal neurosurgery. It is most probable that at the end of such a research rotation, the resident would drop these studies. Such a wrong selection of research is one of the reasons why many young neurosurgeons after years of research and successful publications do not pursue their activities after having received their PhD or title of professor. This is obviously a waste, as such well educated colleagues would rather be qualified to instruct younger trainees. The interest of the trainees should be taken into consideration.

b) Another trainee at the end of his/her second or third year of training sees his/her future commitment in neurooncology and plans to add a neurooncological fellowship after having finished the regular residency programme. After several talks with the programme director the resident uses his/her elective time to work 6 months in an oncology department to learn the theoretical and practical principles of chemotherapy and then participates as the local investigator in a multicenter prospective randomized study on brain tumor therapy. There is no doubt that with appropriate supervision and support, this trainee will learn much from such an experience. With this knowledge the resident will be able to later extend his/her studies and organize projects independently. The authors are convinced that it is one of the important goals of training that at a certain point our trainees become independent and able to organize their own research projects. Such candidates will form the future cadres in academic neurosurgery and also represent the necessary links to the other neuroscience communities [8].

In planning the neurosurgical training programme, when should research begin?

Should it be done at the beginning or even before starting clinical practice, after a certain time of clinical exposure or rather towards the end of the programme?

Probably there is no definite and generally valid answer to this question since one will find individual success stories for all three situations. However, there are convincing arguments that a research rotation should be recommended for the majority of our trainees after at least one, but not later than three or four years of clinical practice in neurosurgery [11]. It seems wise to first observe the clinical abilities and conduct of a young trainee, and then, only in cooperation with him, choose a suitable scientific activity.

The other extreme would be to organize the research period at the end of the residency programme. At that stage, an intelligent resident would know his/her clinical interests and also have the desire to acquire the tools for a more systematic and scientific approach to some unsolved problems. However, this is exactly the time when surgical progress and success are greatest and the resident's interests are focused entirely on improving his/her surgical skills and techniques, and acquiring new surgical methods, etc. There will certainly be a collision of interests. Oftentimes this conflict is resolved based on economic restraints, and in the process some very gifted young colleagues are lost.

If the resident starts with a period of clinical neurosurgery, there are obvious advantages for both teacher and trainee. The arguments for the teacher are: The resident can be observed in his/her clinical and social performance, problem-solving skills, ability to perform under stress, manual dexterity, judgment, etc. The teacher can obtain an idea of his/her talents through observation and regular evaluation by the staff members. He can discuss with the resident his/her abilities and interests and find the best solution for this individual. It should be attempted to define a research area that motivates the trainee for a long time and thus lead to success! The candidate has time to look for a suitable research position in his own or in another department or in another institution. The chosen position should serve the interests of both the resident and those of the department, if one considers eventually a continuing activity in a research project. Hence selection of the topic and of the laboratory is a very important task! It is of no benefit to have the resident participating only as a technician in an ongoing project, he/she should be provided with the skill for independent research and the time to prepare funding of a research rotation, if necessary.

Vice versa, during this initial clinical period, the resident has the chance to examine whether or not his decision for neurosurgery as a career was the right one. He will find out how genuine his commitment to patient management really is and pursue training with more passionate intensity. He learns to deal with the unusual, complex case where he sees that routine alone is not sufficient and that search of literature and reading become necessary. He has the opportunity to observe the work of various specialists and to develop a specific interest, for instance in vascular or spinal problems, etc. With this background he can plan his research rotation more specifically, and may have an opportunity to participate in a research project.

The role of the programme director

From the previous discussion it results that the chairman, the programme director or an assigned tutor has high responsibility in advising the trainee in the above-described decision process and in preparing a research rotation. It is important that the trainee's abilities and interests are discussed with him/her at regular intervals. As soon as the trainee has developed a specific preference for one of the various subspecialty areas, a suitable research topic should be defined. Selection of the research project should be tailored to the individual trainee's clinical interests. The next step would be to choose a well-suited laboratory or another department to best serve the needs of the resident for his research rotation. The staff member in such a system certainly plays an important role in the career planning of the resident [4, 7, 11]. Such a procedure will enhance the success of a research rotation and chances that the trainee will later continue research as part of his career. Furthermore, the rotational proceeding will no doubt prefer clinically oriented research and reduce purely basic research, as defined previously.

Outlook

Today it is knowledge, the product of human intellect, that drives modern society. Knowledge in the form of scientific achievement, know-how and technology are the commodities worth the greatest value. Scientific achievement controls the industry, agriculture, health and the environment as well as all the other important lifelines of modern living [16].

Modern society will flourish from knowledge that is the product of scientific research. Research is not only individual scientific action carried out by small groups of scientists, on subjects that are of interest to their narrow involvement, but also a component of a general concerted scientific action that is launched for the purpose of creating a science-based, science-driven society, to serve sustainable economic competitiveness of nations and the future welfare of their people.

In this vein, the European Community has completed the fifth of such programmes, termed Framework Programmes (FP), focused on raising scientific research capability and quality of the European society. The sixth such programme was initiated in 2002 [3]. The 17.5 billion EURO FP6 will allocate as much as 10.5 billion EURO (about 62%) to the European Research Area (ERA) projects that consist of funding research in 7 fields considered of vital importance in building the future of the European society, based on science and technology [3]. Europe is another such programme that envisages building an information society allowing easy and wide access to information and automation of official applications and technicalities.

Conclusions

Research rotation is an important component in the education of a neurosurgical trainee.

Selection of research should be tailored to the individual trainee's clinical interest.

Avoid studies remote from the individual trainee's clinical interest.

Select a proper laboratory, institution or clinical research programme for the research rotation.

If possible utilize the advantages of cooperation with other institutions.

References

1. Baethmann A, Messmer K (1992) Stellung der klinischen Forschung in der Chirurgie: Konzept der Experimentellen Chirurgie. Chirurg BDC 31 [Suppl] 1: 29–33
2. Baethmann A, Meßmer K (1992) Experimentelle Chirurgie. In: Schweiberer L, Izebicki IR (eds) Akademische Chirurgie. Springer, Berlin Heidelberg New York Tokyo, pp 177–181
3. Commission of the European Union, Communication to the European Parliament concerning the common position of the Council about the 6th Framework Programme of the European Community for research, technological development and dem-

onstration activities contributing to the creation of the European Research Area (2002–2006) and to innovation – 30.01.02

4. Fahlbusch R (1997) From the scientific idea to its realisation. Principles and Strategies in Neurosurgery. Acta Neurochir (Wien) [Suppl] 69: 120–125
5. Hoff JT (1994) Toward better balance. The 1994 presidential address. J Neurosurg 81: 651–655
6. Kanpolat Y (2002) Significance of research, research and publishing in neurosurgery. In: Kanpolat Y (ed) Acta Neurochir (Wien) [Suppl] 83: 1–5
7. Langfitt Th W (1982) Research and training in the neurosurgical sciences, J Neurosurg 57: 733–738
8. Messmer K, Baethmann A (1997) Role of surgical research in the training of neurosurgeons. Acta Neurochir (Wien) [Suppl] 69: 111–115
9. Messmer K (1990) Perspektiven der Experimentellen Chirurgie. Chirurg 61: 248–250
10. Ojeman GA (1985) The role of research rotation in a neurosurgical residency. Neurosurgery 17: 138–139
11. Ojemann RG (1992) Training the neurosurgeon for the twenty-first century. Surg Neurol 37: 167–174
12. Pickard JD. (1995) What is neurosurgical research? Acta Neurochir (Wien) 133: 236–238
13. Reulen HJ, Olteanu-Nerbe V, Steiger HJ (1996) The neurosurgical clinic at the Ludwig-Maximilians-University in Munich. Neurosurgery 39: 1224–1231
14. Reulen HJ (1997) When should residents be exposed to research Acta Neurochir (Wien) [Suppl] 69: 106–110
15. Teasdale GM (1997) Research in Neurosurgical Training. Clinical reviews and trials. Acta Neurochir (Wien) [Suppl] 69: 116–119
16. Yoshimoto T, Tominaga T (1997) Contents and structure of a training rogramme. The Japanese proposal. Acta Neurochir (Wien) [Suppl] 69: 81–83

Correspondence: Prof. Dr. H.-J. Reulen, Schwojer Str. 19, 81249 Munich, Germany. e-mail: Ilona.Anders@med.uni-muenchen

Acta Neurochir (2004) [Suppl] 90: 103–106

Printed in Austria

XIII The accreditation of a training programme

H.-J. Reulen

Department of Neurosurgery, Ludwig-Maximilians-University, Klinikum Großhadern, Munich, Germany

General information

The U.E.M.S. has been active in the field of quality improvement of specialist training in the EU for years and has formulated general guidelines and criteria, published in the European Training Charter for Medical Specialists [1]. In accordance with these guidelines, the UEMS specialist sections develop specific recommendations to set and maintain standards of training, training quality and accreditation of training institutions for the respective speciality. Also they should monitor the contents and the quality of training. This aids harmonization of the training level in the EC. At a national level, training of medical specialists is regulated by the national authorities [1] who set standards in accordance with their national rules and EC legislation as well as in consideration of the UEMS/European Board recommendations.

An important feed-back instrumentation for quality improvement is the visitation and evaluation of training centers, which in future may be coupled with certification and recertification of trainers. At present an external evaluation of the quality of training exists in only a few of the countries of the EU, and the national approaches differ considerably. Therefore, in the UEMS, the need was felt for a harmonisation and encouragement in the field of visitation of training centers. The new *European Charter on Visitation of Training Centers*, 2002 [3] presents the outlines as to organize a national programme for visitation of training centers. The following definitions are made by the UEMS.

Statutory visitation

The responsible national authority is recommended to establish programmes for this purpose as far as these have not been developed yet. Such programmes are increasingly required and tend to become obligatory, as is already the case in several member states of the EU.

Voluntary visitation

Training centers are encouraged to participate in voluntary visitation programmes that award additional quality titles. The UEMS European Boards are active in this field [3].

In Neurosurgery the EANS and the Section of Neurosurgery of the UEMS have entrusted the *Joint Residency Advisory and Accreditation Committee (JRAAC)* with the above task, i.e. the preparatory work for the visitation of European neurosurgical training centres. The visitation programme is now being offered on a voluntary basis for a European recognition of quality of teaching and training in Neurosurgery by the JRAAC. Training centers in Europe are encouraged to participate in such a voluntary visitation programme. Since this European visitation programme follows the highest standards, it should be secured that in the future an accreditation issued by JRAAC as an UEMS/EANS body is recognized by the national authorities.

In the following it is described how the European Accreditation of a Training Programme is organized.

The application

a) The Programme Director must submit the *Programme Application Form* to the Secretary of the JRAAC to arrive sufficiently in advance of a Committee Meeting to allow for proper reviewing. In future a second questionnaire will have to be completed by a representative of the trainees. Applications will be evaluated within the subsequent 6 months. The Programme Application Form can be obtained from the Chairman or the Secretary of JRAAC (presently Prof. T. Trojanowski or Prof. M. Zerah, secretary) or be downloaded from the website of the EANS (www.eans.org/→committees →accreditation form).

b) The Application Form has to be completed in detail by the applying department, describing the personnel, space, technical facilities, clinical experience available, and in particular their Residency Training Programme. The data are necessary for the Evaluation Committee to assess the status of the Residency Programme. The completed Application Form should be sent to one of the addresses given below (see Important addresses).
c) The completed Programme Application Form will be reviewed by at least two members of the JRAAC. The reviewers may request additional material directly from the Programme Director and shall prepare a proposal for the next JRAAC committee meeting.

The site visit

a) If the Training Programme according to the Application Form is in essential compliance with the requirements, the next step will be a *Site Visit* of the applying institution, conducted by two independent visitors (experienced chairpersons) nominated by the JRAAC, and a third national observer/visitor appointed by the applying institution. One member will act as Chairman, another as secretary. The national visitor acts as an observer, may help explain national requirements and peculiarities and should be an experienced member of the respective National Neurosurgical Society. The date of the Site Visit will be arranged between the Programme Director and the site visitors as soon as possible. The Programme Director will receive the necessary information to prepare the time table of the site visit in due time. The site visit will be performed according to the Guidelines of the UEMS Charter on Visitation of Training Centers (2002) [3]. The Site Visit preferably lasts 1 day and serves to see the hospital(s) and department(s) involved in the Training Programme and the educational and scientific environment, holding personal discussions with the Programme Director, the Chairman, the teachers, the trainees, and a representative of the administration. A team discussion with the trainees may be helpful as well. Information given by the trainees must remain confidential. Concluding the visit, a debate with the teaching staff should take place. The time table should include a 30 minute section for the visiting team to formulate its conclusions, conditions and recommendations. Details can be added later when a report is compiled. The center being visited is responsible for the expenses of travel and local accommodation of the site visitors.

   A *report* will be prepared by the site visitors for the subsequent meeting of the JRAAC. The training center should be granted an inspection of the draft of the report to correct any factual errors. The final report will be part of the final decision on the accreditation status of the Programme. Confidentiality regarding the contents of the site visitors' report must strictly be adhered to by the site visitors. All information obtained during the interviews with trainers and trainees remain absolutely confidential.
b) The *accreditation status* as decided by the JRAAC will be reported to the respective Programme Director by formal Letter of Notification. The letter will be forwarded by the Committee Chairman within a reasonable period of time following the committee meeting. On request of the Programme Director, the Dean of the Faculty and/or authorities of the hospital also may receive the official report. Together with the report additional advice and recommendations if necessary will be given to further improve the Training Programme.

The accreditation

The following decisions may be taken by the JRAAC with regard to the accreditation status of a Training Institution and Programme.

Full accreditation (accreditation at full status) may be granted by the JRAAC if the programme has demonstrated to be in full compliance with the Essentials of the European Training Charter [1], the Essentials to Organize a Residency Training Programme [2], and the Charter on Visitation of Training Centers [3]. The Department that has been fully accredited will receive a certificate indicating that the Department and the

Residency Programme fulfil the criteria to meet European Standards of Excellence for Education in Neurosurgery. The certificate is issued by the Chairman and the Secretary of the JRAAC and co-signed by the President of the Section of Neurosurgery of the U.E.M.S. as well as the President of EANS. The accreditation shall be re-assessed after 5 years or within one year after change of a Programme Director.

Provisional accreditation indicates that the programme is basically in line (not in compliance) with the above cited requirements and standards. It is considered as being in a status of development. The Programme Director will be requested to submit a so-called *Progress Report* within one or two years after notification. The Committee shall exactly specify the information to be provided. When a Progress Report is asked for, a specific due date should be included in the request.

Accreditation may be withheld if the programme does not substantially comply with the essentials of the above defined requirements and standards. The Committee will cite those areas in which the reviewed programme does not comply with the standards. A new application can be submitted when the areas indicated are brought in compliance with the requirements and standards.

Accreditation may be discontinued if a programme for some reason is no longer in compliance with the requirements and standards defined above. A new application can be submitted if the requirements are again fulfilled.

Reassessment of a Programme is usually done after 5 years or within one year after change of a Programme Director. For re-assessment the Application Form must be supplemented with the desired data of the calendar year before.

Confidentiality

Information as well as documents acquired during the accreditation process (application form, site visit report) as well as the decisions of the JRAAC remain strictly confidential. The training center that has been visited is entitled to make public the visitation report and the certification. JRAAC will publish the following information about accredited programmes in Acta Neurochirurgica: name and address of the training center, name of the programme director.

Commentary

Some comments shall be made on the present status of the external evaluation and accreditation process.

Experiences gathered so far by the members of JRAAC with regard to the evaluation of a number of departments have shown some interesting aspects: In various European countries individual departments, national neurosurgical societies, or national authorities to some extent have previously developed some form of local or national log-book, an educational programme, etc.; i.e. they established their own requirements and standards for the training of their residents, set up of course in their national language. This made judgement sometimes very difficult for the Review Committee members and thus the role of the national observer during the site visit became very important in explaining the respective background or translating some of the important written material. This diversity at present is typical for the European situation and is taken into account by the JRAAC when making their decision concerning the accreditation status. On the other side it has been observed – although the training programmes of the visited departments generally were on a good or high level – that several "deficits" existed in most of the training centers regarding the formal organisation of a programme in general, the periodic internal progress assessment, the involvement of the resident in all areas of our specialty, and in particular the rotations. The Programme Directors of the visited departments, therefore, received together with the Accreditation of the Training Programme a detailed report specifying those areas where the Review Committee thought were deficits. In addition, a *Progress Report* was asked for after one or two years to show how such deficiencies have been improved.

The Charter on Training of Specialists in the EC and particularly the requirements and standards defined by the UEMS Section of Neurosurgery have been in existence for a few years only and therefore "full compliance" with all the requirements at this time cannot be expected. The new edition of the Charter on Training is being written in summer and fall of 2003 and will probably be available in 2004. It defines more precisely the requirements for a training center as well as the contents of a training programme and thus facilitates their implementation in practice. The national societies or the responsible authorities in a country will now have the responsibility to adapt their requirements and criteria to the UEMS/European Board recommendations. This certainly will lead to a further harmonization of the training in our speciality. At present we still have the great chance of being able to define our standards ourselves, but we have to seize this unique opportunity.

The staff groups of the centers visited so far all expressed that the whole evaluation process, the preparation of the Application Form, the site visit, and the report represented a very positive experience. During the compilation of the extensive material for the Application Form, a new insight into the organisation, the working processes, the strengths and weaknesses etc. of the department could be gained. During the evaluation process and particularly the site visit, they receive friendly advice and support from a group of experienced colleagues in order to complete or upgrade their programmes to be in compliance with the new European standards. Also the trainees regarded the exchange with the site visitors as helpful and they were proud to pass their training in an institution with a quality mark of the EANS and UEMS. With very high probability, intelligent and gifted young colleagues who want to apply for a training position will in future give their preference to institutions with such a quality mark.

Proposal of a time table for visitation

There should be a preliminary meeting (working dinner) with the Chairman, staff members, and the visitation team the evening before the visitation.

The general frame of the department, its collaboration with neighbouring specialities, advantages and deficits can be discussed. Also a general outline of the training programme or specific national requirements may be given.

These are the main sites of interest for the site visitors. Others may be given in the letter to the Programme Director.

Time table (Example)

.... morning conference
.... conference with all staff members
visit of the department
- operating theatre
- intensive care
- outpatient department
- general ward
- neuroradiology/imaging, endovascular techniques,
- radiotherapy or radiosurgery as available
- library and computer facilities
- research facilities
- records

.... interview with trainees (individual and confidential)
.... (the trainees should have their Log-Book available)
.... interview with Medical Director and/or Administrative Director
.... discussion with Chairman and/or Programme Director summary
.... concluding session for the visiting team (30 mins)

References

1. European Training Charter (Charter on Training of Medical Specialists in the EC) U.E.M.S., Secretariat Avenue de al Couronne 20, B – Bruxelles, 1995 – www.uems.be/5charte.htm
2. "Essentials to organize a Residency Training Programme". (1998) Acta Neurochir (Wien) 141: 731–733
3. Charter on Visitation of Training Centres (U.E.M.S., 24 October 1997) – www.uems.be/visitati.htm

Correspondence: Prof. Dr. H.-J. Reulen, Schwojer Str. 19, 81249 Munich, Germany. e-mail: Ilona.Anders@med.uni-muenchen.de

Acta Neurochir (2004) [Suppl] 90: 107–114

Printed in Austria

XIV The European Examination – its present status and potential development

J. Haase

Department of Health Science and Technology, Aalborg University, Aalborg, Denmark

Too much knowledge leads to over activity;
Better to calm the mind.
The more you consider, the greater the loss;
Better to unify the mind
Excessive thinking weakens the will.
The more you know, the more your mind is confused;
A confused mind gives rise to vexation.
Merely reading books is of no lasting value
(Wang Ming)

Historical background

The diversity among the over 30 European countries sending trainees to the EANS training courses is immense. This is based on both cultural factors and differences in priorities of teaching in the different countries. The explosion of medical knowledge poses an additional challenge to all those responsible in setting priorities for what is thought to be the core knowledge in a given training programme. With the concept of developing European standards in neurosurgical training it was evident that an evaluation of both trainers and trainees was necessary. In the US, already in 1940 the first American Board of Neurological Surgeons (ABNS) examination took place. In the UK, the first Intercollegiate Specialty Board Examination was held in 1991, consisting of a multiple-choice questionnaire (MCQ), a clinical examination and 3 vivas [13]. The objective of such examinations is to assure a certain level of theoretical and practical knowledge and to establish safety. While the majority of chairmen believe that such a set of exams is the best available method for probing clinical judgment and knowledge regarding the management of patients, others express a more critical view as for instance stated by John Picard: "Certainly my personal impression, unsubstantiated by any official statistics, is that the level of excellence in the examination does not correlate with subsequent progress, academic innovation, or medical legal complications [13]."

Within the EANS, Mario Brock and Reinder Braakman [5] stimulated the creation of a committee with the aim to develop a European examination of a similar standard as those described above. In 1991, this committee published the recommendations and plans: "Towards a European Board of Neurological Surgery" [5] which was accepted by the EANS Administrative Council. The Examination Committee of the EANS was then founded, chaired by Reinder Braakman. He and subsequently Hans-J. Reulen and Jens Haase were invited guests of the American Board of Neurological Surgeons (ABNS) both at their multiple choice exam (MCQ) and at the oral examinations, and these visits implemented important information in our own EANS setup. Instead of long fruitless discussions on the validity of testing, the European Exam was finally constructed using established features of the ABNS – trying to give all trainees a platform of minimal requirements for being a practicing neurosurgeon in Europe.

The exam was designed with a primary Multiple Choice exam (MCQ) on different main topics and a final oral exam after a trainee had received his/her authorization as a neurosurgeon and having been practicing for 2 years

The updated **guidelines** of the European Exam can be found in addendum 1.

Primary Multiple Choice Examination (MCQ)

In 1992 the first MCQ exam was held with 57 trainees and a passing rate of 33%. After the initial testing of the system, the MCQ exam in 1994 in Turkey attracted 64 participants, the largest number ever! The drop of participants to 34 in 1996 and later down to a mere 12 was unacceptable with regard to the basic ideas of the exam. This led to a critical re-evaluation within the Examination Committee and the Administrative Council (AC) of the EANS. A breakthrough occurred when the Swiss and later the Swedish neurosurgical societies decided to have the MCQ as an integral part of their national specialization in neurosurgery. It was suggested in the AC that this should be extended to all European nations. Finally in 2001, the AC of the EANS decided to make the MCQ an integral part of the EANS Courses with the given consent of the trainees. Since then all 4th and 5th year trainees are taking the MCQ during the last course of their cycle at no cost.

The Examination Committee selects from a pool of questions in various categories around 180 questions for each exam. The pool consists of new and old questions kept strictly confidential.

The questions are prepared by the appointed committee according to the following guidelines: use of simple, "European English" language, with 5 possible answers of which one is correct, possible answers always in alphabetical order.

An example is shown in Table 1.

Table 1.

x) Plateau waves (type A waves of Lundberg)
a: are related to periodic breathing
b: have a low amplitude
c: have no clinical significance
*d: indicate reduced craniospinal compliance
e: usually last less than one minute

The answers are approximately of the same length. The only correct answer is marked with *.

Questions where e.g. 2 out of 5 answers may be correct, as used elsewhere, have not been included so far.

Another set of questions was later introduced being of a yes/no type, an example of which is given in Table 2.

Table 2.

y) The formation rate of cerebrospinal fluid in congenital hydrocephalus is normal
*a: true
b: false

The correct answer is also marked with *.

All previous and newly structured questions are evaluated by an MCQ expert to secure answer quality statistically. A question should not be too simple and provide nearly 100% correct answers, nor should it allow more than one correct answer. Such answers are removed from the question pool.

The exam is taken during a 3-hour-period in two parts, A and B with a short intermission. The trainees cannot remove the original question sheets from the examination room. We have emphasized that cheating during the exam will result in direct demission but have never seen any example of this. Each trainee receives his/her personal results and evaluations together with the average result of the whole group in order to visualize weak points.

An example of the gross evaluation of the 2002 exam is shown in Table 3.

Table 3 presents an overview of the primary exam of August 2002. In part A of the exam 5 questions were found to be invalid and therefore were removed from

Table 3. *European Association of Neurosurgical Societies. European Primary Examination, August, 2002*

Final Result	
Total Examination Valid Questions = 174 of 180 Participants = 46 Sum of Scores = 4693 Mean Score = 102.02 (58.63%) Standard deviation = 18.56	
Neuroanatomy	(17 Questions): 10.76 (63.30%)
Neuropathophysiology	(20 Questions): 11.61 (58.04%)
Neurology	(14 Questions): 8.28 (59.16%)
General skills	(10 Questions): 5.50 (55.00%)
Head Injury	(7 Questions): 3.48 (49.69%)
Brain Tumours	(29 Questions): 17.26 (59.52%)
Cerebrovascular Disease	(30 Questions): 16.72 (55.72%)
Peripheral Nerves	(10 Questions): 6.07 (60.65%)
Spine and Cord	(24 Questions): 13.57 (56.52%)
Neuroradiology	(3 Questions): 2.63 (87.68%)
Pediatric and Congential	(10 Questions): 6.15 (61.52%)
Part A Valid Questions = 90 of 95 Participants = 46 Sum of Scores = 2457 Mean Score = 53.41 (59.35%) Standard deviation = 11.54	
Part B Valid Questions = 84 of 85 Participants = 46 Sum of Scores = 2236 Mean Score = 48.61 (57.87%) Standard deviation = 7.95	

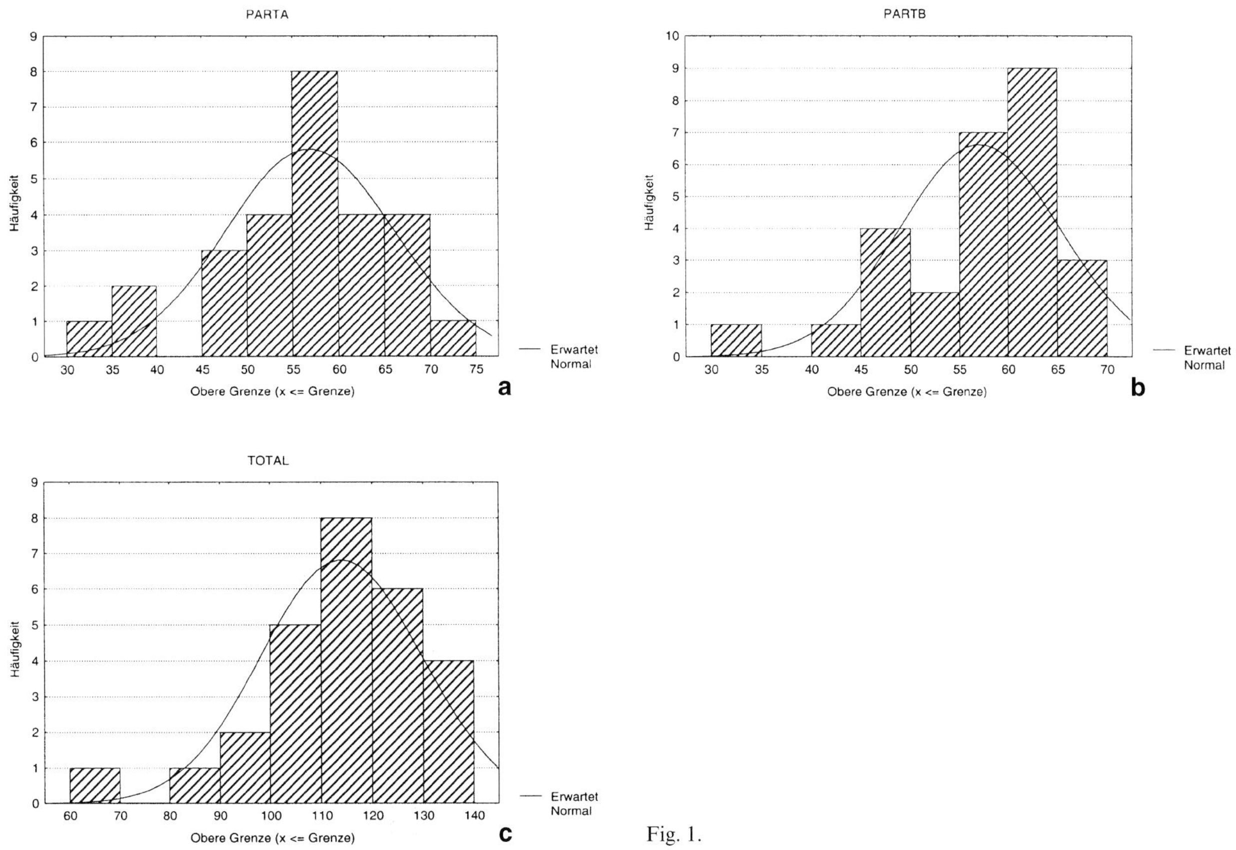

Fig. 1.

the final statistical evaluation. The 46 trainees had a total score of 2457 with a mean score of 53,41 reached by 59,35% of the trainees. Standard deviation from the mean score was 11.54. For part B one question was invalid and removed, mean score was lower (but also lesser number of questions) and was reached by 57.87% of the trainees. Thus there was no difference between the two parts of the exam. Out of 174 questions 6 (3%) were thus removed. For comparison at the ABNS exam of 1994, 23 out of 517 (4%) questions were removed before the final evaluation.

If, for instance, the mean score of part B had been significantly lower, e.g. reached by only 40% of the trainees, we would have evaluated the reasons for this and possibly found that some questions had been inadequately formulated, badly structured or too difficult to answer.

In the ABNS exam 147 out of 220 (67%) US citizens passed the MCQ. Foreign trainees who participate in the ABNS usually have much lower passing scores. As a control group last year, US medical students were tested with the same questions and they usually had a passing rate of 90%!!!

The passing score in our exam is decided upon each year after consultations in the Examination Committee. A fixed, yearly predetermined passing score of e.g. 67% is thus not used, nor do we use a control group. If we look upon a histogram of an exam, the distribution of trainees' results influences the decision (Fig. 1a).

Results of part A. (Number of valid questions were 90; only one candidate had between 70 and 75 correct answers, as seen at right of the abscissa).

Similarly, results of part B are shown in Fig. 1b. 84 valid questions, (3 candidates had between 65 and 70 correct answers).

Results of the whole exam are presented in Figure 1c.

Total number of questions was 174. Four candidates had between 130 and140 correct answers.

The Fig. also demonstrates that the one candidate with a very low score does not statistically influence the fit.

The scores are subjected to the normal Gaussian distribution to test whether distribution is statistically OK. From that we should decide the passing score. Each group of questions is scrutinized carefully.

In conclusion, we try to be fair to the candidates, use simple English words and have an external statistical expert to test the validity of our questions and results. From that evaluation a level for passing score of the exam is determined by the Examination Committee, usually with the mean score −2SD as the cut-off but always with a minimum of 60% correct answers as an absolute minimal requirement.

Secondary oral examination

The oral exam has never become the success we anticipated in the beginning, and in one year – 2001 – it had to be cancelled due to lack of participants. The first oral exam took place in Brussels in 1994 and exams are still carried out at that site. In the following it is important to notice that "EU-countries" and "European countries" are not synonymous. Potential candidates for our Oral Exam must have passed the primary MCQ exam, have a license to practice neurosurgery in *a European country*, must have had one year of practice in neurosurgery as an officially licensed neurosurgeon in *any European country*, and present a log book of operations performed independently after license has been granted. Initially they had to pay a fee of 1.200 DM, later 800 DM and presently 100 Euros [5]. The candidates are tested for 3 × 55 minutes on topics like brain and spinal diseases, peripheral nerves and other issues. Two examiners per session examine them consecutively. If insufficient performance becomes evident on some specific issues during the two first sessions, these topics are covered again in the last part of the exam by two new examiners. We use the American grading system (A, A−, B+, B, B−, C+, C, C−, D and E), "A" being an extraordinary result, "B" the standard, "C" not totally satisfactory and "D" definitely not acceptable for a professional function. If a candidate is given an "E", this candidate has failed the whole exam.

Having passed the oral exam, the candidate receives a **European Certificate** that has no "financial" or "accreditation" value as a license, but can be considered a true award. The best candidate of each year is awarded the "**Braakman Price of Excellence**", a book gift and a certificate.

The number of participants and pass/no pass are found in Table 4.

The total passing rate so far is 75.4%, thus being much higher than the passing rate of the multiple-choice exam.

Table 4. *Results of the oral examinations*

	Number candidates	Passed
1994	12	11
1995	7	6
1996	7	4
1997	6	6
1998	7	5
1999	7	5
2000	6	3
2001	0	0
2002	5	3

All details of the exams including the general rules are now published on the EANS Homepage *www.eans.org* [7]. Applications for exams and notification of results is now organized by the EANS Secretariat, in contrast to previously when the whole work was carried out by the members of the EANS Examination Committee.

Discussion

The problem we face is that the oral examination, in contrast to the written exam, costs money and does not give any legal rights to the candidate who passes, unlike the situation in the UK or in the US. The costs have recently been reduced significantly. Furthermore a candidate, already well integrated in professional life, may risk failing the exam and thus being personally and publicly humiliated [11]. Many potential candidates in the past have used these two reasons for not participating. Another important point is that most European countries today have a compulsory National Board Exam the trainees have to pass at the end of their training. The great disadvantage is that the EANS European oral exam – at least at present – is not equivalent to the National Board Exams and cannot "replace it" – so why take it?

One strategic goal of the EU is to provide a free exchange of persons and services, also within the medical services, as specified in the Directive 93/16 EEC of April 5, 1993. To meet these goals a harmonization of the training of medical specialists, in combination with quality assurance according to given standards will become mandatory. Considering the existing diversities among the countries of the EU in providing quality assurance for recognition as a specialist, a definition of standards with agreed levels of knowledge, practical skills and judgment is urgently needed. The

Table 5. *European Primary Examination September 1997*

Individual results in percent

Participant no.	Result of Part A	Part B	Total
0001-3	62	59	121
0002-6	67	49	116
0003-9	76	64	140
0004-2	65	53	118
0005-5	67	50	117
0006-8	52	43	95
0007-1	61	53	114
0008-4	41	35	76
0009-7	44	42	86
0010-7	46	55	101
0011-0	54	50	104
0012-3	68	55	123
0013-6	63	51	114
0014-9	59	42	101
0015-2	64	58	122
0016-5	49	52	101
0017-8	56	44	100
0018-1	50	45	95
0019-4	64	53	117
0020-4	38	38	76
0021-7	48	38	86
0022-0	43	36	79
0023-3	64	56	120
0024-6	64	57	121

recent enlargement of the EU by additional 10 countries stresses the importance of such a demand. The European Examination must therefore become the accepted and agreed standard which can either be used by the national societies or used as a model when national societies or authorities prefer to organize this on a national level. A critical discussion of the actual European Examination therefore is needed to eventually secure the best possible solution.

Critical remarks to the present multiple choice question exam and the passing rate

As mentioned, the pass/fail score for the primary exam is determined by the Examination Committee, using statistical comparisons. Sixty percent of the correctly answered questions are the minimum number to pass (Table 5). For comparison, in the USA, 64% of correctly answered questions are needed and this percentage is remarkably consistent each year [4]. In the UK, the respective figure is between 64–69%. Despite the slightly higher pass score in the USA and the UK, a

Table 6. *Results of trainee NN ("John Doe"), who failed the exam*

Result of John Doe	DID not PASS	35.1% correct answers
	Your result	Whole group
Total	35.1%	58.6%
Neuroanatomy	41.2%	63.3%
Neuropathophysiology	30.0%	58.0%
Neurology	14.3%	59.2%
General skills	50.0%	55.0%
Head injury	14.3%	49.7%
Brain tumours	48.3%	59.5%
Cerebrovascular disease	36.7%	55.7%
Peripheral nerves	50.0%	60.7%
Spine and cord	25.0%	56.5%
Neuroradiology	100.0%	87.7%
Paediatric and congenital	10.0%	61.5%

higher percentage of trainees pass the exam that means that they in general reach a higher average score as compared to the EU trainees. This may indicate that EU candidates are either less used to this type of exam, have more difficult questions, or have less comprehensive knowledge. This needs to be examined in more detail.

An important factor is the feedback if a resident fails the exam. In Table 6, the results of a candidate NN ("John Doe") are presented.

It is evident that this candidate had major problems in neurology, head injury and paediatrics (congenital disorders) and also remained below the average level of the whole group in most of the other fields. What can he learn from his results? Certainly that his efforts to acquire adequate theoretical ad hoc knowledge did not reach the anticipated levels.

In addition, the demonstration of his weakness in most of the main fields of our speciality may entice him to actively continue studying these topics specifically. He should recognize that his professional performance might be significantly hampered by lack of theoretical knowledge.

A feedback so far does not exist. In future a member of the examination committee or even better his local tutor, after revising the results of table 6, should discuss the results with this candidate and give him/her adequate support. It is therefore important to provide the Programme Director with the performance of his trainees in the individual test categories. Only this allows the Programme Director to take respective measures and adjust the curriculum, if needed.

Issues to be discussed in the future

Recognition or certification as a neurosurgeon should mark the successful completion of a well-structured training curriculum designed to train safe and competent clinical neurosurgeons [8]. The recognition or certification process is a very responsible decision and should be made as objectively as possible. The process itself and its quality have to be scrutinized from time to time! Is the present European examination – which is more or less a copy of the ABNS and the UK examination – adequate or can it be improved in certain aspects?

The written MCQ examination is designed as to test cognitive knowledge about topics related with neurosurgery. Questions include neuroanatomy, neurobiology, neuropathology, neurophysiology, neuroradiology, theoretical clinical skills, intensive care and all categories of clinical neurosurgery. The validity of the individual questions of the various subjects and of the entire examination can be tested and allows standardisation, so that each year's performance can be compared to the previous exams. In the written exam the assessment itself is as objective as possible. The real weakness is the selection of questions, whether they really represent the total amount of knowledge required for a neurosurgeon to make his surgical decisions as safe and correct as possible. How much ad hoc information is needed for instance in anatomy to perform operative procedures in the CNS or the peripheral nerves? How much of this anatomical knowledge must be immediately available in our memory? The question is easily answered for routine and standardised procedures like a lumbar disc or the removal of an acoustic neurinoma. For rare approaches is it appropriate to rely on memory or information systems, books, Internet can be used to update knowledge (clinically relevant knowledge). The MCQ can by no means tell neither the trainee nor the trainer whether the trainee will be a successful neurosurgeon.

The oral examination is thought to be a test of judgement and knowledge (how to use information) regarding the management of cases, including complications, outcome, etc. Typically case scenarios are presented and the candidates with appropriate scans, MRI s, etc. explain how they would manage the case and plan and perform an operation, if indicated. Complications and outcomes are discussed. By having 3 different sessions with 2 examiners each, examiner bias is reduced to a certain extent.

The oral exam in its present form cannot be considered an objective and standardised test because examiners often use personal case material of different difficulty, and this probably varies from examination to examination. Also individual examiners differ in their evaluation of the candidate's performance [8, 13].

Standard scenarios include subarachnoidal haemorrhage-, severe brain trauma-, brain tumours-, intensive care- etc.

In order to reduce this potential bias, more work needs to be done to validate (format) these case scenarios, weigh their difficulties and make the examiner's rating of the candidates' performance more objective. Even taking into account all these potential bias, the oral exam is still the best available technique to assess judgment and clinical performance.

The oral exam is presently offered 2 years after being recognised as a neurosurgeon. Experience so far has shown that this is not acceptable to the majority of candidates for different reasons, as discussed before. The Examination Committee therefore should consider to offer the oral exam at the end of training, which would make it part of a board examination and allow to accept both examinations as equivalent/ supplement to a national examination within the EU.

Some additional instruments may be developed in the future to judge other relevant qualities of a candidate, such as operative skills on the one side and "social intelligence" on the other. The assessment of operative skills may be improved with the recent introduction of minimum and optimum figures for operative procedures required at the end of the 6-year training programme, and particularly by introducing competence levels for each procedure which have to be confirmed by the Programme Director (see "A structured Neurosurgical Training Plan"/K. W. Lindsay and "The new Neurosurgical Training Charter"/ J. Steers, H. J. Reulen, K. W. Lindsay in this book).

This would give the Training Programme Director, who in fact is responsible for training the individual trainee, more responsibility when signing personally the level of competence acquired for each of the demanded operative procedures.

"Social intelligence" is a term comprising different qualities and personal characteristics, which are difficult to measure. Probably the best way is to review (study) the trainees' portfolio with his accomplishments and particularly the assessment forms for each training period signed by the trainers. In the USA, endorsement letters of practicing neurosurgeons in the community where the candidate works are required.

Conclusions

Efforts must be undertaken to convince national societies and authorities that in future a common European standard is mandatory for recognition (accreditation) of a neurosurgeon. The present European Examination can certainly be improved and become a model for such examinations.

Addendum 1. European examination in neurosurgery (Guidelines)

The European Examination in Neurosurgery consists of two parts:

The Primary Examination and the Oral Examination.

Both examinations may be taken on a voluntary basis and are prepared by an Examination Committee consisting of 12 program directors in Neurosurgery from various parts of Europe.

Primary examination

The **Primary examination** is a written, multiple-choice question examination (MCQ) consisting of 150–200 questions covering neurosurgery, neuroanatomy, neuropathophysiology, neuropathology, neurology, neuroradiology, fundamental clinical skills and other disciplines deemed suitable and important by the Committee.

These multiple choice questions have to be answered within 3–4 hours.

The Examination is prepared by the Examination Committee with the assistance of a professional institute for the design and analysis of examinations in medicine. This institute also carries out the analysis.

The examination can be taken once a year in August/September in two separate cities in Europe, either the same city where the annual European Course of Neurosurgery takes place or alternatively Uppsala, Sweden, provided there is a sufficient number of applicants.

The primary examination is open to all residents in accredited neurosurgical programs in Europe, and to all neurosurgeons with a license to practise neurosurgery in a European country.

It is advised that residents should not take the examination before their 4th or 5th year of training, otherwise the examination is generally considered to be too difficult. The level of the examination is intended to be comparable to the level of the Primary Examination of the American Board of Neurosurgery.

Participants agree that they are bound by the Committee's rulings regarding credentials and examination scores.

Candidates, fail or pass, may sit the examination as often as desired either for certification or self-assessment.

Fees are payable in advance and determined by the examination committee. They may vary at the discretion of the committee according to the costs of preparation and analysis.

The completed application form must be filed with the Examination Committee by middle of June.

The candidate must indicate on the form whether the examination is taken for credit toward certification or for self-assessment.

Oral examination

The Oral Examination is the second and final step of the European Examination. A pass at the oral examination is the only way to obtain the European Certificate in Neurosurgery.

Requirements to participate in this examination are:

- Evidence that the Primary European Examination in Neurosurgery has been passed.
- License to practise neurosurgery in any European country. A copy of this license must be submitted for inspection.
- At least one year of practice in neurosurgery as an officially licensed neurosurgeon in any European country.
- A logbook of operations performed independently during a consecutive period of 12 months after the license has been granted.
- Advance payment of the fee payable in Euro as specified on the application form. (approx 100 Euro).

The Examination Committee makes the final decision regarding a candidate's eligibility for the Oral Examination and potential certification after considering all available information pertaining to the entire process of certification.

The examination in the English language consists of three parts, each lasting one hour:

The first hour is dedicated to an oral examination by two examiners on operative neurosurgery of brain and skull.

In the second hour operative neurosurgery of spine, cord and peripheral nerves is covered.

In the third hour topics will be discussed that could not adequately be covered in the first two hours.

The examination is a clinical problem solving and patient management test. It is not a theoretical examination such as the Primary Examination. Case histories are given, and – where appropriate – X-rays, scans, MRI's and other visual aids are shown to augment the presentation and development of cases. Candidates explain verbally how they would proceed to evaluate or manage the cases and to plan and perform the proposed operations, if indicated. 3–6 cases will be discussed per hour.

Each of the three hours is conducted in an interview setting with two examiners, experienced neurosurgeons from a European country.

During these three hours the candidate will, therefore, meet six different European examiners, each of whom will give an independent score.

The final, combined score will be made available by mail within 14 days after the examination.

The best candidate will be awarded the Braakman Diploma of Excellence.

The candidates who pass this examination will receive **The European certificate in neurosurgery**.

This is a document that certifies that the owner has a good level of theoretical and practical knowledge in neurosurgery. It marks a successful completion of standard neurosurgical training.

It is not a license to practise neurosurgery in any European country. The European Certificate – at least at the moment – does not replace any national license. It is an award rather than a license.

If a candidate fails the examination once, a minimum of one year but not more than 3 years should elapse before taking the exam for the second time. A re-examination fee must be paid prior to the examination.

A candidate who has failed the Oral Examination twice must take and pass the Primary Examination again before reentering the certification process and retaking the Oral Examination. An additional year of practice data is also required. The candidate must again pay the examination fee.

Application forms for both examinations can be obtained from the EANS secretariat (stephanie.garfield@virgin.net; registration is possible via the internet *www.eans.org* – What's new – European Examination).

The responsibility for the European Examination rests with the Examination Committee of the EANS.

References

1. ACGM outcome project *www.acgme.org/oome/implement/rsvp.asp*
2. Adams JA (1987) Historical review and appraisal of research on the learning, retention, and Transfer of human motor skills Psychol Bull 101: 41–74
3. American Association of Neurological Surgeons (AANS): *www.neurosurgery.org/aans*
4. American Board of Neurological Surgery (ABNS): *www.abns.org*
5. Braakman R (1977) Experience with the EANS Examinations Acta Neurochir (Wien) [Suppl] 69: 88–92
6. Campbell CM, Parboosingh JT, Gondocz, ST, Babitskaya G, Lindsay E, De Guzman RC, Klain LM (1996) Study of physicians use of a software program to create portfolio of their self-directed learning. Acad Med (Wien) [Suppl] 10: 49–51
7. European Association of Neurological Societies (EANS): *www.eans.org*
8. Haff JT, Eisenberg HM (1997) Assessment of Training Progress and Examinations. Acta Neurochir (Wien) [Suppl] 69: 83–88
9. Kolmos A, Hjalmered JO (1993) Project-organized curricula in relation to basic and applied Science, long and short curricula and distance learning, in: SEFI Cahier nor 4: Project-Organized curricula in Engineering Education, Brussels
10. McDonald JM (1993) Mental readiness and its links to performance excellence in surgery. Kinetek, Ottawa, Canada
11. Palmer J (ed) (1996) Neurosurgery 96, manual of neurosurgery. Churchill, Livingstone, p 13
12. Physician Achievement Review *www.par-program.org/PAR-Inst.htm*
13. Picard J (1997) Experience with the United Kingdom Examinations in neurosurgery Acta Neurochir (Wien) [Suppl] 69: 93–97

Correspondence: Prof. Dr Jens Haase, Department of Health Science and Technology, Aalborg University, Frederik Bajers Verj 7 D-1, 9220 Aalborg Denmark. e-mail: jph@ist.auc.dk

Acta Neurochir (2004) [Suppl] 90: 115–119

Printed in Austria

XV Neurosurgical subspecialization: pros and cons

G. Schackert[1] and **H.-J. Reulen**[2]

[1] Neurosurgical Department, University Clinic TU Dresden, Dresden, Germany
[2] Neurosurgical Department, Ludwig Maximilians University, Munich, Germany

Introduction

In the past decades neurological surgery as a specialty has seen a burst of growth. As a result such multifaceted development has led to the fact that a single neurosurgeon might not be in a position to offer excellent expertise in all aspects of the speciality anymore. Consequently, groups of special interests have been forming in order to clinically and scientifically develop specific aspects of neurological surgery at a high level.

For these reasons, in Germany a committee for 'subspecialization' was formed, embedded in the constitution of the Neurosurgical Academy for Continuing Education (*Neurochirurgische Akademie für Aus-, Fort- und Weiterbildung* – NCAFW). The function of this committee is to critically examine the advantages and risks of developing such subspecialties and acquiring additional qualifications. In this review, we like to present the German concept which may serve as a basis for a broad discussion among the members of the neurosurgical community, not only in Germany but also in other European countries.

As a matter of fact, many departments have already adopted and developed special areas of interest within the specialty of neurosurgery. This movement cannot and must not be decelerated. However, it makes sense to reroute such developments in neurosurgery by means of clear definitions and timely planning. There is an urgent need to draft a general concept that would be feasible for all future subspecialties of neurosurgery, whereas the specific criteria and demands of training institutions and programs for the various subspecialties may be developed in an individual manner.

From postgraduate training to subspecialization (= additional qualification)

Chances and advantages

It seems that the only real possibility to practise and further develop our specialty at a high level of competence in all its breadth is to form special areas of interest and directed training of subspecialists. For instance, a neurosurgeon with an additional qualification in 'spinal neurosurgery' should be in a position to perform complex surgical procedures such as reconstruction of the vertebral body, tumor removal, stabilization etc., and at the same time should closely follow new developments, i.e. application of endoscopic procedures in spinal surgery, etc., in order to constantly widen his knowledge and hone his manual skills and techniques. He should also be in a position to represent his special area from both clinical and scientific aspects and to train younger colleagues of the department. On account of his high competence he should attract patient referrals which, otherwise, the department would not have gained.

Another prominent example is 'neurosurgical oncology' where various special techniques such as neuronavigation, intraoperative ultrasound, electrophysiological monitoring for determining eloquent brain areas (brain mapping), PET, functional MRI, stereotactic biopsy and many more come into play. Some of these techniques are frequently used and have become today's standards, while others are quite difficult and may be applied only to a small patient population. Presently there is a trend to acquire these techniques and to spend more time in treating patients requiring these techniques. If this development towards subspecialization is seen from a positive point of view, it broadens competence, knowledge and scopes of an improved training of specialists in a given department; moreover, such an aspiring attitude increases patient referrals and offers a considerable improvement of professional perspectives.

Risks and dangers

First of all, there exists the danger that the subspecialty will separate from the mother specialty, thus forming a subunit with demarcations of technology and personnel. Such a separation of subspecialties in the sense of complete independence should be avoided by all means. The neurosurgical service should remain one unit. All subspecialties should be part of the neurosurgical department and should not only take part in basic neurosurgical treatment (standard operations), emergency cases and training, but should also recognize these tasks as mandatory. The rights and obligations of a subspecialist should be clearly defined. The involvement of subspecialists in the emergency treatment also means that he keeps his skills in such treatment constantly honed.

It should be ensured that the right to perform a neurosurgical subspecialty procedure is not completely reserved to the acquirer of an additional qualification in that subspecialty. The non-acquirement of an additional qualification should not restrict or exclude performance of standard procedures, especially from the medico-legal point of view. The danger of undermining this principle in a legal trial is present; for example, in trying for compensation of loss, a lawyer may insist on the absence of an additional qualification of the neurosurgeon. Thus the concept of 'standard neurosurgical procedure' should be well defined. One of the decisive aspects in this matter is that training of a specialist in neurosurgery should include all theoretical and practical aspects in such a way that unrestricted practising of all of neurosurgery is permitted through board examination. A thoroughly trained neurosurgeon should be in a position to perform all standard procedures in neurosurgery. With the acquirement of a subspecialization, the surgeon should gain special depth of knowledge and expertise in that particular field of neurosurgery. Thus the task of neurosurgical subspecialization may be defined as the mastery of all conservative and operative standard treatment strategies, the treatment of complex and rare cases, and particularly the performance of highly specific, extraordinary procedures that demand advanced technological and technical dexterity in that area. It would be the task of the subspecialist to represent his area scientifically and develop novel treatment modalities. Successful new concepts offering a considerable improvement and, above all, simplification of an already existing treatment method should be included in the list of standard procedures, and thus be made available to all future neurosurgical trainees during their routine training.

Criteria for additional qualification in a subspecialty area of neurosurgery

In the following, the so-called "Institutional criteria", and the guidelines for training (Training Criteria) in a subspecialty area shall briefly be outlined.

Institutional criteria

For a neurosurgical department it is possible to establish a subspecialty area if the patient traffic within that area has reached a certain magnitude. The required minimum number of patients per year may be defined by a specifically entrusted work group and additionally be confirmed by the respective national neurosurgical societies. Alternatively, this may be organized on a European level. It may be concluded that large departments might tend to have more subspecialties while smaller departments may be limited to a few. For example, the implementation of 'spinal neurosurgery' may be possible in most of the clinics, whereas the necessary number of cases to establish 'pediatric neurosurgery' may be reached by only a small number of departments.

A department that aspires to establish a subspecialty should meet several prerequisites: a pre-defined minimal number of patients in that subspecialty area, related diagnostic departments and technology, interdisciplinary approach (for instance, neurooncology: tumor clinics, regular interdisciplinary meetings, library, access to neuropathology, neuroradiology, radiotherapy etc.). The institutional criteria for other subspecialties such as 'spinal neurosurgery' may look different. Thus, it would be the task of a subspecialist to organize and conduct the interdisciplinary approach according to such specified criteria.

Training criteria

Continuing education and training of a subspecialist demands the development of a training catalogue for that specific field in a similar manner as has been accomplished already in the United States. These catalogues are being developed by specifically charged work groups of a national society. At the European level, this would probably be a work group defined by the Section of Neurosurgery of the UEMS and/or the EANS. The guidelines for institutional criteria and the training catalogue for 'neurooncology', 'spinal neurosurgery', and 'neurosurgical pain management' that have already been published in the organ of the German Society of Neurosurgery *(Mitteilungsblatt)* are typical examples. Minimal duration of an additional qualification should be $1–1\frac{1}{2}$ years after board certification in general neurosurgery, judging from the experience of our colleagues from the USA.

The acquirement of an additional qualification in a neurosurgical subspeciality

When establishing a subspecialty for the first time in a department,

(a) the institution should fulfill the required criteria and
(b) the designated head of the subspecialty area should be able to fulfill the criteria of a subspecialty training catalogue.

When such a subspecialty meets all the institutional criteria and would be chaired by a qualified subspecialist, he may be formally nominated as a trainer for this subspecialty.

How can a specialist be trained to be a subspecialist in the future?

A candidate for subspecialization should have completed his training in neurological surgery. The subspecialization training should be offered only in such institutions where it does not compromise the training of residents in general neurosurgery. Training of residents should always have priority over subspecialization.

In Germany, it has been proposed to recognize a subspecialty qualification (a) by means of confirmation through the trainer and the chairman of the department, and (b) through the *Neurochirurgische Akademie für Aus-, Fort- und Weiterbildung* according to the criteria developed by the respective work groups and presidents of the German Society of Neurosurgery and the German Federation of Neurosurgery. A certification through the German Federal Chamber of Physicians (Deutsche Ärztekammer) or the State Chamber of Physicians has deliberately not been aimed at since adequate practical experience with the recognition criteria must be gathered primarily within the Neurosurgical Academy for Continuing Education. However, we plan to remain in close contact with the Federal Chamber of Physicians regarding these matters. If, at a later stage, the introduction of an additional qualification should be discussed in the Order of Postgraduate Training of the Federal Chamber of Physicians, then subspecialization in neurosurgery might fall under the category of a 'Qualification Upgrading Certificate'.

A list of possible subspecialty areas in neurosurgery

At the present point of time, training catalogues for the following subspecialties are being prepared:

- Neurosurgical oncology (completed)
- Spinal neurosurgery (completed)
- Stereotactic and functional neurosurgery
- Neurosurgical pain therapy (completed)
- Vascular neurosurgery
- Peripheral nerve surgery
- Pediatric neurosurgery.

Subspecialties should be large enough, it should be avoided that individually planned small areas of our specialty isolate themselves and form a 'subsubspecialty'.

Rights and obligations of a subspecialty

Furthermore, internal limitations of rights and obligations have to be described. It has already been mentioned that any tendency towards independence of the subspecialties from neurosurgery should be avoided. The chairman of a department carries final responsibility for the subspecialties practised in his department and has the right and obligation to organize the treatment of patients as well as the conduct of training residents including their rotational assignments. It may be specifically mentioned that the chairman or one of the staff members assigned by him is responsible as the Director of Training Program for student education and training of specialists as well as for the assignment of trainees to a respective subspecialty rotation. However, it is desirable to discuss and agree upon these matters with the respective subspecialist of the department.

The chairman of the department can delegate responsibilities for a subspecialty area to a staff member possessing an appropriate additional qualification. This subspecialist would be completely responsible for the treatment of patients in his area of interest; however, he still remains subordinated to the chairman. This means that diagnostic procedures, indications and operative therapy are administered according to the general guidelines of the department. The introduction of new treatment policies should be discussed prior to accomplishment (also in view of economical reasons). The subspecialist is obliged to train residents during their rotation in the respective area. Furthermore, those candidates showing interest in a subspecialization would be trained by the respective subspecialist. Subspecialization in a specific area also encompasses responsibility for research and development in that particular field. An academic subspecialist would be required to lecture and perform innovative research in his field of specialty.

Subspecialty: "Special Neurosurgical Oncology"[1]

The institutional criteria and the training catalogue of the subspecialty "Special Neurosurgical Oncology" may be presented as an example:

Minimum time for subspecialization amounts to $1\frac{1}{2}$ years. This may be undertaken after completion of neurosurgical training.

Subspecialization requires the following items:

1. Special knowledge, experience and skills in the indication and surgery of 50 supra- or infratentorial tumors and 20 spinal tumors in addition to the numbers required for general training in neurosurgery.
2. Knowledge of indication and surgery of at least 20 biopsies, with the aid of three-dimensional planning systems (e.g. stereotaxy, navigation systems) in addition to the procedures required for general training in neurosurgery.
3. Chemotherapy treatment according to established protocols in 30 patients, including follow-up examinations, awareness and treatment of chemotherapeutical side-effects (e.g. blood counts, prophylaxis of infections, antiemetics, disturbance of lung or kidney function etc.). Knowledge of handling chemotherapeutic drugs.
4. Knowledge and experience in radiosurgical procedures.
5. Participation in clinical scientific projects is recommended. Knowledge of good clinical practice (GCP) is especially demanded in clinical research.
6. Participation in at least two medical education meetings in neurooncology are required.

Institutional criteria for the subspecialty area "Special Neurosurgical Oncology". Medical centers or hospitals have to fulfill the following requirements:

1. Diagnostic institutions and techniques
   - Neuroradiology (incl. MRI)
   - Neuropathology
   - CSF cytology
   - Electrophysiology, incl. intraoperative monitoring
2. Treatment procedures
   - Surgery: routine surgery of supra- and infratentorial tumors and spinal tumors.
   - At least 60 gliomas, 30 meningiomas, and 30 other brain tumors annually, pediatric brain tumors, 25 spinal tumors/year (intra- and extradurally).
   - At least 30 biopsies with the aid of computer assisted three-dimensional planning systems (e.g. stereotaxy, navigational systems).
   - Radiotherapy, eventually in connection with radiosurgery.
   - Tumor outpatient clinic: regular follow-up of neuro-oncology patients.
   - Scientific research: Concomitant scientific research is desirable. Participation in multicenter studies, scientific-clinical work or basic science projects.
3. Neuro-oncological conferences on a regular basis:
   - The treatment of neurooncology patients should be discussed in regular interdisciplinary conferences with neuroradiologists, neurologists, neuropathologists, radio-oncologists and other disciplines that are involved in the tumor center.
4. Department library
   - A library with neuro-oncology literature and updated journals is required.

[1] These guidelines have been developed by a task force led by J.C. Tonn and G. Schackert

The subspecialty "Special Neurosurgical Oncology" can only be established in centers certified for neurosurgical training.

Conclusion

The need for subspecialization in neurological surgery is obvious. The increase in surgical and scientific knowledge, provided by the different subspecialties, will strengthen the field of neurosurgery. Its realization should be unanimously supported by the neurosurgical community.

References (for further reading)

Couldwell WT, Rovit RL (2002) Rethinking neurosurgical subspecialization. Surg Neurol 58(6): 359–363

Dwyer AP, Herkowitz HN, Benzel EC (2002) Controversies in spine: Should there be subspecialty certification in spine surgery? Spine 27(13): 1478–1483

Herkowitz HN, Benzel EC (2002) Controversies in spine: Subspecialty certification should not be a requirement for spine surgery. Spine 27(13): 1478–1483

Mainprize TG, Taylor MD, Rutka JT (2000) Perspectives in pediatric neurosurgery. Childs Nerv Syst 16(10–11): 809–820

[No authors]: Program requirements for residency/fellowship education in neuroendovascular surgery/interventional neuroradiology: special report on graduate medical education: a joint statement by the American Society of Interventional and Therapeutic Neuroradiology, Congress of Neurological Surgeons and American Association of Neurological Surgeons, American Society of Neuroradiology. Neurosurgery 48(2): 457–458

Ostertag C (1997) Subspecialty training in neurosurgery. Acta Neurochir (Wien) [Suppl] 69: 126–129

Piepgras DG (1997) Post residency subspecialty training in neurosurgery – The impact of subspecialty training on organized neurosurgery and resident training – benefits, responsibilities and liabilities. In "Training in Neurosurgery". In: Reulen HJ, Steiger HJ (eds) Acta Neurochir (Wien) [Suppl]

Correspondence: Prof. Dr. med. Gabriele Schackert, Klinik für Neurochirurgie, Universitätsklinikum Carl Gustav Carus der TU Dresden, Fetscherstr. 74, 01307 Dresden, Germany. e-mail: Gabriele.Schackert@uniklinikum-dresden.de

Important addresses

Address of Chairman JRAAC

Prof. Dr. T. Trojanowski
Department of Neurosurgery
University Medical School
Ul. Jaczewskiego 9
20-950 Lublin
Poland
e-mail: *Trojan@asklepios.am.lublin.pl*

Address of Secretary JRAAC

Prof. Dr. M. Zerah
Service de Neurochirurgie
Hôpital Enfant Malades
149, rue de Sèvres
F-75015 Paris
France
e-mail: *michel.zerah@nck.ap-hop-paris.fr*

Application Form for European Accreditation

to be downloaded from the internet:
www.eans.org
What's new: Accreditation of Training Centers

European Logbook

Electronic version can be downloaded from the internet
www.eans.org
What's new: Trainee Logbook
Also see: Information
Printed version is available from
Aesculap AG & Co. KG
P.O. Box 40
D-78501 Tuttlingen
Germany
Attn. Mrs. Eva Streit
e-mail: *eva.streit@aesculap.de*

European Charter on Training

to be downloaded from the internet
www.uems.be
Training/Formation – visitation – UEMS Charter on Visitation of Training

Evaluation Form

to be received from
Klinikum Großhadern
Neurosurgical Department
Co: Mrs. Ilona Anders
Marchioninistr. 15
81377 Munich
Germany
e-mail: *Ilona.Anders@med.uni-muenchen.de*

Author index

Index of keywords

Springer and the Environment

WE AT SPRINGER FIRMLY BELIEVE THAT AN INTERnational science publisher has a special obligation to the environment, and our corporate policies consistently reflect this conviction.

WE ALSO EXPECT OUR BUSINESS PARTNERS – PRINTERS, paper mills, packaging manufacturers, etc. – to commit themselves to using environmentally friendly materials and production processes.

THE PAPER IN THIS BOOK IS MADE FROM NO-CHLORINE pulp and is acid free, in conformance with international standards for paper permanency.